NETWORK SECURITY AND COMMUNICATION ENGINEERING

PROCEEDINGS OF THE 2014 INTERNATIONAL CONFERENCE ON NETWORK SECURITY AND COMMUNICATION ENGINEERING (NSCE 2014), 25–26 DECEMBER 2014, HONG KONG

Network Security and Communication Engineering

Editor

Kennis Chan

Advanced Science and Industry Research Center

CRC Press
Taylor & Francis Group
Boca Raton   London   New York   Leiden

CRC Press is an imprint of the
Taylor & Francis Group, an **informa** business

A BALKEMA BOOK

CRC Press/Balkema is an imprint of the Taylor & Francis Group, an informa business

© 2015 Taylor & Francis Group, London, UK

Typeset by diacriTech, Chennai, India

Published by: CRC Press/Balkema
P.O. Box 11320, 2301 EH Leiden, The Netherlands
e-mail: Pub.NL@taylorandfrancis.com
www.crcpress.com – www.taylorandfrancis.com

ISBN: 978-1-138-02821-0 (Hardback)
ISBN: 978-1-315-68355-3 (eBook)

Table of contents

Grid and cloud computing

Networking algorithms and performance evaluation

Multimedia, signal and image processing

Data mining

Wireless communications and sensor networks

Intelligent computing and communication system

Network Security and Communication Engineering – Chan (Ed.)
© 2015 Taylor & Francis Group, London, ISBN: 978-1-138-02821-0

Preface

The 2014 International Conference on Network Security and Communication Engineering (NSCE2014) was held in Hong Kong on December 25–26, 2014. The purpose of this conference is to deepen the academic exchanges and understandings of network security and communication engineering between scholars, scientists and engineers across all over the world and provide a unique opportunity to exchange information, to present the latest results as well as to review the relevant issues on contemporary research in network security and communication engineering. Young scientists were especially encouraged to attend the conference and to establish international networks with senior scientists.

With the time goes on, network security has become more and more important in all walks of life, from business to politics, the essential value of network security is shown everywhere. The appearance of Internet marked the beginning of the age of communication engineering. And the growing communication engineering becomes more and more depth in people's work and life.

More than 180 scientists and researchers coming from more than 15 countries are interested in this conference. These articles were divided into several sessions, such as network and system security, grid and cloud computing, networking algorithms and performance evaluation, multimedia, signal and image processing, wireless communications and sensor networks and so on. The large number of presented papers emphasizes the considerable academic and keen interest in the conference theme. We wish to thank the authors for their participation and cooperation, which made this conference especially successful.

Finally, we wish to express our warm thanks and appreciation to our committee for their sustained assistance, help and enthusiasm during the preparation of the conference.

We wish all the attendees at NSCE2014 can enjoy a scientific conference in Hong Kong. We really hope that all our participants can exchange useful information and make amazing developments in network security and communication engineering after this conference.

Network Security and Communication Engineering – Chan (Ed.)
© 2015 Taylor & Francis Group, London, ISBN: 978-1-138-02821-0

Organizing committee

General Chair

S.K. Chen, *Altair Engineering Inc., California, USA*
T.R. Vijayaram, *VIT University Chennai Campus, India*

Program Chair

K. Weller, *Science and Engineering Research Center, Hong Kong*

International Scientific Committee

F. Kiani, *Istanbul S.Zaim University, Turkey*
M. AHMAD, *Jamia Millia Islamia, India*
T. Kumar, *Indus University, Pakistani*
D. Dutta, *The University of Burdwan, India*
W.C. Su, *Chang Gung University, Taiwan*
M. Subramanyam, *Anna University, India*
B. Mukherjee, *PDPM Indian Institute of Information Technology, India*
C. Qin, *University of Shanghai for Science and Technology, China*
R.R. Jorge, *Technological University of Ciudad Juarez, Mexico*
X. Lee. Hong Kong Polytechnic University, *Hong Kong*
P. Velayutham, *Madras Christian College, India*
K. Weller, *Science and Engineering Research Center, Hong Kong*
J. Yeh, *Tallinn University of Technology, Estonia*
Y. Wu, *Harbin Engineering University, China*
B. Sarker, *Louisiana State University, USA*
K.S. Rajesh, *Defence University College, India*
M.M. Kim, *Chonbuk National University, Korean*
X. Ma, *University of Science and Technology of China, China*
M.V. Raghavendra, *Adama Science & Technology University, Ethiopia*
J. Ye, *Hunan University of Technology, China*
Z.Y. Jiang, *University of Wollongong, Australian*
V.K. Jain, *Indian Institute of Technology, India*

Network and System Security

Network Security and Communication Engineering – Chan (Ed.)
© *2015 Taylor & Francis Group, London, ISBN: 978-1-138-02821-0*

A privacy amplification protocol against active attacks in information theoretically secret key agreement

Q.H. Wang, X.J. Wang, Q.Y. Lv & X.Y. Ye
College of Communication Engineering, Hangzhou Dianzi University, Hangzhou, China

ABSTRACT: Privacy amplification is an important step in information key agreement that is theoretically secret, and it makes possible two legitimate parties, Alice and Bob who share a partially secret string S that is about an adversary. By communication through a public channel, Eve has partial information to extract a shorter but highly-secret string S' such that Eve's information about S' is negligible. In this paper, we treat the problem of privacy amplification to be secured against active attacks, and propose a privacy amplification protocol that enables communication through the public channel to be authenticated by the partially secret common information between the two legitimate parties. The proposed protocol can make privacy amplification secure against active adversaries with some probability. The relationship between the block size and the success probability of active attack is analyzed and a lower bound on the length of the common string that needs to implement privacy amplification is also given.

1 INTRODUCTION

Privacy Amplification, introduced by Bennett et. al in [1] and generalized in [2], is the technique of shrinking a partially secret string to a highly secret one by public discussion. More precisely, in the privacy amplification phase, two legitimate parties, Alice and Bob, who share a partially secret string S about an opponent Eve has partial information to extract, and a shorter but highly-secret string S' by communication through an public channel such that Eve's information about S' is negligible, i.e., $H(S'/U = u) \geq \log_2^{|S|} - \varepsilon$ holds with very high probability for some small $\varepsilon > 0$, where the random variable U summarizes Eve's complete knowledge about S', and u is the particular value known to Eve [2]. Privacy amplification is an important step in both quantum cryptography and key theoretically secret information agreement [3].

The problem of privacy amplification has been well studied in the passive-adversary model [2]. In this model, it is assumed that Alice and Bob are connected by an authentic public channel, Eve has access to the public channel but it is passive or, equivalent, and she is not able to insert or modify messages without being detected. However, the actual public channel is necessarily an insecure and non-authentic one, and Eve may introduce fraudulent messages or modify messages over the public channel. Consequently, it is necessary to research on the privacy amplification that is secure against active adversaries who are able to insert or modify messages. However, until now, the study results in the field of privacy amplification against active attacks are rare.

The generalized problem of privacy amplification secured against active attacks was studied in [4], [5], [6], [7] and [8]. [4] and [5] treated the case of one-way-transmission protocols for privacy amplification by public discussion over a completely insecure channel. In [5], the partially secret string S is divided into three parts S_I, S_{II}, S_{III} and the first two S_I, S_{II} are used for authentication to transmit a random string over the public channel, while the last part S_{III} is served as the input of an universal hashing. The basic idea is to use the first part of the partially secret string in a first step to authenticate a message containing the description of a function from a suitable class of universal hash functions. In the second step, this hash function is used for privacy amplification, and the remaining string is used as the input to this hash function [7]. Wolf presented a two-way communication protocol in [7] which is an authenticated message in an interactive way. And later, an improvement was made in [8].

In this paper, we investigate the same problem with respect to an active adversary and propose a privacy amplification protocol in which the communication over the public channel is authenticated by the partially secret common information between the two legitimate parties and privacy amplification secured against active adversaries is achieved with some probability.

2 THEORY OF PRIVACY AMPLIFICATION

Privacy amplification technology typically applies a random and public universal hash function to shorten the weak secret key and reduce the amount of information obtained by Eve as much as possible. More precisely, Alice randomly chooses a proper universal hash function G from a universal class and sends its description to Bob over an authentic channel. Then they both compute $S' = G(S)$.

Theorem 1[2]: Assuming that Alice and Bob share a random n-bit string, S, with uniform distribution over $\{0,1\}^n$, U is Eve's knowledge about S over the public discussion, u is a particular value of U observed by Eve. If Eve's Rényi entropy about S, $R(S|U = u)$, is at least c, and Alice and Bob choose $S' = G(S)$ as their secret key, where G is chosen at random from a universal class of hash functions from $\{0,1\}^n$ to $\{0,1\}^k$, Eve's entropy of K given G and V satisfies

$$H(S'|G,U = u) \ge k - \log_2(1 + 2^{k-c}) \ge k - \frac{2^{k-c}}{\ln 2}. \quad (1)$$

Hence, when $k<c$, Eve's entropy of S' is close to the maximum, that is, her information about S', $H(S) - H(S|G,U = u)$, is arbitrarily small.

In the privacy amplification phase, universal hash function plays a very essential role. Alice must choose a proper universal hash function G from a universal class and send its description to Bob effectively and reliably. Otherwise, it is difficult to achieve the extraction of a highly secret string S'. Hence, the communication between Alice and Bob must be authenticated.

3 A PRIVACY AMPLIFICATION PROTOCOL AGAINST ACTIVE ATTACKS

Assuming that Alice and Bob share a partially secret string, $S \in \{0,1\}^n$, S is divided into two parts S_1 and S_2, S_1 is used to authenticate a message decrypting a hash function over the public channel; S_2 is used as the input of a universal hash function for the final privacy amplification. The detailed steps of our proposed protocol are as follows:

1 Assuming that $S_1 \in \{0,1\}^m$ and m can be divided exactly by r, Alice and Bob divide S_1 into blocks of length r, denoted as $\{B_0, B_1, ..., B_{m/r-1}\}$, where $B_i = (B_{i,0}, B_{i,1}, ..., B_{i,r-1})$ and $1 \le i \le \frac{m}{r} - 1$. If there are same blocks among $\{B_0, B_1, ..., B_{m/r-1}\}$, those repeated blocks will be discarded to ensure that all of the data blocks are not repeated. Assuming that the final blocks are $\{B_0, B_1, ..., B_{m/r-l}\}(l \ge 1)$, they are called authentication blocks.

2 Alice randomly chooses a proper universal hash function G from a universal class and wants to send its q-bit description $A=(a_0, a_1, ..., a_{q-1}) \in \{0,1\}^q$ to Bob through public channel. Alice divides A into blocks of length k ($k<r$ and q can be divided exactly by k), denoted as

$$A_0 = \{a_0, a_1, ..., a_{k-1}\}$$
$$A_1 = \{a_k, a_{k+1}, ..., a_{2k-1}\}$$
$$...$$
$$A_{\frac{q}{k}-1} = \{a_{q-k}, a_{q-k+1}, ..., a_{q-1}\}$$

a. Assuming that Alice wants to send k-bit authentication bits $A_i = \{a_{ik}, a_{ik+1}, ..., a_{(i+1)k-1}\}$ ($1 \le i \le \frac{q}{k} - 1$) to Bob through the public channel, she computes $\mu = \sum_{j=0}^{k-1} a_{ik+j} \cdot 2^j$, and transmits $B_{\mu+\beta \cdot 2^k}$ to Bob, where $\beta \ge 0$ is an integer, and it denotes the number of occurrences of μ. In this way, we can assure that each authentication block transmitted by Alice can only be used once.

b. After receiving $B_{\mu+\beta \cdot 2^k}$, Bob verifies whether the received block $B_{\mu+\beta \cdot 2^k}$ matches what he holds. If the block matches, then Bob thinks that the authentication bits sent by Alice is the binary equivalent corresponding to number $((\mu + \beta \cdot 2^k) \mod 2^k \mod 2^k)$. If the block doesn't match, Bob thinks that $B_{\mu+\beta \cdot 2^k}$ is not transmitted by Alice or has been tampered with.

c. Alice and Bob perform Steps (a)~(b) iteratively until all bits are transmitted.

3 Alice and Bob compute $S' = G(S)$.

4 ANALYSIS OF THE PROPOSED PROTOCOL

In the above proposed protocol, by eavesdropping transmitted messages over the open channel, Eve can estimate the error bit rate between her and Alice's (or Bob's) secret strings. When Eve eavesdrops an authentication block B_j, she will compare it with all data blocks she holds to find a data block with minimum number of error bits e_j. After many eavesdropping (assuming that Eve eavesdrops ω authentication blocks), Eve estimates error bit rate

$$\varepsilon_E = \frac{1}{\omega} \sum_{t=1}^{\omega} \frac{e_{j_t}}{r}. \quad (2)$$

4.1 Eve's attack strategy

When Eve intercepts an authentication block, she can adopt the following two attack strategies.

Strategy 1: Eve substitutes this intercepted authentication block directly for another block randomly chosen from all blocks she holds.

Strategy 2: Eve randomly chooses a block, corrects errors randomly according to her estimated bit error

4

rate, and sends the corrected block to Bob, that is, Eve substitutes the intercepted authentication block for her corrected block. As Eve doesn't know the location of the error bits, error correction has a certain probability.

4.2 Performance analysis

Theorem 2: In our proposed protocol, among Eve's n bits information, there are λ bits that are not consistent with the common secret string between Alice and Bob, and the λ bits are uniformly distributed in Eve's n bits information. Then, when $r < \dfrac{n}{\lambda}$, Eve chooses strategy 1, and the maximum success probability of active attack is at least $\dfrac{\lambda}{n}$; when $r \geq \dfrac{n}{\lambda}$, Eve chooses strategy 2, and the success probability of active attack is no more than $1/C_r^{\frac{\lambda}{n} \cdot r}$.

Proof: (1) Assume that Eve chooses strategy 1. As the λ bits are uniformly distributed in Eve's n bits information, when $r < \dfrac{n}{\lambda}$, Eve's data block number $\dfrac{n}{r}$ is greater than the number of error bits λ, so, at least one Eve's data block agrees with the authentication blocks held by Alice and Bob. For Eve, she doesn't know which block is right, so, she can only randomly select one block to perform active attack. Only when Eve picks the right block and the block is not used by Alice and Bob before, can the attack be successful. Hence, the maximum success probability of active attack is at least $1/(\dfrac{n}{\lambda}) = \dfrac{\lambda}{n}$.

(2) Assume that Eve chooses the strategy 2. When $r \geq \dfrac{n}{\lambda}$, there is at least one error in Eve's each data block. After estimating the error bit rate $\varepsilon_E \approx \dfrac{\lambda}{n}$, Eve speculates that she has, on average, about $\varepsilon_E \approx \dfrac{n}{\lambda} \cdot r$ error bits in each of her block. In order to correct those errors, she can only guess those error bits at random, and at most one speculation is correct among all $C_r^{\frac{\lambda}{n} \cdot r}$ speculations. In the same way, only when Eve's guess is right and the block is not used by Alice and Bob before, can the attack be successful. Hence, Eve's success probability of active attack is no more than $1/C_r^{\frac{\lambda}{n} \cdot r}$.

From **Theorem 2**, in order to avoid the case that Eve has larger success probability of active attack when she adopts the first strategy, the length of the authentication blocks should meet $r \geq \dfrac{n}{\lambda}$, and in such case, Eve will adopt strategy 2. When the value of n and λ is certain, the longer the length of the authentication block r,

the lower the success probability of Eve' active attack .

Theorem 3: If we treat Eve's n bit information E as the output of the binary symmetric channel with the bit error rate ε_E when Alice and Bob's n bits information S is the channel input, then, Eve's success probability of active attack is no more than $(1 - \varepsilon_E)^r$ when the strategy 1 is adopted, and Eve's success probability of active attack is no more than $[(1 - \varepsilon_E)^2 + \varepsilon_E^2]^r$ when the strategy 2 is adopted.

Proof: (1) When adopting the strategy 1, Eve randomly chooses a data block from all blocks she holds. As the bit error rate of the binary symmetric channel is ε_E, the probability that Eve's selected data block agrees with that of Alice and Bob is about $(1 - \varepsilon_E)^r$. Hence, the success probability of active attack is no more than $(1 - \varepsilon_E)^r$.

(2) When the strategy 2 is adopted, the probability that an Eve's selected data block contains exactly i errors is $p_i = C_r^i \varepsilon_E^i (1 - \varepsilon_E)^{r-i}$. However, Eve doesn't know how many errors are there in her selected data block, and she can only guess it according to the probability distribution of $\{p_i\}$, so, the probability that Eve guesses right is $\sum_{i=0}^r p_i^2$. Moreover, Eve has to guess the position of the error bits, and the probability that Eve also can guess correctly is $\dfrac{p_i^2}{C_r^i}$. The probability that Eve can correct all errors is $\sum_{i=0}^r \dfrac{p_i^2}{C_r^i} = \sum_{i=0}^r C_r^i \varepsilon_E^{2i}(1 - \varepsilon_E)^{2(r-i)} = ((1 - \varepsilon_E)^2 + \varepsilon_E^2)^r$. Hence, the success probability of active attack is no more than $[(1 - \varepsilon_E)^2 + \varepsilon_E^2]^r$.

Theorem 4[2]: For all positive ε_E and γ, there exists a positive α such that if Alice and Bob share a random u-bit string S_2, which Eve receives through a binary symmetric channel with bit-error probability ε_E, and if they apply privacy amplification with universal hashing to obtain an v-bit string S', where $v = \lfloor (h(\varepsilon_E) - \gamma)u \rfloor$, then for all sufficiently large u, Eve's expected information about S' is at most $2^{-\alpha u}$ bits.

Based on Theorem 4 and the consideration of the common information needed to prevent active attack, we deduce the following theorem.

Theorem 5: Assuming that Alice and Bob share a random n-bit string S, Eve receives it through a binary symmetric channel with bit-error probability ε_E. When the open channel is a non-authentic insecure one and the length of the bit string shared by Alice and Bob is at least $\left\lceil \dfrac{v}{h(\varepsilon_E) - \gamma} \right\rceil + \dfrac{q}{k} \cdot r$ bits, our proposed privacy amplification protocol can resist against active attacks with probability $1 - \max\{(1 - \varepsilon_E)^r, [(1 - \varepsilon_E)^2 + \varepsilon_E^2]^r\}$, and there exists a positive α such that when $\left\lceil \dfrac{v}{h(\varepsilon_E)-\gamma} \right\rceil$ is sufficiently large, Eve's expected information about S' is at most $2^{-\alpha u}$ bits.

Proof: (1) From **Theorem 4**, to extract v-bit highly-secret string S', the length of the input of the universal hash function S_2 should be $u \geq \left\lceil \dfrac{v}{h(\varepsilon_E)-\gamma} \right\rceil$. When u is large enough, Eve's expected information about S' is at most $2^{-\alpha u}$ bits.

According to our proposed privacy amplification protocol, if Alice wants to successfully send k certification bits to Bob, she needs to send r-bit authentication block through the open channel. If Alice wants to successfully transmit the universal hash function G to Bob, she needs to send its q-bit description to Bob through public channel, that is, Alice should transmit $\dfrac{q}{k} \cdot r$ bits through public channel. Hence, the length of S_2 should be at least $\dfrac{q}{k} \cdot r$ bits.

In summary, only when the length of the string S shared between Alice and Bob is at least $\left\lceil \dfrac{v}{h(\varepsilon_E)-\gamma} \right\rceil + \dfrac{q}{k} \cdot r$ bits, can the privacy amplification protocol against active attacks be possible.

Moreover, from **Theorem 3**, under the above conditions, the achieved probability resisting active attacks of our proposed privacy amplification protocol is at most $1 - \max\{(1-\varepsilon_E)^r, [(1-\varepsilon_E)^2 + \varepsilon_E^2]^r\}$.

Note: it is needed that a universal hash function should be transmitted through the public channel, and the common used class of universal hash function for privacy amplification is $h(x) = (a \cdot x)_v$, where $a, x \in \{0,1\}^u$, $h : \{0,1\}^u \rightarrow \{0,1\}^v$, $(a \cdot x)_v 0$ denotes the first v bits of the element $a \cdot x$. Under this condition, to transmit a universal hash function, at least u bits should be transmitted, that is, at least $\dfrac{u}{k} \cdot r$ bits should be transmitted. Hence, in our proposedv protocol, the length of the bit string shared between Alice and Bob should be at least $\left\lceil \dfrac{v}{h(\varepsilon_E)-\gamma} \right\rceil \cdot \left(\dfrac{r}{k}+1\right)$.

Moreover, when $\varepsilon_E \leq 0.5$, the success probability of active attack is no more than $(1-\varepsilon_E)^r$.

5 CONCLUSIONS

In this paper, we treat the generalized problem of privacy amplification secured against active attacks, and propose a privacy amplification protocol that enables communication through public channel to be authenticated by the partially secret common information between the two legitimate parties. It is concluded that the proposed protocol makes privacy amplification secure against active adversaries possible with some probability. The relationship between the block size and the success probability of active attack is analyzed and a lower bound on the length of the common string needed to implement privacy amplification is also given. The proposed method helps to achieve the key theoretically secret information agreement.

ACKNOWLEDGMENTS

This work is supported by National Natural Science Foundation of China (No.61401128), Zhejiang Province Natural Science Foundation (No. LQ14F020010), Project of Zhejiang Provincial Smart City Research Center, Hangzhou Dianzi University and Zhejiang Provincial Key Lab of Data Storage and Transmission Technology, Hangzhou Dianzi University.

REFERENCES

[1] Bennett C.H., Brassard G. and Robert J.M. 1988. Privacy amplification by public discussion, *SIAM Journal on Computing*, Vol. 17, pp. 210–229.
[2] Bennett C.H., Brassard G., Crepeau C. and Maurer U. M. 1995. Generalized privacy amplification, *IEEE Transactions on Information Theory*, Vol. 41, No. 6, pp.1915–1923.
[3] Wang Q., Lv Q., et. al. 2014. A New Bit Pair Iteration Advantage Distillation/Degeneration Protocol in Information Theoretically Secret Key Agreement, *Journal of Computational Information Systems*, Vol.10, No.12, pp.5017–5024.
[4] Maurer U.M. 1997. Information-theoretically secure secret-key agreement by NOT authenticated public discussion, *Advances in Cryptology-EUROCRYPT '97(Lecture Notes in Computer Science)*. Berlin, Germany: Springer-Verlag, Vol. 1233, pp. 209–225.
[5] Maurer U.M. and Wolf S. 1997. Privacy amplification secure against active adversaries, *Advances in Cryptology-CRYPTO'97 (Lecture Notes in Computer Science)*. Berlin, Germany:Springer-Verlag, vol.1294, pp.307–321.
[6] Maurer U.M. and Wolf S. 2003. Secret-key agreement over unauthenticated public channels Part III: Privacy amplification, IEEE *Trans. on Information Theory*, Vol. 49, No. 4, pp. 839–851.
[7] S. Wolf. 1998. Strong security against active attacks in information-theoretic secret-key agreement, *Advances in Cryptology-ASIACRYPT '98 (Lecture Notes in Computer Science)*. Berlin, Germany: Springer-Verlag, vol. 1514, pp. 405–419.
[8] Liu S. and Wang Y. 1999. Privacy amplification against active attacks with strong robustness, *Electronics Letters*, Vol. 35, No. 9, pp. 712–713.

Network Security and Communication Engineering – Chan (Ed.)
© 2015 Taylor & Francis Group, London, ISBN: 978-1-138-02821-0

Zero knowledge proof protocol from Multivariate Public Key Cryptosystems

S.P. Wang, B. Yue & Y.L. Zhang
Xi'an University of Technology, Xi'an, ShaanXi, China

ABSTRACT: A zero-knowledge proof protocol is an interactive method for one party to prove to another that a (usually mathematical) statement is true, without revealing anything other than the veracity of the statement. We present an interactive zero knowledge proof protocol from Multivariate Public Key Cryptosystems (MPKCs), and the protocol satisfies the properties of completeness, soundness and zero knowledge. It is a standard 3-move commit-challenge-response protocol. Our zero knowledge proof protocol is from MPKCs, as MPKCs is one of the promising alternatives which may resist the future quantum computing attacks. Therefore currently our protocol seems to be immune to quantum computing attacks.

1   INTRODUCTION

The idea of Zero Knowledge Proof (ZKP) was first proposed by Goldwasser, Micali and Rackoff [S. Goldwasser, et al. 1989.] in 1989. It is a mutual protocol to solve the problem: the prover demonstrates to the verifier that he has some secret information, but after that the verifier doesn't know what the secret information is. In the verification process, the prover lets out zero information about the secret to the verifier. ZKP can be divided into two basic kinds: interactive and non-interactive zero knowledge proof. Zero knowledge proof protocols are used extensively in the field of information security, such as identity authentication, fair exchange, key agreement, electronic voting and electronic payment system, etc.

The classic public-key cryptosystems are facing security threat due to the emergence of quantum computing, and there is an urgent requirement for new secure public-key cryptosystems in the post-quantum era. Multivariate public key cryptosystem (MPKCs) [Jintai Ding, et al. 2006.] as a new public key cryptosystem is becoming one of the research focuses, and its security is based on a NP-hard problem of solving Multivariate Quadratic (MQ) equations over a finite field. Currently, the quantum computing does not show any advantages in the treatment of the above NP-hard problems. MPKCs with its high computing efficiency are thought to be one of the candidates for security cryptosystem in post-quantum era. MPKCs can be divided into two types: the bipolar systems and the mixed systems. The bipolar systems mainly have MI, HFE, OV, and TTM systems, and contain many signature algorithms, such as SFLASHv2, Rainbow, Quartz and TTS signature algorithm. To research secure zero knowledge proof protocol against

quantum computing is of important theoretical value and practical significance.

In this paper, we present an interactive zero knowledge proof protocol from MPKCs. In our protocol, the prover P is supposed to be proved to the verifier that he or she is the owner of public key $PK = F = L_1 \circ F \circ L_2$ of MPKCs, that is, he or she knows the private key $SK = \{L_1, F, L_2\}$ corresponding to the public key $PK = F = L_1 \circ F \circ L_2$, in a zero knowledge proof style. Our protocol is different from the existing zero knowledge proof protocols from MQ problem. Our scheme is based on a trapdoor function of the public key which is the composition of easily invertible maps, so our scheme is a new type of zero knowledge proof protocols form MPKCs. Our protocol is designed to prove that the prover has the secret key of the public key from MPKCs, while the existing zero-knowledge protocol from MQ problem is designed to prove that the prover has a solution for MQ equation.

2   PRELIMINARIES

2.1   *Interactive zero knowledge proof protocol*

In this section we briefly review the definition of interactive Zero Knowledge Proof (ZKP) protocol. In an interactive ZKP protocols whereby one party, the prover, convinces another party, the verifier, some assertions are true with the remarkable property that the verifier "learns nothing" is other than the fact that the assertion being proven is true.

A ZKP protocol should satisfy the following three properties, **Perfect correctness, Soundness** and **Statistically zero knowledge**.

2.2 Multivariate public key cryptosystems

Here we review some basic facts about the multivariate public key cryptosystems (MPKCs). A MPKCs is a multivariate quadratic polynomial system over a finite field $k=GF(q^l)$. The key idea of MPKCs is to use two invertible affine transformations L_1 on k^m and L_2 on k^n to hide an invertible core map $F : k^n \rightarrow k^m$. There are a lot of different invertible core maps F as we can see this in MI,HFE,OV, TTM, SFLASHv2, Rainbow, Quartz and TTS MPKCs [Jintai Ding, Jason E. Gower & Dieter S. Schmidt. 2006.]. The core map F is at the heart of the MPKCs structure, and the composition of the map in MPKCs can be expressed as follows:

$$\overline{F} = L_1 \circ F \circ L_2 = \left(\overline{f}_1, \cdots, \overline{f}_m \right)$$

$$\overline{f}_i \left(x_1, \cdots, x_n \right) = \sum_{1 \leq j \leq k \leq n} \alpha_i x_j x_k + \sum_{j=1}^n \beta_i x_j + \gamma_i$$

where $1 \leq i \leq m$, and $\alpha_i, \beta_i, r_i \in k$.

The Public Key

1 The finite field $k=GF(q^l)$ including its additive and multiplicative structure;
2 The m polynomials $\overline{f}_1, \cdots, \overline{f}_m \in k[x_1, x_2, ..., x_n]$.

The Private Key

1 The invertible core map $F : k^n \rightarrow k^m$;
2 The two invertible affine transformations L_1 on k^m and L_2 on k^n.

Signature Generation

Let $Y = (y_1, \cdots, y_m) \in k^m$ be a message (or message digests) to be signed. The signer first computes $Y' = L_1^{-1}(Y)$, then computes $X' = F^{-1}(Y')$, and finally computes $X = L_2^{-1}(X') \in k^n$, so $X = (x_1, \cdots, x_n)$ is a signature on message Y.

Signature Verification

To verify that $X = (x_1, \cdots, x_n)$ is a valid signature on the message $Y = (y_1, \cdots, y_m)$, the verifier checks if indeed $\overline{F}(x_1, \cdots, x_n) = (y_1, \cdots, y_m)$. If equality holds, then accept the signature X; otherwise reject it.

3 AN INTERACTIVE ZERO KNOWLEDGE PROOF PROTOCOL FROM MULTIVARIATE PUBLIC KEY CRYPTOSYSTEMS

3.1 Our protocol

The following is our zero knowledge proof protocol from MPKCs. The setup and the parameter of our protocol is as follows

System Parameter Setup: System parameters generation algorithm outputs (k, q, l, m, n), where q, l is the security parameters, $k=GF(q^l)$ is a finite field, m is the number of multivariable equations, and n is the number of the variables.

Private Input of Prover (P): The secret for the Prover is $SK = \{L_1, F, L_2\}$, where $F: k^n \rightarrow k^m$ is an invertible core map [see section 2.2], where L_1 and L_2 are invertible affine transformations on k^m and k^n respectively.

Common Input: the common input for both the Prover and the Verifier is the public key which is $PK = F = L_1 \circ F \circ L_2$, where the notation $\circ$ represents the composition of function.

A **round** for our new interactive zero knowledge proof protocol is described as follows:

Step1. The prover P randomly chooses $u \in k^m$, and sends it to the verifier V;

Step2. The verifier V randomly chooses a Challenge $c \in k^n$, and computes $\overline{F}(c) \in k^m$ and then sends it to the prover P;

Step3. The prover P computes the Response $s = L_2^{-1} \circ F^{-1} \circ L_1^{-1}(u - \overline{F}(c)) \in k^n$ and sends s back to the verifier V;

Step4. The verifier V checks whether or not $\overline{F}(s) = u - \overline{F}(c)$.

If the checking is true, the verifier accepts, otherwise rejects.

The above rounds may run in several times. If all the outputs of the rounds are accepted, the verifier accepts the proof, otherwise the verifier rejects it.

3.2 Security analysis

Theorem 1 (Completeness): The above protocol satisfies the property of perfect completeness.

Theorem 2 (Soundness): The scheme satisfies the property of soundness, in a round, the success cheating probability $\delta = \max\{\eta, \frac{1}{q^{nl}}\}$, where η is the maximum success probability to attacks on the underling MPKCs used in our scheme with computational complexity at least 2^l by using some of the known attacks such as Algebraic Attacks, Linearization Equations Attack, Rank Attack, and Differential Attack etc.

Theorem 3 (Zero Knowledge): The protocol satisfies the property of Zero Knowledge.

Proof. Let us construct an equator EQ to prove the property of the complete zero knowledge of our protocol. EQ can create a proof transcript and its probability distribution is indistinguishable from those in a real proof transcript:

i Selects Response $s_i \in k^n$ at random;
ii Selects Challenge $c_i \in k^n$ at random;
iii Computes Commit $u_i = \overline{F}(s_i) + \overline{F}(c_i) \in k^m$.

Then $\{u_i, c_i, s_i\}$ is a transcript for our protocol. Here only the order of the simulation transcript $\{s_i, c_i, u_i\}$ is different from that of the real transcript $\{u, c, s\}$, so we can translate simulation transcript $\{s_i, c_i, u_i\}$

8

into $\{u_i, c_i, s_i\}$, which is a perfect simulation for the real protocol. Clearly, the elements in the transcript $\{u_i, c_i, s_i\}$ have uniform distributions which are the same as those in a real proof transcript. So our protocol is a zero knowledge proof system.

4 A NON-INTERACTIVE ZERO KNOWLEDGE SIGNATURE SCHEME BASED ON MPKCS

4.1 A knowledge signature scheme based on MPKCs

Parameter Setup: System parameters generation algorithm outputs (k, q, l, m, n, H) ,where q, l is the security parameters, $k = GF(q^l)$ is a finite field, m is the number of multivariable equations, n is the number of the variables, and $H : \{0,1\}^* \to k^n$ is a secure cryptographic hash function with properties of one-way and collision-free.

Key Generation: The public key is $PK = \overline{F}(x_1, \cdots, x_n) = L_1 \circ F \circ L_2(x_1, \cdots, x_n)$, and the private key is $SK = \{L_1, F, L_2\}$,where F is the invertible core map, and both $L_1 \in k^m$ and $L_2 \in k^n$ are invertible affine transformations.

Signature Generation: To get a signature on a message M, the signer does as follows:

1 Chooses $u_i \in k^m$ at random for $i = 1, ..., t$;
2 Computes $c = H(M \parallel PK \parallel u_1 \parallel ... \parallel u_t) \in k^n$;
3 Computes $s_i = L_2^{-1} \circ F^{-1} \circ L_1^{-1}(u_i - \overline{F}(c)) \in k^n$, for $i = 1, ..., t$;
4 The knowledge signature on message M is $\sigma = (c, s_1, ..., s_t)$.

Signature Verification: Any verifier can check knowledge signature $\sigma = (c, s_1, ..., s_t)$ on message M with equation

$$c = H(M \parallel PK \parallel \overline{F}(s_1) + \overline{F}(c) \parallel ... \parallel \overline{F}(s_t) + \overline{F}(c))$$

If the equality holds, the verifier accepts the proof; otherwise, he or she rejects it.

4.2 Security analysis

Theorem 4 (Completeness): The above knowledge signature scheme satisfies the property of completeness.

Proof. If the signature $\sigma = (c, s_1, ..., s_t)$ on a message M is computed correctly according to the Signature Generation algorithm, then it is easy to prove

$$H(M \parallel PK \parallel \overline{F}(s_1) + \overline{F}(c) \parallel ... \parallel \overline{F}(s_t) + \overline{F}(c))$$
$$= H(M \parallel PK \parallel u_1 \parallel ... \parallel u_t)$$
$$= c$$

So our signature scheme satisfies the property of completeness.

4.3 Unforgeability

Theorem 5 (unforgeability): Assume that our knowledge signature scheme is $(\tilde{t}, q_H, q_s, \varepsilon)$ -breakable in the random oracle model, that is, if there is an adversary, it can forge a valid signature with probability ε after q_H hash query , q_s signatures query, and in time $\tilde{t}$. Then there is an algorithm to solve the MQ problem in time $\tilde{t}$ with probability $\dfrac{1}{q_H + q_s}(1 - 1/q^m)\varepsilon$.

5 CONCLUSION

Multivariate public key cryptosystems (MPKCs) potentially could resist the future quantum computing attacks, and it is much more computationally efficient than number theoretic-based systems. Our zero knowledge proof protocol is an attempt to extend the field of the zero knowledge proof protocol from the traditional cryptosystem to MPKCs, and it can be used as the reliable identity authentication scheme in the quantum computing. Our zero knowledge proof protocol is very different from the existing similar protocols which are from MQ problems. The knowledge signature scheme based on MPKCs which is transformed from our zero knowledge proof protocol is in accordance with Fait Shamir heuristic, which provides a new method to obtain knowledge signature for any message, but the provable security of the MPKCs needs further study.

ACKNOWLEDGMENTS

This work is supported by the National Natural Science Foundation of China under grants 61173192, Research Foundation of Education Department of Shaanxi Province of China under grants 12JK0857. Science Plan Projects of beilin bureau of Xi'an of china under grants GX1407.

REFERENCES

[1] S. Goldwasser, S. Micali & C. Rackoff. 1989. The Knowledge Complexity of Interactive Proof-Systems, SIAM Journal on Computing, 18:186–208.
[2] Jintai Ding, Jason E. Gower & Dieter S. Schmidt. 2006. Multivariate Public Key Cryptosystems [M], University of Cincinnati USA: Springer.
[3] Vivien Dubois, Pierre-Alain Fouque1, Adi Shamir & Jacques Stern. 2007. Practical Cryptanalysis of SFLASH, Advances in Cryptology-CRYPTO 2007, LNCS 4622. Berlin: Springer-Verlag, pp. 1–12.
[4] Singh, R.P., Saikia, A. & Sarma, B.K. 2010. Little Dragon Two: An Efficient Multivariate Public Key Cryptosystem, International Journal of Network Security and Its Applications (IJNSA), 2010, 2, 1–10.

[5] Fiat, A., Shamir, A. 1986. How to Prove Yourself: Practical Solutions to Identification and Signature Problems, In: Odlyzko, A.M. (ed.) CRYPTO 1986, LNCS263, Berlin: Springer-Verlag, , pp. 186–194.

[6] Feige, U., Fiat, A., & Shamir, A., Zero-Knowledge Proofs of Identity. J. Cryptology , 1998, 1(2), 77–94.

[7] Bellare, M., Namprempre, C., & Neven, G.2009. Security Proofs for Identity-Based Identification and Signature Schemes, J. Cryptology , 22(1), 1–61.

[8] Jens Groth, Short Non-interactive Zero-Knowledge Proofs, In ASIACRYPT2010, LNCS 6477, Berlin: Springer-Verlag, 2010, pp. 341–358.

[9] N. Courtois.2001. Efficient Zero-Knowledge Authentication Based on a Linear Algebra Problem MinRank, In Proceedings of ASIACRYPT2001, LNCS 2248, Berlin: Springer-Verlag, , pp. 402–421.

[10] Koichi Sakumuto, Taizo Shirai & Harunaga Hiwatari, 2011. Public-Key Identification Schemes based on Multivariate Quadratic Polynomials, Advances in Cryptology – CRYPTO 2011, LNCS 6841, Berlin: Springer-Verlag, pp. 706–723.

[11] Abdalla, M., An, J.H., Bellare, M., & Namprempre, C., From Identification to Signatures via the Fiat-Shamir Transform: Minimizing Assumptions for Security and Forward-Security, In: Knudsen, L. (ed.) EUROCRYPT 2002, LNCS 2332. Berlin: Springer-Verlag, 2002, pp. 418–433.

[12] Sakumoto, K. 2012. Public-Key Identification Schemes Based on Multivariate Cubic Polynomials. In: Fischlin, M., Buchmann, J., Manulis, M. (eds.) PKC 2012. LNCS, vol. 7293, pp. 172–189.

[13] Yang B Y, & Chen J M.2004. TTS: Rank Attacks in Tame-like Multivariate PKCs. [EB/OL]. http://eprint.iacr.org/2004/061.

[14] Nachef. V, Patarin.J & Volte. E. 2012. Zero-Knowledge for Multivariate Polynomials,Progress in Cryptology – LATINCRYPT 2012 ,Lecture Notes in Computer Science Volume 7533,pp 194–213. [EB/OL] http://eprint.iacr.org/2012/239.pdf

Protect the semantic network identity: Attribute-based searchable encryption over routing in named data network

X.H. Jiang & Q.M. Huang

School of Computer & Communication Engineering, University of Science and Technology Beijing

ABSTRACT: Named Data Network (NDN) is a future Internet architecture. It uses semantic network identification. Meanwhile NDN network router caches content's identification which the user requested, so the NDN network has risks in privacy and confidentiality. In this paper, we made use of attribute-based encryption technology and searchable encryption technology to perform routing at full cipher text state, and designed a routing encryption scheme, which allowed a router to match and retrieve Content Store (CS), Forwarding Information Base (FIB), and Pending Interest Tables (PIT) in the cipher text state. The contribution of this work is to dynamically encrypt semantic network identification in routing system of NDN. We built a simulation experimental platform and tested our scheme. Security and performance analysis proved that this program can effectively improve NDN network privacy.

1 INTRODUCTION

With the expanding demand, TCP/IP system has so many problems to deal with. The problems mainly lie in three points. Firstly, network traffic surges faster than the speed of the router performance, and the main source of its traffic is data-intensive computing and multimedia file distribution. Secondly, there isn't a systemic solution for security in current network, and it is basically in a passive response state, such as the expansion of network protocols, encryption, and authentication, and the increase in safety equipment. These measures will reduce the communication efficiency and cause difficult comprehensive coordination. End to end communication model is designed to only provide a secure channel data, but it can't achieve personalized service and security services. Thirdly, internet's terminal morphology has changed a lot, with a significant increase in dynamic and a growing proportion in mobile terminal traffic. Internet of things rapidly increases in the number of low smart terminals. IP address which represents the identity and characterization address, has no strong mobility support. For the above-mentioned issues, the new design of the Internet architecture is able to solve the problem fundamentally. In 2010, the United States established the NSF Future Internet Architecture (FIA) program. FIA has funded four projects which were committed to the future network structures in different researches and design directions, NDN, Mobility First, Nebula and XIA. NDN network's model is content-oriented rather than traditionally host-oriented. It is a content-centric network. NDN uses the contents of keyword to locate the contents.

Although the NDN is seen as the current network problem that can be solved in the future network architectures, yet it also faces many challenges. Because NDN routers in the network cache data to the CS table, an attacker could break through to view router cached content. In order to improve the security of the cached data, content providers need to encrypt data before releasing it, which can keep the contents invisible for the storage system and protect the privacy.

In this paper, we propose a new scheme to solve the NDN network content protection and privacy problems, by combining the attribute-based encryption and searchable encryption techniques to enhance the NDN network privacy during transmission in the untrusted network environment. In the second section, we introduce the related research on NDN network security. In the third section, we focus on NDN security route design. In the fourth section, we conduct security and system analysis.

2 RELATED WORKS

Abdelberi Chaabane, Emiliano De Cristofaro [1] systematically analyzed the privacy problems in CON network, pointing out four main features of CON network as follow: Content identified by name, Content-based routing, Multipath content distribution, and Network Storage. According to these characteristics, the paper pointed out privacy security threats in four aspects, including the cache privacy, content privacy, identity privacy, and private signature. The article first systematically analyzed privacy issues, and then discussed the different attacks, and finally made a detailed description of their impact on users' privacy. Authors proposed counter-measures and tried to find a balance between privacy and performance. Gergely ACS and Mauro Conti [2] suggested that users and content providers needed to make an agreement in which contents are in the scope of protection of privacy, and proposed several techniques to weigh the relationship between privacy and network latency. Meanwhile, the paper presented a formal model to quantify the degree of privacy protection offered by a variety of caching algorithms. Emmanuel A. Massawe [3] proposed the NDN network scalable routing privacy protection protocol (SP-NDN), by the use of multiple Bloom filters to improve the transmission of user data packets privacy interest. Compared with the existing program, the authors proposed a content-related secret key tree based on multicast secret key management protocol, integrated with Bloom filters and multicast secret key to reduce the leakage of user keywords to hinder eavesdroppers to guess users' keywords. Authors try their best to ensure the security of user's interest package, and seek to reduce the extra load of additional information as much as possible during query process. Nonhlanhla Ntuli [4] proposed a method for detecting snoop cache. Model of the system limits users to establish a direct connection to the same router cache, through a combination of formal signature and trust systems to improve the accuracy of detection.

3 SECURE NDN ROUTING WITH ATTRIBUTE-BASED SEARCHABLE ENCRYPTION

3.1 *NDN routing process*

NDN network data structure is divided into two kinds. One is interested package, which is used to route the query. Another is data packet, a data entity. The node model includes three main modules: the forwarding information tables FIB, content repository CS, and pending interested table (PIT). The general NDN routing process is that users' query will generate interested packets. Routers can record the request from the source port and search the local records of FIB. If the router cannot find the record of the request in FIB, it will forward the request to the next router, until the data is found. Each router on the path will keep interested and data packages into PIT and CS for a period of time. Only the first request can access to data source directly, and following request will search the same content request in PIT. If the same content registration records are found, router gets data from the corresponding CS and removes records with same PIT entrance.

3.2 *Network model*

The network model can be simplified as a content provider, user and an untrusted router, which has a certain amount of memory and calculation ability. In this model, the content service provider firstly applied the attribute to generate the corresponding access control tree with different contents, which is used to encrypt content index. Then the contents and index will be published to the network. When user sends a search request, the attribute of users' application and threshold generated by users' searching network identification (keyword group) will be attached. A router stores the encrypted index and the cache content published by network service providers. Router receives user's searching request packet, which contains user's attributes. After the algorithm is processed, the encrypted content does not affect the process of retrieval requests in the router, which does not damage the router and protect privacy.

3.3 *Secure NDN routing scheme*

Denote p, q as prime numbers, and $\mathbb{G}_1, \mathbb{G}_2$ are the multiplication of cyclic group of p, q. Then $e: \mathbb{G}_1 \times \mathbb{G}_1 \rightarrow \mathbb{G}_2$ is a bilinear pairings. If mapping e satisfies the following property [5]: (1) Bilinear (2) Computable.

In $\mathbb{G}_1$, as the security assumptions of cryptography, there are several cryptography difficulties.

Definition 1. Calculate Diffie-Hellman problem (CDH).

Given (g, g^a, g^b), in where $a, b \in \mathbb{Z}_p$, g is the generating element of $\mathbb{G}_1$, calculating g^{ab}.

Definition 2. Judge bilinear Diffie-Hellman problem(DBDH):

Given $(g^a, g^b, g^c, e(g, g)^z)$, where $a, b, c, z \in \mathbb{Z}_p$, judging $e(g, g)^z = e(g, g)^{abc}$.

For DBDH problem, there is no efficient algorithm, so generally considered DBDH is a difficult problem. Based on the above theory, the program can be summarized as the following five basic algorithms: (1) System Initialization. Input system safety factor and attribute set, generating system parameters and the main public secret key. (2) Encryption Index. Input public system parameters, access control tree and keywords to generate the encrypted index for this keyword. (3) Request parameter generation. Input common parameters, user attributes and random safety factor to generate a user access request parameter. (4) GenTrapdoor. Input keywords and user request parameters to generate the threshold. (5) Retrieval. Input the encryption index, access control tree, the user request parameters and thresholds, then output query results by calculation.

3.4 *Construction of secure NDN routing algorithm*

1 System initialization:

It defines a bilinear group $\mathbb{G}$ of prime order ρ with a generator g, and ρ is a prime number. Define a bilinear map $e: \mathbb{G} \times \mathbb{G} \rightarrow \mathbb{G}_1$, which has the properties of bilinearity, non-degeneracy and computability. Content providers use system security parameter λ and properties collection $\mathbb{N} = \{n_1, n_2, n_3, n_4 \cdots\}$ to generate public parameters $\text{Pub} = \langle O, P, Q, A, N_1, N_2, N_3 \cdots N_i \rangle$ and master secret key $\text{Mk} = \langle a, b, c, \lambda, n_1, n_2, n_3 \cdots n_i \rangle$, where $O = g^a$, $P = g^b$, $Q = g^c$, $A = e(g, g)^\lambda$, $N_i = g^{H(n_i)}$, $a, b, c \in \mathbb{Z}_p$. Then the content provider releases Pub and retains Mk. Then the content provider generates access control tree Γ_W for content, $W = \{\omega_1, \omega_2, \omega_3 \cdots \omega_i\}$, and ω_i represents each stage of the network identification (equivalent to the key) of the content, $\Gamma_W = \{\Gamma_{\omega_1}, \Gamma_{\omega_2}, \Gamma_{\omega_3} \cdots \Gamma_{\omega_i}\}$, and Γ_{ω_i} represents the keyword ω_i access control tree, while $\Gamma_{\omega_i}(N) = 1$ represents an attribute set N that meets the keyword ω_i access control tree. Let $\text{attr}(k)$ represent an attribute associated with a node k, and the calculation method of $\Gamma_{\omega_i}(N)$ is that [6] when k is a root node, all its child nodes circularly calculate $\Gamma_{\omega_i}(N)$ value. When k is a leaf node, if and only if $\text{attr}(k) \in N$ that $\Gamma_{\omega_i}(N)$ returns 1, finally through the NAND gate only when meeting the access control tree $\Gamma_W = 1$. When a user first requests content providers, content providers make a judgment according to the user's attribute access control tree. If it meets the tree, then $\lambda = \text{abc}$. If not, λ is random in the collection.

2 Encryption index:

Before the data are released, content providers need to generate encryption index for this content η.

Set $\eta^\wedge = g^{ab}, \eta^{\check{}} = c, \eta_i = N_i^{\frac{a\Gamma_W}{\Gamma_{\omega_i}(N)H(\omega_i)}}$ $(i \in n)$, $\omega_i \in W$ is the key word of content. N' is an attribute set which user should have to access this content. The encryption index generation $\eta = \langle \Gamma_W, \eta^\wedge, \eta^{\check{}}, \{\eta_i\}_{i \in n} \rangle$.

3 Request parameter generation:

The user can obtain property from the authoritative third party and use this and the Pub to generate a request parameter RP for himself. The attribute X which users have is a subset of N. For each attribute, the user selects the random parameters of r_i from the $\mathbb{Z}_p$, $r = \sum_{i=1}^n r_i$, and $M^\wedge = Q^{H(X)}, M_i = P^{\frac{r_i H(X)}{H(n_i)}}, M_f = e(O, P)^{H(X)}, M^{\check{}} = A$, then $\text{RP} = \langle M^{\check{}}, M_f, M^\wedge, \{M_i\}_{i \in n} \rangle$.

4 Gen Trapdoor:

In this process, the user will generate a trapdoor ϕ for the searching content identification ω_i, which is actually the combination of content keyword, using the personal request parameters RP. Set $\phi^\wedge = M^\wedge$, $\phi^{\check{}} = M^{\check{}H(X)}$, $\phi_i = M_i^{H(\omega_i)}$, $\phi_f = M_f^r$, and the Trapdoor is $\phi = \langle \phi^\wedge, \phi_f, \phi^{\check{}}, \{\phi_i\}_{i \in n} \rangle$.

5 Retrieval:

Router stores contents which are published by content providers, encryption index η and access control tree Γ. When the router receives a request from a user, the router will search whether it has the request content in its encryption index η in the cache, according to user's ϕ (verification $w^{\check{}} = w$), and the router needs to verify as follows [7].

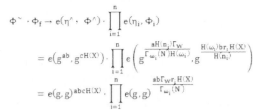

When the user's property set X meets the access control tree Γ_W which is set by the content provider, $\Gamma_W = 1 \cdot \Gamma_{\omega_i}(N^{\check{}}) = 1$, then the formula can be transformed into:

$$e(g, g)^{abcH(X)} \cdot e(g, g)^{abrH(X)}$$
$$= A^{H(X)} \cdot e(O, P)^{rH(X)} = M^{\check{}H(X)} \cdot M_f^r = \phi^{\check{}} \cdot \phi_f.$$

Thus, if the user's property meets the access control tree which is set by content provider, the router returns the correct results to the user.

4 SECURITY AND PERFORMANCE ANALYSIS

First, the program encrypts the content index to realize that the content is opaque for the storage device; NDN network router uses PIT, CS, and FIB t tables to record user access requests and cache data. Because of semantic network identification, the attacker can pry these records in the table to access user's privacy. The program can effectively solve this problem. First, the user sends an access request which is generated and based on users' attributes and keywords of the content to access. In the Router, PIT, CS, and FIB tables store encrypted cipher text data and cipher text index, so even if an attacker compromises the router, because the data recorded is opaque for the router, the attacker cannot snoop into useful information. This solves a major NDN security risk.

Secondly, when generating keyword threshold, the user can use the random parameters to confuse the threshold. The user can use the same keyword but different random numbers in the process, so that even if the same keywords are in different generation process, it can get different threshold. An attacker cannot get a fixed threshold by exploratory polling a keyword. Finally, since the program makes access control tree structure that is closely integrated into the encryption algorithm indexing and retrieval algorithms process, the user who does not meet the access control tree cannot get the correct search results. The design of offensive and defensive game is as follows.

1) Attacker initialization: Assume the attacker owned property set for $\mathcal{B} = \{b_1, b_2, b_3, b_4 \cdots\}$. Because the attacker is an illegal user, its property set $\mathcal{B} \notin N$. However, $\mathcal{B}$ may have the same elements in N, and the attacker may announce that he wanted to challenge attribute set $\psi(\psi \in \mathcal{B})$ and inform the challenger.

2) System initialization: Challenger runs system initialization algorithm, and public arguments Pub is sent to the attacker;

3) Stage 1: An attacker can select a set of attributes $\psi \in \mathcal{B}$ and sensitive keywords $\omega_i{'}$ to query request parameters and use the threshold generation algorithm to generate the appropriate parameters;

4) Challenge: the attacker will generate good threshold and access request parameters which are sent to the router, and the router is calculated with a search algorithm;

5) Stage 2: This phase is the first phase of operation which is repeated. An attacker could use more properties set to challenge the challenger, where $\psi \in \mathcal{B}$;

6) Guess: the attacker outputs the guess for $\omega_i{'}$ guess $\omega_i{'} \in \{0,1\}$. If $\omega_i{'} = \omega_i$, attacker wins. The advantage of the attack is defined as $\mathrm{Adv} = \left| \Pr[\omega_i{'} = \omega_i] - 1/2 \right|$.

Accordingly search algorithm shows that attacker obtains the public parameter Pub which contains g^a and g^b. The encryption index is generated by the content provider, $\eta^\wedge = g^{ab}$. According to the CDH problem, the attacker can't infer the value of $\eta^\wedge$ in polynomial time. So $\mathrm{Adv} = \left| \Pr[\eta^{\wedge'} = \eta^\wedge] - 1/2 \right| < \varepsilon$. At the same time, according to the DBDH hard problem, although the attacker gets O, P, Q, A, he cannot infer whether $\lambda = abc$. When the property attribute sets do not match with content providers' attacker provisions exactly, $\Gamma_{\omega_i} \neq 1$, matching algorithm on both sides are not equal, and so an attacker is failed.

We used the JPBC library to build the simulation experimental platform. In the performance analysis of this part, we first defined some parameters. Set E is the group exponentiation in $\mathbb{G}$, and E_1 is the group exponentiation in $\mathbb{G}_1$. M is the multiplication of $\mathbb{G}$, and P is the generation on the operation of bilinear mapping. In the performance analysis, we mainly considered the above three kinds of operations. The computation complexity of each part of the system is shown in the following table.

Table 1. Computation complexity.

Operation	Computation complexity
System initialization	$P+(n+3)E+E1$
Encryption index	$(n+1)E$
Req. param generation	$(n+1)E+E1$
Gen Trapdoor	$3E$
Retrieval	$(2+n)E1+(1+n)M+E$

The picture below shows the performance of our design in operation efficiency, noting that System initialization and Encryption index are run by content providers. Meanwhile, Request parameter generation and Gen Trapdoor are run by users. Only Retrieval is run in the router, so our scheme doesn't bring too much burden on the router.

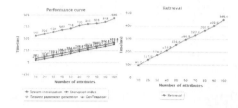

Figure 1. Performance curve.

5 SUMMARY

This paper analyzed the routing process NDN network and its privacy problems. Then a solution was proposed. The program uses access structure with searchable encryption technology and attribute-based encryption technology, effectively solving the problems of NDN network routing information leakage and preventing malicious attackers from snooping router routing table to obtain users' sensitive information. While protecting users' privacy, the program maintains the inherent advantages of semantic-based NDN network and the integrity of NDN routing functions. This program enhances NDN network security in untrusted distributed storage system.

ACKNOWLEDGMENT

This work is supported by the Funding Project for Beijing Excellent Talents Training (No. 2013D009006000002).

REFERENCES

[1] Chaabane A, et al. Privacy in content-oriented networking: Threats and countermeasures[J]. ACM SIGCOMM Computer Comm. Review, 2013.

[2] Acs G, et al. Cache privacy in named-data networking[C].Distributed Computing Systems, IEEE 33rd International Conference on. 2013.

[3] Massawe E A, et al.. A Scalable and Privacy-Preserving Named Data Networking Architecture Based on Bloom Filters[C].Distributed Computing Systems Workshops, IEEE 33rd Inter. Conf. 2013.

[4] Ntuli N, Han S. Detecting router cache snooping in Named Data Networking[C].ICT Convergence (ICTC), 2012 Inter. Conf. on. IEEE, 2012.

[5] Sun W, Yu S, Lou W, et al. Protecting Your Right: Attribute-based Keyword Search with Fine-grained Owner-enforced Search Authorization in the Cloud[C]// IEEE INFOCOM. 2014.

[6] Beimel A. Secure schemes for secret sharing and key distribution[D]. Technion-Israel Institute of technology, Faculty of computer science, 1996.

[7] Boyen X, Waters B. Anonymous hierarchical identity-based encryption (without random oracles)[M]// Advances in Cryptology-CRYPTO 2006.

Network Security and Communication Engineering – Chan (Ed.)
© *2015 Taylor & Francis Group, London, ISBN: 978-1-138-02821-0*

Choosing a method for generating one-time passwords and an information transport technology in the authentication system for ACS

A.Y. Iskhakov, R.V. Meshcheryakov & I.A. Hodashinsky
Tomsk State University of Control Systems and Radioelectronics, Tomsk, Russia

ABSTRACT: Applicability of the QR code and the Near-Field Communication (NFC) technology for transport of authentication data are investigated. The investigation is performed in the framework of developing a two-factor authentication system using a software token stored in a mobile communication device.

KEYWORDS: authentication, identification, OTP, QR code, NFC, tag.

1 INTRODUCTION

In modern access control systems (ACSs), the procedure of authentication when entering a building is generally implemented by means of electronic security checkpoints in the form of turnstiles, threepods, or gates with built-in controllers and proximity card readers (Shelupanov et al. 2009). This allows one to control the access to the building in the most convenient way as well as to log the staff working time.

Proximity cards are widely used for personal authentication. It is a classical example of the second-type authentication, inheriting all drawbacks of one-factor authentication systems.

In 2012, the International Telecommunication Union (UTI) published the statistics concerning the world use of mobile devices. It was reported that mobile communication services were enjoyed by approximately six out of seven million people. Today, the mobile phone has become so ingrained in our everyday lives that we cannot possibly imagine how we could do without it. Therefore, we propose an authentication mechanism that uses the mobile device (smartphone) as a carrier of a personal identifier (user ID).

2 DESIGN OF THE SYSTEM

2.1 *Requirements for the system*

One of the problems with much used proximity cards is that their identifier can easily be compromised. Today, static passwords are no longer considered to provide a reliable protection against intruders. In addition to the static password, authentication systems generally use the one-time password (OTP), which is changed every time when logging into the system and is valid for a limited time (Smith, 2002).

Modern smartphones make it possible to use programmed OTP generators, which allows one to protect the authentication system against compromise of the static identifier (Meshcheryakov, 2007).

As the second authentication factor, we use the protection against unauthorized access to an authenticator application (in the case of the loss/theft of mobile device).

When developing the authentication system for electronic security checkpoints, it is important to take into account the time of authentication. It must be as close as possible to the time of passing through the turnstile with proximity cards. Hence, to transport authentication data, we should choose the technology that is maximally effective in terms of connection time. As an example, we can point to NFC tags and quick response (QR) codes.

These technologies require no transfer of authentication data via public communication channels (Wi-Fi or 3G), which increases the system security and reduces the requirements for the infrastructure of an object being deployed.

2.2 *Authentication scheme*

Using the mobile device, the access subject generates the user ID and the one-time password. This information is read in the user terminal and is then transferred to the authentication system. It is suggested that the authentication system is implemented as a module for modern ACSs. Having processed the information received, the ACS performs a control action (lets the subject through or rejects the request).

As compared to hardware implementations, the proposed method has the following advantages:

1 there is no need for the organization to provide the user with smartcards or expensive hardware tokens;

2 the user does not have to carry around an extra object (token or smartcard);
3 mobile devices are used by the overwhelming majority of people;
4 the frequent use of the smartphone enhances the likelihood that the user will quickly detect a theft (or loss) of the token.

3 TECHNOLOGY OF ONE-TIME PASSWORDS

Today, several OTP authentication methods are used in practice:

- request—response;
- response only;
- time synchronization; and
- event synchronization (Sabanov, 2012).

The request–response method operating in the asynchronous mode implies the bidirectional data communication between the client and the authentication server. Since this imposes heavy restrictions on technical parameters of the smartphone and increases the authentication time, this method is used only in the active communication mode of the NFC technology.

A potential problem with the other methods operating in the synchronous mode is the asynchronous behavior of the OTP token and the server.

Even though a certain time margin rather than a discrete value is used for the authentication procedure in the time-synchronization mode, the clock of the programmed OTP generator can run faster or slower. This can be due to either the intentional change of the device time or the reset because of battery discharge or software failure. As a rule, the users rarely change the time parameters, especially as modern communication devices support the synchronization with NTP servers and, if required, the change of time zones is performed automatically. In the case of full battery discharge, the internal clock of the mobile device is usually fed by a back-up battery.

In the response-only and event-synchronization modes, the authentication failure may result in the "backlog" of the server from the authentication device. As an example, we can consider a situation when the user ran the authenticator application, but for some reasons, didn't use it and shut it down. To handle such situations, alternative OTPs can be generated for several events at a time.

Thus, to implement the programmed OTP generator, we use the time-synchronization mode with the time interval of 10 s. This method is chosen upon analyzing the frequency of possible desynchronizations occurring in actual practice.

4 COMMUNICATION BETWEEN CLIENT DEVICES AND AUTHENTICATION SERVER

4.1 QR code technology

The transfer of authentication data between the mobile device and the reader can be implemented using the quick response (QR) code. In this case, the authentication system is a hardware and software complex consisting of the PC, reader in the form of an IP camera, and software (ACS integration modules) (see Fig. 1).

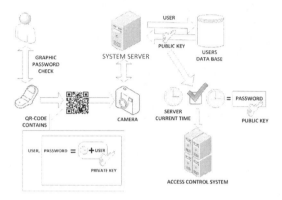

Figure 1. Authentication scheme based on the QR-code.

The QR code is a two-dimensional barcode invented by Denso Wave in 1994.

The QR code has the following advantages over other barcodes:

- it can contain a large amount of digital and text information in any language;
- its print size can be rather small, which places no restrictions on the size of the smartphone display;
- it is quickly recognized by scanners;
- it can be read in any direction (360° of scanning).

According to the ISO/IEC 18004 specification, the QR code has the following limitations on amount of data:

- digits: 7089;
- digits and letters (including the Cyrillic alphabet): 4296;
- binary code: 2953 byte;
- hieroglyphic symbols: 1817.

Thus, the QR code can encrypt more than 2 Kb of text information, which is the acceptable key length for modern cryptographic algorithms.

4.2 NFC technology

Undoubtedly, the near-field communication (NFC) technology is now one of the most popular solutions in the field of wireless communication.

This technology extends the proximity card standard (ISO 14443) to combine the smartcard and the reader into a single device. The NFC-enabled device supports smartcards and readers of the ISO 14443 standard and it can communicate with other NFC-enabled devices; therefore, such a device is compatible with the present infrastructure of proximity cards used in public transport and payment systems. The NFC technology is developed to be used basically in mobile phones.

The NFC is based on inductive coupling; the frequency is 13.56 MHz and the transfer rate is 106 Kb/s (can be increased up to 212 and 424 Kb/s). The signal is subjected to the on-off keying with a depth of 10 or 100% and to the binary phase-shift keying.

When estimating the applicability of this technology for transfer of authentication data, we can point to its following important advantage. In contrast to barcode-based systems and public data transfer channels (Wi-Fi or Bluetooth), the NFC-enabled authentication system has a very short connection time. Instead of execution of conditioning instructions to identify the device, the connection between two NFC-enabled devices is established immediately (in less than one tenth of a second). Moreover, the NFC can function when the device has no power supply (for example, the turned-off phone will allow the user to pass through the electronic turnstile).

When using the NFC technology in the active communication mode, the request–response method of OTP generation is preferable (see Fig. 2).

In this mode, the random question is used by the server for each user request. It is this question that is encrypted by the user and can then be used to verify the authenticity of the access subject.

This scheme allows one to deal with the desynchronization problem; however, in practice, it requires devices with active NFC modules.

5 CONCLUSIONS

Considering the increase in the number of NFC-enabled smartphones (more than 50 models are currently available in the Russian market), we can expect that, within a few years, this technology will become a good alternative to the bank card for paying various services using the mobile phone.

The limited support of NFC modules in the present model range of mobile devices complicates the use of this technology as a single solution to transport of authentication data.

To reach for universality and increase the number of potential consumers, we implemented both the modes of data transfer described above. This allowed us to minimize the limitation on use of mobile devices.

REFERENCES

[1] Shelupanov, A.A., Grusdeva, S.L. & Nachaeva, Yu.S. 2009. *Authentication. Theory and practice of providing access to information resources.* Moscow: Goryachaya liniya - Telekom.
[2] Smith, R.E. 2002. Authentication: from passwords to public keys. Moscow: Vil'yams.
[3] Meshcheryakov, R.V., Davidova E.M. & Savchuk M.V. 2007. Enhanced user authentication methods, Bezopasnost Informatsionnykh Tekhnology 4: 60–68.
[4] Sabanov, A.G. 2012. Basic authentication process. Information protection issues 3: 54–57.

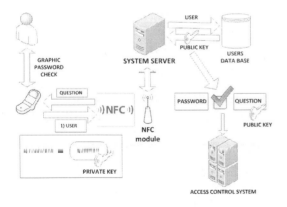

Figure 2. NFC-based authentication scheme.

Network Security and Communication Engineering – Chan (Ed.)
© 2015 Taylor & Francis Group, London, ISBN: 978-1-138-02821-0

Research and design of network security coordinated recovery system

J. Xia, L.T. Jiao, S.S. Zhao & L.T. Li

Department of Automatic Control, Northwestern Polytechnical University, Xi'an, Shanxi, China

ABSTRACT: The article is about subsystem of the network cooperative security system, which contains high-performance, low cost multiple backup disaster recovery systems in different places which can work together with other network security devices. The main design of the system includes monitoring and auditing of the system files, the code of network backup recovery, the backup and recovery for data, the technology design of data synchronization, the management of members and the maintenance, etc. And the experiment proves that the system is effective.

1 INTRODUCTION

It is well known that information and data are regarded as valuable resources. The importance of the information and data is self-evident. For the enterprises in information Age, sound information and data are necessary to make the companies work well, while how to protect the data resource is a big problem. So an effective remote disaster and accident rapid recovery system is very important, which means that even if there is a sudden accident, the data can also be saved completely. Though it is unable to control or foresee a disaster, when a disaster comes, the appropriate preparation can protect the important data and deal with all kinds of possible disasters [1-5]. Therefore, the study of a new disaster recovery system to reduce business investment in disaster areas and further improve the safety and reliability of the data are of great significance and have practically promotional value.

2 SYSTEM ARCHITECTURE DESIGN

2.1 System general design

Network Cooperative Security System (NCSS) is wholly composed of multiple subsystems, and the design and implementation of the whole System are based on the Network dynamic defense model PDRR (Protect / Detect / React / Restore).

Cooperative Disaster Recovery System is a subsystem of Network Cooperative Security System, which belongs to the last link of PDRR. The relationship between the various modules in the overall frame of the network security cooperative defense system is shown in Figure 1.

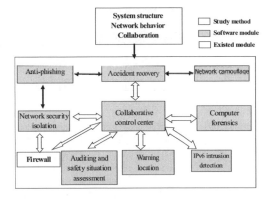

Figure 1. The overall frame of the network security cooperative defense system.

2.2 The architecture of the accident recovery system

Accident automatic recovery system is a subsystem of Network Cooperative Security System, which consists of file information module, monitoring module and control module. It can receive messages from the audit system and coordinate with each other through the communication agents and other hosts in backup alliance, and its structure is shown in Figure 2.

Monitoring module monitors the critical list files by FAM, and if there is any change (modify, delete or add, etc) in the key file or folder, it sends the event message to the control module immediately. File information module includes key files, copy file and file list. The key document library is detected by the monitoring module, and copy file library stores the backup of the key file from other nodes. The core of the control module is to restore the control center, which takes responsibility for making the processing of the various control messages, renewal and maintenance of each list, the digital signature for files and the identity of the host authentication, and auditing system linkage and the accident recovery system between every host to work well together.

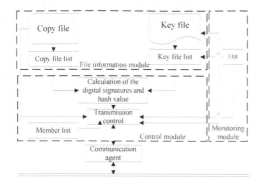

Figure 2. The architecture of the accident recovery system.

The communication agents have the function to control the communication of the file backup recovery and the inspection and maintenance for copy file between each accident recovery system. The inspection and maintenance of multiple host work together to copy their respective key file respectively, and form an accident automatic recovery system alliance. In the alliance each node position is equal, and there is no copy file server, so a single point with failure could not have influence on the whole system. The establishment of the league is based on the identity authentication which is not the connection mode, where the node has no regional restrictions and can exist in a LAN; the node could also be in different ground, which is shown in Figure 3.

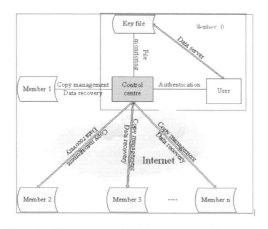

Figure 3. The structure of accident recovery alliance.

2.3 Accident recovery process

In the process of system, the accident recovery system installed on a host puts the key file needed to protect into the real-time monitoring and builds a backup in the alliance. Once the destruction or illegal tampering for the key data or file is found, the accident recovery system can use various means for recovery. The whole process can be represented by the flow chart in Figure 4.

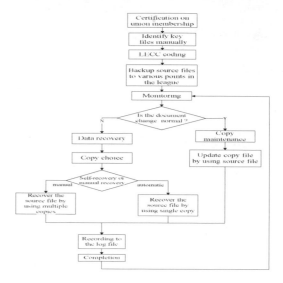

Figure 4. The workflow of the accident rapid recovery system.

3 RESEARCH AND DESIGN ON EACH SUBSYSTEM

The design of subsystems includes the design for the file monitoring and auditing system, code design on network backup recovery, data backup and recovery system, data synchronization technology and the management on members and the maintenance on the replica, etc.

3.1 File monitoring and auditing system

Accident recovery system must first put the key files into the real-time monitoring system, once the monitoring system finds any change, it will make a notice to the audit system by the Agent, then request the audit system to audit the document, and at last decide the next operation according to the results of the auditing. That is to say, the monitoring and the auditing are the premise of the accident recovery system, which make the detection and the judgment for the accident.

Auditing system has the full name: event collaborative audit analysis system, which is the subsystem of the network collaborative security system. It mainly includes the audit of network data and the host data as well as the joint audit for the two. Through real-time recording the information on the network camouflage, IDS and Firewall, the security status of the current system can be grasped accurately; through mutual interaction and collaboratively work with other subsystems, the security incidents of the system could make immediate response [6-8]. The overall structure of the collaborative auditing system is shown in Figure 5.

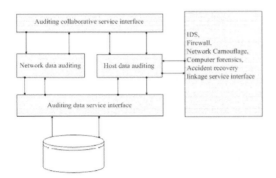

Figure 5. The structure of event collaborative auditing system.

Network collaborative security system realizes the multi-system collaborative auditing mechanism successfully. The audit not only has the effect of simple logging, but also has become an important part of the network security system.

Collaborative auditing system could make auditing for the network and host data. Emergency recovery system is concerned with the important documents. So the relevant auditing with accident recovery system is mainly the security event auditing in the host data audit. The content of the auditing system is shown in Figure 6.

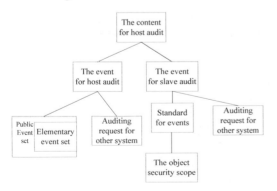

Figure 6. The content of the host audit.

3.2 *The coding technology for network backup recovery*

Due to the unreliability of storage devices when faced with devastation, the long-distance backup is often used to solve the problem of disaster. In the face of a lot of information file, the unreliability of Internet transmission brings a series of problems. Traditional TCP/IP deals with the response retransmission mechanism on the packet losses with time-consuming method and has low reliability, which is hard to guarantee the integrity and availability of the data transmission when recovering the data.

To solve the above problems, the Linear Error Correction Code method is used to solve the data transmission error control problems of emergency recovery system. Each coding block forms a random subset of the original information, and each original packet is distributed to some coded packets. As long as the receiving terminal receives enough (less than total number coding block) coding block, the original data could be got by decoding without acknowledging which package is lost and which package can be received. The receiver does not need to reply packets, and this solves the problems such as packet loss or damage and improves the efficiency and stability of the transmission.

3.3 *Data backup and recovery*

In order to improve the reliability of the backup data, accident recovery system uses the method of long distance multiple backup copies. In current backup transmission method, all copy nodes get data from the original data nodes and use unicast method to transmit, the time required is the sum of all points of transmission time whose efficiency is very low. For vast amounts of data, it takes hours or several days to transmit multiple nodes, which is unacceptable for the users. If the multicast mode [9-11] is used to make backup transmission (the source node transmits data to the copy nodes in parallel), the transmission time is increasingly reduced and the transmission efficiency is greatly improved, while the upstream bandwidth will become the bottleneck of the parallel transmission, which will affect the normal service for the original data nodes and even stop it.

Multi-source fast backup and restore transmission algorithm make full use of the feedback channel bandwidth at various points in alliance and save when we use the linear error correction coding. In backup process, the source data nodes send different encoding packages to each copy node of the source data in parallel; when each copy node receives the code package of the source node, they send different packages to each other, which greatly reduces the workload for the data node and improves the efficiency for backup transmission.

3.4 *Data synchronization*

The data synchronization technology is one of the key technologies of accident recovery system and is the foundation of the fast recovery, fast backup and copy maintenance. Remote synchronization technology is used to share the modified data between the source and the file copy, which can form a new copy when the modified data is sent to each node from the key files, namely fast backup and copy maintenance; it can also use a copy file to synchronize a key file with an unauthorized tampering.

This paper uses the algorithm of the data remote synchronization. The algorithm not only has realized the efficient copy maintenance and real-time recovery in data recovery system but also can be used for differential backup. Accident recovery system has the function of multilevel data backup and the data need to have the ability to restore some point in time ahead. The incremental and differential backup is based on file-level data backup. When there are lots of files with small change or huge amounts of data, the quantity of data needed to transfer is very large. So there are problems in backup and storage efficiency.

3.5 *Management for members and maintenance for replicas*

In order to restore the damaged data quickly and efficiently, the copy data must be saved to multiple nodes in the disaster alliance. Multiple copies store in multiple nodes, and can improve the reliability of the data. But the increase of copies could bring up new problems; for the problem a new plan with effective management for copy, keeping reliability and balance of performance are put forward.

1 Maintenance for replicas: The maintenance for backup alliance includes the maintenance of alliance member list and replica consistency. The member node in union makes "heartbeat" test for other member node by sending encrypted messages regularly and maintains its alliance member list. The detection process between nodes is mutual, and the condition of network load between member nodes can be gotten by detecting. At the same time, the original node calculates the hash function value of key files and then sends it to the corresponding copy file nodes; if the monitoring module does not find changes for the original file, once the hash values of copy file is not the same as the value of the original file, the corresponding measures must be taken to maintain the copy file immediately.

2 Choice of replicas: When the key documents of a node are destroyed, they need to choose one or more backup members to restore it. The monitoring module finds the abnormal change first, then informs the control module; the control module makes a choice for the reachable backup node, then compares the hash value

of each node and chooses a credible and efficient node using the majority rule. If the hash value of the selected replica has no difference in the value of original file, the original file could be recovered; otherwise, it is considered that the backup alliance is destroyed largely, and then the administrator is requested to intervene.

4 EXPERIMENT AND SIMULATION

We apply the experiment to the campus network, and decorate the whole system in this lab, then protect the network and the host in the lab.

The network topology (Figure 7) in lab is similar to the topology when making simulation; and the server also installs RedHat 9.0, on which there is the configuration of the Apache, PIMP and MySQL, which can offer Web server for the whole education network; and also has the configuration of the Wu-ftp server, so it can offer ATP services for the whole lab. The authentication center for distributing X.509 certificate is located in XI'AN Jiaotong University.

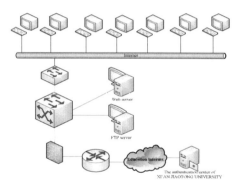

Figure 7. The network topology in lab.

1 The important documents included to the monitoring on the server are:
2 /bin and /usr/bin: The directory stores the command the user used mostly.
3 /boot: This directory stores some of the core files used to boot Linux.
4 /etc: This directory is one of the most important directories in Linux system, which stores various configuration files needed to use.
5 /lib: This directory is used to store system dynamically linked with shared libraries, and almost all of the applications make use of the shared libraries in this directory.
6 /sbin and /usr/sbin: This directory is used to store the system management program which some system administrators use.
7 /WWW: Apache is used to store the web directory.
8 /usr/local/mysql/var: This directory is used to store the MySQL database files.

The host in LAN did not provide other services, so there is no need to monitor the last two mentioned above. The server makes the host in LAN as the replica host; the hosts in LAN backup each other and form the alliance freely according to the principle of the "group peer for each" from the accident recovery system.

Experiment 1 – the destruction of the unknown viruses. A user in LAN receives an e-mail with attachments, and users are curious to open mail attachments, then accident recovery system immediately alarms and displays that ps file viewing the process of /bin being changed; for the first level of the automatic recovery is selected, the changed file can be recovered immediately. Next, the firewall discovers that the host sends messages to the outside without permission, and IDS reports that the host is scanning the other host in the network, auditing system reports that the host has a large number of abnormal processes. The ps command is used to the host, and it has a lot of concurrent processes to scan the other host in the network segment and tries to make the next attack. So only with turning off the internet and killing the attack process, and then accessing the network, IDS and auditing system can have no alarm.

Experiment 2 – huge amounts of data recovery. The Web server can provide the dynamic Web service to the user in education network and get data from the background of the MySQL database through a PHP script. It has proved that it only takes 1 minute and 3 seconds to recover with multi-source backup and restore fast transmission algorithm for a MySQL database file with 200M, PHP script with about 150M, HTML files and other configuration files, and each node in the data backup works, and this improves the utilization rate of the replica and truly realizes the quick recovery for the multiple replicas.

5 SUMMARY

As it can be seen from a lot of simulations and experiments, the time required for data backup and recovery is only about 20% of that in traditional transmission methods and the quantity of data transmission for source data node is only 40% of that in the traditional way. Efficient remote data synchronization algorithm is favorable for the files which have relatively small changes, the smaller the changes, the higher the transmission efficiency becomes. The management plan of the union members and copy files determines the position and the quantity of the replica between each member, and the check for a copy can be completed in 20-40s, which ensures the reliability of the system.

The system has high speed in backup and recovery; which is useful to realize real-time recovery for data without other hardware equipment. So the cost is low and the system can link with other network security equipment; the data recovery (automatic or manual) rate can reach 100%.

REFERENCES

[1] Li Weihua, Jiang Lan. A New Multi-Function System to Deal with Hacker Intrusion[J]. Journal of Northwestern Polytechnical University, Vol.23, No.3,2005.
[2] Wang Shenghang, Ding Wei, etc. White paper of disaster. China Information Support Center of IBM. 2004.
[3] HUANG Zun-guo, REN Jian-yong, HU Guang-ming. On a Skeleton of Rapid Response and Recovery (r-RR) from Info-Sec Incidents [J]. COMPUTER ENGINEERING & SCIENCE. Vol.23,No.6,2001.
[4] Rick Schiesser, "Develop an effective disaster recovery plan", TechRepublic,2002.
[5] H.F.Lipson, DAFisher, "Survivability: A New Technical and Business Perspectiveon Security", Procedings of the New Security Paradigms Workshop, IEEEComputer Society Press, 1999.
[6] Shi Xingjian, Li Weihua, Wang Wenqi, Jiang Weihua. Model approach to fuzzy security audit[J]. Journal of Information and Computation Science, 2004.2.
[7] SHI Xing-jian, LI Wei-hua, WANG Wen-qi. Model approach to security audit domain. Computer Applications[J]. Vol.24, No.10, 2004.
[8] Shi Xingjian, Li Weihua, Wang Wenqi.Security Audit Model Based on Optimized Clustering Algorithm[J]. Computer Engineering and Application. 2005.17.
[9] S.Bhattacharyya, J.F.Kurose, D.Towsley, et al. "Efficient rate-controlled bulk data transfer using multiple multicast groups." In Proc. San Francisco, CA, Apr. 1998.
[10] John W. Byers, Michael Luby, Michae Mitzenmacher. A Digital Fountain Approach to Asynchronous Reliable Multicast. IEEE Journal on Selected Areas Communications-JSAC, vol. 20, no. 8, pp. 1528–1540, 2002.
[11] J. Byers, M. Luby, and M. Mitzenmacher. "Accessing multiple mirror sites in parallel: Using Tornado codes to speed up downloads." In Proc. IEEE INFOCOM, New York, Mar. 1999.

Network Security and Communication Engineering – Chan (Ed.)
© 2015 Taylor & Francis Group, London, ISBN: 978-1-138-02821-0

Mandatory access control scheme based on security label for state power control system

K.L. Gao
China Electric Power Research Institute, Beijing, China

Z.M. Guo
State Grid JIBEI Electric Power Company Limited, Beijing, China

Z.H. Wang
China Electric Power Research Institute, Beijing, China

X.P. Li
State Grid JIBEI Electric Power Company Limited, Beijing, China

ABSTRACT: The access control mechanism for the power control system is proposed in this paper. The scheme takes advantage of the trusted software base and security label system in power control system, which provides operating system kernel-level access control and the application-level access control. The trusted software base first determines whether service requests are allowed. Moreover, only those services permitted by trusted software base can enter the application-level access control system which is based on the power control system and it is difficult to bypass. In addition, the proposed scheme does not affect the operation environment and procedures, original services, deployment, and updates of the power control system. Moreover, the original system does not need to modify any code to support the scheme.

1 INTRODUCTION

With the increasing popularity of computer applications, particularly the development of network and database technology, information security is increasingly becoming a pressing problem, and its fundamental goal is to protect the information from the aspect of confidentiality, integrity, and availability. It increasingly becomes an indispensable tool to use the computer for information collection, processing, storage, analysis, and exchange. Most information processing may be related to national military and political, economic, industrial, and commercial intelligence as well as some private data and other sensitive information. Therefore, information security of the whole system will be threatened if you do not take effective security measures. Once attacked, there will be an incalculable loss to the political, economic, military, and diplomatic intelligence of the country, as well as some user departments and individuals, making information security become increasingly prominent. In recent years, the related researches have even become hotter in IT field. Operating system access control is the key to operating system security control, which takes control of access to system resources based on identification. Access Control [1-3] is one of the most common security

policies for modern operating systems. Discretionary Access Control (DAC) has great disadvantage. DAC is usually based on the assumption that the user itself can ensure credibility of the object. However, such an assumption is often wrong which provides an opportunity for Trojan horse attacks [4-5]. Therefore, Mandatory Access Control has become the best solution for the operating system security. Mandatory Access Control uses the mandatory requirements to prevent the insecure information flow, which can be very effective to prevent Trojan horse attacks.

2 RELATED STUDIES ON ACCESS CONTROL MODEL

Access control is an effective protection method by allowing an authorized body to access to the object of resources and refusing to provide service for unauthorized body [6]. In the database management system of C2, DAC [7-8] is often applied, that is to say, the system administrator (DBA) authorizes users to access to data within the permitted scope. However, DAC cannot guarantee the system security for database applications with high security requirements alone. Therefore, more and more access control methods have been studied and developed, including Mandatory access control (MAC) and RBAC (Role-Based Access Control). Some studies show [9-10] that the secure control of sensitive information flow is implemented by the protection functions based on hardware which is provided by trusted computing platform. Security label, the key technology to mandatory access control, has not been described in detail by former research [1-8]. Therefore, the detailed information and researches about mandatory access control model will be provided.

3 THE MANDATORY ACCESS CONTROL MODEL BASED ON SECURITY LABEL

The so-called mandatory access control [11-15] enforces access control mechanisms to subjects and objects by system. Mandatory access control is determined by security level labels between subjects and objects that meet the relationship of "partial order", which is also called multi-level security model.

3.1 *Security label definition*

Mandatory access control [16-17] grants every subject and object with security level labels corresponding to their identities based on BLP model. The security level label is called "permission label" for the subject, and it is called "sensitivity label" for the object. Both license tag and sensitive tag are called security label or security level because they have the same composition. Here are some key definitions.

Definition 1: Security Level refers to the degree that information is protected by the system.

Definition 2: Sensitivity label is the external representation of **security** level for the object, which is used to judge whether the subject has access rights to the object.

Definition 3: **Permission** level refers to security level of the current subject, which is used to determine the access level to information for the subject.

Definition 4: Permission label is the external representation of the **security** level for the current subject, which is used to judge whether the subject has access right to information.

For multi-level **security** policies [18-20], access control mechanisms should meet the following requirements:

1 Assign each object with a sensitivity label which represents security level for the protected information.
2 Assign each subject with a permission label which is used to specify its credibility. The system determines whether a subject has corresponding access right to an object by permission label and sensitivity label.

3 Ensure all input and output information be correctly marked with sensitivity label which represents its security level. Multi-level security policies are realized by such label mechanisms. It means that it is possible to set up different security levels (category, scope) for classified information, which makes sure that users can only access information permitted.

3.2 *Security label composition*

Security label [21-22] based on multi-level security policies is composed of category and scope. As category reflects a hierarchical relationship, it is also known as hierarchy. Security level is unique, so there is only one category. For example, there exists public U, secret C, confidential S, and confidential TS in Logic-SQL [5] database. Assume Security Level L = {U, C, S, TS} with ascending order: U <C <S <TS. Scope consists of elements without hierarchies, and represents a clear information field. "Without hierarchies" means that there is not a category "greater than" another one. Scope is also a set, for example, assuming C = {Finance Dept., Technical Dept., Sales Dept., and Marketing Dept.} as a multi-scope set. Finance Dept., Technical Dept., Sales Dept., and Marketing Dept. are independent elements of this set.

There exist the following four types of relationships between two security labels A and B:

A dominates B: if and only if security level of A is higher than or equal to the one of B, the scope A contains scope B, that is to say, scope B is a subset of scope A.

B dominates A: if and only if security level of B is higher **than** or equal to the one of A, the scope B contains scope A, that is to say, scope A is a subset of scope B.

A is equal to B: if and only if the security level of A is equal to the one of B. Any element of scope A belongs to scope B, and vice versa. That is to say, A dominates B and B dominates A.

A is independent of B: none of the above conditions meet.

3.3 *Operation conditions of access control mechanism*

With the mandatory access control mechanism based on the security label [23-24], the permission label of a subject must meet the following conditions if a subject wants to access an object:

1 Write right: security level of the subject (hierarchy+ scope) must be dominated by security level of the object (the security level of the object ≥ the security level of the subject, the scope of the object ≥ the scope of the subject).

2 Read right: the security level of the subject (security level + scope) must dominate the security level of the object (the security level of the subject $\geq$ the security level of the object, the scope of the subject $\geq$ the scope of the object). The subject can only access the object if it meets all the above requirements. Generally speaking, the security level of security label is defined by unclassified $\leq$ confidential $\leq$ secret $\leq$ top secret, and access control follows the mechanism of "Bell–LaPadula Model".

4 DESIGN AND IMPLEMENTATION OF THE SCHEME

4.1 Framework of the scheme

A mandatory access control system is proposed based on the operation behaviors of the power control system and functions of the trusted software-based mandatory access system. Trusted software-based mandatory access control system provides kernel-level security access control, which makes sure that the access requests are strictly controlled by security label of subjects and objects. An access request is first judged by access control policies provided by the trusted software. The access request will be refused if it does not meet system policies. If it meets system policies, it will be submitted to the application-level security label system for the second judgment of access control rights. The proposed solution has completely solved problems in the application-level security label system applied in power control system, which can effectively prevent unauthorized access and is difficult to bypass. The solution procedures are as follows.

The trusted software base deployed in the remote agent intercepts the connection request when power control system launches remote service request. Security label of the subject will be extracted from the fixed position in the first package after TCP three handshakes succeed. The security label of the object, corresponding to service name found by the connection port number, will be extracted from the label list. XOR operation is performed between the security label rights value of the subject and the security label rights value of the object stored in memory in advance, which is used to judge whether the connection is permitted. If permitted, the service request will be submitted to the application-level security label system, which will verify whether the service request is allowed. The connection will be established after both access control systems allow the service request. Otherwise, the session will end and the original requester will receive feedback which states the

connection ends. The original requester will receive feedback which states the connection ends if the trusted software base refuses the connection, and the original requester will receive feedback which states refuse of access right if the application-level security label denies it.

Here is an example of how users (user1 with right identity 0x00000001 and users2 with right identity 0x00000000) access the remote browsing service (right identity 0x00000001).

When the user1 accesses remote browsing services, trusted software base will first **extract** user1 identify 0x00000001 after the request is intercepted in the agent. Then, the right identity 0x00000001 of remote browsing service is found out in local server lists. XOR operation is running between these two right identities with result of "0x00000001 XOR 0x00000001 is equal to 0", which means the connection is allowed.

However, when user2 initiates remote browsing request, the **service** will be denied because the XOR operation result between their right identities is equal to 1 of 0x00000000 XOR 0x00000001.

4.2 Determination principles

The following principles are followed during the determination.

1 Other services are not managed by access control policies, except for the remote browsing service and the issued command services.
2 Trusted software-based mandatory access control system only intercepts the first packet and makes access rights determination. Subsequent packages from the request with an allowed tag will be allowed directly without any other rights determination.
3 The trusted software base sends the information that the connection will be closed when the determination fails and the connection session ends.
4 If and only if the connection request passes through both access control systems, it will be allowed. Otherwise, the connection request will be denied.

4.3 Determination references

The following determination references are required for the mandatory access control system based on trusted software base.

1 Local object service label list.
2 The corresponding table between port and service.
3 The offset and format of security label in the request packet.

4.4 Operation procedures of the scheme

The proposed access control mechanism takes advantage of trusted computing technology and security label system in power control systems. The trusted software base provides operating system kernel-level access control, and the power control system provides the application-level access control. The access control with the trusted software base runs in operating system kernel and provides more security strength, which is difficult to bypass. Any service determination with security label must first pass the trusted software base. Moreover, only those services permitted by trusted software base can enter the next determination system of the application-level access control system based on the power control system.

The process of mandatory access control system with trusted software base is as follows (as is shown in figure 1):

1. HMI initiates service access request.
2. The request is intercepted by trusted software base deployed in SCADA.
3. The security label of the subject is extracted from the fixed position of the first packet.
4. The service name is found out in the port service list according to port number of the accessed service.
5. Security label of the object is found out in the object service list according to the service name.
6. XOR operation is performed between the value of the subject security label and the value of the object security label.
7. If the judge succeeds, the packet will be realized. Otherwise, the connection is closed.
8. Subsequent packages from the request will be allowed directly without any other rights determination.

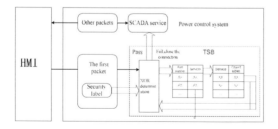

Figure 1. Mandatory access control system based on trusted software base.

Photographs should be with good contrast and preferably in TIFF or EPS format (see Artwork document). Photographic reproductions cut from books or journals, photocopies of photographs and screened photographs are unacceptable. The proceedings will be printed in black only. For this reason, it must avoid the use of color in figures and photographs. Color is also nearly unnecessary for scientific work.

5 CONCLUSION

The proposed scheme provides the access control mechanism with more security strength for the power control system. The scheme takes advantage of the trusted software base and security label system in power control systems, which provides operating system kernel-level access control and the application-level access control. The access control with the trusted software base runs in operating system kernel, which is difficult to bypass. Any service determination with security label must first pass the trusted software base. Moreover, only those services permitted by trusted software base can enter the next determination system of the application-level access control system which is based on the power control system.

REFERENCES

[1] D. Ferraiolo, D. R. Kuhn, and R. Chandramouli. Role-Based Access Control. Artech House, 2003.
[2] D.F. Ferraiolo, R. Sandhu, S. Gavrila, D. R. Kuhn,and R. Chandramouli. Proposed NIST Standard for Role-Based Access Control. ACM Transactions on Information and Systems Security, 4(3), 2001.
[3] R. France, D. Kim, S. Ghosh, and E. Song. AUML - Based Pattern Specification Technique. IEEE Transactions on Software Engineering, 30(3):193–206, 2004.
[4] L. Notargiacomo. Role-Based Access Control In ORACLE7 And Trusted ORACLE7. In Proceedings of the 1st ACM Workshop on Role-Based Access Control, page 17, Gaithersburg, MD, 1995.
[5] S. L. Osborn, R. Sandhu, and Q. Munawer. Configuring Role-Based Access Control to Enforce Mandatory and Discretionary Access Control Policies. ACM Transactions on Information and System Security, 3(2):85–106, 2000.
[6] R. Sandhu and Q. Munawer. How To Do Discretionary Access Control Using Roles. In Proceedings of the 3rdACM Workshop on Role-Based Access Control(RBAC-98), Fairfax, VA, 1998. ACM Press.
[7] R. Sandhu and P. Samarati. Access Control: Principles and Practice. IEEE Communications, 32(9):40–48, September 1994.
[8] M. M. Swift, A. Hopkins, P. Brundrett, C. Van Dyke,P. Garg, S. Chan, M. Goertzel, and G. Jensenworth. Improving the granularity of access control for Windows 2000. ACM Trans. on Information and System Security, 5(4):398–437, November 2002.
[9] H. Yao, H. Hu, B. Huang, and R. Li. Dynamic role and context-based access control for grid applications. In Parallel and Distributed Computing, Applications and

Technologies, 2005. PDCAT 2005. Sixth International Conference on, pages 404–406, Dec. 2005.

[10] T. Zhao and S. Dong. A trust aware grid access control architecture based on abac Networking, Architecture, and Storage, International Conference on, 0:109–115, 2010.

[11] T.F. Lunt, D.E. Denning, et al., The Sea View security model, IEEE Transaction on Software Engineering 16 (6)(1990) 593–607.

[12] S. Jajodia, R. Sandhu, A novel decomposition of multilevel relations into single-level relations, in: Proceedings of IEEE Symposium on Research in Security and Privacy, 1991, pp. 300–315.

[13] J. Melton, A.R. Simon, J. Gray, SQL: 1999 Understanding Relational Language Components, Morgan Kaufmann Publishers, Los Altos, CA, 2001.

[14] P.G. Griffiths, B. Wade, An authorization mechanism for a relational database system, ACM Transaction on Database Systems 1 (3) (1976) 242–255.

[15] P.D. Stachour, B. Thuraisingham, Design of LDV: a multilevel secure relational database management system, IEEE Transaction on Knowledge and Data Engineering 2 (2) (1990) 190–209.

[16] Li N, Mao Z, Chen H. Center for Education and Research in Information Assurance and Security, Technical Report: CERIAS TR 2006–38 host integrity protection through usable non-discretionary access control. West Lafayette, IN, USA: Purdue University, 2007.

[17] C. Turker, M. Gertz, Semantic integrity support in SQL: 1999 and commercial (Object) relational database management systems, The VLDB Journal 10 (4) (2001) 241–269.

[18] J. Abramov, O. Anson, A. Sturm, P. Sho val, Tool support for enforcing security policies on databases, CAiSE Forum 2011 (2011) 41–48.

[19] J. Abramov, A. Sturm, P. Shoval, A Pattern based Approach for Secure Database Design, CAiSE Workshops 2011, LNBIP 83(10), Springer, Berlin, Heidelberg, 2011. pp. 637–651.

[20] A. Ampatzoglou, A. Chatzigeorgiou, Evaluation of object -oriented design patterns in game development, Information and Software Technology 49 (5)(2007) 445–454.

[21] D. Basin, M. Clavel, J. Doser, M. Egea, Automated analysis of security-design models, Information and Software Technology 51 (5) (2009) 815–831.

[22] D. Basin, J. Doser, T. Lodderstedt, Model driven security: from UML models to access control infrastructures, ACM Transactions on Software Engineering and Methodology 15 (1) (2006) 39–91.

[23] L. Chen, M.A. Babar, A systematic review of evaluation of variability management approaches in software product lines, Information and Software Technology 53 (2011) 344–362.

[24] L. Compagna, P.E. Khoury, A. Krausová, F. Massacci, N. Zannone, How to integrate legal requirements into a requirements engineering methodology for the development of security and privacy patterns, Artificial Intelligence and Law 17 (1) (2009) 1–30.

Network Security and Communication Engineering – Chan (Ed.)
© 2015 Taylor & Francis Group, London, ISBN: 978-1-138-02821-0

The complete cyclic structure of the RSA cryptosystem

H. Suzuki, H. Yasuoka & K. Matsushita
Tokyo University of Information Sciences, Chiba, Japan

ABSTRACT: For a plaintext M, a public key set (n, e) of the RSA cryptosystem has an RSA cycle $x : M^{e^x} \equiv M$ (mod n). It is relatively easy to calculate the RSA cycle for a given (n, e) and a plaintext M. However, for all M in $\mathbb{Z}_n$, it is infeasible to calculate all the RSA cycles by exhaustive trials due to huge amount of M. This paper, for the first time, shows an enumerative function ηn,e(x) for the explicit cardinality of plaintexts M in $\mathbb{Z}_n$ each with a given RSA cycle x. Thus, this paper reveals the complete cyclic structure of the RSA cryptosystem for any given key set (n, e).

1 INTRODUCTION

RSA (Rivest-Shamir-Adleman) cryptosystem (Rivest 1978b) appeared at MIT in 1977 as a method of public key encryption and digital signature. RSA encryption and decryption processes are $C = M^e \bmod n$ and $M = C^d \bmod n$, respectively, with a plaintext M, the corresponding ciphertext C, a public modulus key n, a public exponent key e and a private exponent key d where e and d must hold $e \cdot d \equiv 1 (\bmod \lambda (n))$ or $e \cdot d \equiv 1 (\bmod \phi (n))$.

Soon after the RSA cryptosystem appeared, Simmons-Norris (1977) mentioned an attack (now called "cyclic attack") to the system using RSA cycle. A public key set (n, e) of RSA cryptosystem has an RSA cycle $x : M^{e^x} \equiv M(\bmod n)$ or $C^{e^x} \equiv C(\bmod n)$ for each M or C. If one found an x that holds the equation $C^{e^x} \equiv C (\bmod n)$, the person can cryptanalyze the ciphertext C by taking $M = C^{e^{x-1}} \bmod n$ only with the public key set (n, e). Rivest (1978a), an inventor, promptly rejected the attack because the approximate probability of plaintexts that can be broken by the attack is vanishingly small like roughly 10^{-90} with simple conditions:

a. $p - 1$ and $q - 1$ should contain very large prime factors (call them p' and q', respectively),
b. similarly, $p' - 1$ and $q' - 1$ should contain very large prime factors (call them p'' and q''), and primes p', q', p'' and q'' are all larger than 10^{90}.
c. $Ord_{p'q'}(e)$ should be greater than or equal to $p''q''$.

At the time of Rivest (1978a), a suggested modulus size was around 200 digits (665 bits). For the security of RSA, the suggested modulus size is now 617 digits (2048 bits) until year 2030, and 925 digits (3072 bits) after that. The bound 10^{90} in the condition (b) would be 10^{278} for 617 digits modulus, and 10^{416} for 925 digits modulus. The attackable probability 10^{-90} would

be 10^{-278} for 617 digits modulus, and 10^{-416} for 925 digits modulus. The attackable probability would not be increased even when the modulus size inceases. Rivest (1978a) has also mentioned that key generation does not need to worry about these conditions since choosing keys p, q and e at random satisfies the conditions at higher probability. It does not exist, thus far, any deterministic attack to RSA cryptosystem even though many attacks are mentioned in Menezes (1996) & Kaliski (1995). The security of RSA cryptosystem is based on intractability of factoring large integers and is adjustable by changing the size of modulus key $n = pq$. Due to its simplicity of encryption and decryption process equations and its adjustable security, RSA cryptosystem is still one of the useful cryptosystems in real world. Herein after, RSA expresses RSA cryptosystem. All information in this paper do not mean to break RSA yet just express the mathematical complete cyclic structure of RSA.

There are previous works showing RSA's mathematical properties. As an example, Blakley-Borosh (1979) mentioned that nine trivial plaintexts, which consist of $\{0, 1, -1\} \bmod p$ and $\{0, 1, -1\} \bmod q$, cannot be encrypted for any n and e. Katsuno (1983) showed a lower bound $\phi (p - 1)\phi (q - 1)$ for the cardinality of plaintexts M taking the maximum RSA cycle for any given n and e. Hayashi-Kawamura (1990) improved his lower bound for any given n and e to $2\phi(p - 1)\phi(q - 1)$. Suzuki (1988) & Grošek (1994) showed a lower bound $\phi(\lambda(\lambda(n)))$ for the cardinality of encryption exponent keys $e \in \mathbb{Z}^*_{\lambda(n)}$ taking the maximum possible RSA cycle $\lambda(\lambda(n))$ for any n. For these information, the user of secret key d and key generation centre only be able to know $p, q, \phi(p-1)\phi(q-1)$, $\lambda(\lambda(n))$ and $\phi(\lambda(\lambda(n)))$ since getting these numbers from modulus n is no easier than factoring of n.

The RSA structure $M^{e^x} \bmod n$, with a number x of iterated encryption processes, is recognized as a special form fixed b of the double modular exponentiation structure like: $a^{b^c} \bmod n$. The structure is well considered and is mathematically elegant since the plaintext and ciphertext spaces can take all elements in a set $\mathbb{Z}_n$ instead of a restricted set $\mathbb{Z}_n^*$. Rivest (1978a) & Maurer (1995) have analyzed RSA as a double modular exponentiation structure. Rivest (1978a) has mentioned the RSA structure as $M^{e^x \bmod k} \bmod n$ where k denotes the order to which M belongs, modulo n, i.e. $k = Ord_n(M)$. Maurer (1995) has mentioned that RSA cycle for given n, e and M is $Ord_{Ord_n(M)}(e)$. Their descriptions using order modulo a composite are enough to explain the RSA properties but are not enough to count the explicit cardinalities of the RSA properties since the word, "order", is restricted to set $\mathbb{Z}_n^*$.

Analyzing the RSA structure is a kind of analyzing the double modular exponentiation structure. Before analyzing RSA, we survey the previous works regarding the single modular exponentiation structure like: $a^b \bmod n$. Blakley-Blakley (1978, 79) mentioned, in detail, the single modular exponentiation structure using words: order, period and cycle modulo a composite n. The word "cycle" in this paper is the same as their "period." This paper does not deal with their "cycle." A possible order/period/cycle for modulo a composite n (that is a divisor of $\lambda(n)$) is known. Gauss (1986) showed a function $\psi_p(b) = \phi(b)$ for the cardinality of elements in $\mathbb{Z}_n^*$ each with a given order b for any odd prime p where b is a divisor of $\lambda(p)$. Suzuki-Nakamura (1995) extended Gauss' $\psi_p(b)$ to $\psi_n(b)$ with any composites n instead of any odd primes p. We show an example application of the function that $\psi\lambda_{(n)}$ $(\lambda(\lambda(n)))$ expresses the explicit cardinality of encryption exponent keys $e \in \mathbb{Z}_{\lambda(n)}^*$ taking the maximum RSA cycle $\lambda(\lambda(n))$.

As an important point, computing order modulo n for an element in $\mathbb{Z}_n^*$ is feasible, but computing all orders modulo n for all elements in $\mathbb{Z}_n^*$ with large composite n is infeasible. In the RSA structure, same thing occurs as well as single modular exponentiation structure. Computing the RSA cycle for a plaintext in $\mathbb{Z}_n$ is feasible by the user of key d, but computing all RSA cycles for all plaintexts in $\mathbb{Z}_n$ with large composite n is infeasible even if by the user of key d.

However, all cardinalities of plaintexts M in $\mathbb{Z}_n$ each with given RSA cycles x can be counted by an analogy of all cardinalities of elements each with a given order b in $\mathbb{Z}_n^*$. Then, this paper, for the first time, shows an enumerative function $\eta_{n,e}(x)$ for the explicit cardinality of plaintexts M in $\mathbb{Z}_n$ each with a given RSA cycle x. We also show an example application of the function that $\eta_{n,e}(\lambda(\lambda(n)))$ expresses the explicit cardinality of plaintexts M in $\mathbb{Z}_n$ each with the maximum RSA cycle $\lambda(\lambda(n))$. Also, an example of complete RSA cycle distribution table for a given key (n, e) is obtained by $\eta_{n,e}(x)$ for all divisors x of $\lambda(\lambda(n))$.

In Section 2, we denote the notation, known functions and known theorems. In Section 3, we consider RSA cycle and RSA parameters. In successive Sections 4, 5 and 6, we show new explicit enumerative function $\eta_{n,e}(x)$, the computational powers of $\psi_n(b)$ and $\eta_{n,e}(x)$, and a numerical example are mentioned.

A preliminary version of this research first appeared in Suzuki (1997).

2 NOTATION, KNOWN FUNCTIONS AND KNOWN THEOREMS

Here we denote the notation, known functions and known theorems that are used in this paper. All the numbers in this paper are integers except probabilities.

$\mathbb{Z}$: the set of integers.

$\mathbb{N}$: the set of positive integers, e.g. $n \in \mathbb{N} = \mathbb{Z}_{>0}$

Primes : the set of prime numbers, a subset of $\mathbb{N}$, e.g. $p \in Primes$.

$[a, b)$: the set $\{x \in \mathbb{Z} \mid a \leq x < b\}$.

$\mathbb{Z}_n$: the ring of integer modulo n, $\mathbb{Z}_n = \mathbb{Z} / n\mathbb{Z} = [0, n)$.

$\mathbb{Z}_n^*$: the multiplicative group of modulo n, $\mathbb{Z}_n^* = (\mathbb{Z} / n\mathbb{Z})^* := \{x \in \mathbb{Z}_n \mid gcd(x, n) = 1\}$.

$\#\{\cdot\}$: the cardinality or the number of elements in a set $\{\cdot\}$.

$a\backslash b$: a divides b, (in this paper, not "set minus" of group theory).

$lcm(a_1, a_2, \ldots)$: the least common multiple of a_1, a_2, ...

$gcd(a_1, a_2, \ldots)$: the greatest common divisor of a_1, a_2, ...

$Ord_n(a)$: the multiplicative order or the exponent of an element a in $\mathbb{Z}_n^*$,

$$Ord_n(a) := \begin{cases} min\{b \in \mathbb{N} \mid a^b \equiv 1 (\bmod\ n)\}, (n > 1), \\ 1 \quad (n = 1). \end{cases}$$

$Cyc_n(a)$: the multiplicative cycle of an element a in $\mathbb{Z}_n$, if n is square free, $Cyc_n(a) := min\{b \in \mathbb{N} \mid a^{b+1} \equiv a (\bmod\ n)\}$.

$\phi(n)$: Euler's totient function of n or $\#\{\mathbb{Z}_n^*\}$, (*Lemma 2.4*)

$$\phi(n) := \#\{x \in \mathbb{Z}_n \mid gcd(x, n) = 1\},$$
$$= \phi(p_1^{\alpha_1} p_2^{\alpha_2} p_3^{\alpha_3} \ldots),$$
$$= \phi(p_1^{\alpha_1})\phi(p_2^{\alpha_2})\phi(p_3^{\alpha_3})\cdots,$$

where $\begin{cases} n = p_1^{\alpha_1} p_2^{\alpha_2} p_3^{\alpha_3} \cdots, \\ \phi(1) = 1, \\ \phi(p_i^{\alpha_i}) = p_i^{\alpha_i - 1}(p_i - 1). \end{cases}$

$\lambda(n)$: Carmichael's function of n, the universal exponent modulo n or the maximum multiplicative order modulo n, (*Lemma 2.5*)

$$\lambda(n) := max\{x \mid y \in \mathbb{Z}_n^*, x = Ord_n(y)\},$$
$$= \lambda(2^{\alpha_1} p_2^{\alpha_2} p_3^{\alpha_3} \ldots),$$
$$= lcm(\lambda(2^{\alpha_1}), \lambda(p_2^{\alpha_2}), \lambda(p_3^{\alpha_3}), \ldots),$$

where

$$\begin{cases} n = 2^{\alpha_1} p_2^{\alpha_2} p_3^{\alpha_3} \cdots, \\ \lambda(1) = \phi(1) = 1, \\ \lambda(p_1^{\alpha_1}) = \lambda(2^{\alpha_1}) = \begin{cases} \phi(2^{\alpha_1}) = 2^{\alpha_1 - 1} & (\alpha_1 = 1, 2), \\ \dfrac{\phi(2^{\alpha_1})}{2} = 2^{\alpha_1 - 2} & (\alpha_1 \geq 3), \end{cases} \\ \lambda(p_i^{\alpha_i}) = \phi(p_i^{\alpha_i}) = p_i^{\alpha_i - 1}(p_i - 1) & (i \geq 2) \end{cases}$$

$divs(n)$: the cardinality of (positive) divisors of n,

$$divs(n) = \prod_{i=1}^{k} \alpha_i + 1, \qquad \text{where } n = p_1^{\alpha_1} p_2^{\alpha_2} p_3^{\alpha_3} \cdots p_k^{\alpha_k}.$$

$\psi_{p^\alpha}(b)$: the cardinality of elements each with a given order b in $\mathbb{Z}_{p^\alpha}^*$ with an odd prime p,

$$\psi_{p^\alpha}(b) := \#\{x \in \mathbb{Z}_{p^\alpha}^* \mid Ord_{p^\alpha}(x) = b\},$$
$$= \phi(b). \qquad (Theorem\ 2.10)$$

$\theta_p(b)$: the cardinality of elements each with a given cycle b in $\mathbb{Z}_p$ with an odd prime p,

$$\theta_p(b) := \#\{a \in \mathbb{Z}_p \mid a^{b+1} \equiv a (mod\ p)\},$$
$$:= \#\{a \in \mathbb{Z}_p \mid Cyc_p(a) = b\},$$
$$= \begin{cases} \psi_p(b) = \phi(b)(b \neq 1), & (\because \theta_p(b) = \psi_p(b) \text{ for } a \in \mathbb{Z}_p^*) \\ (\phi(1) + 1 = 2(b = 1), & (\text{for } a \in \{0, 1\}). \end{cases}$$

Here we mention known theorems and lemmas that are used in this paper.

Theorem 2.1 (*Chinese Remainder Theorem (Knuth 1981)*). *Let* $n = p_1^{\alpha_1} p_2^{\alpha_2} p_3^{\alpha_3} \cdots p_k^{\alpha_k}$ *be a positive integer. For a set of integers* $u_1, u_2, u_3, \ldots, u_k$, *there is exactly one integer u that satisfies the conditions*

$$0 \leq u < n, \text{ and } (\forall i \in [1, k])[u \equiv u_i (mod\ p_i^{\alpha_i})], \qquad (1)$$

in other words, for a set $\{u_i | i \in [1, k]$, *an element* u_i *in* $\mathbb{Z}_{p_i^{\alpha_i}}\}$, *there is exactly one element u in* $\mathbb{Z}_n$ *that satisfies the conditions*

$$(\forall i \in [1, k])[u \equiv u_j (mod\ p_i^{\alpha_i})]. \qquad (2)$$

Lemma 2.2 (*Isomorphic form of* $\mathbb{Z}_n$). *Let* $n = p_1^{\alpha_1} p_2^{\alpha_2} p_3^{\alpha_2}$ *be a positive integer. Due to Chinese Remainder Theorem, the additive ring* $\mathbb{Z}_n$ *is isomorphic to the direct sum* $\mathbb{Z}_{p_i^{\alpha_i}}$ *as follows:*

$$\mathbb{Z}_n \simeq \mathbb{Z}_{2^{\alpha_1}} \oplus \mathbb{Z}_{p_2^{\alpha_2}} \oplus \mathbb{Z}_{p_3^{\alpha_3}} \oplus \cdots. \qquad (3)$$

Lemma 2.3 (*Isomorphic form of* $\mathbb{Z}_n^*$). *Let* $n = p_1^{\alpha_1} p_2^{\alpha_2} p_3^{\alpha_2}$ *be a positive integer. Due to Chinese Remainder Theorem, the multiplicative group* $\mathbb{Z}_n^*$ *is isomorphic to the cartesian product of* $\mathbb{Z}_{p_i^{\alpha_i}}^*$ *as follows:*

$$\mathbb{Z}_n^* \simeq \mathbb{Z}_{2^{\alpha_1}}^* \times \mathbb{Z}_{p_2^{\alpha_2}}^* \times \mathbb{Z}_{p_3^{\alpha_3}}^*, \times \cdots. \qquad (4)$$

Lemma 2.4 (*The cardinality of* $\mathbb{Z}_n^*$). *Since* $\mathbb{Z}_n^*$ *is expressed by the isomorphic form of the cartesian product of* $\mathbb{Z}_n^*$, *the cardinality of* $\mathbb{Z}_n^*$ *is also expressed as follows:*

$$\begin{aligned} \#\{\mathbb{Z}_n^*\} &= \#\{\mathbb{Z}_{p_1^{\alpha_1}}^*\} \cdot \#\{\mathbb{Z}_{p_2^{\alpha_2}}^*\} \cdot \#\{\mathbb{Z}_{p_3^{\alpha_3}}^*\} \cdots, \\ &= \phi(2^{\alpha_1})\phi(p_2^{\alpha_2})\phi(p_3^{\alpha_3}) \cdots, \\ &= \phi(n). \end{aligned} \qquad (5)$$

Theorem 2.5 (*Fermat-Euler Theorem*). *Let n and a be integers,*

$$a^{\phi(n)+1} \equiv a \ (mod\ n) \ or$$
$$a^{\phi(n)} \equiv 1 \ (mod\ n) \ if \ \boldsymbol{gcd}(a, n) = 1, \qquad (6)$$

where $\phi(n)$ denotes Euler's totient function of n.

Theorem 2.6 (*Generalized Fermat-Euler Theorem (Carmichael 1910 & Singmaster 1966)*). *Let n and a be integers,*

$$a^{\lambda(n)+1} \equiv a \ (mod\ n) \ or$$
$$a^{\lambda(n)} \equiv 1 \ (mod\ n) \ if \ \boldsymbol{gcd}(a, n) = 1, \qquad (7)$$

where $\lambda(n)$ denotes Carmichael's function of n. $\lambda(n)$ is the least exponent which holds the equation for any integer a. Then $\lambda(n)$ is called as the universal exponent modulo n or the maximum multiplicative order modulo n.

Lemma 2.7 *Let n and a be integers. If b is the minimum positive integer that holds*

$$a^{b+1} \equiv a \ (mod\ n) \ or$$
$$a^b \equiv 1 \ (mod\ n) \ if \ \boldsymbol{gcd}(a, n) = 1, \qquad (8)$$

then b denotes the order of an element a modulo n if gcd(a, n) = 1, i.e. $b = Ord_n(a)$, and also b denotes the cycle of an element a modulo n for any integer a, i.e. $b = Cyc_n(a)$.

Lemma 2.8 (*The order of an element in* $\mathbb{Z}_n^*$). *From Lemma 2.3, the order of an element a in* $\mathbb{Z}_n^*$, $Ord_n(a)$, *is calculated as follows:*

$$Ord_n(a) = lcm(Ord_{p_1^{\alpha_1}}(a), Ord_{p_2^{\alpha_2}}(a), Ord_{p_3^{\alpha_3}}(a), \ldots). \qquad (9)$$

Lemma 2.9 (*The cycle of an element in* $\mathbb{Z}_n$). *From Lemma 2.2, the cycle of an element a in* $\mathbb{Z}_n$, $Cyc_n(a)$, *is calculated as follows:*

$$Cyc_n(a) = lcm(Cyc_{p_1^{\alpha_1}}(a), Cyc_{p_2^{\alpha_2}}(a), Cyc_{p_3^{\alpha_3}}(a), \dots). \quad (10)$$

Though Gauss (1986) showed the function $\psi_p(b)$ instead of $\psi_{p^\alpha}(b)$, we show the function $\psi_{p^\alpha}(b)$ as a generalized form of $\psi_p(b)$, and the function $\psi_n(b)$ as a generalized form of $\psi_{p^\alpha}(b)$. Proofs for Theorems 2.10 and 2.11 are in Suzuki (1995).

Theorem 2.10 *(Gauss 1986). Let $\psi_{p^\alpha}(b)$ be the function for the cardinality of elements each with a given order b in $\mathbb{Z}_{p^\alpha}^*$ with an odd prime p, where b is a divisor of $\lambda(p\alpha)$*

$$\psi_{p^\alpha}(b) = \phi(b), \quad if \ b \setminus \lambda(p^\alpha). \quad (11)$$

Theorem 2.11 *(Suzuki 1995) Let $n = p_1^{\alpha_1} p_2^{\alpha_2} p_3^{\alpha_3}$ ($p_1 = 2$) be a positive integer and let $\psi_n(b)$ be the function for the cardinality of elements each with a given order b in $\mathbb{Z}_n^*$ with a composite n.*

$$\psi_n(b) \quad := \quad \#\{x \in \mathbb{Z}_n^* \mid Ord_n(x) = b\},$$
$$= \quad \sum_{\substack{combinations(b_1, b_2, \dots) \\ hold \\ b = lcm(b_1, b_2, b_3, \dots)}} \prod_i \psi_{p_i^{\alpha_i}}(b_i), \quad (12)$$

$$where \begin{cases} b_i \ is \ a \ divisor \ of \ \lambda(p_i^{\alpha_i}), \\ \psi_{p_1^{\alpha_1}}(b_1) = \psi_{2^{\alpha_1}}(b_1) \\ = \begin{cases} 1 & (\alpha_1 = 0), \\ 1 & (b_1 = 1), \\ 1 & (b_1 = 2, \alpha_1 = 2), \\ 3 & (b_1 = 2, \alpha_1 \geq 3), \\ b_1 & (b_1 \geq 4), \end{cases} \\ \psi_{p_i^{\alpha_i}}(b_i) = \phi(b_i) \ (i \geq 2). \end{cases}$$

3 RSA CYCLES AND RSA PARAMETERS

We redefine RSA cryptosystem for explanation in this section.

RSA encryption process:

$$C = M^e \pmod n, \quad (13)$$

Conversely, RSA decryption process:

$$M = C^d \pmod n, \quad (14)$$

are defined for a plaintext M, the corresponding ciphertext C with a public modulus key n, a public exponent key e and a private exponent key d for each user. For security, $n = pq$ is a product of two large distinct primes p and q.

Since there exists order/cyclic property in modulo operation (Theorem 2.5–Lemma 2.9), RSA

cryptosystem with a public key set (n, e) has an **RSA cycle** $x : M^{e^x} \equiv M \pmod n$ or $C^{e^x} \equiv C \pmod n$ for each M or C. As we described in Introduction, this could be a possibility of attack to RSA, but Rivest (1978a) rejected the attack as meaningless. Therefore, we deal with RSA cycle as a mathematical interesting,

The RSA structure $M^{e^x} \mod n$, with a number x of iterated encryption processes, is recognized as a special form fixed b of the double modular exponentiation structure like: $a^{b^c} \mod n$. The structure is well considered and is mathematically elegant since the plaintext and ciphertext spaces can be taken all elements in a set $\mathbb{Z}_n$ instead of a restricted set $\mathbb{Z}_n^*$. Rivest (1978a) & Maurer (1995) have analyzed RSA as a double modular exponentiation structure. Rivest (1978a) has mentioned the RSA structure as $M^{e^x \mod k} \mod n$ where k denotes the order to which M belongs, modulo n, i.e. $k = Ord_n(M)$. Maurer (1995) has mentioned that RSA cycle for given n, e and M is $Ord_{Ord_n(M)}(e)$. Their descriptions using order modulo a composite are enough to explain the RSA properties but are not enough to count explicit numbers of the RSA properties since the word, order, is restricted to set $\mathbb{Z}_n^*$. Hence we introduce a function $RSA\_Cyc_{n,e}(M)$ for the RSA cycle of a plaintext M in $\mathbb{Z}_n$ with a given RSA public key set (n, e),

$$RSA\_Cyc_{n,e}(M)$$
$$:= min\{x \in \mathbb{N} \mid M^{e^x} \equiv M \pmod n\},$$
$$= min\{x \in \mathbb{N} \mid e^x \equiv 1 \pmod{Cyc_n(M)}\}, (\because Lemma\ 2.7) \quad (15)$$
$$= Ord_{Cyc_{n(M)}}(e).$$

The expression $Ord_{Cyc_n(M)}(e)$ instead of $Ord_{Ord_n(M)}(e)$ allows us to deal with all of plaintexts $M \in \mathbb{Z}_n$ which consist of both $gcd(M, n) = 1$ and $gcd(M, n) \neq 1$ elements.

For correctness of RSA, (e, d) holds the conditions:

$$e \cdot d \equiv 1 \pmod{\phi(n)} \ or \ e \cdot d \equiv 1 \pmod{\lambda(n)}. \quad (16)$$

This implies that an e and the corresponding d have same order in modulo $\phi(n)$ and modulo $\lambda(n)$. Similarly, an M and the corresponding C have same order/cycle in modulo n. Therefore, we can say

$$RSA\_Cyc_{n,e}(M) = RSA\_Cyc_{n,d}(M)$$
$$= RSA\_Cyc_{n,e}(C) = RSA\_Cyc_{n,d}(C). \quad (17)$$

Herein after, we can deal only with $RSA\_Cyc_{n,e}(M)$ as the RSA cycle.

For a given (n, e) and a plaintext M, it is easy to calculate the RSA cycle. However, for all M in $\mathbb{Z}_n$, it is infeasible to calculate the RSA cycles by exhaustive trials due to huge amount of M. To know all the RSA cycles of all plaintexts, we need a reduced method to calculate this. Therefore, we need a new explicit enumerative function $\eta_{n,e}(x)$, which is derived in the next section, for the explicit cardinality of plaintexts M in $\mathbb{Z}_n$ each with a given RSA cycle x.

Here, we survey RSA parameters which describe the property of RSA cryptosystem.

n : the public modulus key for each user, a product of two large distinct primes p and q, The size of n is a security measure against to factoring.

e : the public exponent key for each user.

d : the private exponent key for each user.

$\lambda(n)$: This is used for checking the correctness of RSA.

$\phi(n)$: This is also used as well as $\lambda(n)$,

$\lambda(\lambda(n))$: the possible maximum RSA cycle.

$max\{RSA\_Cyc_{n,e}(M)\}$: the actual maximum RSA cycle with given (n, e), a divisor of $\lambda(\lambda(n))$, e is preferable to hold $max\{RSA\_Cyc_{n,e}(M)\} = \lambda(\lambda(n))$.

$2\phi(p-1)\phi(q-1)$: a lower bound of cardinality of plaintexts $M \in \mathbb{Z}_n$ taking $max\{RSA\_Cyc_{n,e}(M)\}$.

$\eta_{n,e}(\lambda(\lambda(n)))$: the explicit cardinality of plaintexts $M \in \mathbb{Z}_n$ taking RSA cycle $\lambda(\lambda(n))$.

$\phi(\lambda(\lambda(n)))$: a lower bound of cardinality of exponent keys $e \in \mathbb{Z}_{\lambda(n)}^*$ taking RSA cycle $\lambda(\lambda(n))$.

$\psi\lambda_{(n)}(\lambda(\lambda(n)))$: the explicit cardinality of exponent keys $e \in \mathbb{Z}_{\lambda(n)}^*$ taking RSA cycle $\lambda(\lambda(n))$.

In addition, we may introduce a security threshold L in RSA cycle against the cyclic attack. The attackable probability by cyclic attack can be expressed as follows:

$$Prob\{M \mid RSA\_Cyc_{n,e}(M) < L\} := \frac{1}{n} \sum_{\substack{x \backslash \lambda(\lambda(n)) \\ and \\ x<L}} \eta_{n,e}(x). \quad (18)$$

4 NEW EXPLICIT ENUMERATIVE FUNCTION: $\eta_{n,e}(\cdot)$

As we mentioned in the previous section, we need a new enumerative function $\eta_{n,e}(x)$ to count the explicit cardinality of plaintexts M each with a given RSA cycle b in $\mathbb{Z}_n$. The function $\eta_{n,e}(x)$ should exist by an analogy of existence of the function to count all cardinalities of elements each with a given order b in $\mathbb{Z}_n^*$. Then, this paper, for the first time, shows an enumerative function $\eta_{n,e}(x)$ for the explicit cardinality of plaintexts M in $\mathbb{Z}_n$ each with a given RSA cycle x.

Theorem 4.12 *Let $\eta_{n,e}(x)$ be the explicit cardinality of plaintexts M each with a given RSA cycle x in $\mathbb{Z}_n$, where x is a divisor of $\lambda(\lambda(n))$.*

$$\eta_{n,e}(X)$$
$$:= \#\{M \in \mathbb{Z}_n \mid RSA\_Cyc_{n,e}(M) = x\},$$
$$= \#\{M \in \mathbb{Z}_n \mid M^{e^x} \equiv M(\bmod n)\}, \quad (19)$$
$$= \sum_{\substack{b\,hold \\ Ord_b(e)=x}} \sum_{\substack{combinations(b_p,b_q) \\ hold \\ lcm(b_p,b_q)=b}} \theta_p(b_p) \cdot \theta_q(b_q).$$

Proof.

$$\eta_{n,e}(x)$$
$$:= \#\{M \in \mathbb{Z}_n \mid RSA\_Cyc_{n,e}(M) = x\},$$
$$= \#\{M \in \mathbb{Z}_n \mid M^{e^x} \equiv M(\bmod n)\},$$
$$= \#\{M \in \mathbb{Z}_n \mid Ord_{Cyc_n(M)}(e) = x\},$$
$$= \#\{M \in \mathbb{Z}_n \mid Cyc_n(M) = b, Ord_b(e) = x\},$$
$$= \sum_{\substack{b\,hold \\ Ord_b(e)=x}} \#\{M \in \mathbb{Z}_n \mid Cyc_n(M) = b\},$$
$$= \sum_{\substack{b\,hold \\ Ord_b(e)=x}} \#\{M \in \mathbb{Z}_n \mid Cyc_p(M) = b_p,$$
$$Cyc_q(M) = b_q, lcm(b_p,b_q) = b\},$$
$$= \sum_{\substack{b\,hold \\ Ord_b(e)=x}} \sum_{\substack{combinations(b_p,b_q) \\ hold \\ lcm(b_p,b_q)=b}} \theta_p(b_p) \cdot \theta_q(b_q).$$

Note: The condition in $\eta_{n,e}(x)$ does not always hold since e is fixed to a value. Therefore, when b that holds $Ord_b(e) = x$ does not exist, the function $\eta_{n,e}(x)$ returns value zero.

5 COMPUTATIONAL POWERS OF $\psi n(b)$ AND $\eta n,e(x)$

The enumerative function $\psi_n(b)$ is for the explicit cardinality of elements $a \in \mathbb{Z}_n^*$ each with a given order b with a composite n. If an exhaustive trial is adopted to calculate this for all a, we need the computational power of $\#\{\mathbb{Z}_n^*\} = \phi(n)$. If we use $\psi_n(b)$ that the computational power of the number of all variations of $\prod_i \psi_{p_i^{\alpha_i}}(b_i)$ is needed, where $n = p_1^{\alpha_1} p_2^{\alpha_2} p_3^{\alpha_2}$. The number of all the variations is just the number of divisors of n, which can be counted by a function $divs(n)$. Therefore the computational power of $\psi_n(b)$ can be expressed by $divs(n)$ and is reduced to be feasible for large integer n.

The enumerative function $\eta_{n,e}(x)$ is for the explicit cardinality of plaintexts $M \in \mathbb{Z}_n$ each with a given RSA cycle x, where x is a divisor of $\lambda(\lambda(n))$. If an exhaustive trial is adopted to calculate this for all M, we need the computational power of $\#\{M \in \mathbb{Z}_n\} = n$. If we use $\eta_{n,e}(x)$, the computational power of the number of all variations of $\theta_p(b_p) \cdot \theta_q(b_q)$ is needed, where b_p denote all divisors of $(p-1)$, and b_q denote all divisors of $(q-1)$. Therefore, the computational power is expressed by $divs(p-1) \times divs(q-1)$ and is reduced to be feasible for large integer n.

6 NUMERICAL EXAMPLE

In this section, we show a numerical example of small integers to see validity of the functions $\eta_{n,e}(\lambda(\lambda(n)))$ and $\psi\lambda_{(n)}(\lambda(\lambda(n)))$. For an example key set ($n = 2773$,

$e = 3$), a part of RSA parameters are calculated as follows:

$$
\begin{aligned}
n = p \cdot q &= 2773\,(= 47 \times 59), \\
e &= 3, \\
d &= 445, \\
\lambda(n) = lcm(p-1, q-1) &= 1334\,(= 2 \times 23 \times 29), \\
\phi(n) = (p-1)(q-1) &= 2668\,(= 2 \times 23 \times 2 \times 29), \\
\lambda(\lambda(n)) &= 308\,(= 11 \times 2^2 \times 7), \\
\boldsymbol{max}\{RSA\_Cyc_{n,e}(M)\} = \lambda(\lambda(n)) &= 308\,(= 11 \times 2^2 \times 7).
\end{aligned}
$$

Using $\eta_{n,e}(\lambda(\lambda(n)))$, we show the explicit cardinality of plaintexts M in $\mathbb{Z}_n$ each with the maximum RSA cycle $\lambda(\lambda(n))$, as follows:

$$
\begin{aligned}
\eta_{2773,3}(308) &:= \#\{M \in \mathbb{Z}_{2773} \mid RSA\_Cyc_{2773,3}(M) = 308\}, \\
&= \#\{M \in \mathbb{Z}_{2773} \mid M^{e^{308}} \equiv M (\mathrm{mod}\, 2773)\}, \\
&= \sum_{\substack{b\, hold \\ Ord_b(3)=308}} \sum_{\substack{combinations(b_p,b_q) \\ hold \\ lcm(b_p,b_q)=b}} \theta_p(b_p) \cdot \theta_q(b_q),
\end{aligned}
$$

$$
\text{where} \left\{
\begin{aligned}
&b \in 1334, 667\} \text{ hold } Ord_b(3) = 308, \\
&combinations\,(b_p, b_q) = \{(46,58),(46,29),(23,58)\} \\
&\text{hold } lcm(b_p, b_q) = b = 1334, \\
&combinations\,(b_p, b_q) = (23,29) \text{ hold} \\
&lcm(b_p, b_q) = b = 667,
\end{aligned}
\right.
$$

$$
\begin{aligned}
&= \theta_p(46) \cdot \theta_q(58) + \theta_p(46) \cdot \theta_q(29) + \theta_p(23) \cdot \theta_q(58) + \theta_p(23) \cdot \theta_q(29), \\
&= \phi(46) \cdot \phi(58) + \phi(46) \cdot \phi(29) + \phi(23) \cdot \phi(58) + \phi(23) \cdot \phi(29), \\
&= 22 \times 28 + 22 \times 28 + 22 \times 28 + 22 \times 28, \\
&= 2464.
\end{aligned}
\tag{20}
$$

In Table 1, we show an experimental result including the same key example. We confirmed, the validity of Theorem 4.12, that all the values of each boxes in Table 1 are the same as the calculation by the $\eta_{n,e}(x)$.

Table 1. A complete RSA cycle distribution table for given keys $n = 2773$ and some e.

RSA cycle	$e = 3$	$e = 11$	$e = 15$	$e = 19$
308	2464	2464	2464	2464
154				
77				
44				
28	168	168	168	168
22		132	132	132
14				
11	132			
7				
4				
2				
1	9	9	9	9
Total	2773	2773	2773	2773

$\longrightarrow$ primitive elements(mod n):1848
order $\frac{\lambda(n)}{2}$(mod n) elements:616
Total:2464

Using $\psi\lambda_{(n)}(\lambda(\lambda(n)))$, we count the cardinality of encryption exponent keys $e \in \mathbb{Z}^*_{\lambda(n)}$ taking the maximum RSA cycle $\lambda(\lambda(n))$ as follows:

$$
\begin{aligned}
\psi_{1334}(308) &:= \#\{x \in \mathbb{Z}^*_{1334} \mid Ord_{1334}(x) = 308\} \\
&= \sum_{\substack{combinations(b_1,b_2,b_3) \\ hold \\ 308=lcm(b_1,b_2,b_3)}} \psi_2(b_1) \cdot \psi_{23}(b_2) \cdot \psi_{29}(b_3),
\end{aligned}
$$

where $combinations\,(b_1, b_2, b_3) \in \{(1,11,28), (1,22,28)\}$
hold $308 = lcm(b_1, b_2, b_3)$

$$
\begin{aligned}
&= \psi_2(1) \cdot \psi_{23}(11) \cdot \psi_{29}(28) + \psi_2(1) \cdot \psi_{23}(22) \cdot \psi_{29}(28), \\
&= \phi(1) \cdot \phi(11) \cdot \phi(28) + \phi(1) \cdot \phi(22) \cdot \phi(28), \\
&= 1 \times 10 \times 12 + 1 \times 10 \times 12, \\
&= 240.
\end{aligned}
\tag{21}
$$

$\psi_{1334}(308) = 240$ is theoretical value of the explicit cardinality of exponent keys $e \in \mathbb{Z}^*_{1334}$ taking the maximum RSA cycle 308. In Table 2, we show all 240 of exponent keys $e \in \mathbb{Z}^*_{1334}$ taking RSA cycle 308. The two lower bound parameters are calculated as:

a lower bound of cardinality of plaintexts $M \in \mathbb{Z}_n$ taking $max\{RSA\_Cyc_{n,e}(M)\}$

$$
2\phi(p-1)\phi(q-1) = 1232(= 2 \times 22 \times 28), \tag{22}
$$

a lower bound of cardinality of exponent keys $e \in \mathbb{Z}^*_{\lambda(n)}$ taking RSA cycle $\lambda(\lambda(n))$

$$
\phi(\lambda(\lambda(n))) = 120(= 10 \times 2 \times 6). \tag{23}
$$

For the exponent keys e, explicit number 240 is double of lower bound 120. In case of an RSA modulus key $n = 2773$, there are two types of RSA cycle distribution patterns in Table 1. In Table 2, $e = 3$ belongs a cyclic group generated by itself 3. $e = 11, 15, 19$ belong a cyclic group generated by 11. We concluded that the exponent keys group determines the type of RSA cycle distribution pattern.

In Table 1, for $e = 3$, within 2773 plaintexts, 2464 plaintexts experimentally take the maximum RSA cycle 308. In this experiment, explicit number 2464 is double of lower bound 1232.

As a consequence of these results, we confirmed the validity of the functions.

7 CONCLUSION

This paper has revealed the complete cyclic structure of the RSA cryptosystem for any given key set (n, e) by showing new enumerative function

Table 2. All 240 primitive elements (mod $\lambda(n) = 1334$). The number of primitive roots, or generators of cyclic group, is $\phi(\lambda(\lambda(n))) = 120$. These can be expressed as Ith power of a generator e, like $e^I \bmod 1334$.

	Cyclic group generated by 3						Cyclic group generated by 11				
I	+1	+3	+5	+7	+9	I	+1	+3	+5	+7	+9
0	$e=3$	27	243		1007	0	$e=11$	1331	971		1307
10		193	403	959	627	10		891	1091	1279	15
20		95	855	1025	1221	20		839	135	327	881
30	317			311	131	30	1215			263	1141
40	1179	1273	785	395		40	659	1033	931	595	
50	1313	1145		699	955	50	375	19		707	171
60	591		1181	1291	947	60	681		205	793	1239
70	519	669	685		791	70	511	467	479		201
80	449	39	351	491	417	80	309	37	475	113	333
90		427	1175	1237		90		1017	329	1123	
100	147	1323		443	1319	100	293	769		1303	251
110	1199	119	1071	301		110	1023	1055	925	1203	
120		653	541	867	1133	120		155	79	221	61
130	859		211	565	1083	130	711		549	1063	559
140	409		1113		775	140	939		1029		727
150	305	77	693	901	105	150	1257	21	1207	641	189
160		501		561	1047	160		433		385	1229
170	85	765		601	73	170	635	797		379	503
180	657	577	1191			180	833	743	525		
190	1139	913	213	583	1245	190	723	773	153	1171	287
200	533		485	363		200	43		1249	387	
210	55	495	453		675	210	569	815	1233		665
220	739	1315	1163	1129	823	220	425	733	649	1157	1261
230		1297	1001	1005	1041	320		1075	677	543	337
240	31	279		1255	623	240	757	885		143	1295
250	271		607	127		250	617		983	217	
260	949	537	831	809	611	260	843	619	195	917	235
270	163			101	909	270	421			1121	907
280	177	259	997		717	280	359	751	159		89
290	1117	715	1099		975	290	97	1065	801		247
300		269	1087	$d=445$	$e=3$	300		1187	889	$d=849$	$e=11$

e.g. for $I = 307$, $d = e^I = 3^{307} = 445 \pmod{1334}$.   e.g. for $I = 307$, $d = e^I = 11^{307} = 849 \pmod{1334}$.

$\eta_{n,e}(x)$ to count the explicit cardinality of plaintexts M in $\mathbb{Z}_n$ each with a given RSA cycle x. We have surveyed RSA parameters including the function $\eta_{n,e}(\lambda(\lambda(n)))$ for the explicit cardinality of plaintexts $M \in \mathbb{Z}_n$ taking the maximum RSA cycle $\lambda(\lambda(n))$, and the function $\psi\lambda(_n)(\lambda(\lambda(n)))$ for the explicit cardinality of exponent keys $e \in \mathbb{Z}^*_{\lambda(n)}$ taking the maximum RSA cycle $\lambda(\lambda(n))$. By deriving the function $\eta_{n,e}(x)$, we have shown a complete RSA cycle distribution table as an example. In the example analysis, we have dealt with small numbers to see validity of functions.

For further research in the subject, we can discuss large numbers since the function can be used to count any large n, e and any RSA cycle x.

REFERENCES

[1] Blakley, B. & Blakley, G.R. 1978, 1979. Security of Number Theoretic Public Key Cryptosystems Against Random Attack I, II and III, Cryptologia, 2(4): 305–321, 3(1): 29–42 & 3(2): 105–118.
[2] Blakley, G.R. & Borosh, I. 1979. Rivest-Shamir-Adleman Public Key Cryptosystem Do not Always Conceal Messages, Comp. & Maths. with Appls., 5: 169–178.
[3] Carmichael, R.D. 1910. Note on A New Number Theory Function, Bulletin of the American Mathematical Society, 16: 232–238.
[4] Gauss, C.F. 1986. Disquisitiones Arithmeticae, English edition of Springer-Verlag. Sections 52–54.
[5] Grošek, O. 1994. Remarks Concerning RSA-Cryptosystem Exponents, Mathematica Slovaca, 44(2): 279–285.

[6] Hayashi, A. & Kawamura, H. 1990. A New Lower Bound concerning the Cyclic Property in the RSA Cryptosystem, Trans. on IEICE, part D-I, J-73D-I: 370–371.

[7] Kaliski, B. & Robshaw M. 1995. The Secure Use of RSA, CryptoBytes, 1(3): 7–13.

[8] Katsuno, K. 1983. A Countermeasure Against a Cryptanalysis of the Rivest-Shamir-Adleman Cryptosystem, Trans. on IECE, part D, J-66D: 963–969.

[9] Knuth, D.E. 1981. The Art of Computer Programming, vol.2: Seminumerical Algorithms, 2/e, Reading, MA: Addison-Wesley. p.271.

[10] Maurer, U. 1995. Fast Generation of Prime Numbers and Secure Public-Key Cryptographic Parameters, Journal of Cryptology, 8(3): 123–155.

[11] Menezes, A.J., van Ooschot, P.C. & Vanstone, S.A. 1996. Handbook of Applied Cryptography, CRC Press.

[12] Rivest, R.L. 1978a. Remarks on a Proposed Cryptanalytic Attack on the M.I.T. Public-Key Cryptosystem, Cryptologia, 2(1): 62–65.

[13] Rivest, R.L., Shamir, A. & Adleman, L. 1978b. A Method for Obtaining Digital Signatures and Public-Key Cryptosystems, Communications of ACM, 21(2): 120–126.

[14] Simmons, G.J. & Norris, M.J. 1977. Preliminary Comments on the M.I.T. Public-Key Cryptosystem, Cryptologia, 1(4): 406–414.

[15] Singmaster D. 1966. A Maximal Generalization of Fermat's Theorem, Mathematics Magazine, 39: 103–107.

[16] Suzuki, H. 1988. Number Theoretic Analysis of RSA Cryptosystem, M.S. Thesis, Nagoya Institute of Technology.

[17] Suzuki, H. & Nakamura, T. 1995. On the Cardinality of Elements each with a Given Order in $\mathbb{Z}_n^*$, Tech. Rep. of IEICE on Information Security, 95(31): 1–6.

[18] Suzuki, H. 1997. A Complete Cyclic Analysis of RSA Cryptosystem, Tech. Rep. of IEICE on Information Security, 97(3): 21–32.

Network Security and Communication Engineering – Chan (Ed.)
© *2015 Taylor & Francis Group, London, ISBN: 978-1-138-02821-0*

A Weighted Secret Sharing Scheme based on the CRT and RSA

X.D. Dong & X.P. Jing

College of Information Engineering Dalian University, Dalian, P.R.China

ABSTRACT: In a Weighted Secret Sharing Scheme (WSSS), each share of a shareholder has a positive weight. The secret can be recovered if the overall weight of shares is equal to or larger than the threshold; but the secret cannot be recovered if the overall weight of shares is smaller than the threshold value. The known WSSS based on the Chinese Remainder Theorem (CRT) needs a security channel which is unpractical. To overcome this problem, we propose to enhance the WSSS based on CRT by incorporating the well-known RSA Cryptosystem. In the proposed WSSS, every shareholder including shareholders having higher weights keeps only one share, participants select their secret shadows by themselves. Also, a secure channel among the dealer and participants is no longer needed. In addition, each participant can check whether another participant provides the true secret shadow or not. The security of the proposed scheme is based on the RSA cryptosystem and intractability of the Discrete Logarithm.

1   INTRODUCTION

In a (t, n)-threshold secret sharing scheme(TSSS), a secret is shared among n participants in such a way that any t (or more) of them can reconstruct the secret while a group of $t - 1$ or fewer can not obtain any information. The idea of a secret sharing scheme was first introduced independently by Shamir (1979) and Blakley (1979). A TSSS has many practical applications, such as opening a bank vault, launching a nuclear, or authenticating an electronic funds transfer. The weighted secret sharing scheme (WSSS) was originally proposed by Shamir (1979). In a WSSS, each share of a shareholder has a positive weight. The secret can be recovered if the overall weight of shares is equal to or larger than the threshold; but the secret cannot be recovered if the overall weight of shares is smaller than the threshold value. In fact, Shamirs (t, n)-threshold secret sharing scheme is a special type of WSSS in which the weight of all shares is the same. There are several threshold secret sharing schemes based on the Chinese remainder theorem(CRT). Mignotte (1983), Asmuth & Bloom (1983) proposed TSSS based on the CRT. Iftene (2005) and Qiong et al. (2005) proposed two CRT-based verified secret share (VSS) schemes. However, Kaya (2008) pointed out that both schemes cannot prevent a corrupted dealer to distribute inconsistent shares to shareholders. They proposed a CRT-based VSS which uses a range proof technique proposed by Benaloh (1987). Recently, Liu et.al (2013) proposed an authenticated group key distribution using the CRT and Guo & Zhao (2013) proposed a quantum secret sharing based on the CRT. Harn et.al (2014) proposed a CRT-based VSS. Dong

(2014) proposed to enhance threshold secret sharing schemes based on CRT by incorporating the well-known RSA cryptosystem. Recently, Harn & Miao (2014) proposed a WSSS based on the CRT. However, their scheme needs a security channel which is unpractical. To overcome this problem, we propose to enhance the WSSS based on CRT by incorporating the well-known RSA Cryptosystem. In the proposed WSSS, every shareholder including shareholders having higher weights keeps only one share, participants select their secret shadows by themselves. Also, a secure channel among the dealer and participants is no longer needed. In addition, each participant can check whether another participant provides the true secret shadow or not. The security of the proposed scheme is based on the RSA cryptosystem and intractability of the Discrete Logarithm.

The rest of this paper is organized as follows. In Section 2 we give some preliminaries about the CRT. A brief review is given in Section 3 about TSSSs based on the CRT. In Section 4 we propose a new WSSS based on the CRT by incorporating the well-known RSA Cryptosystem. Section 5 gives the analysis of the proposed scheme. Finally, concluding remarks are given in Section 6.

2   PRELIMINARIES

Several versions of the CRT have been proposed. The next one is called the general CRT (Fraenkel 1963, Ore 1952).

   Theorem: Let $k > 2$, $p_1 > 2, \dots, p_k \geq 2$, and $b_1, \dots, b_k \in Z$. The system of equations

$$\begin{cases} x \equiv b_1 (\text{mod } p_1) \\ x \equiv b_2 (\text{mod } p_2) \\ \vdots \\ x \equiv b_k (\text{mod } p_k) \end{cases}$$

has solutions in Z if and only if $b_i \equiv b_j \ (\text{mod } (p_i, p_j))$, for all $1 \leq i, j \leq k$. Moreover, if the above system of equations has solutions in Z, then it has a unique solution in $Z[p_1, \cdots, p_k]$, where $[p_1, \cdots, p_k]$ is the least common multiple of $p_1, \cdots, p_k$.

When $(p_i, p_j) = 1$, for all $1 \leq i, j \leq k$, one gets the standard version of the CRT. Garner (1959) has found an efficient algorithm for this case and Fraenkel (1963) has extended it to the general case.

3 BRIEF REVIEW OF MIGNOTTE'S SCHEME, ASMUTH-BLOOM'S SCHEME AND HARN-MIAO'S SCHEME

3.1 Mignotte's threshold secret sharing scheme

Mignotte's threshold secret sharing scheme (Mignotte 1983) uses some special sequences of integers, referred to as the Mignotte sequences. Let n be a positive integer, $n \geq 2$, and $2 \leq t \leq n$. An (t, n)-Mignotte sequence is a sequence of pairwise coprime positive integers $p_1 < p_2 < \cdots < p_n$ such that $\prod_{i=0}^{t-2} p_{n-i} < \prod_{i=1}^{t} p_i$.

Given a publicly known (t, n)-Mignotte sequence, the scheme works as follows:

1 The secret S is chosen as a random integer such that $\prod_{i=0}^{t-2} p_{n-i} < S < \prod_{i=1}^{t} p_i$,
2 The shares I_i are chosen as $I_i \equiv S(\text{mod } p_i)$, for all $1 \leq i \leq n$,
3 Given t distinct shares $I_{i_1}, \cdots, I_{i_t}$, the secret S is recovered, using the CRT, as the unique solution modulo $p_{i_1} \cdots p_{i_t}$ of the system

$$\begin{cases} x \equiv I_{i_1} (\text{mod } p_{i_1}) \\ x \equiv I_{i_2} (\text{mod } p_{i_2}) \\ \vdots \\ x \equiv I_{i_t} (\text{mod } p_{i_t}) \end{cases}$$

3.2 Asmuth-Bloom's threshold secret sharing scheme

This scheme, proposed by Asmuth & Bloom (1983), also uses some special sequences of integers. More exactly, a sequence of pairwise coprime positive integers $p_0, p_1 < p_2 < \cdots < p_n$ is chosen such that $p_0 \prod_{i=0}^{t-2} p_{n-i} < \prod_{i=1}^{t} p_i$.

Given a publicly known Asmuth-Bloom sequence, the scheme works as follows:

1 The secret S is chosen as a random element of the set Z_{p0};
2 The shares I_i are chosen as $I_i = (S + \gamma p_0)(\text{mod } p_i)$, for all $1 \leq i \leq n$, where γ is an arbitrary integer such that $(S + \gamma p_0) \in Z_{p_1 \cdots p_i}$;
3 Given t distinct shares $I_{i_1}, \cdots, I_{i_t}$, the secret S is recovered as $S = x_0 \ (\text{mod } p_0)$, where x_0 is obtained, using the CRT, as the unique solution modulo $p_{i_1} \cdots p_{i_t}$ of the system

$$\begin{cases} x \equiv I_{i_1} (\text{mod } p_{i_1}) \\ x \equiv I_{i_2} (\text{mod } p_{i_2}) \\ \vdots \\ x \equiv I_{i_t} (\text{mod } p_{i_t}) \end{cases}$$

3.3 Harn-Miao's weighted secret sharing scheme

This scheme, proposed by Harn & Miao (2014), consists of three steps: parameters selection, shares generation and secret reconstruction.

1 Parameters selection

The dealer selects an integer p_0 and a sequence of pairwise coprime positive integers, $p_1^1 < p_2^1 < \cdots < p_n^1$ such that $p_0 \prod_{i=0}^{t-2} p_{n-i}^1 < \prod_{i=1}^{t} p_i^1$, and $\gcd(p_0, p_i^1) = 1, i = 1, 2, \cdots, n$ where p_i^1 is the public information associated with each shareholder U_i^1 having his/her share with the minimal weight 1. For this given sequence, the dealer chooses the secret s as an integer in the set Z_{p_0} The dealer selects an integer α such that $\prod_{i=0}^{t-2} p_{n-i}^1 < s + \alpha p_0 < \prod_{i=1}^{t} p_i^1$. For each shareholder U_i^j, having a larger weight $j > 1$, the dealer selects p_i^j satisfying $\prod_{i=t-j+1}^{t} p_i^1 < p_i^j < \prod_{i=n-t+2}^{n-t+(1+j)} p_i^1 \cdot p_i^j$ is the public information associated with shareholder U_i^j. Note that the selected value p_i^j should be relatively coprime to all other public parameters of shareholders and $\gcd(p_0, p_i^j) = 1, \forall i, j$.

2 Share generation

Share for the shareholder U_i^j is generated as $s_i^j \equiv s + \alpha p_0 \bmod p_i^j \cdot s_i^j$ is secretly sent to shareholder U_i^j.

3 Secret reconstruction

Given any subset of distinct shares having overall weight t or larger than t, the secret value $x = s + \alpha p_0$, can be reconstructed by using the standard CRT. Then, the secret s can be recovered by computing $s = x \bmod p_0$.

4 PROPOSED SCHEME

Let $\{P_1, P_2, \cdots, P_n\}$ be a set of participants and D the dealer of the scheme. The scheme needs a bulletin board. Only the dealer D can change and update the information on the bulletin board and other persons can read and download the information from the bulletin board.

4.1 Initialization phase

1 The dealer D chooses two strong primes $p = 2p' + 1$ and $q = 2q' + 1$, where p' and q' are also primes. Both p and q should be so safe that anybody can't factor $N = pq$ efficiently. Then the dealer chooses an integer g such that $1 < g < N$, $(g, N) = 1$ and $(g \pm 1, N) = 1$. Then the order of g is equal to p' q' or $2p'$ q' (Ateniese, Camenisch, Joye & Tsudik 2000). D publishes system information $[g, N]$ on the bulletin board and keeps p and q in secret. Each participant P_j randomly chooses a secret integer s_j from $[2, N]$ as her/his own secret shadow, and computes $R_j = g^{s_j} \pmod{N}$, and then sends R_j to D. D must make sure that R_i and R_k are different when $i \neq k$. If $R_i = R_k$, D asks these participants to choose secret shadows again until $R1, \cdots, R_n$ are different.

2 D chooses the secret integer e, $1 < e < \phi(N) = (p-1)(q-1)$, such that $(e, \phi(N)) = 1$, computes $R_0 = g^e \pmod{N}$ and then uses extended Euclidean algorithm to compute a unique integer h, $1 < h < \phi(N)$, such that $eh \equiv 1 \pmod{\phi(N)}$. D publishes R_0, h on the bulletin board.

3 D selects an integer p_0 and a sequence of pairwise coprime positive integers, $p_1^1 < p_2^1 < \cdots < p_n^1$ such that $p_0 \prod_{i=0}^{t-2} p_{n-i}^1 < \prod_{i=1}^{t} p_i^1$ and gcd $(p_0, p_i^1) = 1, i = 1, 2, \cdots, n$. i.e., the sequence $p_1^1, \cdots, p_n^1$ is a Asmuth and Bloom's sequence. For each shareholder U_i^j, having a larger weight $j > 1$, the dealer selects p_i^j satisfying $\prod_{i=t-j+1}^{t} p_i^1 < p_i^j < \prod_{i=n-t+2}^{n-t+(1+j)} p_i^1$, where $p_{n+1}^1 = p_{n-t+1}^1$ when $j = t$. p_i^j is the public information associated with shareholder P_i^j. Note that the selected value p_i^j should be relatively coprime to all other public parameters of shareholders and gcd $(p_0, p_i^j) = 1, \forall i, j$.

4.2 Divide secret phase

Suppose that $S \in Z_{p0}$ are a secret to be shared. The dealer selects an integer α such that $\prod_{i=0}^{t-2} p_{n-i}^1 < S + \alpha p_0 < \prod_{i=1}^{t} p_i^1$. The dealer D computes $y_{ij} = S + \alpha p_0 \pmod{p_i^j} \oplus R_i^e \pmod{N}$ where $1 \leq j \leq t$ and $\oplus$ denotes the XOR operation, i.e., componentwise addition modulo 2. D publishes triples

(p_i^j, R_i, y_{ij}), where $i = 1, \cdots n$, $1 \leq j \leq t$, on the bulletin board.

4.3 Recover secret phase

Without loss of generality, assume that participants $P_1, P_2, \cdots, P_a$ have the overall weight of shares being equal to or larger than the threshold t and cooperate to reconstruct the secret data S.

1 Each participant P_v, $v = 1, 2, \cdots, a$ downloads public information R_0, h, and uses her/his secret shadow s_v to compute $R_0^{s_v} \pmod{N}$ and then sends it and $R_v = g^{s_v} \pmod{N}$ to the designated combiner.

2 After receiving $R_0^{s_v} \pmod{N}$ and $R_v = g^{s_v} \pmod{N}$, the designated combiner computes $(R_0^{s_v})^h \pmod{N}$, and checks whether $R_0^{hs_v} \equiv R_v \pmod{N}$ is true or not. If $R_0^{hs_v} \not\equiv R_v \pmod{N}$, the designated combiner knows that P_v does not provide her/his true secret shadow s_v.

3 The designated combiner downloads public information (p_i^j, R_i, y_{ij}) on the bulletin board, where $i = 1, \cdots a$, and computes $y_{ij} \oplus R_0^{s_i} \pmod{N} = S + \alpha p0 \pmod{p_i^j} \oplus R_j^e \pmod{N} \oplus R_0^{s_i} \pmod{N} = S + \alpha p0 \pmod{p_i^j}$, where $i = 1, \cdots, a$.

4 The designated combiner uses the standard CRT to solve the system of equations

$$
\begin{cases}
x \equiv y_{1j_1} \oplus R_0^{s_1} (\bmod N)(\bmod p_1^{j_1}) \\
x \equiv y_{2j_2} \oplus R_0^{s_2} (\bmod N)(\bmod p_2^{j_2}) \\
\vdots \\
x \equiv y_{aj_a} \oplus R_0^{s_a} (\bmod N)(\bmod p_a^{j_a})
\end{cases}
$$

and gets the general solutions $S + \alpha p_0 + (p_1^{j_1} \cdots p_a^{j_a})u$, where $u \in Z$. The unique nonnegative solution less than $\prod_{i=1}^{t} p_i^1$ is the secret data $S + \alpha p_0$. In fact, when $j_w > 1$, $\prod_{j=t-j_w+1}^{t} p_j^1 < p_1^{j_w}$. The product $\prod_{j=t-j_w+1}^{t} p_j^1$ consists of j_w elements. So, $(p_1^{j_1} \cdots p_a^{j_a})$ is greater than a product of some p_i^1, which consists of $j_1 + \ldots + j_a$ elements. However, $j_1 + \ldots + j_a \geq t$ and $p_1^1 < p_2^1 < \cdots < p_n^1$ which imply $(p_1^{j_1} \cdots p_a^{j_a}) > \prod_{i=1}^{t} p_i^1$. Then, the secret S can be recovered by computing $S = x_0 \bmod p_0$, where x_0 is the unique nonnegative solution less than $\prod_{i=1}^{t} p_i^1$ the resulted system.

5 ANALYSIS OF THE SCHEME

5.1 Verification analysis

From the Euler Theorem it follows that $g^{\phi(N)} \equiv 1 \pmod{N}$. If P_v isn't a cheater, then $R_0^{hs_v} \equiv g^{ehs_v} \equiv g^{s_v} = R_v$

(mod N) since $eh \equiv 1 (\mathrm{mod}\ \phi(N))$. Otherwise, P_v does not provide her/his true secret shadow.

Remark: If a malicious participant randomly chooses an integer s from the range of 2 to N, then performs the subsequent procedures based on s instead of s_v, she/he can pass the above verification successfully. However, $R_s = g^s (\mathrm{mod}\ N)$ is not equal to any one of R_i in the public information (p_i^j, R_i, y_{ij}) on the bulletin board, where $i = 1, \dots n$. So, the combiner can identify the cheater.

5.2 *Security analysis*

1 Participants whose the overall weight of shares is less than the threshold t can not cooperate to reconstruct the secret data S. In fact, when $j_w > 1, p_i^{j_w} < \prod_{i=n-t+2}^{n-t+(1+j_w)} p_i^1$. The product $\prod_{i=n-t+2}^{n-t+(1+j_w)} p_i^1$ consists of j_w elements. So, $(p_1^{j_1} \cdots p_a^{j_a})$ is less than a product of some p_i^1, which consists of $j1 + \dots + j_a$ elements. However, $j_1 + \dots + j_a < t$ and therefore $j_1 + \dots + j_a \leq t-1$. Thus, $(p_1^{j_1} \cdots p_a^{j_a}) < \prod_{i=0}^{t-2} p_{n-i}^1 < S + \alpha p_0 \prod_{i=1}^{t} p_i^1$. One can only get that $S + \alpha p0 \equiv x0 (\mathrm{mod}\ (p_1^{j_1} \cdots p_a^{j_a}))$, where x_0 is the unique solution modulo $(p_1^{j_1} \cdots p_a^{j_a})$ of the resulted system. So, $S + \alpha p_0 = x_0 + \beta p_1^{j_1} \cdots p_a^{j_a}$, where β such that $\prod_{i=0}^{t-2} p_{n-i}^1 < x_0 + \beta p_1^{j_1} \cdots p_a^{j_a} < \prod_{i=1}^{t} p_i^1$ since $p_0 < \prod_{i=1}^{t} p_i^1 / \prod_{i=0}^{t-2} p_{n-i}^1$. But, there is only one value of β such that $S + \alpha p_0 = x_0 + \beta p_1^{j_1} \cdots p_a^{j_a}$. The successful probability of this approach is smaller than $1/p_0$. Therefore, the secret S can not be recovered by computing $S = x_0 \bmod p_0$.

2 If system attacker personates the dealer to publish a pseudo secret data, she/he has to get the secret number e. Since $R_0 = g^e (\mathrm{mod}\ N)$, she/he is faced with the difficulty in solving the discrete logarithm problem. Another method of getting e is to solve the equation $eh \equiv 1 (\mathrm{mod}\ \phi(N))$. This needs factorization N into a product of primes which is also difficult.

3 In the secret reconstruction phase, each participant only provides a public value and does not have to disclose her/his secret shadow. Anyone who wants to get the participant's secret shadow will be faced with the difficulty in solving the discrete logarithm problem. The reuse of the secret shadow is secure.

4 Sung et.al (2011) proposed new modular exponentiation and CRT recombination algorithms which are secure against all known power and fault attacks.

5.3 *Performance analysis*

There are efficient algorithms for modular exponentiation and CRT recombination (Fraenkel 1963, Garner

1959, Sung, Tae & Dong 2011). The XOR operation is of negligible complexity. What's more, each participant chooses her/his secret shadow by her/himself in the proposed scheme, P_j computes $R_j = g^{s_j} (\mathrm{mod}\ N)$, this also cuts the computation quantity of D. In addition, the system doesn't need a security channel, which also cuts the cost of the system. Therefore the proposed scheme is efficient and practical.

6 CONCLUDING REMARKS

This paper proposes to enhance the weighted secret sharing scheme based on the CRT by incorporating the well-known RSA Cryptosystem. In the proposed scheme, a security channel is no needed for the proposed scheme. Participants select their secret shadows by themselves. Furthermore, once the secret has been reconstructed, it is not required that the dealer redistributes a fresh shadow over a security channel to every participant. The scheme is based on the RSA cryptosystem and intractability of the Discrete Logarithm.

ACKNOWLEDGMENT

This work was supported by the National Natural Sciences Foundation of China under Project Code 10171042 and the Research Project of Liaoning Education Bureau under Project Code L2014490.

REFERENCES

Asmuth, C. A. & J. Bloom (1983). A modular approach to key safeguarding. *IEEE Transactions on Information Theory. 29*, 208–210.

Ateniese G., J.Camenisch, M.Joye & G. Tsudik (2000). A practical and provably secure coalition resistant group signature scheme. in *Proc. of CRYPTO00.*, Santa Barbara, USA, pp. 255–270.

Benaloh (1987). Secret sharing homomorphisms: Keeping shares of a secret secret. in *Crypto '86.LNCS 263*, Springer–Verlag, pp.251–260.

Blakley, G. (1979). Safeguarding Cryptographic Keys. *Proceedings of the National Computer Conference. 48*, 313–317.

Dong, X.D. (2014). A Multi–secret Sharing Scheme Based on the CRT and RSA. *International Journal of Electronics and Information Engineering.* in press.

Fraenkel, A. S. (1963). New proof of the generalized CRT. *Proceedings of American Mathematical Society 14*, 790–791.

Garner, H. C. (1959). The residue number system. *IRE Transactions on Electronic Computers 8*, 140–147.

Guo Y.& Y. Zhao (2013). High–efficient quantum secret sharing based on the Chinese remainder theorem via the orbital angular momentum entanglement analysis. *Quantum Information Processing. 12*, 1125–1139.

L. Harn & F. Miao (2014). Weighted Secret Sharing Based on the Chinese Remainder Theorem. *International Journal of Network Security. 16*, 420–425.

L. Harn, F. Miao & C. C. Chang (2014). Verifiable secret sharing based on the Chinese remainder theorem. *Security and Communication Networks. 7*, 950–957.

Iftene, S. & I. Boureanu (2005). Weighted threshold secret sharing based on the Chinese remainder theorem. *Scientific Annals of the "Al. I. Cuza" University of Iasi Computer Science Section, XVI*, 161–172.

Kaya, K.,A. A. Selcuk (2008). A verifiable secret sharing scheme based on the Chinese Remainder Theorem. in *Advances in Cryptology–Indocrypt'08. LNCS 5365*, Springer–Verlag, pp. 414–425.

Liu, Y.,L. Harn & C. C. Chang (2013). An authenticated group key distribution mechanism using theory of numbers. *International Journal of Communication Systems.* online : 4 JUL 2013, DOI: 10.1002/dac.2569.

Mignotte, M. (1983). How to share a secret. *T.Beth, editor, Cryptography–Proceedings of the Workshop on Cryptography. Burg Feuerstein, 1982, Lecture Notes in Computer Science 149*, 371–375.

Ore, O. (1952). The general CRT. *American Mathematical Monthly. 59*, 365–370.

Qiong, Zhifang, Xiamu & Shenghe (2005). A non-interactive modular verifiable secret sharing scheme. in *Proceedings of ICCCAS 2005: International Conference on Communications, Circuits and Systems.*, IEEE, Los Alamitos, pp. 84–87.

Shamir, A. (1979). How to share a secret. *Communications of the ACM 22*, 612–613.

Sung–Kyoung Kima, Tae Hyun Kima & Dong–Guk Hanb (2011). An efficient CRT–RSA algorithm secure against power and fault attacks. *Journal of Systems and Software. 84*, 1660–1669.

Network Security and Communication Engineering – Chan (Ed.)
© 2015 Taylor & Francis Group, London, ISBN: 978-1-138-02821-0

Study on the method of intrusion detection cluster combined FCM with BP optimized by GA

G.Y. Chen, X.Q. Pan, Y.C. Cao, R.X. Li & N. Sun
School of Information Engineering, Minzu University of China, Beijing, China

ABSTRACT: First, we use Fuzzy C-means Algorithm (FCM) method to cluster the intrusion data, and then we establish a classification model based on the result of FCM clustering and BP neural network. With this model, we can forecast the intrusion data, improve the clustering results, and make the result of FCM clustering more stable. And we optimize the BP neural network with GA algorithm and improve the results in the global scope. The input variables are the intrusion detection data KDDCUP1999 [1], and the output are the results of the intrusion data clustering. Experimental results show that it is an effective network intrusion detection method.

KEYWORDS: Intrusion detection, Fuzzy c-means (FCM), BP neural network (BP), Genetic Algorithm (GA).

1 INTRODUCTION

Intrusion detection is the detection of intrusion behavior, and it is done through the collection and analysis of network behavior, security logs, audit data, information on network and a number of key points in computer systems to check whether the network and system exist behaviors violating security policy and signs of being attacked [2]. Fuzzy C-Means Clustering algorithm is an effective algorithm, and it has been introduced to the intrusion detection. However, fuzzy C-means clustering algorithm itself is sensitive to the initialization value, easy to fall into local extreme point, and not the optimal solution. BP neural network has a strong information processing capability, which can effectively be applied to the classification, clustering, training BP network fuzzy clustering results, and generate new classification model, and then the intrusion data can be predicted with classification model, but BP neural networks also has disadvantages, that is, it is easy to fall into local minimal, and thus take advantage of better global optimization ability of genetic algorithm to optimize the BP network, so that clustering results of intrusion data are greatly improved.

2 FCM, BP, GA ALGORITHM OUTLINE

2.1 *FCM algorithm*

Clustering methods are commonly used in data mining, and it divides physical or abstract objects into several categories. Objects in the same class have high similarity between individuals; Otherwise, objects in different classes have low similarity between individuals. Fuzzy c-means clustering algorithm (FCM) [3] is a clustering algorithm which uses the membership to determine the degree of each element belonging to a class. FCM puts n data vector x_k into c fuzzy classes, and calculates the center of every class so as to minimize the objective function, and the objective function of FCM is shown in equation (1) [2].

$$J = \sum_{i=1}^{n} \sum_{j=1}^{c} \left(u_{ij} \right)^m \| x_j - v_j \|$$ (1)

In equation (1), u_{ij} is the fuzzy membership of individual x belonging to the j-class, m is the fuzzy weighted index, v_j is the cluster centers of j-class, and u and v are calculated as shown in equation (2) and equation (3).

$$u_{ij} = \begin{cases} \left[\sum_{k=1}^{c} \frac{\| x_i - v_j \|^{\frac{2}{m-1}}}{\| x_i - v_k \|^{\frac{2}{m-1}}} \right]^{-1} & \| x_i - v_k \| \neq 0 \\ 1 & \| x_i - v_k \| = 0 且 k = j \\ 0 & \| x_i - v_k \| = 0 且 k \neq j \end{cases}$$ (2)

$$V_j = \frac{\sum_{i=1}^{n} u_{ij}^m x_i}{\sum_{i=1}^{n} u_{ij}^m}$$ (3)

FCM algorithm iterative process is as follows:

Step 1: Given category c, weighted index m.
Step 2: Select the initial cluster centers v.
Step 3: Calculate the fuzzy membership matrix u according to equation (2).
Step 4: Calculate each class center v according to equation (3).
Step 5: Calculate fuzzy clustering target value J according to equation (1), and determine whether it meets the condition, if so, the algorithm terminates; otherwise, return to Step 3.

FCM algorithm eventually gets fuzzy membership matrix u, and determines which category an individual object belongs to according to the position of the largest element in each column of the membership matrix [2].

2.2 BP algorithm

BP neural network is a multilayer feed forward neural network, which contains input layer, hidden layers, and output layer. The hidden layer is the middle layer and may be one layer or multilayer [4]. Input Signal from the input layer is processed by hidden layer and output from the output layer. Neurons status in each layer only affects the neurons state of the next layer. If the output layer is not the expected output, it is transferred to the back-propagation, and according to the forecast error, adjust the value of network weights and thresholds so that the predicted output of neural network constantly approaches the desired output. Figure. 1 is the topology of a neural network.

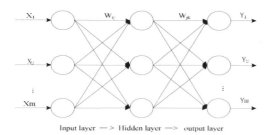

Figure 1. The topology of BP neural network.

In Figure 1, X_1, X_2,...X_n are the input values of BP neural network, Y_1, Y_2, and Y_m are the predicted values of BP neural network, and W_{ij} and W_{jk} are Weights of BP neural network. BP neural network can be regarded as a non-linear function, and the input value and output value of the network are the independent variable and dependent variable of this function. When the number of input node is n and output node is m, the BP neural network expresses mapping relationship of n independent variable to m dependent variable.

BP neural network is one of the most widely used neural network models, and BP network can learn and store a lot of input–output model mapping relationship without prior revealing of mathematical equations describing the mapping relationships.

2.3 GA algorithm

Genetic algorithms simulate the natural genetic mechanisms and biological evolution from a parallel random search optimization method.

It introduced the natural biological evolution principles of "survival of the fittest" to optimize the parameters form an encoded series group. Group of individuals is actually an entity with a characteristic chromosome. Chromosome is a collection of a plurality of genes, so it must first be encoded, and can implement the mapping from phenotype to genotype [5]. And through genetic selection, crossover, and mutation screening of individuals in accordance with the fitness function selected, we can make individuals with good fitness be retained, and individuals with poor fitness be eliminated. The new generation inherits previous generation's information and is also better than the previous generation. This cycle will be repeated, until the condition is satisfied.

The Genetic Algorithm (GA) is a randomized searching global optimization method drawn from biological natural selection and group evolutionary mechanisms. Its main feature is to operate directly on the structure object, and there is no limit to derivation and function continuity. GA has internal implicit parallelism and better global optimization capability. GA uses Probability of Optimization method, and it can automatically get and guide optimized search space, and can adaptively adjust search direction and do not need identified rule.

GA is an iterative process of calculating and its basic arithmetic steps include: coding, initial population, selection operator, crossover operator, mutation operation, evaluation, and decision to stop [6].

3 CLUSTER THE NETWORK INVASION DATA

3.1 Use FCM to cluster the network invasion data

Network intrusion data has a lot of dimensions, but the data between different intrusion categories has minor differences. Therefore, use of FCM to deal with the intrusion data is appropriate and desirable. In this paper, we selected 4500 attack data from KDD CUP99 randomly, using FCM clustering algorithm to process the data, and the results are shown in Table 1.

Table 1. Clustering result of FCM.

Actual \ Clustered	Normal	Probe	DOS	U2R	R2L
Normal	703	432	313	15	100
Probe	107	1226	8	59	697
DOS	35	18	40	2	35
U2R	258	60	0	311	29
R2L	2	0	5	8	37

From Table 1 we can see that FCM, dealing with intrusion detection data classification, and its accuracy rate are not very high. And because the initial cluster centers are randomly selected, differences in each clustering result are quite different, so we need to improve the clustering algorithm.

3.2 Combine FCM with BP to upgrade the result

FCM algorithm clustering results have a close relationship with the selection of the initial cluster centers. If we select a different cluster center, it will give a different result. Therefore, the data processing result by FCM clustering algorithm is unstable. In this paper, we combine FCM and BP neural networks together, so that the stability of clustering results has improved. And we use the improved methods to process the 4500 data selected, and the results are shown in Table 2.

Table 2. Clustering result of FCM combined with BP.

Actual \ Clustered	Normal	Probe	DOS	U2R	R2L
Normal	914	152	227	131	139
Probe	154	1507	28	191	217
DOS	23	8	70	9	20
U2R	201	58	16	349	34
R2L	4	2	7	0	39

From Table 2 we can see that FCM, combined with BP, the accuracy of clustering results have been greatly improved. After many times' tests, the stability of the clustering results has also been greatly improved.

Using BP neural network algorithm to improve FCM algorithm has a significant effect, but the final cluster result is still not very satisfactory. It needs to be further improved.

3.3 FCM and BP which is optimized by GA cluster and the invasion data

3.3.1 BP Neural network optimized by GA

Traditional BP network connection weights learning algorithms are based on gradient descent.

Its drawback is it is easy to fall into local minimum and can not obtain the global optimum [7]. Therefore, we use genetic algorithm to optimize BP neural network's weights and thresholds. Each individual contains a network weights and thresholds, and individual calculates individual fitness values by fitness function. Genetic algorithm uses selection, crossover, and mutation operations to find out the individual with optimal fitness value. BP neural network gets the best initial individual weights and thresholds assignment by genetic algorithm. BP neural network after being trained, predicts the system output. Because the thinking of neural networks is reflected in the weights, therefore, genetic algorithm optimization neural network weights are better able to improve the overall performance of neural networks. The main idea of optimizing neural network using genetic algorithm is to improve the initial weights and node threshold of neural network [8].

Using GA to optimize BP neural network's principle structure is shown in Figure. 2.

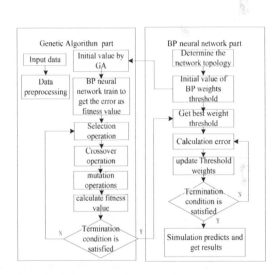

Figure 2. BP process optimized by GA.

Then, we use the optimized BP neural network algorithm to cluster the intrusion data and to improve the cluster process performance of intrusion of data.

3.3.2 Optimization algorithm to cluster the network intrusion data

Intrusion detection cluster uses FCM and BP which optimized with GA. The flowchart of the process is shown in Figure. 3.

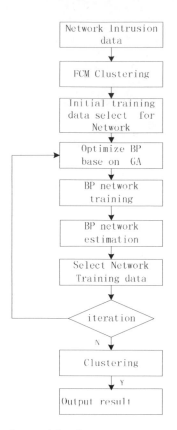

Figure 3. Improved flowchart.

We use the above method to process 4500 data selected, and the results are shown in Table 3.

Table 3. Clustering results of FCM combined with BP Optimized by GA.

Actual \ Clustered	Normal	Probe	DOS	U2R	R2L
Normal	1205	89	79	63	127
Probe	37	1803	0	74	183
DOS	0	7	91	11	21
U2R	88	70	25	439	36
R2L	0	0	5	3	44

From Table 3 we can see that, with respect to Table 2, the accuracy of the results obtained in Table 3 is greatly increased.

4 CONCLUSIONS

FCM has a great influence on results by the selection of the initial cluster centers. BP has strong local searching ability, but it is easy to fall into local minimum. GA is a self-adaptive, global optimization, and implicit parallelism, but the local search capability is not strong. Through the introduction of the foregoing and BP neural network's optimization by genetic algorithms, we combine them with FCM algorithm to deal with intrusion data, making the ability of processing the Network intrusion data significantly improved. And we have improved the accuracy and stability of the intrusion data clustering results.

REFERENCES

[1] http://kdd.ics.http://kdd.ics.uci.edu/ml/databases/kddcup99/kddcup99.html[EB/OL].
[2] Feng S et al. Matlab neural network 30 cases analysis[M].BeiJing: BeiJing Aerospace University Press, 2006.
[3] Bezdek J C. Pattern recognition with fuzzy objective function algorithm[M]. New York : Plenm'. Press, 1981.
[4] Anthony M, Bartlett P L. Neural network learning: Theoretical foundations [M]. Cambridge university press, 2009.
[5] Hertz J. Introduction to the Theory of Neural Computation [M]. MA: Addison-Wesley Press, 1991.
[6] Yao X, Xu Y (2006) Recent advances in evolutionary computation [J]. J Comput Sci Technol 21(1):1–19.
[7] Sutton R S. Two Problems with Back propagation and Other Steepest-Descent Learning Procedures for Networks[C]. Proceedings of 8th Annual Conf. of the Cognitive Science Society. Hillsdale,NJ, 1986: 823–833.
[8] LAM H K.Tuning of the structure and parameters of neural net-work using an improved genetic algorithm [C].Industrial Electronics Society, Denver, CO, USA: IECON'01, 2001.

Network Security and Communication Engineering – Chan (Ed.)
© *2015 Taylor & Francis Group, London, ISBN: 978-1-138-02821-0*

An automatic network protocol reverse engineering method for vulnerability discovery

X.H. Wu, H.T. Ma & D.F. Gu

Science and Technology on Integrated Information System Lab, Institute of Software Chinese Academy of Sciences, Beijing, China

ABSTRACT: With the rapid development of web applications, the efficiency and accuracy of vulnerability exploiting becomes more significant. Fuzzing test is an important method which inputs abnormal data based on network protocol into test system. However, current practice in deriving protocol specifications is mostly manual. To solve this problem, the paper proposes a scheme of automatic network protocol reverse engineering for fuzzing test. The architecture is composed of packet classification, multiple sequences alignment, specific domains recognition, and other key procedures. The result shows that this method has good performance both in the practicability and effectiveness of automatic vulnerabilities discovery, and provides better basis for network security.

1 INTRODUCTION

According to the investigation from VERC, unpatched vulnerabilities in systems are the major causes leading to network security events. Fuzzing is widely applied to discover serious vulnerabilities of network protocols [1] such as DoS, Buffer overflow [2], Integer overflow [3][4], format string and so on. Firstly, "malicious" network inputs to a special application are generated to uncover potential vulnerabilities. Effective mutations are generated based on protocol formats, but protocol specifications are always kept secret and the knowledge of nonpublic "open-source" protocols specifications traditionally is obtained by reverse engineering. Current practice in identifying protocol specifications is mostly manual. Such efforts are painstakingly time consuming and ineffective.

To achieve automatic protocol reverse engineering for fuzzing, the paper proposes a method based on the research of predecessors, in which the key words are extracted and the fuzzing test cases generated will be more effective. It also improves the multiple sequences alignment algorithm slightly. As expected, the result shows that this method has good performance both in the practicability and the effectiveness of automatic vulnerabilities discovery. The paper is organized as follows. Section II highlights the main contributions of the relevant research of network protocol reverse engineering and fuzzing test, and presents our main work. In section III, we describe the network protocol reverse engineering technology based on packet sequences and how to make fuzzing test according to the network protocol in detail. To verify the effectiveness of this method, the experimental results and conclusions are presented in section IV.

2 RESEARCH OF PROTOCOL REVERSE ENGINEERING

Nowadays, vulnerability discovery has attained great achievement such as static analysis, dynamic analysis, patches comparison, fuzzing test, and so on. Network protocol reverse engineering is mainly divided into two groups. One is packet sequence analysis to get the protocol's formats based on network trace; the other is instructions sequence analysis [5][6], which traces the dynamic process of binary file. The latter method is more accurate, but the process requires manual skills. However, analysis based on packet sequence has better performance in automaticity and efficiency. It started from the Protocol Information project [7] led by Beddoe in 2004. The project introduced sequence alignment algorithm in bioinformatics and tried to analyze the packet structure of the target protocol. Weidong Cui et al. proposed a solution for network protocol recognition and automatic recovery called RolePlayer [8], which focused on recognizing the particular fields in the packet. The specific methods are as follows: firstly, it obtains the users' parameters, and recognizes the data fields; and then, it generates the python scripts. Furthermore, system can implement interactions by replacing key word fields. This method is not applicable to complicated protocols. In the following research, they propose a recursion classifying resolution called Discoverer [9]. It firstly sorts out text and binary items, then initially classifies the packets by sequence alignments; it repeats these processes until the number of packets in the subclass is no more than the threshold. The disadvantage of this resolution is that extracting the information about state machine is not considered.

Weiming Li et al. from HUST propose the complete scheme of automatically fuzzing test for network protocol [10], which are domestic pioneers. This scheme firstly extracts packet sequences of the same type by matching types, uses multiple sequences alignment algorithm to distinguish invariant and variable domains, and deduces the text field, binary string field, Unicode code domain, and length domain, and at last extracts the state machine information and generates the SPIKE test cases for fuzzing tests. This scheme is targeted and effective, but does not consider the flags and key words in the packets, which reduces the efficiency and effectiveness in vulnerability exploiting.

3 MODEL DESCRIPTION AND ACHIEVEMENTS

To reverse engineer packet formats automatically for a wide range of protocols, the model is divided into three parts as Fig.1: preprocessing, packet classification, and packet format extraction.

3.1 Preprocessing

This step includes capturing of messages, filtering out the concerned data flow that are received and sent by the target application, reassembling fragment, removing the heading of data packets and so on, and extracting the payload.

Message capturing is the basic step. Taking the most common TCP/IP as the target protocol, we select WinPcap, which has a filter, as the underlying interface. Turn on the hybrid mod, keep listening, and we can distinguish the concerning packets by setting the filtering conditions such as the source and destination IP, port and transmission protocol. Fragment reassembling is usually required when TCP/IP has slices. As for the IP datagram, we can judge from the slices offset and the flags in heading, and for TCP datagram, we determine the slices existing on whether the number of ACK is appropriate.

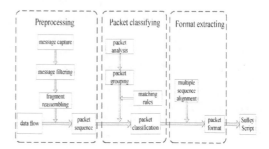

Figure 1. Protocol reverse engineering model.

Extracting payloads is the last part of the preprocessing. It firstly judges whether the data is effective by analyzing the heading and length of the datagram. After that, it removes the invalid or empty data, and extracts the data of application level, forms the packet sequence and marks the direction.

3.2 Packet classification

Packet classification is using the matching rules to divide large number of packets into groups. To improve the effectiveness, this paper makes up a key words dictionary which mainly contains terminologies in IT, abbreviations and common commands of systems and protocols. With further analysis, the dictionary is extensible.

3.2.1 Key words technology

Domain is the element of network protocol, and the common domains are the length, key words, command, and delimiters. The key words domain is always the identification of the protocol, and the values are always constant, which are used to verify the legality. So we introduce the key words to improve the effectiveness of classifying.

The process of identifying key words is as follows: we select the font domain depending on the delimiters and compare the value with key words dictionary. If the value is consistent and reappears in the same position with certainty, this is judged to be a key word domain. Introducing key words into the classifying procedure not only improves the accuracy of packet classification, but also reduces the difficulty in identifying packet format by aligning with key words domain.

3.2.2 Packet classifying based on key words

Packet classifying is clustering in essence. There is no formula accurately measuring the relative distance among packets in the past research, so the traditional clustering algorithm such as K-means is not applicable. This paper gives a packet classifying method applying key words based on the tentative clustering algorithm with the nearest neighbor rule. The main steps are shown as below. Taking the first packet as the clustering center of the first group, as for a new packet, analyze the crucial information firstly such as the type, the direction and so on, and at the same time recognize key words by comparing with the dictionary; and then match it with the packets classified; if it does, just put it into the classification; if not, create a new classification.

3.3 Packet format extraction

There are three kinds of multiple sequence alignment: precise alignment algorithm, progressive alignment algorithm, and iterative alignment algorithm. As packets are various and long, the precise alignment algorithm is time consuming and requires a lot of memory space. The progressive alignment algorithm is based on double sequence alignments and adds new sequences gradually until all the sequences are added. The shortcoming of this algorithm is that different adding orders result in different outcomes. The paper [10] compares the improved progressive alignment algorithm based on the packet length with genetic iterative alignment algorithm, and the former is better in aligning results and efficiency.

Progressive alignment algorithm has three main procedures: 1) calculating the matrix of distance; 2) constructing a guide tree; and 3) progressive aligning according to the tree. The proper order of all the sequences can be determined by measuring the similarity with distance matrix and aligning them with priority. However, the matrix of distance between two sequences is complicated to calculate. Supposing there are n sequences the average length is m, and the complexity of calculating the matrix is up to $O(m^2 n^2)$. As the format of every group is similar, the length difference determines the similarity between two sequences. The paper [10] uses the length difference instead of distance matrix and sorts the packet sequences by length to construct the guide tree. However, the length difference is not the decisive factor to the similarity when the difference value has limits.

The paper makes some improvement: firstly, screen the candidate packets whose length differ by no more than 10%; secondly, calculate the similar value between the candidate packets and the selected packet using double sequence alignments algorithm. This improvement finds a balance between the complexity and the accuracy.

Similarity calculating:

Needleman Wunsch is the classic double sequence global alignment algorithm based on dynamic planning. There are two main procedures in the algorithm,

the first step is calculating the similarity value and forming the matrix; the second step is backtracking to find the optimal alignment according to dynamical planning.

Suppose there are two sequences to be analyzed:

$$A = a_1 a_2 \cdots a_i \cdots a_m, \qquad B = b_1 b_2 \cdots b_j \cdots b_n,$$

then score matrix can be represented as two-dimensional table of columns as in Fig.2. Every cell in the first line and column is initialized to be zero, and the score in every cell is the most similar value.

		a_1	a_2	...	a_i	...	a_m
b_1	0	0	0	0	0	0	0
b_2	0	$M_{1,1}$					
...	0						
b_j	0						
...	0				$M_{i,j}$		
b_n	0						
	0						$M_{m,n}$

Figure 2. Similarity score matrix.

Every element of score matrix is obtained by iteratively computing the state conversion. The formula is as follows:

$$M_{i,j} = max \begin{cases} M_{i-1,j-1} + S_{i,j} \\ M_{i,j-1} + w \\ M_{i-1,j} + w \end{cases}$$

$M_{i,j}$ refers to the matching score of current state. The value is 1 or −1. If matching, it is 1; otherwise, it is −1.

In order to reduce the spaces inserted and accurately align the key information, we make some improvements in the Needleman Wunsch state conversion formula. 1) adjust the weight and change the value to −2, that means taking more off; 2) increase bonus for the continuous chars matching; 3) increase bonus for key words matching. When chars are continuously matched, identify if there are key words in them. If matching, increase the bonus further to highlight the importance. The new state conversion formula is as below:

$$M_{i,j} = max \begin{cases} M_{i-1,j-1} + S_{i,j} + (n-1)b + x \\ M_{i,j-1} + w \\ M_{i-1,j} + w \end{cases}$$

where "n" is the number of continuous matching chars; "b" is the bonus for continuous matching and set to be 2; x is the bonus for key words matching and set to be 10. The improved algorithm effectively reduces fragments and improves the reliability of alignment. As Fig. 3 shows, before optimizing, "Referer:http://192.168.1.1/" is not aligned with; after optimizing, it is accurately recognized.

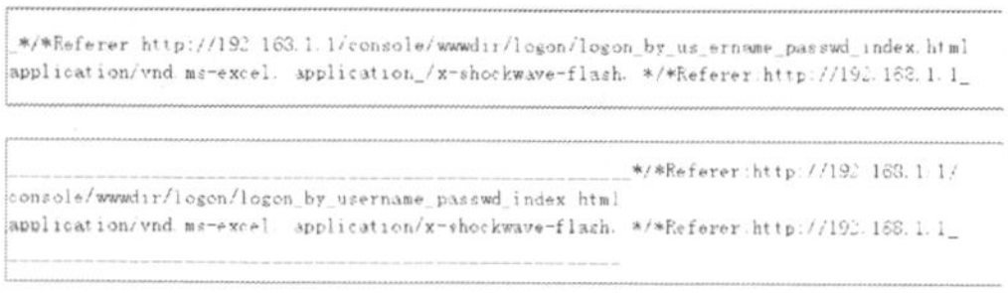

```
*/*Referer http://192.163.1.1/console/wwwdir/logon/logon_by_us_ername_passwd_index.html
application/vnd.ms-excel. application_/x-shockwave-flash. */*Referer http://192.163.1.1_
```

```
                                          */*Referer http://192.163.1.1/
console/wwwdir/logon/logon_by_username_passwd_index.html
application/vnd.ms-excel. application/x-shockwave-flash. */*Referer http://192.168.1.1_
```

Figure 3. Alignment results.

The process of progressive multiple sequences alignment algorithm is as below. Firstly, all packet sequences are ordered by length. Choose the shortest sequence, and find the most similar sequence using the improved method. Then take the two sequences as leaf nodes, use the improved Needleman Wunsch algorithm to align, and the result is viewed as parent node. And as for the left sequences, repeat the above procedures until all the sequences are handled and construct a binary guide tree. After that, start from the root node, recursively apply the saved information to the child nodes and after going through all the nodes, we will get the final results.

3.3.1 *Packet domain reorganization*
Upon the aligned results, the special format of every domain will be recognized.

3.3.1.1 Recognize variable domain and invariant domain
We calculate byte change rate by counting the char in the same position. If the rate is zero, it is invariant domain; otherwise, it is variable domain.

3.3.1.2 Recognize binary domain and text domain
According to the context of the byte, the domain can be identified to be binary or text ones. In the sequence, the text domain usually has the least length value. If the length of the field is less than the least value, it is a binary domain. A lot of tests prove that, when the least value is 4, the type of the packet can be distinguished in most cases.

3.3.1.3 Recognize the delimiters
In the packet sequence of text type, different function domains are separated by delimiters ones, so it is necessary to recognize the delimiters' domains. We scan the chars and match them with the delimiters' dictionary.

3.3.2 *Format output*
We apply the fuzzing framework called Sulley. According to the requirements of fuzzing test, we change the domains recognized into test cases. There are many kinds of data types describing different domains such as s_string, s_delim, s_static, s_random, s_qword, s_double, s_block, s_size, and so on. What is more, it can record all the test cases and crashes, which are critical for discovering the vulnerabilities. It also provides monitoring function for users to watch the progress and suspend or resume the system if necessary.

4 RESULT AND CONCLUSION

We select the classic FTP and HTTP as the target protocols, and two different cases are designed to test the chosen server Warftp and Imgsvr.

- Case 1: Classifying the packet without key words dictionary, compare the improved progressive alignment algorithm with traditional one.
- Case 2: Classifying the packet with key words dictionary and reverse engineer the network protocol's key information using the improved progressive alignment algorithm.

Table 1. Different results.

	without key words		without key words		with key words	
	improved progressive alignment		traditional progressive alignment		improved progressive alignment	
No.	length domain	binary	length domain	binary	length domain	binary
8	0	0	0	0	0	1
10	2	0	2	0	2	7
11	1	0	1	0	2	5
13	0	0	0	0	0	1
15	1	0	0	0	0	1
17	0	0	0	0	0	1
20	1	0	0	0	1	1
sum	5	0	3	0	5	17

From the table, we can see that, as the classifications are not very accurate without key words dictionary technology, the numbers of domains recognized by progressive alignment algorithm are all small, and the improved algorithm is a little better than the traditional one. And key words dictionary helps much to recognize more domains. That means the protocol specifications will be identified more accurately and the fuzzing test cases generated automatically are more targeted.

As for Warftp, five domains are recognized, which are "USER", "PASS", "NLST", "CWD", and "STOR". These domains make up the control scripts for fuzzing test. There are about 5605 test cases generated and three cases make the Warftp crash; As for the Imgsvr, "Host", "connection", "Accept", and "User-Agent" are recognized. There are about 4,217 test cases generated and seven cases make the Imgsvr crash.

The network protocol reverse engineering method for fuzzing test designed in the paper can automatically recognize the format of protocols with high efficiency in fuzzing test, and provides further basis for network security. The future work can be focused on the encryption protocols.

ACKNOWLEDGMENT

This work was supported by the MOST 863 Projects under 2012AA011206.

REFERENCES

[1] Zhiyuan A, Haiyan L. Realization of Buffer Overflow[C]//Information Technology and Applications (IFITA), 2010 International Forum on. IEEE, 2010, 1: 347–349.

[2] Zeng F, Mao L, Chen Z, et al. Mutation-based testing of integer overflow vulnerabilities[C] //Wireless Communications, Networking and Mobile Computing, 2009. WiCom'09. 5th International Conference on. IEEE, 2009: 1–4.

[3] Dietz W, Li P, Regehr J, et al. Understanding integer overflow in C/C++[C]//Proceedings of the 2012 International Conference on Software Engineering. IEEE Press, 2012: 760–770.

[4] Tsankov P, Dashti M T, Basin D. SECFUZZ: Fuzz-testing security protocols[C]//Automation of Software Test (AST), 2012 7th International Workshop on. IEEE, 2012: 1–7.

[5] Portokalidis G, Slowinska A, Bos H. Argos: an emulator for fingerprinting zero-day attacks for advertised honeypots with automatic signature generation[C]// ACM SIGOPS Operating Systems Review. ACM, 2006, 40(4): 15–27.

[6] Newsome J,Song D. Dynamic taint analysis for automatic detection, analysis, and signature generation of exploits on commodity software // Proc of Network and Distributed System Security Symposium. 2005.

[7] BEDDOE M. Protocol information project [EB/OL]. http://www.4tphi.net /~awalters /PI /PI.html.

[8] Cui W, Paxson V, Weaver N C, et al. Protocol-independent adaptive replay of application dialog[C] // Proc of the 13th Annual Network and Distributed System Security Symposium. 2006.

[9] Cui W, Kannan J, Wang H J. Discoverer: Automatic protocol reverse engineering from network traces[C]// Proc of 16th USENIX Security Symposium. 2007: 1–14.

[10] Li Weiming, Zhang Aifang, Liu JianCai, et al. An automatic network protocol fuzz testing and vulnerability discovering method[D]// Chinese Journal of computers.2011,34(2).(in Chinese).

Network Security and Communication Engineering – Chan (Ed.)
© 2015 Taylor & Francis Group, London, ISBN: 978-1-138-02821-0

The new method of security development for web services based on Moving Target Defense (MTD) technologies

M. Styugin
Siberian Federal University, Krasnoyarsk, Russia

ABSTRACT: This article describes the new method of security development for web services based on Moving Target Defense (MTD) technologies. This method addresses the security issue of websites from attackers that employ SQL injection, cross-site scripting, etc. The system interaction is built so that when the adversary tries to investigate the system structure, he obtains ever-increasing complexity of information from the system. These technologies allow us to defend websites from users with malicious intent and to research the behaviors of those attackers. Researching the behaviors of such intruders gives us the opportunity to find new vulnerabilities. This paper provides some examples of these technologies.

1 INTRODUCTION

Providing robust defense for websites is the most challenging task when we are developing an information security system. Any website should always be available to all users. And this is the reason why there is an attack that surfaces from this side. Because websites receive data from the environment, this generates vulnerabilities that might be exploited by attackers. Security risks cannot be preventively calculated by a defender because there are a lot of vulnerabilities that will be discovered in the future. These vulnerabilities might be critical for the system. These vulnerabilities are also critical because an intruder can conduct reconnaissance before his attack and this will provide him with more additional benefits than a defender. There is a common method for mitigating these intruder benefits. This method is called Moving Target Defense (MTD). The MTD method is based on continuous change of the system. Hence, when the attacker conducts reconnaissance, he doesn't obtain current information, so he cannot use it to launch attacks.

The purpose of this paper is to propose security technology for websites based on the MTD methodology.

Almost every attack is virtually preceded in the process of reconnaissance. Attacker tries to implement various exploits based on the information acquired. We can make the reconnaissance process useless or inefficient by introducing dynamic parameters into an information system. The MDT technology has already become some kind of trend in building information security systems. A more detailed description of the technology and its application can be found in publications (Jajodia et al. 2011) and (Jajodia et al. 2013). Practical implementation of those methods has taken place in virtualization (Paulos et al. 2013), software-defined networks (Carvalho & Ford 2014, Jafar Haadi Jafarian & Ehab Al-Shaer 2012), address space layout randomization (ASLR) (J. Li & Sekar 2006.), etc. Similar implementations were used for web–server protection. However, all of them are usually limited to dynamic names of variables, tables in databases, etc.

This paper presents another approach to web-resource protection where the MTD problem is regarded from the point of system research protection. An algorithm for building such protected systems was established by introducing additional parameters.

2 SYSTEM PROTECTION AGAINST RESEARCH

Acquiring new information is researching a system when we interpret information in accordance with hypotheses regarding its structure. That means even during research, some information always precedes actions a priori, otherwise we would not be able to interpret feedback from the system. The following system counter-research principle is based on it: in order to protect a system's functional structure from unauthorized activities, it shall be diversified to such an extent that it should look like chaos to an uninformed researcher when it is difficult to define an unambiguous hypothesis. That problem was thoroughly explored in (Styugin 2014). The present paper does not consider the methodology for system counter-research protection; it will focus on developing an algorithm for modification of structure in an information system which is essentially similar to the MTD technology.

Hence, research protection technologies do not put any restrictions on attacker's actions, but it controls information the attacker uses to make decisions. That solution results in reduction of risk in the whole set of vulnerabilities (even undiscovered ones) because it blocks informative feedback when an attempt to research the system takes place. The technology described in this paper is based on forced complication of a system by introducing "pointless" parameters into it. However, the result we obtain upon its implementation is that the system's structure deviates to unconventional patterns by becoming a complete chaos for an outside observer.

The counter-research technology offered herein is about functional changes in processes in the system. According to some concept of uniqueness, the process is demonstrated on Figure 1.

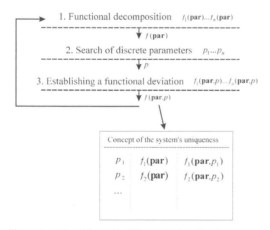

1. Functional decomposition $f_1(\mathbf{par})...f_n(\mathbf{par})$

$f(\mathbf{par})$

2. Search of discrete parameters $p_1...p_n$

p

3. Establishing a functional deviation $f_1(\mathbf{par},p)...f_n(\mathbf{par},p)$

$f(\mathbf{par},p)$

Concept of the system's uniqueness

p_1	$f_1(\mathbf{par})$	$f_1(\mathbf{par},p_1)$
p_2	$f_2(\mathbf{par})$	$f_2(\mathbf{par},p_2)$
...		

Figure 1. Algorithm of building a "unique" system.

At step one system's functional decomposition is performed in order to single out sub-processes for a specific purpose (f_1 (par)...f_n(par)), where par represents any process parameters. For example, access to a database is a sequential authentification: connecting to a database and generating a query. Authentification can be divided into entering a username and a password, etc.

Step two is for specific processes acquired during functional decomposition. We look for possible additional parameters having discrete features ($p_1...p_n$). For example, time in minutes, position of symbols, number of the session, etc.

After establishing discrete parameters, a pointless deviation from the initial process is introduced according to its value ($f_1(\mathbf{par}, p)...f_n(\mathbf{par}, p)$). For example, for entering a password, a misplacement of keyboard symbols is performed depending on their position in a line, and for switching reassignment of ports and host, addresses in the network is performed by the session number, etc. The deviating function should be included in the concept of system's uniqueness. Knowing it we can restore the initial functional structure.

What benefits does it create for us? The initial value of f (par) function is a stereotype scheme and the attacker tries to use it to attack. In case we change it, we do not allow performing standard actions and force the intruder to research. But research is only possible when the attacker correctly establishes the hypothesis (additional parameters) of the new function. Otherwise the attacker will not be able to obtain informative feedback from "the black box". It is quite simple to capture the dependency for one case of an additional parameter $f(\mathrm{par}, p_1)$, and therefore to produce a correct hypothesis. However, it is possible to introduce any large number of parameters into a process $f(\mathrm{par}, p_1, p_2,..., p_n)$, and turn the system into some kind of chaos for potential attackers.

3 WEB-SERVER RESEARCH PROTECTION

When complexifying processes in information systems, it is necessary to be precisely aware of the possible area for such modifications. Many processes such as protocol stack in an operating system, switching principles, etc. cannot be made more complex (or in other words, they are too resource-consuming to complexify them), but in general, additional parameters may be introduced into any process and "a concept of uniqueness" can be made for them. These ideas were used to develop a whole complex of software modules for system research protection (ReflexionWeb project, presently a resident of Skolkovo Innovation Centre). It is more reasonable to protect the most popular vulnerabilities as far as security is concerned so that the attacker would not acquire informative feedback by attacking them. A scenario for such approach is quite simple and operates in the following way: an attacker does not see any "opposing actions" of the system and hence the attacker spends a lot of time "disentangling" the logics of the system. The attacker "ensnares" in the system as it does not take the right actions to achieve the goal without the correct interpretation. Meanwhile, the application can easily register unauthorized activity as it is able to detect actions using stereotyped schemes.

Web-server applications developed up to present time refer to server protection from SQL injections. That is the most popular type of vulnerabilities, thus it is the most reasonable one to be protected from researching. Operation of the two modules will be analyzed below.

The first example is quite easy to implement as it is a function of only one fixed parameter and it does not require adjusting the "uniqueness concept". Its operating principle is similar to the HoneyPot network device. It is known to emulate a nonexistent network being part of a real network and is used as bait for attackers. It deters unauthorized activities and allows researching an intruder on a "secure territory". The HoneyPot may be interpreted as protection from system's research with deviation in a single parameter. In our case, a copy of database structure is similarly created, which is set up as a real one when unauthorized activity is detected.

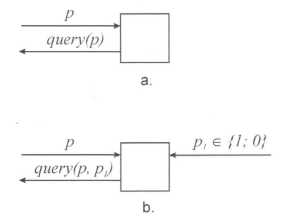

a.

b.

Figure 2.   Database query.

A standard database query is shown in Figure 2a. The p parameter is a database query generated by the system based on the information transferred though *GET* and *POST* arrays. The database management system processes the query and provides a feedback shown as *query(p)*.

We introduce an additional parameter p_1 into that process (Figure 2b). It can have only two values, which are 0 and 1. It obtains a value of 1 when regular expressions in charge of checking *GET* and *POST* arrays detect symbols characteristic for ISS and SQL injection attacks.

Generation of a database query is shown in Figure 3. DB_1 is the initial (original) website database. It is used to create a copy database DB_2 to take (replace) all required information out of it. Besides, there is no risk in deleting and modifying information in the DB. When the module detects attacks of ISS and SQL injections, the p_1 parameter becomes equal to 1 and the system connects to DB_2 database and registers data of the *SERVER* array that contains the attacker's IP-address, browser version, type of operating system, etc.

The attacker does not encounter opposition from the system and takes the same bait as in the HoneyPot system. It needs to acquire informative

feedback to research the system and it is possible when $f(p_1)$ function is found which is in charge of regular expressions.

That module describes the research protection technology in a satisfactory but not in a complete manner and it does not have the uniqueness concept (to be more precise, it is always constant and consists of one parameter), however it is very simple to install and does not require any additional adjustments. The following section reviews an extended database module that allows adjusting the concept of a system's uniqueness.

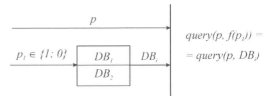

Figure 3.   Database query generation.

4   DATABASE MODULE.

As plugins cannot influence database query structure then all we can do is increase the multitude that is used to generate a query that is database structure and contents. Research protection technology requires introduction of a discrete parameter set $p_1...p_n$ and a system deviating function $f(p_1, ... p_n)$ can be introduced using the parameter set. Since the result shall be a query

$$Query\ (p, f(p_1, ... , p_n)) = query(p, DB_i),$$

then, f function is mapping for a set of databases:

$$f(p_1,...,p_n) \in \left\{ DB_1, DB_2, DB_2^1, DB_2^2,..., DB_2^m \right\}$$

The query generation principle is shown in Figure 4.

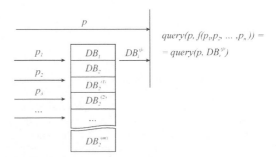

Figure 4.   Database query generation.

Recording databases with two indexes shows the principle of their compilation. DB_1 is the initial (original) database of a website. DB_2 is a copy of the database structure with empty tables like in the previous example. If the uniqueness concept has the rules for the structure's modification depending on parameters found, then the structure of DB_2 changes and a new $DB_2^{(i)}$ state is generated.

The uniqueness concept allows the administrator to adjust the individual system's functional structure. It can be made in an xml-file or any other configuration file. The structure of the concept is shown in Table 1.

Table 1. Concept of a system's uniqueness.

Parameter type	Parameter	Function type	Function
1	DELETE	1	DB_2
2	UNION+LOAD_FILE	3	+1
3	USERS	2	md5(pass)

Parameter type: 1 – SQL language operator, 2 – SQL operator pair, 3 – table names.

Function type: 1 – choice of database, 2 – number of columns, 3 – column values.

For example, we shall review a module's operating principle on Table 1. If DELETE operator is encountered in an SQL-query, system switches to DB_2 database. The second line: if operators UNION and LOAD_FILE are encountered at the same time, then an additional column is added to the table. This may be required to block file download. When the number of columns in the initial table and the merged table does not match then the database indicates an error. And the last line: if USERS table is requested in the text of SQL-injection then the UPDATE operator with the indicated function is performed. In this case a hash function is calculated for each password one more time.

This approach allows an administrator to add any parameters in the websites operation (database query) and create a unique system as far as research protection is concerned.

5 SYSTEM'S OPERATION

Operation of modules can be illustrated with the following example. The first thing that attacker will do when searching for vulnerabilities is trying to find out the presence of checking parameters transferred through GET and POST arrays by the website. To do that, the attacker may enter a single quote symbol or structures like " = 1 and 1 = 1", " = 1 and 1 = 0", etc. All the simple queries will be forwarded to the initial database and the intruder may come to a conclusion that a successful attack is possible. However, further research will take place in the set of $\{DB_1, DB_2, DB_2^1, DB_2^2, \ldots, DB_2^m\}$ and it is not likely to provide the attacker with any sufficient informative feedback. Besides, the module will register the source of unauthorized activity (IP-address, browser version, type of operating system, etc).

Research protection mechanisms do not include blocking vulnerabilities (even though they allow implementing them) as well as MTD technology in general. Research protection technology allows an additional risk reduction for information security systems. Mechanisms provided above are examples of technical implementation for the technology.

Testing such systems and analyzing their operation yielded positive results. When three or more parameters were used in the "uniqueness concept", an attacker spent a fair bit of time performing reconnaissance in the system and finally it refused from further research. Attacker's behavior signatures obtained from that interaction enabled to identify it with 80% accuracy when it attempted to attack other web-servers.

REFERENCES

[1] Jajodia, S. et al. 2011. Moving Target Defense. Creating Asymmetric Uncertainty for Cyber Threats. Series: Advances in Information Security. Springer, 184 p.

[2] Jajodia, S. et al. 2013. Moving Target Defense II. Application of Game Theory and Adversarial Modeling. Series: Advances in Information Security, Vol. 100. 203 p.

[3] Paulos, A., et al. 2013. Moving target defense (MTD) in an adaptive execution environment. *ACM International Conference Proceeding* Series. *8th Annual Cyber Security and Information Intelligence Research Workshop*: Federal Cyber Security R and D Program Thrusts, CSIIRW 2013.

[4] Carvalho, M. & Ford, R. 2014. Moving-target defenses for computer networks. *IEEE Security and Privacy.* Volume 12, Issue 2, March/April 2014, Pages 73–76.

[5] Q.D. JafarHaadiJafarian & Ehab Al-Shaer. 2012. Openflow random host mutation: Transparent moving target defense using software-defined networking. *Proceedings of the 1st Workshop on Hot Topics in Software Defined Networking (HotSDN)*, p. 127–132.

[6] J. Li & R. Sekar. 2006. Address-space randomization for windows systems. *In Preceedings of 2006 Annual Computer Security Applications Conference (ACSAC)*, pp. 329–338.

[7] Styugin M. 2014. Protection against system research. *Cybernetics and Systems*. Volume 45, Issue 4, p. 362–372.

Network Security and Communication Engineering – Chan (Ed.)
© *2015 Taylor & Francis Group, London, ISBN: 978-1-138-02821-0*

CVDE based industrial system dynamic vulnerability assessment A.T. Balkema & G. Westers

X.Q. Zhang & M.J. Chen
East China University of Science and Technology, Shanghai, China

ABSTRACT: Evaluating risk of vulnerability has become particularly important since there are a large number of vulnerabilities being reported every day. In order to adjust the configuration to industrial network security in time and make better prioritization of vulnerability responses, this paper presents Common Vulnerability Dynamic Evaluating method (CVDE), an improving vulnerability rating mechanism on the basis of CVSS, and uses DS evidence theory for vulnerability fusion process. We applied the method on a sample of 186 vulnerabilities for industrial system from NVD, illustrating the validity and flexibility of the mechanism.

KEYWORDS: Industrial system, Vulnerability assessment, DS evidence theory, CVDE.

1 INTRODUCTION

Analyzing the risks caused by system vulnerabilities is core task in industry system security management and assessment [1]. The Common Vulnerability Scoring System (CVSS) now is a common evaluation standard by providing metrics for the vulnerabilities. It includes three groups of metrics (Base, Temporal and Environmental). However, since temporal and environmental metrics which contains dynamic factors are optional in CVSS and the scoring process of them is so subjective that it is difficult to quantify the metric factors [2], most organization just use the base metrics group to evaluate vulnerability. So the industry system cannot be correctly evaluated when real-time changing occurs, such as the vulnerability being mended by patch. Therefore the feasible security strategy cannot be applied to these systems.

In this paper, we propose a Common Vulnerability Dynamic Evaluating method (CVDE), which is an improving CVSS mechanism. Based on the NVD vulnerability database, it can dynamically analyze the vulnerability risk value of some equipment in an industry system by completing CVSS metrics and adapting numerical calculation method. Furthermore, standing on the entire system, DS evidence theory is presented to fuse the risk of vulnerability on separate equipment to get the vulnerability risk of the whole industrial system.

2 BASIC KNOLOWLEDGE OF CVSS AND DS EVDIENCE THEORY

2.1 CVSS (Common Vulnerability Scoring System)

CVSS is an open and free vulnerability evaluation criteria introduced by the National Infrastructure Advisory Council (NIAC) and is now managed by the Forum of Incident Response and Security Teams (FIRST), and has become the industry standard supported by most vendors[3]. CVSS uses three groups of metrics (Base, Temporal and Environmental) to calculate vulnerability scores. The base metric can either be used alone or be in combination with the other two optional metrics. Fig1 shows the three groups and each set of component metrics.

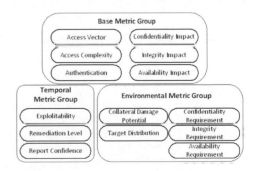

Figure 1. CVSS metrics group.

1 Base Score

$BaseScore=(0.6 \times Impact+0.4 \times Exploitability-1.5) \times f$

$ImpactScore=10.41 \times (1-(1-ConfImpact) \times (1-IntegImpact) (1-AvailImpact))$

$Exploitability = 20 \times AcessVector \times Acess Complexity \times Authentication$

$f= 0$ if $Impact=0$, $f = 1.176$ otherwise

Here, *ImpactScore* and *Exploitability* are calculated by the elements provided by CVSS.

2 Temporal score

$TemporalScore = BaseScore \times Exploitability \times RemediationLevel ReportConfidence$

3 Environmental score

$EnvironmentScore=(AdjustedTemporal +(10-AdjustTemporal) \times CollaterDamagePotential) \times TargetDistribution$

$AdjustedImpact=min(10,10.41 \times (1-(1-Conf Imapct \times ConfReq) \times (1-IntegImpact \times IntegReq) \times (1-AvailImpact AvailReq)))$

Finally, the *Environmental* score is regarded as the final risk score of some vulnerability.

2.2 DS evidence theory

The Evidence Theory, known as Dempster-Shafer (DS) belief function theory, was developed by Dempster and Shafer in [4]. The key concepts are described as below.

1 Basic definition

Definition 1: Assuming is a frame of discernment, and $m: 2^{\Theta} \to [0,1]$ is the basic belief assignment function on set Θ then if it satisfies the conditions $\sum_{A \subseteq \Theta} m(A)=1 \; \forall A \subseteq \Theta, m(A)$ is defined as the basic probability of A, and A is called as the proposition.

Definition 2: Assume is a frame of discernment, and $m: 2^{\Theta} \to [0,1]$ is the basic belief assignment function. Define a function as $Bel: 2^{\Theta} \to [0,1]^{Bel(A)=\sum_{B \subseteq A} m(B)}$, Bel is called belief function on set Θ, which denotes the support degree to A.

2 Combination rule of belief functions

Assume Bel_1 and Bel_2 are the belief functions based on separate evidences on the same frame of discernment Θ; both m_1 and m_2 are the corresponding basic probability assignment; suppose $A_1, A_2, ... A_p$ are the focus elements of m_1 and $B_1, B_2, ... B_q$ are the focus elements of m_2, and $\sum_{A_i \cap B_j = \varnothing} m_1(A_i) m_2(B_j) < 1$, where $i = 1,2,...,p$ $j 1,2,...,q$.

Then the combined basic probability assignment function $m: 2^{\Theta} \to [0,1]$ can be defined as follows:

$$m(A) = \begin{cases} 0 & A=\varnothing \\ \dfrac{\sum_{A_i \cap B_j = A} m_1(A_i) m_2(B_j)}{1 - \sum_{A_i \cap B_j = A} m_1(A_i) m_2(B_j)} & A \neq \varnothing \end{cases} \quad (1)$$

Above is the Dempster combination rule of two belief functions Bel_1 and Bel_2. For a number of functions to combine, apply the same algorithm until all the functions could be combined to the end.

3 CVDE BASED INDUSTRIAL SYSTEM DYNAMIC VULNERABILITY ASSESSMENT

3.1 Industrial system

A typical topological structure of industrial system consists of controlled area, non-controlled area and directorial area. The network system brings some potential danger of its inherent vulnerability. For instance, the storage and identification of data security and the rational structural arrangement of intrusion detection system.

3.2 CVDE mechanism

To meet the popularization and dynamic of the CVSS, its elements and the calculation equation of CVSS should be in continuous perfection. We integrated advantages of security institutions and manufacturers about vulnerability rating method, and proposed an improving rating method CVDE (Common Vulnerability Dynamic Evaluating).

3.3 Base metric group

CVSS has provided the metrics of *Exploitability* group (AV,AC,Au) and *Impact* group (I_C,I_I,I_A) and is defined as the quantitative values. Thinking the object type, access authority and attack vector are also impacting the vulnerability risk, we newly add an *Objectivity* group, which describes the network environment item. The details are described as below:

1 *Object Type*: Object Type evaluates the items affected by vulnerability, which would trigger risk in server, client, or network infrastructure.

• Network infrastructure. The vulnerability affect infrastructure of the industry network.
• Server. Affected servers involve in large amounts of data storage and system information.
• Client. Affected host is a common client, and doesn't involves in important data communications.

2 *Authority*: Authority indicates that what kind of attacker obtains permission. It has three optional items: administrator, consumer or others.

3 *Attack Vector*: Attack Vector means how many objects the vulnerability threatens. According to Microsoft subcommittee [5], we define it as below.

- Network. Data transmission and infrastructure equipment.
- Host. The operating system and core services.
- Application. The application of the internal and external development.
- Account. Identity related equipment or user credentials.

We classify the attack vector to 2 condition, one is the affected items equal or greater than 2, and the other is less than 2. The quantitative values of metrics are designated according to CVSS evaluate method.

Table 1. Metrics and score value of base group.

Metrics	Evaluation	Score
Object Type (OT)	Network infrastructure/ Server/Client	1/0.91/0.61
Authority (PR)	Administrator/ Consumer/Others	1/0.85/0.65
Attack Vector (VE)	>2 /<2	1/0.72

Table 1 shows the updated base metric group. Then the BaseScore (V_B) can be gotten by below equation:

$$V_B = (0.4 \times I + 0.3 \times E + 0.3 \times O - 1.5) \times f(I)$$
$$I = 10.41 \times (1 - (1 - I_C)) \times (1 - I_I) \times (1 - I_A)$$
$$E = 20 \times AV \times AV \times Au, \ O = 10 \times OT \times PR \times VE \quad (2)$$
$$f(I) = \begin{cases} 0 & I = 0 \\ 1.176 & I \neq 0 \end{cases}$$

Here, *I, E, O* represent *Impact, Exploitability* and *Objectivity* score respectively. The value range of *BaseScore* is from 0 to 10.

3.4 Temporal metric group

CVSS has provided temporal metrics and defined quantitative values. The detail is listed in Table2. In order to solve the "not defined" issue, according to Frühwirth's[7] and Frei's[8] research, the items of *Exploitability* and *Remediation Level* of vulnerability in CVSS temporal metrics meets the Pareto distribution and Weibull distribution respectively, it could be calculated by the following equation:

$$TE = 1 - \left(\frac{k}{x}\right)^a \ a = 0.26, k = 0.00161 \quad (3)$$

$$RL = 1 - \exp\left(1 - \frac{x}{\lambda}\right)^k \ \lambda = 0.209, k = 4.04 \quad (4)$$

In the equation, *TE* denotes the Pareto function. *RL* denotes the Weibull function. x is the time span from the vulnerability discovered now. Thinking the risk is changing with the occurrence frequency and distribution potential, we added *Frequency* and *Distribution Potential* items.

1 *Frequency*: Frequency means the frequency which is used for current vulnerability, and applied to measure suffering area in the scope of hosts, and it could be defined by referencing [6].

- High (global threat). The infected hosts reach thousands, and the infected sites come to dozens.
- Medium (several areas threat).The infected hosts reach hundreds.
- Low (local threat). Others.

2 *Distribution Potential*: For any specific target, the propagation of vulnerability could be divided into 4 kinds of circumstances:

- High. Attacks could spread rapidly accompanied by many victims.
- Medium. Attacks could spread rapidly while system has installed the safety equipment.
- Low. Attack could not spread as expected due to limited transmission vector or other issues.
- None. Attacks couldn't spread.

The quantitative values of metrics are designated according to CVSS evaluate method.

Table 2. Metrics and score value of temporal group.

Metrics	Evaluation	Score
Frequency (FR)	Low/Medium/High/ Not defined	0.94/0.97/1/1
Distribution Potential (DP)	Low/Medium/High/ None/ Not defined	0.92/0.96/1/0.9/1

Table 2 shows the updated temporal metric group. The score is defined by referencing[9]. Then the *Temporal* score (V_T) can be gotten by below equation.

$$V_T = V_B \times TE \times RL \times RC \times FR \times DP$$

Here, the range of *Temporal* score is from 0 to 10.

3.5 Environmental metric group

We still use the CVSS scoring method for environmental group, because it has already covered the environmental factors generally.

Ideally, the requirement data can be obtained from an organization's security policy. The environmental score can be gotten by below equation:

$$A_T = V_B' \times TE \times RL \times RX \times FR \times DP$$
$$V_E = (A_T + (10 - A_T) \times CDP) \times TD$$
$$V_B' = (0.4 \times I' + 0.3 \times E + 0.3 \times O - 1.5) \times f(I)$$
$$I' = \min(10, 10.41 \times (1 - (1 - I_C \times CR)$$
$$\times (1 - I_I \times IR) \times (1 - I_A \times AR)))$$

Through CVDE mechanism, we take environmental score as the final vulnerability risk score. Four risk levels are defined in table3.

Table 3.   Risk levels and rating score.

Level	Low	Medium	High	Critical
Range	0.0~2.9	3.0~5.9	6.0~8.9	9.0~10.0

3.6   DS evidence theory for vulnerability fusion

According to DS evidence theory, we define $\Theta = \{safe, unsafe\}$ discernment frame, $safe, unsafe$ represents system safe and unsafe state under vulnerability V_i respectively. The process is as follows.

1 Normalizing the risk value of vulnerability. Assuming there are two vulnerabilities existing in some equipment the risk value of this two vulnerability is V_{E_1} and V_{E_2} respectively, then the normalization processing is as:

$$T_{V_1} = \frac{V_{E_1}}{10} \times 100\%, \quad T_{V_2} = \frac{V_{E_2}}{10} \times 100\%$$

2 Defining the basic belief function. After normalizing, the basic belief function can be calculated as follows:

$$\left. \begin{array}{ll} m_1(safe) = 1 - T_{V_1} & m_1(unsafe) = T_{V_1} \\ m_2(safe) = 1 - T_{V_2} & m_2(unsafe) = T_{V_2} \end{array} \right\} \quad (6)$$

3 Calculating the risk value of the equipment. For the pointed equipment, by combining the belief functions, the risk value T_V of this equipment is:

$$m(unsafe) = \frac{T_{V_1} T_{V_2}}{(1 - T_{V_1})(1 - T_{V_2}) + T_{V_1} T_{V_2}}$$

$$T_V = m(unsafe) \times 100\% \quad (7)$$

For all of the equipment in one industry system, by applying the above three steps repeatedly, we could obtain the risk of the whole equipment.

Apply the same algorithm to combine the equipment risks of one industry system. We could get risk score and level of the system. Thereby, the technical personnel could take appropriate remedial measures so that they can resolve the issues in efficient manner.

4   EXPERIMENT AND ANALYSIS

In order to verify the validation of CVDE method, there are 186 industrial vulnerabilities, which arise from January 1, 2013 to July 30, 2014, are captured from National Vulnerability Database (NVD) by web crawler tool SoMiner 5.1. The description of these vulnerabilities can be got in [10].

Table 4.   Evaluate scores of industrial vulnerabilities.

CVE ID	OT	PR	VE	TE	RL	RC	FR	DP	CDP	TD	AR	IR	CR	CVSS	CVDE	Level
CVE-2013-2781	1.0	1.0	1.0	1.0	0.87	1.0	0.97	0.9	0.3	1.0	1.51	0.5	1.51	*9.995*	*8.314*	*High*
CVE-2014-0757	1.0	0.85	0.72	1.0	1.0	1.0	0.97	0.9	0.1	1.0	1.51	0.5	1.51	*7.124*	*6.231*	*Medium*
CVE-2014-0767	1.0	0.85	0.72	1.0	0.9	1.0	0.97	0.9	0.4	1.0	1.51	0.5	1.51	*7.485*	*7.642*	*High*
CVE-2014-0773	1.0	0.85	0.72	0.9	1.0	1.0	0.97	0.9	0.4	1.0	1.51	0.5	1.51	*7.485*	*7.642*	*High*
CVE-2014-0772	1.0	0.85	0.72	1.0	1.0	1.0	0.97	0.9	0.4	0.75	1.51	0.5	1.51	*4.958*	*5.07*	*Medium*
CVE-2012-4704	1.0	0.85	0.72	1.0	0.87	1.0	0.97	0.9	0.4	1.0	1.51	0.5	1.51	*7.786*	*7.26*	*High*
CVE-2013-0653	1.0	0.85	0.72	0.95	0.87	1.0	0.97	0.9	0.4	0.75	1.51	0.5	1.51	*5.761*	*4.965*	*Low*
CVE-2013-0685	1.0	0.85	0.72	1.0	0.87	1.0	0.97	0.9	0.4	1.0	1.51	0.5	1.51	*7.124*	*7.034*	*High*

......

4.1 Experiment 1

This experiment calculated the risk values of the whole 186 vulnerabilities by CVDE. By analyzing the description of vulnerabilities in NVD and applying the formula in section 3, the quantitative value of each metric can be obtained. Limited by the length of the article, table 4 shows parts of the calculating results. In the table, CVSS score is provided by NVD directly, which just using *Basescore* but omitting *Temporal* and *Environmental* score. The risk level is rated by CVDE score demonstrated in 3.2.

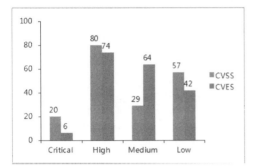

Figure 2.   The scores comparison of CVDE and CVSS.

Fig.2 shows the statistical results of the risk level assessed by CVDE and CVSS respectively on the whole 186 vulnerabilities.It can be seen of , as more environment and temporal factors are considered in CVDE, the number of vulnerabilities which are evaluated as *high* or *medium* occupied about 74.2%,the *critical* is about 3.2% and the *low* is 22.6%. The evaluation result is more in line with normal distribution. Meanwhile, when more vulnerabilities is evaluated as *Critical* level, more swifter response processes which could interrupt production operations must be

triggered to protect industry system, as they will cause more indirect costs for enterprises. But the more accurate evaluation for the vulnerabilities could drop the influence on producing and making these vulnerabilities which can be resolved in an appropriate manner.

4.2 Experiment 2

This experiment is designed to verify the dynamic evaluating character of CVDE method.

In table 5, eight vulnerabilities and their release time, update time, influenced equipment and dynamic points are listed. Since CVDE method considers the temporal and environmental metrics in the evaluating process, it can achieve dynamic assessment effectively. From table 5, it can be seen that the CVDE risk evaluation score changed with the passing of time, the modifying of environment and the developing of patch etc. Then the enterprise could take corresponding right remedial measures to guarantee that its system is running with feasible cost with the change of risk level.

4.3 Experiment 3

We analyzed the vulnerability data of each equipment from the previous samples, and obtained the risk score of each issue through CVDE. The risk score of each issue is listed in table 6.

In this experiment, we simulated an industrial system contains the followed ICS equipment: Siemens SIMATIC, Schneider Electric, Rockwell Automation ControlLogix, Invensys Wonderware, Information Server, GE Proficy CIMPLICITY, CoDeSys Gateway Server. Furthermore, DS evidence theory was applied to fuse the risk of each industrial equipment. For this simulation industrial system the risk score is 6.517, so the risk level is High.

Table 5.   Time changed influence the vulnerability.

CVE_ID	Equipment	Release Time	Update Time	Score1	Score2	Dynamic Point
CVE-2014-0757	3S CoDeSys	2014/2/12	2014/2/12	6.3	5.231	Official fixed. FR and DP reduced.
CVE-2014-0772	Advantech WebAccess	2014/4/10	2014/4/11	5	4.07	Official fixed. DP reduced.
CVE-2012-4704	CoDeSys 'Gateway Server'	2013/2/21	2013/2/26	8.2	7.26	Official fixed. TD reduced.
CVE-2013-0653	GE Proficy CIMPLICITY	2013/5/10	2013/5/27	5.1	5.965	Temporary fixed. CDP increased.
CVE-2013-0685	Invensys Wonderware	2013/1/24	2013/5/27	7.4	7.034	Temporary fixed. FR and DP reduced.
CVE-2013-1499	Schneider Electric Modicon M340	2013/12/10	2013/12/12	5.6	5.813	Unconfirmed.
CVE-2013-1823	GE Proficy HMI	2013/12/2	2013/12/3	7.4	8.114	Workaround. FR increased.
CVE-2013-1745	ABB MicroSCADA	2013/11/21	2013/12/2	6.9	6.147	Official fixed. DP reduced.

......

Table 6. Equipment risk level.

ID	Equipment	Vulnerabilities	Risk Score
VULN_1	Siemens SIMATIC	CVE-2014-2252,CVE-2014-2908 CVE-2014-2250	6.5
VULN_2	Schneider Electric	CVE-2013-2672,CVE-2013-2762 CVE-2013-2761,CVE-2014-0779 CVE-2013-6142	6.1
VULN_3	Rockwell Automation ControlLogix	CVE-2012-6440,CVE-2012-6435 CVE-2012-6441,CVE-2012-6437	7.2
VULN_4	Invensys Wonderware Information Server	CVE-2012-4710,CVE-2012-4709 CVE-2013-0686,CVE-2013-0688 CVE-2013-0685,CVE-2013-0684	7.1
VULN_5	GE Proficy CIMPLICITY	CVE-2012-4689,CVE-2013-0653 CVE-2013-0654,CVE-2013-2785	6.9
VULN_6	CoDeSys Gateway Server	CVE-2012-4708,CVE-2012-4705 CVE-2012-4706,CVE-2012-4704 CVE-2012-4707	5.3

5 CONCLUSION

Based on the analysis of CVSS static vulnerability evaluation method, a CVDE based dynamic vulnerability evaluation method is proposed in this paper. By adding *Object Type, Authority, Attack Vector* items into the Base metrics group and *Frequency* and *Distribution Potential* items into temporal metrics group of CVSS, more static and dynamic factors that influence vulnerability risk level are considered. Especially the calculating functions of metric *Exploitability and Remediation Level* in temporal metrics group are also pointed out to help do dynamic evaluation. At last, DS evidence theory is presented to apply to fuse the vulnerability risk on each of equipment to form the whole vulnerability risk of the entire system. Experiment results show that the CVDE evaluation mechanism is more credible and flexible, which is suitable for dynamic risk assessment. And the assessment result of CVDE can guide the strategy of vulnerability to have more feasible responses in industry system.

REFERENCES

[1] Assad Ali, Pavol Zavarsky, "A Software Application to Analyze Effects of Temporal and Environmental Metrics on Overall CVSS v2 Score", Concordia University College of Alberta, Edmonton, Canada, October 2010.

[2] Pengsu Cheng, Lingyu Wang,"Aggregating CVSS base scores for semantics-rich network security metrics", Proc. 31st International Symposium on Reliable Distributed Systems, Irvine, California, October 8–11, 2012.

[3] M. Keramati, A. Akbari, "An attack graph based metric for security evaluation of computer networks", 6'th International Symposium on Telecommunications (IST'2012), pp. 1094–1098, 2012.

[4] Otman Basir,Xiao hong Yuan, "Engine fault diagnosis based on multisensor information fusion using Dempster-Shafer evidence theory ," Information Fusion, 2007, vol. 8, pp. 379–386.

[5] MicrosoftSecuritySolutions. http://technet.microsoft.com/en-us/solutionaccelerators/ default.aspx.

[6] ThreatSeverityAssessment.http://www.symantee.com/avcenter/threat.severity.html.

[7] C. Frühwirth, "Improving CVSS-based vulnerability prioritization and response with context information," Third International Symposium on Empirical Software Engineering and Measurement, IEEE, 2009, pp. 535–544.

[8] S. Frei, M. May, U. Fiedler, B. Plattner. "Large-Scale Vulnerability Analysis," SIGCOMM' 06 Workshops, 2006, pp. 131–138.

[9] C. Frühwirth, "On Business-Driven IT Security Management and Mismatches between Security Requirements in Firms, Industry Standards and Research Work," Product-Focused Software Process Improvement, 2009, pp. 375–385

[10] National Institute of Standards and Technology (NIST), "National Vulnerability Database (NVD)" Available online at http://nvd.nist.gov/.

Network Security and Communication Engineering – Chan (Ed.)
© *2015 Taylor & Francis Group, London, ISBN: 978-1-138-02821-0*

An improved cloud-based privacy preserving approach for secure PHR access

D. Sangeetha
Department of Information Technology, Madras Institute of Technology, Anna University, Chennai, India

V. Vaidehi
Department of Electronics Engineering, Madras Institute of Technology, Anna University, Chennai, India

ABSTRACT: Patient Health Record (PHR) is comprised of personal and medical information that are accessed by PHR owners, doctors, and researchers. There are several issues like flexible access, security, and privacy when PHR is ported onto the cloud. Due to increasing health issues, it is essential to share the micro data with researchers so that the researchers can analyze the characteristics of the diseases. In this paper, a novel secure framework named Improved Shamir secret sharing Multi Authority Attribute Based Encryption (IS³-MAABE) is proposed which provides privacy of PHR and micro datastored in cloud. This scheme integrates Improved Shamir secret sharing and Multi Authority Attribute Based Encryption to enhance the security of keys that are used for encryption and decryption. To achieve fine grained access control, the users of the PHR are divided into personal and public domain based on the access requirements. To preserve the privacy of shared micro data, an algorithm named Micro Data Aggregation and Deduplication (M-AD) is proposed. This algorithm integrates techniques such as data aggregation and deduplication which inturn reduces the storage cost. From the experimental results, the proposed system is proved to be efficient in terms of time and storage cost.

KEYWORDS: Personal Health Record (PHR), Data security, Data anonymization, Privacy preservation, Access control, Data storage.

1 INTRODUCTION

The successful growth of cloud computing has fascinated various sectors such as industrial companies, government agencies, and healthcare providers to shift enormous applications onto the cloud (Jie Huang *et al*2012). Healthcare application is one of the promising areas of cloud where the patient's Personal Health Record are stored. There are several PHR services that are established to store and share the personal health information. In cloud computing, the PHRs are uploaded to the cloud server to share the records with healthcare experts and individuals of interest. To avoid improper use of PHR that are stored in cloud, the PHRs are encrypted before storing onto the cloud.

Today most of the health institutions share their micro data for research purposes. The hospital shares the medical records with the researchers so that they can analyze the characteristics of diseases. While disclosing these micro data to a third party, the privacy of patients must be preserved. To eliminate the risk of privacy, a method called Anonymization is used (ApekshaSakhareet al2013). It is a process of making data worthless to anyone except the owner. It is a method that transforms the

data in such a way that it prevents the detection of key information (Bhushan Mahajan and Swati Ganar2012). Different types of data anonymization techniques such as generalization and Clustering based anonymization (SangeethaDet al 2014) are used for privacy preservation. Data deduplication is a compression technique which is used to eliminate the duplicated records which reduces the storage cost. In data duplication process, each record is compared with the other records and whenever the record matches, the redundant records are removed. Data aggregation is one of the data mining techniques that searches, gathers, and presents a summarized report.

In this paper, a novel framework is proposed for secure sharing of PHR and the micro data that are stored in cloud. The proposed IS³-MAABE scheme integrates the Improved Shamir's Secret Sharing and Multi Authority Attribute Based Encryption. This scheme is specially designed to secure the encryption and decryption keys that are used to encrypt/decrypt the PHR and to provide proper access to the users of the system based on the access policy. To reduce the storage cost and information loss, the proposed M-AD algorithm integrates data deduplication and aggregation techniques.

This paper is organized as follows. Section 2 presents the proposed IS³-MAABE framework. Implementation and performance analysis is presented in section 3 and 4. Conclusions and future work are eventually presented in section 5.

2 PROPOSED IS³-MAABE FRAMEWORK

The aim of the proposed framework is to provide secure access to the PHR that are stored onto cloud and efficiently share the PHR among users. The users of the system are divided into public and personal domain according to the access requirements. The users in the public domain are those who have access to the data based on professional roles and the users in personal domain are those who are personally associated with the patient. Each domain will have attribute authority governing a subset of users.

2.1 IS³-MAABE architecture

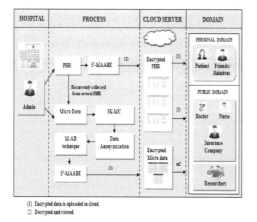

(1) Encrypted data is uploaded in cloud.
(2) Decrypted and viewed.

Figure 1. Architectural diagram of IS³-MAABE framework.

Figure 1 presents the architectural diagram of the proposed framework. In this paper, the PHR is encrypted using the proposed Shamir's Secret Sharing-Multi Authority Attribute Based Encryption (S³-MAABE) scheme before storing onto the cloud. The encrypted PHR's is accessed by the users of the system based on the access structure. The micro data are a collection of data which are recurrently collected from several PHRs that are stored onto the cloud. The admin in the hospital collects and publishes the micro data over a period of time so that the researchers can analyse the characteristics of the diseases. To facilitate analysis, micro data to be shared is clustered

and then anonymized to preserve the privacy of the patient's medical information. Anonymization is done using existing techniques like generalization and suppression. The anonymized micro data contain several repeated records. Therefore, to minimize the information loss and the cost of cloud storage, the anonymized micro data are aggregated and deduplicated. Finally, the micro data are encrypted before storing onto the cloud. The registered users in public and personal domain can access to the micro data after decrypting the data. The decryption of the data is possible only after authenticating the users.

2.2 Improved Shamir's secret sharing - Multi Authority Attribute Based Encryption (IS³-MAABE)

The proposed IS³-MAABE is used to enhance the security of the data that are stored in cloud environment. This scheme presented in Algorithm 1 integrates techniques such as Improved Shamir's Secret Sharing and existing Multi Authority Attribute Based Encryption to enhance the security of keys that are used for encryption/decryption and to provide fine grained access to the users of the system based on the access requirements.

Algorithm 1: Improved Shamir's Secret Sharing

Input:
 Public and Secret keys
Output:
 Complete Key for encryption / decryption
Procedure
If (request ==" key storage")
  { Split the public and secret key based on encryptor's request.
   Store the components of keys at different storage;
   Display user's component of keys to user.}
else if (request == "retrieval of encryption key"){
    Take the no. of split and user's component of the public key from the encryptor;
if the key and user's public key matches
then collect all components of public key
for encryption process;}
else if (request == "retrieval of decryption keys"){
    Take the no. of split and user's component of the public and secret key from the decryptor.
if the key split and user's public and secret key component matches then
collect all components of public and secret key for decryption process;}
End

2.3 Micro Data-Aggregation and Deduplication (M- AD)

The anonymized micro data contain several repeated records. To reduce the cost of cloud storage and information loss, the proposed M-AD algorithm integrates techniques such as data aggregation and de-duplication. The data aggregation is a data mining technique that is used to present a summarized report of the data to achieve specific business objectives. Data deduplication is a data compression technique that eliminates repeated records to reduce the amount of storage space. Each anonymized record is compared with the other records. If similar records are found during comparison then the count is incremented to the corresponding record and then the repeated records are eliminated.

2.4 Encryption/ decryption of the micro data

The micro data are then encrypted using the proposed IS³-MAABE scheme. The admin in the hospital uploads the encrypted micro data onto the cloud server. The number of key split and the first component of the decryption key are communicated to the registered researchers through AA's. The researchers are authenticated before entering the system and can decrypt the encrypted micro data after specifying the number of key split and the first component of the decryption keys. Once the key split and the first component of the key match, then the original secret key that is split into several parts are gathered and used for decrypting the micro data.

3 SYSTEM IMPLEMENTATION

The proposed framework is implemented on Cloud bees environment. It is an online cloud tool that offers platform as a service to manage the developed applications.Figure 3 and 4 show the encryption and decryption of PHR with IS³-MAABE.

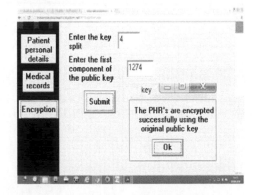

Figure 2. Encryption of PHR using IS³-MAABE.

Before entering the patient details, the keys are generated and split based on the admin's request. The system displays the first component keys. The admin in the hospital is authorized to enter the number of key split and the first component of the public key.

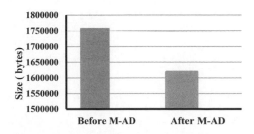

Figure 3. Decryption of PHR using IS³-MAABE.

If the number of key split and first component of the key matches then original public key is gathered and it is used for encrypting the PHR. The encrypted PHRs are stored onto the cloud. When the users of the system login, the system prompts the user to enter into the number of key split, the first component of public and secret key. If they match, then the encrypted PHR is decrypted by using the original secret key. The user views the PHR based on the access privileges. The number of key split, first component of the public and secret key are communicated to Attribute Authorities (AA's) of various domains. The AA's shares it with registered users in their domains.

4 RESULTS AND DISCUSSION

This section explains about the experimental analysis that was performed to prove the effectiveness of our approach. Figure 4 presents the size of the data before and after the M-AD algorithm has been implemented. From the analysis, it was concluded that the proposed M-AD algorithm reduces the cost of cloud storage to a great extent.

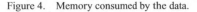

Figure 4. Memory consumed by the data.

The overall performance of the proposed framework is evaluated by comparing existing schemes with the proposed schemes. The comparisons are tabulated in Table 3. From the comparative analysis, it was concluded that the proposed scheme is efficient in terms of time consumption and storage cost.

Table 3. Overall system performance.

	Time taken in milliseconds		
Schemes	Key generation	Encryption	Decryption
KPABE	290	313	128
MAABE	332	336	141
Proposed IS$_3$-MAABE	316	310	110

	Storage space (bytes)		
Before M-AD process	1758942	After M-AD process	1617679

5 CONCLUSION AND FUTURE WORK

The proposed framework addresses the complicated issues such as security and the privacy of the shared PHR in cloud. It provides efficient analysis to the published micro data. The security of the PHR is guaranteed through the proposed IS^3-MAABE scheme. The proposed IS^3-MAABE scheme is used to encrypt the PHR before storing onto the cloud. To facilitate analysis, the micro data are clustered and anonymized. The proposed M-AD algorithm reduces the storage cost of micro data in cloud by integrating techniques such as data aggregation and deduplication. Through the performance analysis, the proposed framework is proved to be efficient.

REFERENCES

[1] ApekshaSakhare and Swati Gana 2013 Anonymization: A Method To Protect Sensitive Data In Cloud. *International Journal of Scientific & Engineering Research*, Volume 4, Issue 5.

[2] Bhushan Mahajan and Swati Ganar 2012 Review Paper on Preserving Confidentiality of Data in Cloud Using Dynamic Anonymization. *International Journal of Computer Science and Network (IJCSN)*, Vol 1, Issue 6.

[3] Bozovic V, Socek D, Steinwandt R, andVil-lanyi V I 2012 Multi-authority attribute-based encryption with honest-but-curious central authority. *International Journal of Computer Mathematics*, vol. 89,pp 3.

[4] Cheung L and Newport C 2007 Provably secure ciphertext policy abe. In *CCS '07: Proceedings of the 14th ACM* conference on Computer and communications security, pages 456–465, New York, NY, USA, ACM.

[5] Elaine Barker, Dennis Branstad, Santosh Chokhani and Miles Smid 2010 Cryptographic Key management workshop summary. *NIST Interagency report 7609 at Computer Security Division, National Institute of Standards and Technology*.

[6] Goyal V, Pandey O, Sahai A, and Waters B 2006 Attribute-based encryption for fine-grained access control of encrypted data. In *Proc. ACM Conf. Computer and Communications Security* (ACM CCS), Alexandria, VA, 2006.

[7] Ibraimi L, Asim M, and Petkovic M 2009 Secure management of personal health records by applying attribute-based encryption. *Technical Report*, University of Twente.

[8] Sangeetha D, Vaidehi V and Logeswari G 2013A Cost Effective Clustering based Anonymizationapproach for Storing PHR's in Cloud. *International Conference on Recent Trends in IT*, MIT, India.

Network Security and Communication Engineering – Chan (Ed.)
© 2015 Taylor & Francis Group, London, ISBN: 978-1-138-02821-0

Measurement of operational risk based on α-stable distribution

W.Q. Wang, Y.N. Sheng & X.D. Liu
Donlinks School of Economics and Management, University of Science and Technology Beijing, Beijing, China

ABSTRACT: Operational risk management plays an indispensable role in modern risk management of commercial banks. Statistics show that between 1987 and 2014, nearly 90% of operational risk loss events took place in Chinese commercial banks, which is related to the "people" factor, and that the sample data demonstrate the characteristic of "sharp peak and heavy tail". This paper adopts the α-stable distribution to better fit the sample data so as to measure the overall operational risk VaR of Chinese commercial banks and evaluate the change of the risk level in recent years. The empirical results show that the overall risk VaR of Chinese commercial banks is in decline year by year. The overall operational risk of Bank of China and Industrial and Commercial Bank of China is higher than the national level.

1 INTRODUCTION

Operational risk is one of the three risks faced by commercial banks. Studies on operational risk in China started relatively late. The China Banking Regulatory Commission formally gave the definition of operational risk in 2005 and proposed 13 relevant suggestions. In late 2009, it published "validation guidelines for measurement of commercial bank capital". Since then, operational risk management in China has moved to a higher criterion, and requirements for the technology applied to measure operational risk are improving as well.

The New Basel Capital Accord put forward loss distribution approach (LDA) on the basis of VaR model to solve the problem of fitting data. LDA calculates the risk capital by estimating the distribution of the operational risk loss events for each product/business type over a period of time. Jobst (2007) believed that for operational risk management, analyzing and forecasting the extreme events remains to be important and the subexponential distribution can be used to fit the loss intensity because of the loss data's "fat tail" characteristic. Feng, Li & Chen (2011) used LDA with left-truncated data to deal with the loss of data integrity, proving that the left truncated LDA model was consistent with data characteristics and it was also robust and sensitive to loss data. Wang et al. (2011) found that LDA method based on bootstrap sampling and piecewise-defined severity distribution (BS - PSD - LDA) performed better than the traditional parametric methods did concerning the lack of loss data and their characteristics of right skewness and fat tail.

LDA takes the loss distribution as a whole into consideration. However, those general methods can't accurately capture the tail of the loss distribution. Medova & Kyriacou (2001), therefore, adopted Extreme Value Theory (EVT) to handle this problem. EVT includes the Peak Over Threshold (POT) model and Block Maxima Method (BMM) model. Moscadelli (2004) found the Generalized Pareto Distribution (GPD) of the POT model can well fit the upper tail of the loss distribution. Yao, Wen & Luan (2013) combined LDA with BMM to measure operational risk of 14 Chinese commercial Banks. The best way of grouping was determined by returns to inspection. Lu & Zhang (2013) measured marginal distributions of several operational risk cells based on the POT model and used multivariate copula functions to depict the dependency of operational risks. It showed that VaR was calculated when considering risk relevance was less than the one calculated by directly summing VaRs of all operational risk cells.

In general, constructing an appropriate measurement model to calculate VaR and capital charge is the main trend of the researches in operational risk management. Existing researches usually choose function with certain shape such as exponential distribution, lognormal distribution, GPD and weibull distribution to fit the loss data. But the corresponding goodness-of-fit is not perfect when they are employed to depict the "sharp peak and fat tail" characteristic of the data. Given the shortcomings of existing researches, this paper selects the α-stable distribution to quantitatively analyze 577 operational risk events of Chinese commercial banks from 1987 to 2014. The α-stable distribution has four parameters that control shape, skewness, scale and position. By adjusting the parameters, it can well fit non-normal distribution, especially the operational risk loss distribution that has "sharp peak and fat tail" characteristic. The parameter estimation is carried out by maximum likelihood estimation method. Based on the above analysis, we discuss the development trend of overall operational risk in Chinese commercial banks.

2 DESCRIPTIVE STATISTICS OF LOSS DATA

We have collected 577 operational risk loss events happening from 1987 to 2014 in Chinese commercial banks with the help of public reports and existing studies. Being in the same social environment and having similar customer groups, all the commercial banks in China are homogeneous to some extent. Therefore, we take Chinese commercial banks as a whole to measure operational risk.

Table 1 shows the frequency and the loss amount proportion of different types of operational risk loss events happening from 1987 to 2014 in Chinese commercial banks. Among all of the seven types of operational risk events, the frequency of internal fraud ranks the highest, accounting for more than 60% of the total, followed by external fraud and clients, products & business practices in that order. But the frequencies related to employment practices and workplace safety and damage to physical assets are very low. Due to the internal fraud, external fraud and clients, products & business practices are relevant to "people", and in this sense, the "people" factor is the dominant factor that triggers the operational risk loss events in Chinese commercial bank. The loss amount of both internal fraud and external fraud events accounts for more than 2/3 of the total, followed by clients, products & business practices and execution, delivery and process management which account for about 10% of the total loss together.

Table 1. Frequency and loss amount proportion of different types of loss events.

Type of loss	Frequency	Loss amount proportion
Internal fraud	63.78%	42.06%
External fraud	17.85%	32.67%
Clients, Products & Business Practices	10.40%	14.51%
Employment Practices and Workplace Safety	0.17%	2.12%
Damage to Physical Assets	0.17%	0.01%
Execution, Delivery & Process Management	6.93%	8.63%
Business disruption and system failures	0.69%	0.0002%

Table 2. Statistical characteristics of sample data. (Unit: ¥ 10 Thousand).

Maximum	Minimum	Standard deviation	Mean	Kurtosis	Skewness
1,010,000	0.01	66,928.1	16,366	114.5	9.31

Although there are events caused by real assets damage and business system failures, the loss amount proportion is close to zero, indicating that the commercial banks in China can better prevent losses of operational risk caused by external causes.

Table 2 offers the statistical characteristics of the sample data. Their kurtosis and skewness are much higher compared with normal distribution, proving the feature of "sharp peak and fat tail".

3 MEASUREMENT MODEL BASED ON α-STABLE DISTRIBUTION

Compared with other distributions such as Student-t distribution and Weibull distribution, α-stable distribution not only allows the fat-tail and asymmetry properties, but also can well handle the tilt phenomenon of the data distribution. Many large data have the characteristics of "sharp peak and fat tail", so it's ideal to fit the data with α-stable distribution. Unary α-stable distribution has four equivalent definitions, and the definition relevant to its characteristic equation is commonly used. If a random variable X obeys an α-stable distribution, X will have characteristic equation which takes the following form:

$$E\left[e^{iuX}\right] = \begin{cases} \exp\left(-|\gamma t|^{\alpha}\left(1-i\beta\left(signt\right)\tan\frac{\pi\alpha}{2}\right)+i\delta t\right), \alpha \neq 1, \\ \exp\left(-\gamma|t|\left(1+i\beta\frac{2}{\pi}\left(signt\right)\ln|t|\right)+i\delta t\right), \alpha = 1. \end{cases} \quad (1)$$

Where α $(0 < \alpha \leq 2)$ is the stability coefficient or shape parameter. The smaller the α is, the longer the tail of the distribution is. β $(-1 < \beta \leq 1)$ is called the skewness parameter, and it decides the tilt of the distribution. The distribution skews to right if $0 < \beta \leq 1$, and the right skewness grows with the increase of β. Oppositely, the distribution skews to left if $-1 < \beta \leq 0$, and the left skewness increases with the decrease of β. δ and γ $(\gamma > 0)$ are position and scale parameters, respectively.

According to different uses of α-stable distribution, there are more than a dozen of parametric forms for it. Among them, two forms: 0 - parametric form and 1 - parametric form are most commonly used. They obey the forms of $S_\alpha(\beta,\delta,\gamma;0)$ and $S_\alpha(\beta,\delta,\gamma;1)$ respectively. Both the two forms only differ in position, and they are the same when $\beta = 0$. Because the 0-parameterized form is continuous with all the values of α, β, δ, γ, it is often used in the calculation and statistical analysis. In this paper, we also use 0-parameterized form when estimating the parameters.

The scale parameter γ indicates the volatility of data under α-stable distribution. According to the property of α-stable distribution (Nolan, 2003), if X is a column of random variables and $X \sim S_\alpha(\beta,\delta,\gamma)$, we have

$$\begin{cases} \dfrac{X-\delta}{\gamma} \sim S_\alpha(\beta,0,1), \alpha \neq 1, \\[2mm] \dfrac{X-\delta-2\beta\gamma \ln\gamma/\pi}{\gamma} \sim S_\alpha(\beta,0,1), \alpha = 1. \end{cases}$$ then

$\dfrac{VaR_{1-c}-\delta}{\gamma} \sim \rho_{1-c}^S$. The VaR under given confidence level $1-c$ can be calculated by (2):

$$VaR_{1-c} = \rho_{1-c}^S \gamma + \delta, \qquad (2)$$

where ρ_{1-c}^S denotes the quantile under confidence level $1-c$ of standard α-stable distribution.

4 EMPIRICAL ANALYSIS

4.1 *"Fat tail" testing*

The main properties of α-stable distribution include fat tail, discontinuity and long-term memory. "Fat tail" refers to the fat tail characteristic of α-stable distribution compared with the normal distribution. "Discontinuity" means that there could be substantial suddenly changes among time series of α-stable distribution. "Long-term memory" refers to the strong correlations between time series with long time interval, and the correlations do not change with time. Because data collected in this paper do not strictly belong to time sequences, we simply focus on the fat tail property. Figure 1 shows that the tail of the sample data has a greater slope compared with the normal distribution. The sample data are gathering at the tail, revealing the "fat tail" property. It is appropriate to construct measurement model based on α-stable distribution.

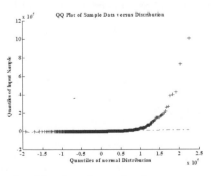

Figure 1. QQ plot of sample data versus normal distribution.

Table 3. Likelihood estimates of α-stable distribution.

Time interval	α	β	γ	δ
		¥ Million		
[1987,2010]	0.522131	1	31.625	7.887
[1987,2011]	0.527313	1	30.414	7
[1987,2012]	0.543237	1	29.6288	5.98
[1987,2013]	0.563374	1	29.6492	5
[1987,2014]	0.56685	1	29.65	5

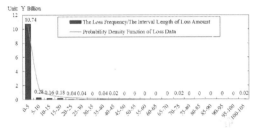

Figure 2. Density function curve of α-stable distribution fitted by loss data.

4.2 *Parameter estimation*

We use maximum likelihood method to estimate the parameters. The adopted tools are Matlab R2012a and Stable4.0. Table 3 lists the likelihood estimates of α-stable distribution. The density function curve drawn according to the α-stable distribution parameter estimates during time interval [1987, 2014] is shown in Figure 2. It's noticeable that the density curve and the loss distribution histogram almost entirely overlap, further proving that the α-stable distribution can well fit the operational risk loss data.

4.3 *Calculation and inspection of VaR*

In this part, we calculate the overall risk VaR of operational risk in Chinese commercial banks by using α-stable distribution, and test the validity of the results through return inspection—failure frequency test proposed by Kupiec (1995). The likelihood ratio LR takes the following form:

$$LR = -2\ln[(1-p_2)^{T-N}p_2^N] + 2\ln[(1-p_1)^{T-N}p_1^N] \qquad (2)$$

Where T is the total number of loss events, N is the number of loss that exceed the VaR with $N \sim B(p,T)$, p_1 is the failure frequency, and p_2 is the significance level. Under null hypothesis $p_1 = p_2$, LR obeys a χ^2 distribution with freedom degree of 1. For significance level 1%, the corresponding critical value is 6.635. If

LR > 6.635, we reject the null hypothesis, proving the VaR to be invalid.

The VaR values and validity test results under 95% and 99% confidence level are shown in Table 4. According to Table 4, all of the *LR* values under 1% and 5% significance level are less than the critical value of 6.635 and 3.841, meaning that the calculations of overall risk VaR are valid.

4.4 *Operational risk measurement of China's top four commercial banks*

We select loss data of Bank of China (BC), Industrial and Commercial Bank of China (ICBC), China Construction Bank (CCB) and Agricultural Bank of China (ABC) from the sample data and calculate their overall operational risk VaR respectively by following the same procedures above after conducting the "fat tail" testing. The results are presented in Table 5, showing that in time interval [1987, 2014], overall operational risk VaR of China's top four commercial banks under 95% and 99% confidence level are effective.

4.5 *The empirical results analysis*

According to Table 4 and Table 5, overall operational risk VaR of Chinese commercial banks calculated by α-stable distribution under 95% and 99% confidence level are valid. When the time interval expands from [1987,2010] to [1987,2014], the overall operational risk VaR shows a trend of decline and drops by around 50%. In time interval [1987,2014], the overall operational risk VaRs of BC and ICBC are higher than the national level, and the VaR of BC ranks the highest followed by ICBC, ABC and CCB in that order.

5 CONCLUSION

This paper uses α-stable distribution to fit operational risk loss data of Chinese commercial banks from 1987 to 2014, and calculates the overall risk VaR.

The empirical results show that the overall operational risk of Chinese commercial banks is decreasing in recent years, indicating that the overall operational risk management in Chinese commercial banks has achieved great success and the management capability is improving year by year. Besides, the capital charge required for covering the overall operational risk is decreasing year by year. The overall risk VaR of BC and ICBC are higher than the national level during time interval [1987,2014]. The two ones should strengthen their operational risk management and reduce the corresponding capital charge.

Table 4. Results of overall risk VaR and validity test.

Time intervals	$VaR_{95\%}$	$LR_{95\%}$	$VaR_{99\%}$	$LR_{99\%}$
	¥ Million			
[1987,2010]	6272.14	2.9230	14,463.3	5.1531
[1987,2011]	5711.48	2.9274	12,139.9	5.5454
[1987,2012]	4736.51	2.7510	9185.9	5.9071
[1987,2013]	3917.94	2.4166	6816.2	3.3004
[1987,2014]	3805.83	2.4732	6513.1	3.3267

Table 5. Overall risk VaR and validity test results of China's top four commercial banks.

China's top four commercial banks	$VaR_{95\%}$	$LR_{95\%}$	$VaR_{99\%}$	$LR_{99\%}$
	¥ Million			
BC	6722.75	1.3847	11,628.5	1.6483
ICBC	6139.81	0.9143	11,299.5	0.9045
CCB	1554.57	2.5137	2439.3	1.3869
ABC	1807.63	2.2237	3288.3	1.3065

The main contribution of this paper is: we adopt α-stable distribution widely used in the field of random signal to better capture the operational risk loss data's "sharp peak and fat tail" characteristics, ensuring the accuracy of calculation and providing new ideas for methods like LDA which needs to fit the loss distribution to measure operational risk.

However, α-stable distribution has no explicit expression of distribution and density function, and for commercial banks, using different models comprehensively according to their own needs is a rational choice.

ACKNOWLEDGMENTS

This work is supported by the Fundamental Re-search Funds for the Central Universities (FRF-TP-14-052A1).

REFERENCES

[1] Feng, J., Li, J. & Chen, J. 2011. Discrimination of Cross-Market Price Manipulations in Stock Index Futures Market: Evidences from Volatility and Liquidity. *Managemen Review* 23:171–176.
[2] Jobst, A . A. 2007. It's all in the data–consistent operational risk measurement and regulation. *Journal of Financial Regulation and Compliance* 15: 423–449.
[3] Kupiec, P. H. 1995. Techniques for verifying the accuracy of risk measurement models. *The Journal of Derivatives* 3.

[4] Lu, J. & Zhang, J. 2013. Measurement of Commercial Bank's Operational Risk Based on Extreme Value Theory and Multivariate Functions. *Chinese Journal of Management Science* 21: 11–19.

[5] Medova, E. A. & Kyriacou, M. N. 2001. *Extremes in operational risk management*: Cambridge University Press.

[6] Moscadelli, M. 2004. The modelling of operational risk: experience with the analysis of the data collected by the Basel Committee. Bank of Italy, Economic Research and International Relations Area.

[7] Nolan, J. P. 2003. *Stable distributions: models for heavy-tailed data*. Swiss: Birkhauser.

[8] Wang, Z., Wang, W., Chen, X., Wang, X. & Zhou, Y. 2012. Application of LDA based on Bootstrap sampling and piecewise-defined severity distribution in operational risk measurement. Journal of Management Sciences in China 15: 58–69.

[9] Yao, F., Wen, H. & Luan, J. 2013. CVaR measurement and operational risk management in commercial banks according to the peak value method of extreme value theory, *Mathematical and Computer Modelling* 58: 15–27.

The expected value management mechanism based on certificate system

X.K. Zheng
China Electric Power Research Institute, Beijing, China

H. Zhang
State Grid JIBEI Electric Power Company Limited, Beijing, China

S. Peng
China Electric Power Research Institute, Beijing, China

L. Song
State Grid JIBEI Electric Power Company Limited, Beijing, China

ABSTRACT: The expected value management system is proposed for the state grid system based on CA certificate system and trusted by computing technology. All software and update packages could be used after going through the developed system, which will help to protect the consistency, trustworthy and security of the installed software and update packages on terminals. The system will control software installation and update in terminals by the regulated and simplified process of software installation, which further enhances the security of the state grid system.

1 INTRODUCTION

With the construction of unified smart grid management system featured with informatization, automatization, and interaction, the system network access becomes more and more complex and challenging. The state power management system processes and stores much more sensitive information of finance, staff, and production material with requirements of cross-system and cross-region resource sharing. Therefore, it is necessary to pay more attention to the security infrastructure in the state power system, such as the authentication system, to make deep study about secure access, secure transmission and secure application, which will secure information of users, terminals, infrastructure network and applications. As an information security framework established on cryptographic techniques, the unified digital certificate system [1-5] is the best solution to meet requirements of authentication, access control, information confidentiality and non-repudiation. It is also the key technology [6-8] to keep applications trusted in the state grid, which covers intranet and internet of the state grid and provides digital certificate services for it.

2 OVERVIEW ABOUT CA

CA [9-12] is an abbreviation for Certificate Authority, usually translated as certification authority or certification center. It is the authority who is responsible for issuing and managing digital certificates and responsible for the legitimacy test in public-key e-commerce system.

Digital certificates [13-15] contain user identity information and all user public keys. The private key of CA can sign the digital certificate to prevent it from faking and tampering by any attacker. The user identity can be confirmed and the certification is accomplished [16] if the certificate is verified to be true and the authentication center issuing the certificate could be trusted. Because CA is the certificate issuing authority [17-19] which is responsible for certificate issuing, certification and the awarded certificate management, security policies, and measures should be developed to verify, identify the user and sign in order to protect the identity of the certificate holder and the ownership of public key.

In order to protect security, authenticity, reliability, integrity, and non-repudiation of the information transferred on the internet, CA is not only going to verify the authenticity of the user identity, but also an authoritative, impartial, unique institution which is responsible for issuing e-commerce security certificate complying with domestic and international security electronic transaction protocol standard, and responsible for the management of digital certificates of all entities involved in online transactions. Therefore, CA is the core technology of electronic transactions.

3 OPERATION PRINCIPLES OF CA CERTIFICATION SYSTEM

The main mechanism applied is encryption algorithm [20-24] during the process of CA certification.

Traditional symmetric key algorithm has advantages of high encryption strength and high speed computation, but security problems from the key transportation and management limit its applications. However, the public key algorithm can solve these problems with a pair of keys which are a private key and a public key. Generally speaking, the public key is publicly available and is used to encrypt, the private key is kept secret personally and used for signing. Secret strength of the algorithm depends on the selected key length [25-28].

If the users want to get a certificate of their own, they should first apply to CA. After CA verifies the identity of the applicant, the applicant will be assigned a public key that will be combined with the applicant's identity information. Then, the authoritative signature will be issued to the applicant.

If a user or an organization would like to identify the authenticity of a certificate, the CA public key will be used to verify the certificate signature. Once validated, the certificate is considered valid. CA has a private key and a certificate (containing the public key) that will be available to any person (any organization). Public online users trust CA by verifying CA's signature.

4 DESIGN OF THE EXPECTED VALUE MANAGEMENT MECHANISM

With certificates technology and trusted computing technology, the expected value management system is developed to check the source consistency and security of the installed software on terminals in the state power system. All software and update packages could be used after going through the expected value management system, which will help to control software installation and update in terminals by the regulated and simplified process of software installation.

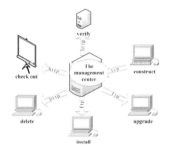

Figure 1. The schematic diagram of expected value management system.

The trusted software management center is used to collect, verify, install, upgrade, view, and delete trusted resource packages, which will help to establish a trusted software library for the state power system gradually.

The resource packages are signed and verified by CA certificate system, which provides the functions of installment, upgrade, and deletion by ftp and http, and view and management by http. The system is designed to keep software installed on terminals in the state power system of being consistent, trusted, and unmodified. The installment upgrade and deletion of software in terminals are simplified by unified policy templates, which ensure that the expected value could be used for a long period once they are calculated and collected.

4.1 Topology of the expected value management system

The expected value management system is mainly composed of CA center, trusted software, terminal, management center, and sub-management center, which provide the following functions.

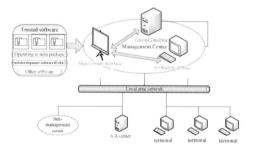

Figure 2. The framework of the expected value management system

CA center runs as CA agencies and the certificate authority. As a trusted third party in transactions, it is responsible for testing the legality of public key, providing functions of generating, updating, and verifying the certificate. Trusted software is the trusted software to be installed. Terminal is a computer with trusted software base which automatically installs software corresponding to template from the management center during the process of the registration or connecting the management center. The management center is the system configuring and managing system resources, security mechanisms and audit information. It provides functions of collecting trusted software and information, generating acquisition package and managing them after CA-signing, configuring templates for terminal, and approving a template and issuing it. Submanagement center signs up to the management center, which synchronizes the template from the management center and downloads the delayed templates.

These function parts operate with the following principles.

1 Management center applies signature public key from the offline CA center, and then saves it.
2 Management center uploads the trusted software and fills in the software information, and collects, analyses, signs, and saves the trusted software.
3 The administrator configures the template or customizes the generic template by management center.
4 When a terminal registers or connects management center, management center will produce a new template by the comparison of the current templates and those installed on terminals, and then issues the template package to the terminals.
5 Terminals install the software package after receiving it.
6 Generic template is issued to sub-management center, and the software package is delayed to issue.

Where, step 1, 2, 3, and 6 will be completed in the management center.

4.2 *Operation procedures*

The generation of CA certificate is as follows:

1 Management Center gets the public key.
2 Management Center initiates a request to the CA Center to request a certificate with the parameter of the public key.
3 After the application is approved, management center saves the CA certificate with its root certificate chain.

The procedures of acquiring the expected value are as follows:

1 Upload software and fill in related information by management interface.
2 Process the uploaded trusted software, including operating system packages, self-developed and software from other venders.
3 Send software and related information to collection terminals, and close the connection after that.
4 After acquiring the expected value and generating installation scripts and the relationship between them, collection terminals send the software package to the management center.
5 Management Center resolves package dependencies, generates the software package templates including program name and version number, and

saves the package template and dependencies to the database
6 Management center signs the software package, generates a new software package with signature files and original software packages and saves it into the trusted database.

Terminal registration procedures are as follows:

1 A terminal applies to register in management center.
2 Management center generates package templates and policies.
3 Management Center issues the CA certificate with the root certificate chain, software packages, and the public key. The connection is closed after that.
4 The terminal verifies CA by the root certificate. If the verification fails, there will display an error message and send an error message to the management center by the heartbeat.
5 The terminal verifies the package signature with the public key and installs the package if the verification is successful.
6 The installation information is sent with the heartbeat after the installation.
7 If the installation is successful on the terminal, the management center will save the template package. Otherwise, there will display an installation error information.

The procedures of modifying the template are as follows:

1 The administrators modify the terminal package template through the management interface.
2 Management center sends notifications about modifying the template to the terminal.
3 Terminal receives the modified notification and initiates an application request.
4 Management center generates a new template by the comparison of the current template with the template installed in the terminal.
5 The following steps are the same as "Terminal registration procedures".

5 CONCLUSIONS

The expected value management system is developed for the state grid system. The system succeeds in ensuring the source consistency and security of the installed software on terminals in the state power system based on certificates technology and trusted computing technology. All software and update packages could be used after going through the expected value management system, which will help control software installation and update in terminals by the regulated and simplified process of software installation.

REFERENCES

[1] Biswas, G.P., 2011. Establishment of authenticated secret session keys using digital signature standard. Information Security Journal: A Global Perspective 20 (1), 9–16.

[2] Diffie, W., Hellman, M., 1976. New directions in cryptology. IEEE Transaction on Information Theory 22, 644–654.

[3] ElGamal, T., 1985. A public-key cryptosystem and a signature scheme based on discrete logarithms. IEEE Transactions on Information Theory 31 (4), 469–472.

[4] Hu, J., Han, F., 2009. A pixel-based scrambling scheme for digital medical images protection. Journal of Network and Computer Applications 32 (4), 788–794.

[5] Levi, A., Caglayan, M.U., Koc, C.K., 2004. Use of nested certificates for efficient, dynamic, and trust preserving public key infrastructure. ACM Transactions on Information and System Security 7 (1), 21–59.

[6] J.K. Liu, J. Baek, W. Susilo, J.Y. Zhou, Certificate-based signature schemes without pairings or random oracles, in: T.C. Wu, et al. (Eds.), Proceedings of the ISC'2008, in: LNCS, vol. 5222, Springer-Verlag, Berlin, 2008, pp. 285–297.

[7] Z.H. Shao, Certificate-based verifiably encrypted signatures from pairings, Information Sciences 178 (2008) 2360–2373.

[8] W. Wu, Y. Mu, W. Susilo, X. Huang, Certificate-based signatures revisited, Journal of Universal Computer Science 15 (8) (2009) 1659–1684.

[9] X. Boyen, Multipurpose identity-based signcryption: a swiss army knife for identity-based cryptography, in: CRYPTO 2003, in: LNCS, vol. 2729, Springer-Verlag, Berlin, 2003, pp. 383–399.

[10] F. Bao, R. Deng, A signcryption scheme with signature directly verifiable by public key, in: Public Key Cryptography, PKC 1998, in: LNCS, vol. 1431, Springer-Verlag, Berlin, 1998, pp. 55–59.

[11] P. Ji, M. Yang, Verifiable short signcryption without random oracle, in: The 2007 International Conference on Wireless Communications, Networking and Mobile Computing, WICOM 2007, IEEE Computer Society, 2007, pp. 2270–2273.

[12] H. Jung, K. Chang, D. Lee, J. Lim, Signcryption schemes with forward secrecy, in: Proceeding of the 2nd International Workshop on Information Security Application—WISA 2001, vol. 2, pp. 403–475.

[13] J. Malone-Lee, W. Mao, Two birds one stone: signcryption using RSA, in: CT-RSA 2003, in: LNCS, vol. 2612, Springer-Verlag, Berlin, 2003, pp. 211–225.

[14] Q. Wu, Y. Mu, W. Susilo, F. Zhang, Efficient signcryption without random oracles, in: ATC 2006, in: LNCS, vol. 4158, Springer-Verlag, Berlin, 2006, pp. 449–458.

[15] J. Zhang, J. Mao, Security analysis of two signature schemes and their improved schemes, in: ICCSA 2007, in: LNCS, vol. 4705, Springer-Verlag, Berlin, 2007, pp. 589–602.

[16] J. Baek, R. Steinfeld, Y. Zheng, Formal proofs for the security of signcryption, in: Public Key Cryptography, PKC 2002, in: LNCS, vol. 2274, Springer-Verlag,Berlin, 2002, pp. 80–98.

[17] M. Seo, K. Kim, Electronic funds transfer protocol using domain-verifiable signcryption scheme, in: JooSeok Song (Ed.), ICISC'99, in: LNCS, vol. 1787, 2000, pp. 269–277.

[18] Y. Mu, V. Varadharajan, Distributed signcryption, in: B. Roy, E. Okamoto (Eds.), INDOCRYPT 2000, in: LNCS, vol. 1977, Springer-Verlag, Berlin, 2000, pp. 155–164.

[19] D. Kwak, S. Moon, Efficient distributed signcryption scheme as group signcryption, in: J. Zhou, M. Yung, Y. Han (Eds.), ACNS 2003, in: LNCS, vol. 2846, Springer-Verlag, Berlin, 2003, pp. 403–417.

[20] I. Gupta, P.K. Saxena, Distributed signcryption from pairings, in: S. Jajodia, C. Mazumdar (Eds.), ICISS 2011, in: LNCS, vol. 7093, 2011, pp. 215–234.

[21] F. Li, Y. Hu, C. Zhang, An identity-based signcryption scheme for multi-domain ad hoc networks, in: J. Katz, M. Yung (Eds.), ACNS 2007, in: LNCS,vol. 4521, Springer-Verlag, Berlin, 2007, pp. 373–384.

[22] S.S. Selvi, S.S. Vivek, D. Shukla, P.R. Chandrasekaran, Efficient and provably secure certificate-less multi-receiver signcryption, in: J. Baek, et al. (Eds.), ProvSec 2008, in: LNCS, vol. 5324, 2008, pp. 52–67.

[23] Z.H. Liu, Y.P. Hu, X.S. Zhang, H. Ma, Certificate-less signcryption scheme in the standard model, Information Sciences 180 (2010) 452–464.

[24] C.K. Li, D.S. Wong, Signcryption from randomness recoverable public key encryption, Information Sciences 180 (2010) 549–559.

[25] C.K. Li, G.M. Yang, D.S. Wong, X.T. Deng, S.S.M. Chow, An efficient signcryption scheme with key privacy, in: EuroPKI 2007, in: LNCS, vol. 4582, Springer-Verlag, Berlin, 2007, pp. 78–93.

[26] C.K. Li, G.M. Yang, D.S. Wong, X.T. Deng, S.S.M. Chow, An efficient signcryption scheme with key privacy and its extension to ring signcryption, Journal of Computer Security 18 (3) (2010) 451–473.

[27] G. Yu, X.X. Ma, Y. Shen, W.B. Han, Provable secure identity based generalized signcryption scheme, Theoretical Computer Science 411 (2010) 3614–3624.

[28] J. Malone-Lee, Identity based signcryption, Cryptology ePrint Archive, Rep. IACR/098, 2002.

Network Security and Communication Engineering – Chan (Ed.)
© 2015 Taylor & Francis Group, London, ISBN: 978-1-138-02821-0

Security defense model of Modbus TCP communication based on Zone/Border rules

W.L. Shang
ShenYang Institute of Automation, Chinese Academy of Science, Shenyang, Liaoning, China

L. Li
ShenYang Ligong University, Shenyang, Liaoning, China

M. Wan & P. Zeng
ShenYang Institute of Automation, Chinese Academy of Science, Shenyang, Liaoning, China

ABSTRACT: To detect the intrusion of advanced industrial virus of industrial control system, design flaws of Modbus TCP Protocol is firstly analyzed in this paper, and a method is proposed, through which the Modbus TCP packet is deeply inspected to deal with the threat from application layer. Furthermore, a general description form of the security rules is proposed, and defense model for Modbus TCP communication in industrial control system or SCADA system is designed, which is based on intrusion detection rules and "white-list" rules. With definition of the minimum set of normal communication between different zones, the system has eliminated exposure greatly. At last, simulation experiments validate that the proposed method is effective and practical.

1 INTRODUCTION

Industrial Control Systems or SCADA systems are widely used in electric power, water plants, petro-chemical, and transportation, aerospace and other fields, which are important parts of the national critical infrastructure and related to the national security strategy. In the early design of industrial control systems, proprietary communication protocols, operating systems and hardware were commonly used, and industrial control systems were isolated from other networks. Therefore, industry control systems are believed to be more secure, and function and physical safety get more attention; information security and network security are less considered.

With the continuous integration of industrialization and informatization, TCP/IP technology, open industrial communication protocols, common hardware, software, and network infrastructure are widely used in the industry communication systems. Industry control system exchanges data with enterprise management information system, or even with the Internet. Thus the closure of the industrial control system is gradually being broken. Industrial control system has many information security and network security flaws that are more vulnerable to hackers, viruses, and penetration attack from hostile forces (Wei. et al. 2013).

Snort2.9 version began to join the Modbus TCP, DNP3 and CIP protocols pre-processor (Xia 2013), which can analyze part of the field of communication protocol and data packet integrity, it can also be combined with the set of intrusion detection rules which is released by the third party agencies, for example, Digital Bond to identify industrial control systems / SCADA system intrusions.

NivColdenberg, Avishai Wool et al. proposed a Modbus TCP communication precise modeling method (Thomas 2013) based on finite state automata. The modeling process according to obtained data packets which grabbed between HMI and PLC data channel, constructed the model based on finite state automata. This method results in a good performance in an actual environment, and it has a low false positive rate. However, the method needs to build on the communication between the HMI and PLC with cyclical characteristics. It needs to be further verified whether it can face the complex industry control system application on the SCADA application.

Dale Peterson et al. have designed a passive method of the security event for DCS system and SCADA system (Goldenberg 2013); this method made up for the lack of records of security incidents and made possible the behavior of audit and fault tracing.

Hadeli, RagnarSchierholz et al. proposed that we can identify abnormal behaviors reliably by the behavior of industrial systems and system description file; they also designed and implemented a framework of generating configuration rules automatically for firewalls and other security facilities configuration rules (Peterson 2013).

WojciechTylman proposed a non-IP protocol processing method based on Snort data acquisition module (DAQ).This method is used for the Modbus RTU communication (Schierholz 2009) and it makes snort to drill fieldbus layer for intrusion detection. What is more, it does not need to add additional program and hardware or modify the Snort code itself.

Current intrusion detection research work can be divided into two categories (Thomas 2013): Misuse detection and Anomaly detection. Misuse detection is based on the method of matching malicious network traffic with known attack signatures. For the reason that the attack signature database of industry control system could not cover all the cases, Misuse detection is easy to generate false negatives.

Anomaly detection is based on the method of identifying abnormal traffic flow of the industrial control system. The fixed communication traffic mode is established by the state model of the system. In comparison with Misuse detection, Anomaly detection is able to detect unknown attacks, and even the disoperation of operators. But Anomaly detection is easy to generate false alarms because it cannot locate error clearly.

The paper is organized as follows: Firstly, security of Modbus TCP is analyzed, and deep packet inspection (DPI) for Modbus TCP is designed. Secondly, we propose a security defense model of industry communication protocol based on Zone/Border rules. At last, simulation experiments are done, which validates that the proposed method is effective and practical.

2 ANALYSIS OF MODBUS TCP SECURITY

2.1 Modbus and Modbus TCP

Modbus is an application layer messaging protocol, located at level 7 of the OSI model and it has become a standard. Its implementations mainly include the achievement based on the serial link (Modbus RTU/ Modbus ASCII), high-speed Token Ring (Modbus Plus), and TCP/IP Technology of Ethernet (Modbus TCP/IP, Modbus TCP for short). As shown in Figure 1, Modbus communication can be connected seamlessly to the same communications system. The role of network gateway is to guarantee conversion of different underlying Modbus communication.

This paper mainly focuses on the communication security which is based on TCP/IP technology of Modbus communication in Ethernet, its request and response packets are encapsulated into the format shown in Figure 2.

MBAP HEADER is the Modbus application protocol header; its purpose is to identify Modbus Application Data Unit. This part mainly includes four parts. They are Transaction Identifier, Protocol Identifier, Length, and Unit Identifier. Function code is a flag of that Modbus client (MASTER)

and indicates server (SLAVE) as to what to do, and it can reflect operating intentions of client to server. The DATA is set by the client according to the difference of function and the specific applications and the server responds correspondingly.

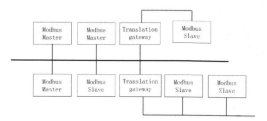

Figure 1.   The topology of Modbus TCP.

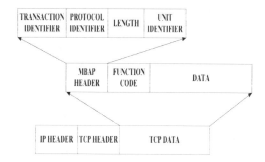

Figure 2.   Modbus TCP packet format.

2.2 Identification/encryption/authorization deficiency

Industrial control protocol, which is widely used today, almost runs in relatively isolated networks environment in the design stage, and there is no data switching with other networks. Therefore, information security and network security factors do not need to focus on consideration, but function safety and physical security should be concerned more (Xia 2013). Although information security needs the existence of industrial control communication protocol system, it is a complex systems engineering to add the industrial communication protocol to the security mechanism. We believe that the security risk of industrial control systems communication protocol will not change from itself in a long situation.

Modbus TCP communication layer is based on standard Ethernet TCP/IP technology, therefore, the traditional attacks on the IT network are also suitable for the industrial control system, and they even cause more harm (Rodrigo 2008). However, the more effective, the more destructive way for control system to build the aggressive behavior to the application layer data. This is mainly due to the defects existing in the design of Modbus protocol application layer, it mainly reflects in identification, authorization, and encryption security mechanism deficiency. Performance of identification deficiency is that only a valid Modbus address and legal function code can establish a Modbus session; Performance of authorization deficiency is that any user can execute arbitrary function without role-based access control mechanism; Performance of encryption deficiency is that address and command are transmitted by plaintext transmission and it is easy to capture and parse.

After obtaining control authority of a system or access authority of a network by a series of penetration attack, an attacker can obtain the information he is interested in by monitoring Modbus TCP traffic in the network, and can construct Modbus TCP packet easily as protocol, what is more, an attacker even can launch an attack on an important controller with the corresponding debugging tool directly. For example, in an industrial control system, the digital quantity may correspond to the coil and holding register of PLC. If a value of a coil of PLC is changed, the switch which should be closed will be likely to open. Obviously it will cause serious consequences in industrial control systems which have extremely stringent and reliable requirements. In another situation, the attacker changes a value in the register, while holding register may correspond to the analog input and output or the important parameter in industry control process, therefore technology cannot reach the standard, or it causes an accident directly.

3 MODBUSTCP DEPTH PARSING METHOD

The above analysis shows that if we want to protect the industrial control system security based on Modbus communication, Modbus packets will need depth parsing in order to detect intrusion and attack. Therefore this section establishes a general model of depth parsing of Modbus message. As shown in Figure 3:

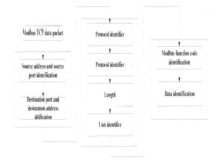

Figure 3. Modbus TCP depth parsing.

Modbus protocol depth parsing model mainly consists of three parts, they are the network layer and the transport layer parsing, Modbus packet header parsing, and Modbus message parsing. Its process is as follows:

Step1: the network layer and the transport layer parsing: This part mainly extracts the source IP address and destination IP address and the source and destination port number and analyzes the information in this part in order to provide basic information for access control in the network layer and transport layer or intrusion detection. Communication network equipments are marked by the IP address and port numbers.

Step2: Modbus packet header parsing: This part mainly explains Protocol identifier, Protocol identifier, Length, and Unit identifier. Protocol identifier provides basis to determine whether it is the Modbus communication or not. The value of the length field represents the unit identifier and the total length of data, by this we can determine the abnormal packets which are constructed artificially. Unit identifier is the address on the Modbus serial link, its length is one byte. Different Modbus devices are addressed according to this field; these different Modbus devices connect with the same protocol conversion gateway and share the same IP address. Therefore, the field is of great significance.

Step3: Modbus message parsing: This part is core and key of Modbus protocol depth parsing, its main duty is to analyze the Modbus function codes which represent the operation intention and to analyze the corresponding data according to different Modbus function codes.

Modbus defines four kinds of data types: they are input coil, coil, input register, and holding register. They correspond to the corresponding variables respectively, such as digital input and output, analog input and output correspond to specific function codes. For example, function code 02 represents reading the input coil operation, its operation object is input coil; function code 03 represents reading the holding register, its operation object is holding resister. Due to the different function codes in protocol, it may operate a single or multiple data objects; hence we need to propose a general Modbus messages analytical mode. As shown in Figure 4:

The process is divided into the following steps:

Step3.1: Parse Modbus Function Code field. If it is not relevant to read and write operation in the four data types of Modbus protocol, the Step2 is executed, otherwise, turn to Step3.

Step3.2: Analyze function codes which may exist and analyze the key fields of corresponding function, then the whole analytical process comes to the end.

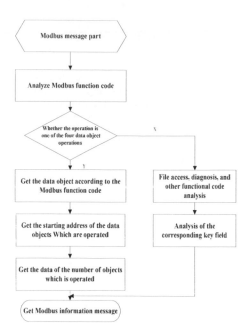

Figure 4. Modbus TCP message parsing.

Step3.3: According to the function code, identify the data objects which are operated.

Step3.4: Analyze the starting address of the data objects which are operated.

Step3.5: Analyze the number information of data objects.

Step3.6: Save the corresponding information of fields. Come to end of the process.

4 E-SECURITY DEFENSE MODEL OF PROTOCOL COMMUCICATION BASED ON ZONE/ENCLAVE RULES

At present, the academic circles and the industry generally believe that the most effective way to ensure the information security and network security of the industrial control system is to build "defense-in-depth" system for industrial system (Fovino 2010). "Defense-in-depth" system contains many aspects, from the perspective of industrial communication protocols depth parsing, and this paper believes that building "defense-in-depth" system of industrial SCADA systems is mainly divided into two core steps: Zone division and Border establishment and security strategy design.

4.1 Zone division and border establishment of industrial control system.

Zone generally has implemented or supported the same functions and has certain assets with the same security requirements. Border exists in different zones, it is the channel of data switching in different zones, the reasonable division and setting of zones and borders are of great significance for ensuring the information security of industry communication system. Generally speaking, Zone division (Eric 2011) can be considered from several dimensions: network connection, control function, data storage, remote access, communication protocol, and critical degree. In this paper, we mainly consider the industrial SCADA system based on the Modbus TCP, therefore, the general method of system zone division and border establishment are simplified as the follow process:

Firstly, the system needs to have a risk assessment and performs important asset identification, and establishes the appropriate documentation.

Secondly, according to the level of an important asset in the enterprise business, the whole system is divided into four layers in accordance with the order from top to bottom, they are business management information layer, data acquisition monitoring network layer, the control unit layer, and the field device layer.

Again, different zones are further divided in each layer according to the functions to be achieved or operating authority. For example, data acquisition monitoring layer has the monitoring picture site which only reads data from the underlying PLC and may have a write operation on the site; this site belongs to different sub-zone levels of data acquisition monitoring layer.

Finally, in order to manage and control the zone communication, in different zones where data communication sets a border and makes the traffic between the two zones via the border.

What need to pointed out is that a division of the sub-region within the region is further characterized for the security needs of different assets within the region. Therefore, finer the "granularity", the more divided the region is and the more nested sub-zones are. In addition, the zone can coincide in logic and physics if we adjust the industrial control system network structure.

4.2 Security strategy design

Border protection among zones may have different security defense technology and equipments, furthermore, the rules description are also different, therefore we first abstract general safety rules modcl:

[Action] [Source Dress] [Source Port] [Destination Address][Destination Port][Specific Protocol Fields]

Action includes:

- Allow: To determine the corresponding data packets is legitimate and releases without any other actions.
- Deny: To determine the corresponding data packets is illegal which discards and triggers an alarm.
- Alarm: To determine the corresponding data packets is abnormal but it releases, and triggers an alarm.

What is more, Specific Protocol Fields indicate the field conditions related to communication protocol. For Modbus TCP communications, the detailed format is as follows:

[Protocol Identifier] [Unit ID] [Function Code][Data]

where: Protocol Identifier refers to the protocol identifier in Modbus packet header; Unit ID refers to unit identifier in Modbus packet header; Function Code refers to Modbus specified function code;

The Data field refers to the relevant Modbus function code. When the Modbus function code read and write corresponding to the four kinds of data or other operations, Data field consisting of the Starting address identified the data type and Number or Sub Function:

[Data] = [StartAddress][Num] OR [SubFunction]

Security defense technology which is used for border may include firewall, IDS, IPS (Sun 2006), etc.

Firewall mainly has access control function for transport layer and network layer, but it lacks support for application layer in industrial protocol. IDS and IPS use schema matching method and take appropriate action to the data packets which meet features.

Judging from deployment, IDS system gets network traffic by bypass monitoring and will not cause the delay effect on network communication. IPS is in the communication link and the illegal communication packages can be intercepted and discarded. Judging from security policy which is reflected from traditional IT information security rules, the security strategy of IDS and IPS mostly take a list of detection rule. If the packet matches, then alarm or discard. While another IPS security policy defines legitimate packet types, then discards any other traffic. These two types of security strategy are shown in Figure 5.

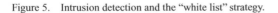

IDS general security policy	IPS whitelist policy
Alert description1;	Allow description1;
Alert description2;	Allow description2;
Alert description3;	Allow description3;
......	......
Alert descriptionN;	Allow descriptionM;
	Deny All Others

Figure 5.   Intrusion detection and the "white list" strategy.

These two security strategies have great limitations for industrial SCADA system. Firstly, the IDS system allows to have a miscarriage, but it is unable to intercept illegal data packet. While as to the "white-list" IPS strategy, if its configuration rules are improper, legitimate packets may be discarded. This is very dangerous to the industrial control system which requires high reliability.

This paper proposed a Modbus TCP communication protection method in industrial control system or SCADA system combining IDS rule with "white-list", and zone division and the border setting together. As shown in Figure 6. Steps are as follows:

Step1: Zone division and border setting of SCADA system, zone division according to industrial control system of the enterprise production information layer, the data acquisition monitoring layer, the control unit layer, and the field device layer. If each zone needs further zone division, and records the partition results with sign in order to further set safe strategy conveniently.

Step2: Set part of "white-list" IPS strategy.

Step2.1: Make clear the minimum set of Modbus TCP communication for data switching between different zones and sub-zones and define this communication as the legitimate communication. The legitimate data defines jointly according to codes in the part of Modbus TCP protocol depth parsing and data whose codes are needed.

Step2.2: Legitimate packet is described by safety rule description language. For example, the Modbus client can use the 01 function code to access the Modbus server address. Safety rules of three coils starting from 0x0000 can be expressed as, [Action:Allow] [Source Address:Modbus_client] [SourcePort:any] [Destnation Address: Modbus _server] [Destnation Port:502] [FunctionCode:01] [StartAddress:0000][Num:03]

Step3: For those that cannot be recognized as legitimatization completely, take IDS strategy to trigger an alarm but not to intercept. This part of the communication may include the communication which can be used by attackers and systems themselves. For example, function code 08 and the sub-function code 01 are reset functions; it cannot be identified as aggressive behavior. Therefore, the best choice is to trigger an alarm timely.

Step4: Default rule setting, that is to say, give up all the other Modbus TCP communication packets which are not in the setting of Step2 and 3.

In fact, the model is not only suitable for Modbus TCP communication security, if we add depth parsing mechanism of other communication protocols,

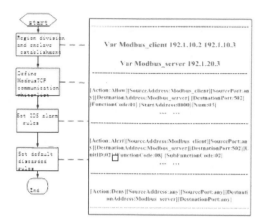

Figure 6. Modbus TCP communication security defense model.

such as MMS, GOOSE, FF-HSE and DNP3, etc, this model can be extended according to the same process to become a normal model.

5 SET THE EXPERIMENTAL ENVIRONMENT

5.1 *Simulation experiment environment and experiment design*

In order to validate Modbus system defense method that this paper supposes, we set up industrial SCADA system simulation environment based on Modbus TCP communication. The topology and structure are shown in Figure 7:

The simulation environment is divided into three layers, the data acquisition monitoring layer, the control unit layer, and virtual field device layer. Data acquisition monitoring layer includes two Modbus clients, they are the monitor screen based on KingSCADA software development and debugging and testing software diagslave of a simulated attack; the control unit layer selects Schneider M340PLC, its CPU model is 2020; virtual field device layer can be implemented in control logic of PLC M340.

Virtual field device layer includes switching values which identify the switch opening and closing, opening analog of three electromagnetic flow valves and the upper and lower limitation of the liquid level and the current level .The logic of PLC limits the current level between the upper level and the lower level. In the real industrial control systems, if the current level is not between the upper and lower limits, it may result in a container explosion or severely affect process quality. Therefore, the attack method design is to modify the holding registers which keep the upper

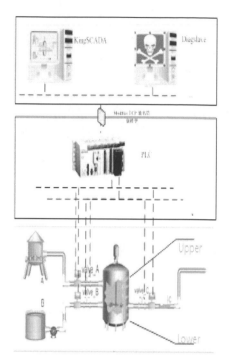

Figure 7. Simulation environment.

level by diagslave software. After the actual liquid level in the container exceeds the real limits, PLC control logic will not change the state of the input and output valves to control the liquid level.

5.2 *Experiment design and results*

The best way to resist the attacks mentioned above is to work normally according to the control logic based on the experimental simulation environment, clear the minimum communication setting required by Modbus client and Modbus server, and well-design border protection strategy. Here the border protection strategy selects the Snort. Security defense strategy is described based on Snort grammatical rules. In order to describe briefly, we still take the same safety rules description manner.

By analyzing the scene, we can realize that the data acquisition software simulated by diagslave only has the rights of reading switching value and analog quantity. While KingSCADA has the rights of reading and writing switching values and analog quantity. The level of the upper and lower limits cannot change frequently, so even if the write operation from KingSCADA should also be identified as abnormal operation and triggers an alarm.

In order to simplify the rules, we assume that there is only has the data subject representing the upper limits and its address is 0x0000. The security rules of the system design process are as followed:

Step1: The system is divided into data acquisition monitoring layer and control unit layer, they respond to Modbus client and server. And the data acquisition monitoring layer is divided into two sub-zones according to which the upper limits can be written.

var Modbus_client [192.168.1.1, 192.168.1.2]
var Modbus_client _1 192.168.1.1
var Modbus_client_2 192.168.1.2
var Modbus_server 192.168.10

Step2: Set "white list" rules, allowing KingSCADA and diagslave to 'read" for the holding register whose address is 0x0000 in PLC.
[Action:Allow][SourceAddress:Modbus_client] [SourcPort:any] [DestinationAddress:Modbus _server][DestinationPort:502] [FunctionCode:03] [StartAddress:0000] [Num:01].

Step3: Set abnormal behavior intrusion detection rules. If KingSCADA has a "write" operation on the PLC, then alarm.
[Action:Alert][SourceAddress: Modbus_client_1] [SourcPort:any] [DestinationAddress:Modbus_server] [DestnationPort:502] [FunctionCode: 06] [Start Address:0000][Num:03]
Set the default rules and intercept all communication which does not appear in the "White-list".
[Action:Deny] [SourceAddress:any] [SourcePort: any] [DestinationAddress:Modbus_server] [Destination Port: any]
The experimental results show that diagslave and the simulation of the attack source and KingSCADA only read 0x0000 holding resister in PLC. At the same time, if the King SCADA modifies the level limits, it will trigger the Snort alarm. This industrial control systems security policy among zones which combines intrusion detection rules with "white list" can greatly eliminate the risk of exposure and guarantee the safe operation of the system.

6   CONCLUSIONS

To solve the problem that it is difficult to detect the intrusion into industrial control system by advanced industrial virus, and design flaws in industrial communication protocols, in this paper we have proposed a method through which industrial communication protocols were deeply parsed. A normal method that describes the security rule is also proposed. Furthermore, this paper designed Modbus TCP zone and border security model based on what combines the intrusion detection rules with "White-list". The core idea of this model is clearly identifying legitimate Modbus TCP communications among zones, and prohibiting all other unnecessary data switching between zones. However, those legal but suspicious traffics will trigger an alarm. The features of the protection strategy are simple, clear and easy spreading. It is very suitable for industrial control systems where communication flow patterns are relatively fixed in network. Validation tests also showed the effectiveness of this method.

However, the Modbus TCP security rule description model is unable to effectively describe the communication behaviors existing in several packets at the same time. Modbus TCP communication protection strategy depended on understanding and grasping of designers to protocols and industrial processes and the understanding deviation will cause the safety rules setting error. In addition, online defensive measures may cause effect on real-time and reliability of industrial control system communication. In the future, we can further study to solve the attack implied in multiple Modbus packets, and real time and reliability of the safety protection strategies on communication system.

REFERENCES

[1] Dale, Peterson & Quickdraw.2009.Generating Security Log Event for Legacy SCADA and Control System Device. Cybersecurity Applications & Thehnology Conference For Homeland Security: 227–229.
[2] Eric,K.2011. Industrial Network Security securing Critical infrastructure Networks for Smart Grid, SCADA, and other Industrial Control System.
[3] Fovino, I.N. & Carcano, A. Modbus/DNP3 State-based Intrusion Detection System. Perth, WA: Advanced Information Networking and Applications: 729–736.
[4] Hadeli,Ragnar Schierholz & Markus Braendle.2009. Leveraging Determinism in Industrial Control Systems for Advanced Anomaly Detection and Reliable Security Configuration. Mallorca: Emerging Technologies & Factory Automation. 1–8.
[5] Niv Goldenberg & Avishai Wool.2013.Accurate Modeling of Modbus/TCP for intrusion detection in SCADA system. Critical Infrastructure protection: 63–75.
[6] Peng, Yong & JIANG, Changqing & Xie Feng. 2012. Industrial control system cyber security research. TsinghuaUniv(Sci&Tech)52(10):1396–1405.
[7] Peter,H. Mauricio, P.2008. Attack taxonomies for the Modbus protocols.Critical Infrastructure protection:37–44.
[8] Sun, Dalin &Jiang Daming.2006.Modbus/TCP protocol safety and its application in industrial monitoring and control system. Journal of Safety Science and Technology2(2):92–95.

[9] Tan, Aiping & Chen, Hao &Wu, Boqiao.2014. Ensemble learning algorithm. Computer Science Network Intrusion Detection Based on SVM: 197–200.

[10] Thomas, H. Morris, Bryan A. Jones, Rayford B.2013a. Deterministic Intrusion Detection Rules for MODBUS Protocols. Hawaii: 46th Hawaii International Conference on System Sciences: 1773–1781.

[11] Rayford Vaughn.2013.A Retrofit Network Intrusion Detection System for MODBUS RTU and ASCII Industrial Control Systems. Hawaii: 2013 46th Hawaii International Conference on System Sciences: 2338 –2345.

[12] Morita, T & Yogo, S .Detection of Cyber-Attacks with Zone Dividing and PCA.

[13] Wei, Qingzhi.2013.Industrial Network Control System Security and Management. Measurement & Control Technology32 (2): 87–92.

[14] Xiong,Qi & Jing, Xiaowei &Zhan Feng.2012. Summary and implications for China of the information security work of the ICS system in the oil and gas industry in America. China Information Security27 (3):80–83.

[15] Xia,Chun-ming & Liu Ta o& Wang Hua-zhong.2013. Industrial Control System Security Analysis. Information Security and Technology: 13–17.

Network Security and Communication Engineering – Chan (Ed.)
© 2015 Taylor & Francis Group, London, ISBN: 978-1-138-02821-0

Permissions abuse detection for android platform based on droidbox

S.Q. Wang, H.T. Ma & X.H. Wu
Science and Technology on Integrated Information System Lab, Institute of Software, Chinese Academy of Sciences, Beijing, China

S.Q. Wang
Wuhan Institute of Post and Telcommunications, Wuhan, China

ABSTRACT: Permission-based security model of Android restricts the applications' behavior from accessing to specific resources, but malicious applications can be invaded by using permission mechanism vulnerabilities. Through the analysis of the permission features of Android applications, we describe a model of permission abuse detection to cover and quantify the categories of all cases logically, and then provide a method in which we combine static permission analysis with dynamic Inter-process Communication (IPC) detection. Permissions abuse behavior of the Android applications can be identified by utilizing the permission to set mapping and taint tracing techniques. Our experiment results indicate that the proposed method can effectively detect permissions abuse in different forms.

1 INTRODUCTION

Android is an open-source operating system based on Linux, mainly developed by Google company [1]. Since the third quarter of 2013, the market share of the Android platform in China has accounted for 59% [2]. However, with the rapid development of smart phones and the popularity of the Android system, malwares and Trojans use the system function defects or code vulnerabilities to steal private information such as account password and credit card numbers in our devices. The permission-based security mechanism is a fundamental policy to defend such attacks and protect users' privacy.

The paper is organized as follows: the first part highlights the main techniques from the relevant literature, and outlines the fundamental theory to previous work which is used in the paper. The second part sets up a model of permission detection analysis in order to cover all the authority abuse. The third part describes the method and technical details such as behavior-based IPC monitoring technology, to analyze the different types of permission abuse situations. The fourth part gives an experimental result aimed at the market common applications, and reveals the comparison in different cases of authority abuse. And the last part provides a discussion on our work's limitations and its future prospects.

2 RELATED WORK

In this section, we present the main contributions related to the malware detection on the Android platform. As the core aspect of taint dynamic analysis, the rationale and implementation of Android Inter-Process Communication (IPC) is particularly introduced. We also discuss the characteristics of the Android permission mechanism and the related techniques used in the permission abuse monitor.

2.1 Android security mechanism

Android platform is developed on the basis of the Linux kernel, so it inherits the structure including the security mechanisms from the Linux operating system [3]. Each application owns a unique UID and a separate sandbox, where the code keeps running independently without interference of other applications. Meanwhile, the program can only visit its own files, which is called file access control. Android has its own specific security mechanisms, including access control, component package, and signature mechanism [4]. The permission mechanism, as one of the core policies, refers whether the application have an access to the specific APIs and restricted resources. By default, when the application has not been granted permission, the system will prevent them from accessing the protected resources to ensure the safety.

However, due to loose permission granting structure in the Android applications, permission will not be removed during the life of the program, once it is granted to the application [5]. In this case, the authorized malicious applications can call the corresponding interfaces and obtain private data in the backstage to achieve a particular purpose. On the other side, the unauthorized ones can exploit transitive permission properties to enable privilege escalation.

Android system requires multiple processes communicating to share resources to complete a task together. Therefore collaborations between applications are supported by Inter-Process Communication (IPC) mechanism. The realization of the IPC is roughly divided into four parts: Binder drive, Service Manager, Service, and Client. Before starting the Android Dalvik machine, the system will start the Service Manager process, open Binder drive, and then inform the relevant drivers in the kernel. The dynamic analysis tool DroidBox referred in this paper, inserts specific code in the Service Manager to detect information flow.

2.2 Detection method

For malware detection, there are various means to obtain the detailed knowledge of application's characteristics. According to the execution time, behavior-based detection methods are divided into static detection and dynamic detection [6].

2.2.1 Static detection

Static detection [7] is a quick, inexpensive approach to search malicious characteristics or bad code segments without executing them in an application. It's a common way used in a preliminary analysis, when a suspicious application is firstly analyzed to detect any obvious security threats. The techniques combine reverse engineering with binary techniques, including decryption, decompilation, and pattern matching. The comparison between behavioral characteristics and the known malicious patterns is to determine whether the application is malicious software.

Adrienne [8] et al. proposed a static analysis tool, Stowaway, that applies static detection on the collected sample applications, then maps the permission with each operation. The target of Stowaway is to investigate over-privilege permissions, so it builds an API mapping in the Android emulator to determine the permissions required to interact with system APIs, through the analysis of function module and execution path.

Nowadays, with the code encryption confusion techniques, the number of malicious code patterns gradually increases, and the extraction of characteristic sequence becomes more difficult. Due to the

nature, static detection has the superiority of high Speed and low power consumption [9]. Meanwhile the primary weakness is that without malicious code patterns known in advance, it is impossible to detect new malware behavior successfully.

2.2.2 Dynamic detection

Researchers primarily use dynamic analysis in system call tracing or taint tracking. A lot of work has been done in monitoring sensitive data and providing approaches to protect users' privacy on phones in dynamic way.

Blasing et al. [10] proposed an Android Application Sandbox (AASandbox) in 2010, which can execute automatically without human interaction. This system performs a static pre-check, followed by a dynamic analysis to save logs of system calls. In the same year, Enck et al. [11] presented TaintDroid, a customized Android platform that tracks the flow of sensitive privacy data. In this paper, DroidBox [12], the open source dynamic analysis tool based on the TaintDroid, runs on the specific Android emulator rather than the real machine like TaintDroid, and it has more functions such as monitoring network file IO, cryptographic operations, the behavior of the SMS and telephone.

To present some work of the static and dynamic work used in identifying malware in privilege aspect and discuss the pros and cons, we set up a model of permission abuse detection, and proposed a method which can cover most cases of such malicious utilization, furthermore tracking the transfer of sensitive data which is the characteristic and innovation of this article.

3 MODELING

In this section, we put forward a detection model for Android permission scheme. This model is based on the API callings and privilege request to cover the most types of permission abuse detection as far as possible.

At present, the Android platform has the weakness in the permission's distribution and utilization. The application has to be authorized with all the permissions that are requested before the setup for successful installation. Once installed, the use of permission is out of control and the transmissions are opaque to the user. There is no way to detect and determine the flow of the sensitive information. This is the common issue existing in the permission mechanism. What is more, an Android application is archived in a Zip-compatible format, which generally compiled codes and binary resources. The package also accompanies a file called AndroidManifest.xml, to list a set of default permissions. For the comprehensive analysis of the authorization-related information, we present the modeling work in this subsection.

According to the mode of occupation, the types of permission are divided into two sets. One is the applied permissions called P_d, the other is required permissions called P_r. The elements of these sets called $p_i (i \in N^*)$ have been stipulated and declared in the Android system itself, such as

P_0: "android.permission.SEND_SMS";
P_1: "android.permission.INTERNET ".

There is another set of permissions referring to the sensitive information which is monitored by DroidBox such as GPS or Contracts called P_o.

When $P_d \subset P_r$, some functions cannot be carried out effectively. So this situation which will not lead to security issues is ignored in this paper.

Generally speaking, each member of the permission sets corresponds to an API [13], and the statements in the APK determine the specific system interface that application can call invoking. The definite corresponding relationship between permission and peculiar API is shown in Table 1.

Table 1. The correspondence of APIs and permissions.

LOCATION_ GPS	GPS Positioning Information	android.permission. ACCESS_FINE_ LOCATION
CONTACTS	Contact Person Data	android.permission. READ_CONTENTS
SMS MMS	Information of SMS & MMS	android.permission. READ_SMS android.permission. RECEIVE_SMS
INTERNET	Networked Information	android.permission. INTERNET
ACCOUNT	Account Information	android.permission. GET_ACCOUNT

3.1 Privilege disparity

We discuss the element of the set: P_m, $P_m \in P_d - P_r$ when $P_r \subset P_d$.

Privilege disparity means that the extra permission has been applied, but without which the application can also function well. For example, an alarm clock app is asked for the privilege to read the contacts or make calls. According to the characteristics of coarse grained Android permission referred above, we have to allow the request of all permissions for the successful installation. However, to the common sense, we can judge that this kind of application is not reasonable.

There is another situation, some applications have complex functions, and the artificial judgment is inadvisable. Therefore, once authorized, the use of permission is out of control and the transmissions are unknown to the user. For instance, If the authority GPS, INTERNET belongs to P_d, but do not belong to P_r, as an application with INTERNET permission has access to the outside server and is likely to send the private information over the network, an attacker can take advantage of buffer overflow vulnerabilities, to get you the location of the positioning information without your permission, and are transmitted through the INTERNET access. In order to guarantee the security of information, we must eliminate the "Privilege Disparity".

3.2 Privilege information leakage

We discuss the element of the set: P_m, $P_m \in P_r$ when $P_r \subset P_d$.

Although the permission is actually required for the accurate execution, it has possibilities to maliciously utilize the applied privilege. For instance, the authority SMS can send the private data on unclear purpose, while the ability of sending message is needed to function well. This situation is referred as the "privilege information leakage".

3.3 Privilege escalation

The Android permission mechanism takes one single application into consideration. However, the loose and open permission definition between component processes leads to the privilege escalation [14]. The principle is shown in Figure 1.

Hypothetically, P_1 is the permission needed to access the component 1 of Application 3 (Comp31). According to the safety of the Android platform design, Comp11 have no access to Comp31 under the premise of lack of P_1. On the other side, the scenario of Comp21 owning the privilege given by permission P_1 but without reasonable protection indicates that any component of other Apps can access to Comp21. That is to say, Comp11 can obtain the Comp31 sensitive data protected by the P_1 via Comp21. However such access paths are not consistent with the intention of the system design obviously. In this way, the malicious applications that exploit transitive permission properties enable privilege escalation.

For instance, App1 should not be allowed to use Internet to download online or access Content Provider for local pictures, while it is permissible to send intent to or receive data from other process access. In this case, the sensitivity of the picture is determined by the ability that App2 has corresponding permissions and can be accessed to. Similarly, privilege escalation attacks can be used to infiltrate servers with personal and malicious content, when a benign App uploads information to the servers; the malicious app uses the unprotected interfaces of the server to download private content.

The aim of this model portion is to monitor the real-time communication between Apps and detect the inter-process communications against a set of pre-defined security analysis.

4 APPROACH

In this section, the details are provided of the detection of permission abuse combined with the static permission analysis with dynamic IPC monitoring by taint transmission. Our approach is indicated in Figure 2.

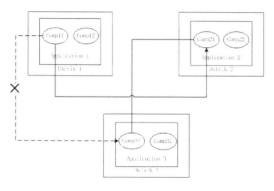

Figure 1. Principle of privilege escalation.

4.1 Static permission analysis

The package of android application has a filename extension called .apk, which compresses various resources. The AndroidManifest.xml, one of the package components, used to declare the basic information of the application, includes package name, basic components, application permissions and version information. We can get the applied permissions through the reverse analysis of the configuration file. In this paper, we parse out AndroidManifest.xml from the given APK package, and extract the string from the label called user-permission which indicates the permission request. To detect the Privilege disparity, we use the mapping of Android API calls to permissions to determine if the set of permissions requested by an application is reasonable. According to the tag, we can distinguish which permission is fragile in front of the malicious attack. These vulnerable permissions will be focused on and tracked by the taint method.

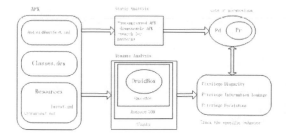

Figure 2. The architecture of the method of permission abuse.

4.2 IPC dynamic analysis

In this paper, the dynamic analysis tool DroidBox, is used in the detection of information flow through inter-process communication (IPC). It stores 32-bit variables on the adjacent location of sensitive source as a taint label. Each label corresponds to one taint type, namely allowing 32 different taint markings. DroidBox modifies Android platform, identifies privacy sensitive sources and instruments taint tags within the operating system. It defines 24 labels to trace private data at the sensitive APIs such as IMEI or IMSI code, location information, short message, phone records, and the contact information. By adding a taint tag to database files, all information read from the file will be automatically tainted. In the runtime process, taint tags propagate between applications through the data's interception, assembling, encryption, and transmission.

According to feature of IPC mechanism, DroidBox implements the taint propagation as follows. The modified platform first changes C++ Parcel Message object to store each corresponding label at the framework layer, and defines two methods of interface joint parameters respectively and retrieves parcel existing stain label, and then modifies the Java parcel object to get and set the Java variable stain label which means adding taint tag at the ServiceManager source, but as a result, C++ JNI interface parameters can be accessed by stain label, and finally modifies a Java interface. This completes the detection of IPC.

On the basis of the static analysis, as DroidBox implements privacy hook placement at several sensitive APIs, the dynamic detection precisely monitors runtime access of user data exposure by applications, and dives deep into the permission abuse.

To monitor dangerous components referring in static analysis, DroidBox detects the component of data flow, to see whether the relevant data is sent out through Inter-Process Communication.

5 EXPERIMENT AND EVALUATION

In this section, we select 30 downloadable popular applications from the third-party market and present the experiments to evaluate the performance based on DroidBox detection methods described above.

5.1 Environment setup

As the detection experiment is based on the DroidBox, therefore the test environment is almost the same as it is. The specific configuration is as shown in Table 2:

Table 2. Test environment.

CPU	Intel(R)Core(TM) i7-3770 CPU @3.40GHz
Memory	8.00GB
OS	Ubuntu 13.10
Android Release	Android 4.1.1_r1
Emulator Memory	1024M

5.2 Experimental results

Case 1: The experiment testing results show that 56% of applications are over-privileged and have extra permissions. 67% of applications have at least one extra permission. With the method proposed in our paper, the verified malicious use of different permissions is shown in Figure 3.

Case 2: The static analysis concluded that there are 45% of frail components which may cause privilege escalation. Figure 4 represents the way and concrete data of transferring sensitive information of such components.

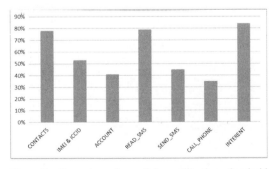

Figure 3. Percentages for malicious utilization of android permission.

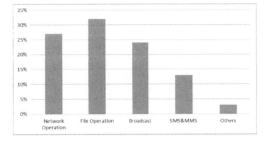

Figure 4. Percentages for data transmission through IPC.

6 CONCLUSION

Through the analysis of the Android platform malware behavior characteristic and contrast of the Android platform current malware detection technology, we point out the shortage of the current detection technology of permission abuse. At present, most of the detection scheme cannot fully cover various circumstances of permission abuse. In this paper the model of permission abuse detection is proposed. With the combination method of static permission check before installation and runtime of the IPC monitoring, the existence rights abuse can be determined. Good results in detection coverage have been achieved from the test on actual applications.

In the future work, we will continue to improve the method of permission abuse detection such as monitoring for higher accuracy and efficiency on malware identification technologies. Large analysis and tests will be conducted as it provides the way to recognize behavior pattern and detects malicious behavior, which may be utilized for prospective automated detection.

ACKNOWLEDGMENT

This work was supported by the National Science and Technology Major Project under 2012ZX01039-004, the 863 Projects under 2012AA011206 and the NSFC under 61202218.

REFERENCES

[1] http://en.wikipedia.org/wiki/Android_%28operating _system%29.
[2] http://www.techspot.com/news/57228-google-shows-off-new-version-of-android-announces-1-billion-active-monthly-users.html.
[3] A. P. Felt, etc, 2011, Android permissions demystified, in Proceedings the 18th ACM conference on Computer and communications security, Chicago, Illinois, USA, 2011, pp. 627–638.

[4] Au, K., Zhou, B., Huang, Z., Gill, P., Lie, D. Short Paper: A Look at SmartPhone Permission Models. In Proceedings of the ACM Workshop on Security and Privacy in Mobile Devices (SPSM), 2011.

[5] Enck W, Octeau D, McDaniel P, et al. A Study of Android Application Security[C]//USENIX security symposium. 2011.

[6] JACOB G, DEBAR H, FILIOL E. Behavioral detection of mal-ware: from a survey towards an established taxonomy [J]. ComputVirol,2008,4:251–266.

[7] Shin w,Kwak S,Klyomoto S,et al.A small but non-negligible flaw in the Android pemminssion scheme. IEEE intenational Symposium on Policies for Disitributed Systems and Net-works,2010:109–110.

[8] A. P. Felt, etc, 2011, Android permissions demystified, in Proceedings the 18th ACM conference on Computer and communications security,Chicago, Illinois, USA, 2011, pp. 627–638.

[9] Moser A, Kruegel C, Kirda E. Limits of static analysis for malware detection[C]//Computer Security Applications Conference, 2007. ACSAC 2007. Twenty-Third Annual. IEEE, 2007: 421–430.

[10] Blasing T, Batyuk L, Schmidt A D, et al. An android application sandbox system for suspicious software detection[C]//Malicious and Unwanted Software (MALWARE), 2010 5th International Conference on. IEEE, 2010: 55–62.

[11] Enck W, Gilbert P, Chun B, et al. TaintDroid: An Information-flow Tracking System for Realtime Privacy Monitoring on Smartphones[C]//Proc. of OSDI'10. Vancouver, Canada: [s. n.],2010.

[12] DroidboxProject[EB/OL].http://code.google.com/p/droidbox/,2013.

[13] Hackborn, D. Re: List of private / hidden / system APIs? http://groups.google.com/group/android-dcvclopers/msg/a9248b18cba59f5a.

[14] L. Davi, A. Dmitrienko, A.-R. Sadeghi and M. Winandy. Privilege escalation attacks on android, In Proceedings of the 13th international conference on Information security, Boca Raton, FL, USA, 2010.

Network Security and Communication Engineering – Chan (Ed.)
© 2015 Taylor & Francis Group, London, ISBN: 978-1-138-02821-0

DroidMonitor: A high-level programming model for dynamic API monitoring on android

D. Liang
Shanghai Jiao Tong University, Shanghai, China

R. Chen
Institute of Command Information System of PLA University of Science and Technology, Nanjing, China

H.Y. Sun
Shanghai Jiao Tong University, Shanghai, China

ABSTRACT: The popularity of Android smart phone makes Android smart phone one of the hottest platforms for mobile applications. At the same time, the users are facing dangers brought by malwares. The variability and complexity of malware patterns require analysts to develop new analysis in a convenient and flexible way. Bytecode instrumentation is an efficient way to apply either dynamic analysis or monitor for Android applications. In this paper, we propose a high-level, convenient, flexible and extensible programming model dedicated for API inspection and monitor using. We apply instrumentation in an elegant way that analysts can design their analysis in Java program without the need of knowledge of low-level bytecode.

1 INTRODUCTION

Android is the clear leader in operating systems for smartphones and tablets [1, 5]. Its open source environment and unrestricted application market have attracted many Android app developers. However, Android is also the most targeted platform for attackers at the same time. Although Android has a permission system where users are notified of the permissions and able to cancel the installation, it's still not fine-grained enough. The users are not aware of how these permissions granted to apps are actually used, what resources they access, when these resources are used and where these resources flow to.

To deal with this problem, analyses of app runtime behaviors of API usage are of great importance. Many researches have been done on either static analysis or dynamic analysis. Most of the static analyses take the application source code or binaries as well as some related configuration (In Android, the AndroidManifest.xml) as input, and try to detect malicious behaviors by diagnosing the dataflow out of the program. However, static analyses face challenges like code obfuscation as well as dynamic execution [4]. Dynamic analyses are also used for malware detection. The direct way of dynamic analysis is by instrumenting the app's bytecode. However, it requires that the analysts have quite a lot of knowledge of bytecode to develop instrumentation-based analyses and there is no mature framework dedicated

for monitoring Android APIs. In addition, some framework requires modifying the Android system [6] or using virtualization technique [7] to achieve dynamic inspection. These system are not easy to build and are not flexible to extend.

In this paper, we propose a programming model for analysts to conveniently develop a new dynamic analysis based on Java bytecode instrumentation, and it enjoys the flexibility and extensibility at the same time. We use dex2jar to bridge the gap between Java bytecode and Android Dalvik bytecode. We extend DiSL[8], an existing mature instrumentation framework for Java bytecode, adding a new feature where analysts can define snippet for targeted APIs and define observation and monitoring logic. Our extension takes place in an elegant way that isolates monitoring logic from instrumentation.

2 BACKGROUND

In this section, we introduce the background context of our work. We first give a brief introduction to bytecode instrumentation. And then we introduce some prerequisite knowledge of Android platform.

2.1 Bytecode instrumentation

Bytecode instrumentation is an important technique for dynamic program analysis.

Applications and some system services on Android take Java as its programming language. Before being deployed onto Android system, the programs are compiled first into Java bytecode (.class files), and then are transformed from the Java bytecode into Dalvik bytecode. DVM (Dalvik Virtual Machine) and JVM bytecode are different in several aspects, e.g., instruction set that JVM is stack-based while DVM is register-based.

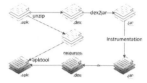

Figure 1. Instrumentation workflow, the gray output stands for the instrumented version.

Although there are several tools targeted at DVM bytecode, e.g., smali/baksmali, there is no mature framework manipulating directly on DVM bytecode. To deal with this problem, we choose to parse Dalvik bytecode back to JVM bytecode, and use DiSL, an open source domain-specific language for Java bytecode instrumentation. We choose DiSL, because it provides flexible instrumentation ability with high-level language, and ability to extend. It is built on the low-level bytecode manipulation library ASM and provides high-level domain-specific programming language. Since Android uses different class loading procedures, we only take DiSL as an offline instrumentation tool and our extension to DiSL doesn't follow its original isolation principle in order to realize control of target application.

2.2 The Android platform

Android application and services written in Java are composed of App components, which are the essential building blocks. Each component is a different entry point. Multiple entries bring difficulty for analysis. There are four types of components: Activities, Services, Content providers and Broadcast receivers. Each type serves a distinct purpose and has a distinct lifecycle.

Analysis to Android application can be categorized into inner component analysis and inter component analysis. Inner component analyses, as the name says, they focus only on the code of a single component. Many traditional analyses can be taken for diagnosing the behaviors inside a single component. Components of one or more applications can connect via Android IPC mechanisms, e.g. When an application wants to get contacts information, it needs to call the content provider API to fetch the contacts information. The IPC calls are in a very loosely coupled fashion which make monitor program execution statically difficult. At the same time, it points out that monitoring the APIs where IPC calls happen is the key point for analyzing app's potential malicious behavior.

3 OVERVIEW

In this section, we give a brief overview of the work flow of our framework for making a new analysis from scratch. You will get more details in the next section.

As is shown in figure 1, the input file is the Android application package file, and output is instrumented apk file. To create a new analysis monitoring APIs, the instrumentation follows the following steps:

1 Select the APIs of interest for analysis;
2 Write guard Class to select exact API in (1);
3 Create initial wrapper methods in a specification class where each method with "@Monitor" annotation corresponds to a target API method and map each method to corresponding guard Class.
4 Define analysis logic in wrapper methods.
5 Run the DroidMonitor and get instrumented apk file as output.

We illustrate with a sample of our model, there are many programming features brought by DiSL which we won't cover in this paper. We will give some more details in the next section.

As is shown in figure 2, assume the situation to observe two apis of *TargetClass* in packaget *test*. *api0* is a public static method which takes Integer and String Object as parameters and *api1* is another public non-static method. We show in figure 2, how analysts can write in Java code when there are two snippets dedicated for monitoring two target APIs.

```
package test;
public class TargetClass {
    public static String api0(int a, String b){...}
    public void api1(int a, String b){...}
}
public class TargetClassApi0Guard {
    @GuardMethod
    public static boolean guard (StaticContext sc) {
        return sc.isMethod("test.TargetClass",
        "api0", "ILjava/lang/String;");
    }
}
public class TargetClassApi1Guard {
    @GuardMethod
    public static boolean guard (StaticContext sc) {
        if(!sc.isSelfOrChildClass("test.TargetClass"))
            retun false;
        else
            return sc.isVirtualMethod(
            "api1", "ILjava/lang/String;");
    }
}
```

Figure 2. Target class as well as the sample guard classes which are used to locate the bytecode of interest.

Then they need to define Guard class to restrict the analysis scope. Guard is a feature in DiSL to give accurate control of bytecode instrumentation. As is shown in figure 2, the guard class uses the *StaticContext* to retrieve static information to match whether it's the target APIs.

Finally the analysts just need to define one method named as *"path_classname_methodname"* with the Monitor annotation for each target API. As is shown in figure 3, they can write their logic before real call to the target API, and these methods will be woven into a global class *APIWrapper*.

Each method annotated with *"Monitor"* contains the logic how the analysis deals with monitoring.

Figure 3. Sample analysis code-skeleton in Java programming.

4 IMPLEMENTATION

In this section, we introduce the implementation details of DroidMonitor. First we start with the concept of APIWrapper, illustrated with bytecode example, and then we introduce the high-level programming model we designed to describe the analysis login in Java programming language. Finally, we show some corner cases of how we deal with the object oriented features of Java language.

4.1 *APIWrapper*

Monitoring sensitive APIs is the key to monitoring the behavior of an application. To achieve this, we introduce the idea of APIWrapper, which is a global class for the observed application and contains all wrapped observed APIs. We realize the functionality based on modifying the invoke- family bytecode on targeted APIs to replace corresponding method in the global APIWrapper class with invocation.

4.1.1 *JVM bytecode stack*
As is mentioned before, we instrument on the JVM bytecode parsed from DVM bytecode. The JVM bytecode includes four opcodes for method invocation:

"invokestatic", "invokespecial", "invokevirtual" and "invokeinterface". The detailed document for these four kinds of bytecode can be found on Java document. In short, bytecode of "invokestatic" represents invocation of a static method, while the others represent none-static method invocations. Java bytecode uses stack-based structure, where method arguments are pushed onto the stack before method invocations. None-static method invocation needs to push an extra reference of the caller object before pushing the arguments as is shown in figure 4. It's clear how static invocation replacement works. The insight lies in the replacement of none-static invocation. The stack of none-static invocation includes instance reference plus arguments, which compose the parameters for the wrapped method in APIWrapper class. In this case, the invocation to none-static methods can be perfectly transformed into static method invocation.

Original	Instrumented
load parameters	<=> load parameters
invokestatic TargetClass.api0	<=> invokestatic APIWrapper.test_TargetClass_api0
load instance	<=> load instance
load parameters	<=> load parameters
invokevirtual TargetClass.api1	<=> invokestatic APIWrapper.test_TargetClass_api1

Figure 4. Invocation replacement via bytecode instrumentation for sample API TargetClass.api0 and TargetClass.api1.

4.1.2 *Instrumentation point choices*
To wrapper the target API, there are several options: 1. instrument at the callee side that is directly inside the target API; 2. instrument at the caller side where target API is called. The first way is not proper for Android situation here, because most of the targeted APIs to observe are inside the Android system library, which is shared among all Android processes. Instrumentation directly on the library need not only to root the device, it may also influence behaviors of all the app processes that are not expected to be observed. In this paper, we choose the second option.

Instead of instrumenting bytecode just before or after the target invoke- bytecode, we have made an innovative design that we add an extra *APIWrapper* class containing static wrapper methods corresponding to each target API and replace target API invocations with that of APIWrapper. This brings benefits including observation of the API usage as well as the ability to control or even block the API usage. It can be isolated from the invocation point. This paves the way for high-level programming model.

In summary, we add a global-visible class *APIWrapper* with wrapper methods for each target API, and replace call to the target APIs to corresponding wrapper methods. Monitoring and controlling logic can be done in the wrapper methods.

4.2 High-level programming model

It's a challenge for analysts to enforce the policies of analysis if they are not familiar with bytecode. To bridge the gap between low-level bytecode and high-level programming Java code, we provide analysts with a way to write Java for their analysis logic.

We extend DiSL making use of its ability of bytecode manipulation, to support a new Annotation "Monitor". The instrumentation logic will be written in Java including snippets for instrumentation.

4.3 Object oriented programming features support

DroidMonitor perfectly supports features of Object Oriented programming like polymorphism and inheritance. As shown in figure 2, with the help of static context provided by DiSL, the guard class can filter invocations correctly. This also works for interface matching. The analysts are able to observe a base class or interface API and gain control over all invocations to child method invocations or interface implementation. None-public members are also supported in DroidMonitor with Java reflection, which makes private or protected call outside TargetClass possible, and makes our model sound.

5 RELATED WORK

There are many researches about analysis on Android. Several techniques have been applied towards the growing number of malware.

Some researches use static analysis to analyze the behavior reflected by the application code. A. P. Felt builds Stowaway [2], a tool for detecting over-privilege in Android application. It diagnoses on application bytecode and maps the API to permissions to check whether the application follows least privilege principle. RiskRanker [3] designed by M. Grace statically classifies applications into multiple security risk categories. Static analysis faces challenges in Android's multiple processes runtime. DroidMonitor makes use of static context for instrumentation.

There are also some dynamic analyses, which focus on monitoring of Android application at runtime. TaintDroid [6] and DroidScope [7] are the most famous ones. However, they are difficult to deploy on real device. Similarly, some other researches, like CopperDroid [9], are designed for offline detection of malware. However, they are often hard to extend for flexible analysis. It is hard for analysts to make new analyses on their demand for lack of interfaces.

6 CONCLUSION

To conclude, DroidMonitor is a highly extensible framework for repackaging Android apps. Analysts can define their logic of monitoring APIs for target apps in high-level programming language Java and deploy instrumented application for observation. DroidMonitor paves the way for a highly simplified way for API monitoring on Android.

ACKNOWLEDGMENTS

This research is supported by National Science and Technology Major Project of Science and Technology of China under grant no. 2012zx03006-002.

REFERENCES

[1] M. Bierma, E. Gustafson, J. Erickson, D. Fritz, Y. R. Choe, "Andlantis: Large-scale Android Dynamic Analysis," IEEE CS Security and Privacy Workshops (SPW), 2014.

[2] A. P. Felt, E. Chin, S. Hanna, D. Song, and D.Wagner. "Android permissions demystified,".Proc. of ACM Conference on Computer and Communications Security (CCS), pages 627–638, 2011.

[3] M. Grace, Y. Zhou, Q. Zhang, S. Zou, and X. Jiang. "Riskranker: scalable and accurate zero-day android malware detection," Proc. of International Conference on Mobile Systems, Applications, and Services (MOBISYS), pages 281–294, 2012.

[4] D. Arp, M. Spreitzenbarth, M. Hubner, H. Gascon, and K. Rieck, "Drebin: Efficient and Explainable Detection of Android Malware in Your Pocket," Proceedings of the 20th Annual Network & Distributed System Security Symposium (NDSS), 2014.

[5] Y. Zhou and X. Jiang. "Dissecting android malware: Characterization and evolution," Proceedings of the 2012 IEEE Symposium on Security and Privacy, SP '12, pages 95–109, 2012.

[6] W. Enck, P. Gilbert, B. gon Chun, L. P. Cox, J. Jung, P. McDaniel, and A. Sheth, "Taintdroid: An information-flow tracking system for realtime privacy monitoring on smartphones," Proc. of USENIX Symposium on Operating Systems Design and Implementation (OSDI), 2010, pp. 393–407.

[7] L.-K. Yan and H. Yin, "Droidscope: Seamlessly reconstructing os and dalvik semantic views for dynamic android malware analysis," in Proc. of USENIX Security Symposium, 2012.

[8] L. Marek, S. Kell, Y. Zheng, L. Bulej,W. Binder, P. Tuma, D. Ansaloni, A. Sarimbekov, and A. Sewe. ShadowVM: Robust and comprehensive dynamic program analysis for the Java platform. In Proc. 12th Intl. Conf. on Generative Programming: Concepts and Experiences, GPCE'13, pages 105–114. ACM, 2013.

[9] A. Reina, A. Fattori, and L. Cavallaro. A system call-centric analysis and stimulation technique to automatically reconstruct android malware behaviors. In Proc. of European Workshop on System Security (EUROSEC), April 2013.

Application of a security protocol to wireless multiple access systems

Ajung Kim
Sejong University, Seoul, Korea

ABSTRACT: Security protocols applicable to wireless communication are proposed to establish a security sublayer in a physical layer as opposed to conventional cryptographic systems which are entirely based on computational complexity. Implementation in wireless multiple access systems is addressed, and system modeling and performance are analyzed for frequency-division-duplexing based code division multiple access systems. This scheme devises a new way of attaining secure communications without entirely relying on mathematical complexity.

1 INTRODUCTION

1.1 *Type area*

In wireless communication networks in which signals are distributed to users with an antenna over a wide span of area, security is an issue of importance. There is every possibility of the signal being intercepted by an eavesdropper and of jammed by a malicious intruder. Furthermore, the advent of the quantum computer [Landahl2003] with the ability of massive parallelism has been a threat to conventional cryptosystems. Such increasing demands for protecting wireless links against passive eavesdropping or active tampering has prompted researchers to devise new cryptosystems not merely relying on computational complexity.

In this paper, a key agreement protocol exploiting physical entities of networks is proposed. This security protocol is applicable to a physical layer in wireless communication networks and its implementation using interference in multiple access systems is suggested. Interference modeling is carried out in section II and system performance according to the protocol is analyzed for frequency-division-duplexing(FDD) based code division multiple access (CDMA) systems and system designs based on analysis are provided in section III with concluding remarks in section IV.

2 INTERFERENCE MODELING

A major issue in cryptosystems is key establishment, distribution, and management. A key agreement protocol exploiting physical entities can be considered, which is based on various system noises and interference. As multiple access interference from other channels is dominantly influential in multiple access systems, system interference modeling is necessary for performance evaluation.

On a forward transmission link in FDD-CDMA systems, the same cell interference (intracell interference) is dominant, which is generated by aggravation of orthogonality of Walshcode through multipaths. Suppose that S is the data signal power received by a mobile station and $S_{o,R}$ is the total signal power arrived at the remote station (RS) from a corresponding base station (BS) which includes signals in pilot channels, overhead channels and other user traffic channels as well as data channels . Power ratios β and α are defined as

$$\beta_l = S_{0,l}/S_{0,R} \quad , \qquad \alpha = S/S_{0,R} \qquad (1)$$

with $S_{o,l}$, the received signal power via the i-th path out of L multipaths and S, the received data power.

Then the signal to interference and noise ratio (SINR) is approximated by

$$\text{SINR} \approx \frac{S}{I_{SC}} = \frac{\alpha S_{0,R}}{I_{SC}} = \sum_{l=1}^{L} \frac{\alpha \beta_l}{1-\beta_l} = \frac{\alpha}{1-\zeta} \qquad (2)$$

The same cell interference is expressed by $I_{SC} = (1-\zeta)S_{0,R}$ with an orthogonality factor ζ from which a degree of orthogonality deterioration due to multipath propagation can be inferred by

$$\zeta = 1 - \frac{1}{\displaystyle\sum_{l=1}^{L} \frac{\beta_l}{1-\beta_l}} \qquad (3)$$

Other factors affecting SNIR include thermal noise of a receiver I_N, other cell interference (intercell interference) I_{OS}, and adjacent channel interference I_{OF} as

$$\text{SINR} = \frac{S}{I_N + I_{SC} + I_{OC} + I_{OF}}.$$

Thermal noise is determined by the reference noise temperature To and a receiver's noise figure F as $I_N = kT_0 F$.

with Bolzman's constant $k = 1.38 \cdot 10^{-23} \, \text{J} / \, ^\circ\text{K}$.

Intercell interference received at the n-th RS is

$$I_{OC,n} = \sum_{m=1} L_{n,m} \eta_{n,m} S_{m,T} \qquad (4)$$

where $S_{m,T}$ is a transmitted power from a BS in the m-th cell and $L_{n,m}$ and $\eta_{n,m}$ are a path loss and a log-normal shield between the n-th RS and the m-th BS, respectively.

Adjacent channel interference is interference from other frequency systems employing nonideal filters and nonlinear active devices, which is caused by out-of-band emission, the receiver's nonideal blocking, and intermodulation. Interference signal power due to out-of-band emission, I_s, is expressed by

$$I_S = P_T \delta_S \qquad \delta_S = \int_{BW_R} S_{TX}(f) df \bigg/ BW_R \qquad (5)$$

P_T, $S_{TX}(f)$ denote the signal power and the spectrum density function of signals, transmitted from an unwanted transmitter, and BW_R is the bandwidth of a receiver's filter. d is the interference ratio for P_T. Interference signal power due to receiver's nonideal filtering, I_B, is similarly obtained as

$$I_B = P_T \delta_B \qquad \delta_B = \int_{BW_T} S_{RX}(f) df \bigg/ BW_T \qquad (6)$$

where $S_{Rx}(f)$ denotes the spectrum density function of the receiver's filter. BW_T is a bandwidth of the interference signal.

Lastly, interference power arriving at the n-th RS due to adjacent frequency services on forward transmissions is approximated by

$$I_{OF,n} = \breve{\rho}_0 L_{n,0} \eta_{n,0} S_{0',T} + \sum_{m=1} \breve{\rho}_m L_{n,m} \eta_{n,m} S_{m,T} \qquad (7)$$

$S_{0',T}$ is the signal power transmitted by a BS with an adjacent frequency band and the same location as the wanted BS. ρ_0 and $L_{n,0}$, $\eta_{n,0}$ are an adjacent channel interference ratio (ACIR), path loss and log-normal shadowing associated with the aforementioned BS, respectively. The subscripts n,m in the second term refer to the n-th RS and the m-th BS.

3 PRINCIPLES AND PERFORMANCE

While interference limits the performance of multiple access systems, interference can be conversely exploited in key agreement in cryptosystems. In secret key distribution between two stations, the transmitting BS S plays the role of a trust authority and transmits a sequence of signals susceptible to interference and noise to station B. An eavesdropping RS, E, might try to detect each bit of the signals with optimal decision region divisions [2], whereas B only accepts signals above a certain threshold signal level θ and discards the unreliable bits falling in the erroneous region below the threshold. B then tells S via a public channel which bits were accepted as a key string without telling the values of his bits, which makes an agreed key string between S and B after post processes. Then, Eve ends up having a higher error rate and, moreover, a measured bit stream that is uncorrelated with B's, because the interference and noise of E's receiver is uncorrelated with that of B's.

Suppose station B measures each symbol at a threshold θ with the following decision rule for QPSK(quardrature phase shift keying) modulation and coherent measurement; the signal is accepted if $|r| > \theta$, and is discarded if $|r| < \theta$ for a received signal vector r with an amplitude $|r|$. Consider an intercell interference-limited system based on FDD-CDMA. With the *SINR* analysis on forward transmission and the relationship between *SINR* and bit energy-to-noise spectral density with Shannon bandwidth, the error rate in the B's key string is calculated in terms of relative threshold $h=\theta/S^{1/2}$

$$P_e^B = \frac{1}{4} erfc(\sqrt{\frac{\alpha^B}{1-\zeta^B}}(h+1)) \qquad (8)$$

where the complimentary error function is $erfc(u) = \frac{2}{\sqrt{\pi}} \int_u^\infty e^{-r^2} dr$. In an eavesdropping attempt, station E makes a decision according to the maximum a posteriori (MAP) decision rule[Van Trees1968]. Then the error probability of station E is obtained as

$$P_e^E = \frac{1}{4} erfc(\sqrt{\frac{\alpha^E}{1-\zeta^E}}) \qquad (9)$$

Whereas B can accept the reliable signals not falling within the erroneous region, E has to measure every signal with no choice of rejecting the erroneous signals that lead to a higher error rate in the signal sequence accepted and shared between the legitimate stations, S and B. E can obtain neither the correct key string, nor the same key string that B has Moreover,

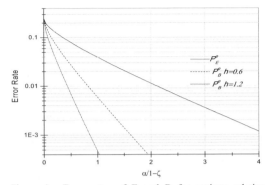

Figure 1. Error rates of E and B for various relative threshold values h=0.6 and h=1.2 in an intracell interference-limited system.

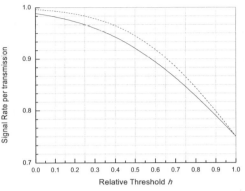

Figure 2. Signal rate per transmission vs. relative threshold value h for $a/1-\zeta =2$ (solid line) and $a/1-\zeta =3$(dashed line).

even Eve's induction of the threshold value does not compromise security at all.

Our scheme is not unconditionally secure, and neither are quantum cryptosystems[Cho2008] not using a single photon per pulse, as E can reduce the error rate with technological advantages over legitimate users. Nevertheless, the measurement results of E's bit sequence are uncorrelated with those of B's.

E's attempt to amplify the signal is futile, as taking several samples reduces the signal to the level where the noise effects are more significant, and amplifying disturbs the states and so introduces noise large enough to nullify the advantage of eavesdropping on strong signals.

Fig.1 shows error rates of station E and station B with this scheme as a function of $a/1-\zeta$, for various values of θ. Station B takes the value of θ that makes P_e^B considerably small.

Considering that station B discards data below the threshold value, the signal rate of key generating per use is

$$R_s = 1 - \frac{1}{4} erfc(\sqrt{\frac{\alpha^B}{1-\zeta^B}}\,|h-1|) \qquad [h=\theta/\sqrt{S}<1] \qquad (10)$$

Fig. 2 shows signal rate per transmission as a function of relative threshold value h for $a/1-\zeta =2$ and 3.

A signal with a low SINR introduces a high error rate of station E and a high threshold value induces a low error rate of station B. For security purposes, it is mandatory to utilize weak signals so as to be vulnerable to the noise factors. The trade-off of the high security achieved by weak signals or higher threshold values is the sensitivity to attenuation and interference, or a reduced signal rate.

Partial mutual information between S and E, which is to be hashed out at privacy amplification

[Bennett1992], can be obtained with eq. (9) as
$$I_{SE} = 1 + P_e^E \log_2 P_e^E + (1-P_e^E)\log_2(1-P_e^E).$$

Although mutual information, which will be hashed out at privacy amplification, is one of the criteria for the system design of cryptosystems, mutual information has only asymptotic significance for noisy systems. We are subsequently unconcerned with possible privacy amplification here and design our cryptographic system according to cryptographic constraints. Counter measurements against various possible attacks can set constraints in system design. With a key trial scheme, E may make a copy of the transmitted signals, measure them and attempt to guess a correct key string by arbitrarily changing the value of n bits out of her $|k|$ bit key string in a possible known plain text attack. Suppose E has the capability of trying N_E key strings, which yields a probability of success P_c^E. For α given a and ζ with BPSK, this determines a maximum signal energy E allowed between S and B. From the constraints, it turns out that P_c^E is smaller than 10^{-37} with a key length $|k|\sim10^3$ for stream ciphers. For $|k|\sim50$ and $P_e^E \sim 0.1$, some hashing for privacy amplification is necessary to make P_c^E smaller. Instead, we propose using a simple mod 2 sum of 10 different bits each to get a key bit from a 500 bit string. With such a key string of $|k|\sim50$, we have $P_e^E \sim 0.45$ and $P_c^E \sim 5.63\times10^{-8}$ for $N_E \sim 2^{20}$. System parameters are to be determined for a given tolerable value of P_e^B. B should take a threshold value in consideration of the number of stations N and the tradeoff between his error rate and the signal rate. A larger value of θ will decrease the signal rate as well as the error rate. In practice, since a reconciliation procedure such as parity checks is required to eliminate possible errors in the key string between S and B, P_e^B is to be set with this overhead cost in mind. The system parameter values are quite encouraging for practical applications. At microwave frequencies including mobile cellular transmission [Mishra2004], a signal data rate R_0 can be in the gigabit range so that the key distribution signal rate R will be $R \sim O(1)$ Mbaud/sec. Even after overhead costs such as reconciliation by a different bit string are taken into account, the resulting actual rate R may still compare favorably to the well-known public-key methods based on computational complexity.

S can distribute a sequence of arbitrary signals to a group of stations and each station measures the signals and accepts the values above a threshold of its own. Then the signals that each station accepts as a key are different from each other, which make private keys of each station's own and enable simultaneous distribution of respective unique keys to multiple users. Though it seems in eq. (10) that the decrease in ζ or the increase in M, the maximum number of unwanted BS transmitters, increases R by pushing signals beyond a given threshold value, it also increases the error rate, P_e^B by exacerbating the SINR. For a given tolerable value of P_e^B, θ_M must be adjusted to a higher value as M and $1-\zeta$ increase and the signal rates are not simply increased with the system size, but must be evaluated and compared with different threshold values for a given P_e^B. In general case, the adverse effect is dominant as M increases, which makes the overall data rate decrease.

Unlike quantum cryptography, this cryptosystem enables signal amplification and implementations over a long distance or cascaded links are feasible.

4 CONCLUSION

This paper proposed security protocols applicable to physical layers of multiple access systems in wireless communications. Whereas interference limits performance, the proposed protocol conversely exploits it for key agreement. We analyzed performance for the FDD-CDMA system with QPSK modulation and coherent measurement. The principle is valid for other modulations including amplitude shift keying, but PSK that needs less signal power for a specified level spacing is favorable in terms of the power budget and security. The system parameter values are promising for practical applications in microwave communications and other networks. It provides a security sublayer to a physical layer as opposed to conventional cryptosystems based on computational complexity. The implementation is easier and requires simpler processing at error correction and private amplification. In addition, this cryptosystem allows signal amplification and implementing in long haul communications, as opposed to quantum cryptosystems. The model and estimates here are applicable to a broad area of wire/wireless communication systems and constitute a new approach by establishing a robust physical security layer in practice.

This work was supported by the National Research Foundation (NRF-20110010691).

REFERENCES

[1] Bennett C. H., *et al.*, 1992, Experimental quantum cryptography, *J. of Cryptology,* vol. 5: 3 28.
[2] Cho A., 2008, Cryptography - Quantum network set to send uncrackable secrets, *Science,* vol. 322: 32–33.
[3] Landahl A. J., 2003.A shortcut through time - The path to the quantum computer, *Science,* vol. 300: 1509–1509.
[4] Mishra A., *et al.*, 2004, Proactive Key Distribution Using Neighbor Graphs, *IEEE Wireless Communications*: 26–36.
[5] Van Trees H. L., 1968, *Detection, Estimation and Modulation Theory.* Wiley.

Network Security and Communication Engineering – Chan (Ed.)
© 2015 Taylor & Francis Group, London, ISBN: 978-1-138-02821-0

Improved energy detector against Malicious Primary User Emulation Attack in cognitive radio network

J.X. Yang, Y.B. Chen, G.Y. Yang, M.X. Zhao & X.J. Dong
College of Electrical and Information Engineering, Yunnan Minzu University

ABSTRACT: The improved energy detector based on cooperative spectrum sensing is proposed to defend Malicious Primary User Emulation Attack (MPUEA) in cognitive radio network. In the scheme, the improved energy detector and optimal weighting algorithm are used to replace the conventional energy detector and Maximum Ratio Fusion (MRC) algorithm respectively. It maximizes the probability of detection when the probability of false alarm is a fixed value. It is shown by simulations that the Primary User (PU) can improve the performance of the detection by increasing the transmission power, and minimize influence of MPUEA on the system in the new scheme.

1 INTRODUCTION

Cognitive radio technology is proposed to solve the problem of low utilization rate of spectrum. With the rapid development of cognitive radio technology, security threat appears. The primary user emulation attack is a common attack in the security threat [1]. It is divided into selfish and malicious attacks. The selfish attack is actualized by a pair of users, who communicate by sending out a signal similar to the primary user's signal in this band. The malicious attack is caused by the attacker who sends out signal continuously to prevent other users' access. One of the important issues in Cognitive Radio Network is to distinguish between the primary user signal and the attacker signal. Yunfei-Chen proposed the improved energy detector [2][3]. It can improve the performance of the detection, compared with the traditional energy detector. The cooperative spectrum sensing sends the sensory information to the fusion center [4] [5] to determine the presence of primary user through a fusion criterion. Hard fusion and soft fusion are the main methods to fuse the sensory information. The soft fusion is better than the hard fusion in fusion effect. The soft fusion is used in the scheme. Due to the impact of MPUEA, the detection performance of system declines sharply. It cannot be improved by increasing the transmission power of PU. The optimal weighing coefficients are obtained by maximizing the probability of detection and minimizing influence of MPUEA on the system.

The second part of this paper describes the model in the presence of the primary user emulation attack in cognitive radio networks. The third part is about the scheme of improved energy detector against malicious primary user emulation attack. Simulation results and performance analysis are given in the fourth part of the paper. The fifth part of this paper summarizes the work of this paper. The last part is acknowledgement.

2 SYSTEM MODEL

In this model, we assume that there is a primary user, K secondary user in the CRN.

The signal received by the kth secondary user at the tth time instant is

$$y_k(t) = \begin{cases} h_{ek}(t)x_e(t) + n_k(t) & H_0 \\ h_{pk}(t)x_p(t) + h_{ek}(t)x_e(t) + n_k(t) & H_1 \end{cases} \quad (1)$$

H_0 corresponding to the primary user signal is absent and H_1 corresponding to primary user signal is present. Where $y_k(t)$ is the signal of the kth secondary user. $h_{pk}(t)$ denotes the channel gain between primary user and kth secondary user and $h_{ek}(t)$ denotes the channel gain between attacker and kth secondary user. $x_p(t)$ and $x_e(t)$ are the signals transmitted by primary user and attacker. $n_k(t)$ is the additive white Gaussian noise at the kth secondary user with zero mean and variance σ_n^2. Soft fusion is used in this paper. In the cooperative manner, the signals received by secondary users are weighed by some coefficients $w_k, (k = 1, 2...K)$ and converged to a fusion center where a final decision which depends on the absence or presence of the primary user is made. The combined signal in the fusion center is

$$y(t) = \sum_{k=1}^{K} w_k y_k(t) \tag{2}$$

The statistic of improve energy detector for decision of the presence of the PU is

$$Y = \sum_{t=1}^{N} |y(t)|^P \tag{3}$$

3 THE SCHEME AGAINST MALICIOUS PRIMARY USER EMULATION ATTACK

In cognitive radio networks, false alarm probability P_f and detection probability P_d are defined as

$$P_f = p_r\left(Y \geq \beta | H_0\right) \tag{4}$$

$$P_d = p_r\left(Y \geq \beta | H_1\right) \tag{5}$$

Where β is a detection threshold.

In our system, as the signals of attacker and primary are similar, the distribution of the attacker's signal is similar with primary user. The attacker's signal and primary user's signal both follow the Gaussian distribution. We defined

$$x_p(t) \sim N\left(0, \sigma_p^2\right) \tag{6}$$

$$x_e(t) \sim N\left(0, \sigma_e^2\right) \tag{7}$$

The channel gain $h_{pk}(t)$ and $h_{ek}(t)$ are fixed values. $y(t)$ also follows the Gaussian distribution.

$$y(t) \sim \begin{cases} N\left(0, \sigma_0^2\right), & H_0 \\ N\left(0, \sigma_1^2\right), & H_1 \end{cases} \tag{8}$$

Where $\sigma_0^2 = \sigma_e^2 \left| \sum_{k=1}^{K} w_k h_{ek} \right|^2 + \sum_{k=1}^{K} |w_k|^2 \sigma_n^2$ (9)

$$\sigma_1^2 = \sigma_e^2 \left| \sum_{k=1}^{K} w_k h_{ek} \right|^2 + \sigma_p^2 \left| \sum_{k=1}^{K} w_k h_{pk} \right|^2 + \sum_{k-1}^{K} |w_k|^2 \sigma_n^2 \tag{10}$$

Given

$$h_e = \left[h_{e1}(t), h_{e2}(t), \cdots, h_{ek}(t) \right]^T$$

$$h_p = \left[h_{p1}(t), h_{p2}(t), \cdots, h_{pk}(t) \right]^T$$

$$w = \left[w_1, w_2, \cdots, w_K \right].$$

σ_0^2 and σ_1^2 can be denoted

$$\sigma_0^2 = \sigma_e^2 w h_e h_e^H w^H + \sigma_n^2 w w^H \tag{11}$$

$$\sigma_1^2 = \sigma_e^2 w h_e h_e^H w^H + \sigma_p^2 w h_p h_p^H w^H + \sigma_n^2 w w^H \tag{12}$$

where T and H are the transpose and the conjugate transpose respectively.

According [2], the mean and variance of Y under hypothesis H_0 are

$$E\{Y|H_0\} = N \frac{2^{P/2}}{\sqrt{\pi}} \Gamma\left(\frac{P+1}{2}\right) \sigma_0^P \tag{13}$$

$$Var\{Y|H_0\} = N \left(\frac{2^P \Gamma\left(\frac{2P+1}{2}\right)}{\sqrt{\pi}} - \frac{2^P}{\pi} \Gamma^2\left(\frac{P+1}{2}\right) \right) \sigma_0^{2P} \tag{14}$$

where $\Gamma(a) = \int_0^{+\infty} e^{-t} t^{a-1} \, dt$ is the gamma function.

The mean and variance of Y under hypothesis H_1 are

$$E\{Y|H_1\} = N \frac{2^{P/2}}{\sqrt{\pi}} \Gamma\left(\frac{P+1}{2}\right) \sigma_1^P \tag{15}$$

$$Var\{Y|H_1\} = N \left(\frac{2^P \Gamma\left(\frac{2P+1}{2}\right)}{\sqrt{\pi}} - \frac{2^P}{\pi} \Gamma^2\left(\frac{P+1}{2}\right) \right) \sigma_1^{2P} \tag{16}$$

We assume that the probability density function (PDF) of Y is the PDF of gamma distribution and the cumulative distribution function (CDF) is the CDF of gamma distribution.

Under hypothesis H_0, the shape parameter k_0 and scale parameter θ_0 of gamma cumulative distribution function are [2]

$$k_0 = \frac{E^2\{Y|H_0\}}{Var\{Y|H_0\}} = N \frac{\Gamma^2\left(\frac{P+1}{2}\right)}{\Gamma\left(\frac{2P+1}{2}\right)\sqrt{\pi} - \Gamma^2\left(\frac{P+1}{2}\right)} \tag{17}$$

$$\theta_0 = \frac{Var\{Y|H_0\}}{E\{Y|H_0\}} = \frac{2^{P/2}}{N} \frac{\sqrt{\pi}\Gamma\left(\frac{2P+1}{2}\right) - \Gamma^2\left(\frac{P+1}{2}\right)}{\Gamma\left(\frac{P+1}{2}\right)\sqrt{\pi}} \sigma_0^P \quad (18)$$

Using (17) and (18) in (4), the P_f can be determined as

$$P_f = \frac{\Gamma\left(k_0, \frac{\beta}{\theta_0}\right)}{\Gamma(k_0)} \quad (19)$$

where $\Gamma(a,x) = \int_x^{+\infty} e^{-t} t^{a-1} dx$ is the upper incomplete gamma function.

Under hypothesis H_1, the shape parameter k_1 and scale parameter θ_1 of gamma cumulative distribution function are

$$k_1 = \frac{E^2\{Y|H_1\}}{Var\{Y|H_1\}} = N \frac{\Gamma^2\left(\frac{P+1}{2}\right)}{\Gamma\left(\frac{2P+1}{2}\right)\sqrt{\pi} - \Gamma^2\left(\frac{P+1}{2}\right)} \quad (20)$$

$$\theta_1 = \frac{Var\{Y|H_1\}}{E\{Y|H_1\}} = \frac{2^{P/2}}{N} \frac{\sqrt{\pi}\Gamma\left(\frac{2P+1}{2}\right) - \Gamma^2\left(\frac{P+1}{2}\right)}{\Gamma\left(\frac{P+1}{2}\right)\sqrt{\pi}} \sigma_1^P \quad (21)$$

Using (20) and (21) in (5), the P_d can be determined as

$$P_d = \frac{\Gamma\left(k_1, \frac{\beta}{\theta_1}\right)}{\Gamma(k_1)} \quad (22)$$

For a given P, $P_f = \varepsilon$, $\varepsilon \in (0,1)$. The decision threshold is

$$\beta = \Gamma^{-1}(k_0, \varepsilon)\theta_0 \quad (23)$$

$$P_d = \Gamma\left(k_1, \Gamma^{-1}(k_0, \varepsilon)\frac{\theta_0}{\theta_1}\right) = \Gamma\left(k_1, \Gamma^{-1}(k_0, \varepsilon)\frac{\sigma_0^P}{\sigma_1^P}\right) \quad (24)$$

According to the character of upper incomplete gamma function, to maximize P_d is equivalent to minimize $\frac{\sigma_0^P}{\sigma_1^P}$. To maximize $\frac{\sigma_0^P}{\sigma_1^P}$ is equivalent to minimize $\frac{\sigma_0^2}{\sigma_1^2}$. By some calculations, the optimized W is

$$w = \left[\left(\sigma_e^2 h_e h_e^H + \sigma_n^2 I\right)^{-1} h_p\right]^H \quad (25)$$

There is a maximum of P_d for each P. The optimal P can be found to make the value of P_d maximal.

$$P_{d\max} = \Gamma\left(k_1', \Gamma^{-1}(k_0, \varepsilon)\left(\frac{1}{1 + \sigma_p^2 h_p \left(\sigma_e^2 h_e h_e^H + \sigma_n^2 I\right)^{-1} h_p^H}\right)^{P_{opt}/2}\right) \quad (26)$$

4 SIMULATION

Above all, the accuracy of using gamma distribution to approximate the distribution of Y is verified.

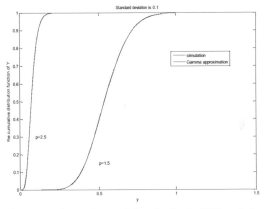

Figure 1. Comparison of the simulated CDF and the gamma CDF for Y when standard deviation is 0.1.

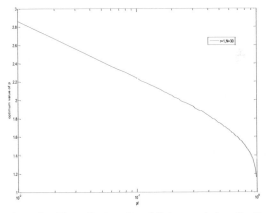

Figure 2. The optimum value of P that maximizes the P_d vs. P_f.

Figure 1 compares the cumulative distribution of the simulation value of Y by using the gamma distribution approximate Y. Under hypothesis H_0, the use of gamma distribution to approximate the distribution of Y is very close to the cumulative distribution of the simulation value of Y.

In the scheme there are a primary user, a malicious primary user emulation attack and four secondary users in the CRN. Given the number of samples $N = 30$ in a detection interval $n_k(t) \sim N(0, \sigma_n^2)$ $x_p(t) \sim N(0, \sigma_p^2)$ $x_e(t) \sim N(0, \sigma_e^2)$.

where $\sigma_p^2 = \sigma_e^2$, $\sigma_n^2 = 1$, $r = \dfrac{\sigma_p^2}{\sigma_n^2}$.

Figure 2 displays the optimal P when $r = 1$. The optimal value of P is got by making P_d maximum when P_f is determinate. The most of P is not equal to 2. That is to say, the improved energy detector can improve the system performance more than the traditional energy detector.

In Figure 3, let $r = 0.5$. It displays the receiver operating characteristic (ROC) curves of our optimal fusion scheme when MPUEA and conventional MRC are absent and present respectively. We find that the detection performance decreases largely due to the impact of MPUEA. The scheme has not made the detection performance improved, compared with conventional MRC. Then we try to increase the transmit power of primary user. In Figure 4, $r = 100$ (the transmit power of primary user increased 200 times). Our scheme makes detection performance improved obviously. The result can attain the standard of 802.22 (P_f from 0.01 to 0.1, P_d from 0.9 to 0.99).

5 CONCLUSION

Due to the impact of MPUEA, the detection performance of system declines sharply. It cannot be improved by increasing the transmission power of PU. The optimal weighting coefficients are obtained by maximizing the probability of detection and minimizing influence of MPUEA on the system. Our scheme can reduce the effects of MPUEA on the performance of the system more effectively than the conventional MRC.

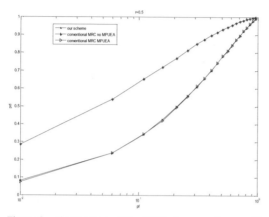

Figure 3. Comparison of the ROCs for our scheme with MPUEA , the conventional MRC when MPUEA is absent and the conventional MRC when MPUEA is present. $r = 0.5$.

ACKNOWLEDGMENT

This work was partly supported by National Natural Science Foundation of China (NO.61261022) and innovation team project of Yunnan University of Nationalities. Yuebin-Chen is the corresponding author.

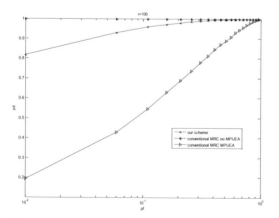

Figure 4. Comparison of the ROCS for our scheme with MPUEA , the conventional MRC when MPUEA is absent and the conventional MRC when MPUEA is present. $r = 100$.

REFERENCES

[1] Attar, A .2012.Survey of Security Challenges in Cognitive Radio Networks: Solutions and Future Research Directions.Proceedings Of The Ieee, Vol.100 (12), pp.3172–3186.
[2] Yunfei.C.2010.Improved energy detector for random signals in Gaussian noise IEEE Transactions On Wireless Communications,Vol.9 (2), pp.558–563, Feb.
[3] Urkowitz.H.1967.Energy detection of unknown deterministic signals.Proceedings of the IEEE, Vol.55 (4), pp.523–531.
[4] Letaief .K. B.2009.Cooperative communications for cognitive radio.Proc. IEEE, vol. 97, no. 5, pp. 878–893.
[5] Yunfei. C.2006.SER of selection diversity MFSK with channel estimation errors.IEEE Transactions On Wireless Communications, vol. 5, pp. 1920–1929.

Network Security and Communication Engineering – Chan (Ed.)
© 2015 Taylor & Francis Group, London, ISBN: 978-1-138-02821-0

Implementing variable length Pseudo Random Number Generator (PRNG) with fixed high frequency (1.44 GHZ) via Vertix-7 FPGA family

Q.A. Al-Haija
Department of Electrical Engineering, King Faisal University, Al-Ahsa, Saudi Arabia

N.A. Jebril
Department of Computer Science, King Faisal University, Al-Ahsa, Saudi Arabia

A. Al-Shua'ibi
Department of Electrical Engineering, King Faisal University, Al-Ahsa, Saudi Arabia

ABSTRACT: In this paper, we are implementing fast Pseudo Random Number Generator (PRNG) based on Linear Feedback Shift Register (LFSR) method with variable datapath sizes (8 bit –to- 1024 bit). The design was synthesized by using Xilinx Virtex 7 chip family with target device XC7VH290T-2-HCG1155 in terms of maximum frequency and area of the FPGA design. As a result, a fixed maximum frequency of 1436.678MHz for different datapaths has been achieved. Thus, the results conformed to a linear relationship between area and bit length. Consequently, the obtained results are attractive for many embedded system applications such as cryptographic algorithm design.

KEYWORDS: Computer arithmetic, fpg a design, hardware synthesize, PRNG algorithm, LFSR.

1   INTRODUCTION

The vast evolution of digital design techniques by means of new configurable logic blocks (CLBs) urged the computer design researcher to navigate through the digital arithmetic field, which plays an important role in the design of digital processors and embedded systems. Digital arithmetic encompasses the study of representations, algorithms for operations on numbers, implementations of arithmetic units in hardware and their use in general-purpose and application specific systems [1, 10].

Digital arithmetic is considered a middleware layer in the design of different number theory algorithms which has hastily grown over the past decade due to its substantial need in the areas of cryptography and coding theory.

The focus of this paper will be on Pseudo Random Number Generator (PRNG) based on Linear Feedback Shift Register (LFSR) method which is well-known iterative algorithm to generate random numbers for different applications. PRNGs are important in practice for their speed and reproducibility in number generation [2]. A pseudo-random number generator, or PRNG, is a random number generator that produces a sequence of values based on a seed and a current state. Given the same seed, a PRNG will always output the same sequence of values [3]. It contributes to systems including decision-making and probability testing as well as cryptographic systems such as RSA [11].

One way to build PRNG is by applying LFSRs based exclusive-OR gates. In LFSR, the random number repeat itself after 2n-1 clock cycles (where n is the number of bits in LFSR). A standard polynomial function (for n = 8): X8 + X7 + X6 + X4 + X2 + 1 is used to generate random numbers, for example; a 4-bit LFSR (shown in figure 1) uses 4 D-Flip-flops and XOR gates where each D-Flip-flop uses asynchronous reset that is independent of clock. LFSR is easy to be implemented in hardware as multiple LFSR's are often combined to achieve better security.

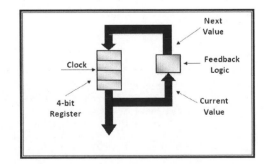

Figure 1.   4-bit LFSR.

The main contribution of this paper is to provide an efficient variable length FPGA design of PRNG by using Xilinx Vertix-7 FPGA Kit Family via VHDL which can be integrated smoothly into cryptosystems. The comparison with other existed designs shows that the proposed coprocessor design has a throughput efficiency of even two or more times faster than other designs.

The rest of the paper is organized as follows: first, a mathematical review of PRNG Based LFSR is presented. Second, we provide a full description of the simulation environment. Then, a full hardware implementation and its performance evaluation presented are followed by the conclusions and recommendations.

2 MATHEMATICAL BACKGROUND

Typical PRNG algorithm can be defined as [4]:

> INPUT: (Key, Seed)
> OUTPUT: random_data, (Key', Seed')
> random_data = F (Key, Seed)
> Key' = F (Key, Seed+1)
> Seed' = F (Key', Seed)
> Return random_data
> Where F is a cryptographic function of n bits.

The mathematical model of PRNG based LFSR is shown in figure 2. It can be derived as follows:

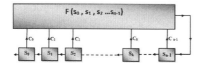

Figure 2. Typical LFSR of length n (n-stage).

We first consider the case that f is a linear function, i.e.

$$f(\vec{S}) = \sum_{i=0}^{n-1}(Ci * Si)$$

The output of this LFSR is determined by the initial values s_0, s_1... s_{n-1} and the linear recursion relationship:

$$S_{k+n} = \sum_{i=0}^{n-1}(C_i * S_{k+i}) , \ k \geq 0$$

Where $c_n = 1$ by definition.
The notion of LFSR can also be represented in matrix equation formula as follows:

$$\begin{bmatrix} X_0(t+1) \\ X_1(t+1) \\ \vdots \\ X_{n-2}(t+1) \\ X_{n-1}(t+1) \end{bmatrix} = \begin{bmatrix} 0 & 1 & 0 & \cdots & 0 & 0 \\ 0 & 0 & 1 & \cdots & 0 & 0 \\ & & \vdots & & & \\ 0 & 1 & 0 & \cdots & 0 & 1 \\ 1 & h_1 & h_2 & \cdots & h_{n-2} & h_{n-1} \end{bmatrix} \begin{bmatrix} X_0(t) \\ X_1(t) \\ \vdots \\ X_{n-2}(t) \\ X_{n-1}(t) \end{bmatrix}$$

Where X (t + 1) = Ts X (t) & Ts: companion matrix.

3 SIMULATION ENVIRONMENT

This work aims to implement and verify a variable length PRNG hardware unit by using Field Programmable Gate Array technique (FPGA) and Xilinx chip technology. FPGA [5] is an integrated circuit made from semiconductors based on a matrix of CLBs that might exceed 2000 pins and billions of transistors

In order to build up the FPGA unit of PRNG, the following simulation specification were involved to accomplish the design:

- HDL Language: VHDL [6], an acronym for Very High Speed Integrated Circuit Hardware Description Language which is a programming language used to describe a logic circuit by function, data flow behavior, or structure. Unlike other programming languages (such c or Pascal) which are designed as a general-purpose language, VHDL aims to modeling or documenting electronics systems.
- HDL Simulator: ModelSim 10.1 [7], an intelligent, user friendly, easy-to-use graphical user interface with TCL interface and considered as a typical environment to describe the hardware using VHDL.
- HDL Synthesizer [8]: Xilinx ISE Design Suite 14.2, a complete platform to build, verify and synthesize the hardware design. The design can be evaluated in terms of both area and performance.
- FPGA Chip Technology: Xilinx Virtrex 7/ VC707with target Device (xc7vh290t-2-hcg1155) [9]. Such FPGAs are optimized for advanced systems requiring the highest performance and highest bandwidth connectivity. Figure 3 shows the target device.

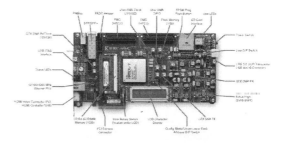

Figure 3. Target FPGA Chip Technology: Xilinx Virtrex 7.

- Datapath size range: (8- 1024 bits), our study is considered to synthesize the design for different bit lengths: 8-, 16-, 32-, 64-, 128-, 256-, 512-, 1024-bits.

- Total REAL time to Xst completion: 9.00 seconds, which is the real time required by simulation platform processor to navigate through the critical path delay.
- Total memory usage is 424268 kilobytes. This forms the amount of real memory required by simulation platform PC to synthesize the design.
- Optimization Goal: Speed. The focus of the study was to enhance the timming issues (Maximum frequency and Minimum delay).
- Simulation Platform: High Performance PC Specifications with Core i7 processor/quad-core [3.4GHz, 8MB Shared Cache] and 16GB Memory (DDR3 of 1600MHz).

4 PROPOSED PRNG DESIGN & EVALUATION

The top view diagram of the proposed PRNG is shown in figure 4 where x is the bit length. Figure 5 shows the internal digital circuit design for PRNG of 1024 bit, which involves D-Flip-flop and XOR gate for each bit of the register.

The simulation results were generated by using Xilinx Synthesizer ISE tools. The target chip technology Xilinx Virtex 7 (xc7vh290t-2-hcg1155. They were applied to the VHDL code that we implemented for PRNG-Unit.

The obtained results showed similar frequency behavior as we change the datapath size where a maximum frequency has been recorded for the different datapath size (from 8-bit – up to 1024-bit) as 1436.678 MHz. The same tendency has been noticed for the minimum delay period as its related to the frequency, which recorded a total 0.587ns, spent as 0.236ns logic, 0.351ns route (40.2% logic, 59.8% route). The area of the design were increasing linearly as we increase the precision (bit length), e.g. the 1024- PRNG needs 1024 D-Flip-Flop and 1024 XOR gate.

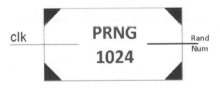

Figure 4. Top level block diagram of PRNG.

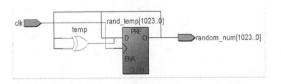

Figure 5. Digital circuit design of PRNG based LFSR.

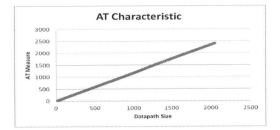

Figure 6. Digital circuit design of PRNG based LFSR.

Thus (Area * Time) characteristic can be assumed as a measurement that combines area and time. AT measure for the design can be calculated as:

$AT = (X + Y) * 0.587$ ns where X is the number of D-Flip-Flop and Y is the number of XOR gates. As $X=Y$ then:

$AT = 1.174 * X$ (shown in the figure 6).

5 CONCLUSIONS

In this paper, we proposed a fast FPGA implementation for PRNG unit based on linear feedback shift register by using Xilinx Virtex 7 device technology that improves the computation process. The performance is studied in terms of both timing issues (maximum frequency (MHz)) and area of the design (number of logic elements). Eventually, a fixed maximum frequency of 1436.678MHz for different datapaths has been achieved. Also, the results show a linear relationship between area and precision.

REFERENCES

[1] Milos D. ERCEGOVAC (2004), Tomas LANG, "Digital Arithmetic," Morgan Kaufmann Publishers, by Elsevier Science (USA), Vol1, Ch2.
[2] Wade Trappe (2006), "Introduction to Cryptography with Coding Theory", 2nd edition, Pearson Education. Inc, pp 10–11, 76–78, 81–82.
[3] LavaRnd (2007), " Terms & Definitions: pseudo-random-number-generator".
[4] NIST Special Publication 800–90(2007), Recommendation for Random Number Generation Using Deterministic Random Bit Generators, Elaine Barker and John Kelsey, National Institute of Standards and Technology.
[5] Clive Maxfield (2004) "The Design Warrior's Guide to FPGAs: Devices, Tools and Flows", Mentor Graphics Corporation and Xilinx, Inc., Elsevier.
[6] D. L. Perry (2002), " VHDL: Programming by Example", Fourth ed., The McGraw-Hill Companies.
[7] Mentor Graphics Corporation (2009), "ModelSim® Tutorial".

[8] X. Corporation (2008), Xilinx ISE 10 Tutorial: A Tutorial on Using the Xilinx ISE Software to Create FPGA Designs for the XESS XSA Boards, XESS Corp.

[9] Xilinx Inc. (2014), "VIRTEX®-7 FPGAS: Optimized for Highest System Performance and Capacity".

[10] W. Stein (2011)," Elementary Number Theory: Primes, Congruences, and Secrets", Vol I.

[11] Q. Abu Al-Haija, M. Smadi, M. Jaffri & A. Shua'ibi (2014), "Efficient FPGA Implementation of RSA Coprocessor Using Scalable Modules", ninth International Conference on Future Networks and Communications (FNC), by Elsevier, Ontario, Canada.

Network Security and Communication Engineering – Chan (Ed.)
© 2015 Taylor & Francis Group, London, ISBN: 978-1-138-02821-0

The software trusted measurement with expected value

Z.H. Wang & L.B. Wu
China Electric Power Research Institute, China

Z.M. Guo & X.P. Li
State Grid JIBEI Electric Power Company Limited, Beijing, China

ABSTRACT: The trusted measurement mechanism for software or programs in the state gird system is proposed based on the trusted computing technology. The method will calculate hash values of the executable programs, dynamic libraries and kernel modules before they are loaded. The results are used to compare the expected value to verify whether they are allowed to load or run. Simulation experiment is designed to test the efficacy of the method.

KEYWORDS: Trusted measurement, Expected value, Trusted computing.

1 INTRODUCTION

In the traditional sense, the software trust mainly contains two aspects: security and dependability[1]. There have been many different statements and studies about the conception of trust. From the perspective of IT system, ISO/IEC 15408 standard suggests that the behavior of a trusted components, operation or process is predictable under any operating conditions and can resist the damage caused by application software, virus and the certain physical interference[2]; Trusted Computing Group think that if an entity always act according to the expecting set goals, it can be called trusted[3]. From the perspective of network behavior, it is suggested that the trusted network should be that the behavior and its results of the network system can be expected, behavior state can be monitored, behavior results can be assessed, and abnormal behavior can be controlled[4]. From the perspective of the user experience, Microsoft proposes that trusted computing is a safe and reliable computation which can be achieved at any time, and the degree of trust on computers from humans, just like the freedom and security which we can feel when we use the electric power system or a telephone[5]; The trustworthy is also defined as that if a software system's behavior is always in line with expectations, it can be called trustworthy[6]. Different expressions about the concept of trustworthy shows people's cognition of trustworthy at different times from different perspective and also represents the different research purposes and application domain.

Trusted computer system builds the trust chain of the system to ensure that the system is reliable by extending the trust from the trusted root to the whole system through trusted measurement. Therefore, trusted measurement is the key technology to build the trusted computer system. In order to protect the reliable operation of computer system, it is necessary to measure the system with both static measurements and dynamic measurements based on the definition of trustworthy which trusted computing group is put forward as that behavior under a given target can be expected. It is more in line with the basic requirements of trusted computing to research the system's trusted measures from the perspective of considering the trustworthy of entity's behavior.

It is a hot spot of trustworthy software to study software reliability measure problem from software behaviors[7-10]. Trusted software behavior refers to that the running behavior can be monitored, behavior results can be assessed and abnormal behavior can be controlled. The key question is the analysis and research of the trusted evidence reflecting software running, extracting the evidence as expectations of software behavior and considering the software operation which need to be interacted with the environment by input and output. The attackers often use the input data to damage the trust of the software behavior.

2 TRUSTED EVIDENCE

The trusted evidence are the facts evaluating the trust of the software[11-13]. In the past, when they investigated or evaluated software, the trusted evidences people were concerned about are usually divided into three aspects:

1 Evidence in the development phase: how to obtain the evidences of software entities that conform to the setting goal in the development process through the standardization of the design, production and management processes;
2 Evidence in the submit stage: it is the trusted evidence after the submit stage, which mainly obtain through analysis, testing and verified tools;
3 Evidence in the application stage: it consists of the extensive degree of software usage, the users feedback information and the trust of the software provider, etc.

Assuming that the compilation system is absolutely authentic, if a person fully and correctly understands and knows the software's source code, that is, he has a good knowledge about the behavior and expectations of the software, the person will absolutely trust the software. However, software systems are more and more complex, even if the software maker can't fully and correctly recognize and understand them. Moreover, the desired trust of the software is objective and universal, rather than one individual's subjective results. So establishing the trust of software on a person's cognition and understanding of the software has two problems:

People's cognitive problems and trust delivery problems [14-17]. That is how to ensure the accuracy and adequacy of the software maker's cognition and understanding and how to deliver the trust for the software from a person or one group to other people. The ideal solution is that the expectation on the software is formally verified. That is to say, formal system refers to the software source code, or the conversion between the system and software source code is mechanized. Moreover, the verification process is mechanized. These mechanisms ensure the consistency between the formal system and the source code and the trust of the formal system, which solves the problem of human cognitive ability and trust delivery. However, there are still many barriers to overcome to reach such goal. Because of the non-decidability of the software's correctness, the machine inspection is mechanical rather than automated.

3 TRUSTED EVIDENCE MODEL BASED ON AUTHENTICATION

Trust for the software, or the understanding and awareness of the software, is an increased process, which embody in people's description for different forms of software and the different concrete degree and a variety of implicit or explicit validation process based on the description. Classifying and collecting trusted evidence according to the different methods of description and validation helps to solve the lack of traditional trusted evidence's classification and collection methods [18-20]. It gives the trusted evidence model which classifies the trusted evidence based on the different description and verification and contains several commonly used software verification way, corresponding to different forms of description in the software system. That is to say, the trusted evidence model will classify trusted evidences with the process of "System Description→Verification→Expectation", such as "Formal System Model→Formal Validation →Formal Expectations", "System Graphical Model→Non-formal Verification→Expectation", etc. Each approach for the authentication includes a description of the software system, the mapping relationship with other description and prospective validation based on the description.

Software quality is usually broken down into various features and sub-features when ruling and evaluating it. Accordingly, trusted evaluation also needs multi-scale quantitative index system.

First of all, the trusted evidence model based on authentication [21-22] is advantageous to the open source software developer personality and is also conducive to trusted evolution of software. Secondly, the trusted evidence model based on authentication decompose the overall trust of the software to trust of multiple different expectations, while the evidence also correspond to the respective expected in line with the thinking of divide and rule about processing complex issues. Moreover, the evidence in the evidence model itself is reflection of the logical thinking of software creator or improvements about software and its expectation. Therefore, the trusted evidence model based on authentication complies with human cognitive habits of the software, is convenient for people to analysis, recognize and understand, and helps establish trust based on awareness and understanding.

4 DESIGN AND IMPLEMENTATION

Trusted measurement mainly controls software before it runs by measuring the execution permission, source and trustworthy. Trusted measurement based on the expected value is a process to measure the integrity of the objects by the measurement agent with preset measurement policies and proper algorithms. As is shown in fig. 1, the main task of the trusted measurement are as follows:

• Determining the expected assertion, which properly describe the desired results of measured object, namely, its integrity.
• Defining policies, which is to define feasible measurement methods to verify the integrity of the measured object.

- Evaluating the measurement, which is to evaluate the creditability of the measured objects by measurement results.
- The trusted measurement will measure the executable program in the system, the dynamic libraries and kernel modules to protect their integrity by calculating their hash value and comparing the expected value.

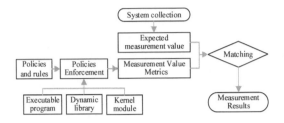

Figure 1.   The trusted measurement system structure.

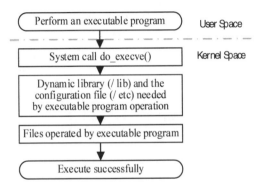

Figure 2.   The flow chart of program execution.

4.1 *Program execution process analysis*

All operating systems are Linux system in the state grid control system. Therefore, the study is focused on Linux system. Program is stored in a disk in the form of an executable file in Linux system. The executable file includes the executive function's target codes and the data used by these functions. Many functions are service routines used by all programs and their target codes are in a particular file so-called "library". A library function code can be copied to the executable file statically or connected to the process in the runtime.

As is shown in fig. 2, if a user wants to display files in the current directory, what he needs is to type the external command /bin/ls under the shell prompt. The user can command shell to create a new process to invoke a system called execve(), which pass the full path of the ls executable file name as a parameter, just like the /bin/ls in this case. Sys_execve() service routine finds the corresponding files, checks executable format and modifies the current process of the execution context according to stored information. Sys_execve() copies the executable file path name into a new distributed page frame and calls do_execve() function with the given parameters as the pointer to the page frame, the pointer of the pointer array, and the location of the kernel mode stack storing the contents in the user mode registers.

4.2 *Measurement object and method*

Hash operation is applied to measure the code of measurement object, including the executable program, the dynamic library and the kernel module. The hash value is compared with the expected value to verify whether the integrity of the measured object is destroyed. The measurement time depends on the size of the measured file. When the executable programs, dynamic libraries and kernel modules are firstly loaded, the measurement agent will be activated. There are two mechanisms to ensure the measurement agent is not loaded repeatedly, which makes the measurement time shorter. One is the system buffer mechanism by which the system stores files recently opened in the cache so that it will not be opened when reloading, which will not invoke the measurement agent. The other mechanism is the trusted software base which makes the measurement agent label the measured files and these files will not be measured again if they are verified to be measured.

As is shown in fig. 3, the measurement point of the executable program is set in the hook function security_bprm_check() provided by LSM mechanism. When the executable program starts, its related information is intercepted. Hash algorithm is called to calculate its hash value and the results are compared with the expected values to verify whether it is destroyed.

The measurement point of dynamic libraries is set in the hook function sys_init_module () provided by LSM mechanism. When the dynamic library is loaded, it will be mapped into the system memory by mapping functions, including dynamic libraries for program startup and loaded during the program operation. The measurement is unnecessary to reload if the dynamic library has already run in the memory. Otherwise, the information about the dynamic library is intercepted when it is loading. Hash algorithm is called to calculate its hash value and the results are compared with the expected values to verify whether running program is tampered. When the program exits, some of dynamic libraries will be released. However, some will still reside in the system memory providing services for other programs. Therefore, these dynamic libraries don't need to be measured then these programs restart.

The measurement point of the kernel module measurement is set in function of sys_init_module () because LSM mechanism doesn't provide appropriate measurement hook function. There are two possibilities to be addressed for the kernel module measurement. First, the necessary modules for the operating system will be loaded when Linux system starts up. Second, the necessary modules will be loaded by the operating system dynamically when needed, for example, U-disk is used. Linux kernel does not differentiate these two cases, and the kernel module is loaded by calling the same system call of sys_init_module (). Therefore, the information about code and length of the kernel module will be intercepted and sent to the measurement agent, which will calculate its hash value and compare it with the expected value to verify whether it is manipulated.

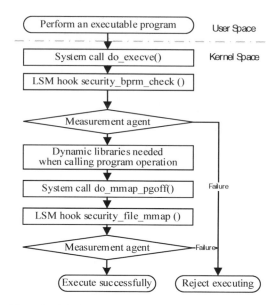

Figure 3. The flow chart of the trusted measurement.

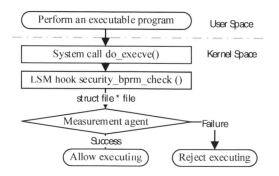

Figure 4. The flow chart of the executable program loading.

4.3 Measurement procedures

The measurement procedures of the executable programs, dynamic libraries and kernel modules are analyzed as follows.

The executable program loading is shown in fig.4. The system application do_execve() is first called when the executable program is loaded in Linux. It will pass the executable file name, the number of parameters and the values of the parameters to the kernel and encapsulate these parameters in struct linux_binprm. LSM mechanism adds hook function security_bprm_check () where the system calls with the parameter of the encapsulated struct linux_binprm. The struct file * file already encapsulated in struct linux_binprm is sent to the trusted agent by the hook function, which will calculate the hash value of the executable program to compare it with the expected value. The executive programs will be allow to run if the compared result is the same. Otherwise, they will refuse to run.

The dynamic libraries loading is shown in fig. 5. The system function do_mmap_pgoff() will be first called when the dynamic library is loaded in Linux that will pass the parameters of the dynamic kernel code to the kernel. LSM mechanism adds hook function security_file_mmap () with the parameter of struct file * file that will be sent to the trusted agent. The agent will calculate the hash value of the executable program to compare the expected value. The dynamic libraries will be allowed to be loaded if the compared result is the same. Otherwise, they will refuse to load.

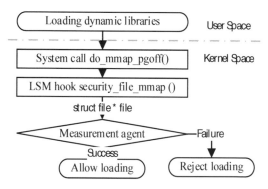

Figure 5. The flow chart of the dynamic libraries loading.

The kernel modules loading is shown in fig. 6. When the kernel modules are loaded in Linux, the system function sys_init_module() will be called, passing the parameters of the kernel code segment and the module size to the kernel. Because LSM mechanism doesn't provide appropriate measure hook function, sys_init_module() will be proposed to be intercepted. It will pass the loading information of the kernel modules to the trusted agent that calculates its hash values to compare the expected value. The kernel modules will be allow to be loaded if the compared result is the same. Otherwise, they will refuse to load.

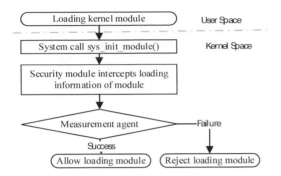

Figure 6.   The flow chart of the kernel module loading.

5   THE PROPOSED MODEL

Assuming that $X = \{x_1, x_2.......x_n\}$ is a set of executable programs, dynamic libraries and kernel modules, and x is one of them. $Se : x \rightarrow [0,1]$ is defined as the state evaluation function of elements in the set with the following evaluation levels. The level of Distrusted meets the requirements of $0 \leq Se(x) \leq E_0$, and the level of Trusted meets the requirements of $E_0 \leq Se(x)$. Where, $0 \leq E_0 \leq 1$.

Suppose that $E\pi = \{E_1, E_2.......E_n\}$ is the minimum trusted value for the application set $AP = \{ap_1, ap_2.......ap_n\}$ allowed to run. If the trusted value of any element ap_i is less than the corresponding minimum trusted value, it will be defined as "Distrusted".

6   EXPERIMENT RESULTS

The simulation is performed by NetLogo, with experiment environment of Intel Core dual-core 2.36G, 4G memory, win7 OS. The detailed parameters are in table 1.

Table 1.   The parameters of the simulation experiment.

Parameter	Initial value	Parameter interpretation
Network environment parameters		
N	200	Number of entities in the network environment
M	600	Snapshot number of the group members
Algorithm parameters		
E_0	0.6	Status unbelievable threshol
t	1000S	System running time

As is shown in fig. 7, the proportion of the distrusted entities versus all entities decreased with the time.

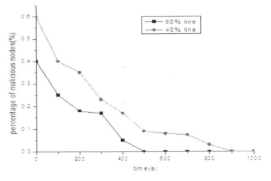

Figure 7. The detection efficiency of distrusted entities.

7   CONCLUSION

The software trusted measurement mechanism is developed to control programs running in the state grid system with the expected values and trusted computing technology. The proposed method is able to control the execution permission, source and reliability of software or programs. The executable programs, dynamic libraries and kernel modules are measured before they are loaded and executed, which will keep their integrity. The simulation experiment shows that the proposed mechanism can detect untrusted entities with great efficiency, including the executable programs, dynamic libraries and kernel modules. Moreover, the method is applied in state grid system with good performance.

REFERENCES

[1] Algırdas A, Jean-Claude L, Brian R, et al. Basic concepts and taxonomy of dependable and secure computing[J]. IEEE Trans Dependable Secure, 2004, 1(1): 11–33.
[2] The International Organization for Standardization. ISO/IEC15408-1 Information technology -security techniques-evaluation criteria for IT security-Part 1: Introduction and general model[S/OL]. (2005-11-14) [2009-12-05]. http://www.clusit.it/whitepapers/iso15408-1.pdf.
[3] Trusted Computing Group. TCG specification architecture overview[EB/OL].(2007-03-28). http: //sparrow.ece.cmu.edu/group/731-s07/readings/TCG_1_3_Architecture_Overview.pdf.
[4] Lin Chuang. Research on trustworthy networks[J]. Chinese Journal of Computers, 2005, 28(5):751–758.

[5] Gates B. Trustworthy computing[EB/OL]. (2002-01-15)[2009-12-06]. http://www.wired.com/news/business/0,1367,49826,00.html.

[6] Wang Huaimin, Tang Yangbin, Yin Gang, et al. Trustworthiness principle of network software[J]. Science in Chinese: Series E, 2006, 36(10):1156–1169.

[7] Mollering G. The nature of trust:From Georg Simmel to a theory of expectation, interpretation and suspension[J]. Sociology, 2001, 35: 403–420.

[8] Lewis J D, Weigert A. Trust as a social reality[J]. Social Forces, 1985, 63(4): 967–985.

[9] Lewicki R J, Bunker B B. Trust in relationships: A model of trust development and decline[M]//Bunker B B, Rubin J Z.Conflict, Cooperation, and Justice. San Francisco: Jossey-Bass.1995.

[10] Yang Fuqing, Mei Hong. Development of software engineering: Co-operative efforts from academia, government and industry[C]//Proceedings of International Conference Software Engineering(ICSE2006), ACM 2006.

[11] Cao Donggang, Yang Fuqing, Mei Hong. Development of software engineering: A research perspective[J]. Journal of Computer Science and Technology, 2006, 21(9): 682–696.

[12] Liu Ke, Shan Zhiguang, Wang Ji, et al. Overview on major research plan of trustworthy software[J]. Chinese Science Funds, 2008(3): 145–151.

[13] ISO 9126-1 2001 Standard for software engineering-product quality-Part 1-quality model[S].

[14] Josang A, Ismail R, Boyd C. A survey of trust and reputation systems for online service provision[J]. Decision Support Systems, 2007, 43(2).

[15] Siani P, Marco C M, Stephen C. Analysis of trust properties and related impact of trusted platforms[EB/OL]. (2005-03-18). http://www.hpl.hp.com/techreports/2005/HPL-2005-55.pdf.

[16] LOSCOCCO P A, WILSON P W, PENDERGRASS J A, et al. Linux kernel integrity measurement using contextual inspection [C]//Proceedings of the 2007ACM Workshop on Scalable Trusted Computing. New York: ACM, 2007: 21–29.

[17] THOBER M, PENDERGRASS J A, McDONELL C D. Improving coherency of runtime integrity measurement[C]//Proceedings of the 3rd ACM Workshop on Scalable Trusted Computing. New York, NY: ACM, 2008: 51–60.

[18] AZAB A M, PENG Ning, SEZER E C, et al. HIMA: a hypervisor-based integrity measurement agent [C]//Proceedings of the 2009 Annual Computer Security Application Conference. Honolulu, HI: IEEE, 2010: 461–470.

[19] JAEGER T, SAILER T, SHANKAR U. PRIMA: policy-reduced integrity measurement architecture[C]//Proceedings of the Eleventh ACM Symposium On Access Control Models and Technologies. New York, NY: ACM, 2006: 19–28.

[20] CAMENISCH J. Better privacy for trusted computing platforms[J]. Lecture Notes in Computer Science, 2004, 3193: 73–88.

[21] PEARSON S. Trusted Computing Platforms, the Next Security Solution[R]. Bristol UK: Trusted E-Services Laboratory, HP Laboratories, 2002.

[22] XU Ziyao, HE Yeping, DENG Lingli. An integrity assurance mechanism for run-time programs [J]. Lecture Notes in Computer Science, 2009, 5487: 389–405.

Network Security and Communication Engineering – Chan (Ed.)
© 2015 Taylor & Francis Group, London, ISBN: 978-1-138-02821-0

The security protection system design for the state grid with trusted computing technology

B.L. Yang & B.H. Zhao
China Electric Power Research Institute, Beijing, China

H. Zhang
State Grid JIBEI Electric Power Company Limited, Beijing, China

ABSTRACT: The security protection system for the state grid control system is proposed with trusted computing technology, which are applied to keep integrity of the system with functions of trusted measurement, trusted report, and trusted storage providing. The state grid control system is defined into different security regions in accordance with the business processes and different types of security equipment and products are suggested to apply among different security regions. The proposed system can meet technical requirements of information system which is classified as protection of level four. The simulation experiment is performed and results show that it could ensure that the state grid control system is secure and trusted.

KEYWORDS: Trusted computing, Security protection system, Trusted measurement, Security region.

1 INTRODUCTION

With the rapid development of computer and network communication, the attack strength is rising quickly [1-5]. Among them, the attacks on industrial control have become a new focus. According to figures released by the national information sharing platform, in recent years, there have been more and more malicious software and codes to steal sensitive information and gather intelligence by APT, which have usually run for many years before they are found. According to CNCERT report, a large number of hosts in the country are infected with the Trojan program with an APT characteristic, involving a number of government institutions, important information systems, and key enterprises. Because of the importance of these servers, repairing its loopholes need to be very cautious, and it takes a long time to make the development of patches and utility processes, whereas the emergence of new vulnerabilities is much faster than the speed of patch development, which results in the accumulation of more and more loopholes. The state grid control system is an important industrial infrastructure. Currently, it applies defense mechanisms of anti-virus software, intrusion detection and other security equipment, which operate in the way of the virus database characteristic value comparison. However, they are helpless for attacks with features of unpredictability, specific targets, and non-reusablity. Such peripheral "blocking" methods cannot prevent man in the middle of attack, tampering, receiving,

reproducing, and other types of network threats in the state grid control system [6-9].

2 RELATED TECHNOLOGIES

2.1 *Trusted computing technology*

In 1999, trusted computing platform Alliance TCPA[10-15] (Trusted Computing Platform Alliance) was made up of Intel, HP, IBM, and Microsoft IT company which were founded, with the aim of promoting a new generation of safe, trusted hardware platform specifications and development. In 2003, TCPA organization was reorganized as the Trusted Computing Group (Trusted Computing Group, TCG), which not only emphasized trusted hardware platform, but also increased concerns about the trust of the software. TCG has developed a range of trusted computing platform architecture, basic functions, commands, interfaces, and specification, which provides a baseline reference for the realization of the trusted computing platform. Its specifications include the TCG main specification, TPM specification, PC and trusted software stack specifications. TCG also published the relevant evaluation standards to evaluate whether the trusted computing equipment meets the TCG specifications.

The core of trusted computing [16-19] technology is the trusted chain and trusted roots. TCG Trusted Platform Module (TPM) specification is trusted computing platform hardware root of trust, which is a security chip providing protected storage security and cryptographic operation ability. TPM is embedded into the motherboard and connected to the CPU through the external bus.

Trusted Cryptography Module (TCM) is a basic continuation of the design structure of the TPM, developed by China independently supporting the domestic SM-series , asymmetric , and symmetric algorithms. Another useful specification is the draft standard of Trusted Platform Control Module (TPCM), which proposes that TPCM operate as an active device compared with trusted platform module as a passive device. Prior to CPU startup, TPCM runs as a trusted root to build the trust chain in computers. Trusted measurement root (Root of Trust for Measurement, RTM) is designed inside TPCM to provide stronger physical protection for it. Such design can enhance security of trusted measurement root by preventing BIOS (Basic Input Output System) tampering. TPCM provides the similar functions as TPM does. However, the CRTM of TPM is in BIOS, outreaching the protection scope of TPM. Therefore, TPCM specification proposes a new computer framework, which changes the traditional boot sequence. As the root trust chain, TPCM will play a key role in protecting the system security by controlling the system boot, I/O interface, and system configuration.

Faced with emerging security threats and attacks, more and more conventional protection measures are developed and deployed, but unsatisfactory [20-22]. In fact, most of the existing defense systems are designed for general computing environment with the goal of the easy deployment, easy operation, and greater compatibility. Therefore, they cannot provide the "tailored" protection system for a specific information system. The trusted computing platform could solve these problems effectively by starting from the bottom of the PC Terminal and building a trust chain. Trusted computing technology will help build a trusted computing system taking the CPU, motherboard, system software, application software, and network into account.

2.2 Access control technology

Access control technologies are used to address how to access system resources. Appropriate access control can prevent unauthorized users getting data intentionally or unintentionally. Access control means include user identification code, password, login control, resource authorization (Such as user profiles, resource profiles, and control list), authorization verification, logging, auditing, and so on [23-34].

Access control technologies are classified into six groups, including defensive, detective, corrective, managed, technical, and operational control. Among them, the defensive control is used to

prevent the occurrence of adverse events; Detective control is used for the detection of adverse events; Corrective control is used for correction of adverse events; Managed control is used for the development, maintenance and use of the management system, including decisions of system policies, procedures, behavior norms, roles and obligations, individual functions and personnel security decisions; Technical control is used to provide automatic hardware and software protection for the IT systems and applications, which is widely applied in technology systems and applications; Operational control is used to protect the operating system and application's routine procedures and mechanisms, which are mainly involved in people use and operation of the security method, and is likely to have an effect on system and application environment.

3 THE SECURITY PROTECTION SYSTEM

The security protection system of the state grid control system is studied with the model of "Triple protection layers" managed by "one management center". The security protection system is proposed to meet technical requirements of information system classified protection of level four. The system consists of trusted computing environment, trusted boundary, trusted communication network, and trusted security management center.

3.1 Trusted computing environment

Trusted computing environment is responsible for information generation, storage, processing, transmission, and security, consisting of trusted chips, trusted hardware, operating systems, and trusted software base.

The key of trusted computing environment is the trusted software base. Trusted software base consists of hardware resources based on trusted computing platform, providing trusted computation services and protecting the system software and the environment through the trusted measurement and certification. Trusted software base runs between the trusted chip and the application system, as a bridge for application systems to access trusted chip. The application system uses security functions provided by the trusted chip through calling the interface of trusted software base.

Trusted computing environment has functions of trusted measurement, trusted storage and trusted report. The trusted measurement is used to keep the operating system environment secure and trusted, the trusted storage is used to protect the critical data and sensitive information of operating system, and the trusted report is used to provide security reports of the operating system, which keeps the computing environment secure and trusted as a whole.

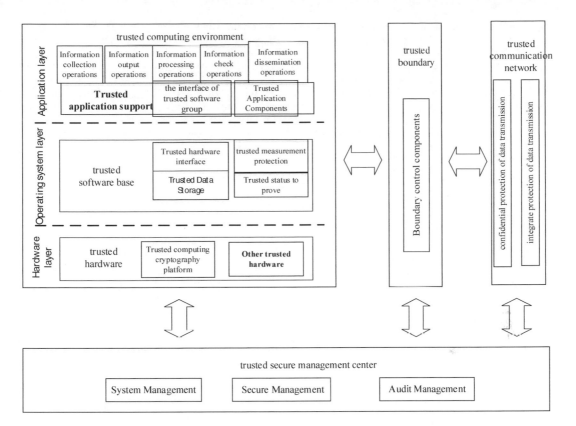

Figure 1. The state grid security protection system.

3.2 *Trusted boundary*

Trusted boundary connects trusted computing environment and trusted communication network, and enforces security policies. Trusted boundary performs security check and control by regional access control and packet filtering technology, which ensure that the information in and out of the boundary is controlled, preventing unauthorized access. At the same time, it also provides functions of trusted audit, illegal behaviors finding and disposal.

3.3 *Trusted communication network*

Trusted Communication Network is a group of relevant devices which help transfer the information and enforce security policies. Trusted communication network provides security by the confidential protection and the integrity protection of data transmission, which ensures that network communications cannot be intercepted and tampered to keep information secure during the transmission. The trusted network connection mechanism is applied to keep devices accessing network secure and trusted, which will

prevent unauthorized connection. Trusted communication network can provide security audit, deal with illegal behaviors and provide alarm.

3.4 *Trusted security management center*

Trusted Security Management Center is the security component, uniformly managing trusted computing environment, boundary, communication network, and applications. It is the central hub which helps the users achieve security policy management, organization management, operation management and technology framework. Trusted Security Management Center can effectively integrate security policy management, organization management, operations management and technology framework.

3.5 *Regional security access control*

The security region definition is the premise of security access control. The state grid control system is defined into different security regions in accordance with the business processes, which are further defined into sub-security regions. Different types of security equipment and products are deployed to provide functions of isolation and access control among different security regions. Defining security regions help make the network structure more clear and the protection more focused, which provides a structured security protection system for the state grid.

Based on data stream and information security requirements, the state grid control system is defined into three security regions, which are security zone I, II, III, respectively.

1 Security zone I is real-time control area and the core area of secondary power dispatch system. Any real-time monitoring systems or systems with the capabilities of real-time monitoring are defined as security zone I. Its users are dispatchers and running operators, and the real-time requirement is in the level of milliseconds. This area communicates with the province control system by SPDnet, including the real-time closed-loop control system (SCADA/EMS), wide area phase measurement system (WAMS), secure automatic control system, and protection setting work station with functions of modifying given value and the distance cast back.

2 Security zone II is non-controlled production area. The systems without production controlling functions and the parts out of system control belong to the security zone II. Its users are running program staff. This area communicates with the province control system by SPDnet, including Water Scheduling Automation System, DTS, electricity trading system, electric energy metering systems, assessment systems, relay protection, and fault recorder information management system without functions of modifying given value and the distance cast back.

3 The security zone III is production management area which provides Web services, including Lightning monitoring systems, weather information, daily / Morning, DMIS, etc.

4 REGIONAL SHARED KEY METHOD

A polynomial $A(x)$ on a finite field $F_p[x]$ is defined on Security Management Center Key Management Server.

$$A(x) = \prod_{i=1}^{m}(x - H(sk_i, z)) \qquad (1)$$

where, m represents the number of members in the region U, sk_i is the permanent key of the members in the group which will be distributed to all the members in the area. $H(x, y): Z_p^* \times \{0,1\}^n \to Z_p^*$ is cryptographic security hash function, z is a random integer from $\{0,1\}^n$. $A(x)$ is called Access Control Polynomial. Just as the equation (1), $A(x)$ takes the legitimate user's key $H(sk_i, z)$ as the root. When taking legitimate customer key, $A(x)$ is equal to zero. Otherwise, $A(x)$ is a random value.

The key management server randomly selects area shared key K for region U, the calculating polynomial:

$$P(x) = A(x) + K \qquad (2)$$

In the end, $(z, P(x))$ is made public. With this public information, any member U_i can get the group key:

$$K = P(H(sk_i, z)) \qquad (3)$$

For any member U_r out of U, $P(H(sk_i, z))$ is a random value. Therefore, U_r can't get the key K. This key management mechanisms can ensure that the guest virtual machine can extract the key from $P(x)$ only if user's SID_i belongs to $A(x)$.

5 EXPERIMENT RESULTS

The anti-attack capability of the proposed security protection system is tested by NetLogo, with experiment environment of Intel Core dual-core 2.36G, 4G memory, win7 OS.

Table 1. Simulation parameters.

	Meaning
N	300 initial node
M	Infection when the virus first broke out in the number of virtual machines
Average-node-degree	Number of other virtual machines per node interaction
Time	Simulation system uptime
$\beta\%$	Groups in the percentage of malicious virtual machine
Recovery chance	The probability of return to normal after the node is infected

Because the proposed security protection system mainly prevents malicious attacks from outside, the successful detection rate of malicious attacks is used to verify its efficiency. Given the period of Δt, malicious attacks $b(t)$ in the system, and successful detection $a(t)$, $\beta\%$ can be defined as follows:

$$\beta\% = \frac{a(t)}{b(t)} \qquad (4)$$

118

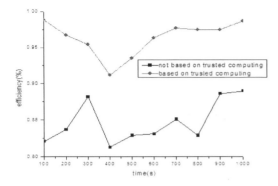

Figure 2. The anti-attack capability of the proposed security protection system.

6 CONCLUSION

The security protection system is developed for the state grid control system with trusted computing technology and access control technology. The proposed security protection system is implemented with the model of "Triple protection layers" managed by "one management center", which meets technical requirements of information system classified protection of level four. The state grid control system is defined into different security regions in accordance with the business processes. Different types of security equipment and products are deployed to provide functions of isolation and access control among different security regions, which provides a structured security protection system for the state grid. Trusted computing technologies are applied to provide trusted root and establish trusted chain, which help keep integrity of the system with trusted measurement, trusted report and trusted storage. The experiment results show that the proposed security protection system is able to prevent malicious attacks, ensuring the state grid control system secure and trusted.

REFERENCES

[1] Geir Hallingstad, Martin Gilje Jaatun and Ronny Windvik. Firewall Technology. 2001.

[2] Steven M. Bellovin. Security problems in the TCP/IP protocol suite. Computer Communications Review, 19(2):32–48, April 1989.

[3] D. Brent Chapman. Network (in) security through IP packet filtering. In Proceeding of the Third USENIX Unix Security Symposium. Pages 63–76, Baltimore, MD. September 1992.

[4] Wiliam R. Cheswick. The design of security internet gateway. In Proceedings of the Summer USENIX Conference, Anaheim, CA, June 1990.

[5] Thomas H. Ptacek and Timothy N. Newsham. Insertion, evasion, and denial of service; Eluding network intrusion detection. Technical Report, Suite 330. 1201 5th Street S.W. Calgary, Alberta, Canada, 2R-0Y6, 1998.

[6] Vem Paxson. Bro: A system for detecting network intruders in real-time, Proceeding of the Seventh USENIX Security Symposium, pages 31–51, 1998.

[7] M. Handley, C. Kreibich, and V. Paxson. Network intrusion detection; Evasion, traffic normalization, and end-to-end protocol semantics. Proceeding of the USENIX Security Symposium, pages 115–131, 2001.

[8] Trusted Computing Group. TCG Specification Architecture Overview Specification Revision 1.2. https://www.tmstedcomputinggroup.org. Apr 2004.

[9] J. A. Halderman, Seth D. Schoen, et al. Lest We Remember: Cold Boot Attacks on Encryption Keys[C]. 17th USENIX Security Symposium.

[10] Evan R. Sparks. A Security Assessment of Trusted Platform Modules. Computer Science Technical Report[DB/OL]. http://www.cs.dartmouth.edu/pkilab/sparks/.

[11] E. Brickell,J. Camenisch, L. Chen. Direct Anonymous Attestation, Oct. 2004 [EB/OL]. http://eprint.iacr.org/2004/205.

[12] S.Goldwasser,S. Micali,C. Racko. The Knowledge Complexity of Interactive Proofs. SIAM J. Comput., 1989, 18(1); 186–208.

[13] L. Chen, R. Landfermann, H. L. ohr, M. Rohe, A. Sadeghi, C_ StUble. A Protocol for Property-Based Attestation. Proceedings of the first ACM workshop on Scalable trusted computing, Alexandria, Virginia, USA: ACM Press, Nov. 2006, 7–16.

[14] Trusted Computing Group. Trusted Network Connect Architecture for Interoperability [DB/OL]. Http://www.trustedcomputinggroup.org/tnc/.

[15] TCG, Trusted Network Connect Architecture for Interoperability, Revision 1.3 [S], 2008.

[16] D.E. Bell, J. Leonard, LaPadula. Security computer systems; Mathematical foundations. Technical Report 2547, MITRE Corporation, Bedford, MA, 1973.

[17] D. E. Bell, J. Leonard, LaPadula. Security Computer System: Unified Exposition and Multics Interpretation. MTR-2997, MITRE Corp., Bedford, MA, March, 1976.

[18] K. Biba. Integrity considerations for security computer systems. Technical Report 76–372, U. S. Air Force Electronic Systems Division, 1977.

[19] R. S. Sandhu. Lattice-Based Access Control Models. IEEE Computer, 26(11):9–19, Nov 1993.

[20] A. C. Myers, A. Sabelfeld, S. Zdancewic. Enforcing robust declassification. 17th IEEE Computer Security Foundations Workshop, 172–186, 2004.

[21] R. S. Sandhu, E. Coyne, et al. Role-Based Access Control Models. IEEE Computer, IEEE Press, 29(2):38–47, 1996.

[22] F. C. B. David, J. N. Michael. The Chinese Wall Security Policy. Proceedings of IEEE Symposium on Security and Privacy, 206–214, 1989.

[23] D. D. Clark, D. R. Wilson. A Comparison of Commercial and Military Computer Security Policy. Proceedings of the IEEE Symposium on Security and Privacy, Oakland, CA, 184–194, 1987.

[24] L. Badger, D. F. Swme, et al. Practical domain and type enforcement for UNIX. Proceedings of IEEE Symposium on Security and Privacy, 66–77, 1995.

Grid and Cloud Computing

Network Security and Communication Engineering – Chan (Ed.)
© 2015 Taylor & Francis Group, London, ISBN: 978-1-138-02821-0

Cloud storage security

T. Galibus
Belarusian State University, Minsk, Republic of Belarus

H.E.R.M. Vissia
Byelex Multimedia Products BV, Oud Gastel, The Netherlands

ABSTRACT: Information security is one of the key problems in distributed cloud services. Providing secure access control mechanisms is an important requirement to any secure cloud solution. The absence of a control zone for mobile device access in a dynamic service model was taken into consideration while we dealt with this problem. Therefore, we propose to use attribute-based encryption, as a comprehensive access control solution for cloud storage including user accountability and key revocation. Previous proposed solutions of this problem are mainly technical and require additional semi-trustable proxy-servers.

To address this limitation, we propose to introduce special parameters for an adaptive attribute-based encryption algorithm. These parameters allow us to guarantee the reliable operation and protection of the system in the presence of malicious users, or in the case of unintentional information leakage (loss of a mobile device). In addition, we propose a simplified procedure for generating encryption attributes. This procedure is sufficient to provide access control in the majority of cloud services. The proposed modifications allow us to implement applicable access control systems in cloud services.

1 INTRODUCTION

When enterprises or organizations move to a cloud infrastructure or switch to a distributed system paradigm, various new problems arise. In fact, the ability to access data from anywhere and from any device is a real issue for the Chief Security Officer.

2 PROBLEM STATEMENT AND SOLUTION

Today most of the protected cloud services only offer a simple private secure data encryption solution (i.e. encryption via server- or client side). Just having this type of encryption is not enough when an enterprise wants to implement the paradigm of "access from anywhere and from any device". Such a distributed system requires a more sophisticated security system.

2.1 *Manageable access policy*

When there is a distributed infrastructure, it is important that all user access is properly configured, so that the user only has access to the files he or she needs. No unauthorized access should be possible. When an employee leaves the organization, the data should stay in the organization and not fall in the hands of people outside of the organization. The system should be able to change certain propertiesproactively. For example, if a device is stolen, the manager should be able to revoke permissions of the user of the device by changing the secret key(s) involved.

The key points:

1 The secure **setup** of the access policy: once the data is encrypted, only authorised users can get access to it.
2 The ability to secure data when the user is **deleted** from the system. That is to say, it becomes unauthorised (user revocation).
3 The safe modification of access rights and policies: it should not make the data vulnerable (user accountability).

2.2 *Full encryption*

By using an access policy, we can enforce the fact that all sensitive information is encrypted when using the high-end protection algorithms based on the strongest cryptographic functions. The encryption is implemented via the normal access policy.

Features of encryption:

1 Public key encryption.
2 Encryption is coordinated along the defined access policy. (i.e. your secret key gives you rights just for your portion of the information and not more)
3 On-the-fly encryption. (No data is sent from the server without prior encryption. This ensures the protection.)

2.3 Mobile and wireless access security

These days, a big security risk is posed on mobile and wireless networks. Therefore, it is necessary to pay specific attention to the protection of the data sent to the public unmanageable mobile devices (like tablets, notebooks and mobile phones).

The client is setup in such a way that it does not allow viewing encrypted data without a prior connection to the server to ensure the user has the valid key.

The following security should be in place:

1 A secure client side application to view the encrypted data.
2 Key verification.
3 Key expiry period.

2.4 Requirements of the users of cloud services [8]

- High-end security
- Managed security
- Automated maintenance and support
- Low cost for all Internet-capable user devices
- Individual security settings regardless of operating system and user device
- Easy use and minimal overhead with the highest level of security
- Security everywhere: your data is not only secure on your premises, but also when you are on the road
- All the data is encrypted and sent via the secure system – no one can intercept your data
- Easy and stress-free use of mobile devices
- Compliance system for manager liability (e.g. file sharing or data theft)
- Freedom from costly consultants (employ administration or create an in-house system)

3 CONCLUSION AND SOLUTION

One of the biggest challenges in the field of distributed networking is to provide the ultimate protection for sensitive data. Now the "security problem" requires a fast-working and effective solution on mobile and wireless devices, which are difficult to manage.

One of the best ways to restrict unauthorised access to shared data is to use a strong cryptographic algorithm [1] [3] [5]. Once the access policy is set and the secret keys are generated and distributed, the security of the system can be guaranteed.

The most promising way to control the access is based on attribute-based encryption (ABE) [2], which allows not only to encrypt the data but also to configure the authorised users. In other words, in such encryption system the secret user keys depend on additional attributes. Generally, this encryption method requires a complicated implementation [2] [4] [7].

Additional features of secure ABE are [8]:

- Data confidentiality
- Fine-grained access control
- Scalability
- User accountability
- User revocation
- Collusion resistant

The two unique concepts of our approach to cloud ABE:

1 **Simplified group hierarchy:** In our model, the large-scale cloud storage has a limited number of user access groups so that each user and each shared file belongs to a specific set of groups. This assumption allows us to propose a simplified key generation for attribute-based access. The advantage of this approach is that the simplified system can be enhanced with other properties like user revocation and accountability, at lower cost.
2 **Mathematical approach to key revocation:** We introduce additional key parameters that allow us to enhance the adaptive properties of a system, i.e. user accountability and key revocation. These parameters allow us to change and re-encrypt the part of confidential data easily without compromising other user data when the revocation is needed. Previous solutions of the problem are mainly technical and require additional semi-trustable proxy-servers [8] [9]. We propose a low-cost solution based on the modification of the base algorithm. We plan to include the verification of the user authorisation key into the ABE in order to give a better solution to revocation problem without re-encryption.

4 DESCRIPTION OF THE SYSTEM

In order to initialize the system of attribute-based encryption based on the simple selective principle we need the following:

1 The set of attributes:

$t_1, t_2, \ldots, t_n \in Z_q$ where q is prime

The set of attributes corresponds to the set of the access group identifications:

$Group1 \rightarrow t_1$
$Group2 \rightarrow t_2$
...
$Groupn \rightarrow t_n$

2   The data M, or the hash-value of the open text.
3   The set of user attributes: $\{t_i\}_U$

The set of attributes of the encrypted text $\{t_i\}_M$
The access rule of the selective scheme is as follows:
If at least one attribute in the set $\{t_i\}_U$ is equal to the attribute in the set $\{t_i\}_M$, the corresponding user U can decrypt the text M.
This access rule is enough for the modeling of a large-scale cloud storage security policy based on group access. So, the general complex structure of the ABE scheme can be simplified with this access control rules.
The encryption system can be implemented as follows:

Initialization
1   G is the group with the generator g.
2   The secret master-key which is stored on server and is accessible only to the administrator.

$MK = (t_1,...,t_n, y)$

3   Public key which is evaluated according to the master-key and is used to access the encrypted data:

$PK = \left(g^{t_1},...,g^{t_n}, Y = e(g,g)^y \right)$, where e(g,g) is the bilinear pairing.

Key generation
The secret user key D is generation based on the user U attribute set:

$\{t_i\}_U \rightarrow D = \{D_i = g^{yw/t_i}\}$

This key is sent to the user by the administrator. It depends on the master-key parameters and additional parameter $w \in Z_q$. This parameter serves for the D_i modification when the public keys are not changed.

Encryption
The encrypted text E consists of the encrypted message along with the attributes and the public key set$\{E_i\}$.
The additional parameter $s \in Z_q$ serves for text re-encryption so that the secret user key set D_i should not be changed:

$E = Me(g,g)^{ysw}, \ \{E_i = g^{t_i s}\}_{\forall i \in \{t_i\}_M}$

Decryption
In order to access the value of M one must evaluate $Y^{sw} = e(g,g)^{ysw}$:

$M = E / Y^{sw}$

The user evaluates $e(g,g)^{ysw}$ using the secret key D_i which corresponds to the attribute t_i and the public key E_i:

$$e(g,g)^{ysw} = e(E_i, D_i) = e\left(g^{\frac{yw}{t_i}}, g^{t_i s} \right) = e(g,g)^{ysw}$$

4.1   *Common scenario description*

4.1.1   *The user is added to the system, the file is added to the storage and the file is shared for the user*

i.   When the user U is added to the system, he gets the set of attributes (at least one attribute when he is not added to any previous group):

$\{t_i\}_U = \{t_1\} \rightarrow NewGroup$

Generally, the set of attributes depends on the user access rights.

ii.   The secret key is generated according to the attribute set:

$\{t_i\}_U \rightarrow D = \{D_1 = g^{yw/t_1}\}$

iii.   When the file M is added, it gets the set of attributes (at least one attribute for the user U):

$\{t_i\}_M = \{t_1\}$

iv.   The text is encrypted and the public key is published:

$E = Me(g,g)^{ysw}, \ \{E_1 = g^{t_1 s}\}$

4.1.2   *The user is added to the group where the file was shared*

i.   The value corresponding to Groupk is added to the user attribute set:

$\{t_i\}_U = \{t_i\}_U \cup \{t_k\} \rightarrow Groupk$

ii.   The value corresponding to Groupk is added to the user secret key set:

$D = D \cup \{D_k = g^{yw/t_k}\}$

4.1.3   *The user is deleted from the group where the file was shared*

i.   The value corresponding to Groupk is removed from the attribute set:

$\{t_i\}_U = \{t_i\}_U \setminus \{t_k\} \rightarrow Groupk$

ii.   The value corresponding to Groupk is removed from the secret key set:

$D = D \setminus \{D_k = g^{yw/t_k}\}$

iii. The text is re-encrypted with the new parameter w' in order to protect the text from the access with the old secret key value

$$E = Me(g,g)^{ysw'}$$

iv. The secret keys of all users able to access M are modified

$$D_i = (D_i)^{w'/w}$$

4.1.4 The new file is shared for the group

i. The value corresponding to Groupk is added to the attribute set:

$$\{t_i\}_M = \{t_i\}_M \cup \{t_k\} \rightarrow Groupk$$

ii. The value corresponding to Groupk is added to the public key set:

$$E = E \cup \{E_k = g^{t_k s}\}$$

4.1.5 The user is deleted from the system

i. Task 3 should be repeated for all user groups.

5 FINAL THOUGHTS

Our encryption system gives a practical solution to the problem of data security in distributed networks i.e. providing effective access control to sensitive data. The algorithm provides additional adaptive parameters: once the user is deleted from the system, his secret key is not only automatically removed but also loses validity. Other properties include re-encryption, access delegation, key expiry period and collaborative access configuration.

REFERENCES

[1] Armbrust, M., Fox, A., Griffith, R., Joseph, A.D., Katz, R., Konwinski, A., Lee, G., Patterson, D., Rabkin, A., Stoica, I. &Zaharia, M., A view of cloud computing. *Communications of the ACM*, 53, pp. 50–58, 2010.

[2] Bethencourt, J., Sahai, A. & Waters, B.,Ciphertext-policy attribute-based encryption. *Proceedings of IEEE Symposium on Security and Privacy*, pp. 321–334, 2007.

[3] Dikaiakos, M.D., Katsaros, D., Mehra, P., Pallis, G. &Vakali, A.,Cloud computing: Distributed internet computing for it and scientific research. *IEEE Internet Computing*, **13,** pp. 10 –13, 2009.

[4] Goyal, V., Pandey, O., Sahai, A. & Waters B., Attribute-based encryption for fine-grained access control of encrypted data. *Proceedings of the 13th ACM conference on Computer and communications security*, pp. 89–98, 2007.

[5] Kamara, S. &Lauter, K., Cryptographic cloud storage. *Proceedings of the 14th international conference on Financial cryptography and data security*, pp. 136–149, 2010.

[6] Liu, Q., Tan, C.C., Wu, J. & Wang, G., Reliable re-encryption in unreliable clouds. *Proceedings of the IEEE Global Telecommunications Conference*, pp. 1–5, 2011.

[7] Sahai, A. & Waters, B., Fuzzy identity based encryption. *Advances in Cryptology: EUROCRYPT*, **3494,** pp. 457–473, 2005.

[8] Secucloud solutions for small and medium-sized companies. Online.secucloud.com/solutions/sccucloud-for-small-and-medium-sized-companies-eng/.

[9] Yu, S., Wang, C., Ren K. & Lou, W., Achieving secure, scalable, and fine-grained data access control in cloud computing. *Proceedings of IEEE INFOCOM*, pp. 534–542, 2010.

[10] Yu, S., Wang, C., Ren, K. & Lou W., Attribute based data sharing with attribute revocation. *Proceedings of the 5th ACM Symposium on Information, Computer and Communications Security*, pp. 261–270, 2010.

Network Security and Communication Engineering – Chan (Ed.)
© 2015 Taylor & Francis Group, London, ISBN: 978-1-138-02821-0

An elasticity evaluation model for cloud storage system

X.N. Zhao, Z.H. Li & X. Zhang
School of Computer Science and Technology, Northwestern Polytechnical University, Xi'an, China

ABSTRACT: Cloud storage system is applied widely to different kinds of applications nowadays, and it is becoming more and more important to evaluate whether a cloud storage system is appropriate. However, due to the high economic and time costs of testing process and the system infrastructure's complexity, there are still no widely accepted benchmarks for cloud storage elasticity evaluation. In this paper, a new evaluating model for the system's elasticity at the IaaS level is proposed. It defines metrics in 3 dimensions, combining with the application type to set the weight of each metric from different dimensions. At last, we discuss this model's effectiveness and evaluating method through experiment and analysis.

KEYWORDS: Elasticity Evaluation, Cloud Storage, Extreme Scalability, Scaling Flexibility.

1 INTRODUCTION

Cost-efficiency and elasticity of data storage have become a key requirement on storage solutions. Cloud storage solutions usually promise high elasticity and low cost. However, it is hard to get both benefits from one system obviously. Therefore, users need an elasticity evaluating system to help them evaluate and compare the numerous cloud products for the best fit.

On the other hand, there are still some challenges to evaluate a cloud storage system's elasticity. For example, cloud storage system usually has a distributed architecture; it has potential availability problems and limited performance benefits. The complex system architecture makes it hard to evaluate. Secondly, cloud storage system has a large scale, and its capacity is at least in the petabyte-scale. For example, our current research project (863 Program) aims to evaluate an exabyte-scale, it is even hard to find a real system in such scale to evaluate its elasticity. Thirdly, as far as we know, even though there are a lot of research works associated with cloud system evaluation, few works focus on cloud storage system evaluation, especially on elasticity evaluation, and it is seldom discussed on either benchmark or evaluation model. At last, it is almost impossible to measure a cloud storage system with traditional method because of unacceptable labor cost and economic cost.

In order to evaluate a cloud system, performance modeling and performance test are the primary work [1]; some works try to discuss performance test process and method [2]. Consistency is an important factor for cloud storage system's availability evaluation. Research[3][4][5] discuss how to guarantee or test the consistency of a cloud storage and evaluate this factor within the Amazon's S3 system or other object storage system, and paper [6] tries to use a benchmark to evaluate the cloud storage system's consistency. Except modeling and test, benchmark is another central topic in cloud evaluation. There are some good works on it. Yahoo proposes a benchmark named YCSB [7] which focuses on evaluating the basic performance of several NoSQL DataBase including HBase, Voldemort, Cassandra and MongoDB which are applied in the cloud storage environment, and CMU's PDL laboratory use its extended version YCSB++ [8] to support complex features and optimizations. In fact, there are some other researches focusing on object storage system, such as Agarwal et al[9] However, all works have a demerit, that is, they usually only take the view from client-centric or server-centric, while no research can give the consideration to both parts and no benchmark is provided for cloud storage elasticity evaluation.

About elasticity, in paper [10], the authors compare various definitions about cloud elasticity, and in their own definition they mainly focus on the challenges of resource allocation. Paper [11] mainly discusses some technologies to improve the cloud database's elasticity and automaticity in order to make the database more cloud-friendly. Moreover, in [12] the authors provide an approach for measuring the elasticity of a cloud, but they only define the elasticity in physics and benchmark of the cloud computing performance. Researches [13] [14][15] all discuss the elasticity at special levels of a cloud system, but none of them did adequate research on storage scalability and its quantitative measurement.

In order to solve these problems, we propose an evaluation model considering both requirements of users and providers for a cloud system at the level of IaaS. The rest of this paper is organized as follows. Section 2 proposes an evaluation model by analyzing all the facts which can affect the cloud system's elasticity, and discusses its metrics in detail. Section 3 introduces how we design the experimental process and analyze the validation by the experiment result. Finally, we conclude this paper and present future work of this study.

2 EVALUATION METRICS AND MODEL

For a cloud system, as we know, it provides levels of service usually, such as IaaS, PaaS, SaaS and so on. In this paper, we discuss the service quality at the IaaS level.

2.1 Analysis and selection on elasticity metrics

When building an evaluating system, the first step is that we must make certain of who cares about the evaluating result, and the second step is to decide which metrics to choose. So before choosing the metrics, we need to make certain of what can affect the special metrics' values.

1 Different views

When we want to judge a cloud storage system, it implies at least two sides of viewpoint, client-centric and service-centric. For the service providers, they prefer to classify the users into some levels, and provide levels of service quality in order to improve the system resources' utility and limit the TCO. For the end user, getting higher service performance with less money is the first need. So, for different roles, the metrics have different priority.

2 Important factors

A cloud system includes network service, computing service and storage service at IaaS level. They provide the system service as a whole, and these 3 parts affect each other, so elasticity can't be evaluated separately. Meanwhile, elasticity is a system level property, so its metrics are affected by both software and hardware. A cloud storage system has its limit capacity at the physical level and it can't provide a performance effective service without a well designed resource allocating policy.

2.2 Elasticity evaluating model

We propose an evaluating system which has three dimensions of metrics to measure. As listed in table 1, they are extreme scalability, scaling flexibility and

scaling level. And we use these metrics to measure the SUT's (System Under Test) elasticity quantitatively. Scaling level is used to evaluate the SUT's elasticity from the view of its architecture.

Table 1. List of main elasticity metrics.

CLASSIFI-CATION	METRICS	DESCRIPTION
scaling level	device level scalability	The precondition of testing and verifying the SUT's extreme scalability. Focus on functions and relationships among the levels.
	file system level scalability	
	system hardware level scalability	
	data center level scalability	
system extreme scalability	extreme capacity reachability	To verify the SUT's maximum capacity is reachable. To measure the system's function is normally useful under the max scale.
	system availability	
system scaling flexibility	online scalability	To test the SUT's performance, service status during scaling.
	scaling efficiency and quality	

2.2.1 Scaling level

It can be considered with 4 levels. For a specific SUT, it may include some of them due to its specific architecture.

1 Storage devices.

It can be a hard disk (attached to a server directly), disk arrays, NAS devices, or an object storage device even a hybrid storage. We use Cap_{Dx} to describe device x's capacity.

2 System hardware.

We use H_y to describe one hardware in the SUT which is connected with storage devices (may be data-manage server, may be a switch or other connector device).

3 File system.

Usually, a cloud storage system chooses the distributed file system, and different file system can manage different scale of files in quantity and size.

4 Data center.

It is hard to organize an exabyte-scale cloud storage system, so usually one SUT includes several data centers. We use S_{DCj} to record data center j's maximum salability.

2.2.2 Extreme scalability

The extreme scalability comes from some components listed above. So, we can calculate a specific

SUT's maximum capacity with equation (1) at physical level.

$$S_{SUT} = \Sigma_j S_{DCj}, \quad S_{DCj} = \Sigma_{x,y} R(H_y, Cap_{Dx}) \quad (1)$$

$R(H_y, Cap_{Dx})$ is used to get the maximum capacity for each isolated hardware which is connected to storage devices. We discuss its capacity at the logical level with equation (2).

$$S_{SUT} = \Sigma_j Cap_{DCj}(N_f, F_s) \quad (2)$$

$Cap_{DCj}(N_f, F_s)$ indicates data center j's logical maximum capacity, and it is a function of maximum amount of files N_f and acceptable file size F_s. Normally, $Cap_{DCj} = N_f \times F_s$, but sometimes it maybe different on a special file system depending on its design.

2.2.3 Scaling flexibility

Scaling flexibility can reflect the cloud storage's elasticity in practical application environment from the user's view better. It must be evaluated more comprehensive including time, space, capacity, cost and other dimensions. We list the whole evaluating objective in table 2. All the metrics from various dimensions in the model constitute the cloud storage system's evaluating system.

Table 2. List of system scaling flexibility metrics.

CLASSIFI-CATION	METRICS	DESCRIPTION
Vertical Scalability	scaling-up	nodes' service ability improved
	scaling-down	nodes' service ability declined
Horizontal Scalability	scaling-out	nodes increased in the system
	scaling-in	scale of system shrinked
Scaling Cost	time cost	time duration of the scaling process
	economic cost	money spent during scaling
Efficiency and Quality	performance earnings	ratio of performance change after scaling (improvement or decline)
	efficiency earnings	average performance earning on unit capacity change
	scaling stability	service stability during scaling process

- Vertical scalability meter

One node i in a cloud storage system, its vertical scalability can be described as:

$$S_{Ver_i} = \Delta Q_i / \Delta Cap_i, \quad S_{Ver} = \frac{1}{n}\Sigma_{i=1}^{n} S_{Ver_i} \quad (3)$$

In equation (3), ΔCap_i represents the capacity variation of node i in the system before and after scaling, and ΔQ_i represents the service quality variation of node i, and $Q_i = IOPS/r$. Here, r means the average response time. S_{ver} is the overall scalability of a SUT, and it is used to measure the SUT's whole scalability quantitively.

- Horizontal scalability meter

It can reflect the system scaling efficiency when the node amount is changed.

$$S_{Hor} = (\Sigma_{i=1}^{n} \Delta Q_i)/(\Sigma_{i=1}^{n} \Delta Cap_i) \quad (4)$$

With ΔQ_i and ΔCap_i mean the same metrics in equation (3), n represents nodes number increased or decreased in the SUT.

- Scaling cost meter

$$C_{SUT} = \Delta T \times \Delta Cap / \Delta Cst_e \quad (5)$$

Here, ΔT means the time cost during the system is scaling, and ΔCst_e is the total economic cost of the scaling process and C_{SUT} is the total scaling cost of a SUT.

- Efficiency and quality meter

Quantitative test of the system's scaling flexibility can be finished by the models introduced above of table 2. In this part, we will discuss how to evaluate and compare some systems' scalability through measuring the efficiency and quality of their scaling process. In a word, scalability must be in a reasonable range which means the users or providers can reach their usage objectives with finite cost and in an acceptable time interval.

1 Performance is the primary problem concerned for both users and providers, here we use performance earning to describe this character, and its computing equation is:

$$PE_{SUT} = P_a - P_b \quad (6)$$

P_a indicates the system performance after scaling, and P_b is the value before scaling. P is a function of IO throughput, utilization of CPU and bandwidth, delay and other parameters.

2 Efficiency earning represents the relationship between performance variation and capacity variation in one scaling procedure. It is

$$E_i = PE_i / \Delta Cap_i \qquad (7)$$

In equation (7), i represents a special node in the SUT, and it means node i's scaling efficiency earning can be measured by its performance earning of unit capacity. So, we get an overall model to measure the SUT's scaling efficiency and quality.

$$EQ_{SUT} = \Sigma_i (w_i \times (PE_i / C_i)) \qquad (8)$$

Here, i represents one scaling node in the SUT, and it may be a provider or a user, wi is the weight of node i, PE_i is node i's performance earning and C_i is the scaling cost of node i. A node's scaling cost can be computed by equation (5).

Certainly, there are many factors affect the scalability, so it is insufficient to evaluate a cloud storage system in reality, so it is better to import some restricted condition, such as location restriction, device type restriction, architecture restriction etc.

3 EXPERIMENTAL AND DISCUSSION

Cloud storage system elasticity need to be evaluated from the system level, as discussed in section 1, because of the system scale and testing cost, and it is impossible to measure all metrics under a real system. On the other hand, some metrics discussed in section 2 can be calculated based on analyzing the characteristics and limitations on the system's architecture and configuration without any actual measuring. In this section, we will discuss how to design a test platform to measure and evaluate a SUT with the least scale and fewest testing cost, then introduce how to measure extreme elasticity with a capacity simulator after discuss evaluating some metrics by analyzing.

3.1 Evaluating method and steps

As shown in Fig. 1, we propose a semi-simulating system as the reference architecture to evaluate a system's elasticity.

It includes 3 types of components, namely the full-real data center, semi-simulating data center and full-simulating data center. Components in a full-real data center are all physical devices, and all the components and their connecting form are from the design of SUT, and we can test the basic performance information and some metrics of scaling flexibility under this type of component. A semi-simulating data center has real components and simulating components. The real parts is organized same to the real data center but with fewer servers, storage and other devices.

With the help of a simulator which can provide larger capacity storage in the semi-simulating data center, we can measure the extreme salability factor of some basic components in the SUT. We can also test and

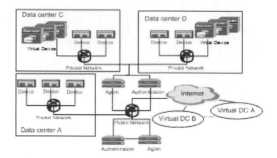

Figure 1. Cloud storage system elasticity testing reference architecture.

verify the extreme scalability of a SUT with this reference architecture.

In fact, the provider cares much about the extreme scalability than the client. Therefore, we design a test process from the provider's view, and the extreme scalability testing basic steps are as follows:

1 Analyze the architecture of the SUT, and evaluate all its nodes' management ability and storage capacity.
2 Test how many data each MDS (meta data server) can manage.
3 According to the critical path's processing ability, conclude the maximum amount of MDS in a SUT.
4 Test the additional bandwidth cost and additional capacity cost ratio in a MDS.
5 Analyze and conclude the SUT's extreme scalability on capacity considering the additional cost information tested in step 4).

The purpose of scaling flexibility evaluation is to evaluate the availability of the system during its scaling process, and the metrics we proposed in the evaluating model mainly focus on performance, time and cost. So, we can use the conventional method to get all the information on them, and we need not design the test steps specifically.

3.2 Measure and analyze

With considering the metrics of scaling level, we can conclude that the value of maximum capacity of a SUT can be calculated by analyzing it without testing any objects. Obviously, device level and file system level salability are the basic parts.

1 Device level extreme scalability. The addressing range of the IO protocol can decide the max capacity at the device level in theory without considering the limitation of max bandwidth, max port amounts and other designed parameters of the device.

2 File system level extreme scalability. We use a tool named dmsetup to test whether BWFS (Blue Whale File System)[16] can scale to the capacity of 10PB level under the Linux OS. As shown in the Fig. 2, the design of virtual storage of our test is described.

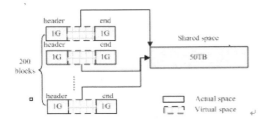

Figure 2.   Simulated storage's configuration.

This virtual storage is designed to simulate 200 LUNs whose capacity is 50TB. Each LUN has three parts, which are header, rail and virtual segment. Header and rail are created by actual space, and virtual segment is virtual space. All the 200 LUNs share one segment on actual storage space with the capacity of 50TB. For the file system, each LUN is a disk with 50TB capacity, so this design can simulate the capacity of 50TB×200 = 10PB, and it only uses 2GB×200 + 50TB = 50.4TB actual storage space. As a result, we can save 99.5% storage. As listed in table 3, when we simulate 1EB level storage, we need to simulate 20000 LUNs and save more than 99.99% actual storage.

Table 3.   Storage saving ratio.

REQUEST	STORAGE	ACTUAL
10PB	50.4TB	99.5%
1EB	90TB	99.99%

As a result, we can save 99.5% storage space. As listed in table III, when we simulate 1EB level storage, we need to simulate 20000 LUNs and save more than 99.99% actual storage space.

Certainly, we confirmed the availability of this simulated system by measuring its IO performance. But, because the IO requests are responded directly by the simulator without reading or writing the actual storage, it is meaningless to discuss the test result in detail. On the other hand, there is no disk product can reach the capacity of 50TB at present, so in this experiment, we only measured the extreme scalability of BWFS instead of cloud storage system.

4   CONCLUSION AND FUTURE WORK

In this paper, we presented a new modeling and measuring approach of cloud storage system elasticity. After analyzing the main factors which affect the cloud storage system, we propose a three-dimension evaluating system and discuss the elasticity evaluating model in detail. In our simulating test environment, it verifies that the model and measure approach is useful and effective with capacity simulator for extreme scalability evaluation.

ACKNOWLEDGMENTS

This work is supported by a grant from the Northwestern Polytechnical University Fundamental Research Foundation No.JC20120209, National High Technology Research and Development Program of China (863 Program) No. 2013AA01A215 and the National Natural Science Foundation of China under Grant No.61033007.

REFERENCES

[1] Z. Yan, L. Mingliang, Z. Jidong, M. Xiaosong, and C. Wenguang, "Cloud versus in-house cluster: Evaluating amazon cluster compute instances for running mpi applications," in Proceedings of International Conference for High Performance Computing, Networking, Storage and Analysis (SC), 2011, Conference Proceedings, pp. 1– 10.

[2] D. Shue, M. J. Freedman, and A. Shaikh, "Performance isolation and fairness for multi-tenant cloud storage," in Proceedings of the 10th USENIX conference on Operating Systems Design and Implementation. 2387914: USENIX Association, Conference Proceedings, pp. 349–362.

[3] T. Kraska, M. Hentschel, G. Alonso, and D. Kossmann, "Consistency rationing in the cloud: pay only when it matters," Proceedings of the VLDB Endowment, vol. 2, no. 1, pp. 253–264, 2009.

[4] W. Golab, M. R. Rahman, A. AuYoung, K. Keeton, and X. Li, "Eventually consistent: not what you were expecting?" Communications of the ACM, vol. 57, no. 3, pp. 38–44, 2014.

[5] D. Bermbach and S. Tai, "Eventual consistency: How soon is eventual? an evaluation of amazon s3's consistency behavior," in Proceedings of the 6th Workshop on Middleware for Service Oriented Computing. 2093186: ACM, Conference Proceedings, pp. 1–6.

[6] W. Golab, M. R. Rahman, A. A. Young, K. Keeton, J. J. Wylie, and I. Gupta, "Client-centric benchmarking of eventual consistency for cloud storage systems," in Proceedings of the 4th annual Symposium on Cloud

Computing. 525935: ACM, Conference Proceedings, pp. 1–2.

[7] B. F. Cooper, A. Silberstein, E. Tam, R. Ramakrishnan, and R. Sears, "Benchmarking cloud serving systems with ycsb," in Proceedings of the 1st ACM symposium on Cloud computing. ACM, Conference Proceedings, pp. 143–154.

[8] S. Patil, M. Polte, K. Ren, W. Tantisiriroj, L. Xiao, J. L´opez, G. Gibson, A. Fuchs, and B. Rinaldi, "Ycsb++: Benchmarking and performance debugging advanced features in scalable table stores," in Proceedings of the 2nd ACM Symposium on Cloud Computing.

[9] D. Agarwal and S. K. Prasad, "Azurebench: Benchmarking the storage services of the azure cloud platform," in Proceedings of IEEE 26th International Parallel and Distributed Processing Symposium Workshops & PhD Forum (IPDPSW), 2012, Conference Proceedings, pp. 1048–1057.

[10] N. R. Herbst, S. Kounev, and R. Reussner, "Elasticity in cloud computing: What it is, and what it is not," in Proceedings of the 10th International Conference on Autonomic Computing (ICAC 13). San Jose, CA: USENIX, 2013, pp. 23–27.

[11] D. Agrawal, A. Abbadi, S. Das, and A. Elmore, Database Scalability, Elasticity, and Autonomy in the Cloud, ser. Lecture Notes in Computer Science. Springer Berlin Heidelberg, 2011, vol. 6587, book section 2, pp. 2–15.

[12] D. M. Shawky and A. F. Ali, "Defining a measure of cloud computing elasticity," in Proceedings of the 1st International Conference on Systems and Computer Science (ICSCS), 2012, Conference Proceedings, pp. 1–5.

[13] B. Suleiman, "Elasticity economics of cloud-based applications," in Proceedings of 2012 IEEE Ninth International Conference on Services Computing (SCC), 2012, Conference Proceedings, pp. 694–695.

[14] J. Gao, P. Pattabhiraman, B. Xiaoying, and W. T. Tsai, "Saas performance and scalability evaluation in clouds," in 2011 IEEE 6th International Symposium on Service Oriented System Engineering (SOSE), Conference Proceedings, pp. 61–71.

[15] E. Stefanov, M. van Dijk, A. Juels, and A. Oprea, "Iris: A scalable cloud file system with efficient integrity checks," in Proceedings of the 28th Annual Computer Security Applications Conference. ACM, 2012, pp. 229–238.

[16] Liu, Z., Meng, X., Xu, L.: Lock management in blue whale file system. In: Proceedings of the 2nd International Conference on Interaction Sciences: Information Technology, Culture and Human, ACM 1142–1147.

Exploring a framework of a cloud-assisted peer-to-peer live streaming

L.Z. Cui, G.H. Li, L.L. Sun & N. Lu
College of Computer Science and Software Engineering, Shenzhen University, Shenzhen, China

ABSTRACT: Although P2P has been the main solution for live streaming distribution, the dynamic restricts the performance. Cloud computing is a new promising solution, which could be introduced as a supplement for P2P. However, for seeking a balance between transmission performance and deployment cost, there has been no mature and integrated solution so far. In this paper, we design a cloud-assisted P2P live streaming system by combing two state-of-art video distribution technologies: cloud computing and P2P. We introduce a two layer framework, including the cloud layer and P2P layer. As for the two respective layers, we propose the corresponding formation and evolution method. The experiment results show that our system can outperform two classical P2P live streaming systems, in terms of the transmission performance and the reduction of cross-region traffic.

1 INTRODUCTION

With the rapid development of Internet, more and more people watch the live television online through live streaming instead of traditional TV. The solution for the live content distribution has obtained a great success, including CDN [1-2] and P2P [3]. CDN is first used to deliver web pages, and then is introduced to streaming distribution [4-5]. The drawbacks of CDN include high renting cost and limited storage. The P2P solution could reduce the deployment cost and provide good scalability. However, since the dynamic is obvious in P2P, the performance can't be guaranteed [6]. Cloud computing provides an alternative method for streaming distribution, which can provide reliable and huge computing capability, storage and bandwidth [7]. The cost is based on the practical usage. It is a good direction on combining cloud computing and P2P in order to seek a balance between transmission performance and deployment cost [8]. As for this direction, there has been no mature and integrated solution so far.

In this paper, we propose a novel cloud-assisted P2P live streaming system, through simultaneously combing two state-of-art video distribution technologies: cloud computing and P2P. The main contributions of this paper can be described as follows. First, we introduce a two-layer framework, including the cloud layer and P2P layer. In the cloud layer, we propose a hybrid tree formation method, including the inter-region tree and intra-region tree. Second, we propose a formation method of the inter-region that organizes each root cloud server of each region into a binary tree as an optimal creation. We also propose a formation method of the intra-region tree that organizes the cloud servers in the same region into a ULF tree structure. Third, as for the P2P layer, we introduce a new metric called *Buffer Value* to evaluate the status of

users' buffer map and propose a formation method based on this metric to improve the transmission QoS and deal with the content availability problem. Finally, through the extensive simulation compared with two practical P2P live streaming systems, CoolStreaming [9] and P2P-locality, our system can achieve better transmission performance and smaller cross-region traffic.

The rest of the paper is organized as follows. Section 2 introduces the framework of our system. In Section 3, we discuss the detailed design of the two layers. The experiments and results are presented in Section 4. Finally, Section 5 concludes our work.

2 THE FRAMEWORK OF A CLOUD-ASSISTED P2P LIVE STREAMING SYSTEM

In this section, the framework of a cloud-assisted P2P live streaming system is presented.

Fig. 1 shows the framework of a cloud-assisted P2P streaming system, which consists of the cloud layer and P2P layer. In the cloud layer, there are dynamically rented cloud servers of different data centers, belonging to geographically different regions. As a user wants to join the live streaming session, it can contact the tracker server and the tracker server will arrange a proper cloud server to it. The chosen cloud server will be in charge of the new user's joining process. After joining, the user will get content from both the cloud servers and the other users in the same P2P swam. The cloud layer also adjusts the number of rented cloud servers and the geographical distribution. In the P2P layer, the users could not only organize the topology and exchange content as the traditional P2P streaming systems, but also seek help from the cloud servers in their region as necessary.

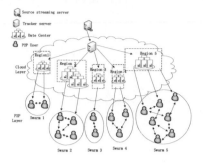

Figure 1. The framework of our cloud-assisted P2P live streaming system.

3 THE DESIGN OF A CLOUD-ASSISTED P2P STREAMING SYSTEM

3.1 *The cloud layer formation and evolution*

Although the tree structure is well known for its delivery low latency, the maintenance overhead is so high that the performance degrades significantly, due to the dynamic of users in P2P system. In our design, since the cloud server is stable, unlike the dynamic P2P user, we adopt a hybrid tree structure for the cloud layer, including the inter-region tree and intra-region tree.

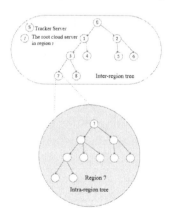

Figure 2. The hybrid tree structure for the cloud layer.

The inter-region tree is a binary tree, and we use tracker server S to be the root of the tree and responsible for the tree formation. Each cloud server in the interior part of the inter-region tree is the root cloud server of a certain region. In current Internet, the latency between different regions (maybe cross ISP) is larger than that within the same region. The streaming latency will accumulate along the tree path, which will degrade the performance of the leaf region. Server S builds the tree as the criterion that makes the sum of the distance between each cloud and servers

S along the path in the tree. The distance means the transmission latency, which of any two regions can be measured by Ping message. Eq. 1 gives the calculation formula of the distance. Di denotes the distance of root cloud server i and server S and parent(i) means the parent server of server i in the tree. The formation of the inter-region tree can be formulated to an optimal problem, described as Eq. 2. This optimization is a classic optimal problem, which can be solved by the dynamic programming or greedy algorithm. We don't describe the details due to space limitation.

$$D_i = D_{i,parent(i)} + D_{parent(i)} \tag{1}$$

$$\min \sum_{i=1}^{n} D_i \tag{2}$$

When a new region is rented, which means the cloud servers are rented in this new region, server S will find a right leaf position in the inter-region tree for the root cloud server of the new region by solving the optimal problem. When all the cloud servers of a certain region leave, server S will be the root cloud server from the tree and adjust the tree structure by solving the optimal problem.

As for the intra-region tree, the root cloud server of each region will organize the servers with its region as the up-left first (ULF) tree structure. In ULF tree, the upload capacity of an upper level node is larger than that of a lower level node and the upload capacity of a left node is larger than that of a right node in the same level. The main idea of our tree topology formation is that the cloud server with larger upload capacity should be placed closer to the root cloud server within the region so as to decrease the height of the tree and thus reduce the delay of lower cloud servers. When a cloud server is newly rented, it will first contact the root cloud server in the same region, and the root cloud server will be responsible to help the newly rented server to the region tree as ULF tree structure. If a root cloud server leaves, the tracker server will select another cloud server as the root and repair the tree structure of this region.

3.2 *P2P layer formation and evolution*

The main design goal of the P2P layer formation is to improve the transmission QoS and enlarge the content diversity so as to provide more mutual assistance among P2P users. When a new user joins the streaming session, it will first contact the server of the same region. The server will then randomly choose some users of this region with enough available upload bandwidth and send the information of these chosen users to the newly joined user, including IP address, buffer map and available upload bandwidth. The new user will then determine the neighbors from the candidate neighbors, and exchange the streaming content with them by a peer-to-peer scheduling strategy. Since we don't focus on the scheduling approach in this paper, we use the traditional rarest-first scheduling strategy.

To improve the transmission QoS and deal with the content availability problem that may exist, the server chooses the candidate neighbors and the user determines the neighbors according to the status of users' buffer map. A server chooses the users with the approximate playback point and high buffer level as the candidate neighbors for the new user. The reason is that these candidate neighbors could provide the streaming content for the new user as much as possible.

For a user i, we introduce a new metric Buffer Value, denoted by BVij, to evaluate the status of its candidate neighbor j. The calculation formula is described as Eq. 3 in Fig. 3, where the value of b_{ik} is 1 or 0.

$$BV_{ij} = \sum_{k=1}^{L} b_{ik} \oplus b_{jk} \qquad (3)$$

When the user i has the block k in its playback buffer, b_{ik} is 1. Otherwise, b_{ik} is 0. L is the length of each user's playback buffer. When a user i determines its neighbors from its candidate neighbors, it first calculates the BV of each candidate neighbor and then sorts the candidate neighbors in descending order by the value of BV. The user will choose the top Nnbr candidate neighbors as its neighbors.

Due to the dynamic of the overlay structure and network transmission fluctuation, the user may not get enough content in its playback buffer. In this case, the frame skipping will happen and the user will not smoothly play the streaming content, which will degrade the user's experience. To avoid the decrease of the performance, when the user finds out the content of its playback buffer below a threshold, the user will request more proper neighbors from its server to maintain the good streaming quality. When the user could still not obtain proper neighbors from the server via several attempts, the user will directly download the streaming content from the server until it finds the proper neighbors.

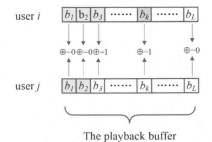

The playback buffer

Figure 3. The playback buffer for the calculation of *BV*.

In the P2P layer, there is the obvious dynamic of the overlay structure since the users could arbitrarily join and leave at any time, even without any hints. The dynamic not only degrades the stability of the overlay topology, but also decreases the transmission performance. To deal with the dynamics, each user will periodically send a keep-alive message to the server. If the sever doesn't receive the keep-alive message from a certain user, it will check whether that user is still in the overlay. When a server decides to quit the cloud layer, it will first transfer all its users to another server of a certain different region, and then stop and leave. To get new neighbors from the new region, the transferred users will execute the same process as they join the overlay.

4 EVALUATION

4.1 Simulation setup

To validate and evaluate our system, we develop a discrete event-driven simulator and conduct several extensive simulations. In our simulation, we introduce 6 regions to simulate the geographical data centers in real world and each user will be randomly assigned to a region with a probability distribution. To simulate the dynamic of users, the users will join and departure 100 person-times during each simulator run. Our system is compared with CoolStreaming and P2P-locality. CoolStreaming is a classic P2P live streaming system with mesh topology and rarest-first scheduling algorithm. P2P-locality adds the locality-awareness based on CoolStreaming.

4.2 Simulation results

A metric LocalityRatio is defined as Eq. 4 to evaluate the reduction of cross-region traffic, which is the ratio of cross-region packets to total packets.

$$LocalityRatio = \frac{CrossRegionPackets}{TotalPackets} \qquad (4)$$

Fig. 4 shows the LocalityRatio of the three systems. Since CoolStreaming doesn't consider the locality factor, its cross-region packets are much higher than that of the other two systems. Although both our system and P2P-locality system are locality-aware, our system gets the assistance of the cloud servers from the intra-region and our P2P layer formation considers the buffer status, which makes the LocalityRatio of our system much lower.

The playback continuity is defined as the fraction of the blocks that could be successfully received before their playback deadlines in each hour. Fig. 5 shows the playback continuity of the three systems. Our system could achieve a higher playback continuity. The reason is that the users could deal with the

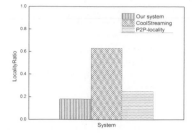

Figure 4. The locality ratio of the three systems.

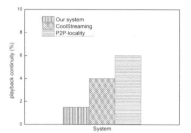

Figure 5. The playback continuity of the three systems.

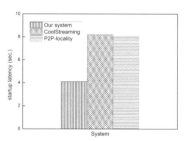

Figure 6. The startup latency of the three systems.

dynamic in time and get the content from the cloud server in the worst case.

The startup latency is the time that a user spends in receiving enough blocks to start playback. Fig. 6 shows the startup latency of the three systems. Our system could achieve a lower startup latency. The reasons can be explained as follows. On one hand, the users gets the content from the cloud servers with enough upload capacity in the same region. On the other hand, in our P2P layer formation, the new user could get the neighbors with the closer play buffer.

5 CONCLUSION

In this paper, to solve the challenges from the diverse and dynamic demand of live streaming users, we propose a cloud-assisted P2P streaming system, the cloud computing systems. We introduce a two-layer framework, namely cloud layer and P2P layer. We propose the detailed design of the formation and evolution for both cloud layer and P2P layer. Through extensive simulation, the effectiveness and efficiency of our system have been validated. In the future, we would like to analyze the rented strategy of the cloud servers. Besides, we will conduct our experiments in real environment via Planet-Lab.

ACKNOWLEDGMENTS

This work was supported in part by National Natural Science Foundation of China (Grant no. 61402294), Guangdong Natural Science Foundation (Grant no. S2013040012895), Foundation for Distinguished Young Talents in Higher Education of Guangdong, China (Grant no. 2013LYM_0076), Major Fundamental Research Project in the Science and Technology Plan of Shenzhen (Grant no. JCYJ20130329102017840 and JCYJ20130329102032059).

REFERENCES

[1] Pallis G, Vakali A. Insight and perspectives for content delivery networks. Commun. ACM, 2006, 49(1):101–106.

[2] Pathan A m K, Buyya R. A Taxonomy and Survey of Content Delivery Networks. Grid Computing and Distributed Systems GRIDS Laboratory University of Melbourne Parkville Australia, 2006, 148:1–44.

[3] Gummadi P K, Saroiu S, Gribble S D. A measurement study of Napster and Gnutella as examples of peer-to-peer file sharing systems. SIGCOMM Comput. Commun. Rev., 2002, 32(1):82–82.

[4] Fortino G, Russo W, Mastroianni C, et al. CDN-Supported Collaborative Media Streaming Control. Multimedia, IEEE, 2007, 14(2):60–71.

[5] Kontothanassis L, Sitaraman R, Wein J, et al. A transport layer for live streaming in a content delivery network. Proceedings of the IEEE, 2004, 92(9):1408–1419.

[6] Y. Huang, TZJ. Fu, DM. Chiu, JCS. Lui, C. Huang. Challenges, design and analysis of a large-scale P2P-VoD System. In Proceedings of ACM SIGCOMM, August 2008.

[7] Hayes B: Cloud computing. Communications of the ACM, 2008, 51(7): 9–11.

[8] Li H, Zhong L, Liu J, Li B and Xu K: Cost-effective partial migration of VoD services to content clouds. In Proceedings of IEEE Cloud Computing, July 2011.

[9] Zhang X, Liu J, Li B, et al. CoolStreaming/DONet: a data-driven overlay network for peer-to-peer live media streaming. Proceedings of INFOCOM 2005.

Network Security and Communication Engineering – Chan (Ed.)
© *2015 Taylor & Francis Group, London, ISBN: 978-1-138-02821-0*

Privacy-preserving data sharing with anonymity in the cloud

Z.M. Zhu & R. Jiang
School of Information Science and Engineering, Southeast University, Nanjing, China

H.F. Kong
The Third Research Institute of the Ministry of Public Security, Shanghai, China

ABSTRACT: Security issues have always been the bottleneck of the development of cloud computing. Recently, identity security has been an important part of the security issues and it is necessary to achieve anonymity for the protection of user identities. In addition, because of the frequent change of the membership, sharing data while providing privacy-preserving in the cloud is still a challenging issue, especially for an untrusted cloud due to the collusion attack. In this paper, we propose a privacy-preserving and anonymous data sharing scheme for dynamic members. Firstly, we propose a secure approach for key distribution without any secure communication channels. Secondly, our scheme can achieve fine-grained access control, any user in the group can anonymously use the source in the cloud and unauthorized users cannot access the cloud. Finally, we can protect the scheme from collusion attack, which means that revoked users cannot get the original data file even if they conspire with the untrusted cloud.

1 INTRODUCTION

In cloud computing, cloud service providers offer an abstraction of infinite storage space for clients to host data [1]. However, security concerns become the main constraint as we now outsource the storage of data, which is possibly sensitive, to cloud providers.

Yu et al. [2] exploited and combined techniques of key policy attribute-based encryption [3], proxy re-encryption and lazy re-encryption to achieve fine-grained data access control without disclosing data contents. However, the single-owner manner may hinder the implementation of applications.

Lu et al. [4] proposed a secure provenance scheme by leveraging group signatures and cipher text-policy attribute-based encryption techniques [5]. However, the revocation is not supported in this scheme.

Liu et al. [6] presented a secure multi-owner data sharing scheme, named Mona. However, the scheme will easily suffer from the collusion attack, which can lead to disclosing other secrets of legitimate members.

Zhou et al. [7] presented a secure access control scheme on encrypted data in cloud storage by invoking role-based encryption technique. Unfortunately, the verifications between entities are not concerned, and the scheme easily suffers from collusion attack.

Ruj et al. [8] proposed an anonymous scheme with decentralized access control by leveraging attribute-based encryption and attribute-based signature scheme. However, the communication channels in this scheme are assumed to be secure. To solve these problems, we propose a secure data sharing scheme, which can achieve secure key distribution and data sharing for dynamic group. The main contributions of our scheme include:

1 We provide a secure way for key distribution without any secure communication channels. The users can securely obtain their private keys from group manager without any Certificate Authorities.
2 We propose a secure data sharing scheme which can be protected from collusion attack. The revoked users are unable to get the original data files once they are revoked even if they conspire with the untrusted cloud.
3 Our scheme is able to support dynamic groups efficiently, and the private parameters of the users do not need to be recomputed and updated.
4 Our scheme can achieve anonymity. The legal users can anonymously utilize the cloud servers to share data with others.

The remainder of the paper proceeds as follows. In section 2, we describe the system model. Our proposed scheme is presented in detail in section 3, followed by the security analysis in section 4. Finally, the conclusion is made in section 5.

2 SYSTEM MODEL

As illustrated in figure 1, the system model consists of three different entities: the cloud, a group manager and a large number of group members.

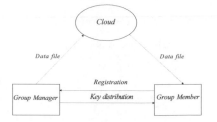

Figure 1. System model.

The cloud, maintained by the cloud service providers, provides storage space. However, the cloud will try to learn the content of the stored data.

Group manager takes charge of system parameters generation, user registration, and user revocation. We assume that the group manager is fully trusted by the other parties.

Group members are a set of registered users that will store their own data into the cloud and share them with others. When a user wants to become a group member, he needs to pay for his registration.

3 THE PROPOSED SCHEME

3.1 Preliminaries

Let G_1 and G_2 be multiplicative cyclic groups of the same prime order p. Let $e : G_1 \times G_2 \rightarrow G_T$ denote a bilinear map constructed with the following properties:

1 Bilinear: For   all $a, b \in Z_p^*$ and $P \in G_1, Q \in G_2, e\left(P^a, Q^b\right) = e(P, Q)^{ab}$.
2 Nondegenerate: $e(P, Q) \neq 1$.
3 Isomorphic: $\psi(Q) = P$ is a computable isomorphic function from G_2 to G_1.
4 Computable: There is an efficient algorithm to compute $e(P, Q)$ for any $P \in G_1, Q \in G_2$.

3.2 Scheme description

Our scheme includes system initialization, user registration, file upload, user revocation, file download and anonymity and provenance.

3.2.1 System initialization

The group manager takes charge of this operation. Firstly, he generates a bilinear map group system $S = \left(p, G_1, G_2, G_T, e(\cdot, \cdot)\right)$. Then he randomly selects a element $Q \in G_1$ and a number $\gamma \in Z_p^*$, and computes $W = Q^\gamma$. At last, the group manager publishes the parameters $\left(S, Q, H, H_1, Enc(\ )\right)$, where H is

hash function: $\{0,1\}^* \rightarrow Z_p^*$, H_1 is hash function: $\{0,1\}^* \rightarrow G_1$, and $Enc(\ )$ is a symmetric encryption algorithm. Besides, the group manager will keep the parameter γ as the secret master key.

3.2.2 User registration

This operation is performed by user and the group manager, which is illustrated in figure 2.

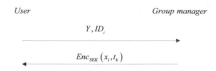

Figure 2. User registration.

When a user i with his identity ID_i comes to register, he randomly selects a number $\xi \in Z_p^*$ and computes $Y = Q^\xi$. Then he sends Y along with his identity ID_i to the group manager.

Having received the message, the group manager performs the following operations:

1 Selecting a random number $x_i \in Z_p^*$ and computing the session key $SEK = Y^\gamma$. Then
2 Sending the encrypted message $Enc_{SEK}\left(x_i, t_k\right)$ to user, where t_k is the time stamp.
3 Adding $\left(ID_i, x_i\right)$ into the group user list, which is stored in his local space. In addition, the group manager checks the group user list and constructs the polynomial function $f_p(x) = \prod_{j=1}^m \left(x - x_j\right) = \sum_{i=0}^m a_i x^i \pmod{p}$ and the exponential function $\{W_0, ..., W_m\} = \left\{G^{a_0}, ..., G^{a_m}\right\}$, where $G \in G_1$.
4 Selecting a random number $s \in Z_p^*$ and computing $PK = P^s$. Then the group manager constructs $EM = \{s \cdot W_0, W_1 \cdots W_m\}$ and the group manager publishes PK and PK, EM will be stored in his local server.

After receiving the message from the group manager, first of all, the user computes the session key $SEK = W^\xi$ and decrypts the message with this session key. Then the user i gets his private parameter x_i, which will be used for file download. Then the user becomes a group member.

3.2.3 File upload

As is illustrated in figure 3, file upload is performed by group member, group manager and the cloud.

For the group member, first of all, he sends a request to the server of the group manager and then the server sends back to the group member. Then the group member recovers the secret number from with his private parameter as the following equation:

$$s \cdot W_0 \cdot \prod_{j=1}^{m} (W_j)^{x_i^j} = s \qquad (1)$$

Then he uses this secret number to encrypt the plaintext M and constructs message $EF = (ID_{group}, ID_{data}, C, t)$, where $C = Enc_s(M)$ and t is a time stamp.

Then the group member sends $\{EF = (ID_{group}, ID_{data}, C, t), sig(EF)\}$ to the cloud, where $sig(EF)$ is the signature of message EF and is computed as the following equation:

$$sig(EF) = P^{\frac{1}{s+H(EF)}} \qquad (2)$$

On receiving the message, the cloud verifies the signature by checking the following equation:

$$e(sig(EF), PK \cdot P^{H(EF)}) \overset{?}{=} e(P, P) \qquad (3)$$

Finally, the data files will be stored in the cloud after successful verification.

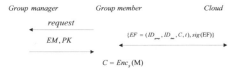

Figure 3. File upload.

3.2.4 *User revocation*

User revocation is performed by the group manager. When a user i with identity ID_i is revoked, the group manager performs the following operations:

1 Removing user i from the group user list in the local storage space.
2 Checking the new group user list, suppose that there are n legal group members in the list. According to the list, the group manager then constructs the polynomial function $f_p(x) = \prod_{j=1}^{n}(x - x_j) = \sum_{i=0}^{n} b_i x^i \pmod{p}$ and the new exponential function $\{W_0, ..., W_n\} = \{G^{b_0}, ..., G^{b_n}\}$, where $G \in G_1$.
3 Selecting a random number $s^* \in Z_p^*$ and computing $PK = P^s$. Then the group manager constructs new $EM = \{s \cdot W_0, W_1 \cdots W_m\}$. And the group manager publishes new PK and new PK, EM will be stored in his local server.

3.2.5 *File download*

This operation is performed by group manager, group member and the cloud, which is illustrated in figure 4. First of all, the group member recovers the secret number s, which is similar to what the group member performs in the phase of file upload. Then the group member sends the request $RE = (ID_{group}, ID_{data}, t_r)$

along with the signature to the cloud, where t_r is the time stamp and the signature is computed as the following equation:

$$sig(RE) = P^{\frac{1}{s+H(RE)}} \qquad (4)$$

On receiving the message, first of all, the cloud verifies the validity of the signature by checking the following equation:

$$e(sig(RE), PK \cdot P^{H(RE)}) \overset{?}{=} e(P, P) \qquad (5)$$

Then the cloud files the corresponding data $\{EF = (ID_{group}, ID_{data}, C, t), sig(EF)\}$ to the group member after successful verifications.

Finally, the group member decrypts the data by leveraging the symmetric encryption key s, which is recovered in the above operations.

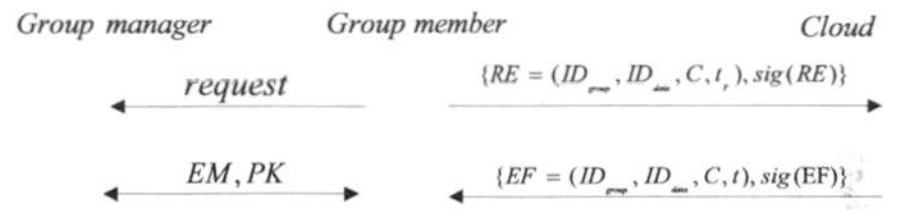

Group manager	Group member	Cloud
← request		$\{RE = (ID_{group}, ID_{data}, C, t_r), sig(RE)\}$ →
← EM, PK		$\{EF = (ID_{group}, ID_{data}, C, t), sig(EF)\}$ →

Figure 4. File download.

3.2.6 *Anonymity*

Any user in the group can anonymously access the cloud for the stored data files with the help of the signature $sig()$. In the phase of file download, the cloud can verify whether the user is a legal user with the help of the public parameter PK, which is generated by the group manager. Meanwhile, the cloud is incapable of learning the real identity of the user through the signature.

4 SECURITY ANALYSIS

We prove the security of our scheme in terms of key distribution, anonymity and data confidentiality.

Theorem 1 (Key Distribution). In our scheme, the communication entities can securely distribute the private keys for users.

Proof. As illustrated in Figure 2, the user can decrypt the message and obtain the private parameter x_i because of the following equation:

$$SEK = W^\xi = Q^{\gamma\xi} = Q^{\xi\gamma} = Y^\xi \qquad (6)$$

If an attacker wants to get the session key, he needs to obtain γ, in other words, given $Q, W = Q^\gamma$, he needs to compute γ. However, this contradicts with the basic Diffie-Hellman problem (BDHP) assumption. Therefore, only the user and the group manager can obtain the correct session key.

Theorem 2 (Anonymity). Benefited from the group signature $sig()$, our scheme can achieve anonymity.

Proof. In the phase of file upload, the group members send message to the cloud along with the signature sig(EF). The cloud verifies the validity of the signature by checking the equation (3). The correctness of above verification equation is based on the following relation:

$$e(\text{sig(EF)}, \text{PK} \cdot \text{P}^{H(\text{EF})}) = e(\text{P}^{\frac{1}{s+H(\text{RE})}}, \text{PK} \cdot \text{P}^{H(\text{EF})})$$

$$= e(\text{P}^{\frac{1}{s+H(\text{RE})}}, \text{P}^s \cdot \text{P}^{H(\text{EF})}) \quad (7)$$

$$= e(\text{P}, \text{P})$$

Therefore, the group member can anonymously upload data files. If an attacker wants to access the cloud, he needs to get the legal secret number, which means that given $P, PK = P^s$ he can compute s, this contradicts with the BDHP assumption.

Theorem 3 (Data Confidentiality). Our scheme can protect data confidentiality, the cloud is unable to learn the content of the stored files $EF = (\text{ID}_{group}, \text{ID}_{data}, C, t)$, even under the collusion with the revoked users.

When a user is revoked, the group manager constructs the new polynomial function $f_p(x) = \prod_{j=1}^{n}(x - x_j) = \sum_{i=0}^{n} b_i x^i \pmod{p}$ and the new exponential function $\{W_0, ..., W_n\} = \{G^{b_0}, ..., G^{b_n}\}$ for existing legal users. From the polynomial function, we can obtain that when $x = V_i$, $\forall i \in [1, n]$, then

$$b_0 + b_1 x + b_2 x^2 + ... + b_n x^n = \sum_{i=0}^{n} b_i x^i = f_p(x) = \prod_{j=1}^{m}(x - x_j)$$

$$= \prod_{j=1}^{m}(x_i - x_j) \quad \forall i \in [1, n] \quad (8)$$

$$= 0$$

And for other values of x, the value of the polynomial function is not zero.

In addition, the group manager constructs $EK = \{s \cdot W_0, W_1 \cdots W_n\}$ and sends it to the cloud with a signature so that others cannot fake it. Only the legal users can recover the re-encryption parameter due to the following equation:

$$s \cdot W_0 \cdot \prod_{j=1}^{n}(W_j)^{x_i^j} = s \cdot G^{b_0} \cdot \prod_{j=1}^{n}(G)^{b_j x_i^j}$$

$$= s \cdot G^{b_0 + b_1 x_i + b_2 x_i^2 + ... + b_n x_i^n} \quad (9)$$

$$= s \cdot G^{f_p(x_i)} = s$$

The revoked user cannot recover the re-encryption key due to the equation $s \cdot G^{f_p(x_i)} \neq s$. Suppose that the revoked user can compute the re-encryption key, which means that given $s \cdot W_0$, he can compute s. Obviously, this contradicts with the BDHP assumption.

5 CONCLUSION

In this paper, we design a secure anonymous data sharing scheme for dynamic groups in the cloud. In our scheme, the users can securely obtain their private parameter from group manager without secure communication channels. Also, our scheme is able to support dynamic groups efficiently, when a new user joins in the group or a user is revoked from the group, the private parameters of the other users do not need to be recomputed and updated. Moreover, our scheme can achieve secure user revocation, the revoked users can not be able to decrypt the encrypted data files and obtain the plain text once they are revoked even if they conspire with the untrusted cloud.

ACKNOWLEDGMENTS

This work is supported by the National Natural Science Foundation of China (NO.61202448), and the Key Laboratory Program of Information Network Security of Ministry of Public Security (No.C14610).

REFERENCES

[1] M. Armbrust, A. Fox, R. Griffith, A. D. Joseph, R. Katz, A. Konwinski, G. Lee, D. Patterson, A. Rabkin, I. Stoica, and M. Zaharia. "A View of Cloud Computing," Comm. ACM, vol. 53, no. 4, pp. 50–58, Apr.2010.

[2] Shucheng Yu, Cong Wang, Kui Ren, and Weijing Lou, "Achieving Secure, Scalable, and Fine-grained Data Access Control in Cloud Computing," Proc. ACM Symp. Information, Computer and Comm. Security, pp. 282–292, 2010.

[3] V. Goyal, O. Pandey, A. Sahai, and B. Waters, "Attribute-Based Encryption for Fine-Grained Access Control of Encrypted Data," Proc. ACM Conf. Computer and Comm. Security (CCS), pp. 89–98, 2006.

[4] R. Lu, X. Lin, X. Liang, and X. Shen, "Secure Provenance: The Essential of Bread and Butter of Data Forensics in Cloud Computing," Proc. ACM Symp. Information, Computer and Comm. Security, pp. 282–292, 2010.

[5] Brent Waters, "Ciphertext-Policy Attribute-Based Encryption: An Expressive, Efficient, and Provably Secure Realization," Proceedings of the 14th International Conference on Practice and Theory in Public Key Cryptography, pp. 53–70, 2011.

[6] Xuefeng Liu, Yuqing Zhang, Boyang Wang, and Jingbo Yang, "Mona: Secure Multi-Owner Data Sharing for Dynamic Groups in the Cloud," IEEE Transactions on Parallel and Distributed Systems, vol. 24, no. 6, pp. 1182–1191, June 2013.

[7] Lan Zhou, Vijay Varadharajan, and Michael Hitchens, "Achieving Secure Role-Based Access Control on Encrypted Data in Cloud Storage," IEEE Transactions on Information Forensics and Security, vol. 8, no. 12, pp. 1947–1960, December 2013.

[8] Ruj, Sushmita, Stojmenovic, Milos, Nayak, Amiya: Decentralized access control with anonymous authentication of data stored in clouds. IEEE Transactions on Parallel and Distributed Systems, v 25, n 2, 384 394, 2014.

Network Security and Communication Engineering – Chan (Ed.)
© 2015 Taylor & Francis Group, London, ISBN: 978-1-138-02821-0

An E-learning system based on cloud computing

S.S. Cheng, Z.G. Xiong & X.M. Zhang
Hubei College of Engineering, Academy of Computer Science, China

ABSTRACT: Cloud computing is becoming an adoptable technology for many of the organizations with its dynamic scalability and usage of virtualized resources as a service through the Internet. Cloud computing is growing rapidly, with applications in almost any areas, including education. At present, most of conventional education forms are becoming not suitable for requirements of social progress and educational development and are unable to catch up with the changes of learning demand in time. This paper describes the architecture of cloud computing platform by combining the features of E-Learning. The authors has tried to introduce cloud computing to e-learning, built an e-learning cloud, and made an active research and exploration for it from the following aspects: architecture, construction method and external interface with the model.

KEYWORDS: Cloud computing, E-learning, cloud based E-learning, architecture.

1   INTRODUCTION

Cloud computing is a virtualized image based on Internet technology which has become a great solution to provide a flexible, scalable, on-demand and dynamic computing infrastructure for many applications. It is a significant technology trend which has a potential to reshape information technology processes and IT marketplace as effective cost. Cloud computing offers the sharing of resources that includes software, storage, data, applications, infrastructure and business processes in the IT marketplace to contest elastic dem and supply.

The emergence of new-fangled generation E-learning systems brings diverse teaching and learning environments for the facilitator and the students. The emerging and elastic cloud computing technology is a special kind of Web service which improves the performance of educational management system in terms of resource sharing and also provides efficient teaching and learning mechanism for different range of students from different environment. Cloud computing provides a platform to share resources in terms of scalable and flexible infrastructures, application development platforms, middleware and business enterprises. It depicts a new enrichment, consumption and delivery model of IT services using various internet protocols at remote computing sites.

With the Rapid growth of contemporary knowledge society there is an enriched demand for the E-learning object models, when the educational information and resource availability is growing drastically. Most of the educational institutions recognize the requirement of adopting new technologies such as E-learning techniques to satisfy the stipulation of the age. Using an amalgamate approach based on the Cloud computing gives the solution to attain the agility and ease access to technology in an affordable manner at institution level.

2   LITERATURE REVIEW

2.1   *Cloud computing*

According to the official NIST (National Institute of Standards and Technology) definition, "cloud computing is a model for enabling ubiquitous, convenient, on-demand network access to a shared pool of configurable computing resources (e.g., networks, servers, storage, applications and services) that can be rapidly provisioned and released with minimal management effort or service provider interaction." The NIST definition lists five essential characteristics of cloud computing: on-demand self-service, broad network access, resource pooling, rapid elasticity or expansion, and measured service. It also lists three "service models" (software, platform and infrastructure), and four "deployment models" (private, community, public and hybrid) that together categorize ways to deliver cloud services. The definition is intended to serve as a means for broad comparisons of cloud services and deployment strategies, and to provide a baseline for discussion from what is cloud computing to how to best use cloud computing.

2.2 E-Learning framework

The E-Learning Framework is a service-oriented factoring of the core services required to support e-Learning applications, portals and other user agents. Each service defined by the Framework is envisaged as being provided as a networked service within an organization, typically by using either Web Services or a REST-style HTTP protocol.

The ultimate aim of the Framework is, for each identified service, to be able to refer to an open specification or standard that can be used to implement the service and also to be able to provide open-source implementation toolkits such as Java and C# code libraries to assist developers.

The intention is not to provide a blueprint for an open-source solution, but to facilitate the integration of commercial, home-grown and open source components and applications within institutions and regional federations, by agreeing common service definitions, data models and protocols.

The Framework began life within JISC as a way of making sense of its funded development activities within the learning and teaching space and to focus future efforts.

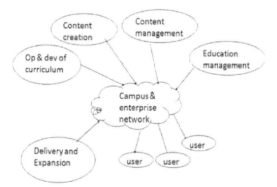

Figure 1.   E-learning framework.

3   CLOUD BASED E-LEARNING ARCHITECTURE

On the other hand, E-learning cloud is a migration of cloud computing technology in the field of e-learning, which is a future e-learning infrastructure, including all the necessary hardware and software computing resources engaging in e-learning. After these computing resources are virtualized, they can be afforded in the form of services for educational institutions, students and businesses to rent computing resources. E-learning cloud architecture is shown in Fig.2[22][23].

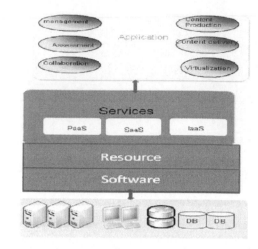

Figure 2.   E-learning cloud architecture.

The proposed e- learning cloud architecture can be divided into the following layers: Infrastructure layer as a dynamic and scalable physical host pool, software resource layer that offers a unified interface for e-learning developers, resource management layer that achieves loose coupling of software and hardware resources, service layer that contains three levels of services (software as a service, platform as a service and infrastructure as a service), application layer that provides with content production, content delivery, virtual laboratory, collaborative learning, assessment and management features.

A. Infrastructure layer is composed of information infrastructure and teaching resources. Information infrastructure contains Internet/Intranet, system software, information management system and some common software and hardware; teaching resources is accumulated mainly in traditional teaching model and distributed in different departments and domains. This layer is located in the lowest level of cloud service middleware, the basic computing power like physical memory, CPU, memory is provided by the layer. Through the use of virtualization technology, physical server, storage and network form virtualization group for being called by upper software platform. The physical host pool is dynamic and scalable, new physical host can be added in order to enhance physical computing power for cloud middleware services. The following Fig. 3 depicts this in a clearer view.

B. Software resource layer mainly is composed by operating system and middleware. Through middleware technology, a variety of software resources are integrated to provide a unified interface for software developers, so they can easily develop a lot of applications based on software resources and embed them into the cloud, making them available for cloud computing users.

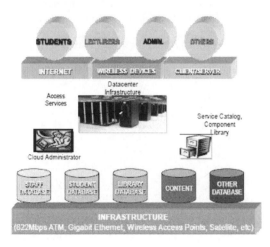

Figure 3. Proposed infrastructure layer in an expandable view.

C. Resource management layer is the key to achieve loose coupling of software resources and hardware resources. Through integration of virtualization and cloud computing scheduling strategy, on-demand free flow and distribution of software over various hardware resources can be achieved.

D. Service layer has three levels of services namely, SaaS (Software as a service), Paas (Platform as a service),IaaS (Infrastructure as a service). In SaaS, cloud computing service is provided to customers. As it is different from traditional software, users use software via the Internet, without needing a one-time purchase for software and hardware, and to maintain and upgrade, simply paying a monthly fee.

E. Application layer is the specific applications of integration of the teaching resources in the cloud computing model, including interactive courses and sharing the teaching resources. The interactive programs are mainly for the teachers, according to the learners and teaching needs, taken full advantage o f the underlying information resources after finishing being made, and the course content as well as the progress may at any time adjust according to the feedback, and can be more effectiveness than traditional teaching. Sharing of teaching resources includes teaching material resources, teaching information resources (such as digital

libraries, information centers), as well as the full sharing of human resources. This layer mainly consists of content production, educational objectives, content delivery technology, assessment and management component [24].

4   EXPECTED BENEFIT FROM THE ARCHITECTURE

The intended advantages derived from the proposed architecture are as follows:

There are numerous advantages when the e-learning is implemented with the cloud computing technology, they are:

A. Low cost: E-Learning users need not to have high end configured computers to run the e-learning applications. They can run the applications from cloud through their PC, mobile phones, tablet PC which have minimum configuration with internet connectivity. Since the data is created and accessed in the cloud, the user need not to spend more money for large memory for data storage in local machines. Organizations also need to pay per use, so it's cheaper and need to pay only for the space they need. This advantage can be visualised from the following Fig. 4 which illustrates the connectivity tier of the proposed architecture.

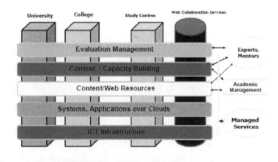

Figure 4.   Connectivity scenario of the institutions in the proposed architecture.

B. Improved performance: Since the cloud based e-learning applications have most of the applications and processes in cloud, client machines do not create problems on performance when they are working.

C. Instant software updates: Since the cloud based application for e-learning runs with the cloud power, the software's are automatically updated in cloud source. So, e-learners always get updated instantly.

D. Improved document format compatibility: Since some file formats and fonts do not open properly in some PCs/mobile phones, the cloud powered e-learning applications do not have to worry about those kinds of problems. As the cloud based e-learning applications open the file from cloud.

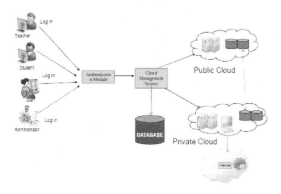

Figure 5.   Interactive mode of the proposed architecture.

E. Benefits for students: Students get more advantages through cloud based e-learning. They can take online courses, attend the online exams, get feedback about the courses from instructors, and send their projects and assignments through online to their teachers. The interactive mode of the proposed architecture is furnished in the Fig. 5.
F. Benefits for teachers: Teachers also get numerous benefits over cloud based e-learning. Teachers are able to prepare online tests for students, deal and create better content resources for students through content management, assess the tests, homework, projects taken by students, send the feedback and communicate with students through online forums.
G. Data security: A very big concern is related to the data security because both the software and the data are located on remote servers that can crash or disappear without any additional warnings. Even if it seems not very reasonable, the cloud computing provides some major security benefits for individuals and companies that are using/developing e-learning solutions.

5   CONCLUSIONS

E-learning is an Internet-based learning process, by using Internet technology to design, implement, select, manage, support and extend learning. Cloud computing is a recently developed and advanced Internet-based computing model. By combination of cloud computing and e-learning, building cloud-based

e-learning system opens up new ideas for the further development of e-learning[5]. In this paper, we introduced cloud computing to e-learning, built an e-learning cloud, and made an active research and exploration for it from the following aspects:[9] its work mode, architecture, construction method, external interface with the business model, challenges and solutions etc.

Our results suggest that the introduction of cloud computing into e-learning is feasible and it can greatly improve the efficiency of investment and the power of management, which can make elearning system development into a virtuous circle and achieve a win-win situation for suppliers and customers.

ACKNOWLEDGMENTS

This paper is supported by Humanities Social Sciences Project (Project Number: 14Q092) of Education Department of Hubei Province, National Natural Science Foundation (Project Number: 61370092), The natural science foundation of Hubei province (Project Number: 2014CFB188).

REFERENCES

[1] F. Jian, "Cloud computing based distance education outlook", China electronic education, 2009.10, Totally 273, pp.39–42.
[2] R. Hua, "Teaching Information System Based on Cloud Computing",Computer and Telecommunications, 2010.02, pp. 42–43.
[3] Y. Juan, S. Yi-xiang, "The Initial Idea of New Learning Society whichBased on Cloud Computing", Modern Educational Technology, Vol.20,No.1, 2010, pp.14–17.
[4] T. Jian, F. Lijian, G. Tao, "Cloud computing-based Design of NetworkTeaching System", Journal of TaiYuan Urban Vocational college, Mar.2010, pp.159–160.
[5] Y. Zhongze, "The basic principles of cloud computing and its impact on education", Satellite TV and Broadband Multimedia, 2010.6, pp.67–70.
[6] W. Xiaomei, J. Xiaoqiang, "Cloud computing on the Impact of Higher Education", Science & Technology Information, 2010.10, pp.397–398.
[7] Z. Zhong-ping, L. Hui-cheng, "The Development and Exploring of E- Learning System on Campus Network", Journal of Shanxi Teacher's University (Natural Science Edition), Vol.18, No.1, Mar.2004, pp.36–40.
[8] W. Jianmin, "Campus Network's E-learning Mode", New Curriculum Research, 2007.08, pp.84–86.
[9] Y. Wei, Y. Rong, "Research of an E-learning System Model Based onAgent", Computer Engineering and Applications, Nov. 2004, pp.156–158.
[10] A. Gladun, J. Rogushina, F. Garcı́a-Sanchez, R. Martı́nez-Be´jar, J.Toma´s Ferna´ndez-Breis, "An application of intelligent techniques and semantic

web technologies in e-learning environments", Expert Systems with Applications 36, 2009, 922–1931.

[11] Y. Li, S. Yang, J. Jiang, M. Shi, "Build grid-enabled large-scale collaboration environment in e-learning grid", Expert Systems with pplications 31,2006, 742–754.

[12] Z. Chengyun, "Cloud Security: The security risks of cloud computing,models and strategies", Programmer, May.2010, pp.71–73.

[13] B. Hayes, "Cloud computing," Comm. Acm, vol. 51, no. 7, pp. 9– 11,2008.

[14] E. Tuncay, "Effective use of Cloud computing in educational institutions," Procedia Social Behavioral Sciences, p. 938–942, 2010.

[15] R. Buyya, C.S. Yeo & S.Venugopal, "Market-oriented Cloud computing: Vision, hype, and reality of delivering IT services as computing utilities," 10th Ieee Int. Conf. High Performance Comput.Comm., p. 5–13, 2009.

[16] M. Lijun, W.K. Chan & T.H. Tse, "A tale of Clouds: Paradigm comparisons and some thoughts on research issues," Ieee Asia-pasific Services Comput. Conf., Apscca08, pp. 464–469, 2008.

[17] K. Praveena& T. Betsy, "Application of Cloud Computing in Academia," Iup J. Syst. Management, vol. 7, no. 3, pp. 50–54, 2009.

[18] K.Λ. Delic & J.A. Riley, "Enterprise Knowledge Clouds," Next Generation Km Syst. Int. Conf. Inform.,

Process, Knowledge Management, Cancun, Mexico, pp. 49–53, 2009.

[19] J. A. Méndcz and E. J. González, "Implementing Motivational Features in Reactive Blended Learning: Application to anIntroductory Control Engineering Course", IEEE Transactions on Education, Volume: PP, Issue: 99, 2011.

[20] S. Ouf, M. Nasr, and Y. Helmy, An Enhanced E-Learning Ecosystem Based on an Integration between Cloud Computing and Web2.0", Proc.IEEE International Symposium on Signal Processing and InformationTechnology (ISSPIT), pages 48–55, 2011.

[21] D. Chandran and S. Kempegowda, Hybrid E-learning Platform based on Cloud Architecture Model: A Proposal", Proc. International Conference on Signal and Image Processing (ICSIP), pages 534–537, 2010.

[22] L. Huanying, "Value and understanding for cloud computing based on middleware", Programmer, 2010.05. pp.68,69.

[23] F. feng, "Cloud-based IT infrastructure of next-generation telecom F",Mobile Communications, 2010, No. 8, pp.76–79.

[24] H. Xin-ping, Z. Zhi-mei, D. Jian, "Medical Informatization Based on Cloud Computing Concepts and Techniques", Journal of Medical Informatics, 2010, Vol.31, No.3, pp.6–9.

Digital forensic investigation of dropbox cloud storage service

A.C. Ko & W.T. Zaw
University of Computer Studies, Mandalay, Myanmar

ABSTRACT: Nowadays, cloud storage services, a type of IaaS, are becoming a popular business paradigm for managing documents anytime and anywhere. It is possible for malicious users to abuse cloud storage services and the number of crime on them has increased rapidly. The investigation of these cloud storage services presents new challenges for the digital forensics community. Therefore procedures for forensic investigation of cloud storage services are necessary. This paper presents a process model for digital forensic investigation of cloud storage services and describes some important artifacts in PCs. Popular cloud storage service Dropbox is used as a case study; we documented a series of digital forensic that they created experiments with the aim of providing forensic practitioners to undertake the cloud storage forensics.

1 INTRODUCTION

In recent years, cloud computing has become popular as a cost-effective and efficient computing paradigm. Cloud storage is a popular option for users to store their data and access it via a range of internet connected devices. Cloud storage services, a type of IaaS, provide users with storage space. There are a range of cloud storage hosting providers, and many offer free cloud storage services, such as Dropbox, Microsoft® SkyDrive®, and Google Drive. ICTs, such as personal computers, laptops, smartphones and tablets, are fundamental to modern society and open the door to increased productivity, faster communication capabilities, and immeasurable convenience.

However, it also changes the way criminals conduct their activities, and vulnerabilities in ICT infrastructure are fertile grounds for criminal exploitation. With the development of the cloud storage, the number of crime on them has increased rapidly [1, 2]. Therefore cloud storage has been identified as an emerging challenge to digital forensic researchers and practitioners. This paper will discuss the need for cloud storage forensics and presents the procedures for forensic investigation of cloud storage services. It will also attempt to discover what evidence can be gathered from Dropbox, including evidence that is located on the computer(s) with Dropbox installed on them as well as evidence that can be gathered from the web portal. Organization of the paper is as follows: Dropbox cloud storage and research questions for digital forensics are described in Section 2. Cloud Storage Forensics Framework is outlined in Section 3. Dropbox forensic investigation is discussed in Section 4. Section 5 draws conclusions from the work conducted.

2 DROPBOX CLOUD STORAGE FORENSIC AND RESEARCH QUESTIONS

2.1 *Dropbox cloud storage*

Dropbox [3] is the most frequently used and popular cloud storage service used by over 50 million people in the world that allows users to backup files to the internet and to share them with other people. It has applications that run on Windows®, Mac, Linux, iPhone, Android and Blackberry. The Dropbox application creates artifacts on a system that may provide pertinent information. The Dropbox servers store many useful logs in regards to account history and a user's file history. Obtaining these artifacts and log files could provide an investigator with valuable evidence. Digital Forensics is a branch of forensic science that is used to encompass the recovery and investigation of data in digital devices. Conventional digital forensic methods are insufficient for investigating cloud storage [4].

2.2 *Research questions for dropbox forensic*

- What artifacts are created during the Dropbox installation process?
- What artifacts are left behind after Dropbox is uninstalled?
- What artifacts are created in VM hard drives, database files and log files of Web Browsers during the accessing VMs?
- What artifacts are left behind in VM hard drives and Web Browsers after uploading and downloading files?
- What artifacts are left behind after Anti-Forensic software is used?

3 FORENSIC FRAMEWORK

Our cloud storage forensic framework is based on the National Institute of Standards and Technology [5]. The framework is iterative, and a practitioner can start one or more iterative processes, while the overall investigation progresses. Our cloud storage forensic framework (Fig 1) is comprised of five phases; preparation, preservation, collection, examination and analysis, and documentation and presentation.

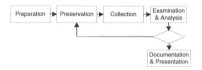

Figure 1. Cloud storage forensic framework.

Preparation: This phase is concerned with preparation of tools, techniques, training, equipment acquisition, search warrants and monitoring authorisation and management support.

Preservation: This phase is concerned with isolation, securing and preserving the state of physical and digital evidence.

Collection: This phase is concerned with recording the physical scene and duplicating digital evidence by using standardized and accepted procedures.

Examination and analysis: This phase is concerned with the examination and analysis of digital evidence. Examination is an in-depth systematic search of evidence relating to the suspected crime and focuses on identifying and locating potential evidence. Analysis determines importance and probative value to the case of the examined product.

Documentation and Presentation: This phase is concerned with complete and accurate reporting steps of findings and the results of the analysis of the digital evidence examination. Documentation is an ongoing process throughout the examination.

4 FORENSIC INVESTIGATION

4.1 *Preparation phase*

To gather the artifacts in relation to the use of Dropbox created in Pcs, we created 18 VMs as shown in Fig 2. We examine and analyze a variety of circumstances of user accessing Dropbox by Windows client software and different Web browsers.

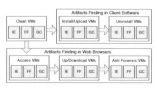

Figure 2. VMs creation for experiment.

Testing environment and summary configurations of VMs are described in Table 1 and 2.

Table 1. Testing environment.

Environment	System Specification
Hosts	Intel® Core™ i7- 2600 CPU @ 3.40GHz, 4GB RAM, 1TB HD, 1Gigabit Ethernet
VMs	1024MB RAM, 50GB, HardDisk (SCSI)
Software	VMware®Workstation 9.0, Windows 7, CCleaner 4.19, Dropbox Client 2.10.30
Web Browsers	Internet Explorer 9, Mozilla FireFox 33.0, Google Chrome 38.0
EDRM File Format Dataset	consists of 381 files covering 200 file formats [6]. Key files: AFTER.doc, LEAR.pdf, piñata.png, Getting Started.pdf

Table 2. Summary configuration of VMs.

VMs	Software Components
3 Clean VMs	Win 8, Browsers (IE,FF,GC)
3 Install/Upload VMs	Win 8, Browsers(IE,FF,GC), Dataset, Dropbox client S/W
3 Uninstall VMs	
3 Access VMs	Win 8, Browsers (IE,FF,GC), Data Set
3 Up/Download VMs	
3 Anti-Forensic VMs	Win 8, Browsers (IE,FF,GC), Dataset, CClearner

4.2 *Preservation phase*

A digital forensic investigation is the ability to conduct analysis on a forensic copy, rather than interact with or alter the original source. At the preservation stage, the physical items are examined and details documented, such as hard drive manufacturer, serial numbers, model numbers, and other information. Data are copied in a forensic manner and verified with an MD5 and/or SHA hash algorithm using write-protection and creating a bit-for-bit copy.

4.3 *Collection phase*

Secure collection of evidence is important to guarantee the evidential integrity and security of information. In order to do analysis, we collect the digital evidence by using disk imaging tool. In this research, a forensic copy of virtual hard drive and each memory file are created by using AccessData FTK Imager [8]. MD5 hash value of each file is calculated and verified with each forensic copy.

4.4 *Examination and analysis phase*

In this phase, forensic copies of the VM hard drives, memory captures were examined and analyzed by using forensic tools [8]. Dropbox cloud storage is a

Web-based service and Internet browsing history is important for forensics investigation. In this research, we examined and analyzed the artifacts of Windows client software and Web browsers stored in PCs. The artifacts found in that VMs are listed in Table 3.

Table 3. Important paths of dropbox client software.

Folder/File	Paths
Installe Folder	"%AppData%\Dropbox
Sync Folder	C:\Users\<username>\Dropbox
Default file	Getting Started.pdf .dropbox (DROPBOX File)
Link Files	C:\Users\<username>\Desktop\ Dropbox.lnk C:\Users\<username>\Links\Dropbox. lnk
Executable & libraries	C:\Users\<username>\AppData\ Roaming\ Dropbox\bin
Prefetch Files	C:\Windows\Prefetch\DROPBOX N.N.NN.EXE-NNNNNNNN.pf C:\Windows\Prefetch\DROPBOX N.N.NN.EXE-NNNNNNNN.pf
Database Files	%AppData%\Dropbox\host.dbx, host. db, unlink.db %AppData%\Dropbox\instance1\ aggregation.dbx, config.dbx, photo.dbx, unlink.db deleted.dbx, sigstore.dbx, filecache.dbx, notifications.dbx,

To examine the result of user uninstalling the client software, we created Uninstall VMs and did the experiment. We found that Dropbox sync folder and file contents remained on the hard drive and was not affected. Only host.dbx file remain in "%AppData%\ Dropbox" and other files are removed. All database files are removed form "%AppData%\Dropbox\ instance1\". The majority of Dropbox's configuration and user info are stored in SQLite database files. Dropbox® Decryptor from Magnet Forensics Tools [9] and SQLite DB Browser [10] are used to decrypt the Dropbox encrypted SQLite databases and to view the contents of the DB file. The important artifacts of Dropbox databases are shown in Table 4.

Table 4. Important artifacts of dropbox databases.

config.dbx

Host_id	Signature value of the Host's ID
Email	DropBox user's email
Displayname	PC Display Name
Userdiaplayname	Dropbox User's name
Dropbox_path	Location of Dropbox User folder

filecache.dbx (file_journal)

id	File's ID assign by DropBox
Server_path	The path of the stored file
Local_filename	The name of the stored file
Local_size	The size of the stored file
Local_mtime	File modification time(UNIX Time)
Local_ctime	File creation time (UNIX Time)
User_id	Dropbox User ID

sigstore.dbx

hash	The hash value of the stored file
size	The size of the stored file

deleted.dbx (File Table)

cache_path	Path of cache file
origin_path	Path of deleted filename in sync folder
Date_added	Deleted date
Size	The size of the stored file
mtime	modification time

The information contain in config.dbx, filecache.dbx, sigstore.dbx and delete.dbx are shown in Fig 3, 4, 5 and 6.

Figure 3. Config.dbx.

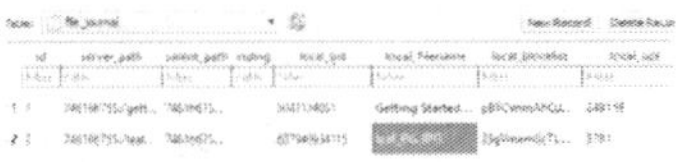

Figure 4. filecache.dbx

Figure 5. sigstore.dbx.

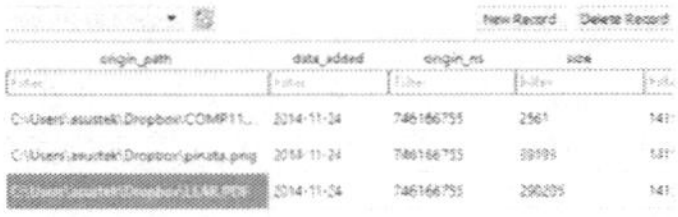

Figure 6. deleted.dbx.

To view the user names and passwords stored by Web browsers (Fig. 7), we used PasswordFox, password recovery tool [11].

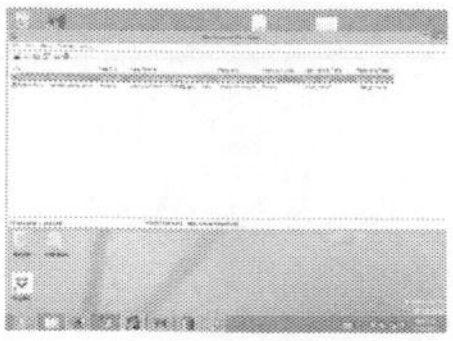

Figure 7.   User name and password.

Figure 8.   Sample RegRipper output of NTUSER.DAT

We examined and analyzed the registry files of all VMS, which was created for doing experiment, using Windows Registry data extraction and correlation tool, RegRipper [12]. There were references to the Dropbox URL, Dropbox Software files and folders, Dropbox default files and EDRM File Format Dataset test files

Table 5.   Important files and paths of web browsers.

Mozilla Firefox 33.0.2

Data	Path
Cache	%LocalAppData%\Mozilla\Firefox\ profile\xxxxx.default\cache2\entries
History	%AppData%\Mozilla\Firefox\profile\ xxxxx.default\places.sqlite %AppData%\Mozilla\Firefox\profile\ xxxxx.default\formhistory.sqlite
Cookie	%AppData%\Mozilla\Firefox\profile\ xxxxx.default\cookies.sqlite %AppData%\Mozilla\Firefox\profile\ xxxxx.default\ permissions.sqlite

IE 9.10.9200.16384

Cache	%LocalAppData%\Microsoft\Windows\ TemporaryInternet Files\Low
History	%LocalAppData%\Microsoft\ Internet Explorer
Cookie	%LocalAppData%\Microsoft\Windows\ cookies

Google Chrome 38.0.2125.111 m

Cache	%LocalAppData%\Google\Chrome\ user data\default\cache
History	%LocalAppData%\Google\Chrome\ user data\default\history
Cookie	%LocalAppData%\Google\Chrome\ user data\default\cookie

in Install/Upload VMs, Access VMs and Download VMs but not in Clean VMs. Dropbox username is not provided in registry file. NTUSER.dat file is a registry file which contains a unique documents folder, custom settings, desktop properties and browsing history. Sample RegRipper output is listed in Fig 8.

From analysis of the 3 anti-forensic VMs, we observed that the references to the file names were removed when all options were used for CCleaner [7].

4.5   *Documentation and presentation phase*

According to the experiment, we found that a variety of data remnants were located when user used Dropbox to access and store data. The important files, paths, artifacts of Dropbox cloud storage service are documented as shown in above tables. This information enables a practitioner to conduct forensic analysis and will assist to determine if Dropbox client software has been used.

5   CONCLUSION

Cloud storage services are becoming more prevalent, and not just for businesses – end- and home-users are taking advantage of opportunities to automate backups, make files available offline or from any computer, share files and photos, and so on. It is possible for malicious users to abuse cloud storage services and the number of crime on them has increased rapidly. In this paper, cloud storage forensic framework is presented and popular cloud storage service Dropbox is used as a case study. We conducted the experiment on Dropbox cloud storage and highlighted a lot of useful information that can be found by analyzing artifacts left by Cloud Storage clients. This methodology is helpful in the investigation of cloud storage services.

REFERENCES

[1] D. Quick, B. Martini and R. Choo, (2013), Cloud Storage Forensics, 1st Edition, ISBN: 9780124199705, Syngress.
[2] H.Chung, J. Park, S. Lee, & C.Kang, (2012), Digital forensic investigation of cloud storage services. *Digital investigation*, 9(2), 81–95.
[3] B. Martini and K-KR Choo, (2012), An integrated conceptual digital forensic framework for cloud computing. Elsevier- Digital Investigation, volume 9(2), pp. 71–80, 2012.
[4] Dropbox: www.dropbox.com
[5] K. Kent, S. Chevalier, T. Grance, & H. Dang, (2006), Guide to Integrating Forensic Techniques into Incident Response. SP800-86, Gaithersburg. U.S. Department of Commerce.
[6] EDRM File Format Data Set: http://www.edrm.net/ resources/data-sets/edrm-file-format-data-set
[7] CCleaner:http://www.piriform.com/ccleaner
[8] FTKImager:http://accessdata.com/solutions/ digital-forensics/forensic-toolkit-ftk
[9] DropboxDecryptor:http://www.magnetforensics.com
[10] SQLiteDB Browser: http://sqlitebrowser.org
[11] Password Recovery: http://www.nirsoft.net
[12] RegRipper: http://regripper.wordpress.com

Network Security and Communication Engineering – Chan Chan (Ed.)
© 2015 Taylor & Francis Group, London, ISBN: 978-1-138-02821-0

Cloud-based material management system for small-medium size printing companies in China

P. Fang
Changjiang Professional College of Electromechanical Engineering, Wuhan, China

Y.B. Zhang
Department of Mechanical Engineering, Beijing Institute of Graphic Communication, Beijing, China

ABSTRACT: Material management is one of the most important factors affecting printing companies' profits in China. More and more printing companies need innovational methods to improve printing management in order to keep their competitive edge. But in China, most of small and medium size printing companies have difficulties in material management. We analyze material management problems in printing companies and present a cloud-based material management to manage printing material effectively and efficiently. We focus on managing raw material within printing company to simplify the prototype system. Business process management method is used to identify material management related processes. Business process model and notation are adapted to model and express identified processes. A printing material management prototype system is designed and deployed, with software as a service model. Results show that this model could reduce cost of printing management and help printing companies manage the raw material effectively and efficiently. Much work will be focused on extending the scope of material management in the future.

KEYWORDS: Material management; business process management; cloud computing; software as a service; printing.

1   INTRODUCTION

Competition among printing companies is fiercer than ever in China. On one hand, the costs of printing have been increasing, especially the cost of paper and manual work. The large number of printing companies is another reason for the increasing competition [1]. For the past decade, the price for the same printing order remains almost the same but the manual cost has doubled, which means that the profit for printing companies has been deduced largely. On the other hand, with the rapid development of digital devices and network, especially the Internet and mobile devices, more and more information is available in digital format instead of paper-printed format. This reduces the requirement of traditional printing.

Innovation is the key for printing plants to survive current competition. There are more than 100,000 printing companies and 95 percent of them are private-owned ones [2]. Big companies have adopted new information technology to enhance their competitive edge. These companies have used enterprise resource planning (ERP) systems to manage all their information [3], including customer information, employee, order, finance, and so on. More than 95 percent printing companies in China have used computers in their daily work according to our surveys. More than 80 percent companies in Beijing have used ERP to manage their information.

However, information technologies have not achieved expected results in most printing companies. Most printing companies just replace the daily paper working documents with digital ones. Workers think that it is more convenient to send and copy digital documents than paper ones. At the same time, they face problems with digital documents, such as outdated documents, inconsistence among related documents and version controlling problem.

Printing material is the most valuable part of a printing order. Paper, ink, plate,s and other printing related material cost more than 70 percent of entire cost of an order. Effective and efficient management of printing material will reduce waste and will help printing company reduce cost and increase competitive edge.

In this paper, we will provide a cloud-based printing material management system framework. With this framework, printing plants can manage their material effectively without extra investment in hardware and software system.

2 MATERIALS MANAGEMENT

Materials management is a part of business logistics and refers to overseeing the location and movement of physical items or products. There are three main elements associated with such management, namely spare parts, quality control, and inventory management [4].

Inventory management is the accurate tracking of all materials in a company's inventory. A company has typically purchased these items from other suppliers. There are three possible areas of loss that can be reduced through effective inventory management: shrinkage, misplacement, and short shipments.

Shrinkage is a general materials management term used to describe the loss of materials once they have reached a company. This type of loss is due to theft or damage. Loss through misplacement is most commonly found in very large organizations or warehouses. Material is received by the shipper and then frequently moved to another location by the distribution staff. However, if it is moved to the wrong location, it can get lost and counted as never having been received.

Short shipments occur when the quantity received is less than the number on the packing slip. This must be identified and corrected as soon as possible, preferably before the shipper receives the package. The more time passes before it is realized, the greater is the risk of a supplier insisting that the product was shipped correctly, and the loss occurred within the customer's warehouse.

Materials management is important in large manufacturing and distribution environments, such as warehouses, where there are multiple parts, locations, and significant money invested in these items.

3 LITERATURE VIEW

More and more researchers have realized that information technology will be the key to success of traditional printing and press industry. With the increment of digital information and smart devices, traditional paper-printed documents are decreasing. Traditional printing and press companies have faced the fact of declining business. Information technologies have been used in many aspects of printing, e.g. improving printing quality and producing management.

In China, most printing companies still focus on printing related devices. About 95 percent of Chinese printing companies are private ones. If a printing device goes out of work, the ongoing work has to be stopped. The owner of the company can see the direct effect. On the other hand, lack of finance support is one characteristic of a private company. It is difficult to calculate the return of the investment on information technology. It is natural for private company owners to focus on the printing related devices.

Enterprise Resource Management (ERP) system is one of the most popular systems for make-to-order companies [5]. At present, there are 50 percent of printing companies which have adopted ERP. There are more than 100 printing ERP systems in China. The prices of ERP range from less than 10,000 RMB to more than 1,000,000 RMB.

Job Definition Format (JDF) specification gets more focused recently. JDF is drafted by The International Cooperation for the Integration of Processes in Prepress, Press, and Post-press Organizations (CIP4). The mission of CIP4 is to foster the adoption of process automation in the printing industry. JDF specification has three main functions including providing a single common language that supports the lifecycle of a print job, providing a command and control language for devices on the shop floor and providing a flexible methodology for constructing workflows and providing the command, control, and configuration of plant automation and job production [6].

JDF can be used in printing ERP systems. Some researchers discuss JDF-based ERP system [7]. This effort helps to combine job information and production information and bring benefits for precious management.

Material management is an isolated system in most small-medium size printing companies. Printing companies have realized that material management is the key factor for business success. Material information was written on paper during the past few decades. At present, with the common use of personal computers, most of printing companies adopt material management system or electronic documents such as office excel or word documents. Material information management system helps printing companies manage materials conveniently. Even for printing companies who use office documents to manage material, it is more convenient to search and find material information with electronic documents than with paper ones.

To solve the problems mentioned above, we focus on material management. We design a printing material management. With cloud computing technology, we provide the system as a service so that printing companies can use the system without investment on the software system and the hardware running the system as a traditional one. Printing companies do not need IT stuff to maintain the system.

4 METHODS

4.1 Business process management

Business process management (BPM) is a systematic approach to making an organization's workflow more effective, more efficient, and more capable of adapting to an ever-changing environment [8]. A business process is an activity or set of activities that will accomplish a specific organizational goal.

BPM is often a point of connection within a company between the line-of-business (LOB) and the IT department. Business Process Execution Language (BPEL) and Business Process Management Notation (BPMN) were both created to facilitate communication between IT and the LOB. Both languages are easy to read and learn, so that business people can quickly learn to use them and design processes. Both BPEL and BPMN adhere to the basic rules of programming, so that processes designed in either language are easy for developers to translate into hard code.

4.2 *Business process model and notation*

Business Process Model and Notation (BPMN) is a method of illustrating business processes in the form of a diagram similar to a flowchart. BPMN was originally conceived and developed by the Business Process Management Initiative (BPMI). It is currently maintained by the Object Management Group (OMG).

BPMN provides a standard and easy-to-read way to define and analyze public and private business processes [9]. BPMN provides a standard notation that is readily understandable by management personnel, analysts, and developers [10]. The original intent of BPMN was to help bridge communication gaps that often exist between various departments within an organization or enterprise. BPMN can also help to ensure that XML (Extensible Markup Language) documents designed for the execution of diverse business processes can be visualized with a common notation.

A standard Business Process Model and Notation (BPMN) will provide businesses with the capability of understanding their internal business procedures in a graphical notation and will give organizations the ability to communicate these procedures in a standard manner. Furthermore, the graphical notation will facilitate the understanding of the performance collaborations and business transactions between the organizations. This will ensure that businesses will understand themselves and participants in their business and will enable organizations to adjust to new internal and B2B business circumstances quickly.

4.3 *Software-as-a-service*

Software as a Service (SaaS) model of cloud computing technology is used to deploy the printing management system. With SaaS model, printing companies can gain access to the printing material management system from anywhere at any time with a web browser if internet connection is available [11]. Printing companies do not need to maintain the hardware and running system.

Cloud computing is a general term for anything that involves delivering hosted services over the Internet. These services are broadly divided into three categories: Infrastructure-as-a-Service (IaaS), Platform-as-a-Service (PaaS) and Software-as-a-Service (SaaS). The name cloud computing was inspired by the cloud symbol that is often used to represent the Internet in flowcharts and diagrams.

A cloud service has three distinct characteristics that differentiate it from traditional hosting. It is sold on demand, typically by the minute or the hour; it is elastic–a user can have as much or as little of a service as they want at any given time; and the service is fully managed by the provider (the consumer needs nothing but a personal computer and Internet access). Significant innovations in virtualization and distributed computing, as well as improved access to high-speed Internet and a weak economy, have accelerated interest in cloud computing.

A cloud can be private or public. A public cloud sells services to anyone on the Internet. Currently, Amazon Web Services is the largest public cloud provider. A private cloud is a proprietary network or a data center that supplies hosted services to a limited number of people. When a service provider uses public cloud resources to create their private cloud, the result is called a virtual private cloud. Private or public, the goal of cloud computing is to provide easy, scalable access to computing resources and IT services.

In the software-as-a-service cloud model, the vendor supplies the hardware infrastructure, the software product and interacts with the user through a front-end portal. SaaS is a very broad market. Services can be anything from web-based email to inventory control and database processing. Because the service provider hosts both the application and the data, the end user is free to use the service from anywhere.

While on-premise business process management (BPM) has been the norm for most enterprises, advances in cloud computing have leaded to increased interest in on-demand, software as a service (SaaS) offerings.

5 A CASE STUDY

We set two principles when we designed the cloud-based printing material management system. Efficiency was the first principle. We did not start from scratch. We would not build the business process engine and would not build the Platform as a service by ourselves. Focusing on core printing material business processes was another principle. There are many different processes for even a small size printing company. It was impossible to design all related process in a short time.

There are two important processes related with material management. One is material procurement process and the other is printing order management process. All material is used during one of prepress, press, and post press processes. The need for any related material is processing during the execution of an ordering as shown in Fig .1.

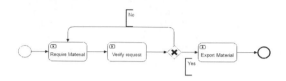

Figure 1.   Material consuming process.

The material procurement and material quality checking process is shown in Fig .2.

Figure 2.   Material procurement process.

An open source business process engine named Activiti is used as business process engine. Activiti is implemented in Java programming and provides flexible application programming interfaces.

A free and restricted Platform as a service named Cloudbees is used as PaaS to deploy our designed printing management system. Cloudbees mainly support different web servers for Java programming language. It reduces the learning cures for cloud computing for java web developers. We designed a printing material management web system with java language and deployed the system on Cloudbees.

Our designed printing management system provided as a material management service for BIGC printing factory. With explicitly expressed material management processes, users could understand the processes better. Material information could be updated instantly. Employees could access to the system from anywhere at any time from personal computer, smart mobile phone, or tablet as long as internet access is available. The printing factory could reduce cost of system maintenance.

6   CONCLUSION

We designed a printing material management system and deployed software as a service model. Two core processes related to printing material management were defined with business process management method. The two processes explicitly expressed the material management activities and logics and helped the printing company to understand material management processes better. The processes would provide a base for future material management processes optimization. With cloud computing technology, it would be possible for different sized printing companies to adopt information technology so as to improve material management efficiency. The value of this paper is that it provides a reference for other printing management applications.

REFERENCES

[1] He mingxin, "Printing companies should be leading group going aboard," China Publishing Journal, , vol. 6, June. 2012,pp.19–21.
[2] Ren zhicheng, "Innovation is the key for printing industry development," China Pbulishing Journal, vol. 4, April. 2011,pp.27–30.
[3] SUN Liang. "Dyeing and Printing Company Production ERP System Design," Journal of Donghua University(Natural Science),vol. 36, DEC. 2010. pp.708–712.
[4] M Christopher.Logistics and supply chain management,Pearson UK,2012.
[5] Aslan, B., M. Stevenson, and L.C. Hendry, Enterprise Resource Planning systems: An assessment of applicability to Make-To-Order companies. Computers in Industry, vol. 63,2012.pp. 692–705.
[6] JDF Specification. http://www.cip4.org/document_archive/download_request.php?did=2591.
[7] Yun Fei Zhong, Xiao Qi Peng, Xi Yu Xiao, Qiao Li Zhang, Qian Kun Cheng, "A Preliminary Study of JDF-Based ERP System in the Printing Enterprises,"Advanced Materials Research, vol. 102–104,2010 pp.851–855.
[8] Bruce Silver ,BPMN Method and Style, 2nd Edition. Cody-Cassidy Press,2011.
[9] Chinosi, M. and A. Trombetta, BPMN: An introduction to the standard. Computer Standards & Interfaces, vol.34,2012. pp. 124–134.
[10] Alexander Benlian,Thomas Hess,Peter Buxmann, Software-as-a-Service,Springer.2010.
[11] Chinosi, M. and A. Trombetta, BPMN: An introduction to the standard. Computer Standards & Interfaces, vol.34,2012. pp. 124–134.

Network Security and Communication Engineering – Chan (Ed.)
© 2015 Taylor & Francis Group, London, ISBN: 978-1-138-02821-0

Mobile cloud platform for big data analytic

N.W. Win & T. Thein
University of Computer Studies, Yangon, Myanmar

ABSTRACT: Big data technologies are to extract value from very large data volumes of a variety of data by enabling analysis. With the fast deployment of cloud computing to mobile devices, big data analysis is shifting from personal computer to mobile devices. For this stream of change of technology, a new mobile cloud platform which can offer seamless connectivity and high query processing performance is needed. This paper presents a mobile cloud platform for accessing big data in the cloud through mobile devices. RESTful web service technology is applied to meet seamless connectivity between mobile device and cloud storage.MapReduceTransformation Model is applied to provide high query processing performance. The query processing performance of proposed platform is evaluated and compared with the performance of other traditional high level query languages, such as Pig, Hive, and Jaql. Comparative results show that our platform can provide better query processing performance.

KEYWORDS: Big data analytic, HadoopMapReduce, Mobile cloud, RESTfulweb service.

1 INTRODUCTION

With the fast deployment of cloud computing with mobile device, big data analyticsmoves from personal desktop computer to mobile devices including Tablet, Smart Phones and so on.In smart devices market, Android is the most widely used mobile Operating System which is based on the Linux Kernel and developed by Google [9]. The well-known technology for accessing data from cloud storage by using a mobile web service is Mobile Cloud Computing.

Recently, this mobile cloud concept has been investigated as a powerful and more importantly, an energy-efficient platform to support massively parallelizable applications. The potential for integrating a mobile cloud platform with the existing cloud computing architecture to form a hybrid system for Big Data has also been the focus of significant recent research.

Mobile web services allow deploying, discovering and executing of web services in a mobile communication environment using standard protocol. Web service can be classified into two main categories: RESTful and SOAP-based web services.

In big data analytics, Hadoop is becoming the core technology to solve the business problem for large organizations with cloud storage. A commonly used architecture for Hadoop consists of client machines and clusters of loosely coupled commodity servers that serve as the HDFS distributed data storage and MapReduce distributed data processing.

MapReduce is the programming model for data processing with a simplified and abstracted basic operation of Map and Reduce, and many complicated problems can be solved. Programmers who are not familiar with parallel programs can easily perform parallel processing for data. It supports high throughput by using many computers. As the core technology of the Hadoop is the MapReduce parallel processing model, all of the high level query languages (HLQLs) that run on Hadoop are the MapReduce based query languages.

A number of HLQLs have been constructed on top of Hadoop to provide more abstract query facilities than using the low-level Hadoop Java based API directly. Pig, Hive, and JAQL are all important HLQLs. Programs written in these languages are compiled into a sequence of MapReduce jobs; to be executed in the HadoopMapReduce environment. Apache Hive is an open-source data warehousing solution built on top of Hadoop and it provides an SQL dialect, called Hive Query Language (HiveQL) for querying data stored in a Hadoop cluster. Apache Pig provides an engine for executing data flows in parallel on Hadoop. It includes a language, PigLatin, for expressing these data flows. Jaql is a declarative scripting language for analyzing large semi-structured datasets in parallel using Hadoop's MapReduce framework. It consists of a scripting language and a compiler, as well as a runtime component for Hadoop.

There are many types of existing big data analytic platforms [1, 5, 7] for large scale data. Most of them are based on MapReduce, distributed file system.

Our mobile cloud platform for big data analysis provides a solution to reduce the query processing time of traditional query languages by using a MapReduce Transformation Model. To achieve the seamless connectivity between mobile and cloud storage, we used RESTful web service technology. By using this platform, users send a request from their mobile device and get back the results without noticeable amount of time.

2 MOBILE CLOUD COMPUTING AND BIG DATA ANALYTIC CONCEPT

2.1 *Mobile cloud computing*

Mobile cloud computing (MCC) at its simplest, refers to an infrastructure where both the data storage and data processing happen outside of the mobile device. Mobile cloud applications move the computing power and data storage away from the mobile devices into powerful and centralized computing platforms located in the clouds, which are then accessed over the wireless connection based on a thin native client. Improving data storage capacity and processing power: it enables mobile users to store/access large data in the cloud and helps to reduce the running cost for compute intensive applications [4].

2.2 *Big data analytics*

Big data analytics requires massive performance and scalability - common problems that old platforms cannot scale to big data volumes, load data too slowly, respond to queries too slowly, lack processing capacity for analytics and cannot handle concurrent mixed workloads [3].

There are two main techniques for analyzing big data: store and analyze, and analyze and store [7]. The store and analyze integrates source data into a consolidated data store before it is analyzed. Analyze and store technique analyzes data as it flows through business processes, across networks, and between systems.

2.3 *Hadoop distributed file system and MapReduce*

The Hadoop distributed file system (HDFS) is designed to store very large datasets reliably and all servers are fully connected and communicate with each other using transmission control protocol (TCP) based protocols. HadoopMapReduce is a software framework for easily writing applications which process vast amounts of data in parallel on large clusters of commodity hardware in a reliable, fault-tolerant manner. The framework sorts the outputs of the map, which are then input to the reduce tasks. Typically, both the input and the output of the jobs are stored in a file

system. The framework takes care of scheduling tasks, monitoring them and re-executing the failed tasks [8].

2.4 *RESTful-based web service*

REST stands for Representational State Transfer: it is a resource oriented technology and it is defined by Fielding in [2] as an architectural style that consists of a set of design criteria that define the proper way for using web standards such as HTTP and URIs. Although REST is originally defined in the context of the web, it is becoming a common implementation technology for developing web services. RESTful web services are implemented with web standards (HTTP, XML and URI) and REST principles. REST principles include addressability, uniformity, connectivity and stateless. RESTful web services are based on uniform interface used to define specific operations that operate on URL resources.

3 MOBILE CLOUD PLATFORM ARCHTECTURE FOR BIG DATA ANALYTIC

Our mobile cloud platform consists of a web service layer, application layer, processing layer and storage layer.

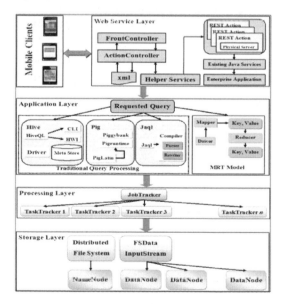

Figure 1. Mobile cloud platform architecture.

The web service layer containsRESTful web service which isidentified by logical URLs. After that it sends the HTTP request to the cloud storage. The application layer organizes with MRT (MapReduce Transformation Model) and it receives the user's request query from web service layer and works out the analytical process with processing layer and storage layer. The processing layer contains JobTracker and TaskTracker nodes and which works as MapReduce framework. The Storage layer consists of NameNode and DataNode. The DataNode clusters are used to store the big data.

The mobile client application invokes the service with query by using Java-based or any other client as long as it is able to support the HTTP methods. After invoking the service, utility services required by the framework uses the Front Controller for centralized request processing and uses the RESTServiceServlet for processing the input query. It supports common HTTP methods like GET, PUT, POST and DELETE. After that, RESTActionController manages the core functionality of loading the services and framework configuration, validation of request and mapping the requests with configured REST actions and executing the actions. This RESTActionController uses RESTConfiguration for loading and caching the framework configuration as well as the various REST services configurations at run-time. The REST action mapping that specified in the configuration file is stored in RESTMapping component and it consists of the URI called by the client and the action class which does the processing. After processing RESTMapping, the next component is ActionContext and it encapsulates all the features required for execution of the REST action. It assists developers in providing request and response handling features so that the developer has to only code the actual business logic implementation. It hides the protocol specific request and response objects from the Action component and hence allows independent testing. It also provides a handle for the XML Binding Service so that Java business objects can be easily converted to XML and vice-versa based on the configured XML Binding API. The RESTActionController configures this component dynamically and provides it to the Action component. The XML Binding component encapsulates the Java XML Binding mechanism and provides a uniform interface to convert Java business object to XML and vice-versa. It allows configuration of any XML binding mechanism. In this platform, querying process takes place on cloud backend and corresponding results are sent back to the mobile user.

In the MapReduce transformation model, user request string that is taken by the web service is compiled in Mapper and Reducer. The Mapper class runs mapping process and produce key, value output and this output are passed to Reducer class as its input. The Reducer class runs reducing process and produces the final output. This output is sent back to the user by using a web service. The work within the MapReduce transformation model is applied data from Hadoop cluster.

The traditional analytic method takes a large amount of time to receive the output result. To improve the query processing performance, reducing query execution time focuses on this platform.

4 PERFORMANCE EVALUATION

4.1 Experiment environment

We implemented the mobile platform for big data analysis and evaluated on different Operating Systems and different high level query languages. To build a storage cluster, we created 12 VMs for NameNode, Secondary NameNode, DataNode, JobTracker and TaskTracker. The specifications of devices and necessary software component used in mobile cloud infrastructure, and data set used in MapReduceprocessing are described in table 1.

Table 1. Experiment parameters.

Parameters	Specification
OS	- Ubuntu 12.04 Linux, - Red Hat Enterprise Linux 6.4
Host Specification	Intel ® Core i7-2600 CPU @ 3.40GHz, Intel ® Core i7-3770 CPU @ 3.40GHz, 8GB Memory, 1TB Hard Disk
VMs Specification	1GB RAM, 50 GB Hard Disk
Mobile Device Specification	Huawei G730-U00, Android OS version 4.2.2 (Jelly Bean), Quad-core 1.3 GHz Cortex-A7, 4GB internal memory
Software Component	- Hadoop 1.1.2 - Hive 0.9.0, Pig 0.12.1, Jaql 0.5.1
Data Set	US census dataset [6] - 114 GB 47 population tables, 14 housing tables, 10 population tables

4.2 Evaluation and results discussion

The query processing time of our MapReduce Model is compared with query processing time of other query languages.Sample queries of HiveQL, PigLatin, and Jaql for query processing are shown in Figure 2.

The traditional big data analytic platform performs analytics directly over HadoopMapReduce framework. All the queries for analytics are executed as Map and Reduce jobs over big data placed into HDFS file system. HadoopMapReduce processes these data with user friendly query languages such as HiveQL, PigLatin and Jaql to get analytical results.

In our big data analytic platform, we use MRT model and processthe data that are stored in HDFS on commodity servers to extract information. At this time, we record the query execution time of MapReduce process.

The **HiveQL** (Hive Query Language) is
hive> create table population (ID int, FILEID string, STUSAB string, CHARITER string, CIFSN string, LOGRECNO string, POPCOUNT int) row format delimited fields terminated by '\,' stored as textfile;
hive> load data inpath '/user/root/Rec250000.csv' overwrite into table population;
hive> select STUSAB, sum(POPCOUNT) from population group by STUSAB;

The **PigLatin** is
grunt> population = load '/user/root/families.csv' using PigStorage(',') as (ID: int, FILEID:
chararray, STUSAB: chararray, CHARITER: chararray, CIFSN: chararray, LOGRECNO: chararray, POPCOUNT: int);
grunt>grouped = group population by STUSAB;
grunt> result = foreach grouped generate group, SUM(population.POPCOUNT);
grunt> dump result;

The **Jaql** is
jaql>$ population = read(del("/user/root/families.csv", {schema: schema { ID: long, FILEID: string, STUSAB: string, CHARITER: string, CIFSN: string, LOGRECNO: string, POPCOUNT: long} }));
jaql> $population -> group by $STUSAB={$.STUSAB} into {$STUSAB, total:sum($[*].POPCOUNT)};

Figure2.   Sample queries of high level query languages on families table.

Figure 3 shows that the query processing time of four querying methods on both Ubuntu and Red Hat OS with varied workloads. As a result of experiments, we can conclude that different numbers of records are used and the proposed big data platform with MapReduce Transformation Model provides better execution time for querying data than other query languages.

Because, other query languages take a large amount of time to be transformedinto the MapReduce model at runtime. For a querying point of view, we can conclude that MapReduce Transformation Model of proposed platform is better than other query languages on

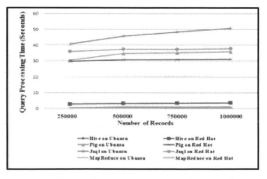

Figure 3.   Comparison of query processing time between Ubuntu and Red Hat OS with different workloads.

both OS. From the operating system point of view, we can also conclude that the Red Hat OS is more convenient than Ubuntu OS for this proposed platform. This proposed platform developed the predefined process that transforms the query into MapReduce programming model. So, it reduces the extra time to transform query language into the MapReduce model and it makes a better performance in querying data.

5   CONCLUSION

Big data analysis is the process of examining big data to uncover hidden patterns, unknown correlations and other useful information which can be used to make better decisions. Moreover, the convergence of mobile computing and cloud computing into a single platform has become an efficient platform for big data analysis. In this paper, we implemented a new mobile cloud platform for big data analysis. This platform operates with RESTful web service technology to provide seamless connectivity between mobile device and cloud storage. To improve the query performance, we developed a MapReduce Transformation Model to transform users' requests into MapReduce form. The analytical result wastransferred to the mobile by using RESTful web service technology. As a result, performance evaluations wereconducted to prove that the proposed platform provides three times faster than other high query languages in both operating systems.

REFERENCES

[1] A. Nandeshwar, August 26, 2013. "Tableau Data Visualization Cookbook", Packt Publishing Ltd.,.
[2] C.Ahn and Y.Nah, 7–9 June 2010 "Design of Location-based Web Service Framework for Context-Aware Applications in Ubiquitous Environments," in Proceeding of International Conference on Sensor Networks, Ubiquitous, and Trustworthy Computing (SUTC), Newport Beach, CA, USA, , pp.426–433.
[3] C.White, "Using Big Data For Smarter Decision Making", BI Research, July 2011.
[4] Hong T.Dinh, Chonho Lee, Dusit Niyato, and Ping Wang," Survey of Mobile Cloud Computing: Architecture, Applications, and Approaches".
[5] "HP Advanced Information Services for the Vertica Analytics latform".
[6] http://www2.census.gov/census_2010/04-summary_File_2.
[7] Kyar Nyo Aye and Thandar Thein, 2014"A Comparison of Big Data Analytics Approaches Based on Hadoop MapReduce",12th International Conference on Computer Application, Yangon.
[8] O' Reilly, Tom White, 27 January, 2012, "The Hadoop Definitive Guide, 3rd Edition".
[9] Phandroid.com, 3 September, 2013, "Android device activation numbers reach 1 billion worldwide".

Network Security and Communication Engineering – Chan (Ed.)
© 2015 Taylor & Francis Group, London, ISBN: 978-1-138-02821-0

Latency-aware interrupt rate control for network I/O virtualization

Z.Y. Qian

School of Software, Shanghai Jiaotong University, Shanghai, China

ABSTRACT: Virtualization technologies are playing an increasingly important role in today's new data centers. The virtualized network resources are becoming the main factor in cloud computing environments, and many researches take efforts to increase the network I/O virtualization performance. Single Root I/O Virtualization (SR-IOV) is a popular technology inherited from direct I/O, and it's able to approve nearly bare metal performance. However, previous researches on SR-IOV all focused on improving throughput and reducing CPU utilization, but had no work on the latency limitation requirement of different applications. In this paper, we analyze the relationship among one VM's network throughput, latency and CPU utilization by several experiments in SR-IOV environment. Based on our analysis, we propose the design of a latency-aware interrupt rate control method to provide high network throughput while ensuring the limited latency of different applications.

1 INTRODUCTION

Today as cloud platforms and new data centers are creating a new level of infrastructures, virtualization technologies are playing an increasingly important role in the cloud computing environments. Virtualization technologies allow a physical device to be elasticly shared between multiple virtual devices, which can be easily used and managed to run multiple applications. This higher level of abstraction provides the agility required by application portability, and reduces cost by increasing infrastructure utilization. Most applications running on the cloud platforms need to use network I/O resources, which require a good performance and high available network connection. Therefore, network I/O virtualization becomes one of the dominant factors in cloud computing environments. However, there are some performance problems in the process of virtualizing network resources (Menon et al. 2005), which motivate several new techniques to achieve high network throughput and reduce overhead.

At present, several I/O virtualization techniques have been developed to resolve these performance problems. Some techniques are software-based, such as device emulation (Landau et al. 2011) and para-virtualized driver domain model (Fraser et al. 2004). Others are hardware-based, such as the self-virtualized devices (Raj et al. 2007) and direct I/O technology (Borden et al. 1989). Recent research indicates that hardware-based approaches can provide better performance than software-based approaches, due to the cut of redundant data copy overhead (Cherkasova et al. 2005).

Single Root I/O Virtualization (SR-IOV) (Dong et al. 2008) is one of the hardware-based I/O virtualization techniques inherited form direct I/O, which shares the NIC device natively between VMs and offloads the data copy overhead. Previous researches have improved the SR-IOV networking performance in many virtualization platforms such as Xen (Barham et al. 2003) and KVM (Kivity 2007), and shows that the overall throughput is close to bare metal in high performance networking environments (Tian et al. 2012). However, these researches all focus on improving throughput and reducing CPU utilization, but don't care about the increased I/O latency. Furthermore, the different requirements of applications also haven't been discussed in these researches.

In this paper, we first evaluate some experiments in SR-IOV environment, and analyze the relationship between one VM's throughput, latency and CPU utilization. Then we propose the design of a latency-aware interrupt rate control method to provide high network throughput while ensuring the limited latency of different applications.

The rest of the paper is organized as follows. In section II, we introduce the background of this paper, including SR-IOV and related studies. Section III shows our evaluation of the experiments, and analyzes the tradeoff between network throughput and latency. In section IV we introduce our design of the latency-aware interrupt rate control method. In the end, we conclude this paper in Section VII.

2 BACKGROUND

In this section, we first give an overview of SR-IOV. Then, we introduce previous studies on network I/O virtualization performance.

2.1 SR-IOV overview

SR-IOV is presented by the PCI-SIG to address sharing of I/O devices in a standard way. SR-IOV is improved from direct device assignment technology, which provides very fast I/O speed but prevents the sharing of I/O devices. The goal of SR-IOV is to bypass the involvement of Virtual Machine Monitor (VMM) during data movement, and it adopts Intel® VT-d technology to offload memory protection and addresses translation. An SR-IOV device has one or more Physical Functions (PF), each PF can support multiple Virtual Functions (VF). Each VF has independent resources such as memory space, interrupts, and DMA streams. A VF can be assigned to a VM, and the VM can access the VF directly and thus transmits or receives packets without intervention of the VMM.

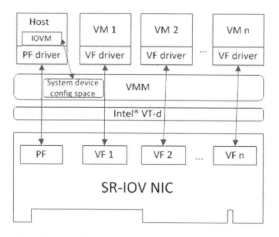

Figure 1. Architecture of SR-IOV.

Figure 1 shows the architecture of SR-IOV. SR-IOV devices have three software components, master driver (also called physical function drive, or PF driver), VF drivers, and SR-IOV manager (IOVM). The PF driver works in the host system and it manages the integrated function of an SR-IOV device and is responsible for configuring shared resources. A VF driver in the guest system works as a normal network driver, it can only access its own dedicated resources. The VF driver is responsible for direct data movement between the operating system and VF. The IOVM runs in the host system and helps VMM to configure all VFs in the system device configuration space, each

with its own configuration space completes with Base Address Registers (BARs). Then VMM can assign a VF to a VM by mapping the VF configuration space to the VM configuration space. Once assigned, the VF can be used as a normal PCI function in the VM.

2.2 Related work

Shivam et al. (2001) firstly proposed a zero-copy transfer layer, called Ethernet Message Passing (or EMP). EMP emulates the entire network protocol stack on the NIC hardware to achieve a very satisfactory performance. Liu et al. (2006) showed that VMM data copy was the major impact of the network I/O virtualization performance. And they provided a new VMM-bypass model, which allowed VMs to complete I/O operations without the VMM's involvement. This performed well but with much higher CPU utilization.

Ram et al. (2009) proposed Virtual Machine Device Queues (VMDq) that helped offload data packets sorted from the VMM to the network. VMDq provided individual send and receive queues for each VM, which doubled the throughput on a virtualized platform.

Dong et al. (2010) discussed a framework for implementing SR-IOV support in the Xen environment. The result showed that SR-IOV can achieve line rate throughput with adaptive interrupt coalescing.

Gordon et al. (2012) proposed Exit Less Interrupts (ELI), which allowed the VMs to handle their own interrupts, and removed the VMM and host system from the interrupt handling path. ELI reduced the overhead of interrupt handling and allowed VM to reach nearly bare metal performance, but suffered from limitation of CPU cores.

Huang et al. (2012) proposed Adaptive Interrupt Rate Control (AIRC) to reduce CPU overhead caused by excessive interrupts. And they also proposed a Multi-threaded Network Driver (MTND) which allowed SR-IOV to make full use of CPU resources in the multi-core environments. They showed that AIRC reduced CPU overhead by half and MTND improved a lot of network performance.

3 ANALYZE

In this section, we perform several experiments to evaluate the relationship among one VM's throughput, latency and CPU utilization based on SR-IOV NIC.

3.1 Experimental setup

The experimental environment consists of two physical machines, which are directly connected by a 10Gbit Ethernet fiber. One machine works as the client to generate network workload and the other works as the server to host the VMs. Each machine has the same hardware configuration: one Intel Westmere dual-socket (6 cores/12 threads) with Intel VT-d supported Xeon X5670 CPUs running at 2.93Ghz, 16GB memory, and Intel 82599 10Gbit NIC. We use the Xen 4.1.2 as the hypervisor, both host and guest setups run RHEL6.1 with Linux 2.6.32. We patch the Linux kernel with multi thread NAPI driver, which allows VM to make full use of SR-IOV in the multi-core environment. We use the ixgbe-3.4.24 as the Host NIC PF driver and the ixgbevf-2.2.0 as the Guest NIC VF driver.

In our experiments, we evaluate network throughput by measuring with netperf benchmark [17]. We run a netperf server in each VM, and open a TCP stream connection from client machine to measure the receive side performance. The netperf client is configured to send 1472-byte packets. In the server machine, we use driver parameter Interrupt Throttle Rate (*ITR*) to control how many interrupts each VF can generate per second. The average latency of each VM is calculated by 1 / *ITR*.

3.2 *Evaluation*

We first analyze the effect of SR-IOV's interrupt throttle rate. We configure one VM with 2 VCPUs, and test the throughput and CPU utilization with interrupt rate of 1000, 2000, 4000, 8000, 16000 per second. Figure 2 shows that the VM's network throughput and CPU utilization increase in line with interrupt rate, which means it should be a tradeoff between latency and network throughput to optimize the resource utilization.

Then we evaluate network throughput performance with increased interrupt rate when the CPU utilization is 100%. We configure one VM with one VCPU, and set the interrupt rate from 1000 to 80000

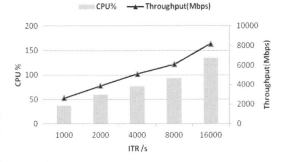

Figure 2.   Interrupt rate impact on one VM's performance.

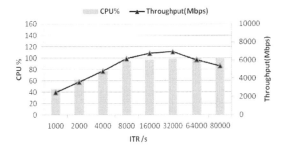

Figure 3.   Network throughput with fully CPU utilization.

per second. Figure 3 shows that the VM's CPU resource is almost fully used when the interrupt rate is 8000, and the throughput is 6 Gbps. Then the CPU utilization maintains 100% when the interrupt rate increases. The VM's throughput achieves the highest when the interrupt rate is 32000, and it will decline when the interrupt rate keep increasing. This is because the CPU takes too much effort to handle the interrupts and has no time to compute.

In conclusion, we need to work out the optimized interrupt rate with bounds of CPU utilization and latency limitation.

4   DESIGN

In this section, we present our design of latency-aware interrupt rate control method.

There are a large number of applications running on cloud platforms, and each has particular latency requirement and CPU workload. It is hard to determine a unified interrupt rate for all the applications. On the contrary, applications can easily define their own maximum latency limitation. Then we can decide one VM's interrupt rate by its maximum latency limitation and the real time CPU utilization.

In our method, we allow each VM to define its own maximum latency limitation dynamically, and work out a minimum interrupt rate, called *ITRdef*. An interrupt rate control manager works in the host system, and reconfigures each VM's *ITR* at set intervals.

In each loop, the manager firstly calculates the VM's free CPU resource from its real time CPU utilization, saved as *CPUfree*. Then the manager checks *CPUfree* with predefined upper-bound and lower-bound. Based on fig. 3, the feasible lower-bound is 10%. If *CPUfree* is smaller than the lower-bound and ITR is bigger than *ITRdef*, the manager will reduce *ITR* with an offset. If *CPUfree* exceeds the upper-bound and the *ITR* is smaller than *ITRmax*, the manager will increase *ITR* with an offset. After that, the manager will sleep for several seconds to wait for the CPU utilization stable. The manager will go on the loops until *CPUfree* is in the bounds. One complete loop is showed as Algorithm 1.

Algorithm I: Latency-aware interrupt rate control

---

loop

    $CPU_{free} \leftarrow getFreeCPU(\ )$

    if $CPU_{free} < L$ **and** $ITR > ITR_{def}$ **then**

        $ITR\ ITR - offset$

    else if $CPU_{free} > U$ **and** $ITR < ITR_{max}$ **then**

        $ITR\ ITR + offset$

    else

        loop break

    end if

        waitSeconds(\ )

end loop

To be more clear, the symbols in Algorithm 1 are listed in Table 1.

Table 1. Algorithm symbols.

Symbol	Meaning
$CPUfree$	The free CPU resource of one VM
ITR	Interrupt Throttle Rate of one VM
$ITRdef$	Minimum interrupt rate defined by VM
$ITRmax$	Predefined maximum interrupt rate (base on NIC device)
U	Predefined upper-bound of free CPU resource
L	Predefined lower-bound of free CPU resource

5 CONCLUSION

This paper focuses on different applications' latency requirement in the SR-IOV based network I/O virtualization environment. In this paper, we first analyze the relationship of one VM's throughput and latency. Then we introduce our design of a latency-aware interrupt rate control method, which allows VM to decide its latency limitation and reconfigures the interrupt rate dynamically according to VM's CPU utilization.

REFERENCES

[1] Barham, P. Dragovic, B. Fraser, K. Hand, S. Harris, T. Ho, A. Neugebauer, R. Pratt, I. & Warfield, A. 2003. Xen and the art of virtualization. *The nineteenth ACM symposium on Operating systems principles*: 164–177.

[2] Borden, T. L. Hennessy, J. P. & Rymarczyk, J. W. 1989. Multiple operating systems on one processor complex. *IBM Systems Journal* vol. 28 (no. 1): 104 –123.

[3] Cherkasova, L. & Gardner, R. 2005. Measuring cpu overhead for i/o processing in the xen virtual machine monitor. *The annual conference on USENIX Annual Technical Conference*: 24–24.

[4] Dong, Y. Yu, Z. & Rose, G. 2008. SR-IOV networking in Xen: architecture, design and implementation. *USENIX Workshop on I/O Virtualization.*

[5] Dong, Y. Yang, X. Li, X. Li, J. Tian, K. & Guan, H. 2010. High performance network virtualization with sr-iov. *IEEE 16th International Symposium on High Performance Computer Architecture*: 1 –10.

[6] Fraser, K. Hand, Neugebauer, S. R. Pratt, I. Warfield, A. & Williams, M. 2004. Safe Hardware Access with the Xen virtual machine monitor. *1st Workshop on Operating System and Architectural Support for the on demand IT Infrastructure.*

[7] Gordon, A. Amit, N. Har'El, N. Ben-Yehuda, M. Landau, A. Schuster, A. & Tsafrir, D. 2012. Eli: bare-metal performance for i/o virtualization. *The seventeenth international conference on Architectural Support for Programming Languages and Operating Systems*: 411–422.

[8] Huang, Z. Ma, R. Li, J. Chang, Z. & Guan, H. 2012. Adaptive and Scalable Optimizations for High Performance SR-IOV. *IEEE International Conference on Cluster Computing.*

[9] Kivity, A. Kamay, Y. Laor, D. Lublin, U. & Liguori, A. 2007. Kvm : The linux virtual machine monitor. *the Linux Symposium.*

[10] Landau, A. Ben-yehuda M. & Gordon A. 2011. SplitX: Split guest/hypervisor execution on multi-core. *USENIX Workshop on I/O Virtualization.*

[11] Liu, J. Huang, W. Abali, B. & Panda, D. K. 2006. High performance vmm-bypass i/o in virtual machines. *The annual conference on USENIX Annual Technical Conference*: 3–3.

[12] Menon, A. Santos, J. R. Turner, Y. Janakiraman, G. J. & Zwaenepoel, W. 2005. Diagnosing performance overheads in the xen virtual machine environment. *The 1st ACM/USENIX international conference on Virtual execution environments*: 13–23.

[13] Raj, H. & Schwan, K. 2007. High performance and scalable I/O virtualization via self-virtualized devices. *International Symposium on High Performance Distributed Computer.*

[14] Ram, K. K. Santos, J. R. Turner, Y. Cox, A. L. & Rixner, S. 2009. Achieving 10 gb/s using safe and transparent network interface virtualization. *The ACM SIGPLAN/SIGOPS international conference on Virtual execution environments*: 61–70.

[15] Shivam, P. Wyckoff, P. & Panda, D. 2001. Emp: Zero-copy os-bypass nic-driven gigabit ethernet message passing. *Supercomputing, ACM/IEEE 2001 Conference:* 49–49.

[16] Tian, K. Dong, Y. Mi, X. & Guan, H. 2012. sebp: Event based polling for efficient i/o virtualization. *IEEE International Conference on Cluster Computing.*

Network Security and Communication Engineering – Chan (Ed.)
© *2015 Taylor & Francis Group, London, ISBN: 978-1-138-02821-0*

Research of a high power factor mobile phone charger based on L6561 and TinySwitch

D.J. Chen
Zhejiang Wanli University, Ningbo Zhejiang, China

ABSTRACT: Currently, the issue of power factor correction for small power charger is rarely considered, and due to a large number of which, the effect on the grid cannot be neglected. In this paper, a mobile charger circuit is proposed based on L6561 and TinySwitch with high power factor. Pre-stage of the circuit is designed based on L6561 to obtain high power factor, and a DC/DC converter is used as post-stage which is established by TinySwitch to get the required charger voltage. Experimental results show that the proposed charger circuit could work properly in the range of 85 to 265V AC input with high power factor.

1 INTRODUCTION

According to the statistics of MIIT (Ministry of Industry and Information Technology), the number of China's mobile phone users amounts more than 1.1 billion. With the popularity of mobile phones, mobile phone charger has become one of the commonly used appliances. Although the power of each phone charger is small, but considering such large number of chargers, the impact on the grid cannot be overlooked.

When scholars conduct researches on a variety of chargers, they are more considering about charging methods (Xiaowei et al. 2001, Bing 2005, Yanbing et al. 2014, Min & Yujie 2012), or on a variety of new input energy (Zhengyu 2004, Wenying et al. 2014). Except some large power chargers (Ping & Xiao 2010), there are few research results about low-power charger, such as mobile phone chargers in this paper. Therefore, this paper will research on a mobile phone charger based on L6561 and TinySwitch. Besides the function of conventional chargers, this charger could achieve high power factor and improve its impact on the grid.

2 DESIGN OF CHARGER CIRCUIT

2.1 *Pre-stage of the circuit*

Figure 1 shows the pre-stage circuit of the mobile phone charger based on L6561. The input voltage Vin is a wide range of AC voltage with a value from 85 to 265V in order to adapt to the global standard. The output voltage of this stage VC5 is the voltage across the electrolytic capacitor C5 and a stable DC voltage. In this stage, the input current must be controlled into sinusoid to achieve high power factor. In addition, a

constant output voltage is for another control purpose. Therefore, a Boost circuit is always used as the main topology which includes transformer T1, diode D1, and MOSFET M1, as it is shown in Figure 1.

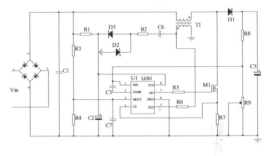

Figure 1. Pre-stage circuit diagram of mobile phone charger.

The power factor correction principle of L6561 can be explained as follows. In Figure 1, pin 4 of the chip works as the current sampling pin, and the control purpose of the chip is to lead the average value of the current to be sinusoidal. Pin 3 of the chip measures the input voltage which is a half sine waveform. This sampling voltage is multiplied by the chip internal voltage waveform (the output signal of internal amplifier with a constant value), and the result is used as the reference signal for pin 4. Therefore, the input current of the circuit can be controlled into a sine wave. (Zhipeng et al. 2004, Danjiang et al. 2011)

In Figure 1, high-frequency filter capacitor C1 can reduce the high-frequency switching noise from the inductor current. The worst situation of this noise happens at the minimum peak value of input voltage. Usually, it is required that a circuit can suppress the maximum frequency ripple voltage at the rate of input voltage between 1% and10%, and the variable r represents the coefficient (r=0.01 to 0.1). The value of capacitor C1 can be calculated by the reference Equation 1.

$$C_1 = \frac{I_{rms}}{2\pi \cdot r \cdot f_{sw} \cdot V_{irms(min)}} \qquad (1)$$

here, $V_{irms(min)}$ is the minimum input RMS voltage which is 85V in this paper. $I_{rms} = P_i/V_{irms(min)}$ is the maximum RMS input current. P_i is the input power of the circuit, which is equal to the output power if efficiency is not considered. f_{sw} is the switching frequency of L6561, which will change as the input voltage changes.

The value of output capacitor C5 is related with the output voltage, output power, and desired voltage ripple. When selecting a low ESR capacitor, its value is

$$C_5 = \frac{I_o}{4\pi \cdot f \cdot \Delta V_o} = \frac{P_o}{4\pi \cdot f \cdot V_o \cdot \Delta V_o} \qquad (2)$$

here, f is the frequency of input voltage Vin, and ΔV_o is typically selected from 1% to 5% of the output voltage.

When we design the boost inductance (the primary inductance of the transformer T1 in Fig.1) many parameters must be brought into consideration, and there are different methods which can be used to calculate it. In addition, it must be ensured that when the inductance value is determined, the minimum system toggle frequency will be greater than the maximum internal starter frequency to make the system work correctly. The minimum switching frequency f_{sw} occurs at the top of the line voltage, while the maximum one occurs at zero crossing of the line voltage. The minimum system toggle frequency may occur in the maximum or minimum line voltage, and the inductance values are defined as the follows assuming unity power factor for the circuit.

$$L = \frac{V_{irms}^2 \cdot (V_o - \sqrt{2} \cdot V_{irms})}{2 \cdot f_{sw(min)} \cdot P_i \cdot V_o} \qquad (3)$$

here, V_{irms} is the minimum or the maximum RMS value of the input voltage, and both of which are substituted into Equation 3 and takes a smaller result as boost inductance value.

The design of other peripheral parameters of L6561 refers to the official manual or some relevant literature (Zhipeng et al. 2004), so it is not repeated here.

2.2 Post-stage of the circuit

As previously mentioned, the output voltage of the pre-stage circuit is VC5 which crosses the electrolytic capacitor C5. This DC voltage is a constant voltage

with the value of about 400V if the input AC voltage is designed to a wide range mentioned in this paper. However, the charging voltage of mobile phone is usually around 5V, so a DC/DC converter must be designed here to step down the input voltage. Here, TinySwitch (Wu et al. 2003, Yihua et al. 2010, Zhiwei et al. 2013, Chunlei et al. 2008) is adopted as the main controller and flyback converter as the main topology.

TinySwitch is a series of low-power single-chip switching power supply chip developed by Power Integrations. The third generation TinySwitch-III is widely used in chargers or adapters of mobile phones, digital cameras, and other low-power appliances (Zhiwei et al. 2013, Chunlei et al. 2008). Tiny Switch-III incorporates a 700 V power MOSFET, oscillator, high voltage which switch current source, current limit (user selectable), and thermal shutdown circuitry. In addition, the current limit circuit senses the current in the power MOSFET. When this current exceeds the internal threshold, the power MOSFET is turned off for the remainder of the cycle. The limit current can be adjusted by a ceramic capacitor connected to BYPASS/MULTI-FUNCTION (BP/M) pin of TinySwitch-III (Chunlei et al. 2008).

1 A 0.1 μF BP/M pin capacitor will select a standard current limit for the application of enclosed adapter.
2 A 1 μF BP/M pin capacitor will select a lower current limit, and reduce the RMS current flowing through the device. As a result, it will improve efficiency and be more suitable for high device temperature requirements design.
3 A 10 μF BP/M pin capacitor will select a higher current limit. At a temperature permitting the peak output power of the device or a continuous power will increase.

TinySwitch-III series chips contain seven products from TNY274 to TNY280. TNY276 is selected as the control chip according to the power of mobile phone charger and the circuit diagram is shown in Fig. 2. Input voltage of this stage is the output voltage of the pre-stage VC5. The main topology flyback converter is composed of transformer T2, internal integrated MOSFET of TNY276, output rectifier diode D5. The clamp circuit is made up of D4, R10, and C6 to suppress voltage spikes caused by transformer leakage inductance to ensure the safety of internal MOSFET. The capacitor C7 which is connected to pin 1 of the chip determines the limit current described above. Feedback adjustment circuit is composed by the optocoupler and TL431, where R11 is the optocoupler current limiting resistor, and R13 and R14 divider ratio can be set up and adjust the output voltage to reach high voltage accuracy. In addition, an RC circuit connected between the cathode and the reference voltage terminal of TL431 constitutes a proportional-derivative controller in order to improve the static and dynamic response of output voltage.

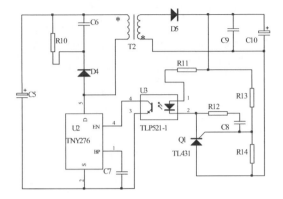

Figure 2. Post-stage circuit diagram of mobile phone charger.

Since the PI company provides an aided design tools PI Expert for its products, development time of switching power supply will be greatly reduced and some optimal design can be achieved easily too (Gang et al. 2007).

3 EXPERIMENTAL RESULTS

According to the above theory and the circuit diagram Figures 1-2, a mobile phone charger experimental circuit with power factor correction function is designed and made with the input voltage 85 ~ 265V, the output voltage 5V, and the output power 10W. Figures 3-4 show the input voltage and current experimental waveforms under 90V input voltage and 220V input voltage.

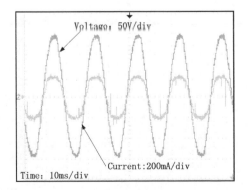

Figure 3. Experimental waveforms under 90V input voltage.

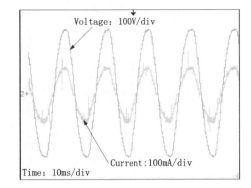

Figure 4. Experimental waveforms under 220V input voltage.

Under different values of input AC voltage, the actual output voltage of the experiment circuit is shown in Tab.1 with full load. It can be seen that the circuit has a high voltage regulation within the set range of the input voltage.

Table 1. Output voltage with different input AC voltage.

V_{AC}/V	90	150	220	250
V_o/V	4.995	4.997	4.998	5.003

In the case of 220V AC input voltage, Table 2 shows the output voltage when the load changes.

Table 2. Output voltage with different load.

Output current/A	100%I_o	50%I_o	10%I_o
V_o/V	4.998	5.001	5.014

4 CONCLUSION

This paper presents a mobile phone charger with power factor correction function. After the theoretical design, an experiments circuit is carried out to verify the theory. The results show that this charger not only has the power factor correction circuit function, but also can achieve high voltage regulation and load regulation within a given range of input AC voltage.

ACKNOWLEDGMENT

The project is supported by Project of Education Department of Zhejiang Province(Y201431817), Science and Technology Innovation Team of Ningbo Science and Technology Bureau (2013B82009).

REFERENCES

[1] Bing L. 2005. Design of smart charger for maintenance-free lead-acid battery based on UC3906. *Mechanical Engineer* (11):94–95.

[2] Chunlei Y. & Honghao G. & Yanfeng C. et al. 2008. Application of TinySwitch-III in small switch mode power supply. *The World of Power Supply* (1):64–67,70.

[3] Danjiang C. & Wei Z. & Zhihong X. et al. 2011. Design of a switching power supply with super wide voltage input range. *Electrical Automation* 33(3):65–67.

[4] Gang Z. & Zhigang L. & Daiwei Y. et al. 2007. Design of a multiple output flyback switching mode power supply based on TOPSwitch and PI Expert. *Power Supply Technologies and Applications* 10(2):1–4.

[5] Min L. & Yujie W. 2012. Research and implementation of intelligent and power-saving charger for mobile phone. *Mechanical Engineering & Automation* (1):143–145.

[6] Ping L. & Xiao S. 2010. Research on coal intelligent charging with high power factor input for gel batteries. *Coal Mine Machinery* 31(8):176–178.

[7] Wenying H. & Mingjun Z. & Wanli S. et al. 2014. Optimum design of solar power charger. *Chinese Journal of Power Sources* 38(1):85–86.

[8] Wu N. & Yongzhen C. & Yuhong K. 2003. Characteristic analysis and application of the TinySwitch-II. *Journal of Liaoning Institute of Technology* 23(2):11–12.

[9] Xiaowei L. & Mingming Z. & Longrui Z. 2001. Studies on chargers for lead-acid batteries used for electric bikes. *Chinese Labat Man* (2):27–29.

[10] Yanbing Y. & Hui L. & Xiaodan L. 2014. Development of microcontroller-based double closed-loop smart charger. *Chinese Journal of Power Sources* 38(2):307–309.

[11] Yihua H. & Yanlin L. & Shiwen L. et al. 2010. Design of a switching mode power supply with dual iso-lated outputs based on TinySwitch-II. *Telecom Power Technology* 27(2):1–4.

[12] Zhengyu L. 2004. Charger of Solar energy for Mobile telephone Battery. *Journal of Zhangzhou Normal University (Natural Science)* 17(4):41–44.

[13] Zhipeng W. & Shenggui T. & Chunhua T. 2004. Design of APFC circuit operated in critical current mode based on L6561. *Telecom Power Technologies* 21(4): 9–15.

[14] Zhiwei H. & Zhenning G. & Feifei Y. 2013. Design of LED driving power supply based on TinySwitch-III. *Telecom Power Technology* 30(2):10–12.

Network Security and Communication Engineering – Chan (Ed.)
© 2015 Taylor & Francis Group, London, ISBN: 978-1-138-02821-0

Research on text classification based on inverted-index in cloud computing data center

G.Y. Yang & L. Chen
College of Information, Liaoning University, Shenyang Liaoning, China

L. Chen & B. Meng
Computing & Information Center, Liaoning University, Shenyang, Liaoning, China

ABSTRACT: The text classification which is based on inverted-index is presented according to dimensional distribution of the text vector sloppy, in order to improve the feature selection various methods of computing speed and text classification algorithms such as KNN and Bayes. The algorithm can build the structure of inverted-index tree which may filter algorithm horizontal dimension, then get small data set quickly. The experiments indicate that under such a massive text classification environment, the processing speed of this method is still faster than that of the traditional one although it may lose the accuracy.

KEYWORDS: Text classification, KNN algorithm, Inverted index, Feature weighing.

1 INTRODUCTION

With the development of Internet, especially cloud computing database center, it becomes increasingly difficult for the traditional algorithm of dimensional calculation to meet the requirements which is due to the high-dimensional text vector and other swelling text data such as web pages. As arrangement of text data and an important way of organization, text classification can decrease dimension by the effectiveness of the feature selection, which can not only reduce the cost of system and run-time, but also can improve the accuracy of classification.

There are two ways to improve the speed of classification. The first one is feature selection which may both improve the accuracy and speed of classification, such as the TFIDF, document frequency, information gain, mutual information, CHI, expect cross entropy, etc [4]. We will adopt TFIDF and make some improvement in it. The second one is category selection. Classifying in similar category may increase the speed of classification, and within-class means classification is one of the most important one. Therefore, text classification based on inverted index is proposed in order to improve the speed of classification, meet various methods of calculation and it is the text vector dimension sloppy we have mentioned before. Text classification is suitable for Bayes [1] algorithm, KNN [10] algorithm, and so on.

2 RELATED WORK

The text vector space model can be used to calculate the text weights for document D, a document d can be seen as a set of documents d n-dimensional vector, $\bar{d} =< d_1, d_2, ..., d_n >$ Text vector weights are calculated by *TFIDF* algorithm which can reflect the feature word space of the document set, and each component corresponds to a term weight of the size $S(i)$:

$$S(i) = TF(t_i, doc) * IDF(t_i) = tf(t_i, doc) \times \log(D/DF(t_i)) \tag{1}$$

where $tf(doc, t_i)$ is the frequency of feature words t_i that appear in the document; D is the total number of documents, $DF(t_i)$ is the number of feature words in the document, $IDF(t_i)$ reflects the ability of distinguishing of the feather word. Document vector is normalized, and the formula[3] is as follows:

$$TFIDF(t_i, doc) = \frac{tf(t_i, doc) \times \log(N/n_{t_i} + 0.01)}{\sqrt{\sum_j \left[tf(t_j, doc) \times \log(N/n_{t_j} + 0.01) \right]^2}} \tag{2}$$

The cosine of the angle between the two document vectors is used to calculate the similarity of documents, and the formula is as follows:

$$sim(d_i, d_i) = \cos(d_i, d_i)$$

$$= \frac{\sum_{k=1}^{n}(t_{i,k} \times t_{j,k})}{\sqrt{\sum_{k=1}^{n}(t_{i,k})^2} \times \sqrt{\sum_{k=1}^{n}(t_{j,k})^2}} \qquad (3)$$

3 DESIGN OF STORAGE STRUCTURE BASED ON THE INVERTED-INDEX

3.1 Dictionary structure

From Figure 1, Term is a type of varchar. *t_id* which is 32 bit fixed length and is the unique number to term. *idf* is the inverse document frequency, 32 bit, which can be used to calculate the cosine of the angle. The entire space of the dictionary share is about 8MB. Under this circumstance, we can get the ideal speed as the result of the achievement of dictionary permanent memory and the storage way which is called Hash graph method that is adopted in the memory of dictionary.

3.2 Index structure of inverted chart

Both positive displacement index and inverted index [8,9] are the index structures of KNN algorithm of text classification. When filtering the data, we use the structure of inverted index and can increase the speed of computation, especially in the process of calculating the cosine of the angle.

Storage structure of inverted chart is shown in Figure 2 and the description is as follows:

(1) Number 1 in Figure 1 is called *RootInfo* which is a representation of the structure of root node which is derived from the whole set of data in the entire inverted index. Both *docSum* the number of documents and *sizeSum* the total number of all the words are in the set of data.

Number 2 is called *ClassInfo* which is a representation of the node structure of category in data set. *cnid* is a label of category, *dcount* is the total number of documents and *tsum* is the number of all words in some category.

Number 3 is called *TermInfo*, which is a node structure of all feature word in some category. *tcount* is the word in category, *t_id* is the number of times shown in every document and *tfavgval* is the average of *tf* in some category.

Number 4 is the node structure of a feature word (*DocInfo*) which is shown in all kinds of documents in some category. *tf* is the showing frequency of word *t_id* in this document, *docid* is the document number of block starting, and *doc_offset* is offset of the document number compared to the last document that can compress the storage space of index.

(2) Block is a continuous storage area of the disk with a fixed size. When the free space of need within

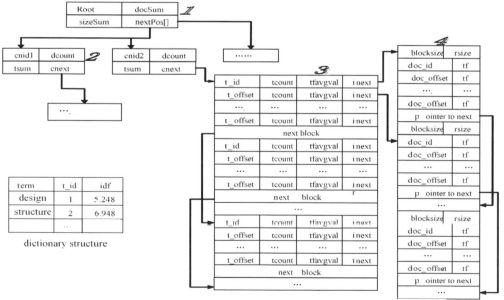

the structure of inverted-index

Figure 1. The structure of inverted-index and dictionary structure.

the storage area inverted chart exceeds the fixed size, the division and addition are needed. The block size, the block length, the block free space, and the address of block are all within the head of block.

The structure of inverted-index can establish inverted-order based on each category of data set. We can get smaller data from the big one by classifying the massive data and then the classification of huge amounts of data is the same as the number of category which can be computed. In order to increase the speed of computation, calculation of every classification of categories can be used during the KNN algorithm. But it may become difficult to maintain the whole capacity of invert-index due to the increased number of it though each index becomes smaller.

3.3 Insertion of inverted-chart

The step of the algorithm is as follows:

3.3.1 Algorithm

The establishment Inverted-index.

Input: D: set of text files within classified label. DS: set of dictionary structure. Output: the Inverted-index tree.

3.3.2 Method

First, the structure of inverted-index tree is established by the following steps:(a) In order to get docSum (the total number of documents), sizeSum (the size of all words within the data set), dcount (the number of every category), tsum (the size of the word of every category), we need to scan the set D. (b) Establishing the RootInfo structure of the root node derived from the inverted-index tree and structure information docSum and sizeSum. (c) Establishing the ClassInfo structure of node information of category and internal information dcount and tsum, then connecting it with RootInfo structure of root node. (d) Establishing TermInfo information of word node and then connecting it with ClassInfo structure of category node respectively. (e) Reading each file of the set of data D, and the information of DocInfo structure of node can be generated by II_create (II_tree,D) and connecting it with TermInfo structure respectively. (f) Return to step (e) if there has been some unread file in the data set, otherwise go to next step. (g) Returning to the root node of the inverted-index, we can get the structure of inverted-index, which is the tree of inverted-index.

Second, the step of generating II_create (II_tree, D) of information of DocInfo structure is as follows: (a) Reading document X in the data set D. (b) Getting cnid which is the category of docid in document X. (c) Getting the structure of ClassInfo information

based on the category of docid. (d) Reading the word t_id in document docid. (e) Getting the information of TermInfo of t_id under the structure of ClassInfo information. (f) The structure of information which is generated by calculating the value of tf in docid document being as follows: store the information of <t_id,<docid,tf>> in the structure of DocInfo. (g) Under the node of category, connecting the structure of TermInfo information of the word node to the structure of DocInfo respectively. (h) Changing the tfavgval of word within the category under the structure of TermInfo information and the tcount which is the total number of the words in the category. (i) If there are some words in the document docid unprocessed, returning to step (d), or going to next step. (j) If there are some documents unread in data set D, returning to step (a), otherwise the procedure is over.

3.4 Inverted-index deletion

The inverted-index deletion is the inverse process of operation for insertion, and the step of algorithm is as follows:

3.4.1 Algorithm

Inverted-index deletion. Input: D: document within text set with class labels. II_tree: inverted-index tree. Output: A deleted inverted-index tree.

3.4.2 Method

The step of deleting II_delete (II_tree, D) the information of DocInfo structure is as follows: (a) Reading document X in the data set D. (b) Getting cnid which is the category of docid in document X. (c) Getting the structure of ClassInfo information based on the category of docid. (d) Repeat. (e) Getting the information of TermInfo of t_id under the structure of ClassInfo information. (f) Getting the structure of DocInfo information, under TermInfo information structure. (g) If docid is outside of DocInfo information structure, returning to step (e), otherwise deleting docid under the structure of DocInfo information. (h) Changing the tfavgval of word within the category under the structure of TermInfo information and the tcount which is the total number of the words in the category. (i) Changing the value of dcount and tsum under ClassInfo. (j) Changing the value of docSum and sizeSum under the root content RootInfo. (k) Until: the TermInfo information structure of the word within ClassInfo can be read.

3.5 Inverted-index update

Updating of inverted-chart can be transformed into the deletion of inverted-chart and the insertion of inverted-chart. Using the batch method, the update strategy can reduce IO effectively by the operation of deletion and the process of insertion after an accumulation of a certain amount. After the insertion of inverted-chart, deletion of inverted-chart and the update of inverted-chart of the three steps, and the inverted-index model can not only meet the existing need of classification algorithm of KNN and Bayes, but also be an incremental text classification algorithm that is based on the inverted-index.

3.6 *Strategy of inverted-index pruning*

The ultimate purpose of pruning strategy is to reduce the unnecessary calculation and cost. Feature selection and category selection are usually used in pruning strategy. Those are common methods such as information gain, mutual information, CHI, expect cross-entropy for feature selection which can make it more achievable based on the inverted-index. Category selection is a kind of algorithm of text classification where KNN Or Bayes algorithm are used within K categories which can be obtained from the calculation of cosine of the angle which is within each category and the estimate of data.

Pruning algorithm of feature selection is a kind of horizontally pruning algorithm which can reduce the number of feather words in each category. But the algorithm of category selection is a kind of vertically pruning algorithm which may cut several categories within the data set and the categories which are close to the samples can be left. So it can be more efficient for algorithm of text classification when the two methods are combined and work together. And various kinds of algorithms of improvement within the text classification can be achieved by the inverted-index.

4 EXPERIMENTAL ANALYSIS

4.1 *Experimental data sets*

In order to verify the effectiveness of amendment based on text classification of inverted-index, we use the international corpus 20NewGroup to text it. Under the circumstance of an average distribution of sample, we regard the problem of the single label as an example to make an experiment. The capacity of 80MB data set can contain 15,529 documents and 20 categories, and each category is about nearly 800 documents. We verify the function of classification of KNN algorithm by 10 cross-validations.

Stopping word removal, stemming treatment and classifier which are all unified within the program of preprocessing which will use the KNN algorithm when calculating. 31,306 features are the result of data processing of 20NewsGroup.

4.2 *Analysis of experimental results*

Comparing the function of the two methods according to the data set above, we get:

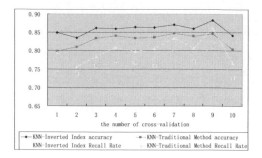

Figure 2. The chart of accurate comparison.

The inverted-index method is superior to the traditional one because both the accuracy of classification and the recall rate [2] of the new method are all higher than those of the traditional one, which are 2% and 4% higher respectively. The result of the chart of time comparison has been omitted, due to the space limitations, the time of traditional calculation classification is about 700s, but the time of inverted-calculation classification is only 400s. Compared to the traditional method, the time of inverted classification can be more stable and has less fluctuation.

5 CONCLUSION

According to the current segmentation of word and hardware software environment and new generation data center, text classification based on inverted-index is proposed. The experiment shows that the inverted-index can not only increase the efficiency of algorithm but also can reduce the time of classification and at the same time increase the accuracy of classification.

REFERENCES

[1] Liangxiao Jiang, Research on Native Bayes Classifiers and Its Improved Algorithms[D].China University of Geosciences,2009.
[2] ZHANG Feng-qin, WANG Lei ,Text feature generation algorithm based on clustering weighted,2013,30:146–148.
[3] YANG Jianwu, CHEN Xiaoou. A Similariry Search Algorithm for Text Based on Inverted-index Computer Engineering.2003,31(5):1–3.
[4] Yan Jun, Liu Ning, Zhang Benyu.OCFS: Optimal orthogonal Centroid Feature Selection for test categorization [M]. Brazi:l SIGIR, 2005.

[5] DENG Pan,LIU Gong-shen, Effective storage of inverted index. Computer Engineering and Application,2008,44(31):149–152.

[6] Zhang Hai-long, WANG Lian-zhi, Automatic text categorization feature selection methods research[J], Computer Engineering and Design,2006,27:20–24.

[7] Mlademnic, D., Grobenik,M, Feature Selection for unbalance Class distribution and Native Bayes [A].Proceedings of the Sixteenth International Conference on Machine Learning[C],Bled: Morgan Kaufmann,1999:258–267.

[8] Putz S.Using a relational database for an inverted text index,SSL91-20[R].Xerox PARC,1999.

[9] Ames T J.XML in an adaptive framework for instrument control[C]//2004 IEEE Aerospace Conference Proceedings,America:IEEE,2004:6–13.

[10] Masand B,Lino G,Waltz D, classifying NewS Stories Using Memory Based Reasoning[C]//15th ACM SIGIR Conference.1992:59–65.

Network Security and Communication Engineering – Chan (Ed.)
© *2015 Taylor & Francis Group, London, ISBN: 978-1-138-02821-0*

Rehabilitation for hearing-impaired children pronunciation system based on cloud application architectures

L.J. Shi
College of Electronic Information and Engineering, Changchun University, Changchun, China
College of Opto-electronic Engineering Changchun University of Science and Technology, Changchun, China

J. Zhao
College of Computer Science and Technology Changchun University, Changchun, China
College of Instrumentation and Electrical Engineering, JiLin University, Changchun, China

H.W. Qin
College of Electronic Information and Engineering, Changchun University, Changchun, China

ABSTRACT: Recent deaf children pronunciation rehabilitation system is limited by equipment and environment, and the rehabilitation needs deaf children training in the fixed training center. But most of the deaf children have no conditions and energy to be trained in training centers, so they are unable to enjoy the technology, few of whom really accept rehabilitation. The training system based on cloud is presented to solve the problem. Framework is built on the cloud data, and the mobile terminal data are uploaded to the cloud platform, and then the processed data go back to the terminal, which can get rid of the fixed place restrictions on the training, so training of deaf children can be carried out at any time and any places. At the same time, training centers can analyze the stored data, and then get the recovery conditions of the deaf children.

1 INTRODUCTION

3D talking head [1-3] is an important branch of visual speech technology. It is a new visual speech technology which overcomes traditional shortcomings of speech synthesis technology. In recent years, many scholars recognize that it is important to apply 3D talking head technology in human–computer interaction to resolve speech therapy. 3D talking head mainly focuses on the parameters acquisition and the parameters driven of vocal organs. The different methods of the parameters acquisition produce the bionic [4] method based on simulating the composition of musculoskeletal and other organs, and it can also produce the data analysis represented by image acquisition and processing [5]. Olov Engwall and Preben Wik [6] first proposed that there should be an interactive conversation act as a virtual language teacher in computer-assisted language teaching system and the virtual teacher is applied to adult and children language learning as well as the hearing-impaired children or patients who cannot speak native language to learn the training.

At present, there are two ways of pronunciation rehabilitation for hearing-impaired children. One is teacher one-to-one teaching method and another one is computer-aided teaching method. Problems that need to be solved in computer-aided teaching method are how to understand pronunciation training process expressed by the computer for children who cannot hear, how to assess the accuracy of pronunciation of hearing impaired children for computer, and whether the movements process of the vocal organs of hearing-impaired children conform to the normal pronunciation process. A special human–computer interaction model should be established to solve these problems. This paper describes the special human–computer interaction model; this model can solve interaction between hearing-impaired children and the computer. Based on this model, this paper gives an introduction of cloud computing framework, and hearing-impaired children can get rid of a fixed place of training restrictions. In recent years, with the rapid development of large-scale Internet application in a new generation, such as social networks, e-commerce, online video and other digital cities, these emerging applications have the characteristic of large amount of data storage, and business growth speed. In order to solve the above problems, this paper puts forward the concept of cloud computing where the pronunciation hearing impaired children rehabilitation system is applied. Cloud computing is to utilize the Internet to achieve on-demand. Whenever and wherever possible, it is convenient to access to the shared resource pool (such as computing facilities, storage devices, applications, etc.) computing model. Computer resource service is an important manifestation of cloud computing, it has data center management, large-scale data processing, application deployment issues such as the user screen. With cloud computing, the user can quickly apply or release the resources of its business, according to the load. [7] So hearing-impaired children can be trained anywhere and anytime.

2 SPECIAL HUMAN–COMPUTER INTERACTION MODEL STRUCTURE AND KEY TECHNOLOGY

Special interaction model frame structure is initially proposed as is shown in Figure 1. Pronunciation rehabilitation of hearing impaired children is guided through video information of the system, including pronunciation process training, breathing process training during pronunciation, and tongue movement training during pronunciations, which are simulated into realistic 3D head portrait and animation by specific systems respectively.

Problems of human-computer interaction model and interactive technology oriented to the pronunciation rehabilitation of the hearing-impaired children were: 1) The establishment of special interactive model; 2) The pronunciation simulation of the realistic 3D head portrait; 3) The establishment of respiratory training model; 4) The establishment of tongue training model; 5) The barriers of obtaining access to parameters of multiple vocal organs and the integration of parameters of multiple vocal organs were problems that need to be solved. Accordingly, studies were made in this paper in the following aspects:

2.1 *Study on realistic 3D head portrait technology and 3D information acquisition of vocal organs*

The parameters of each vocal organ were extracted depending on different equipments and technologies, and the parameters were stored in the corresponding parameters bank. The following methods were adopted:

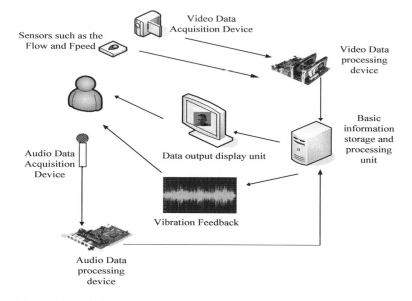

Figure 1.   Special interaction model between hearing-impaired children and computer.

2.1.1 *The establishment of translucent 3D head portrait model*
Realistic 3D head portrait model is created with 3D facial information scanned by 3D face scanner by using 3D software. The hearing impaired children can learn pronunciation process by visual feedback of the pronunciation process of 3D translucent virtual man in the display.

2.1.2 *The 3D dynamic information collection of the pronunciation process of facial vocal organs*
The speaker's audio data (natural continuous speech) and facial three-dimensional dynamic data (acquiring the dynamic deformation data of the vocal organs) are acquired synchronously by using audio acquisition devices and 3D motion capture system. The deformation data of the speaker's facial vocal organs are

obtained through the real-time data collection of the speaker's front side and orthogonal side and the collection of continuous pronunciation movements. 3D dynamic information of moving facial feature points is acquired by using 3D motion capture system at the rate of 20 frames per second.

2.1.3 The establishment of vocal organs model of tongue and palate and the 3D dynamic information collection

Take tongue for an example. Reference pictures of the tongue in different positions are manifested through the use of two-dimension or profile of the central head of three-dimensional tongue contour, and different measurements shows different points of view. The dynamic three-dimensional information of the tongue and palate in the pronunciation process obtained by the electropalatography is stored in the database of vocal organs.

2.1.4 Algorithm study on integration and drive of vocal organs of realistic 3D head portrait model

The synthesis algorithm suitable for the realistic 3D head pronunciation is proposed, synthesizing 3D information of the pronunciation process of each vocal organ acquired by independent equipment. The simulation of realistic 3D head pronunciation is completed finally by driving vocal organs by using data from 3D database.

2.2 Study on realistic 3D head portrait technology and 3D information acquisition of vocal organs

Respiratory training model is an important component of language rehabilitation training, and the base of learning pronunciation methods, skills, and developing good pronunciation habits. The key of the breathing method training is the air flow, and air flow control is a very important part of the pronunciation. Traditional breathing training approach is deep breath training by teachers with interesting breathing training performed simultaneously.

2.3 The establishment of special human–computer interaction model in tongue training

Feedback model of respiration training applying sensor technology and 3D reconstruction technology is established. The tongue movement trail is fed back in the display according to tongue information of the hearing-impaired children in pronunciation rehabilitation by sensors. It is compared with the required trail to guide the tongue training.

3 THE REHABILITATION OF CHILDREN BASED ON THE CLOUD COMPUTING PLATFORM WITH HEARING IMPAIRMENT PRONUNCIATION SYSTEM

Cloud computing platform can be divided into three categories: Data-storage-based cloud storage platform type, Data-processing-based cloud computing-based platform, and both comprehensive cloud computing platform. The main data storage based on cloud storage platform is the rehabilitation process data for pronunciation hearing impaired children. The main computing platform for data analysis and the analysis of the results return to the user.

3.1 Construction and cloud storage software and hardware platform for intelligent network data transmission system platform to build

From the short-range wireless data transmission perspective, the data collection terminal information transmit wirelessly to the cloud. For data processing in the cloud, the results will be returned to the terminal, and the terminal displays evaluation results. So it is easy to operate, and it just needs to have the sensor terminal. And wireless data transmission device can carry and install easily.

3.2 Establish training assessment system based on cloud computing framework pronunciation

Pronunciation training system by hearing-impaired children pronunciation guides and establishes a communication channel such as cloud computing by building an intelligent network platform for data transmission software and hardware systems, and the data collected from pronunciation training transmission via a wireless communication channel is processed in the central processing unit. Pronunciation level is assessed by the trainer. And the calculation results are fed back to the system via a communication channel training, as is shown in Figure 1.

3.3 Rehabilitation of hearing impaired children data management system based on cloud computing

Cloud computing needs distribution, massive data processing, and analysis. Therefore, data management technology must be able to manage large amounts of data based on cloud computing and the data are massive, [8] heterogeneous, non deterministic characteristics, which needs to use the data management technology with effective analysis and processing of the massive data and information. The construction of distributed data storage is highly

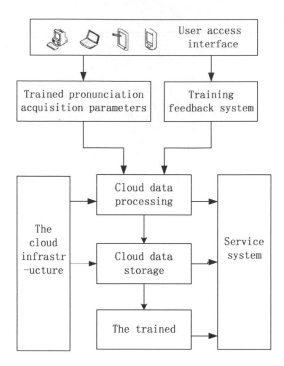

Figure 2.   Hearing-impaired children rehabilitation training system of cloud computing framework.

available and has scalable systems. The current cloud computing includes data management technol [9], BigTable [10], MapReduce [11] data management technology and Amazon Dynamo. [12] Data in the cloud characteristics [is mainly manifested in the following aspects12]: mass/ heterogeneity and non deterministic. Therefore, the data processing of deaf children rehabilitation training system is relatively complex.

4   CONCLUSION

Special human–computer interaction model of the session by the above method to the 3D-talking head is proposed. Framework is built on the cloud data, and the mobile terminal data are uploaded to the cloud platform. Then the processed data is back to the terminal, which can get rid of the fixed place restrictions on the training, so training of deaf children can be carried out at any time and any place. While the training center can store the data analysis, it needs to understand the situation of the hearing-impaired children's rehabilitation.

ACKNOWLEDGMENTS

This work was supported in part by China Postdoctoral Science Foundation under Grant No. 2013M541298; The Fund of Ministry of Education of China No.Z2012062; Jilin Province Science and Technology Development Plan under Grant No. 201101088, 20130522162JH; JiLin Province Education Development Plan under Grant Nos. 2013278, 2013285,2011223. Changchun Administration of Science and Technology Plan under Grant No.13GH03.

REFERENCES

[1] DIX A, Finlay J, Gregory D A. 2003. Human computer interaction [M].*3rd Edition. Prentice Hall*:25–40.
[2] Zhao Jian , Shi Lijuan, Du Qinsheng, Wang Lirong:2013.Special Human-Computer Interaction Model Oriented to Hearing Impaired Children[J], *Journal of Convergence Information Technology,* Vol. 8, No. 3, pp. 503–511.
[3] Zhao Jian,Shi Lijuan,Fan Qinyin,Lin Jun,Du Qinsheng, Wang Lirong, Zhao Peng. 2013.Breath Training for Hearing Impaired Hearing Chinldre Based on Computational Fluid Dynamics[C] *In Proceeding of AEIT2013*, Dalian, China, December 11–13, pp. 506–509,.
[4] R.Wilhelms-Tricarico1995.Physiological model ing of speech production: Methods for modeling of soft-tissue articulators.J.Acoustics Soc Amer.97(5):3085–3098.
[5] T.Guiard-Marigny.Adjoudani, and C. Benoit. 1994.A 3D model of the lips.*In Proceedings of the 2nd ETRW on Speech Synthesis*, New Plat, NY,.
[6] O. Engwall, .Combining MRI.2003.EMA & EPG in a threedimensional tongue model[J]. *Speech Communication*, vol. 41/2-3, ,303–329.
[7] LUO Jun-zhou, JIN Jia-hui, SONG Ai-bo, DONG Fang, July 2011.Cloud computing: architecture and key technologies, *Journal on Communications*. Vol.32 No.7 pp:3–19.
[8] Liu Zheng Wei,Wen Zhong Ling. 2012.Cloud Computing and Cloud Data Management Technology, *Journal of Computer Research and Development*, 49(Suppl.):26–31.
[9] Ghemawat Sanjay,Gobioff Howard,Leung Shun Tak. 2003.The Google file system,Proco of 19th ACM SOSP New YORK [2010-12-29].http:www.cs.vu.n.ralf?Map. Reduce?paper. pdf.
[10] Chang Fay,Dean Jeffrey,Ghemawat Sanjay. Bigtable:2006. Adistributed storage system for structured data??Proc of the 7th USENIX Symp on OSD Berkeley Berkeley USENIX.
[11] Lammel Ralf .Googles MapReduce 2010. programming modelRevisited. http:?? Userpages .uni-koblenz. de? laemmel? MapReduce? papeer. pdf.
[12] Li Qiao ZHENG Xiao. 2011.Research Survey of Cloud Computing,*Computer Science*, Vol.38.No.4 Apr. pp:33–37.

Networking Algorithms and Performance Evaluation

Self-learning method for industrial firewall rules based on hash algorithm

Y.Q. Lei
University of Chinese Academy of Sciences

W.L. Shang, M. Wan & P. Zeng
Shenyang institute of Automation, Chinese Academy of Science, Shenyang, Liaoning, China
Key Laboratory of Networked Control Systems, Chinese Academy of Sciences, Shenyang, Liaoning, China

ABSTRACT: In order to realize intrusion detection of the new unknown industrial virus, a self-learning algorithm is needed to study the information model and behavior characteristics of network communication in industrial control system, automatically generating a comprehensive rules white-list for industrial firewall detection. This paper proposes a self-learning algorithm based on statistical analysis of log rules, it dynamically generates and updates rules, using hash algorithm to add up the number of data flow, and resulting in a new filtering rules to provide intelligent self-learning ability for industrial firewall. The experimental results show that, compared with traditional IT firewall, the industrial firewall can detect a variety of illegal access, and automatically update and generate a new industry with the firewall rules.

1 INTRODUCTION

1.1 *Background*

In 2010, the "Stuxnet" attacked Iran's nuclear facilities, which made the industrial control system security up to the national strategic security. Although traditional IT security technology is relatively mature, it can't be directly used for industrial control systems, one of the main reasons is that the industrial control system uses a proprietary protocol specification (such as Modbus TCP, DNP3, IEC61870-5-104, MMS, GOOSE etc.), while the form of protocol specifications is achieved to be diverse (Peng et al. 2012, Vinay et al. 2006).

The protocol security issues in industrial control system can be divided into two categories: one is caused by the design and description of the protocol itself, and the other is caused by the achievement. The main proprietary communication protocol or statute vulnerability in the industrial control systems can be used as a breakthrough to attack industrial terminal equipment. Thus, industrial controllers for intrusion detection technology should also be able to analyze proprietary protocols of industrial control system.

From the perspective of intrusion detection, traditional TCP/IP network makes communication with the dynamic and unpredictable behavior. Different from traditional TCP/IP network, industrial control system communicates network with the characteristics of "Status limited" and "limited behavior" (Zhu&Sastry 2010, Niv&Avishai 2013)."Status limited" refers to communicating with regularity and

stability, namely traffic rules. "Limited behavior" refers to fixed behavioral characteristics and predictable behavior patterns, thus simplifying the description of the model. On specific communications equipment, it usually needs to repeat the limited operations.

1.2 *Related research*

In self-learning method for industrial firewall rules, there are scholars from different levels to carry out related research. Zhao & Zou (2009) for the factors that affect firewall performance and adaptive capacity. The paper introduces new filter rules reference model and more general adaptive algorithm, by using a time-based global information collection and statistical weight calculation, considering the recent network traffic and historical accumulation of data, to achieve a dynamic generation of personal firewall filtering rules. However, the model is based on expert experience to pre-set weight factor, but with different personal experiences the weighting factors are quite different, and it is difficult to set. Ren et al. (2006) the frequency of the use of filter rules within a certain time for statistical analysis, and dynamically adjust the relative order of the rules in the rule list according to the analysis results. So the most frequently used rules at the top of the rule list, which can reduce the packet rules match time, improve the performance of the firewall. However, the statistical time parameter T is determined and over simplified, it is difficult to deal with complex environmental requirements of industrial systems. Yun et al. (2013) analyzing DNP3 protocol communication behavior between master and slave in SCADA System, establish based white-list rule set which is based on the sub-frame mode. The use of sub-frame method represents the operating characteristics of application layer data and the arrival time interval of packets. Log law studies using statistical methods, generating rule set by self-learning algorithm, and security experts analyze packet framing regularity to extract white-list rules.

1.3 *Contents*

Traditional host-based firewall role in network layer, only in accordance with pre-configured filter rules to monitor and analyze datagram, and can only provide a single-level, static network security, which was set by network security experts based on experience and existing knowledge, it is difficult to adapt to ever-changing network (Qiu 2006, Hao 2010).

In response to these problems, this paper proposes a new idea, through machine learning methods, learning behavioral characteristics of industrial control systems and regularity of communication networks. It is expressed as rules or model and automatically generates white-list rules, providing theoretical and technical support for intrusion detection system.

2 SELF-LEARNING METHOD EMBODIMENTS

2.1 *Architecture*

Install Ubantu 10.04 system in the VMware Worksation, using Libpcap (Rao et al. 2011, Liu 2010) to capture packets, it can not only improve the efficiency of packet capture, but also enhance the portability of the system in favor of secondary development. After the industrial firewall started, packet capture module started capturing packets flowing through the network and preprocessing the captured data packet. Data packets change to txt log files after pretreatment and the self-learning module learns the rules from the txt log files. The basic schematic of statistical analysis based on the log rule self-learning algorithm is shown in Figure 1.

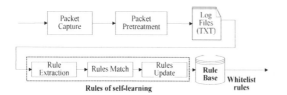

Figure 1. Statistical analysis based on the log rule self-learning algorithm.

Statistical analysis based on the log rule self-learning algorithm mainly consists of the following modules:

1 Packet capture and decoding module: This module is responsible for crawling all packets flowing through the network, capture the traffic date, and decoding the link-layer packet.
2 Packet pre-processing module: This module is responsible for normalizing the data packet, recombinant IP fragmentation, TCP stream reassembly, etc.
3 Rule self-learning module: Rules self-learning module to search and match rules, learn and automatically generate the new rules.

2.2 *Packet capture module*

Figure 2 shows a standard Libpcap library-based application flow.

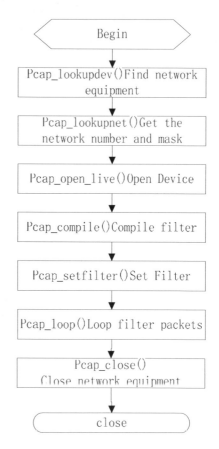

Figure 2. Libpcap library-based application flow.

The NIC is responsible for data transceivers work, which has four operating modes: broadcast mode, multicast mode, direct mode, and mixed mode. To capture all the data flowing through network for analysis, it should be set to promiscuous mode. To achieve packet capture module can use Libpcap software under Linux systems (Gao &Xie 2004), which provides a rich API function that can help users quickly develop packet capture program. Programs written using Libpcap free to cross-platform are used so powerful that it can capture and send raw packets from the NIC, filter and store packets.

2.3 *Packet decoding module*

Use pcap_loop() function to capture network packets, decoding the packet header information for analysis by layer, the specific process is as follows:

1 Firstly, by Ethernet frame header information packet analysis is done to get the source MAC address, destination MAC address and the network layer packet encapsulation type. As for the IP packet type value 0x0800, 0x0806 for ARP request/response packet type value, 0x8035 for RARP request/response packages.
2 After parsing the data link layer header information, we then analyze the network layer header. Assuming the IP packet, we analyze the important information of IP header, obtain the source IP address, destination IP address, and the identifier field of upper layer protocol and so on. Determine transport layer protocol type by the identifier field value, such as the field value of 1 indicates the ICMP protocol data, 6 for TCP protocol data, and 17 for UDP protocol data.
3 Analyze the transport layer header. Such as TCP protocol, it obtains the destination port number and source port number from the first two bytes, and further obtains the upper layer protocol information, such as 21 for the FTP protocol, 80 for the HTTP protocol, 23 for the Telnet protocol, etc.
4 By extracting the header information of the link layer, network layer, transport layer protocol, and then the payload data packet is analyzed to obtain the data of this packet.

2.4 *Packet preprocessing module*

Used libnids (Zheng et al. 2004, Chen & Chang 2005) to complete the packet preprocessing, specifically including IP fragment reassembly and TCP data stream reduction function. IP fragment reassembly is an important work of the IP layer. We define an ip_ callback () function, and then call nids register ip () function to register and receive the correct IP packets, which means no fragmentation has completed the head of IP fragment reassembly.

To receive the TCP stream data which is exchanged must define the callback function tcp callback (), to achieve the TCP protocol analysis and call nids register tcp () function to register and achieve the TCP stream reduction.

3 RULES SELF-LEARNING METHOD

Used statistical analysis to optimize firewall filter rules based on the assumption that the packet filtering feature has some continuity in a certain period of time. After real-time statistics, the most frequently used visits path rules are extracted out as patterns, in order to achieve lower rules match time, and improve the performance of the firewall. The so-called self-learning refers to that it automatically generates a new rule according to the network administrator for rules processing method and the actual rules of the network model, so that it is better to meet the needs of the actual network. Statistical analysis of the rule-based self-learning, by extracting data rules, searching and matching with rule set to learn and automatically generate the new rules.

3.1 *Algorithm flow*

Rules self-learning method is based on statistical analysis model. Used libpcap to capture packets, after pre-processing the data to generate a log file, and then extract keyword segments (source address, destination address, action type, etc.), and according to the statistics of visits between the source host and destination host within a certain time interval, forming a rule set. Determine whether the user's request can be accessed by the established threshold condition, thus achieving self-learning.

A typical industrial firewall path rules can be expressed such as:-sip 192.168.12.1 -dip 192.168.123.4 -p Modbus/TCP -sport 1024:65535 -dport 502 -j drop. Sip represents the source address, dip indicates the destination address, p represents the protocol, mainly for Modbus/TCP protocol, sport is the source port, dport indicates the destination port, j is represented as the action, usually to "accept " or "drop ", secure access path means the action of firewall "accept ".

Through the analysis of secure access paths between source host and destination host, to achieve self-learning safe firewall path rules, you can use the following steps:

Step 1: Extract all access paths from logs;
Step 2: Statistics of visit times within a certain time interval for each access path (respectively in days / weeks / months for access time interval);

Step 3: Calculate the safety factor of each path based on statistical statistic results;

Step 4: Set different thresholds depending on the time interval;

Step 5: The safety factor is compared with the set threshold, and the administrator determine whether to put the path into the default security access list, thus forming a white list rules.

Secure access factor of each path can be expressed as: S="*SafetyVisits*"/"*TotalVisits*"

secure access factor greater, indicating that the path has greater probability of crossing the firewall, reflecting how legal the path is.

3.2 *Establish rules*

The packet store in the form of txt text file after pretreatment, in the phase of establishing rules, reads the contents of text file to extract quintuple stream information. The so-called stream refers to the data packet set which has certain attributes and life cycle, quintuple refers to the source IP address, destination IP address, source port number, destination port number, and protocol. This paper describes the flow of a quintuple flow using CRC20 (Chen et al. 2008) algorithm to calculate the hash value of quintuple, according to hash value of the data packets to put them into different streams, so as to achieve the purpose of the packet diversion. After the split, the quintuple information packet is stored in the corresponding flow table, thus completing the statistical data stream. The statistics are compared with threshold value, if it is greater than the threshold value, then default it for the safe access and interact with administrators, reminding whether to add the quintuple to security list and depending on the result to automatically generate rules.

3.3 *Searching and matching rules*

The action after matching rules (Hadziomanovic et al. 2012) contains accept, log, and drop. Accept means through the firewall; log record network traffic but not alarm; drop means to discard packets that are in violation of the rules.

Rule matching workflow is shown in Figure 3.

According to quintuple information, calculate the corresponding hash value, and then look for the hash table content, if the quintuple information already exists in the hash table, then it does not need match filtering rules, directly do data statistics and accept/drop the packet; if not, then record the quintuple information, and initialize a new connection to add it into the hash table.

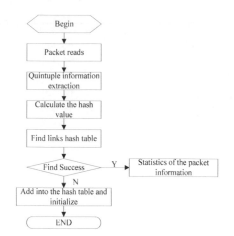

Figure 3.   Rule matching workflow.

4   ASSESSMENT TEST

Structure experimental environment as shown in Figure 4, using Linux development platform, industrial system firewall is based on the white-list policy (Stouffer et al. 2011, Chen 2012), which means the visit does not belong to the white-list rule which is regarded as illegal access. Thus, the experimental simulation of a small industrial environment, the IP address of 210.72.141.91 to generate abnormal data which comes from legitimate users' new behavior that is not in the white-list rule, IP address of 210.72.141.86 generates abnormal data belonging to illegal users, IP address of 210.72.141.168 generates valid data, and the path rules have been in the white- list rule base.

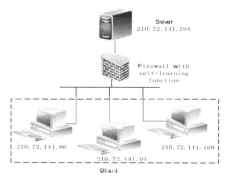

Figure 4.   Structure experimental environments.

Using packet capture and analysis method, getting the recent week of log information given by industrial firewall with self-learning function or not separately, IP address of 210.72.141.168, 210.72.141.91, and 210.72.141.86 correspond to path 1, path 2, and path 3 respectively. In order to do better comparative analysis of the data, setting the total visit number of each path at a daily basis are the same, through the hash statistical analysis, getting safety visits number of each path this week is shown in Figure 5,6, and7 respectively.

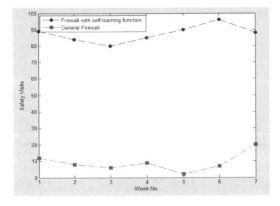

Figure 5.   Safety visits number of path 1 this week.

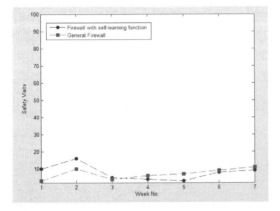

Figure 6.   Safety visits number of path 2 this week.

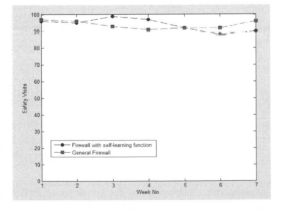

Figure 7.   Safety visits number of path 3 this week.

According to the formula mentioned in section 2.1 to obtain secure access factor of each paths and convert it to secure access, probabilities histogram is shown in Figure 8.

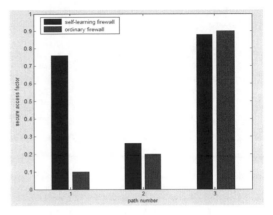

Figure 8.   Secure access probabilities histogram.

Applying the algorithm of self-learning rules processes each access path rules, setting the threshold at 0.72 according to the actual situation. By experimental results what can be seen in Figure 8, the path 1 and 3 will be the default safety path for firewalls that have the self-learning ability, rather than only considered path 3 legitimate. Moreover, by blocking the path 2 successfully, indicating self-learning firewall like ordinary fire wall, with the ability to block malicious data, it fully explains that the self-learning firewall can do intelligent learning, and adapt to changes in user demand, and promptly update the rules, with strong ability to learn.

The experimental result shows that compared with traditional IT firewall, the industrial firewall is capable of preventing unauthorized access, it can automatically update and generate the firewall rules and has good real-time performance and safety.

5   CONCLUSIONS

This paper presents a statistical analysis of log-based rule self-learning algorithm to identify industrial firewall white-list rule through learning the communication path rule. Experimental results show that the proposed self-learning firewalls with a strong security and learning ability, compared with ordinary firewalls, can effectively respond to changes in demand and update the rule base real-time.

Although the proposed algorithm can be self-learning path rules, generate new path rules, but how to improve the availability, reduce their risk of the rules, and increase the reliability of threshold setting is for future research. At the same time, the rule does not include the value of industrial application layer protocol data and other parameters, which needs further research.

REFERENCES

[1] B. Zhu & S. Sastry. 2010. SCADA-Specific intrusion detection/prevention systems: a survey and taxonomy. *In Proceedings of the First Workshop on Secure Control Systems (SCS'10)*. Stockholm, Sweden.

[2] Chen, Y. & Lu, X. & Sun, Z. 2008.Research hashing algorithm for flow management. *Computer Engineering and Science* 30(4):26–29.

[3] Chen, Y. 2012. *Linux Network Security Technology and Implementation*. Beijing: Tsinghua University Press.

[4] Chen, Z. & Chang, J. 2005.Libpcap and Libnids based network intrusion detection system (NIDS) Design and Implementation. *Fujian computer* 5(4):56–57.

[5] D. Hadziomanovic& D. Bolzoni& P.H. Hartel. 2012. A Log Mining Approach for Process Monitoring in SCADA. *International Journal of Information Security*11(4): 231–251.

[6] Gao, G. &Xie, Z. 2004. Network Intrusion Detection System packet capture and packet parsing. *Qiqihar University* 20(3):47–50.

[7] Hao, Y. 2010.*Research and implementation of deep packet inspection firewall host*. Chengdu: University of Electronic Science and Technology.

[8] J. H. Yun & S. H. Jeon & K. H. Kim & Woo-Nyon. 2013. Burst-based anomaly detection on the DNP3 protocol.

[9] K. Stouffer & J. Falco & K. Scarfone. 2011. Guide to Industrial Control Systems (ICS) security. *National Institute of Standards and Technology (NIST) Special Publication* 23(9):800–820.

[10] Liu, H. 2010. *Network Security Technology*. Beijing: Mechanical Industry Press.

[11] Niv, Goldenberg &Avishai, Wool. 2013. Accurate modeling of Modbus/TCP for intrusion detection in SCADA systems. *International Journal of Critical Infrastructure Protection* 6(2): 63–75.

[12] Peng, Y. &Jing, C.2012.Industrial Control System Information Security Research. *Tsinghua University (Natural Science)* 52(10): 1396–1408.

[13] Qiu, Z. 2006. *Intelligent firewall filtering rules to learn and Optimization*. Harbin: Harbin Institute of Technology.

[14] Rao, M. & Du, Z. & Han, Q. & Li, Q. 2011.Embedded Linux Firewall Design and Implementation. *Intelligent Computer and Applications* 1(3):35–39.

[15] Ren, A. & Yang, S. & Li, H. 2006. An optimization matching method of firewall rules based on statistical analysis. *Computer Engineering and Applications* 4(9):162–164.

[16] Vinay, M. Igure & Sean, A. Laughter& Ronald D. 2006.Williams. Security issues in SCADA networks. *Computers & Security* 25: 498–506.

[17] Zhao, Y.& Zou, H. 2009. Adaptive optimization strategy based personal firewall. *Computer Engineering*35(10): 109–111.

[18] Zheng, L. & Zhang, H. & Liu, W. &Huang, W. & Fang, S. 2004. Research and Implementation of Linux Network Firewall IP fragment reassembly algorithm. *Computer Engineering and Applications*3(5):153–155.

[9] *International Journal of Control and Automation* 6(2): 313–324.

Network Security and Communication Engineering – Chan (Ed.)
© 2015 Taylor & Francis Group, London, ISBN: 978-1-138-02821-0

Intrusion detectors design with self-configuring multi-objective genetic algorithm

E.A. Sopov & I.A. Panfilov
Siberian State Aerospace University, Krasnoyarsk, Russia

ABSTRACT: This work focuses on the problem of the automated design of efficient network intrusion detectors. The SVM-based classifier with the problem attributes (features) selection is proposed. The problem of designing the classifier is viewed as a two-criteria optimization problem of maximizing the detector accuracy and minimizing the number of features. The self-configuring multi-strategy and multi-objective co-evolutionary genetic algorithm is used for the Pareto set approximation. Finally, the results of numerical experiments are proposed for the PROBE attack and e-mail spam detection problems.

1 INTRODUCTION

Global network activity has become an essential part of daily life. Almost every communication and computer device has the internet access. Moreover, many complex electronic domestic appliances are also connected to the global network to receive functional instructions, to report operation failures and so on. At the same time, the number of network threats and attacks of various types is constantly increasing.

There exist many intrusion detection systems (IDS) that are widely used in network security systems (Patcha 2007). Unfortunately, conventional security software requires a lot of human effort to identify and resolve threats. Most IDS use heuristics, which are based on statistics from past attacks and the activity presuming. Each new essential incident results in new statistical analysis and IDS adaptation or total rebuild. Thus, there is an actual challenge to design efficient techniques that are able to deal with complex data and automatically form and tune the decision-making algorithms for the IDS.

Many practical applications of IDS are based on the intelligent information technologies approach. In particular, data-mining, web mining and web-usage mining are widely used. In the field of network, anomaly and intrusion detection classification-based techniques are reasonably popular. There exist neural networks, fuzzy logic classifiers, nature-inspired algorithms and others (AmalrajVictoire 2011, Bukhtoyarov 2012). The classification-based techniques demonstrate a good ability to generalize; even if the given data are noisy and have missing values. At the same time, the classifier designing and tuning are complex problems themselves, comparable to the original problem.

In this work, the support vector machine (SVM) approach is applied to the classification problem

solved in the IDS. The problem statement is enhanced to provide the minimization of problem features that leads to the selection of significant features for the given data. The SVM classifiers are automatically configured with a novel multi-objective genetic algorithm. The final decision about intrusion is made by using an ensemble of classifiers.

The rest of the paper is organized as follows. Section 2 describes the techniques. Section 3 states the problem in a formal way and describes proposed methods. Section 4 demonstrates the approach effectiveness at solving benchmark problems from the area of network security. In the conclusion the results and further research are discussed.

2 RELATED WORK

Classification problems are the problems of identifying that which is of predefined categories and is a new instance belong. Many intrusion detection problems can be formulated in the same way. Usually problems with only two classes are considered (for example, legitimate or illegitimate connections, spam or not spam and so on). As previously mentioned, intelligent information technologies are very popular. In particular, many researches use artificial neural networks.

In this work, the SVM-based classifier is used (Vapnik 1982). The efficiency of SVM-based classifiers has been established and shown in various studies. Many researches note that the SVM shows better results than artificial neural networks for many complex classification problems.

The SVM can be characterized as a supervised learning algorithm capable of solving linear and non-linear binary classification problems. A training set contains m patterns $\{(x_i, y_i)\}$, $i=1,...,m$, where

$x_i \in X \subseteq R_n$ is an input vector and $y_i \in \{-1,+1\}$ its corresponding binary class label. The main idea of the SVM-classification is to separate examples by means of a maximal margin hyper-plane (Cristianini & Shawe-Taylor 2000). To construct the SVM classifier, one has to minimize the norm of the weight vector w under the constraint that the training patterns of each class reside on the opposite sides of the separating surface (see Equation 1).

$$\|\mathbf{w}\| \to \min$$
$$y_i \left(< \mathbf{w}, x_i > + b \right) \geq 1,\, i = \overline{1, m} \tag{1}$$

where $< \ldots >$ is a dot product, w is the weight vector and b is a shift parameter. The dot product can be substituded with another non-linear operation (so-called kernel function), but choosing the proper kernel is also not a trivial problem.

Thus, the most discriminating hyper-plane can be constructed by solving the constrained optimization problem. A new instance x is classified by applying the following classifier (see Equation 2):

$$y(x) = sign\left(< \mathbf{w}^*, x > + b^* \right) \tag{2}$$

where w^* and b^* are determined by (1).

Real-world classification problems may be associated with instances presented by a great number of features. The increased dimensionality of data makes the testing and training of classification methods very difficult. Moreover, it is usually a priori unknown whose features are useful for the problem solving and whose features contain redundant or even irrelevant information. Such a problem is called feature-induced over fitting.

Feature selection is a process that selects a subset of original features that provides better performance for the given classification problem solved. Training with the reduced set of attributes decreases computation time and complexity, and makes the patterns easier to understand (Karegowda et al. 2010).

An approach for feature selection that uses the classification method itself to measure the importance of the feature set is called a wrapper model. Wrapper methods generally result in better performance than other selection methods because the feature selection process is optimized for the classification algorithm to be used. However, wrapper methods are too expensive as they lead to a complex search problem.

Thus, we can formalize two objectives for the optimization problem of the SVM-classifier design: classification accuracy and minimization of features. In general, the objectives are in conflict. This means that better accuracy requires more information (more features are used) and, minimizing features lead to lower accuracy.

Such optimization problem is called multi-objective and it leads to find a set of effective (compromise, non-dominant, or Pareto-optimal) solutions. It is obvious that the Pareto set contains very different solutions with various combinations of the accuracy value and the number of features. As they all represent a balance between generalization and accuracy, it seems to be a good idea to construct a collective decision.

There exist many heuristics for calculating the ensemble output, including simple averaging, weighted averaging, majority voting and ranking (Wiering & Hassel t2008). In (Popov et al. 2012) a new scheme, based on the estimation of the local effectiveness of the ensemble members, was proposed. All studies show that the ensemble provides better results on average than any member does on average.

3 SVM CLASSIFIER DESIGN

We formulate the given problem in a formal way. The first objective $f_1(x)$ is the standard classification accuracy (see Equation 3). The second objective $f_2(x)$ corresponds to the number of features, which are used in the model building.

$$f_1(X) = \frac{CP}{TP} \cdot 100\% \tag{3}$$

where TP is a total number of patterns of the dataset, CP is a number of correctly classified patterns; X is a certain SVM using a specific number of features.

The multi-objective optimization problem is as follows (see Equation 4):

$$\begin{cases} f_1(X) \to \max \\ f_2(X) \to \min \end{cases} \tag{4}$$

As previously mentioned, the objectives are in conflict, so we need to deal with the concept of Pareto optimality. In the field of multi-objective optimization, the evolutionary optimization approach is known as a very effective one for many complex multi-objective problems (Deb 2001). One more reason to use the stochastic search is that objectives are algorithmically evaluated. The population-based search also provides a better representation of the final approximation of the Pareto set.

In this work, we use an efficient approach called the self-configuring co-evolutionary multi-objective genetic algorithm (SelfCOMOGA). The SelfCOMOGA was introduced in (Ivanov & Sopov 2013) and its effectiveness was demonstrated with real-world problems in (Ivanov & Sopov 2014).

The key feature of the Self COMOGA is a parallel evaluation (so-called co-evolution) of many multi-objective strategies. The total population is initially divided into disjoint sub-populations of equal size. The portion of the population is called the computational resource. Each subpopulation corresponds to a certain multi-objective search strategy (algorithm) with its own parameter values and evolves independently. After a period, the performance of individual algorithms is estimated and the computational resource is redistributed. Algorithms with better performance increase their population size. Finally, random migrations of the best solutions are presented to equate start positions of algorithms for the run with the next period.

Such a co-evolution technique eliminates the necessity to define an appropriate algorithm for the problem as the choice of the best algorithm is performed automatically during the run. Moreover, the Self COMOGA performs better than the average individual techniques on.

The Self COMOGA is a binary genetic algorithm. We apply a straight forward representation for the candidate solutions. The chromosome consists of two parts. The first one corresponds to the SVM model and represents the values of the weight vector and the shift parameter. The second part defines the attributes (features), which are included into the SVM model. If a certain feature is included, the corresponding value of the weight vector is used while estimating the model performance. Otherwise, the value of the weight vector is ignored.

The list of multi-objective strategies is included in the Self COMOGA, its parameter values and general control parameters are discussed in detail in (Ivanov & Sopov 2013) and are not a focus of this work.

The majority voting is used to build the collective decision for the ensemble of the Pareto-optimal SVM.

4 EXPERIMENTAL RESULTS

The performance of the proposed technique is evaluated to solve the following problems in the field of network security.

The first problem corresponds to the detection of PROBE attacks. The goal is to build a predictive model (a classifier) capable of distinguishing between "bad" connections, called intrusions or attacks, and "good" normal connections. Attacks fall into four main categories: DOS (denial-of-service, e.g. syn flood), R2L (unauthorized access from a remote machine, e.g. guessing password), U2R (unauthorized access to local super user (root) privileges, e.g., various "buffer overflow" attacks) and probing (surveillance and other probing, e.g., port scanning). The total number of attack types in the dataset is 38. All data can be found in the Machine Learning Repository of the UC Irvine KDD archive (Frank 2010).

The dataset contains 4,000,000 instances. The total number of attributes is 42, but in (Stolfo et al. 2000) the following attributes are proposed as the most relevant to the problem solved: 1, 3, 5, 8, 33, 35, 37, and 40.

To build the classifier, all patterns relevant to PROBE attacks are marked as referring to the first class, the others are marked as referring to the second class. The results are compared with other approaches collected in (Malik 2011) and (Semenkin et al. 2013) (see Table 1).

Table 1. Performance comparison for the PROBE attack detection problem.

Approach	Accuracy (detection rate, %)	False positive rate, %
PSO-RF	99.92	0.029
SelfCGP+ANN2	99.79	0.027
Random Forest	99.80	0.100
SelfCGP+ANN1	98.78	0.097
Bagging	99.60	0.100
PART (C4.5)	99.60	0.100
NBTree	99.60	0.100
Jrip	99.50	0.100
Proposed approach	**99.48**	**0.078**
Ensemble with majority voting	99.41	0.043
Ensemble with weighted averaging	99.17	0.078
Ensemble with simple averaging	99.18	0.122
BayesNet	98.50	1.000
SMO (SVM)	84.30	3.800
Logistic	84.30	3.400

The second problem is the e-mail spam detection. The dataset contains 4,600 instances (e-mails) of two classes: spam and non-spam. Each instance is presented with 57 quantitative attributes, which are extracted from e-mail content. All data can also be found in the Machine Learning Repository of the UC Irvine KDD archive (Frank 2010). The results are compared with other approaches collected in (Dimitrakakis 2006) and (Semenkin et al. 2013) (see Table 2).

Table 2. Performance comparison for the SPAM detection problem.

Approach	Accuracy, %
SelfCGP+ANN2	94.96
Proposed approach	**94.81**
Ensemble with majority voting	94.77
Ensemble with weighted averaging	94.67
Ensemble with simple averaging	94,57
SelfCGP+ANN1	94,57
Boost	93.52
RL	92.59
MLP	91.67

One realization of the SVM ensemble is presented in Table 3. All members of the ensemble are the Pareto optimal solution for the previously mentioned optimization problem.

Table 3. Ensemble structure for the SPAM detection problem.

Ensemble	Accuracy, %	Number of features
SVM1	92.17	23 of 54
SVM2	90.32	35 of 54
SVM3	78.02	16 of 54
SVM4	92.38	30 of 54
SVM5	63.33	12 of 54
SVM6	67.01	11 of 54
SVM7	89.87	33 of 54
SVM8	57.11	16 of 54
SVM9	85.34	33 of 54
SVM10	93.05	37 of 54

As we can see, the proposed approach demonstrates sufficiently high results. It yields to the best techniques, but out performs the average results (which are still sufficient). The proposed approach has important advantages. First of all, the results are obtained in an automated way. We also obtain not only a decision making model (namely the SVM), but a set of models with different numbers of features. The models with a small number of features can be easily analyzed to be included in future solutions.

5 CONCLUSIONS

The work presents an original approach to the automated design of efficient network intrusion detectors. The detector in the form of a classifier is built by using the ensemble of the SVM. We have stated the multi-objective optimization problem of designing the SVM with high accuracy and minimum features. We have applied the proposed approach to two real-world problems: the PROBE attack and the SPAM detection. The experimental results demonstrate high performance even with such a simple core technique (the linear SVM).

In further works, we intend to configure the kernel of the SVM to obtain more complex and non-linear classification models. The proposed approach should be tested with more network security problems.

ACKNOWLEDGMENTS

Research is performed with the financial support of the Ministry of Education and Science of the Russian Federation within the federal R&D programme (project RFMEFI57414X0037).The author expresses his gratitude to Mr. Ashley Whitfield for his efforts to improve the text of this article.

REFERENCES

[1] Amalraj Victoire, T.& Sakthivel, M. 2011. A Refined Differential Evolution Algorithm Based Fuzzy Classifier for Intrusion Detection. European Journal of Scientific Research, vol.65, No. 2: 246–259.

[2] Bukhtoyarov, V., Semenkin, E. 2012. Neural networks ensemble approach for detecting attacks in computer networks. Evolutionary Computation (CEC), 2012 IEEE Congress: 3401–3406.

[3] Cristianini, N. & Shawe-Taylor, J. 2000. An introduction to support vector machines and other kernel-based learning methods. Cambridge: Cambridge University Press.

[4] Deb, K. 2001. Multi-objective optimization using evolutionary algorithms. John Wiley & Sons, Ltd. England.

[5] Dimitrakakis, C. & Bengio, S. 2006. Online Policy Adaptation for Ensemble Classifiers. IDIAP Research Report 03–69.

[6] Frank, A. & Asuncion, A. 2010. UCI Machine Learning Repository: http://archive.ics.uci.edu/ml. Irvine, CA: University of California, School of Information and Computer Science.

[7] Ivanov, I. & Sopov, E. 2013. Investigation of the self-configured coevolutionary algorithm for complex multi-objective optimization problem solving. Control systems and information technologies, 1.1 (51): 141–145.

[8] Ivanov, I. & Sopov, E. 2014. Design Efficient Technologies for Context Image Analysis in Dialog HCI using Self-configuring Novelty Search Genetic Algorithm. In proc. of the 11th International Conference of Informatics in Control, Automation and Robotics, Vol.2: 832–839.

[9] Karegowda, A.G., Jayaram, M.A., Manjunath, A.S. 2010. Feature Subset Selection Problem using Wrapper Approach in Supervised Learning. International Journal of Computer Applications (0975 – 8887) Volume 1, No. 7: 13–17.

[10] Malik, A.J., Shahzad, W., Khan, F.A. 2011. Binary PSO and random forests algorithm for PROBE attacks detection in a network. IEEE Congress on Evolutionary Computation: 662–668.

[11] Patcha, A. & Park, J-M. 2007. An Overview of Anomaly Detection Techniques: Existing Solutions and Latest Technological Trends. Computer Networks.

[12] Popov, E.A., Semenkina, M.E., Lipinskiy, L.V. 2012. Evolutionary algorithm for automatic generation of neural network based noise suppression systems. Vestnik. Bulletin of Siberian State Aerospace University. Vol. 4 (44): 79–82.

[13] Semenkin, E., Semenkina M., Panfilov, I. 2013. Neural network ensembles design with self-configuring genetic programming algorithm for solving computer security problems. Advances in Intelligent Systems and Computing. - Volume 189 AISC, 2013: 25–32.

[14] Stolfo, S., Fan, W., Lee, W., Prodromidis, A., Chan, P. 2000. Cost-based Modeling for Fraud and Intrusion Detection: Results from the JAM Project. In Proceedings of the 2000 DARPA In-formation Survivability Conference and Exposition (DISCEX '00): 130–144.

[15] Vapnik, V.N. 1982. Estimation of Dependences Based on Empirical Data. New York: Springer.

[16] Wiering, M.A. & Hasselt, H.V. 2008. Ensemble algorithms in reinforcement learning. IEEE Trans. Syst. Man Cybern. Part B: Cybern., vol. 38, no. 4: 930 –936.

Network Security and Communication Engineering – Chan (Ed.)
© *2015 Taylor & Francis Group, London, ISBN: 978-1-138-02821-0*

A novel EM algorithm for stable optimum

Y. Xiao, G.R. Xuan & Z.Q. Yang
Department of Computer Science, Tongji University, Shanghai, China

Y. Q. Shi
Department of ECE, New Jersey Institute of Technology, Newark, New Jersey, USA

ABSTRACT: The main contribution of this paper is to propose a novel EM algorithm that utilizes ideas of EM algorithm and maximum-entropy uniform distribution to find a stable optimum in an uncomplicated way. The conventional EM algorithms may suffer from the following two problems: first, it may converge to an undetermined local maximum; second, the algorithm may suffer from singularity. The novel EM algorithm is deterministic that the solution is determined solely by the initial condition. In our novel EM algorithm, a stable and optimal solution can be obtained by using a uniform distribution instead of special initial condition. In addition, a positive perturbation scheme is adopted to avoid singularity. Experimental results have demonstrated that the novel EM is uncomplicated and effective for stable optimum compared with some prior arts.

1 INTRODUCTION

The Expectation Maximization (EM) algorithm is one of the most popular methods for obtaining the maximum likelihood estimate. One of its convenient properties is that it guarantees an increase in the likelihood function in every iteration (Dempster 1977). Moreover, since EM operates on the log-scale, it is analytically simple and numerically stable for distributions that belong to the exponential family such as Gaussian. However, the conventional EM algorithm has some drawbacks. First, it may converge to a local optimum of the likelihood function. Second, the algorithm may suffer from singularity. A great deal of efforts for EM algorithm to obtain the global optimal solution has been made, including the Deterministic Annealing Algorithm (Rose 1998) and the Minimum Message Length (Figueiredo 2002). Our primary goal is to achieve stable optimal solutions in an uncomplicated way when the conventional EM attains locally optimal solutions and other efforts to obtain the globally optimal solutions are rather complicated.

In this paper, a novel EM algorithm is proposed. First, we use the distribution instead of parameter as initial condition so as to easily manipulate the EM iteration. Second, the uniform distribution instead of peak-shape distribution is used to provide the most random initialization. That is, the proposed novel EM algorithm has been devised based on the use of an uniform distribution (having maximum entropy) as initial condition that can achieve global optimality. In addition, a positive perturbation scheme is proposed in our algorithm to avoid suffering from singularity. The proposed scheme has been shown effective in the experimental works.

The rest of the paper is organized as follows. The motivation, diagram, and flowchart are introduced in Section 2. We apply our method to several image data sets, experimental results of the novel EM are reported in Section 3. Conclusions are drawn in Section 4.

We first present some notations which are necessary for formulas in this paper:

C *number of components of Gaussian, i = 1,…,C*
N *number of Samples (Pixels)*
Y_j *sample vector j = 1,…, N*
K *number of feature (Gray levels), k = 1,…, K*
P_i *Priori probability*
M_i *Mean vector*
Σ_i *Covariance matrix*
λ *or* λ' *Simplified form of parameters before or after estimation where* λ *or* λ' *means [P_i, M_i, Σ_i]*
w_{ik} *Weight distribution*
X_k *Feature vector of sample vector Y_j*
$h(X_k)$ *Histogram with feature X_k*
$q_i(X_k)$ *or* $q_i(Y_j)$ *Posteriori distribution for X_k or Y_j respectively*
q *Simplified form of posteriori distribution*
L *Likelihood function*
$N(x_k| m_i, \sigma_i^2)$ *or* $N(X_k|M_i, \Sigma_i)$ *one or multidimension Gaussian Distribution*

2 CONVENTIONAL EM ALGORITHM

The Expectation maximization (EM) algorithm is an iterative procedure designed to produce maximum likelihood estimates in incomplete data problems. It has been mathematically proved that the likelihood will increase during each iteration but may not always produce the global optimizer of the likelihood.

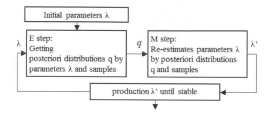

Figure 1. Block diagram of conventional EM.

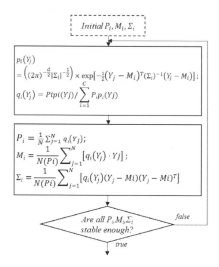

Figure 2. Flowchart of conventional EM.

The block diagram of conventional EM for Gaussian Mixture Model (GMM) is shown in Figure 1. The EM iteratively produce estimates to the maximum likelihood estimate (MLE) of the parameters λ, denoted by λ^*, by maximizing the log-likelihood function. In Figure 1, the parameters λ of GMM is set as its initial condition, the iteration will begin with getting conditional distribution and posteriori distributions q and samples in E step. The parameters λ will be re-estimated by posteriori distributions q and samples in M step. It successively alternates between the E step and M step until convergence while the difference or relative difference of successive log-likelihood values fall below a specified tolerance.

The flowchart of conventional EM is shown in Figure 2. The solution generated by EM is determined solely by the initial condition. And depending on its starting value, EM is a locally converging algorithm, which may not produce the global optimizer of the likelihood function. It may suffer from local maximum and singularity.

3 MOTIVATION OF THE NOVEL EM ALGORITHM

First, we use the distribution instead of parameter as the initial condition of EM to control the iterative process more easily. In the conventional EM for GMM, we will set the initial condition for all the parameters: mean vectors, covariance matrices, and prior probabilities. Differently, if we use the distribution as the initial condition, we will only need to set up an initial distribution. And, the number of distributions is equal to the number of GMM components, and it will be quite simple.

Second, we use the uniform distribution as the initial condition, this will provide us with the most random initialization, and hence with the maximum entropy among various distributions that can be used in the algorithm initialization. This is most important because it solves the issue of "local optimum".

4 NOVEL EM ALGORITHM

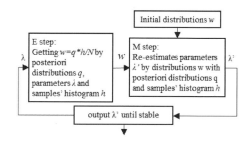

Figure 3. Block diagram of proposed novel EM.

The novel EM is based on a realization of uniform distribution which is known to have maximum entropy. The block diagram of the novel EM is shown in Figure 3. The likelihood by histogram h(Xk) is defined as:

$$L = \sum_{i=1}^{N} ln\big(p(Y_i|\lambda)\big) = \sum_{k=1}^{K} h(X_k)ln(p(X_k|\lambda)),$$

were Mixed distribution: $p(X_k|\lambda) = \sum_{i=1}^{C} P_i p_i(X_k)$ and Gaussian distribution: $p_i(X_k) = N(X_k|M_i, \Sigma_i)$
The novel EM's flowchart is shown in Figure 4.

We choose to work with the histogram instead of the samples in order to simplify the iterative computation greatly (Xuan 2001). The formulas in Figure 4 are different from those in Figure 2, i.e., the novel EM works with the histogram h(Xk), while the EM works with sample Yj. Figure 4 however still corresponds to Figure 3. The proposed novel EM algorithm, on the one hand, utilizes ideas of EM algorithm and maximum-entropy uniform distribution to find a stable and globally solution in an uncomplicated and practical way. On the other hand, the features generated satisfy both uniqueness and similarity requirements for image retrieval, authentication and some forensic tasks.

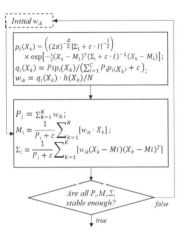

Figure 4.   Flowchart of proposed novel EM.

5   POSITIVE PERTURBATION SCHEME

The E step and M step of EM algorithm are repeated as necessary. Each iteration is guaranteed to increase the log-likelihood and the algorithm is guaranteed to converge to a local maximum of the likelihood function. Even though the EM algorithm will converge to a (possibly local) maximum, the algorithm may still suffer from the singularity problem, such as the boundary overflow.

Table 1.   Comparison of the novel and standard EM.

Novel EM proposed	Conventional standard EM
Uniform distribution used as random initialization. It converges to a global optimal solution.	Parameters λ used as random initialization. It may converge to a local maximum.
Adding a very small perturbation to avoid singularities.	Algorithm may suffer from singularity.

The proposed novel EM algorithm solves the issue of singularity by adding a small quantity (e.g., $\varepsilon=10-20$ or other small numbers), which is referred to as a Positive Perturbation Scheme. In this way, the reliable computation can be guaranteed. Adding a small quantity usually takes place for the following situations, e.g., division by a denominator, logarithm, and the matrix inverse. Because these measures are adopted, the summand in iteration will always not be a negative number.

The comparison of the novel EM and standard EM of GMM is shown in Table 1.

6   EXPERIMENTAL WORKS

The experimental works have been presented in Figures 5-8 to illustrate the proposed novel EM algorithm for comparison between using the uniform distribution (b) and peak-shape distributions (c-f) as the initial condition. The sample images are selected from each category from the test dataset (Wang 2003), which contains 1,000 generic images in 10 categories, with 100 images in each category. The uniform distributions are wik = rand(1,256)/256. The peak-shape distributions are with different means and fixed standard deviations: wik = $N(Xk|m,\sigma i2)$; i=1,2; k=(1:256); m=60,100,150,210; σ1=5, σ2=15. The values of likelihood L for five images under different initial conditions are shown in Table 2.

Since the uniform distribution is a fixed specific realization of probability distribution, the solution is surely deterministic. By comparison, other

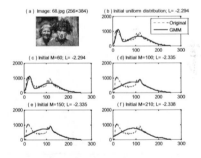

Figure 5.   Novel EM of Image 68.

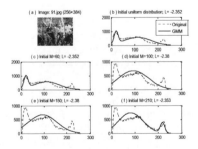

Figure 6.   Novel EM of Image 91.

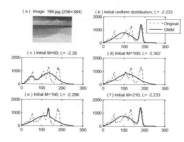

Figure 7. Novel EM of Image 199.

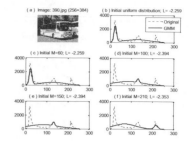

Figure 8. Novel EM of Image 390.

solutions are determined by the different peak-shape distributions.

The likelihood L achieved with a uniform distribution as initialization are always equal to the largest one among the maximum likelihood values that can be achieved with the different peak-shape distributions as initialization. The results in Table 2 confirm that for all the images the novel EM could find the global optimal solution every single time. When the peak-shape distributions are used as starting values, it is unlikely that they all yield the exact same solution. Thus, we consider a solution as approximately equal to the global optimum if it achieves the maximum of all the likelihood values with different start values. It has been demonstrated in our experimental works that this novel EM algorithm using a realization of uniform distribution as initial condition can achieve the global maximum for natural images.

Table 2. Likelihood L for images with different initialization.

Image No.	Uniform Distribution	Mean M of peak distribution			
		60	100	150	210
68	−2.294	−2.294	−2.335	−2.335	−2.338
91	−2.352	−2.352	−2.38	−2.38	−2.353
199	−2.233	−2.28	−2.302	−2.296	−2.233
390	−2.259	−2.259	−2.394	−2.394	−2.353

7 CONCLUSION

In this paper we introduce a new novel EM algorithm to find stable optimal solution. Using a random uniform distribution (known to have maximum entropy) as the initial condition, the novel EM converges iteratively to an optimal solution. Furthermore, the estimation process is repeatable, stable, and uncomplicated. For every time, the novel EM algorithm produces stably the optimal solution which could achieve the maximum of all the likelihood values obtained by different initial conditions. A positive perturbation scheme is also proposed to avoid

boundary overflow, often occurring with the conventional EM algorithms.

In our Experiments, the novel EM algorithm is used for digital images and the discussion is made with respect to Gaussian Mixture Model (GMM). The histogram has been used in order to speed up computation involved in the novel EM algorithm. Numerical experiments indicate the ability for obtaining the stable optimal solution in an uncomplicated way over solutions obtained with some prior arts, including the Deterministic Annealing Algorithm and the Minimum Message Length.

ACKNOWLEDGMENTS

This work is supported by "National Fundamental Research Funds for the Central Universities" 2012KJ001.

REFERENCES

[1] Bilmes J. A. 1998. A Gentle Tutorial of the EM Algorithm and its Application to Parameter Estimation for Gaussian Mixed and Hidden Markov Models. U.C. Berkeley. TR-97–021.

[2] Dempster A., Laird N. & Rubin.D.1977. Maximum Likelihood from Incomplete Data via the EM Algorithm. *Journal of the Royal Statistical Society B* 39:1 22.

[3] Figueiredo M. A.T. & Jain A. K. 2002. Unsupervised learning of finite mixture models. *IEEE Transactions on Pattern Analysis and Machine Intelligence* 24:381–396.

[4] Li J. and Wang J. Z. 2003.Automatic Linguistic Indexing of Pictures by a Statistical Modeling Approach. *IEEE Transactions on Pattern Analysis and Machine Intelligence*, 25(9):1075–1088.

[5] Rose K. 1998. Deterministic annealing for clustering, compression, classification, regression, and related optimization problems. *Proceedings of the IEEE* 86 (11): 2210–2239.

[6] Xuan G., Zhang W.& Chai P. 2001. EM algorithm of Gaussian mixture model and hidden Markov model. *IEEE International Conference on Image Processing (ICIP2001).* vol.1:145–148.Thessaloniki, Greece.

Network Security and Communication Engineering – Chan (Ed.)
© 2015 Taylor & Francis Group, London, ISBN: 978-1-138-02821-0

Survey and analysis of digital watermarking algorithms on HEVC video coding standard

Ali A. Elrowayati & M.F.L. Abdullah
Faculty of Electrical and Electronic Engineering, Universiti Tun Hussein Onn Malaysia

ABSTRACT: The High Efficiency Video Coding (HEVC) is the newest video coding scheme which was issued in January 2013 by the ITU-T and ISO/IEC. It has the advantage of reducing the bit rate by as much as 50 % when compared to H.264 while maintaining the same visual quality. In the last decade, authentication and copyright protection methodologies have become one of the key items required in order to protect video contents by embedding within an efficient video codec. Thus the objective of this paper is to review and investigate the applicability of digital watermarking algorithms in this new Standard. The results of this study provide motivation to achieve higher embedding capacity and higher compression performance for HEVC compared to H.264/AVC especially, for low bit-rate coding.

1 INTRODUCTION

In the last decade, video compression standards have been improved to provide more compressed data along with better visual quality resolution. As a result, providing a security method with higher efficiency coding is a priority demand. Video digital watermarking integrated with these standards is a technology that can serve this purpose. A large number of video watermarking techniques have been proposed to embed copyright marks and other information in digital video, which can later be extracted or detected. Digital video is a sequence of still images or frames. However, video watermarking techniques need to meet other requirements than those required in image watermarking schemes, such as video coding technologies, redundancy between sequences of frames and some special video attacks (Bhowmik and Abhayaratne, 2012). In addition, video watermarking techniques that use coding structure techniques as a basic component in the watermarking schemes are the primary target for the goal of integrating watermarking and compression to reduce the complexity of video watermarking processing. Therefore, integrating watermarking requires an optimal implementation of the watermarking framework at the encoder to reach that goal.

Despite the success of these video codes such as MJPEG2000, MPEG-2 and H.264 for video coding and video watermarking, a new standard known as HEVC with higher compression efficiency, is poised to replace them. As HEVC digital video becomes more prevalent, the industry will need appropriate copyright protection and authentication methods. The new HEVC video coding scheme was issued in January 2013 by the ITU-T and ISO/IEC (Sullivan et al., 2012). It has the advantage of reducing the bitrate in the range of 50 % when compared to H.264 while maintaining the same visual quality. In the last decade, authentication and copyright protection methodology has become a key factor in order to protect video content by embedding within an efficient video codec, so the development of new algorithms is required to integrate robust watermarking techniques for different profiles of HEVC.

2 OVERVIEW ON HEVC VIDEO CODING

The main architecture of HEVC is based on the same architecture as in the H.264 standard. Both methods utilize the following steps: prediction, transformation, quantization of prediction residues, and entropy coding (Correa et al., 2014). The significant compression ratio improvement of HEVC compared to H.264 is achieved by adding new efficiency coding tools such as an adaptive loop filter, sample adaptive offset, and a motion merge technique (Cai et al., 2012).

The core unit of HEVC is the Coding Tree Unit (CTU) instead of the macro-block in H.264, which is divided into smaller Coding Units (CUs) based on the frame partitioning structures. Furthermore, the CU splits into different block sizes, ranging from 4 x 4 to 64 x 64 (Zhao et al., 2013). A generic block diagram of the HEVC encoder is shown in Figure 1.

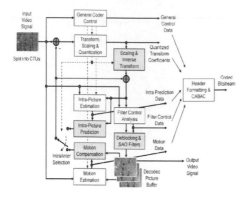

Figure 1. HEVC encoder block diagram (Sullivan et al., 2012).

3 DIGITAL VIDEO WATERMARKING

3.1 Overview on digital watermarking

Digital watermarking has been presented as a key to the problem of copyright protection of multimedia data in the next generation networks (Shi et al., 2010). Essentially, nowadays with a wide range of video applications, Multimedia transmissions over wireless channels and the Internet requires that multimedia resources should be protected. Thus, video coding integrated with digital watermarking techniques can be efficiently employed to achieve this goal. However, digital video watermarking has many aspects which must be considered during the design of any digital video watermark schemes such as imperceptibility, robustness, capacity, complexity, synchronization, bit-rate control and error drift.

Generally, a watermarking system consists of two main components: a watermark embedding unit and a watermark detection unit as shown in Figure 2. The embedding unit adds the watermark component to the host data. The output of the embedding unit is the watermarked data. Attackers intend to do one of the following illegal actions such as modification, copying, destroying or removing the watermark from the host data.

Figure 2. Main units of watermarking.

3.2 Video watermarking techniques

According to the work of watermarking embedding, video watermarking techniques are divided into three main groups or classifications, namely spatial domain, frequency domain and bit-stream domain as shown in Figure 3. Each group applies different methods. In this sub section a brief review is given of the current video watermarking techniques based on their watermarking domain. First, spatial domain watermarking techniques embed the watermark by modifying the pixel positions or pixel values of the host video. The main advantages of using this technique are the low time complexities and simplicity of implementation. However, generally these techniques have some disadvantages in providing robustness and meeting imperceptibility requirements (Paul, 2011). Second, frequency transform domain which used to overcome the main disadvantages of the spatial domain. Further, analysis of the bands in a frequency domain is a prerequisite to enhance watermark robustness and imperceptibility. However, these techniques generally have some disadvantages in terms of complexity. A number of transforms are used to transfer from the spatial to frequency domain, such as the discrete Fourier transform, discrete cosine transform, discrete wavelet transform, and hybrid transforms (Pereira et al., 1999), (Gaobo et al., 2006), (Wu et al., 2011). Finally, bit-streams domain Watermarking techniques embed the watermark into the compressed encoded bit streams. The advantage of this technique the computational cost is low. However, the disadvantage is that the compressed bit-rate constrains the size of the watermarked data. The strength of the embedded watermark is limited by the error rate of the video decoder and the embedding strategy is constrained by the coding standard and the video compression method (Langelaar et al., 2000) , (Fridrich et al., 2004).

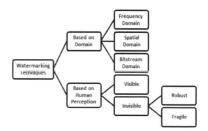

Figure 3. Classification of watermarking techniques.

4 VIDEO WATERMARKING IN A HEVC BITSTREAMS

Video watermarking methods that use the coding structure technique as a basic component in watermarking schemes are primarily targeted by the goal of integrating watermarking and compression standards. This is to reduce the complexity of the video watermarking processing and this integrated watermarking requires an optimal implementation of the watermarking framework for the encoder to reach that goal.

4.1 Special features of the HEVC

The watermarking methods was developed based on previous video coding, such as H.264/AVC cannot be applied directly to HEVC. This is because HEVC applies two types of transforms which are different from what are used in H.264/AVC so the watermarking methods must be modified to apply the HEVC standard. Moreover, as HEVC digital video becomes more prevalent, the industry will need appropriate copyright protection and authentication methods in the near future. So the development of new methods is required to integrate robust watermarking techniques for different profiles of HEVC.

4.2 Watermarking techniques and algorithms of the HEVC standard

To facilitate the classification of the different algorithms, we can classify them based on the place where the watermarking occurs during the video compression. The four options are: before encoding, after quantization, during entropy coding and during encoding process. The figure 4 illustrates a classification of watermarking techniques based on HEVC structure.

Option 1
- Before encoding: Spaial domain techniques which based on the coding tree unit structure.

Option 2
- After quantization: Frequency domain techniques which based on the transform unit structure and quantization parametrers.

Option 3
- During entropy coding: Bitstream domain techniques which based on the entropy unit structure.

Option 4
- During encoding process: Spacial techniques which based on motion vectors modification.

Figure 4. Classification of watermarking techniques based on HEVC coding.

Option 1: before encoding we find video watermarking algorithms such as spread spectrum can be applicable on HEVC by modifying these algorithms based on the coding tree unit structure. However, no works on the HEVC can be found in the literature according to this option.

Option 2: after quantization phase, one may modify the quantized integer DCT/DST coefficients according to the quantization parameter chosen by the user. We can cite the clever proposal of (Chang et al., 2014) developing a new technique to categorize blocks based on certain intra prediction mode combinations of neighbouring blocks and then the quantized coefficients determine and divide into perturbation groups for each I-frame without any error propagation in the HEVC streams. The proposed technique embeds bits in 8×8, 16×16, and 32×32 transform blocks of the HEVC. The experimental results for the PSNR (Peak Signal to Noise Ratio) and the embedding capacity for different bit-rates are given in Figure 5 for comparison between those techniques. With regard to the bitrate, the proposed technique achieves higher embedding capacity and higher compression performance compared to H.264/AVC (Ma et al., 2010) for low bit-rate video error free coding.

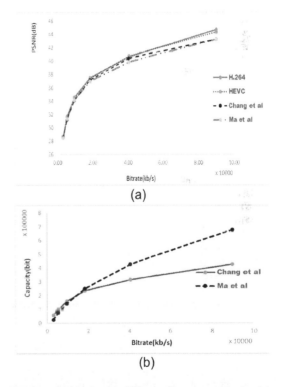

(a)

(b)

Figure 5. Objective data hiding performance evaluation (a) PSNR vs. bitrate and (b) capacity vs. bitrate plots for tested techniques on video test named PeopleOnStreet with size 2560x1600.

Option 3: one may achieve the watermarking during the entropy coding phase. This type of algorithm necessitates to enter structure of CABAC and the bit capacity is obviously very low and algorithms are not robust. In addition, no works on the HEVC can be found in the literature according to this option.

Finally, there are other techniques based on motion vectors can be applicable on the HEVC. For example, the authors (Zhang et al., 2001) proposed a video watermarking method based on this technique for H.264 standard. The motion vector value is optimized to select the best value to embed the watermark information. Then, different watermarks are embedded in the different parts of the motion vector. This technique enhances the robustness of the watermark scheme and might be applicable to HEVC as a future work.

5 CONCLUSION

In this paper, recent developments have been reviewed in the area of watermarking techniques and algorithms for a new Standard HEVC. As mentioned above, the HEVC is a new standard which is still an open field of research in the area of video watermarking. The results of this study provide motivation for future works to modify and apply existing digital watermarking methods to HEVC and achieve a higher embedding capacity and higher compression performance compared to H.264/AVC for low bit-rate video coding.

ACKNOWLEDGMENT

This work is partly supported from the Universiti Tun Hussein Onn Malaysia (UTHM), Department of Communication Engineering. Johor, Malaysia

REFERENCES

[1] BHOWMIK, D. & ABHAYARATNE, C. 2012. 2D+ t wavelet domain video watermarking. *Advances in Multimedia,* 2012, 6.
[2] CAI, Q., SONG, L., LI, G. & LING, N. Lossy and Lossless Intra Coding Performance Evaluation: HEVC, H.264/AVC, JPEG 2000 and JPEG LS. *In:* QI CAI, L. S. A. G. L., NAM LING, ed. Signal & Information Processing Association Annual Summit and Conference (APSIPA ASC), 2012 Asia-Pacific, 2012. IEEE, 1–9.
[3] CHANG, P.-C., CHUNG, K.-L., CHEN, J.-J., LIN, C.-H. & LIN, T.-J. 2014. A DCT/DST-based error

[4] CORREA, G., ASSUNCAO, P., AGOSTINI, L. & DA SILVA CRUZ, L. A. 2014. Complexity scalability for real-time HEVC encoders. *Journal of Real-Time Image Processing,* 1–16.
[5] FRIDRICH, J., GOLJAN, M., CHEN, Q. & PATHAK, V. Lossless data embedding with file size preservation. Electronic Imaging 2004, 2004. International Society for Optics and Photonics, 354–365.
[6] GAOBO, Y., XINGMING, S. & XIAOJING, W. A genetic algorithm based video watermarking in the DWT domain. Computational Intelligence and Security, 2006 International Conference on, 2006. IEEE, 1209–1212.
[7] LANGELAAR, G. C., SETYAWAN, I. & LAGENDIJK, R. L. 2000. A State-of-the-Art Overview. *IEEE Signal processing magazine.*
[8] MA, X., LI, Z., TU, H. & ZHANG, B. 2010. A data hiding algorithm for H. 264/AVC video streams without intra-frame distortion drift. *Circuits and Systems for Video Technology, IEEE Transactions on,* 20, 1320–1330.
[9] PAUL, R. T. 2011. Review of robust video watermarking techniques. *IJCA Special Issue on Computational Science,* 3, 90–95.
[10] PEREIRA, S., RUANAIDH, J. J., DEGUILLAUME, F., CSURKA, G. & PUN, T. Template based recovery of Fourier-based watermarks using log-polar and log-log maps. Multimedia Computing and Systems, 1999. IEEE International Conference on, 1999. IEEE, 870–874.
[11] SHI, F., LIU, S., YAO, H., LIU, Y. & ZHANG, S. 2010. Scalable and credible video watermarking towards scalable video coding. *Advances in Multimedia Information Processing-PCM 2010.* Springer.
[12] SULLIVAN, G. J., OHM, J., HAN, W.-J. & WIEGAND, T. 2012. Overview of the high efficiency video coding (HEVC) standard. *Circuits and Systems for Video Technology, IEEE Transactions on,* 22, 1649–1668.
[13] WU, C., ZHENG, Y., IP, W., CHAN, C., YUNG, K. & LU, Z. 2011. A flexible H. 264/AVC compressed video watermarking scheme using particle swarm optimization based dither modulation. *AEU-International Journal of Electronics and Communications,* 65, 27–36.
[14] ZHANG, J., LI, J. & ZHANG, L. Video watermark technique in motion vector. Computer Graphics and Image Processing, 2001 Proceedings of XIV Brazilian Symposium on, 2001. IEEE, 179–182.
[15] ZHAO, T., WANG, Z. & KWONG, S. 2013. Flexible mode selection and complexity allocation in high efficiency video coding.

propagation-free data hiding algorithm for HEVC intra-coded frames. *Journal of Visual Communication and Image Representation,* 25, 239–253.

Network Security and Communication Engineering – Chan (Ed.)
© 2015 Taylor & Francis Group, London, ISBN: 978-1-138-02821-0

An improved SRDI imaging algorithm

B.P. Wang
Science and Technology on UAV Laboratory, Northwestern Polytechnical University, Xi'an, P.R. China

X.T. Wang, Y. Fang, Y.L. Lan & H. Wang
School of Electronics and Information, Northwestern Polytechnical University, Xi'an, P.R. China

ABSTRACT: Making full use of the characteristics of space debris, high-speed spinning around its main axis, Single Range Doppler Interference (SRDI) algorithm can be used for ISAR imaging of space debris with small dimension. But the imaging result of SRDI algorithm is easily influenced by the quality of time-frequency spectrum. To improve the imaging performance and robustness, an improved SRDI algorithm is proposed in this paper. By extracting and fitting for the time-frequency curve, the quality of time-frequency spectrum is greatly improved. Then, after an integral along the sine curves, the two dimension (2D) image is reconstructed. The experimental results show the effectiveness of the proposed algorithm.

KEYWORDS: Spinning target, SRDI algorithm, Time-frequency spectrum, Integral.

1 INTRODUCTION

There exists a lot of space debris in the space, such as a large number of meteorites, the abandoned rockets, the invalid spacecraft, etc. These objects move in high speed (usually 8 km/s). They make collisions with spacecraft, which directly changes the external performance of the spacecraft, resulting in breakdown of spacecraft and bringing serious consequences. Therefore, it is an important task to detect and recognize the space debris. It is possible to protect spacecraft against small debris of less than 1 cm by an appropriately designed shield, and those larger than 10 cm have been constantly monitored by the existing network of radar and optical sensors around the world. But the main target of our observations of space debris is debris in the size region of 1-10 cm. The present ISAR technique cannot be applied to the imaging of small debris of 1-10 cm size.

Space debris usually spins in high speed around its major axis. It makes Doppler modulation for the echo, which has a very big difference from conventional moving target's Doppler modulation. So the traditional ISAR imaging method based on the turntable model is invalid for the high-speed spinning target. We usually adopt a method - SRDI [1]. The algorithm makes use of the fact that space debris usually follows simple spinning motion around their major axis and the imaging of space debris can be realized. By analyzing Doppler spectrum which migrates a cycle and two-dimensional image of the target can be got. But the algorithm needs to translate echo data from a single range unit into time-frequency domain and perform integral along a fixed curve. But it has complex operations and large amount of calculations. Imaging effect of SRDI algorithm is affected by the quality of time-frequency spectrum and cannot achieve the ideal resolution. In order to improve the imaging quality of SRDI algorithm, this paper puts forward an improved SRDI imaging algorithm. The algorithm firstly extracts signal time-frequency spectrum, performs elaboration and increases amplitude for the extracted spectrum, then we use trigonometric function to do fitting for extracted time-frequency curve, finally we get the two-dimensional imaging of target with integral along the fitting curves, which can improve the imaging resolution.

2 SRDI IMAGING ALGORITHM

We assume that target has the constant spinning angular velocity and the translational motion has been fully removed by current motion compensation completely during the coherent processing interval (CPI) [2]. The target spins in high speed. During the CPI, target often spins several cycles. The rotational angle of target relative to the radar usually is only a few degrees, so its effects can be ignored. Figure 1 is the motion model of space high-speed target which is built in the Cartesian coordinate system *XOY*. We take

O as the rotational center of target and take the direction of radar sight line as *Y* axis. Assuming that rotational angular velocity of target is ω_s, we estimate it by autocorrelation method in actual situation [1, 3].

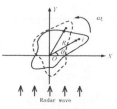

Figure 1. High-speed spinning target model.

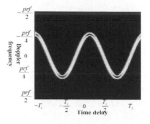

Figure 2. Time-frequency analysis of the spinning signal.

Under the far field condition, the instantaneous range of a scattering point is given by:

$$R_i(t_m) = R_0 + R_i \sin(\theta_i + \omega_s t_m) \qquad (1)$$

where R_0 is the constant part, (R_i, θ_i) presents the polar coordinates of scattering point i and the sampling time (slow time) is t_m. After down conversion to the baseband, the echo can be written as:

$$s_i(f, t_m) = A_i \exp\left[-\frac{j4\pi f}{c}(R_0 + R_i \sin(\omega_s t_m + \theta_i))\right] \qquad (2)$$

where A_i denotes scattering intensity, c presents the speed of light and radar's signal frequency is f. When radar transmits narrow band signals, after range compression echo envelope will be compressed into a certain range unit. Range pulse compressed signal is expressed as:

$$s_i(r, t_m) = A_i \text{sinc}\left[\frac{2\pi B_r}{c}(r - R_0)\right] \exp\left[-j4\pi f_c R_i \sin(\theta_i + \omega_s t_m)/c\right] \qquad (3)$$

where f_c denotes signal carrier frequency, and B_r denotes the signal bandwidth. The instantaneous phase is defined as $\vartheta(t_m) = 4\pi f_c R_i \sin(\theta_i + \omega_s t_m)/c$. The instantaneous Doppler in the azimuth direction can be written as follows:

$$f_d(t_m) = -\frac{d}{d\tau_m}\left[\frac{\vartheta(t_m)}{2\pi}\right] = -\frac{2R_i \omega_s \cos(\theta_i + \omega_s t_m)}{\lambda} \qquad (4)$$

where $\lambda = c/f_c$ is the signal wavelength. When the CPI is not less than one period, the full Doppler

bandwidth is proportional to R_i as $B_d = 4R_i \omega_s/\lambda$. The Doppler frequency caused by the target's spinning angular velocity makes rotational target on the signal echo produce sinusoidal phase modulation [4], [5]. For space spinning target, the traditional ISAR imaging method is no longer available.

Make a short-time Fourier transform for (3) to get the two-dimensional time-frequency spectrum which is:

$$S(t_m, f_d) = \int s_i(R_0, \tau) w(\tau - t_m) \exp(-j2\pi \cdot \tau f_d) d\tau \qquad (5)$$

where $w(\tau) = 1, (-\Delta\tau \leq \tau \leq \Delta\tau)$ denotes the time window function, make $\tau - t_m = t$, (5) becomes:

$$S(t_m, f_d) = A_i \int_{-\Delta\tau}^{\Delta\tau} \exp\left[-j \cdot \vartheta(t + t_m)\right] \cdot \exp\left[-j2\pi(t + t_m) f_d\right] dt \qquad (6)$$

When the window length is short enough, we can take the following approximation for instantaneous phase $\vartheta(t + t_m)$:

$$\vartheta(t + t_m) \approx \vartheta(t_m) + \vartheta'(t_m) \cdot t = \vartheta(t_m) - 2\pi f_d(t_m) \cdot t \qquad (7)$$

Take (7) into (6), the plural expression of time-frequency distribution is:

$$S(t_m, f_d) = A_i \cdot \exp\left[-j \cdot \vartheta(t_m)\right] \cdot \exp\left[-j2\pi \cdot t_m f_d\right] \\ \cdot 2\Delta\tau \cdot \text{sinc}\left[2\pi\Delta\tau(f_d - f_d(t_m))\right] \qquad (8)$$

Figure 2 shows the magnitude distribution. Single range Doppler interference (SRDI) [4] algorithm makes use of the magnitude distribution to estimate the shape of spinning space debris. With an integral along the sine curves, the two dimension (2D) image is reconstructed.

3 AN IMPROVED SRDI IMAGING METHOD

Combining theory with experiment analysis, we find that the imaging effect of SRDI algorithm is affected by the quality of time-frequency spectrum and it cannot achieve the ideal resolution. The quality of reconstructed image has a certain relationship with the resolution of signal time-frequency spectrum. The higher the quality of the time-frequency spectrum is, the better the effect of the reconstruction image is. In order to improve the imaging quality of SRDI algorithm, this paper puts forward an improved SRDI imaging algorithm. The algorithm firstly extracts signal time-frequency spectrum, does the elaboration and increases amplitude for the extracted spectrum, then we can use trigonometric function to do fitting for extracted time-frequency curve, finally the target's two-dimensional imaging is gotten with integral along the fitting curve. We introduce the algorithm in a detail by a simulation example next. The simulation parameters are as follows: the radar's carrier frequency is $2\ GHz$, its wavelength is $\lambda = 0.15\ m$, and its bandwidth is $20\ MHz$. The point simulation model is shown in Figure 3 (f), and three points are located at $(\lambda, 0), (3\lambda, \pi/2), (3\lambda, \pi)$ respectively.

Because the signal that radar transmits is narrow band, radar envelope will be compressed into the same range unit after range pulse compression for the echo signal. We can get signal time-frequency spectrum diagram which is shown in Figure 3 (a) via the short-time Fourier transform for lateral echo data of range unit. We can see that time-frequency resolution of the signal is poorer. The time-frequency curves of different scattering points have mutual interference, which results in breakpoints and has an influence on the follow-up image processing. So we need to deal with curve of the time-frequency spectrum according to the following steps;

We use low-pass filter to smooth the slice of time-frequency spectrum at different time [6]. Noise peaks in the frequency domain are filtered and the signal components peaks are kept. The time slice sequences before and after filtering are shown in Figure 3 (c) and (d) respectively;

On the basis of step (2), we can do sliding window detection for time slice sequences and find the sequence fragment whose sequence maximum is located in the center of the window in window scope. We can increase its amplitude. For other data segments, weighted inhibition need to be done in amplitude. We can change the length of the window L and weighted value ω to control the clarity of Doppler frequency curve in the process of extracting the time-frequency spectrum curve. We can use the following criterion function to find the parameter optimal value:

$$\min_{L,w} MSE(x,y) = \min_{L,w} \left\| \hat{f}(x,y) - f(x,y) \right\|_2^2 \qquad (9)$$

where $MSE(x,y)$ denotes the mean square error, $\hat{f}(x,y)$ denotes the extracted time-frequency spectrum diagram and $f(x,y)$ denotes the primitive time-frequency diagram. The extracted time-frequency spectrum curve is shown in Figure 3 (b). In order to improve the calculative efficiency, we choose 3 as the window value and 0 as the weighted value;

According to the derivation above, we conclude that the time-frequency spectrum curve of spinning target is sine form. So we can fit extracted time-frequency curve in step (3) in accordance with the trigonometric function and get smooth and coherent time-frequency curve, then according to the original time-frequency spectrum assign a value for the fitting curve, which is shown in Figure 3 (e).We use moving least square method to fit curve. The method, which introduces concept of influence area that the nodes of data point

within limits has influence on the fitting function value, and improves least square method. The idea of moving least square method is that selecting suitable weighted function and working out the fitting function instead of the least square fitting curve whose idea is that looking $f(x)$ to minimize ε^2 which denotes the sum of the error square between the right function and measured value.

$$\varepsilon^2 = \sum_{i=0}^{N} [y_i - f(\mathbf{x}_i)]^2 \qquad (10)$$

The algorithm flow is that the area firstly should be segmented, each sectional point should be circled and we can determine the influence area size of the grid point and form fitting curve by connecting grid points. It has higher fitting precision and better fitting smoothness [7].

For the sine wave of a single cycle, we use 5 order polynomial or 7 order polynomial to fit error and achieve the desired effect. But when the cycle of sine waveform is more, we need to use higher order polynomial to do fitting. Here we can adjust it according to what we need.

Then two-dimensional image of the target can be got with integral according to the fitting time-frequency curve in the step (4).

At the same time, the imaging effects of SRDI algorithm and improved algorithm are shown in Figure 3 (g) and Figure 3 (h) respectively. Because the improved algorithm improves the quality of the time-frequency curve by extracting and fitting time-frequency curve, we can see that the imaging effect of improved algorithm is more ideal and the imaging resolution is higher.

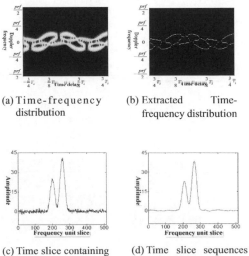

(a) Time-frequency distribution

(b) Extracted Time-frequency distribution

(c) Time slice containing noise

(d) Time slice sequences after filtering

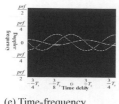

(e) Time-frequency
distribution after fitting

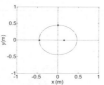

(f) Point target model

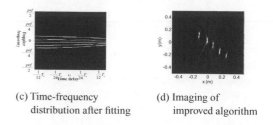

(c) Time-frequency
distribution after fitting

(d) Imaging of
improved algorithm

Figure 4.   Contrast from imaging effect of measured data.

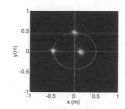

(g) SRDI algorithm
imaging effect

(h) Improved algorithm
imaging effect

Figure 3.   Curve fitted and imaging effect contrast of simulation data.

4   EXPERIMENT AND PERFORMANCE ANALYSIS

We use microwave dark room to test actual measurement data of five metal ball targets. Where the signal carrier frequency is 18 *GHz*, bandwidth is 5 *MHz*, observation angle is 60°and step is 0.02°. The time-frequency distribution of the echo measured data of the five metal balls is presented in Figure 4 (a), from which we can find that the quality of time-frequency spectrum is poor. The imaging result of SRDI algorithm is shown in Figure 4 (b). Due to the influence of time-frequency spectrum's quality, imaging focusing result of SRDI algorithm is poor. The fitting time-frequency curve diagram is shown in Figure 4 (c). We can find that the fitting time-frequency curve is clearer. The imaging result of improved algorithm is shown in Figure 4 (d). By contrasting the imaging result of SRDI algorithm with the imaging result of improved algorithm, we can know that the focusing effect of improved algorithm is superior to SRDI algorithm obviously.

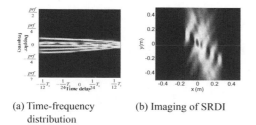

(a) Time-frequency
distribution

(b) Imaging of SRDI

5   CONCLUSION

Based on the fact that space debris spins around its major axis, SRDI algorithm analyses Doppler spectrum which migrates a cycle. The two-dimensional image of the target can be got with integral along the time-frequency curve. Because of the poor quality of time-frequency curve, it has an influence on the imaging effect. By extracting and fitting the time-frequency curve, we get imaging result of better focusing effect with integral along the fitting curve. We use measured and simulated data to compare improved algorithm with SRDI algorithm. We find that the improved algorithm is better to a certain extent. The experimental simulation results show that the validity and accuracy of method in this paper.

ACKNOWLEDGMENTS

This work is supported by the National Natural Science Foundation of China (61472324), Aerospace Science and Technology Innovation Fund (CASC201105), and graduate starting seed fund Of Northwestern Polytechnical University (Z2014131).

REFERENCES

Sato T. Shape estimation of space debris using single-range Doppler interferometry [J]. *IEEE Trans. on Geoscience and Remote Sensing*, 1999, 37(2): 1000–1005.

M.-D. Xing, Z. Bao. Study of Translational Motion Compensation and Instantaneous Imaging of ISAR Maneuvering Target [J]. *ACTA ELECTRONICA SINICA*, 2001, 29 (6):733 – 737.

M.D. Xing,Z. Bao, .Radar imaging technology[M]. Beijing: *Electronic Industry Press*, 2005: 230–254.

Chen V C and Qian S. Joint time-frequency transform for radar time Doppler imaging [J]. *IEEE Trans. on Aerosp. Electron. Syst.* 1998, 34(2): 486–499.

J Li, H Ling. Application of adaptive chirplet representation for ISAR feature extraction from targets with rotating parts [J]. *IEE Proc2Radar Sonar Navig*, 2003 ,150 (4):284 – 291.

X.-F Ding, J.-M.Hu, ect. The Research on Doppler-only Imaging Algorithm of the Mid-course Missile Object[J],*Journal of Electronics & Information Technology*,2009,31(12): 2864–2868.

Q.II. Zeng, D.T. Lu. Curve and surface fitting based on moving least square methods. *Journal of Engineering Graphics*, 2004,25(1):84–89.

Network Security and Communication Engineering – Chan (Ed.)
© *2015 Taylor & Francis Group, London, ISBN: 978-1-138-02821-0*

Register one-to-one mapping strategy in Dynamic Binary Translation

Y.N. Cui, J.M. Pang, Z. Shan & F. Yue
State Key Laboratory of Mathematical Engineering and Advanced Computing, Zhengzhou, China

ABSTRACT: Dynamic Binary Translation (DBT) is an important technology to cross-platform software transplantation such as enhancing the compatibility of Alpha. However, there exist some factors that impede its performance: high translation overhead and translated code of low quality. In this work, we take advantage of host registers, using register mapping strategy to promote the performance of DBT. By mapping registers from guest Instruction-Set Architectures (ISA) to host machines, it could off-load the overhead caused by load and store operations of memory. By simplifying the rules of the intermediate code generated, it could lower the number of intermediate code and improve the quality of translated code. Form X86 to ALPHA on SPEC 2006, the code expansion reduces 32.86% and the performance of the optimized translator is improved by 7.55%.

KEYWORDS: Dynamic binary translation, Overhead, Registers mapping, Performance optimization.

1 INTRODUCTION

Dynamic binary translation [1, 2] could emulate an executable program in an instruction-set architecture (ISA) on a host machine with a different ISA. Dynamic binary translators are gaining importance because it is widely applied with system virtualization, mobile platform, and security monitoring. However, there exist some factors that impede the performance of a DBT: translation overhead and the quality of the translated code [3].

Register that is an information storage unit is an important part of modern computer architecture. Register usage policy could greatly affect the running time of a program. It is a benefit method to promote efficiency of DBT running by adopting right registers management strategy. We would like to use register mapping strategy to promote the performance of DBT. As ALPHA has 32 registers, the number of registers could meet the need of register mapping. The DBT could thus take advantage of the registers of the host machine from X86 to ALPHA. Translation overhead could be off-loaded and the host code could be simplified.

In this work, we took advantage of host machine registers, using register one-to-one mapping strategy to promote the performance of DBT. We used QEMU [4, 5], an efficient DBT system, to emulate and translate binary application from guest machine X86 to host machine ALPHA. By using register mapping strategy to QEMU, we successfully addressed the goal of low translation overhead and the good translated code quality on the SPEC2006.

The main contributions of this paper are as follows:

a. Based on process-level QEMU, we designed the register one-to-one mapping strategy from X86 to ALPHA, making 8 general-purpose registers of X86 one-to-one mapped to the registers of ALPHA, and off-loaded the translation overhead.
b. The intermediate code generation rules were simplified, the code expansion was reduced and the performance of binary translator was promoted.

2 RESEARCH ON DYNAMIC BINARY TRANSLATION

2.1 Overview of dynamic binary translator

QEMU is a typical dynamic binary translator (DBT) that could enable system-level virtualization and process-level emulation. System-level virtualization is used for simulating the operation system and process-level emulation is used for emulating application. This article is mainly based on the process-level emulation.

Fig. 1 illustrates the organization of QEMU. QEMU loads guest code into host machine memory, and then divides the code into several basic blocks (TB), which is the basic translation unit. QEMU's key translation engineer is called Tiny Code Generator (TCG) [6], which provides a serial of Instruction translation operation. When a TB is fetched, TCG disassembles and translates the TB into intermediate code. Then, the intermediate code is improved with an optimization method: register liveness analysis [7].

Dead intermediate code deletion is also done during the optimization. Finally, the intermediate code is translated into the host code that will be stored into cache and executed by DBT. Although the whole translation process includes the optimization, the host code quality is not as good as it should be. It is because there are often many redundant loads and store operations left in the code.

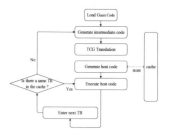

Figure 1. The basic structure of QEMU.

2.2 Register management mechanism of dynamic binary translator

QEMU has three types of temporary variables (shown in Table 1): *Global, Local, and Temp.*

QEMU defines global variable *env* (as shown in Fig. 2), which is used to store the content of guest

Table 1. The types of QEMU temporary variables.

Temporary variables type	Action scopes
Global	Throughout the translation process
Local temp	In the function of the intermediate code generated
Temp	In a Translation Block

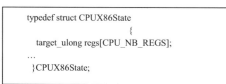

Figure 2. The structure of env.

registers. The variable *env* ensures the consistency of guest registers between different TBs. However, the method with host machine memory simulating guest registers leads to emit many redundant loads and store operations on the host memory. We analyze addition instruction of guest code as an example.

As shown in Fig. 3, firstly the immediate variable 0x1 is placed in *Temp* variable *tmp0*, and then *tmp0* is

stored into the *env->regs[0]*, which simulates the register *EAX* of X86 in the host memory. This completes the translation of *MOVL $1, %eax*. Similarly, *EBX* is also stored into the related memory of ALPHA. Finally, QEMU fetches two additions from host machine memory and completes the addition operation. The final result is stored into the memory again. 3 guest instructions produce 8 intermediate operations, which include 5 *loads* and *store* operations. The number of *loads* and *store* operations accounts for 62.5% of the total amount of the host code.

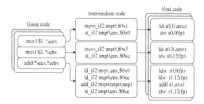

Figure 3. The translation process of addition instruction.

It shows that the register management mechanism, using host machine memory to simulate guest machine registers, produces many redundant *loads* and *store* operations on memory. This could lower the translation overhead and lower the performance of DBT.

3 DESIGN OF REGISTER ONE-TO-ONE MAPPING STRATEGY

The goal of this work is to design a new register management strategy that could lower translation overhead and emit high-quality host codes. We adopt a registers one-to-one mapping approach to achieve the goal.

There are eight general purpose registers under X86, including *EAX, EBX, ECX, EDX, EBX, ESP, EBP, ESI, and EDI.* That the number of ALPHA registers reaches 32 provides a strong hardware conditions to achieve registers mapping from X86 to ALPHA. We select eight host machine registers *R6* to *R13*, corresponding to the X86 general-purpose registers, as mapped registers. Registers mapping relationship is shown in Fig. 4.

Generally, QMUE has to save and restore the contexts when the control switches between the guest code translation and the execution in the dispatcher. QEMU uses global variable *env* to save and restore guest registers during such context switching. Because register mapping doesn't need to use env during translating a TB, we add two functions: *env_to_regs()* and *regs_to_env()*. The function *env_to_regs()* loads *env* into host machine registers before the host code is executed, and *regs_to_env()* stores host machine registers into *env* after the host code is executed. The two functions ensure the correctness

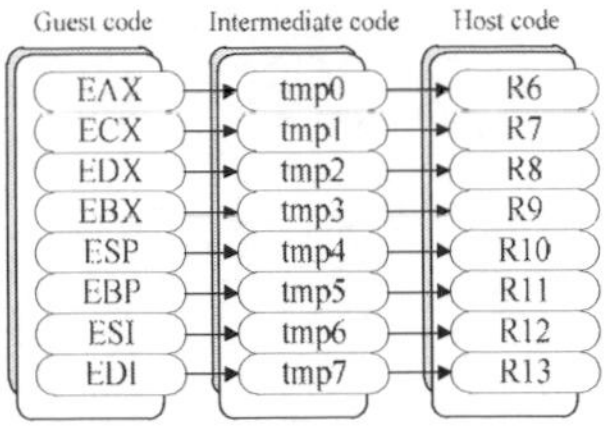

Figure 4. Mapping relationship from X86 to ALPHA.

of emulation when QEMU uses register one-to-one mapping strategy.

Further we try to simplify the intermediate code. Because register mapping doesn't use host memory to load and store guest registers, the loads and store operations can be eliminated as the redundant instructions. We also take the additional instruction for example. Obviously, the *Temp* variable *tmp0* corresponds to *EAX* and the *Temp* variable *tmp3* corresponds to *EBX*, so loads and store operations could be deleted. The result (as shown in Fig. 5), the number of the intermediate code and host code is the same as the guest code, and shows that the code expansion is reduced. It load-offs the translation overhead and promotes the quality of the host code.

Figure 5. Translation process of addition instruction simplified.

By using register mapping strategy, we achieve the goal of registers one-to-one mapping from X86 to ALPHA. When executing the host code, QEMU does not need to use the host memory to load and store the guest registers, and the redundant loads and store operations can be eliminated.

4 PERFORMANCE EVALUATION

In this section, we present the code expansion test and the performance of the register mapping strategy based on process-level QEMU.

4.1 Experimental setup

All experience tests are performed on a system with one ALPHA processer and 4 GB main memory. The operation system is 64-bit Linux with kernel version 2.6.28.10. The 13 SPEC2006 benchmark suite is tested. The 13 benchmarks are compiled with GCC 4.4.0 for X86 guest ISA. QEMU version is 0.10.1.

4.2 Test of the code expansion

By comparing the account of the host code before and after optimized QEMU, we evaluate the effect of simplifying the code. The result is shown in Table 2.

Table 2. Results of SPEC2006 Test.

	Original	Optimized
perlbench	7291679	5004513
bzip2	131057	88894
Gcc	2325734	1612154
Gobmk	240883	163464
Hmmer	151804	102587
Sjeng	171460	118416
libquantum	99208	66218
h264ref	390071	264155
Omnetpp	390495	255672
Astar	169702	112702
xalancbmk	1742634	1092506
Specrand	56677	37593

The effect of simplifying the code is represented by

$$\eta = 1 - N_{op} / N_{ori} \times 100\%$$

where N_{ori} is the number of original QEMU host code, and N_{op} is the number of optimized QEMU host code. The η is the code simplifying rate (CSR). Fig.6 shows the result of comparing CSRs after the 13 benchmarks are separately emulated by original and optimized QEMU. The highest CSR from xalancbmk is 37.13% and the lowest CSR from gcc also reaches 30.6%. According to the results, it is easy to see that the average of CSR is 32.86%. Moreover, we can see register one-to-one mapping strategy is effective to simplify the intermediate code.

4.3 Performance of the register mapping strategy

We test the DBT performance of the register one-to-one mapping strategy by adopting the time of application running. Table 3 illustrates the overall performance results of X86 to ALPHA. The performance is represented by

$$\mu = 1 - T_{op} / T_{ori} \times 100\%$$

where T_{ori} is the time of SPEC2006 running on original QEMU and T_{op} is the time of optimized QEMU. The μ is the DBT performance promoted rate (PPR). The result (shown in table 3) shows that the PPR is promoted by 7.55% using register mapping strategy.

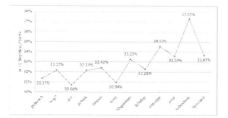

Figure 6. The translation process of addition instruction.

Table 3. Results of SPEC2006 test.

SPEC2006	T_{ori}	T_{op}	μ
Perlbench	3min17.381s	3min1.117s	8.24%
bzip2	3min55.064s	3min37.505s	7.47%
gcc	5min10.771s	4min52.187s	5.98%
gobmk	0min18.668s	0min17.666s	5.37%
hmmer	1min17.994s	1min11.747s	8.01%
sjeng	11min33.538s	10min51.232s	6.10%
libquantum	0min5.966s	0min5.646s	5.36%
h264ref	73min56.714s	66min57.445s	9.45%
omnetpp	4min1.289s	4min39.174s	7.34%
astar	36min6.878s	31min31.251s	12.72%
xalancbmk	2min28.301s	2min16.051s	8.26%
specrand	0min1.302s	0min1.220s	6.32%

5 RELATED WORK

Dynamic binary translation is widely used in many fields: system virtualization [8], mobile platform [9], and security monitoring [10].

A.L Liang et al. [11] had a detailed analysis on registers management of QEMU. However, this paper didn't propose an optimization method to improve QEMU. A very related work is [12] which also uses the host registers. The authors map guest registers EAX, ESP, and EBP to host registers s4, s5, and s6 from X86 to GONSON. The test result shows that the system start time is faster. However, the optimized method which only uses 3 registers is interfered by original register management strategy and the correction of translation is affected.

Y.H Wen et al. [13] proposed the strategy of the specific registers mapping and register function cutting out. This maps the specific registers to the host registers from PowerPC to ALPHA. Because of complexity of register function cutting out, the optimization brings some translation overhead to affect DBT performance.

6 CONCLUSION

In this paper, we presented the register mapping strategy, which is an effective optimization method of QEMU. By one-to-one mapping X86 registers to

ALPHA registers, the register mapping strategy has reduced the number of loads and store operations on memory and simplified the rules of the intermediate code generation. We showed the strategy could effectively off-load the translation overhead and promote the quality of the host code.

REFERENCES

[1] Gschwind M, Altman E R, Sathaye S, et al. Dynamic and Transparent Binary Translation[J].IEEE Computer, 2000, 33(3):54–59.
[2] J.H LI, X.N MA, C.Q ZHU. Dynamic binary Translation and Optimization [J]. Journal of Computer Research and Development, 2007, 44(1):161–168.
[3] D.Y Hong, C.C Hsu, P.C Yew, et al, HQEMU: A Multithreaded and Retarget Dynamic Binary Translator on Multicores[C]. International Symposium on Code Generation and Optimization, 2012:104–113.
[4] Fabrice B. QEMU: The Open Source Processor Emulator [EB/OL].http://Bellard.Free.fr/QEMU/zbout.html.
[5] Choi J W, Nam B G. Development of High Performance Space Processor Emulator Based on QEMU-Open Source Dynamic Translator [C]. Proceedings of the 12th International Conference on Control, Automation and Systems, IEEE, 2012:300–304.
[6] QEMU. http: // wiki. qemu.org/Documentation/TCG
[7] Probst M, Krall A, Scholz B. Register Liveness Analysis for Optimizing Dynamic Binary Translation [C]. 2002 9th Working Conference on Reverse Engineering (WCRE), IEEE, 2002: 35–44.
[8] Zhang S, Jiang L, Zhang X, et al. Hardware Software Co-design of Pipelined Instruction Decoder in System Emulation [C]. 2014 4th IEEE International Conference on Software Engineering and Service Science (ICSESS), IEEE, 2013: 149–153.
[9] C. Wang, S.L Hu, Ho-seop Kim, StarDBT: An Efficient Multi-platform Dynamic Binary Translation System [J]. Computer Science. 2007:4–15.
[10] Andrew Henderson, Aravind Prakash, Lok Kwong Yan, et al. Make It Work, Make It Right, Make It Fast: Building a Platform-Neutral Whole-System Dynamic Binary Analysis Platform [C]. The 2014 International Symposium on Software Testing and Analysis (ISSTA). 2014: 248–258.
[11] A.L LIANG, H.B GUAN, Z.X LI. A Research on Register Mapping Strategies of QEMU [C]. Second International Symposium on Intelligence Computation and Applications, 2007:168–172.
[12] S.S CAI, Q. LIU, J. WANG, et al. Optimization of Binary Translationor Based on GODSON CPU [J]. Computer Engineering, 2009, 35(7).
[13] Y.H WEN, D.G TANG, F.B QI. Register Mapping and Register Function Cutting out Implementation in Binary Translation.

Network Security and Communication Engineering – Chan (Ed.)
© *2015 Taylor & Francis Group, London, ISBN: 978-1-138-02821-0*

Fuzzy classification and implementation methods for text advertisements in Weibo

X.H. Xu & X.Y. Yang
Beijing Key Laboratory of Network Systems and Network Culture,
School of Digital Media & Design Arts, Beijing University of Posts and Telecommunications, Beijing, China

ABSTRACT: The establishment of classification index system of advertisements in Weibo, which is a newly developed short-text media in China, is briefly discussed first. After that, the fuzzy method and procedures used in classification are introduced. Several important techniques were explored including the knowledge expression, the hierarchical structure of knowledge library, the pre-processor, and the forward inference engine. The classification of text advertisements in Weibo was implemented. Some specific texts collected from Weibo were analyzed and classified for certification, and the results were effective and efficient compared with manual classification.

1 INTRODUCTION

Weibo is a newly developed social media in China. It mainly consists of less than 140 words along with some other multimedia resources, and is posted on network for users to obtain, disseminate, and share information. Since it is accessible to all kinds of users, Weibo is now being flooded and contaminated with advertisements, which has great influence on users' experience and could also cause safety problems. Therefore, it is necessary to explore the classification index system of text advertisements in Weibo, considering the contents of text, based on the usage laws and regulations in China, especially on the Internet. By researching the classification method and its implementation techniques, advertisements in Weibo can be analyzed and classified, through which we can filter certain kinds of advertisements, and also lay the foundation for subsequent research.

2 ADS CLASSIFICATION INDEX SYSTEM

As its contents are edited and posted by users themselves, lots of advertisements published by regular users or company users are now flooding into Weibo. For now, some draft rules of handling waste implementing information were published. For example, Sina defined "excessive implementing information which damage user experience" as wasted. In this draft, along with waste information and malicious actions, there were three kinds of information being identified as implementation actions that impacted the normal order of Weibo or damaged the users'

experience, which would be imposed gag or silence as penalty.

However, when it comes to the classification of advertisements in Weibo, these draft rules are vague and does not have the maneuverability. After a large number of text ads and their catalog were analyzed, based on "Information System Security Protection Grading Guide" published by China's Ministry of Public Security and other extensive research data of internet cultural security, a classification index system of text advertisements in Weibo was initially built. The system classified Weibo advertisements with some certain index as shown in Figure 1.

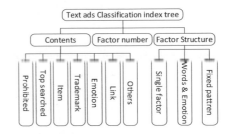

Figure 1.   Ads classification index tree.

3 FUZZY METHODS AND PROCEDURES OF CLASSIFICATION

Advertisements in Weibo is a relative and qualitative fuzzy concept. It is difficult to have a quantitatively accurate represent of classification index as the contents and formats of advertisements are variable, which makes the classification a multi-factor and multi-level fuzzy classification.

A hierarchical factor-based fuzzy classification method was used in our classification, that is, to classify from the bottom to the top, as well as from a single indicator to multi-indicator considered.

1 The classification index domain of the classification object is determined, and then divided into n subsets.

$$U = \overset{n}{\underset{i=1}{U}} u_i \tag{1}$$

where $u_i = \left\{ u_{i1}, u_{i2}, \cdots, u_{ip_i} \right\} (i = 1, 2, \cdots, n)$

2 To each classification index subsets u_i, perform the single level and single indicator fuzzy transformation. Classification level domain $V = \left\{ v_1, v_2, \cdots, v_q \right\}$ is given, where q is better to be an odd number so that we can have a middle level.

To quantify each indicator in u_i, the membership of a certain ad to all categories classified by single indicator is determined. The value of membership function can be determined by using expert scoring method. So we can have the single indicator fuzzy matrix R_i of u_i.

The fuzzy weight vector of every indicators in u_i is determined as

$$A = (a_{i1}, a_{i2}, \cdots, a_{ip_i})$$

Before performing the synthesis process, A_i needs to be normalized. Ensure

$$\sum_{r=1}^{p_i} a_{ir} = 1 \tag{2}$$

Performing fuzzy transformation to A_i and R_i gives us B_i, the single level classification result vector of u_i, which is

$$\begin{cases} A_i \circ R_i = B_i \\ B_i = (b_{i1}, b_{i2}, \cdots, b_{iiu}) \end{cases} (i = 1, 2, \cdots, s) \tag{3}$$

The fuzzy weight vector of comprehensive index $u_i (i = 1, 2, \cdots, s)$ is given as follow:

$$A - (a_1, a_2, \cdots, a_s)$$

3 Multi-level fuzzy classification. Repeat steps 1~2 using u_i as the comprehensive index and B_i as the single indicator classification result. We can have Multi-level fuzzy classification model.

4 Analysis of fuzzy classification results. The weighted average method is used to process fuzzy classification result vectors. The basic idea of this process is firstly to get the rank of classification result vector according to its position, which is to use $1, 2, \cdots, q$ to represent every category, and then secondly to get the weighted sum of each components' rank of result vector, thus we can have the relative position of classified object, written as

$$E = \sum_{j=1}^{q} b_j^k j \Big/ \sum_{j=1}^{q} b_j^k \tag{4}$$

where k is a coefficient to be determined for controlling the impact of larger b_j. k is often set as 1 or 2, so we can sort several classified objects in the order of E.

4 KEY TECHNIQUES IN CLASSIFICATION

4.1 Knowledge expression

4.1.1 Production rule

Key words, emotions, other factors, and the factor structure were identified and summarized as knowledge, based on the afore mentioned classification index system. The knowledge was expressed as production rule with a basic structure below:

IF <A> THEN <B>

in which, A is the production prerequisite, and B is a set of conclusions. Both A and B can generously be expressions composed by text, numbers, and logical operators AND, OR, and NOT.

Here in our research, the prerequisite A led by IF of production rule was the classification index of different levels, and conclusion B led by THEN was the corresponding membership function.

When single indicator was used in classification, some simple rules with ordinary form would be generated. Such as the Single Factor in Fig.1, four of its specific indicators were empty, single emotion, single link as well as single keyword, respectively meaning nothing, only emotions or links, and meaningless keywords accumulation. These factors can be matched with sub-bases to get different results. With factors above, some of the rules can be described as:

IF (text=empty) THEN (1, 0, 0, 0, 0, 0)

IF (single factor=emotion OR link) THEN (1, 0, 0, 0, 0, 0)

In which, value of the membership (1, 0, 0, 0, 0, 0) means it is classified into the first category as the malignant ads, and so on.

Simple rules above can only classify texts with a primary structure, and the contents that can be identified by these rules mostly belonged to malignant ads. Meanwhile, indicators such as Fixed Patterns can generate more complex rules depending on different common patterns used in ads. Some of rules produced like this are listed below.

IF (trademark>0 AND other keywords='game' OR 'download' OR 'new edition') THEN (0,0,1,0,0,0)

Rules can be added by using Matlab visual editing tool for fuzzy rules, or directly added in the classification program.

4.1.2 Frame structure

Corresponding to the hierarchical structure of classification index tree, the multiple nested frame structure, such as {Frame {Sub-Frame {Slot {Side {Sub-Side {Value…}}}}}}, was used for the knowledge expression to manage the production rules. The framework was composed by different levels. Its basic logical structure can be expressed as:

```
{Frame{Slot 1{Side 1{Value 1,
                Value 2,
                    …
                Value m}
            Side 2{Value …}
      Slot 2{…}
      …
      Slot n{…}
}}
```

In this structure, location frame was for the classified object, such as certain Weibo text ad. Slot was for the classification index, like the factor content. Side was for classification sub-index, such as prohibited words. Value was for classification indicator like some certain prohibited words like "naked". Compatible with the index, one side can have lots of values where production rules would be stored. Multiple nested linear table was used to store the contents of framework inside the computer.

This frame-rule mixed expression of knowledge can greatly simplify the workload of the inference engine, reduce the search volume compared with a simple rule system, shorten the reasoning time, and improve the efficiency of solving.

4.2 Hierarchical structure of knowledge base

A comprehensive knowledge base was necessary for the classification. Hierarchical tree structure, as shown in Figure 2, was used to build knowledge base, considering the contents and structure of classification index system.

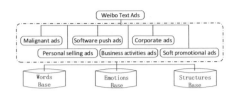

(a) Top level of knowledge base structure

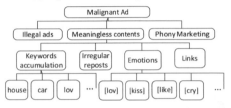

(b) Bottom level of classification knowledge base structure

Figure 2.   Hierarchical structure of the knowledge base.

The storage, management, and maintenance of knowledge can be easily achieved within this structure, through which the knowledge would also be conveniently used in inference.

4.3 Preprocess and inference

4.3.1 Pre-processor

Demanded by the classification knowledge, feature factors in a text advertisement must be extracted to compose the fact library. The pre-processor we used in our research is described as follows.

For an original text advertisement, pre-processor firstly filtered the contents of feature factors in it to compose the Feature Factor Sets. Then the number of all kinds of factors was counted. A text composed purely by factors, which is structurally corresponding with the original text, was extracted. The results consisted with the fact library, and would be matched with the prerequisite of rules.

An example is given here. A text advertisement in Weibo we collected was "Cushlon foam and a full-length Max Air Unit [gift] the lightweight mesh upper smooth and comfortable…NIKE best breathable sport shoes-Air Max [video] http://t.cn/ zTXZzaE 10 years of legend [heart] NOT EVER, BUT NOW {group 70% off promotion}[gift]former price 880 now only 368 with free shipping! item link>>>http://t.cn/8sNbxQt".

The information extracted after pre-processing is listed in Table 1.

Table 1. The Pre-processing result of one text ad.

Content	Words and category	frequency
Keywords	Cushlon/Air unit/ shoe(item)	2
	NIKE/nike/ Max (trademark)	2
	group (activities)	1
	promotion/discount/ price/%off(discount)	1
	Free shipping (selling)	1
	link(other keyword)	1
	legend(phony)	1
Emotions	[video]	1
	[heart]	1
	[gift]	2
Links	http://t.cn/zTXZzaE	1
	http://t.cn/8sNbxQt	1
Pure factors text	Cushilon foam, air unit, Max, [gift], shoes, NIKE, shoes, Max, [video], http://t.cn/ zTXZzaE, myth, [heart], group, discount, promotion, [gift], price, free shipping, item link, http://t.cn/8sNbxQt	

4.3.2 Inference engine

Fitted with the knowledge expression, forward inference engine was applied in our research. By matching the fact library with the prerequisite of rules, the values of membership function can be determined through the inference engine.

Indicators in different levels of index system were filled in the hierarchical structure of knowledge base. The slot and side that would be used according to the levels of classification were firstly determined by the inference engine, and then the rules needed in solving were searched. Contents in fact library was matched with the prerequisite of the rules in the rule library one by one. The membership function contained in the conclusion of rules successfully matched would be stored as one of the row elements of membership matrix.

5 IMPLEMENTATION

The collected text advertisements in Weibo were classified with the methods above. To space limitations, the advertisement text in 4.3 is given next as an example of the general classification.

Given the classification categories domain V

$V =\{$Malignant ads (M), personal selling ads (S), business activities ads (A), software push ads (SW), corporate ads (C), soft promotional ads (P)$\}$

Main Indicators was chosen, according to the contents referred to, to build classification indicators domain U. The general indicators domain used in classification of different text ads was transcendental, while different kinds of sub-sets of indicators could also be generally used in classification of the same type of ads (such as all the text from a certain user).

The fixed patterns used, which are also common and efficient for the rest classifications, are listed below. The classification result is shown in Table 2.

Pattern I: item$\geq$1 AND trademark$\geq$0 AND keyword='free shipping' OR 'discount'

Pattern II: keyword='address' OR 'download' AND link

Pattern III: keyword +emotion>2 AND link

Table 2. The classification result of one Weibo ad.

Categories	M	S	A	SW	C	P
prohibited	0.03	0.52	0.21	0.02	0.2	0.02
Hot words	0.12	0.35	0.24	0.02	0.24	0.03
Item	0.20	0.23	0.32	0.03	0.12	0.10
Trademark	0.18	0.21	0.34	0.05	0.10	0.12
Emotion	0.32	0.43	0.12	0.01	0.08	0.05
Link	0.13	0.44	0.09	0.26	0.05	0.03
Onefactor	0.27	0.24	0.18	0.08	0.18	0.05
W-Emo-W	0.29	0.32	0.16	0.12	0.08	0.03
Pattern I	0.06	0.39	0.16	0.28	0.02	0.09
Pattern II	0.03	0.27	0.33	0.31	0.03	0.04
Pattern III	0.1	0.22	0.39	0.14	0.12	0.03

However, when data in Table 2 were used as the fuzzy matrix R, the value of each component of weight vector was too small to affect the classification results since there were too many categories. We divided the data into 3 groups in the form of 4-4-3, as the final three are relatively independent from the front 8. Fuzzy transformation was performed respectively to have

$$R = \begin{bmatrix} 0.13 & 0.33 & 0.28 & 0.03 & 0.17 & 0.06 \\ 0.25 & 0.36 & 0.14 & 0.12 & 0.10 & 0.03 \\ 0.06 & 0.29 & 0.29 & 0.24 & 0.06 & 0.06 \end{bmatrix}$$

Delphi method was used to determine the weights of each indicators. Here, we use the weight vector

$A = (0.2, 0.3, 0.5)$

Fuzzy transformation was performed to have

$$B = (0.13, 0.33, 0.24, 0.16, 0.09, 0.05)$$

Weighted average method was used to analyze and process the fuzzy classification results. Set the value of each components of the resulting vector according to their positions, that is, 1 for malignant ads, 2 for personal selling ads, and so on. The final classification result was calculated as Equation (4).

$$E = \frac{1 \times 0.13^2 + 2 \times 0.33^2 + 3 \times 0.24^2 + 4 \times 0.16^2 + 5 \times 0.09^2 + 6 \times 0.05^2}{0.13^2 + 0.33^2 + 0.24^2 + 0.16^2 + 0.09^2 + 0.05^2} \approx 2.57$$

The result $E \approx 2.57$ showed that the classification result of this example was between category 2 and 3, and slightly partial to the 3rd, which meant it is probably a personal selling ad or a business activities ad, though more likely to be the latter. However, it could be easily identified literally to both of those two categories. As for which one it actually belonged to, the answer would be relative to the user who actually had posted it, and that is the poster attribute. It was rather hard to tell that on this text alone.

In order to solve the problem above, the subsequent completion of classification on commercial users based on the classification results of advertisements is now proceeding, to which our research now has built a great foundation.

6 CONCLUSIONS

With methods and procedures above, several text advertisements in Weibo were classified for certification, and the results were considered reasonable and acceptable. The methods and implementation were effective and efficient.

REFERENCES

[1] Wang Yan, Yang Wen-Yang, Zhang Yi etc. Research on the Evaluation Index System of the Recommended Information in the Network Cultural Security of China. *Journal of information*. 2008,27(5):64–66. DOI:10.3969/j.issn.10021965. 2008.05.020.

[2] Zhang Xu. Explore of Weibo Security and its solution. *Tribune of Social Sciences in Xinjiang*.2013,(2):6668,80. DOI:10.3 969/j.issn.16714741.2013.02.015.

[3] Yang, Christopher C. Ng, Tobun D.. Terrorism and Crime Related Weblog Social Network: Link, Content Analysis and Information Visualization. *Intelligence and Security Informatics*. 2007 IEEE; New Brunswick,NJ.2007:55–58.

[4] Xu Xiaohui, Tong Bingshu. Fuzzy Evaluation and Implementation Techniques of Product Structure Design for Assembly. *Transection of the Chinese Society for Agriculture Machinery*. 2007,38(2):129133. DOI:1 0.3969/j.issn.10001298. 2007.02.033.

[5] Deng Bing-Na. Spam Recognition Research in Blog Comment. *HEBEI University*. 2011.DOI:10.7666/d. d154574.

[6] PRC Ministry of Public Security of Information. *Systems Security protection level grading guidelines (Trial Version)*. 2005,12.

Optimization of depth-first search for smart grid

X.Z Hou & H.L Sun
State Grid Chongqing Electric Power Research Institute, Chongqing, China

S.Q Liu, Y.G Li, C.X. Yang & J.T Cai
Beijing Techrefine Technology Co., Ltd., Beijing, China

ABSTRACT: With the development of information technology, communication technology applications will be diversified in various industries. Application of communication technology also needs to be improved and optimized according to the characteristics of industry. It can serve the industry better. The application of wireless sensor network is taken into account in the centralized meter reading system of State Grid in communication technology. Its characteristics include multi-node, multi-hop, connectivity, and irregular distribution as shown in figure 1. This network is required to be intelligent, stable, fast, multi-hop, self-healing, and so on. We need research as to how the nodes are arranged, how to build mathematical models, how to choose the algorithm, and how to optimize the algorithm. In this paper, we analyze the distribution of nodes and build a mathematical model to choose an algorithm and optimize the algorithm.

1 INTRODUCTION

With the development of information technology, the Wireless Sensor Network is used in various fields because of its high transmission speed, high information integration and easy installation. In different application fields, it is combined with the characteristics of the industry and its technology is optimized. Wireless sensor network has the characteristics of intelligent, stable, and self-healing. It can be applied in different industries.

The China State Grid was convinced by the advantages of wireless sensor network in data transmission. They use it in centralized meter reading system; it is called the micro power wireless communication.

If we want to research the centralized meter reading system network, we must research the characteristics of the country network collecting system network first, and build a mathematical model for research.

In this paper, we research the characteristics of China network collecting system network first and build a mathematical model from the study on mathematical model algorithm to prove the advantages of the proposed algorithm.

2 THE CHARACTERISTICS OF THE CENTRALIZED METER READING SYSTEM OF NETWORK

In the centralized meter reading system for the national grid, the application needs are considered. The characteristics of sensor network self-organizing multi hop (micro power wireless communication) include:

- There are a plurality of nodes in space (less than 1000);
- The distribution of nodes is irregular; the hops of path is maximum of 7;
- Communication depends on the RF signal quality between the nodes;
- The nodes can communicate with each other.

There is one and only one master node in the space which can be called the root or source.

According to the characteristics above and in the application of centralized meter reading system, we make an assumption on the number of nodes and hop a space node distribution of relative extreme, and list the distribution of several representative studies. There are: node equilibrium distribution, the concentrated distribution for the end of master node, and the concentration distribution for the front of master node.

Table 1. Node distribution of relative extreme.

Hops\Total	128	254	508	1016
Equilibrium distribution of nodes				
1	18	36	73	145
2	18	36	73	145
3	18	36	73	145
4	18	36	73	145
5	18	36	73	145
6	18	36	73	145
7	18	36	73	145
Concentration distribution of network terminal (similar binary tree)				
1	1	2	4	8
2	2	4	8	16
3	4	8	16	32
4	8	16	32	64
5	16	32	64	128
6	32	64	128	256
7	64	128	256	512
Concentration distribution of network front (similar binary tree)				
1	64	128	256	512
2	32	64	128	256
3	16	32	64	128
4	8	16	32	64
5	4	8	16	32
6	2	4	8	16
7	1	2	4	8

In Table 1, that is the node distribution of extreme condition. However, in practical application, the most communication is 3-4 jumps in the network. In Figure 1, search algorithm can start from the simplified mathematical model. This is a simplified mathematical model, starting from the main node V0. The nodes are named according to the hierarchy, as shown in figure 1.

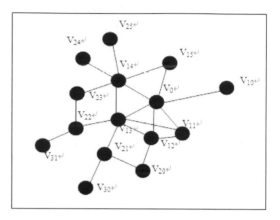

Figure 1. Simplified mathematical model of network.

As shown in figure 1, the connection between the nodes proves that nodes can communicate with each other. There is Adjacency List in the RAM of nodes; subscript serial number is sequential in Adjacency List; Adjacency List is node information that can be charged which contains a node address and the node communication parameters.

3 THE MATHEMATICAL MODEL OF NETWORK

According to the characteristics of centralized meter reading system of state grid, and the node distribution representative, the network for meter reading system is more like the data structure in Figure 1. It is connected with no direction and is a more complex data structure than the proportional table and tree. Fig (Graph Theory) is a graph composed of a plurality of given points and joint points of the line. This pattern is usually used to describe certain relationship between certain things. Points are used to represent things; the line connecting two points is used to express the underlying relationship between two things.

When we research the questions about the Figure 1, we need to consider each vertex information in Figure 1. In Figure 1, the process of each vertex is visited, which is the search. In Figure 1, the search algorithm is the basis for communication between the nodes, the topological sort and the critical path calculation.

4 THE ROUTING ALGORITHM OF NETWORK

Usually, the search method of graph has two algorithms: the depth-first search and the breadth-first search. The depth-first search is similar to the pre-order search of the tree. The characteristics of the algorithm are searching the depth direction of the network and searching the next layer first; the breadth-first search is similar to the method of the tree's searching according to the level. The characteristic of the breadth-first search is searching the same layer first.

In the searching process, both the depth-first search and the breadth-first search will be visited more than once in order to avoid the same node. Each node which has been visited must be marked. So we set up an auxiliary array visited [0..n-1] for each node in the network. The initial value of the array is set to "false" or zero, if a node V_i is visited, then the visited[i] is set to "true". In this way, we calculate the path from V_0 to V_{30} and V_{31}. In Figure 1, the breadth-first searches first divide all nodes into multilayer, and then does the calculation. The results are as shown below:

Table 2. The graph depth-first and breadth-first solving 1.

Mode	BFS	DFS
V_{30}		
Solving	V_0-V_{13}-V_{21}-V_{30}	V_0-V_{11}-V_{12}-V_{13}-V_{21}-V_{30}
search points	13	14
V_{31}		
Solving	V_0-V_{13}-V_{21}-V_{30}	V_0-V_{11}-V_{12}-V_{13}-V_{21}-V_{30}
search points	14	9

The results have the following characteristics: the breadth-first search is easy to calculate the shortest path, but the method will take more time and memory; the depth-first search calculates the approximate solution, the calculation time is not fixed due to the number of neighbor nodes, but it takes up less memory.

No matter which algorithm you choose, in order to maintain robustness of the network, multiple paths need to be stored to keep the network self-healing capability. According to this requirement, we need to optimize the existing search algorithms. It is necessary to modify the qualifications while searching. The auxiliary array aforementioned visited [0.n-1] scopes are modified to include only the neighbors during the search starting node. In this way, we calculate the path from V_0 to V_{30} and V_{31}. In Figure 1, the Breadth-First Search first divides all nodes into multilayer, and then does the calculation. The results are as shown below:

Multiple paths can be found in limited conditions for the depth-first search was changed. One of the paths must be the same as the results for the breadth-first search. Since the breadth-first search algorithm needs to do the layered operation, nodes of the same layer are access restrictions. So only 1 path is calculated. If you need to search more paths, mechanism of hierarchical processing needs to be broken and more memory will be occupied if we change the mechanism.

Which routing algorithm will be used is determined by the size of the RAM memory used in routing algorithm, the time complexity of the routing algorithm mainly affects the time of the network formation.

Depth-first recursive algorithms are easier to implement for a connected graph is concerned. It often only needs to call once transversal. After weighing the pros and cons of the two algorithms, the depth-first algorithm is more applicable to State Grid reading system.

Table 3. The graph depth-first and breadth-first solving 2.

Mode	BFS	DFS
V_{30}		
Solving	V_0-V_{13}-V_{21}-V_{30}	V_0-V_{11}-V_{12}-V_{13}-V_{21}-V_{30} V_0-V_{11}-V_{12}-V_{20}-V_{21}-V_{30} V_0-V_{11}-V_{13}-V_{21}-V_{30} $\ldots\ldots$ V_0-V_{13}-V_{21}-V_{30} $\ldots\ldots$ V_0-V_{15}-V_{14}-V_{23}-V_{22}-V_{13}-V_{11}-V_{12}-V_{20}-V_{21}-V_{30} (22 paths)
search points	13	182
V_{31}		
Solving	V_0-V_{13}-V_{21}-V_{30}	V_0-V_{11}-V_{12}-V_{13}-V_{21}-V_{30} V_0-V_{11}-V_{12}-V_{13}-V_{14}-V_{23}-V_{22}-V_{31} V_0-V_{11}-V_{12}-V_{13}-V_{22}-V_{31} V_0-V_{11}-V_{12}-V_{20}-V_{21}-V_{13}-V_{14}-V_{23}-V_{22}-V_{31} $\ldots\ldots$ V_0-V_{13}-V_{22}-V_{31} $\ldots\ldots$ V_0-V_{15}-V_{14}-V_{23}-V_{22}-V_{31} (16 paths)
search points	14	180

5 PATH OPTIMIZATION ALGORITHM-BASED ON DEPTH-FIRST SEARCH

In order to keep the self-healing ability of the network, it often keeps multiple paths. In general, communication on the shortest path will spend the least time; communication is more reliable on the strongest signal path. The path which is self-healing is used as a backup path. This can be optimized for search:

- Limit the depth of the search process for the solution to 7-hop, you can refine your search efficiency and reduce invalid path.
- The weight value is calculated from the hops and the signal strength.
- The router algorithm calculates four optimal paths and uses the weight value, an old path will be replaced by the better one in the router calculating process.

- All the paths have different key nodes, so if one of the key node fails, it still can communicate through other paths.
- The search depth reduces to 7 hops.

The depth means the hops in the path, more hops will reduce the communication success rate of path; more hops also increased transmission time. Considering the actual application situation, 7-hops is sufficient to meet the centralized meter reading system of State Grid. For example in Table 3, the V_{30} has a path: V_0-V_{15}-V_{14}-V_{23}-V_{22}-V_{13}-V_{11}-V_{12}-V_{20}-V_{21}-V_{30}, its hops is 10. For a four layer network, the path has no advantage in the many paths. Reducing the hops to 7 will optimize search efficiency and reduce invalid path.

Weight value of path is calculated by the signal strength and hops.

Signal strength is an important factor which influences the success rate of communication. The poor signal strength will increase the error rate in communication. Select the minimum signal strength whose error rate is 0 in communication as the reference value. The path with signal strength below the reference value will decrease the priority.

The weight of path will consider the following factors:

- Record the path whose signal strength is better than the reference value;
- Find the path which has minimum number of hops in the records which are calculated above and calculate the key nodes;

The key node is a node in all paths. If this node is at fault, all paths will fail. So to calculate the key nodes is an important manifestation of network self-healing ability.

As it is shown in the following table, a list of key nodes in the V_{30} for all solutions is:

Table 4. Key-node list of V_{30}.

NO.	Key-node	Paths/total paths	Description
1	V_{21}	22/22	the absolute key node
2	V_{13}	20/22	the secondary key node
3	V_{20}	13/22	the secondary key node

* The absolute key node: all paths are used.

The path which is self-healing is used as a backup path. The backup path doesn't consider the absolute key node. For example, in Table 4, the secondary key nodes, V13 is included in most paths, so the 4 paths storage may include this node. To ensure the node failure will not cause the network fault, a path without the secondary key node of V13 is necessary.

Select the strongest signal quality as routing evaluation standard.

One path includes multi hops. The hop which has the least signal strength influences the stability of the path. For example, if a path to the V0-V13-V21-V30 signal intensity (acting on the edge) are (-90db) (-85db) (-80db), then the path quality is the lowest value (-90db).

Table 5. Results of depth-first optimized.

Mode	V_{30}	V_{31}
	DFS	DFS
Solving	V_0-V_{13}-V_{21}-V_{30}	V_0-V_{13}-V_{22}-V_{30}
	V_0-V_{12}-V_{20}-V_{21}-V_{30}	V_0-V_{14}-V_{23}-V_{22}-V_{31}
	V_0-V_{14}-V_{13}-V_{21}-V_{30}	V_0-V_{14}-V_{13}-V_{22}-V_{31}
	V_0-V_{11}-V_{12}-V_{20}-V_{21}	V_0-V_{12}-V_{13}-V_{22}-V_{31}-V_{30}
	(4 paths)	(4 paths)
search points	99	89

In the above results, we reduce the number of search nodes, so the search time is reduced. The path which is calculated has less hops and better signal strength. The depth-first search algorithm achieved good results in the centralized meter reading system in practice.

6 CONCLUSION

To research of micro power wireless automatic meter reading system, the main difficulty is to analyse the characteristics of node distribution, to build the mathematical model of network, optimize route searching algorithm, and make the wireless network be stable, efficient, intelligent, and robust.

In this paper, we analyzed the distribution characteristics of nodes using the graph mathematical model and the graph searching algorithm to calculate all the nodes. We believe that the depth-first search algorithm can meet the application.

REFERENCES

[1] Peter Brass, 2014, Advanced Data Structures, England: Cambridge University Press.
[2] Chapman and Hall, 2007, Dinesh Mehta and Sartaj Sahni Handbook of Data Structures and Applications, Boca Raton: CRC Press.
[3] Niklaus Wirth, 1985, Algorithms and Data Structures, Washington, D.C.:Prentice Hall Press.
[4] Diane Zak, 2011, Introduction to programming with c++, copyright 2011 Cengage Learning Asia Pte Ltd.
[5] Wei-Min YAN and Wei-Min WU, 2007, Fundamentals Of Structural Data In C, Beijing: Tsinghua University Press.

Network Security and Communication Engineering – Chan (Ed.)
© *2015 Taylor & Francis Group, London, ISBN: 978-1-138-02821-0*

Operational risk measurement based on POT-Copula

Enhemende
Network Management Center, Hohhot Vocational College, Inner Mongolia, China

ABSTRACT: According to the sample data of the operational risks of the Chinese commercial banks during 2003 to 2013, this paper constructs a risk measurement model based on POT-Copula. Using Hill graph and the excess mean function diagram determining the model of threshold, it measures the overall value at risk by using HKKP to estimate the tail parameters of the POT model and calculates the value at risk of the conditional risk of the inside and outside deception of Chinese commercial banks' operational risk, and finally employs t-Copula function to connect the marginal distribution of the inside and outside deception in the operational risk.

1 INTRODUCTION

Recently there is a worldwide increase in the operational risk frequency of the commercial banks, leading the bank industry and the academic field to pay great attention to the operational risk management. Measurement of operational risk is the core of the operational risk management work of commercial banks. Therefore, many financial institutions are sparing no effort to develop the tools and methods for operational risk measurement. The Basel committee formally divided the calculating method of operational risk into Basic Index Approach (BIA), Standard Approach (SA), and Advanced Measuring Approach (AMA) in the New Capital Accord issued in 2014. However, the practice has proved that the BIA and SA have lots of deficiencies. So the recent researches on the calculating of operational risk have been focusing on the typical AMA, giving rise to a quantity of operational risk measurement models. Peaks over threshold (POT), among all these methods, becomes an important approach for its extraordinary data sensitivity and expansibility. POT methods have many applications on measuring operational risk. Song and Turk [1] pointed out that when lacking the data and its heavy-tailed characteristic, POT methods can easily and reliably measure, predict, and manage the relative risk elements. Bin and Wu [2] used POT to simulate the loss distribution in the operational risks of Chinese commercial banks and diagnosed the simulation, and tested the feasibility of the operational risk by using POT methods.

Despite all these, there still exist many weaknesses in POT. Most of the researches use SME to make sure of the threshold value [3, 4]. Some of the researches also use Hill diagram to assist [5]. Lu [6] believes that SME and Hill graph rely on the objective judgments of the pictures, which may have many errors. Hill estimation always has errors. Only when the sample capacity reaches infinite can the figure reach zero. Based on the deficiencies, HKKP offers a reasonable way to determine the threshold. Not only can it overcome the high precision of the threshold requirements, but it prevents excessive dependence on the parameter hypothesis. HKKP is unbiased though it is with simple sample. In addition, in most previous research, researchers usually did simple summary sums of the operation risk of each category, ignoring the risk correlation between them; hence it probably causes the deviation by the risk measurement results.

Aiming at overcoming the shortage of current study on operational risk measurement, this paper uses the HKKP method to estimate the threshold of POT, and employs the Copula function to connect various types of operational risk's marginal distribution, and constructs an operational risk measurement model for the whole operational risk based on POT-Copula. It makes an empirical study on the operational risk of Chinese commercial banks, hoping to provide a meaningful reference for the operational risk measurement of commercial banks and corresponding capital allocation.

2 CONSTRUCTION OF THE RISK MEASUREMENT MODEL BASED ON POT-COPULA

2.1 POT model

Suppose a sequence of independent random variables with the same distribution subjects to the profit and loss distribution F and define F_u as excess of loss distribution beyond the observed value of a certain large threshold u, according to conditional probability formula:

$$F_u(x) = \Pr(X - u \le x | X > u) = \frac{F(x+u) - F(u)}{1 - F(u)}, x \ge 0,$$

$$F(x) = [1 - F(u)]F_u(z) + F(u).$$ (1)

for a threshold large enough u, F_u converges to generalized Pareto Distribution $G_{\lambda,\eta}$:

$$G_{\lambda,\eta} = \begin{cases} 1 - (1 + \lambda \frac{x}{\eta})^{\frac{-1}{\lambda}}, \lambda \ne 0, \\ 1 - \exp(-\frac{x}{\eta}), \lambda = 0, \end{cases}$$ (2)

η is scale parameter, $\eta > 0$, λ is the tail parameters, $\lambda \in R$, when $\lambda > 0, x \ge 0$; when, $\lambda < 0, 0 \le x \le -\frac{\eta}{\lambda}$.

when $\lambda \to 0$, $G_{\lambda,\eta}(x)$ is called Paretol type, a typical exponential distribution. When $\lambda > 0$ and, $\lambda < 0$ $G_{\lambda,\eta}(x)$ distribution corresponds to Pareto II and Pareto III distribution. Pareto II distribution is a heavy-tailed distribution that is widely applied in time order, which is consistent with the thick tail characteristics of asset loss distribution. The bigger the tail parameter λ is, the more obvious the fat-tail characteristic is. In determining the threshold u, $\frac{n - N_u}{n}$ is used to approach $F(u)$. Uniting formula (1) and (2), we have:

$$\hat{F}(x) = \begin{cases} 1 - \frac{N_u}{n} \exp \frac{x-u}{\hat{\eta}}, & \lambda = 0 \\ 1 - \frac{N_u}{n} (1 + \hat{\lambda} \frac{x-u}{\hat{\eta}})^{-\frac{1}{\lambda}}, & \lambda \ne 0 \end{cases}$$ (3)

This formula is based on the general Pareto distribution's POT model. Application of POT model to carry on the risk measure needs the reasonable estimation of threshold u and model parameter λ and η. According to the estimation methods of threshold u from Danielsson et al. [7], Gomesm and Martins [8], Sample Mean Exceedance and Hill diagram, technically speaking, all rely on the subjective judgment of the graph.

This paper will use HKKP estimate to overcome high accuracy requirements of threshold value, and in the case of insufficient data, we can still get unbiased estimation. After getting POT model parameter estimation results, we can use this formula to calculate VaR and CVaR of risk:

$$\begin{cases} VaR = F^{-1}(1 - \alpha) = u + \frac{\eta}{\lambda}((\frac{n}{N_u}\alpha)^{-\lambda} - 1), \\ CVaR = \frac{VaR}{1 - \lambda} + \frac{\eta - \lambda u}{1 - \lambda}, \end{cases}$$ (4)

$N_u = \sum_{i=2}^{n} i$, which is the number of samples exceeding the threshold value.

2.2 Copula function

Copulas connect theory was first put forward by the French statisticians Sklar in 1959 [9]. He pointed out that the joint distribution can be decomposed into a marginal distribution and a copulas connect function that describes the correlation between variables, which connects multiple random variable marginal distribution and the formation of a joint distribution function. Binary Copula function $C(u, v)$ is the function defined $I^2 = [0,1] \times [0,1]$, meeting the following conditions:

1 $C(u, v)$ has solutions and is 2d progressive increasing.
2 To any $u, v \in I = [0,1]$, $C(u, v)$'s marginal distribution $C_U(u)$ and $C_V(v)$ separately meet $C_U(u) = C(u, 1) = u$, $C_V(v) = C(1, v) = v$

In short, Copula is a function that associates joint distribution of multidimensional random variable with one-dimensional marginal distributions. In terms of binary Copula function, if we make $H(x, y)$ the binary joint distribution function that has $F(x)$ and $G(y)$ as the marginal distribution functions, there will be a Copula function named $C(u,v)$. And we have $H(x, y) = C(F(x), G(y))$. $F(x)$ and $G(y)$ are continuous functions, then $C(u,v)$ can be confirmed uniquely. Otherwise, $C(u,v)$ can be confirmed uniquely on $RanF \times RanG$. On the other hand, if $F(x)$ and $G(y)$ are one-dimensional marginal distributions and $C(u,v)$ is a Copula function, $H(x, y)$ which is defined by the formula (1), is a binary joint distribution function with $F(x)$ and $G(y)$ as marginal distributions.

2.3 Risk measurement model building

Using POT-copulas model for modeling commercial Banks operating risk of low frequency high risk loss events, the main processes are as follows:

1 Using the sample average excess function to select threshold of u, and exceed the threshold excess sample data were determined.
2 Using the generalized pareto distribution GDP fitting excess sample data, and using maximum likelihood estimation POT model parameters.
3 Type (4) is used to calculate operational risk of loss value of VaR and conditional value at risk of CVaR.
4 Using t-copulas to connect connection consists of internal fraud and external fraud combination of operation risk

216

3 OPERATIONAL RISK MEASUREMENT

3.1 Data collection and analysis

This paper collected China's commercial Banks operating risk events of 499 loss data which is from Chinese court net, the China banking regulatory commission, China's financial network, the national audit office, Sohu News, HeXunWang, and Netease news from 1993 to 2003. According to our statistical results, internal and external fraud take the highest rates in operational risk loss events, which is 50.8% and 21.4%, so this passage chooses internal and external fraud which have high frequency to happen and make a lot of damage as the range of study. First, we test the normality of these two types of operational risk loss events with MATLAB.

Table 1. Internal fraud and external fraud data distribution characteristics analysis table.

	Maxi-mum Value	Min-imum value	Mean value	SD	Skew-ness	Kur-tosis
Internal fraud	399400	0.01	11102	35078	6.95	65.90
External fraud	1010000	1	26143	109689	7.42	63.61

According to table 1, we can see that China's operational risk event losses vary greatly and that the lowest loss is 100 Yuan while the highest is 10.1 billion Yuan; two kinds of loss events of skewness and kurtosis are different from standard normal distribution, which suggests that the two types of operational risk loss amount do not obey the normal distribution. According to table 1, 499 cases of loss amount distribution can be seen and the Chinese commercial bank operation risk loss event has the characteristics of the spike thick tail.

3.2 Estimating tail parameter with HKKP

Firstly, we should construct a set of order statistics, and we sort the operational risk loss data from the lowest to the highest. Then, we pick k order statistics which are the highest. The research result of Huisman et al. [11] shows that when k is between $n/3$-$n/2$, the robust estimates can be acquired. Therefore, the integers from 1 to $n/2$ will be checked one by one in this essay to get the accurate k. And then we will use least square method to get the error sum of squares of every statistic. Then we will take the partial derivatives to α_1 and finally get the estimate value $\hat{\alpha}_1$. The final result is that the HKKP estimate of internal fraud losses is $\hat{\alpha}_1 = 0.874$ and that tail parameter is $\lambda = 1/\hat{\alpha}_1 = 1.144$.

Also, the HKKP estimate of external fraud losses is $\hat{\alpha}_1 = 2.37$ and tail parameter is $\lambda = 1/\hat{\alpha}_1 = 0.422$.

3.3 Determining the threshold value and parameter estimate

The core of building extreme model is to choose the right threshold value. And too large or too small threshold value will produce biased estimate parameter. And this will make the measurements to be questioned. This essay will comprehensively use Hill estimation and mean excess function to get the threshold value which has a high fitting degree. Due to limited space, the pictures are not shown in the paper. But from the pictures we can know that the threshold of intern fraud is 51,000,000 and that of extern fraud is 89,000,000. Then, with maximum likelihood estimation, we can know that the scale parameter of intern fraud is 37.558 and that of extern fraud is 579.318. And the excess numbers are 67 and 30.

3.4 The calculation of CVaR value

After the confirmation of the threshold value, tail parameter and scale parameter, we substitute all the estimates of parameter into formula (4). And we respectively calculate the two types of operation risk corresponding to VaR and CVaR value. The results are shown in table 2.

Table 2. The CVaR value of different confidence level when the event is not relevant.

Confidence Interval	0.99	0.975	0.95	0.9
The Intern Fraud CVaR	5007.9	2989.2	1956.7	885.9
The Extern Fraud CVaR	11606.4	6632.8	3678.0	1773.8
Total CVaR	16614.3	9622.0	5634.7	2659.7

Table 2 shows that the CVaR value will grow with the increase of confidence level in both intern fraud and extern fraud and that the level of sensitivity to CVaR value of intern fraud is higher than that of extern fraud. The former can be obtained by the increase of probability distribution function, and the latter shows that the fat tail of intern fraud is more significant than that of extern fraud in operation risk in China.

We further choose Student-t Copula function to calculate the total CVaR of two kinds of risks that may be correlated. We first get the correlation coefficient is 0.3103. And then calculate the total CVaR of operation risk with software R to compare with the former result. The results are shown in table 3.

Table 3. The comparison of CVaR when events are independent and relevant in different confidence levels.

Confidence level	99%	97.5%	95%	90%
CVaR when events are independent	16614.29	9621.99	5634.73	2659.68
CVaR when events are relevant	15394.76	8813.167	4873.512	2157.53
Reduced ratio	7.34%	8.41%	13.51%	18.88%

Table 3 shows that the CVaR of the relevant events are higher than that of independent events. And the level of the decrease of CVaR will be increased with the decrease of confidence level. A reasonable explanation of this is that when the confidence level reaches a certain level, the times of the operation loss will be quite low.

4 CONCLUSION

A risk measurement model which is based on POT-Copula is built in this paper. And we use it to study the measurement of operation risk in Chinese Commercial Banks. Firstly, we analysis the situation of the operation risk in Chinese Commercial Banks according to the collected data from 1993 to 2013 and find that the risk is mainly made up of intern and extern fraud. Besides, the slant tail phenomenon is obvious. Then we use HKKP to estimate the tail parameter of operation risk and confirm the threshold value of POT with hill diagram and mean excess functional diagram. At this point, we calculate the VaR and CVaR value of intern and extern fraud. Considering that there might be correlation between intern and extern fraud, we connect the marginal distribution of two kinds of risks with Copula function and calculate the VaR and CVaR value of total risk.

The result shows that estimating tail parameter with HKKP can avoid the partial tail of data. The total operation risk made up of intern and extern fraud is obviously less than the simple sum of the two kinds of risks. The operation risk value which is calculated by the model in this paper provides reference for commercial bank to conduct relevant capital allocation.

REFERENCES

[1] Turk A B. Quantitative operational risk management[M]. Ginancarlo N Advance in risk management Sciyo, 2010.
[2] Bing Jiancheng, Wu Jun. The Fitting and Diagnosis of Chinese Commercial Banks' operational risk loss [J]. Investment Research, 2013(3): 20–32.
[3] Zhang Wen, Zhang Yishan. An Empirical Study on the Measurement of Operational Risk of Commercial Bank which Applies Extreme Value Theory[J]. The South Finacial, 2007(2): 12–14.
[4] Gao Lijun, Li Jianping, Xu Weixuan. The Extreme Value Estimation of the Operational Risk of Commercial Bank Based on the POT Method [J]. Operations Research and Management, 2007, 16(1): 112–117.
[5] Wu Hengyu, Zhao Ping. Operational Risk Measurement of Chinese Commercial Banks——Research Based on Extreme Value Theory [J]. Journal of Shanxi University of Finance and Economics, 2009(8): 109–115.
[6] Lu Jing, Zhang Jia. Research on Operational Risk Measurement of Commercial Bank Based on Extreme Value Theory and the Multiple Copula Function [J]. Chinese Journal of Management Science, 2013, 21(3): 11–19.
[7] Danielson K G, Baribault H, Holmes D F, et al. Targeted disruption of decorin leads to abnormal collagen fibril morphology and skin fragility[J]. The Journal of cell biology, 1997,136(3):729–743.
[8] Gomes M I, Martins M J. Generalizations of the Hill estimator–asymptotic versus finite sample behaviour[J]. Journal of Statistical Planning and Inference, 2001,93(1):161–180.
[9] Sklar A. Functions de Repartition àn Dimensions et Luers Marges[J]. Publications de Institut de Statistique Universite de Paris, 1959, (8): 229–231.
[10] Huisman R, Koedijk K G, Kool C J M, et al. Tail-index estimates in small samples[J]. Journal of Business & Economic Statistics, 2001,19(2):208–216.

Network Security and Communication Engineering – Chan (Ed.)
© 2015 Taylor & Francis Group, London, ISBN: 978-1-138-02821-0

Power allocation based on average BER for two-relay three-hop cooperative communications

X.F. Lan, J. Zhang & Y.X. Ma
Shanghai Normal University, Shanghai, China

ABSTRACT: As a virtual Multi-Input Multi-Output (MIMO), cooperative communication allows users to obtain a space diversity gain without increasing the number of user's antennas. This paper focuses on a two-relay three-hop cooperative diversity system model, which adopts AF and MRC manners. A power allocation algorithm is proposed which aims to minimize Average Bit Error Rate (ABER) of destination node under the condition of total power constraint. Simulation results show the power allocation scheme reduces ABER at low signal to noise ratio (SNR), so as to make the communication quality of cell-edge users better.

KEYWORDS: Cooperative communication, ABER, AF, power allocation.

1 INTRODUCTION

In cellular mobile communication systems, low SNR of a cell-edge user occurs because of base station coverage and inter-cell interference. For the cell-edge user, reasonable combination with cooperative relay can effectively make up base-station coverage and bring diversity gain without changing antenna configuration.

The basic idea of cooperative communication is that each single-antenna user is considered as a relay node which forwards information for each other. Common forward manners are amplify-forward (AF) and decode- forward (DF). In AF manner, relay node only amplifies the power of its received signal from source, and then destination decodes its received signal from the relay. In DF manner, relay node decodes its received signal from source and then transmits to the destination.

Concerning with power allocation in cooperative diversity system, most literatures consider system capacity or outage probability as objective function to find the optimal power allocation solution. In [1]-[3], capacity maximization is adopted as objective function for two-relay three-hop AF diversity system. In [4]-[5], BER is as objective function and optimal power allocation algorithms are proposed for two-node two-hop DF diversity system.

This paper focuses on optimal power allocation problem in a cooperative diversity system by using AF manner. We adopt two-relay three-hop cooperative system model. A power allocation model and algorithm based on minimizing average bit error rate (ABER) is presented. Simulation results show that the algorithm reduces ABER at low SNR and improves the performance of cell edge users.

The paper is organized as follows. In the first section, the system model is presented, and signal to noise ratio is analyzed. The second section is the analysis of ABER using MRC. In the third section, an optimal power allocation algorithm for minimizing ABER is presented. In the fourth section, simulation results are given. The last is the conclusion.

2 SYSTEM MODEL

Consider two-relay communication models as shown in Fig. 1 and Fig. 2. Both models contain four nodes: source S, destination D, and cooperative relays R_1 and R_2. These dual-relay systems can be divided into two types of transmission: two-hop and three-hop transmission.

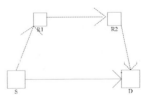

Figure 1. Two-relay three-hop cooperative diversity model.

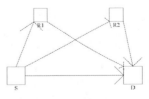

Figure 2. Two-relay two-hop cooperative diversity model.

In the first scenario, the relay nodes use AF mode. To ensure the signals of nodes are not interacted on each other, the system adopts time division multiple access (TDMA). The signal from source is transmitted in three time slots. In the first time slot, source S sends signals to node R_1 and destination D. In the second, relay node Rs_1 amplifies the received signal and then sends the signal to node R_2. In the third, R_2 receives the signal from R_1, amplifies it and then sends to destination D. Destination D combines the two received signal of the first and the third time slot by using MRC. In the second scenario, signal transmission is divided into two time slots. In the first time slot, source S sends signals to the node R_1, R_2, and destination D. In the second time slot, R_1 and R_2 amplify the received signal's power and send the signal to destination D. Destination D combines three-path signals by using MRC.

Assume that all channels are flat Rayleigh fading channels. Use h_{SD}, h_{SI}, h_{12}, and h_{2D} represent channel gains of S to D, S to R_1, R_1 to R_2, and R_2 to D, respectively. Let p_s, p_1, and p_2 be transmit power of S, R_1, and R_2, respectively. Let x represent the signal of source S. y_1 and y_2 denote the received signal of R_1 and R_2. We take y as the received signal of D by using MRC. The three received signals are expressed as follows:

$$y = \alpha_o(\sqrt{p_s}\,xh_{SD} + n_{SD}) + \alpha_1(\beta_2 y_2 h_{2D} + n_{2D}) \qquad (1)$$

$$y_1 = \sqrt{p_s}\,xh_{s1} + n_{s1} \qquad (2)$$

$$y_2 = \beta_1 y_1 h_{12} + n_{12} \qquad (3)$$

where β_1 and β_2 represent amplification gain of

R_1 and R_2[8], respectively. $\beta_1 = \sqrt{\dfrac{p_1}{P_s\,|\,h_{s1}\,|^2 + \sigma^2}}$,

$\beta_2 = \sqrt{\dfrac{p_2}{P_1\,|\,h_{12}\,|^2 + \sigma^2}}$. n_{SD}, n_{S1}, n_{12}, and n_{2D} denote

complex Gaussian white noises, whose means are zero and variances are σ_2. α_0 and α_1 represent weight factors of MRC, which can be written as

$$\alpha_0 = \frac{\sqrt{p_s}}{\sigma^2}\,h_{SD}^* \qquad (4)$$

$$\alpha_1 = \frac{\sqrt{p_s}\,\beta_1\beta_2 h_{s1}^* h_{12}^* h_{2D}^*}{(\beta_1^2\beta_2^2\,|\,h_{12}\,|^2|\,h_{12}\,|^2 + \beta_2^2\,|\,h_{2D}\,|^2 + 1)\sigma^2} \qquad (5)$$

In such a AF system, when relay node amplifies the power of signal, the power of noise is also amplified. Power allocation is affected by total transmission power, channel gain, and the SNR of each link, which is analyzed as follows.

Assume channels h_{SD}, h_{S1}, h_{12}, and h_{2D} are complex Gaussian random variables whose means are zero and variances are σ^2_{SD}, σ^2_{S1}, σ^2_{12}, and σ^2_{2D}, respectively. Let $\overline{\gamma}_0$ be the average SNR of direct path, $\overline{\gamma}_0 = \dfrac{\sigma^2_{SD}p_S}{\sigma^2}$. $\overline{\gamma}_1$ represents the average SNR for the cooperative paths, which is expressed as

$$\overline{\gamma}_1 \approx \left(\frac{1}{\overline{\gamma}_{S1}} + \frac{1}{\overline{\gamma}_{12}} + \frac{1}{\overline{\gamma}_{2D}}\right)^{-1} \qquad (6)$$

where $\overline{\gamma}_{s1} = \dfrac{\sigma^2_{s1}p_S}{\sigma^2}$, $\overline{\gamma}_{12} = \dfrac{\sigma^2_{12}p_1}{\sigma^2}$ and $\overline{\gamma}_{2D} = \dfrac{\sigma^2_{2D}p_2}{\sigma^2}$.

According to MRC, output average SNR is the sum of the average SNR on various path signals of the receiver. That is, let $\overline{\gamma}$ be the average SNR of destination D using MRC, then

$$\overline{\gamma} = \overline{\gamma}_0 + \overline{\gamma}_1 \qquad (7)$$

3 AVERAGE BER ANALYSIS

To obtain power optimization solution when we take ABER as objective, the ABER theoretical express of the two-relay three-relay cooperative diversity system is extremely necessary. According to moment generating function (MGF) presented in literature [6], ABER can be expressed as the following equation when MRC and BPSK modulation are adopted.

$$P_e = \frac{1}{\pi}\int_0^{\pi/2}\prod_{m=0}^{1} M_{rm}\left(\frac{-1}{\sin^2\theta}\right)d\theta \qquad (8)$$

Where $M_{rm}(.)$ is MGF, which is defined as

$$M_{rm}(s) = \int_0^{\infty} p_{rm}(\gamma)e^{s\gamma}d\gamma \qquad (9)$$

where $p_{\gamma m}(\gamma)$ is the probability density function of γ_m. Because all the channels are Rayleigh channels, according to the literature [7], the relationship of MGF is

$$M_{rm}\left(\frac{-1}{\sin^2\theta}\right) = \left(\frac{-1}{\sin^2\theta}\overline{\gamma}\right)^{-1} \qquad (10)$$

220

Taking the formula (10) into equation (8),we can obtain approximate upper limit expression of ABER as shown in the formula (11).

$$p_e \approx \frac{3}{8}(\bar{\gamma}_0 + \bar{\gamma}_1)^{-1} =$$

$$\frac{3\sigma^2}{8}(\frac{1}{p_S^2 p_{SD}^2 p_{S1}^2} + \frac{1}{p_S p_1 \sigma_{SD}^2 \sigma_{12}^2} + \frac{1}{p_S p_2 \sigma_{SD}^2 \sigma_{2D}^2}) \tag{11}$$

4 OPTIMAL POWER ALLOCATION ALGORITHM

The total system power is assumed to be constant. The minimum ABER is taken as the optimization criterion to decide the optimal power among source S relay R_1 and R_2, which is described as

$$\min_{p_S, p_1, p_2} \quad P_e$$
$$s.t. p_s + p_1 + p_2 = p \tag{12}$$

This optimization problem can be transformed into mathematical expression by Lagrange multiplier method. By taking its partial derivatives to p_1, p_2, and p_s respectively along with the combination of the constraint, we can obtain the optimal solution which is

$$p_S = \{\frac{A-4B+\sqrt{A^2+8AB}}{4(A-B)} p (A \neq B时), \frac{2}{3} p \ (A=B时)\} \tag{13}$$

$$p_1 = \frac{\sigma_{2D}}{\sigma_{2D}+\sigma_{12}}(p-p_S), p_2 = p - p_1 - p_S$$

Where $A = (\sigma_{2D}+\sigma_{12})^2 \sigma_{12}\sigma_{S1}^2$, $B = \sigma_{2D}^2 \sigma_{12}^3$.

5 SIMULATION RESULTS AND ANALYSIS

Simulation conditions are as follows. Channel state information of each node is known. Assume that all channels are flat Rayleigh fading channels. The transmission power of each node is limited by the total power. Namely, $p_s + p_1 + p_2 = p$. We adopt the normalized power, i.e. p=1.

From Fig. 3, we can observe that the ABER of the proposed algorithm in the two-relay three-hop cooperative diversity system is better than that of the two-node two-hop system. When the SNR is higher than 6dB, there is little BER difference between the proposed algorithm and reference BER value. When the signal to noise ratio is lower than 6dB, the BER value of the proposed algorithm is better than the reference value. The result indicates that when the channel state is not in an ideal situation, three-hop two-node

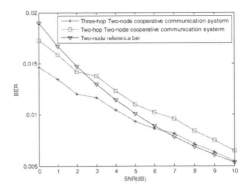

Figure 3. BER versus SNR.

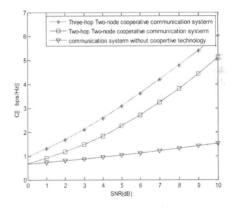

Figure 4. System capacity versus SNR.

cooperative system can improve the communication quality of a cell-edge user.

From Fig. 4, it can be observed that using cooperative communication may improve the throughput of the entire system. The two-node three-hop cooperative diversity system is superior to two-node two-hop system. The result also shows that the proposed power allocation algorithm in the three-hop two-node cooperative diversity system improves the system throughput.

6 CONCLUSION

For a two-relay three-hop AF cooperative diversity system, a power allocation algorithm is proposed to minimize ABER under total power constraint. Simulation results show that in low SNR environment, the algorithm makes the performances of BER and system throughput better than two-node two-hop system. The results also mean that the method may improve cell-edge user communication quality. Further research will be done to analyze the power allocation method when a system model of multiple cell-edge users is adopted.

ACKNOWLEDGMENT

This work has been supported by Natural Science Foundation of China (61101209), Natural Science Foundation of Shanghai (11ZR1426600), the Program of Shanghai Normal University (DZL126).

REFERENCES

[1] Min Lee;Seong Keun Oh. Joint Power Allocation for Coordinated Two-Point Diversity Transmission Under a Coordinated Power Constraint in Rayleigh Fading Channels.

[2] Kezhi Wang;Yunfei Chen,Alouini, M.-S.,Feng Xu. "BER and Optimal Power Allocation for Amplify-and-Forward Relaying Using Pilot-Aided Maximum Likelihood Estimation" Transactions on Communications IEEE, 2014: 3462–3475.

[3] Chin-Liang Wang;Ting-Nan Cho;Kai-Jie Yang,On Power Allocation and Relay Selection for a Two-Way Amplify-and-Forward Relaying System, Communications, Transactions on IEEE,2013: 3146–3155.

[4] Tera, Akhil Dutt ; Gurrala, Kiran Kumar; Das, Susmita,Power allocation for AF cooperative relaying using particle swarm optimization, Green Computing Communication and Electrical Engineering (ICGCCEE) IEEE, 2014:1–4.

[5] Liang Han;Shihai Shao;Ying Shen;Chaojin Qing;Youxi Tang,Outage Probability and Power Allocation for Amplify-and-Forward Cooperative Relaying Systems with Correlated Shadowing, Vehicular Technology Conference (VTC Fall) IEEE, 2013 : 1–5.

[6] Zhipeng Wang;Kunqi Guo;Lixin Sun ; Shilou Jia,A power allocation scheme in two-hop multi-node cooperative communication system, Measurement, Information and Control (ICMIC), 2013 International Conference on 2013 : 439–442.

[7] Ng, C.T.K.;Goldsmith, A.J.The impact of CSI and power allocation on relay channel capacity and cooperation strategies, Wireless Transactions on Communications IEEE,2008: 5380–5389.

[8] Xiaojun Yuan; Tao Yang; Collings, I.B., "Multiple-Input Multiple-Output Two-Way Relaying: A Space-Division Approach," Transactions on Information Theory IEEE, vol.59, no.10, pp.6421,6440, Oct. 2013.

Network Security and Communication Engineering – Chan (Ed.)
© *2015 Taylor & Francis Group, London, ISBN: 978-1-138-02821-0*

The comprehensive carrying capacity assessment of lands in Jing-Jin-Ji region based on projection pursuit approach

X.D. Liu, S. Chen, C. Liu & X. Gao
Donlinks School of Economics and Management, University of Science and Technology Beijing, Beijing, China

ABSTRACT: There exists a serious problem of subjectivity in the evaluation index weight empowerment for the comprehensive carrying capacity of lands in previous literature, so this paper constructs a comprehensive carrying capacity evaluation model of lands based on projection pursuit approach of particle swarm optimization to solve the problem, and employs the model to assess the lands in Jing-Jin-Ji region. Projection and pursuit approach can project high-dimensional data onto one-dimension, and then can find the optimal projection direction through particle swarm optimization, so that the low-dimensional data can effectively reflect the characteristics of high-dimensional data, avoiding the human factors of the index weight empowerment. The results show that this model can effectively solve the problem of subjectivity of indicators empowerment and can make a scientific assessment of comprehensive carrying capacity of lands in Jing-Jin-Ji region, and can also provide an important reference for comprehensive utilization and rational exploitation of land resources in this area.

1 INTRODUCTION

The outbreak of the global resources and environmental crisis in 1960s and 1970s caused a series of problems. Since then, people have begun to pay attention to land capacity. Nevertheless, the early concept of land carrying capacity only considered the capacity of food production. In the 21st century, the connotation and denotation are extended. For one thing, the scope of land expands, from the initial arable land for food production to a broader sense of the land. For another, the scope of bearing capacity increases, including not only the population, but also a variety of human activities and their social and economic scale and intensity as well as strategic thinking for sustainable development and harmonious society. At the same time, analyzing technologies and methods of land capacity also tends to diversify, but they contain deficiencies. For instance, the subjectivity of AHPindex weight is too large; the rules of genetic algorithm is complex and its astringency is general; the ecological footprint method does not consider technical factors.

To make up for the deficiencies of the previous studies, we combine with projection pursuit and particle swarm to build up an evaluation program for evaluating comprehensive carrying capacity of the land. The improvements of the program include the following three aspects. First, it takes four aspects into account to build up a comprehensive carrying capacity's index system that adapts to social

development. Second, it selects projection pursuit model to evaluate, effectively avoiding the problem that the determination of the evaluation index weight is too subjective. Third, according to the principles of genetic algorithms, it adds adequate variation into PSO, which reduces the possibility of falling into local optimum and increases the accuracy and efficiency of the algorithm. On the basis of constructing the evaluation program, the paper focuses on the evaluation of comprehensive carrying capacity of the land of Jing-Jin-Ji area, compares the differences between Beijing, Tianjin, Hebei, Jing-Jin-Ji and nationwide respectively, and at last, proposes some suggestions, hoping to provide valuable references for rational plan and use of Jing-Jin-Ji land.

2 INTRODUCTION OF ALGORITHM PRINCIPLE

2.1 *Projection pursuit model*

The method is used to analyze and process high-dimensional observation data. The basic idea of it is that high-dimensional data is projected onto low-dimensional (1 to 3-dimensional) subspace to search for the structure or feature projection to reflect the original high-dimensional data.

Supposing the multi-index sample set of projection pursuit problem is $\{x(i,j)|i=1,\ldots,m,j=1,\ldots,n\}$, of which m is the number of samples, and n is the number of indicators. The steps to establish PPC model are as follows:

2.1.1 Data preprocessing

To normalize the sample evaluation index set for indicators that the bigger they are, the better they are:

$$x^* (i,j) = \left(x_{max}(j) - x(i,j) \right) / \left(x_{max}(j) - x_{min}(j) \right) \quad (1)$$

And for indicators that are the smaller the better:

$$x^* (i,j) = \left(x(i,j) - x_{min}(j) \right) / \left(x_{max}(j) - x_{min}(j) \right) \quad (2)$$

Among them, $x_{max}(j)$ and $x_{min}(j)$ are maximum values and minimum values of indicators respectively.

2.1.2 Constructing projection index function

Supposing $A(j)$ as projection direction vector, the projection value of the sample i in this direction is:

$$Z(i) = \sum_{j=1} A(j) X(i,j) \quad (3)$$

That is, to construct a projection index function $Q(A)$ as the basis for determining the projection direction optimization is that when the index reaches a maximum value, the optimal projection direction is considered to be found.

The projection index function is set as:

$$Q(A) = S_Z D_Z \quad (4)$$

In (4), S_Z represents the scatter degree inter-variety, indicated as the standard deviation of $Z(i)$, and D_Z represents the density inner-variety, indicated as the local density of $Z(i)$. That means:

$$S_Z = \left\{ \sum_{i=1}^{m} \left[Z(i) - \overline{Z} \right]^2 / (m-1) \right\}^{\frac{1}{2}} \quad (5)$$

$$D_Z = \sum_{i=1}^{m} \sum_{j=1}^{n} \left(R - r_{ij} \right) * I \left(R - r_{ij} \right) \quad (6)$$

Z is the average of the sequence $\{Z(i)|i=1\sim m\}$. R is the partial width parameter. When the point spacing value r_{ij} is no more than R, we calculate according to the inner-variety formula; otherwise, we use this formula: $r=|Z(i)-Z(j)|$. Signum function $I(R-r_{ij})$ is unit step function. When $R-r_{ij} =0$, its value is 1; when $R-r_{ij}<=0$, its value is zero.

2.2 Estimating optimal projection direction

The question of estimating optimal projection direction can be transferred as the optimization question:

$$\begin{cases} maxQ(A) = SzDz \\ \|a\| = \sum_{j=1}^{m} a_j^2 = 1 \end{cases} \quad (7)$$

The absolute value of each component of the optimal projection direction reflects the degree of the influences of carrying capacity of lands to the general goal and sub-goals, namely, the weights.

2.3 PSO algorithm

PSO is used to solve optimization problem. PSO has both the whole and the local searching abilities, with simple and practicable parameter adjustment, fast convergence rate and wide versatility.

Supposing that within a D-dimension searching space, the group $X=(X_1,X_2,...,X_n)$ consists of n particles, and the ith particle is represented as $X=(x_{i1},x_{i2},...,x_{iD})$, referring to its position in the D-dimension searching space. The fitness value of each particle's position X_i can be calculated according to the object function. The ith particle's speed is $V=(V_{i1},V_{i2},...,V_{iD})$, and its $Pbest$ is $P=(P_{i1},P_{i2},...,P_{iD})^T$ and the $Gbest$ is $P_g=(P_{g1},P_{g2},...,P_{gD})^T$.

Particles change their speeds and positions by changing the $Pbest$ and the $Gbest$. The formula is as follows (8) (9):

$$V_{id}^{k+1} = wV_{id}^k + c_1 r_1 \left(P_{id}^k - X_{id}^k \right) + c_2 r_2 \left(P_{gd}^k - X_{gd}^k \right) \quad (8)$$

$$X_{id}^{k+1} = X_{id}^k + V_{id}^{k+1} \quad (9)$$

of which, w is inertia weight; $d=1,2,...,D$; $i=1,2,...n$; k represents the current iterations; V_{id} refers to particle's speed; c_1 and c_2 are non-negative constants, which are called Speed Factors and r_1 and r_2 are random numbers within [0,1].

The iteration formulas of $Pbest$ and of $Gbest$ are as follows:

$$Pbest_{id}^{k+1} = \begin{cases} s_{id}^{k+1} & Q_{id}^{k+1} \geq Q \left(Pbest_{id}^k \right) \\ Pbest_{id}^k & Q_{id}^{k+1} \leq Q \left(Pbest_{id}^k \right) \end{cases} \quad (10)$$

$$Gbest^{k+1} = s_{max}^{k+1} \quad (11)$$

The steps of PSO are as follows. First, initialize the particles' positions and speeds. Second, calculate the particles' fitness values and decide $Pbest$ and $Gbest$. Third, change particles' positions and speeds, and accordingly, $Pbest$ and $Gbest$ are changed. Iterate for the given times. For this study, the fitness function is $Q(A)=S_Z D_Z$, and the fitness value is function value.

PSO has satisfactory convergence and global searching ability, but it also has disadvantages of low convergence precision caused by local extreme and of having difficulty in converging to a global optimum. By referring to the idea of mutation in genetic algorithm, we add some mutating operation to PSO algorithm. That is, we re-initiate some variables with certain probability. Through the mutating operation, the group searching space, which diminishes continually during the procedure of iterations, is extended. That makes it possible for particles to leave the best positions where they have been searched and to reach to a wider space. At the same time, the diversity of group is maintained, which increases the possibility for the algorithm to find out a better value.

2.4 Projection pursuit model of PSO

Combine PSO with the projection pursuit model to assess the comprehensive carrying capacity of lands in Jing-Jin-Ji region. The process is as follows:

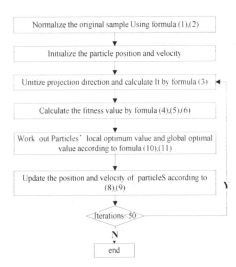

Figure 1. Algorithm flowchart.

When the process ends, the particle's position is the optimal projection direction, and accordingly, the projected value is got.

3 EMPIRICAL ANALYSIS

Jing-Jin-Ji region is in the heartland of the Bohai Sea region of Northeast Asia, and also is the largest and liveliest economic area in North China. Jing-Jin-Ji region here includes Beijing, Tianjin, and Hebei Province. With the development of regional economy, the intensity of the land resource utilization in Jing-Jin-Ji grows gradually, which leads to the decrease of the carrying capacity of the lands. This section will assess the comprehensive carrying capacity of the lands in Jing-Jin-Ji region with the help of Projection Pursuit Model of PSO.

3.1 Construction of index system

The paper constructs an assessment index system which contains four aspects: resources, society, economy, and ecological environment. And the number of group particles in the assessment model is 5 (table I). For each of them, there are 4 first sub-goals, which are goals of the land resources utilization, goal of social development, goal of economic technology, goal of economic technology, and goal of ecological environment. These five sub-goals include 5, 7, 12, and 6 second sub-goals respectively, 30 in total. It is noted that the second sub-goals contribute to the assessment of comprehensive land carrying capacity negatively as well as positively. Thus, we divide them into positive indexes and negative indexes. In addition, due to the fact that part of the data cannot be directly obtained, we get them indirectly through regressively analyzing the existing data.

3.2 Assessment of analyzing results

We realize the model's specific algorithm by using MATLAB. According to the conclusions of the relevant literature, we ultimately determine the model parameter values as follows: learning factor: $c_1 = 2$, $c_2 = 2$; iterations = 50; population size = 400; inertia weight = 0.9965; adaptive mutation rate = 0.9. The final assessment results are shown in Table 1.

The final evaluation results show that in the aspect of the use of land resources, the level in Jing-Jin-Ji is about twice higher than the national average level. Therefore, in the scope of the whole country, it is in the dominant position in terms of the use of land resources.

As for the social development, the level in Jing-Jin-Ji is much lower than the average level of China. It is mainly due to the high population density of the region. In the internal view from Jing-Jin-Ji, because of its high population density and low per capita arable land, Beijing is after Tianjin and Hebei, but its urbanization rate is still the highest among them. Therefore, in reality, Beijing can bring out its strengths to make up for its weaknesses. Low urbanization rate of Hebei is an important factor in restricting its land carrying capacity. As a major agricultural province, Hebei can improve the development of industry while developing agriculture stably to keep its economic growth.

As for the economic technology, the level in Jing-Jin-Ji is higher than the national average level. In the internal view from Jing-Jin-Ji, being the economic, political, and cultural center, Beijing is on the top in economy and technology. Tianjin, being one of China's four municipalities, northern economic center and international port city, has a thriving economy and advanced technologies as well. As for Hebei, its economic technology development is far behind Beijing and Tianjin, which severely restricts the utilization of its advanced land resource. It can learn the advanced science technology from Beijing and Tianjin and improve agricultural industry at the same time.

Tianjin, as the largest industrial city of North China, should pay attention to the ecological environment, increase environment protection and greening work, or fight against the environment-polluted enterprises from the source.

Table 1. Comprehensive carrying capacity assessment results of each area.

Regions	Sub-goal 1 Assessment of the land resources utilization	rank	Sub-goal 2 Assessment of social development	rank	Sub-goal 3 Assessment of economic technology	rank	Sub-goal 4 Assessment of ecological environment	rank	General goal Assessment of the comprehensive land carrying capacity	rank
Beijing	1.35683	2	0.21511	5	2.37229	2	1.12952	2	2.51012	2
Hebei	1.31375	3	0.989	2	1.14749	4	0.97537	3	1.56261	4
Tianjin	2.17326	1	0.33562	4	2.38060	1	0.36221	5	2.77650	1
Jing-Jin-Ji	1.29346	4	0.75008	3	1.40811	3	0.92986	4	1.79727	3
China	0.74398	5	1.42500	1	0.92258	5	1.40202	1	1.04666	5

From the perspective of the general goal, the level of Jing-Jin-Ji is higher than national average level. Therefore, society development and environment protection in the region of Jing-Jin-Ji should be given high priority, especially in terms of population and forestation.

4 CONCLUSION

The paper builds up a projection pursuit evaluation model based on PSO to evaluate the comprehensive carrying capacity of land in Jing-Jin-Ji area. The greatest advantage of the evaluation model lies in its ability to effectively solve the problem of subjectivity of indicators empowerment and make a scientific assessment, which provides a rational reference for comprehensive utilization and rational exploitation of land resources. At the same time, in view of the reality that the data for each indicator does not meet the Gaussian distribution and indicators are high dimension, the model can effectively find the structure and characteristics of high-dimensional indicators. In addition, the introduction of some mutating operations into PSO method reduces the likelihood of falling into a local optimum and improves the efficiency of finding the global optimum.

ACKNOWLEDGMENTS

This work was supported by the Fundamental Research Funds for the Central Universities (FRF-TP-14-052A1).

REFERENCES

[1] Chen,G.Z.&Wang,J.Q, Xie H M. 2008.Application of particle swarm optimization for solving optimization problem of projection pursuitmodeling.*Computer Simulation*8: 159–161.
[2] Fu,S.F.Zhang P.& Jiang J L. 2012.Evaluation of land rsources carrying capacity of development zone based on planning environmentimpact assessment.*Chinese Journal of Applied Ecology*2: 459–467.
[3] Gössling,S.Hansson,C.B,&Hörstmeier O, et al.2002. Ecological footprint analysis as a tool to assess tourism sustainability.*Ecological Economics*2–3(43):199–211.
[4] Hoekstra,A.Y.2009. Human appropriation of natural capital: A comparison of ecological footprint and water footprint analysis.*EcologicalEconomics*7(68): 1963–1974.
[5] Huang,H. 2011.Multi-objective genetics algorithm for land use structure optimization.*Journal of Mountain Science*6:695–700.
[6] Huijbregts,M.A. J.Hellweg,S.&Frischknecht R, et al. 2008.Ecological footprint accounting in the life cycle assessment of products.*Ecological Economics*4(64):798–807.
[7] Ma,S.F.He,J.H.&Yu,Y. 2010.Model of urban land-use spatial optimization based on particle swarm optimization algorithm.*Transactions of the Chinese Society of Agricultural Engineering*9: 321–326.
[8] Monfreda,C. 2004.Wackernagel M, Deumling D. Establishing national natural capital accounts based on detailed ecological footprint and biological capacity assessments.*Land Use Policy*3(21):231–246.
[9] Siche,R.Pereira,L.&Agostinho,F.et al. 2010.Convergence of ecological footprint and emergy analysis as a sustainability indicator of countries: Peru as case study. *Communications in Nonlinear Science and Numerical Simulation*10(15):3182–3192.

Network Security and Communication Engineering – Chan (Ed.)
© 2015 Taylor & Francis Group, London, ISBN: 978-1-138-02821-0

A study about LS-SVM's application in RFID-based indoor positioning

J. Xue, L.T. Li, J.W. Xu & L.T. Jiao
School of Automation, Northwestern Polytechnical University, Xi'an, China

ABSTRACT: This paper studies the indoor positioning technology based on RFID technology and selects Time Difference of Arrival (TDOA) as the basic location algorithm. The paper introduces the theory of Least Squares Support Vector Machine (LS-SVM) to optimize the TDOA algorithm since TDOA algorithm does not have the ability of learning and cannot change as environment changes. Finally, with the help of the nano Track development kit, the paper has completed indoor positioning system design and experiment. Experiments show that the optimized algorithm can achieve higher precision.

KEYWORDS: RFID, the indoor positioning technology, time difference of arrival (TDOA), least squares support vector machine (LS-SVM).

1 INTRODUCTION

At present, RFID technology is widely used in industrial production and life. Because RFID is a non-contact automatic identification technology, it can automatically identify target objects and access to relevant data through radio frequency signal, and can identify work without manual intervention through wireless real-time remote reading, and it has a faster recognition and longer service life. It can not only recognize the dynamic object, but also identify multiple tags at the same time. The RFID technology has been widely used.

Due to the lack of a complete standard system in frequency RFID technology and open frequency range, it makes the RFID indoor positioning technology only stay in the experimental stage, and the use of RFID technology at the present stage is mainly in terms of automatic identification, so it is necessary to deepen the study of RFID indoor positioning technology. Indoor positioning technology is very practical which has a large space to develop, and it has a quite wide range of applications, especially in complex environments, such as libraries, gymnasiums, underground garages, warehouses, and so on. At present the commonly used indoor positioning technologies mainly include the following: the positioning technology based on the technology of ultrasonic [1], the technology based on the infrared technology [2], UWB location technology [3], and RFID positioning technology (WLAN, ZigBee) [4] et al. Because RFID technology applied to indoor location has the characteristic of high feasibility and positioning accuracy, so this paper researches and designs an indoor positioning system based on RFID technology to meet the needs of the indoor personnel accurate positioning and statistics.

2 THE INDOOR POSITIONING ALGORITHMS

At present the popular RFID localization algorithms mainly contain RSSI algorithm [5], the time of arrival (TOA) location algorithm [6], angle of arrival (AOA) algorithm [7], TDOA algorithm [8], and so on. RSSI algorithm can be easily affected by the weather, obstacles or mobility, resulting in imprecise distance, thereby positioning accuracy is not high. Requirements station in time of arrival TOA positioning algorithm precisely synchronizes with time and technology, otherwise the positioning accuracy will greatly reduce. When there is an error in microseconds between base stations, it will produce 300 meters of the theoretical error on the distance. Angle of arrival AOA algorithms is to use multiple antenna arrays to measure the signal from the mobile station of arrival angle, and a measured value of DOA is to make the target location of the mobile station to measure DOA by drawing a straight line. If there are at least two DOA values from two different locations on the two antennas, then the target position of the mobile station must be from the two antennas on the intersection of two straight lines. Under normal circumstances, only the use of multiple measurements of DOA can provide redundant information to achieve the goal of improving positioning accuracy. Time difference of arrival TDOA algorithm is determined by measuring the target mobile station to send the signal arrival time difference to the plurality of base stations and receive the target mobile station positioning, and it has the characteristics of a simple hardware requirement and accurate positioning. In order to improve the accuracy of positioning, this paper selects TDOA as the basis of location algorithm, then makes a optimization in view of the lack of TDOA.

3 THE OPTIMIZATION OF TDOA ALGORITHM

Standard TDOA algorithm has no memory and learning ability, and it will not change with changes of the environment, which is the biggest deficiency of the TDOA algorithm. And in practical applications it performs particularly. For example, in indoor positioning, the interior could store a large number of furniture, furnishings and decorative items. But the TDOA algorithm could only output according to the actual value, which sometimes makes the error high. The error still exists when the label is installed on the human body, but there are laws to follow when people move. For example, people do not coincide exactly with walls and other fixed objects. In order to improve the positioning accuracy, it is necessary to introduce new algorithms. The paper refers to the SVM to optimize the output of the TDOA algorithm.

3.1 The principle of the SVW

SVM method is a new machine learning method based on the statistical learning theory, which is given by Vapnik et al in the early 1990s. SVM evolved from the optimal classification face in linearly separable case, and it can be illustrated from the Figures 3-2. For points in one dimensional space, two-dimensional space in a straight line, plane in 3d space, as well as the hyperplane in high dimensional space, solid and hollow points in the figure represent two classes of samples, H represents the classification hyperplane between them; H1 and H2 respectively represents the nearest sample from the classification surface and the hyperplane parallel to the classification surface, and the distance between them is called classification interval.

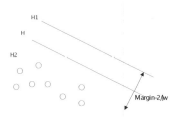

Figure 1. The optimal classification surface.

3.2 The principle of the LS-SVM

Least squares that support vector machine (LS-SVM) is a new development of support vector machine (SVM) in recent years, and the LS - SVM training process also follows the rule of structural risk minimization criterion; unlike support vector machine (SVM) for the choice of slack variables, LS-SVM changes the inequality constraints into equality constraints and the empirical risk is changed from a power into quadratic deviation, and finally the problem of solving quadratic programming is transformed into solving linear equations, which is avoiding the insensitive loss function, reducing the computational complexity and improving the operation speed.

Set D={(xk,yk)|k=1,2,...,N}, where the $x_k \in R_n$ is the input, $y_k \in R_n$ is the output analog. In power space w the LS-SVM classification problems can be described as follows:

$$\min \phi(a,b,e) = \frac{1}{2}w^T + \frac{1}{2}\gamma \sum_k \qquad (1)$$

The constraint is the follow:

$$y_k[w \; \varphi(x_k)+] = 1 - {}_k$$
$$k = 1, \cdots, N \qquad (2)$$

The Lagrange function is set:

$$(\,,b,e,\alpha) = \phi(w,b,e) - \alpha_k \sum \alpha_k y_k[w \; \varphi(x_k)+b]-1+e_k \qquad (3)$$

Where the Lagrange multipliers $a_k \in R$. Optimized for the above formula, the partial derivative of w, b, e_k, a_k is equal to zero.

$$\begin{cases} \frac{\delta L}{\delta w} = 0 \rightarrow w = \sum_{k=1}^{N} \alpha_k y_k \varphi(x_k) \\ \frac{\delta L}{\delta b} = 0 \rightarrow \sum_{k=1}^{N} \alpha_k y_k = 0 \\ \frac{\delta L}{\delta e_k} = 0 \rightarrow \alpha_k = \gamma e_k \\ \frac{\delta L}{\delta e_k} = 0 \rightarrow y_k[w^T \varphi(x_k)+b]-1+e_k = 0 \end{cases} \qquad (4)$$

The above formula can be turned into solving the matrix equation below:

$$\begin{vmatrix} I & 0 & 0 & -Z^T \\ 0 & 0 & 0 & -Y^T \\ 0 & 0 & \gamma I & -I \\ Z & Y & I & 0 \end{vmatrix} \begin{bmatrix} w \\ b \\ e \\ \alpha \end{bmatrix} = \begin{bmatrix} 0 \\ 0 \\ 0 \\ I \end{bmatrix} \qquad (5)$$

That is:

$$
\begin{bmatrix} 0 & -Y^T \\ Y & ZZ^T + \gamma^{-1}I \end{bmatrix} \begin{bmatrix} b \\ \alpha \end{bmatrix} = \begin{bmatrix} 0 \\ I \end{bmatrix}
\tag{6}
$$

here $\quad Z = \left[\varphi(x_1)^T y_1, \varphi(x_2)^T y_2, \cdots \varphi(x_N)^T y_N \right]^T$,

$Y = \left[y_1, y_2, \cdots, y_N \right], I = [1,1,\cdots,1]^T$,

$e = \left[e_1, e_2, \cdots, e_N \right]^T, \alpha = \left[\alpha_1, \alpha_2, \cdots, \alpha_N \right].$

At the same time, substitute the Mercer condition into the $\Omega = ZZ^T$, and then it can be obtained:

$$
\Omega_{kl} = y_k y_l \phi(x_k)^T \phi(x_l) = y_k y_l \Psi(x_k, x_l)
\tag{7}
$$

Therefore, the decomposition of the formula (1) can be obtained by solving formula (6) and (7). LS-SVM classification decision function is:

$$
y(x) = \text{sgn} \left[\sum_{k=1}^{N} \alpha_k y_k \Psi(x, x_k) + b \right]
\tag{8}
$$

Among them, $\psi(,;,)$ is the kernel function, and the purpose is to extract features from the original space, and map the samples in the original space as a vector in high-dimensional feature space in order to solve the original linear inseparable problem.

3.3 *Optimization steps for TDOA*

The main purpose for the Location prediction based on least squares support vector machine is trying to find a function to determine the relationship between the future value and the values in the past, that is to say, the prediction makes use of past and present observations to estimate the future value, which is actually based on a hypothesis. Namely there is a certain functional relation between the future value and the past value. The prediction steps are as follows:

1 Selection of the sample. The premise condition for the forecast of the least squares support vector machine (SVM) is that it needs enough and high accuracy sample. In this paper the training sample is the imprecision area that TDOA algorithm outputs, and the inaccuracy location in this area may be caused by the environment or the system clock. In the imprecise region, we choose 20 points as the training samples.

2 The optimization of regional judgment. Optimization methods are mainly system location determination method and zoning laws. First, the

system determination method is determined by the TDOA algorithm. The TDOA algorithm shows that three stations can determine a coordinate system. When the number of base stations is 4, four positioning coordinates can be determined. And the four coordinates are got ultimately through the centroid algorithm, so when the four coordinate are different, the output result is not accurate, thus it can be used as optimization decision basis. Second, system imprecise area can be obtained through on-the-spot test; the coordinates of inaccurate area can be got through sampling from the inaccurate area; when the test coordinates are in inaccurate area, use LSSVM algorithm to modify the coordinates. This paper uses the second method. In the imprecise region it selects 20 points as the training samples for testing.

3 Select the appropriate kernel function and the corresponding parameter for least squares support vector machines by using training samples to train the model and to forecast the future position coordinates. This paper selects the RBF with Gaussian as kernel function. The penalty factor C is set to 500, the formal parameter r is set to 10, and square broadband parameter $\sigma 2$ is set to 0.4.

4 The steps for TDOA algorithm optimization using LS-SVM is as follows:

Step 1: Initialize, determine the location and sequence of the anchor node and ensure clock synchronization of the anchor nodes;

Step 2: Anchor node, in turn, launches the ranging signal;

Step 3: The test node receives the pulse signal of anchor nodes and gets the arrival time;

Step 4: To solve the signal propagation time;

Step 5: To transmit the distance measured to the server;

Step 6: Use TDOA algorithm to calculate the location coordinates;

Step 7: Use the LS-SVM algorithm to optimize the result of the step 6;

Step 8: The output of the optimization results.

4 THE EXPERIMENTAL SIMULATION

4.1 *The experiment platform*

Hardware is nanoTrack development kit, and the development kit is mainly designed to meet the RFID range and communication and satisfy the requirement of experiment. The development kit includes five of the same development board and a signal receiver, by an external power supply in the form of power or a notebook USB. By using a laptop to display and process data, all test equipments are shown in Figure 4-1.

Figure 2. The experiment equipments.

Development board is mainly the following modules: micro controller module, data transceiver module, serial communication module, antenna, and power module. Among them, the micro controller is to complete data processing, communication and storage function; The data transceiver module is to complete the generation of linear frequency modulation signal, sending and receiving; A serial port communication module is used for a master station and processing server communication interactive; Antenna is used for sending and receiving linear frequency modulation signal in space; The power supply is the hardware circuit of power supply for the whole positioning system. The following is the data transmission figure.

Figure 3. Data transmission figure.

First, the computer sends the request location instructions to the reception/transmitter through the USB cable. Then receiving/transmitter sends the instructions to tier 1 station, and next tier 1 station sends location instructions to the secondary stations through RFID. At the same time the distance measurement is carried out on the label by tier 1 station and feedback with RFID to receiving/transmitter. The secondary stations begin to do the distance measurement to label and return the ranging command to tier 1 station when they receive the locate command. At last, tier 1 station returns the ranging data to station signal receiving/transmitter and the signal receiving/transmitter sends the data through the USB cable.

4.2 LS - SVM for TDOA optimization experiment

During the experiment, we have selected the four stations and number of each base station. First we measure the actual distance to each base station, and then

set the coordinates of each base station. The coordinates of the label are obtained by TODA algorithm. The accuracy of base stations will directly affect the label location coordinates. 4-3 is monitoring interface figure. For the convenience of analysis, first of all, the output of measuring distance by TDOA algorithm converting coordinates are shown in the figure. As is shown in Figure 4-4, red "+" stand for actual coordinates and blue "*" stands for the result of measurement.

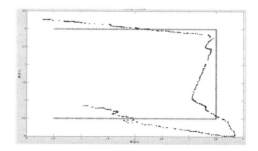

Figure 4. Monitoring interface figure.

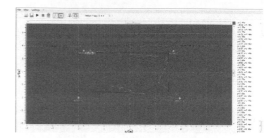

Figure 5. The real values and the measured values.

The Figure 6 is the result that is got after using LSSVM for TDOA algorithm to optimize. Red "+" stands for actual coordinates, blue "*" stands for the result of measurement and green "△" stands for the LSSVM algorithm of TDOA algorithm results after optimization. Obviously, the result after optimized by LS-SVM algorithm is closer to the actual coordinates, which can achieve higher precision.

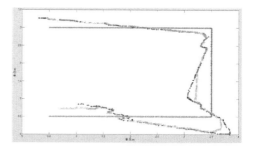

Figure 6. The result after optimized by LS-SVM.

5 CONCLUSION

This paper has studied the RFID positioning technology and introduced the mechanism of least squares support vector theory to optimize TDOA algorithm. As a result, more accurate results are obtained. Indoor personnel precise positioning system based on RFID technology has combined with RFID technology and personnel management, realized the monitoring in the whole process of the indoor personnel management, enriched the management, and improved work efficiency and management quality.

REFERENCES

[1] Li Guohua. Research and Development of 3-D Positioning System Based on the Ultrasound Technology[J]. Computer Measurement & Control. 2005. 13.

[2] Lu Qi. An IndoorO rientation System based on Singlechips [J]. Microprocessors. Apr,2006.

[3] Win M Z, Dardari D, Molisch A F, et al. History and applications of UWB[J] .Proceedings of the IEEE, 2009, 97 (2) : 198–204.

[4] AMORIM S R, ALVESPAD. Enhancing the efficiency of active RFID-based indoor location systems[C]//Wireless Communications and Networking Conference.[S.l.]: IEEE, 2009: 1–6.

[5] BOUETM,SANTOSALD.RFID tags: positioning principlesandlocalizationtechniques[EB/OL][2012-02-12]. http: // ieeexplore. Ieee. org /xpls /abs_all. jsp? arnumber = 4812905.

[6] M.K.Steven. Fundamentals of Statictical Signal Processing [M].Estimation Theory,2003.

[7] ZHU Juan, ZHOU Shang-wei, MA Qi-ping. An indoor localization algorithm based RFID.Microcomputer Information [J]. 2009 (23) :160–162.

[8] Han Xia-lin. Indoor location Algorithm Based on RIFD Technology and Its Improvement[J]. Computer Engineering. 2008, 34 (22):266–267.

Network Security and Communication Engineering – Chan (Ed.)
© *2015 Taylor & Francis Group, London, ISBN: 978-1-138-02821-0*

Paragraph reconstruction for postscript documents with complex layout

M.N. Zhang, M.Y. Li, W. Wang & Z. Yu
Zhejiang Provincial Key Laboratory of Service Robot, College of Computer Science, Zhejiang University, Hangzhou, China

ABSTRACT: Electronic publishing has become an irreversible trend in publishing industry. An important step in electronic publishing is to convert a scanned copy or a digital image of a document into editable texts, which is very important to the people with visual impairment, who access the information by means of screen-readers. Existing methods for document parsing and paragraph reconstruction only work on PostScript documents with a simple and standard layout. These techniques are severely limited when applied to documents with complex layout and irregular typesetting. To address this issue, we propose a rectangle segmenting approach to make paragraph reconstruction for Postscript documents with complex layout.

1   INTRODUCTION

Recently, electronic publishing has become an important branch in publishing industry. Many software applications could convert the traditional paper-based magazines into digital magazines publications. A widely used software OCR could convert the scanning or photographing images into editable texts. However, the paragraph structure lost in OCR reduces the interaction capability for the reconstructed document. This is especially difficult for people with visual impairment who read electronic documents with the help of a screen reader. Without paragraph structure, the screen reader can only output all the words in a document sequentially, which reduce their efficiency in accessing specific information in a document and limit possible interaction with the information in a document.

To improve interaction, paragraph reconstruction is an essential step in converting the paper documents to electronic documents. It means users can selectively read paragraphs according to their needs, or generate document summary based on the paragraph structure.

The typeset and layout of a document directly affect the degree of difficulty and accuracy of paragraph structure reconstruction (Xu Y. et al. 2007). Existing works for paragraph reconstruction from a scanned document or a digital image will need to extract words and their two-dimensional coordinates and then determine the layout by comparing the coordinates of the neighboring words with a preset threshold (Wu Y. M. et al. 2010). These simple methods, however, do not work for documents with complex layout. In this paper, we propose a new paragraph reconstruction approach rectangle segmenting for documents with complex layout and typesetting.

By retaining the character sequence and coordinates for the internal text blocks, we map all the characters in a document to a one-dimensional linear space to obtain the initial paragraph reconstruction. Then we partition the indented text blocks resulted from inserted pictures or other objects. After that we merge the adjacent text blocks from the same paragraph if they meet some certain threshold, thereby improving the ability to resolve complex typesetting.

While the passage structure of the document is lost in the conversion process, which leads to the interaction between the user and the document more single, making users unable to read a paragraph based on their interests.

The document reconstructed from the scanned copy using our method will effectively retain paragraph structure and hence will enable better information interaction for the document.

2   PROPOSED APPROACH

In this section, we'll discuss our rectangle segmenting approach in detail. We first extract useful information from the document such as block coordinates, character coordinates, character width and height, the number of pages, coordinates of pictures, etc. Then we map all the characters in a document to a one-dimensional linear space to obtain the initial paragraph reconstruction. After that we segment the indented text block resulted from inserted pictures or other objects into several rectangle text blocks. Below shows the general steps of our proposed algorithm:

- Map all the characters in a document to a one-dimensional linear space to obtain the initial paragraph reconstruction;
- Get the segmenting point;
- Segment the irregular text block resulted into several rectangles;
- Merge paragraphs if they meet some certain threshold.

2.1 Detailed descriptions of our approach

After getting the characters and the corresponding two-dimensional coordinates etc., we use them for linear mapping to make sure that the distance of adjacent characters within a line is much smaller than the distance of characters between different lines in Figure 1.

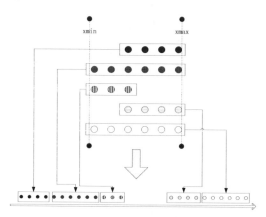

Figure 1. Linearization mapping example.

Therefore we can obtain the initial paragraph reconstruction by comparing the distance between new coordinates.

Algorithm 1: Linear mapping.

INPUT: matrix of characters' position;
OUTPUT: the one-dimensional coordinate new;

1. Calculate the left edge $xmin$, the right edge $xmax$ of the text block;
2. Initialize the line number $linenum = 0$;
3. For $i=2$ to n
4. If the ordinate difference between the i-th and $(i-1)$-th characters is bigger than height, $linenum + +$;
5. Calculate the distance between i-th character and the left edge as x_i;
6. $new_i = linenum \times (xmax - xmin) + x_i$
7. EndFor;
8. Return new;

If the shape of the text block is rectangular, we can use Algorithm 1 to make paragraph structure construction. While most shapes of the complex documents are not rectangular due to inserted pictures or other objects. If we only use Algorithm 1, we would get some errors in Figure 2. The asterisk '*' is the paragraph header and we can find all the real headers are marked correctly, but mark some wrong headers at the same time, which are surrounded by the green dashes.

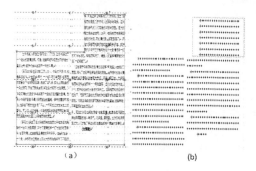

(a)   (b)

Figure 2. Paragraph reconstruction result by Algorithm 1 on complex documents.

To address these issues, this paper proposes a method to segment the complex irregular shapes into rectangles and make sure the number of the rectangles is smallest.

Given a character i ($1<i<n$) in one text block of n characters, we use Algorithm 1 to calculate the number of the headers up_i before characters j, ($1<j\leq i$) and the number of the headers $down_i$ of the rest characters j, ($i<j\leq n$). Then the total number of headers after segmenting is

$$sum_{seg} = up_i + down_i - b \qquad (1)$$

We can also get the number of the headers before segmenting sum_i by ALG1. Then we get the contribution of character i in reducing the headers:

$$red_i = sum_i - (up_i + down_i - b) \qquad (2)$$

We set $b = 1$ when the last paragraph of the up text and the first paragraph of the $down$ text belong to a paragraph, otherwise we set $b = 0$.

In order to achieve the goal of segmenting the irregular shapes into rectangles and the smallest number, we consider every character as segmenting point and calculate the corresponding reduction of headers and choose the largest reduction as a segmenting point. We continue to segment the text block until the new segmenting point will not cause reduction in the total number of headers. After segmenting the text block by reduction the number of headers, we use coordinates of inserted pictures to get other segmenting points. The algorithm is Algorithm 2:

Algorithm 2: Segmenting.

INPUT: new coordinates of the text matrix;
OUTPUT: segmenting points set *tag* and the number *count*;

1. Initialize $tag_n = \{0,\ldots0\}$, $tag_1 = 0$, $tag_2 = n$;
2. While the total number of the header is no longer reduced;
3. FOR $i=1$ to $n-1$
4.   Use ALG1 to calculate sum_i, up_i, $down_i$;
5.   If meet the condition of paragraph merge
6. $redu_i = sum_i - (up_i + down_i - 1)$
7.   Else
8. $redu_i = sum_i - (up_i + down_i)$
9.   EndIf
10.   Add i into set *tag*, $count + +$
11. EndWhile
12. Get other segment points by images and add them to tag and increase the corresponding number.

After getting the segmenting points, we can segment the irregular shape into several rectangle text blocks as in Figure 3. Then we can make paragraph reconstruction in each rectangle text blocks by Algorithm 1. We also need to merge the last paragraph of up text and the first paragraph of down text if they belong to the same paragraph.

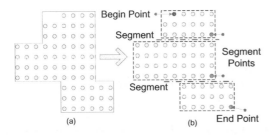

Figure 3.   Irregular shape text box segmentation result example.

3 EXPERIMENT

Dataset: We collected 28 documents of 2058 paragraphs. To show the effectiveness of our approach, firstly we marked all the paragraph headers manually and make them the real headers. Then we compared the headers obtained by our algorithm and the real headers to validate the effectiveness of our algorithm.

We extracted the text block information including character coordinates, images information and so on by the literatures (Zhang Z. W. et al. 2008, Zhu M. et al. 2010). We only demonstrated the results of 2 documents by our approach due to limited space as Figure 4 and Figure 5. The points (red and black) denote the characters of the documents, the red dots are the headers marked by our algorithm. The blue squares and green squares denote the upper left and the lower right coordinates of the inserted images.

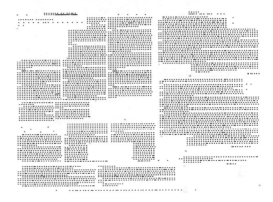

Figure 4.   Paragraph reconstruction result example 1.

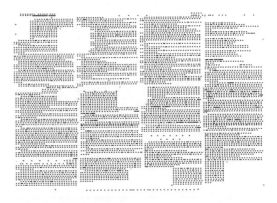

Figure 5.   Paragraph reconstruction result example 2.

Evaluation metric: We use 3 ordinary metrics to value our results, which are:

$$Precision = \frac{|S_{find} \cap S_{real}|}{|S_{find}|} \quad (3)$$

$$Recall = \frac{|S_{find} \cap S_{real}|}{|S_{real}|} \quad (4)$$

$$F1 = \frac{2 \times Precision \times Recall}{Precision + Recall} \quad (5)$$

where $|S|$ denotes the number of the set S, S_{find} denotes the set of headers by our algorithm, S_{real} denotes the real headers. Therefore $S_{find} \cap S_{real}$ denotes the set of correct headers found by our algorithm. The results are shown in Figure 6-8.The black dashed line in each figure is the average value.

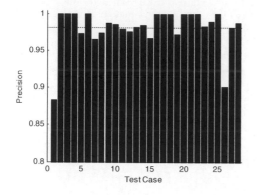

Figure 6.   Precision result on all test data.

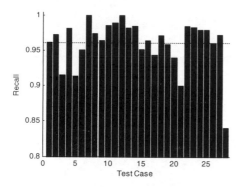

Figure 7.   Recall result on all test data.

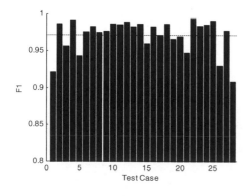

Figure 8.   F1 result on all test data.

4   CONCLUSION AND FUTURE WORK

In this paper, we proposed a rectangular segmentation approach. We addressed the problems of reconstruction of complex layout of Postscript documents. Experimental results indicated that our rectangular segmentation approach did mark all the real headers of the documents.

ACKNOWLEDGMENT

This work is supported by National Key Technology R&D Program (Grant No. 2012BAI34B01).

REFERENCES

[1] Wu, Y. M. &Zhu M. & Luo M. C. 2010. Algorithm of extracting and restructuring text block paragraph of post-script document. *Journal of Computer Application*,vol. 12, pp. 27–30.
[2] Xu, Y. & Song R. 2007.Encyclopedia Text Topic Segmentation Based on CRF. *Computer Engineering*, vol. 27, no. 12, pp. 16–18.
[3] Zhang Z. W. & Kong F. R. 2008.Extracting Mathematical Expressions from Scientific and Technical Documents in PostScript Format.*Journal of Computer Application*. vol. 25, no. 11, pp. 157–158.
[4] Zhu M.2010. Research and Application of Technologies of PostScript File Parse. *South China University of Technology*.

High performance load balancer for restful web service

X.D. Wu & J.CH. Zhong
The Third Research Institute of Ministry of Public Security, Shanghai, China

X.Y. Li
State Grid Yingkou Electric Power Supply Company, Liaoning, China

ABSTRACT: Along with the fast development of internet application and rapid growth of restful application, the current load balancer can hardly satisfy the need of such workload. On the base of the characteristics of restful web service, a dynamic load balancer method is proposed. This method takes the weight of different restful web services into consideration, which dynamically deals with the maximum capability of the server rather than statically estimate it. As a result, the balancer can be able to judge the workload of the servers and make balance of them effectively. It has been implemented on the optimized Linux system. Finally, the result of the experiments shows that this system presents a higher performance than other similar ones.

1 INTRODUCTION

With the development of the internet application, the types of application grow very fast. Moreover, with the growth of users, the load of the servers increases rapidly even to the degree that one server can't bear the whole workload any more. In order to cope with this situation, the REST framework model has been proposed. This framework is based on the use of the standard protocol and independent of states, such as the HTTP and only uses four verbs to realize all the applications. Considering those characteristics of simplicity and convenience, it develops very fast. To deal with the problem of the enormous servers' load, the typical technology achievements, such as cluster computing and cloud computing, have been proposed, which use a lot of machine to deal with mass service requests. Under this situation, a load balancer is necessary to assign the service request to each server to get the best performance.

The concept of REST was initially proposed in 2000 by Roy Fielding in the paper of "Architectural Styles and the Design of Network-based Software Architectures"[1]. The REST defines a set of architectural principles. Based on those principles, developers can design web service system that is focused on system resources, including providing service for clients speaking different languages and how to use the HTTP to process and transmit resource. Considering the amount of website services, REST has become the main Web service design model in recent years and has been broadly put into use [2] [3][4]. Since it is very convenient to use, REST has widely replaced the design of interface which is based on SOAP and WSDL. Normally, REST is easy to use partly because it is based on the existing popular protocols, such as HTTP, URI, XML, and HTML.

In order to guarantee the performance of the load balancing system, using the high performance hardware is the basis, for example, an Intel Xeon processor, a high speed PCI-e data bus, a high performance network card, etc. Especially, when it is necessary, the specialized hardware should be adopted. Secondly, it is crucial to use a high processing performance of network operating system. In this respect, the UNIX or the Linux operating system can be applied, and even make use of a real-time operating system. Finally, we need to use a good load balancing algorithm to achieve load distribution. Making decision about the program should be based on comprehensive consideration, since it involves various factors such as cost, time, the difficulty of development and its maintenance complexity, etc. In this way, this paper has adopted the general X86 platform development and the Linux operating system, with in-depth optimization.

Based on guaranteeing the performance of the load balancer itself, it is important to ensure the load is balanced among all servers. That is to say, the working target of load balancing algorithm is to assign clients' requests to the appropriate server in the shortest response time. To measure the load balancing algorithm, the main criterion is the average waiting time of the user, which is the time that the request stays in the system including the process of researching the inlet system, processing and returning the result. In order to achieve this goal, under normal circumstances, we need to let user's request to be processed by the strongest server. This server refers to instantaneous processing capacity, which is a dynamic index rather than a static one. To achieve this object, a measuring index is required to select the most appropriate server to process this request. It can facilitate in preventing server overload and improve the server performance.

The traditional methods, RR (Round Robin), WRR (Weighted Round Robin), LC (Least Connection), WLC (Weighted Least Connection), use static information and only suited for small scale static web page system [5]. However, with the development of application, most of them need to be backed up by database. Even for website service, it also uses dynamic page (such as JavaScript) technology. Actually, it shows a big difference in the process time comparing static and dynamic page [6]. Since the traditional load balancing technology is not suitable to deal with the present problems, a lot of dynamic methods have been proposed. The basic thought of them is to compute the dynamic performance of the real server, then assign requests to the lightest load server. Stochastic Hill Climbing uses central site to gather the other's sites load information [7]; the literature [8] uses Ant Colony Optimization Algorithm based method; the literature [9] uses Auction-based method.

This paper will first optimize operation system, then propose an algorithm suited for restful service, at last do an experiment to get the real performance.

2 OPTIMIZE OPERATION SYSTEM

We use Linux to implement the load balancer. In order to get higher performance, Linux kernel futures have been applied to optimizing system's network performance, mainly RPS (Receive Packet Steering), XPS (Transmit Packet Steering), IRQ affinity. Network card multi-queue technology creates multi-queue by the support of network interface card or software simulation, and assign IRQ for every queue, bind every different queue to a fixed CPU through IRQ. Specifically compute the packet hash according to for parameter: source IP, source port, destination IP, destination port, then assign the same hash code packet to the same queue. Picture 1 gives a four queues sample.

Additionally, in order to improve the cache hit rate, we need to bind one receive queue and one transmit queue to one CPU, where the receiver queue and the transmit queue process the same hash packets. Figure 2 shows how to get the IRQ for receive and transmit queues.

After getting queue's IRQ, select one CPU to process this queue, modify/proc/irq/*/smp_affinity to bind them to same CPU, where * is the IRQ number. Figure 3 shows the modification result.

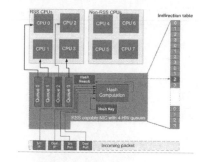

Figure 1. Example of multi queue.

Figure 2. Receive and transmit queue's IRQ.

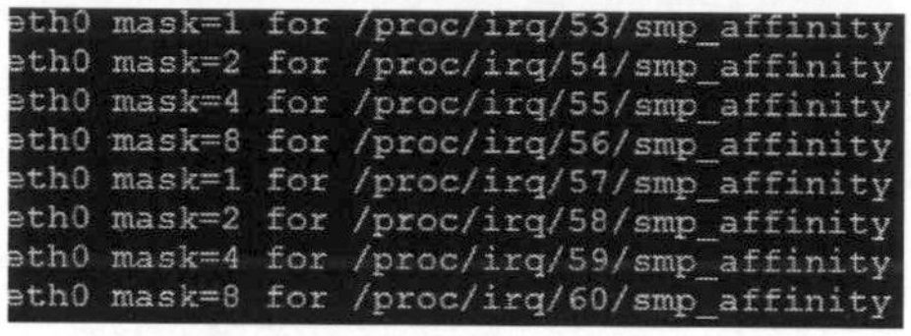

Figure 3. Bind result.

3 LOAD BALANCE ALGORITHM

On the base of other algorithms, a new method which is suitable for the restful service propose has been proposed. It assigns the workload to different restful servers by measuring the capacity of them. That is to say, this design is able to compute their dynamic load, them assign request to the relatively idle server.

3.1 Definition

S_i : the ith real server $i \in \{1, n\}$
L_i : the ith server's instant load
N_i : the ith server's capability to process restful requests in unit time
Assign there has m Restful services :
ε_j : weight of the jth restful service, $j \in \{1, m\}$
N_{ij} : the ith server's capability to process the jth restful requests in unit time

3.2 Determine parameter

For N_{ij} can be determined by experiment or can be deduced by comparison such as CPU memory and so on. Figure 4 shows typical N_{ij}. From Figure 4 we know that for restful service N_{ij} first ascend with the time, then for a point of time it will drop, then shock around one value. This value is N_{ij}.

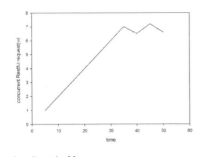

Figure 4. Sample N_{ij}.

For average response time, as shown in figure 5, with the growth of requests, it first grows slowly, but after a point of time, it grows very rapidly. This phenomenon is called "Denies of Service". It is adopted by system software to prevent system halt. To ensure normal service, it must keep the server from "Denies of Service".

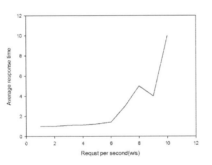

Figure 5. Sample average response time.

After we get N_{ij}, we must evaluate the weight of every restful service. The principle is consider service's feature, give the most important service the biggest weight, and let $\sum_{j=0}^{j<m} \varepsilon_j = 1$, then compute

$N_i = \sum_{j=0}^{j<m} \varepsilon_j * N_{ij}$ $i \in \{1, n\}$. After that, we can compute the instant load for every real server, let $n_1\, n_2, ..., n_m$ is the instant restful request , then the weighted request, is $N_i' = \sum_{j=0}^{j<m} \varepsilon_j * n_{ij}$, the instant load is $L_i = \dfrac{N_i'}{N_i}$, then for ideal load balance status, it has $L_1 = L_2 = ... = L_n$.

3.3 Load assign

First get every server's load L_i, if no load it is zero. Because it is hard to estimate server's load after assign request, the assign procedure will assign request to the lightest load server.

While a request comes, first judge whether the system is in critical status. The critical status can be decided by the user, typically is it's L_i near to 1. If all of the servers are in critical status, give a warning and deny this request. If not, select some lightest load server to assign. After selecting the destination server set use probability method to get the final one. For example, for a 100 nodes cluster, select 10 lightest load server, then

1 Generate a random number between 0~100;

2 Compute every 10 server's assign section, let $L_i' = 1 - L_i$, then the low limit of section of the ith server is: $100 * \dfrac{\sum_{j=1}^{j<i} L_j'}{\sum_{j=1}^{j<10} L_j'}$ and the high limit of section is $100 * \dfrac{\sum_{j=1}^{j<i+1} L_j'}{\sum_{j=1}^{j<10} L_j'}$.

3 Judge which server's section contains the random, then assign this request to this server.

4 At last, update L_i, end one assign.

Using this method, the lightest load server have the biggest probability to deal with the request, at the mean time get the better load balance effect because the probability method can assure every assign is an independent incident.

4 EXPERIMENT

The following figure shows the experiment network connection.

The hardware configuration of the servers is shown in Table 1.

The software we use is optimized RedHat Enterprise Linux (RHEL) 6.3, and use the load balanced algorithm proposed by this paper. Test instrument, we use Ixia Xcellon-Flex. For convenient define one resource (URI) and 3 restful services: Get, Post, and Put. For restful service's parameter, gives 0.3, 0.5, and 0.2., use Ixia get server's capability.

At last show the load balance's test result, as following figure shows.

According to the figure, the max concurrent request is 9,500,000 and the request per second is about 76,000, the average response time is 0.15 ms. As a comparison, show the test result which uses WRR algorithm in Figure 8. From the test result we know that though the max concurrent request is 9,500,000, but the request per second is about 28,000, the average response time is 0.4 ms.

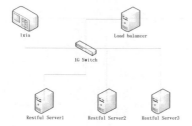

Figure 6. Network sketch map.

Table 1. Server's configuration.

	CPU	memory	NIC
Load Balancer	Xeon E3-1275	64G	Intel(R) PRO/1000
Server1	Core i3-2120	32G	Intel(R) PRO/1000
Server2	Core i7-2600	32G	Intel(R) PRO/1000
Server3	Xeon E3-1225	32G	Intel(R) PRO/1000

Table 2. Server's N_i.

	Get(w/s)	Post(w/s)	Put(w/s)	N_i
Server1	12.6	5.5	6.3	7.79
Server2	13.2	5.9	6.7	8.25
Server3	12.3	5.1	5.5	7.04

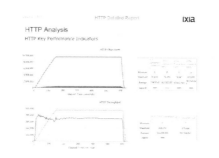

Figure 7. Our algorithm test result.

The test result shows that the load balance is a high performance load balance and our method is effective.

5 CONCLUSION

According to the feature of restful service, combination with load balance we propose the method to implement high performance load balance for restful service, which solves the problem encountered in

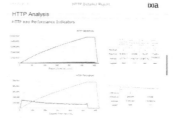

Figure 8. WRR algorithm test result.

actual work. And it also gives valuable reference to solve similar problem.

ACKNOWLEDGMENTS

This paper is supported by the National High Technology Research and Development Program of China (No. 2012AA01A403).

REFERENCES

[1] Roy Thomas Fielding, Architectural Styles and the Design of Network-based Software Architectures 2000.
[2] Lathem, J., Gomadam,K., Sheth,A.P, SA-REST and (S)mashups : Adding Semantics to RESTful Services Semantic Computing, 2007. ICSC 2007. 469 – 476.
[3] Jason H. Christensen Using RESTful web-services and cloud computing to create next generation mobile applications,OOPSLA '09 Proceedings of the 24th ACM SIGPLAN conference companion,ACM New York, NY, USA,2009, 627–634.
[4] Min Choi and Namgi Kim; A REST Web Service for Simulation of Global Income Tax Deduction and Year-end Tax Settlement; International Journal of Software Engineering and Its Applications, Vol.7, No.6 (2013), 399–410.
[5] Casslicchio Emiliano,Tucci Salvatore. Static and Dymanic scheduling algorithm for scalable Web server farm. In:Proceedings of the IEEE 9th Euromicro Workshop on Parallel and Distributed Processing, 2001, 369–376.
[6] Iyengar Arun, MacNair Ed, Nguyen Thao. An analysis of Web server performance. In: Proceedings of Global Telecommunications Conference, 1997, 3:1943–1947.
[7] Mondal B, Dasgupta K, Dutta P. Load balanceing in cloud computing using stochastic hill climbing – a soft computing. C3IT-2012: Proceedings of the 2nd Intertional Conference on Computer, Communication, Control and Information Technology. New York:Elsevier, 2012. 783–789.
[8] Chen Z, Resource allocation for cloud computing base on ant colony optimization algorithm. Journal of Qingdao University of Science and Technology: Natural Science Edition, 2012. 33(6):619–623.
[9] Huu T T, Tham C K. An auction-based resource allocation model for green cloud computing, Proceeding of the 2013 IEEE International Conference on Cloud Engineering. Piscataway:IEEE, 2013:269–278.

Network Security and Communication Engineering – Chan (Ed.)
© 2015 Taylor & Francis Group, London, ISBN: 978-1-138-02821-0

A two-level mixed service polling model with batch arrival

Z. Guan & W.H. Qian
School of Information Science & Technology, Yunnan University, Kunming, Yunnan, China

Y.Q. Wang
State Grid Shanghai Changxing Electric Power Supply Company, Shanghai, China

ABSTRACT: We present a single-server and two-level mixed service polling systems with two queue types, one center queue and N normal queue. Customers arrive at the system according to a batch process. The center queue will be successively served with exhaustive discipline after each normal queue being served with parallel 1-limited discipline. In the first level, server visits between the center queue and the normal queue. In the second level, normal queues are polled by a cyclic order. We propose an imbedded Markov chain framework to drive the closed-form expressions for the mean queue length and average waiting time. Numerical examples demonstrate that theoretical and simulation results are identical and the new system efficiently differentiates priorities.

1 INTRODUCTION

A polling model is a queue system which consists of N queues, which are attended by a single server by visiting the queues in a cyclic order [1-3].

Many studies in the literature assume that customers arrive at queues follow the Poisson process, because it is tractable for performance analysis. In many applications of polling system, this assumption is appropriate, such as in computer-communication systems, and information packets arrive at the termination with the Poisson process. Yet this might in some cases, not be a realistic assumption, e.g., in bursty systems like voice and video, transportation systems, and some production systems, events might occur simultaneously. We study the system with Markova Batch arrival where the number of customers arriving at queue in each slot is governed by a Markov chain.

The traditional routing mechanism is the cyclic server routing. Boon et al. developed a polling model with N queues providing K priority levels in each queue [4], where priority levels are distinguished by the order in which customers are served in each queue. In the other hand, priorities between different queues can be also distinguished by setting different service disciplines for customers in the queues. Hwang and Chang proposed a polling system with general service order and mixed service discipline [5], L. Qiang et al. has provided an exact mean cycle time of a N queues mixed service polling model [6].

We have discussed a parallel two-level poling system with Poisson arrival in ref [7-9]; in the present paper we extend the result to a more normal Batch arrival model (BTLP). The special feature of the model considered in the present paper are: (1) Customers arrive at queues according to a Markov batch process; (2) By adapting the polling order and the service discipline for a queue type, we distinguish two priority levels; (3) For this model, we derive the probability generating function of the joint queue length distribution at polling epochs. (4) Based on these results, we achieve the exact closed-form expression of mean queue length and mean waiting time. Specially, the unknowns in these equations are all first moments of random variables or could be calculated from the first moments of random variables and, thus, no correlation terms are required.

2 MODEL DESCRIPTION

In this section, we study a batch arrival polling system consists of $N+1$ ($N>2$) infinite-buffer queues and a single sever. The server visits and serves the customers in each queue in a two-level logical cyclic order.

2.1 Preliminary

We index the N normal nodes by i, $i=1,2,\ldots,N$ as $(Q_1, Q_2,\ldots,Q_N)$, in which customers are served in parallel limited mechanism [10], and a high priority central queue by h as Q_h, served in exhaustive scheme.

Assuming that the arrival of customers in queue j ($j=1,2,\ldots,N,h$) waiting for service to meet the independent Batch process with arrival rate $p\lambda_j$, the interval time between each batch follows the independent geometric distribution. Customers arrive at queue j ($j=1,\ldots,N,h$) in batch with Geom-probability p at each time slot, and the arrival rate in each batch is λ_j. The generating function of arrival process in queue j is $A_j^*(z_j)= p_j+p_jA_j(z_j)$, with the variance of $\sigma_{\lambda j}^2= A_j^{*\prime\prime}(1)+ p\lambda_j-p\lambda_j^2$.

The service time of a customer at each queue is independent of each other. The mean value is β_i, and their generating function is $B_j(z_j)$, with the variance of $\sigma_{\beta j}^2= B_j''(1)+ \beta_j-\beta_j^2$. The sever which visits queues is governed by a two level routing. In the first polling level, the server polls between the high priority queue Q_h and a normal queue; in the second level, for each time after the exhaustive service at Q_h, one normal queue is visited in a cyclic order, i.e., the sever routing in this model is $1\to h\to\ldots\to i\to h\to i+1\to\ldots\to h\to N$. We specified exhaustive discipline for the central queue and 1-limited discipline for normal queues, so that, the entire customers in the central queue could be served in the last serve round, while who in normal might need several round when it has more than one customer.

The switch-over times is a random variable with mean value γ_i and generating function $R_i(z_i)$, with the variance of $\sigma_{rj}^2= R_j''(1)+ r_j-r_j^2$. $i=1,\ldots,N$. Otherwise, when the sever polls a queue at times with customers in its storage, the server will provide service, and switch polls the following immediately without switching over time, when it has finished the current serve.

Besides, we make the assumption for BTLP model which is as follow: customers arrive at central queue Q_h at the condition that $p_h\lambda_h$ is served in exhaustive mechanism. The generating function of the service time is $F_h^*(z_h)=p_h+p_h F_h(z_h)$, $F_h(z_h)= A_h(B_h(z_h F_h(z_h)))$, is the corresponding generation function in a Poisson arrival system. Each queue has enough storage so that no customer is lost under the first-come-first-serve rule.

2.2 Generation function

The buffer occupancy method is used to determine the moments of the steady-state queue length, waiting time, and delay. We determine the relationships between the set of buffer occupancy variables by deriving $G(\mathbf{z})$, the generating function for the number of customers present at polling instants. Upon the assumption outlined in the previous paragraph, it is possible to define that a Markov processes to describe status of nodes which are as follow:

Assuming that the sever begins the service of Q_i at t_n, defining a random variable $\xi_i(n)$ as the number of customers in storage at queue i at time t_n. Then the status of the entire polling model at time t_n can be

represented as $\{\xi_1(n),\ldots,\xi_N(n),\xi_h(n)\}$. While $\xi_i(n^*)$ is the number of customers in storage at node h at time t_n^*, at which the sever begins to provide service to Q_h, and the status of the entire polling model at time t_n^* which can be represented as $\{\xi_1(n^*),\ldots,\xi_N(n^*),\xi_h(n^*)\}$. For $\sum\rho_i+\rho_h<1$ ($\rho=p\lambda\beta$), we will assume the queues are stable. Then, the probability distribution is defined as

$$\lim_{n\to\infty}P\Big[\xi_j(n)=x_j; j=1,\cdots,N,h\Big]=\pi_i\left(x_1,\cdots,x_N,x_h\right)$$

$$\lim_{n\to\infty}p[\xi_j(n^*)=y_j; j=1,\ldots,N,h]=\pi_{ih}\left(y_1,\ldots,y_N,y_h\right)$$

and the generation functions at t_n^* and t_{n+1}

$$G_i\left(z_1,\cdots,z_N,z_h\right)=\sum_{x_1=0}^{\infty}\cdots\sum_{x_N=0}^{\infty}\sum_{x_h=0}^{\infty}z_1^{x_1}\ldots z_N^{x_N}z_h^{x_h}\pi_i\left(x_1,\ldots,x_N,x_h\right)$$

$$G_{ih}\left(z_1,\cdots,z_N,z_h\right)=\sum_{y_1=0}^{\infty}\cdots\sum_{y_N=0}^{\infty}\sum_{y_h=0}^{\infty}z_1^{y_1}\ldots z_N^{y_N}z_h^{y_h}\pi_{ih}\left(y_1,\ldots,y_N,y_h\right)$$

According to the proposed mechanism, the system variables have the following equations.

$$\begin{cases}\xi_j(n+1)=\xi_j(n^*)+p_j\eta_j(v_h), & j\neq h\\ \xi_j(n+1)=0, & j=h\\ \xi_j(n^*)=\xi_j(n)+p_j\eta_j(v_i), & j\neq i, \quad \xi_i(n)\neq 0\\ \xi_j(n^*)=\xi_i(n)+p_i\eta_i(v_i)-1, & j=i, \quad \xi_i(n)\neq 0\\ \xi_j(n^*)=\xi_j(n)+p_j\mu_j(u_i), & j\neq i, \quad \xi_i(n)=0\\ \xi_j(n^*)=p_i\mu_i(u_i), & j=i, \quad \xi_i(n)=0\end{cases}\quad(1)$$

$v_j(n)$ is the service time in Q_j, $\eta_k(v_j)$ is the number of arrivals to Q_k during service at Q_j with probability p. $u_i(n)$ is the switchover time from Q_i to node Q_h, and $p_k\mu_k(u_j)$ is the number of arrivals to Q_k during $u_i(n)$ with probability p. ($j,k=1,2,\ldots,N,h$).

We define generation function of the two categorizes queues as follow.

The generating function for the number of customers present at polling instants t_n^* is

$$G_{ih}(z_1,z_2,\cdots,z_N,z_h)=\lim_{n\to\infty}E\left[\prod_{j=1}^{N}z_j^{\xi_j(n^*)}z_h^{\xi_h(n^*)}\right]$$

$$=\frac{1}{z_i}B_i\left(A_h^*(z_h)\prod_{j=1}^{N}A_j^*(z_j)\right)G_i\left(z_1,\cdots,z_i,\cdots,z_N,z_h\right)$$

$$-G_i\left(z_1,\cdots,z_i,\cdots,z_N,z_h\right)\Big|_{z_i=0}\Bigg]\qquad(2)$$

$$+R_i\left(A_h^*(z_h)\prod_{j=1}^{N}A_j^*(z_j)\right)G_i\left(z_1,\cdots,z_i,\cdots,z_N,z_h\right)\Big|_{z_i=0}$$

Where $G_i(z_1,...,z_N,z_h)$ is the generating function for the number of data packets present at polling instants of Q_i, abbreviated as $G_i(\mathbf{z})$. $G_{ih}(z_1,...,z_N,z_h)$ is the generating function for the number of data packets presented at polling instants of central queue Q_h, abbreviated as $G_{ih}(\mathbf{z})$.

The generating function for the number of customers present at polling instants t_{n+1} is

$$G_{i+1}(z_1,z_2,\cdots,z_N,z_h)=\lim_{n\to\infty}E\left[\prod_{j=1}^{N}z_j^{\xi_j^{(n+1)}}z_h^{\xi_h^{(n+1)}}\right]$$
$$=G_{ih}\left(z_1,z_2,\cdots z_N,B_h\left(\prod_{j=1}^{N}A_j^*(z_j)F_h^*\left(\prod_{j=1}^{N}A_j^*(z_j)\right)\right)\right) \tag{3}$$

3 PERFORMANCE ANALYSIS

3.1 Derivative of generation functions

We assume the first derivative of generation functions:

$$g_i(j)=\lim_{z_1,\cdots,z_N,z_h\to1}\frac{\partial G_i(z)}{\partial z_j},$$

$$g_{i0}(j)=\lim_{z_1,\cdots,z_N,z_h\to1}\frac{\partial G_i(z)|_{z_i=0}}{\partial z_j}, \text{ and}$$

$$g_{ih}(j)=\lim_{z_1,\cdots,z_N,z_h\to1}\frac{\partial G_i(z)}{\partial z_j}, j=1,2,\cdots,N,h.$$

Taking it with Eq.(2) and Eq.(3), we have:

$$1-G_i\left(1,\cdots,z_i,1,\cdots1\right)\Big|_{z_i=0}=\frac{Np\lambda\gamma}{1-\rho_h-Np+N\lambda\gamma} \tag{4}$$

$$g_{ih}(h)=\frac{p\lambda_h\gamma(1-\rho_h)}{1-\rho_h-Np+Np\lambda\gamma} \tag{5}$$

The second first derivatives of generation functions are defined as:

$$g_i(j,k)=\lim_{z_1,\cdots,z_N,z_h\to1}\frac{\partial^2 G_i(z)}{\partial z_j\partial z_k},$$

$$g_{i0}(j,k)=\lim_{z_1,\cdots,z_N,z_h\to1}\frac{\partial^2 G_i(z)|_{z_i=0}}{\partial z_j\partial z_k}, \text{ and}$$

$$g_{ih}(j,k)=\lim_{z_1,\cdots,z_N,z_h\to1}\frac{\partial^2 G_{ih}(z)}{\partial z_j\partial z_k},$$

where $i=1,2,...,N$; $j,k=1,2,...,N,h$.

Assuming that the normal queues are symmetric, ie. $\lambda_i=\lambda_j$, $\beta_i=\beta_j$, $p_i=p_j$, $i,j=1,2,...,N$, by substituting the second derivative of Eq.(2) and ßEq.(3), give the expressions of $g_{ih}(h,h)$ and $g_i(i)$:

$$g_{ih}(h,h)=\left[p_hA_h^*(1)\beta+(p_h\lambda_h)^2B^*(1)-(p_h\lambda_h)^2R^*(1)-p_hA_h^*(1)\gamma\right] \tag{6}$$
$$\frac{Np\lambda\gamma}{1-Np-\rho_h+Np\lambda\gamma}+(p_h\lambda_h)^2R^*(1)+p_hA_h^*(1)\gamma$$

$$g_i(i)=\frac{1-\rho_h}{2(1-Np-\rho_h)}\left\{N\left(\gamma pA^*(1)+(p\lambda)^2R^*(1)\right)+\frac{2N\rho_h}{1-\rho_h}\left((p\lambda)^2R^*(1)+(p\lambda)^2r\right)+\frac{Np\rho_h^2}{(1-\rho_h)^2}\left(\gamma pA^*(1)+(p\lambda)^2R^*(1)\right)\right.$$

$$+\frac{Np\lambda}{1-Np-\rho_h+Np\lambda\gamma}\left[N\gamma(\beta-\gamma)pA^*(1)-N(p\lambda)^2\gamma R^*(1)+\frac{1}{1-\rho_h}\left(2N(p\lambda)^2\gamma p_h(\beta-\gamma)-2N(p\lambda)^2\gamma(1+\rho_h)R^*(1)\right.\right.$$

$$+2\gamma(1-\rho-\rho_h)+p\lambda\gamma p_h(1+\rho_h)-(N-1)p\lambda\gamma^2+2(N-1)p\lambda\gamma(\gamma-\beta))+\frac{p\lambda}{(1-\rho_h)^2}\left(Np\lambda\gamma B^*(1)\right.$$

$$\left.\left.\left.-Np\lambda\gamma p_h^2R^*(1)+\gamma\beta_h^2(\rho_h+Np\lambda(\beta-\gamma))p_hA_h^*(1)+p_h\lambda_h\gamma B_h^*(1)\right)\right]\right\} \tag{7}$$

3.2 Mean queue length and average waiting time

The mean queue length $E(L_j)$ of queue j ($j=1,...,N,h$) is defined as the number of jobs in the buffer between two successive arrivals of the server at this queue. Let the mean number of customers in node j at time t_n/t_n^* be defined as $g_i(j)/g_{ih}(j)$ when Q_i/Q_j is polled by the center server. Then, it is given by $E(L_i)=g_i(i)$, and in like manner $E(L_h)=g_{ih}(h)$.

Mean waiting time is the time from which a customer enters the storage of Q_i to which it is served. As the normal queues are served in the 1-limited service, by adapting an approach in ref [9], the mean waiting time in Q_i could be calculated as:

$$E(w_i)=\frac{1}{p\lambda C}g_i(i)-\frac{1}{p\lambda} \tag{8}$$

Taking with Eq. (7), it is easy to obtain the closed form expression of mean waiting time in normal queue.

The high-priority node Q_h is served in the exhaustive scheme, under the theory in reference [11], we could get the expression of the average waiting time which is as follow:

$$w_h=\frac{g_{ih}(h,h)}{2p_h\lambda_hg_{ih}(h)}-\frac{p_hA_h^*(1)}{2(p\lambda_h)^2(1+\rho_h)}+\frac{p_h\lambda_hB_h^*(1)}{2(1-\rho_h)} \tag{9}$$

243

Where, $g_{ih}(h)$ and $g_{ih}(h,h)$ could be acquired from Eq.(5) and Eq.(6).

4 NUMERICAL ANALYSIS

To assess the accuracy of the expression and the efficiency of the BTLP, we have performed numerical experiments to test the accuracy of the exception for different values of the workload of the system. We should consider a ten queues-one high priority queue and nine normal queues polling system with the arrival rate λ, service time β, and switch over time γ when it is necessary.

In this example we illustrate the accuracy of theoretical analysis and the model performance with superiors in the high priority queue. "o" and "*" respect the simulation results. We simulate cases with $p=p_h=1$ and $p=0 \& p_h=0$ in Fig.1 and Fig. 2.

Fig.1 shows the queue length and the mean waiting time for BTLP. The analytical results coincide with the simulation results. The x-axis shows the number of workload of the model ($\Sigma\rho < 1$) growing with the increase of the customer arrive rate λ; the y-axis shows the mean queue length (number of customer) in Fig.1 (a) while mean waiting time (slot) in Fig.1 (b). We can clearly see that with the growth of G, both the mean queue length and the mean waiting time is increasing distinctly in Q_i, while the performances in Q_h are much better, both queue and mean waiting time are much lower than normal queues, and the growth in Q_h with G presents much more smoothly.

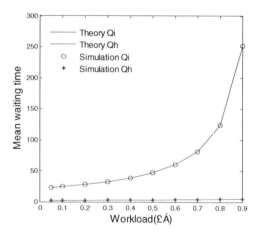

Figure 1. Average waiting time in BTLP ($\beta=\beta_h=10$, $\gamma=5$, $N=9$, $p=p_h=1$).

The geom-probability p_j could be used to describe the state of Q_j. Fig. 2 shows the case of $p_i=1$ ($i=1,2,\ldots,N$), $p_h=0$, i.e. the central queue works in a sleep state with no customer, and then all the customers arrive at the normal queues. Consequently, the model works as an N queues classical exhaustive polling system with parallel switch over process.

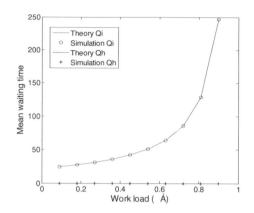

Figure 2. Average waiting time in BTLP ($\beta=\beta_h=10$, $\gamma=5$, $N=9$, $p=1$, $p_h=0$).

5 CONCLUSION AND EXTENSIONS

In this paper, we have studied a two-level mixed service polling with batch arrival (BTLP) model, in which customers arrive at each queue according to a batch process with Geom-probability p_j, and queues are provided with a distinguished priority service. The exact closed-form expression of mean queue length and mean waiting time are achieved by using the Markov chain and generation function framework. The vast prior literature study on waiting time of priority polling system are numerical analysis or expressed as the solution of a set of linear equations, however, we get the closed from expressions of mean queue length and average waiting time, which are exactly coincided with the computer simulation results. As we have shown, the system fulfills requirements of multi-priority. High level queue has a superior performance in queue length and delay, and low level queues still worked stably. Both of them are more effective than those in non-priority systems. Furthermore, Slightly more general batch arrival processes are discussed in BLTP, the analyzed result could be promoted to several polling systems in special cases, for example, two-level mix service polling system [7] when $p=p_h=1$; Gemo/G/1 system when $p=0$; 1-limited parallel system [10] when $p_h=0 \& p=1$.

ACKNOWLEDGMENT

This work was supported by a Grant from the National Science Foundation of China (no. 61463051 and no. 61461054), the National Science Foundation of Yunnan Province (no.2012FD002), and the Science Foundation of Yunnan Provincial Department (no.2014Z010).

REFERENCES

[1] H. Levy, M. Sidi, Polling systems: applications, modeling and optimization, IEEE Transactions on Communications, vol.38, no.10, pp.1750–1760, 1990.

[2] H. Takagi, Queueing analysis of polling models: an update, Stochastic Analysis of Computer and Communication Systems, H. Takagi (ed.), North-Holland, Amsterdam, pp.267–318, 1990.

[3] M.A.A. Boona, R.D. van der Mei, E.M.M. Winands. Applications of polling systems. Surveys in Operations Research and Management Science vol.16, no.2, pp. 67–82, 2011.

[4] M.A.A. Boon, et al., A polling model with multiple priority levels, Performance Evaluation, vol.67, no.6, pp.468–484, 2010.

[5] L.C. Hwang, An exact analysis of an asymmetric polling system with mixed service discipline and general service order, Computer Communication, vol.20, no.1, pp.1293–1299, 1997.

[6] L. Qiang, Z. Zhongzhao, Z. Naitong, Mean Cyclic Time of Queueing Priority Station Polling System, Journal of China Institute of Communications, vol.20, no.2, pp. 86–91, 1999.

[7] Y. Zhijun, D. Hongwei, and C. Chuanlong. Research on E(x) characteristics of two-class polling system of exhaustive-gated service. Acta electronica sinica, vol.42, no.4, pp.774–778, 2014.

[8] Q. L. Liu, D. F. Zhao & D. M. Zhou. An analytic model for enhancing IEEE 802.11 point coordination function media access control protocol, European transactions on telecommunications, vol. 22, no.2011, 332–338.

[9] G. Zheng, Z. Dongfeng, and Z. Yifan. A Discrecte Time Two-level Mixed Service Parallel Polling Model. Journal of electronics (China), vol. 29, no.1/2, pp.103–110, 2012.

[10] Y. Zhijun, Z. Dongfeng. Polling strategy for wireless multimedia LANs. *Tsinghua Science and Technology*, vol.11, no.5, pp.606–610, 2006.

[11] D. F. Zhao , S. M. Zhao. Analysis of a polling model with exhaustive service. Acta. Electronica Sinica, vol.22, no.5, pp. 102–107, 1994.

Network Security and Communication Engineering – Chan (Ed.)
© 2015 Taylor & Francis Group, London, ISBN: 978-1-138-02821-0

Separation and reception of sub-chip multipath components

X.G. Zhang, Y. Guo & Y.J. Sun
School of Information and Electrical Engineering, China University of Mining and Technology, Xuzhou, China

L.L. Li
China Mobile group Jiangsu Co., Ltd. Xuzhou branch, China

ABSTRACT: The traditional RAKE receiver is able to identify and use the multipath information of more than one relative delay chip. In practice, especially in circumstance of intensive buildings, indoor, underground coal mine, many path which time delay is smaller than time of one chip Direct affect Performance of the system. Based on this, the paper discusses matched filter, matrix and MUSIC three kinds delay algorithm, the simulation results show that: the three algorithms can achieve multipath estimate within one chip, and under the same circumstances MUSIC is better.

KEYWORDS: CDMA, Multipath separation in sub-chip, RAKE.

1 INTRODUCTION

RAKE receiver makes the diversity reception of CDMA come true, including multipath separation and combination technique and so on. The traditional RAKE receiver is able to identify and use the multipath information of more than one relative delay chip. In practice, especially in circumstance of intensive buildings, indoor, underground coal mine, relative delay is common at a multipath chip. Therefore, it is of great practical significance to study the impact of relative delay on multipath signal of chips and the effective detection method. The multipath separation whose relative delay is more than one chip include serial correlation method, matched filtering method and inverse filter[1-4] and so on. But these methods could not tell the difference between latency in a chip in the multipath. Many algorithm research have been approached aiming at multipath separation whose delay time is less than one chip, for example, algorithm based on minimal L1 norm[5], algorithm based on MUSIC[6]. Algorithm based on minimal L1 norm is usually used in sparse channel, for example, in water. The algorithm based on the matrix and MUSIC are commonly used in general channel. The paper analyzes and emulates two kinds of chip time delay estimation algorithm based on matrix and MUSIC and compares with matched filtering method. At last, the paper realized on Separation and RAKE receive of CDMA sub-chip multipath components accordingly combining MRC and EGC.

2 SEPARATION OF SUB-CHIP MULTIPATH COMPONENTS

Basic idea of Matrix method: The decoding chip multipath problem can approximately be transformed to solve a system of linear equations based on the derived the analytical expression of the related solution to enlarge the output ignoring the noise. Considering the particularity of dispreading, every sampling value consists two parts: the phase component and the out of phase component, so the sample output can be expressed as array.

$$R_p = \{ \ \dots, \ a^a_{p,q}(t_p - qT_c)e^{j\phi^a_{p,q}} = \sum_{l=1}^{L^a_p} a_m s(t_p - (q+n_l)T_c)\beta^{a,l}_p e^{j\phi^a_l}, \ \dots \ \} \tag{1}$$

$$p \in \{1,2,\dots,M\} \qquad q \in \{1,2,\dots,N\}$$

The phase component can be written as:

$$a^s_{p,q}(t_p - qT_c)e^{j\phi^s_{p,q}} = \sum_{l=1}^{L^s_p} a_m s(t_p - qT_c)\beta^{s,l}_p e^{j\phi^s_l} \tag{2}$$

And the unusual phase component can be written as:

$$a^a_{p,q}(t_p - qT_c)e^{j\phi^a_{p,q}} = \sum_{l=1}^{L^a_p} a_m s_l(t_p - qT_c)\beta^{a,l}_p e^{j\phi^a_l} \tag{3}$$

$$a^a_{p,q}(t_p - qT_c)e^{j\phi^a_{p,q}} = \sum_{l=1}^{L^a_p} a_m s(t_p - (q+n_l)T_c)\beta^{a,l}_p e^{j\phi^a_l} \tag{4}$$

on account of formula (4) and formula (3), formula (1) can be written as:

$$R_p = \left\{ ..., \sum_{l=1}^{L_p^s} a_m s(t_p - qT_c)\beta_p^{s,l} e^{j\phi_i^s} + \right.$$

$$\left. \sum_{l=1}^{L_p^a} a_m(t_p - (q+n_l)T_c)e^{j\phi_p^{a,l}} e^{j\phi_i^a}, \cdots \right\} \tag{5}$$

The interrelated de-spreading operations is multiplied in order and sum in formula (5) with pseudo-random pulse sequence in which the initial phase is t_p, and the interval is T_c. Assume that the contribution of the adjacent symbols is zero.

On the basis of the double value autocorrelation feature in the pseudo-random sequence, can show:

$$Y_p = a_m \{ N \sum_{l=1}^{P} X_l O(-\frac{M-l+1}{T_c})e^{j\phi_l}$$

$$-\sum_{l=p}^{M} \frac{M-l+p}{T_c}e^{j\phi_l} \} \qquad p \in \{1,2,...,M\} \tag{6}$$

of which, function $O(t)$ represents the output of the receive filter, when the output pulse output from the waveform shaping filter input the receive filter.

The output samples of interrelated de-spreadings shown in formula (6) can construct equations shown in formula (7) in which the unknowns is the multipath amplitude and the phase.

$$Y \approx AX \tag{7}$$

$$Y \approx \begin{bmatrix} Y_1 \\ Y_2 \\ \vdots \\ Y_M \end{bmatrix} X \approx \begin{bmatrix} X_1 e^{j\phi_1} \\ X_2 e^{j\phi_2} \\ \vdots \\ X_M e^{j\phi_M} \end{bmatrix} \tag{8}$$

$$A \approx \begin{bmatrix} NO(-\frac{M}{T_c}) & -O(-\frac{M-1}{T_c}) & \cdots & -O(-\frac{1}{T_c}) \\ NO(-\frac{M}{T_c}) & NO(-\frac{M-1}{T_c}) & \cdots & -O(-\frac{1}{T_c}) \\ & & \vdots & \\ NO(-\frac{M}{T_c}) & NO(-\frac{M-1}{T_c}) & \cdots & NO(-\frac{1}{T_c}) \end{bmatrix} \tag{9}$$

$$= \begin{bmatrix} 31 & -1 & -1 & -1 & -1 & -1 & -1 & -1 \\ 31 & 31 & -1 & -1 & -1 & -1 & -1 & -1 \\ 31 & 31 & 31 & -1 & -1 & -1 & -1 & -1 \\ 31 & 31 & 31 & 31 & -1 & -1 & -1 & -1 \\ 31 & 31 & 31 & 31 & 31 & -1 & -1 & -1 \\ 31 & 31 & 31 & 31 & 31 & 31 & -1 & -1 \\ 31 & 31 & 31 & 31 & 31 & 31 & 31 & -1 \\ 31 & 31 & 31 & 31 & 31 & 31 & 31 & 31 \end{bmatrix}$$

Assume that the output of baseband filter is sampled at tomes the chip rate, the Cycle time of the chip is 31. The coefficient matrix of the Linear equations (7) is (9).

Basic idea of MUSIC algorithm: Discrete measurement data can be got by L values of $H(f)$ which is the frequency response under the same frequency interval sampling. Consider the impact of Gaussian white noise, frequency response of sampled discrete channel is :

$$x(l) = H(f_l) + \omega(l) = \sum_{l=0}^{L_p-1} \alpha_k e^{-j2\pi(f_0 + l\Delta f)\tau_k} + \omega(l) \tag{10}$$

It can be written as formula (11), which is a vector matrix

$$X = H + W = Va + W \tag{11}$$

The MUSIC algorithm in here is broken down the matrix X in the formula (11).

$$R_{xx} = E\{XX^H\} = VAV^H + \sigma_\omega^2 I \tag{12}$$

A L-dimensional space containing the signal vector X can be divided into two orthogonal subspaces Which is by noise eigenvectors and feature vector signal it can be divided into noise subspace and signal subspace. Projection matrix of the noise subspace is:

$$P_\omega = Q_\omega (Q_\omega^H Q_\omega)^{-1} Q_\omega^H = Q_\omega Q_\omega^H \tag{13}$$

In which $Q_\omega = \begin{bmatrix} q_{L_p} & q_{L_p+1} & \cdots & q_{L-1} \end{bmatrix}$, $L_p \leq k \leq L-1$, q_k is the noise eigenvectors, vector $v(\tau_k)$ must exist in the signal space. Defining

$$P_\omega v(\tau_k) = 0 \tag{14}$$

The multipath delay τ_k can be received by seeking the MUSIC pseudospectra, that is to say let $S_{MUSIC}(\tau)$ be the maximum.

$$S_{MUSIC}(\tau) = \frac{1}{|P_\omega v(\tau)|^2} = \frac{1}{v^H(\tau)P_w v(\tau)}$$

$$= \frac{1}{|Q_\omega^H v(\tau)|^2} = \frac{1}{\sum_{k=L_p}^{L-1} |q_k^H v(\tau)|^2} \tag{15}$$

3 ANALYSIS OF SIMULATION RESULT

3.1 *Delay estimation of CDMA codes in a chip*

Simulation parameters are shown in Table 1, Delay estimation of CDMA codes in a chip is shown in Figure 1. As Figure 1 shows, matched filter cannot distinguish the delay in each chip, while the MUSIC algorithm and the Matrix can be used to estimate the delay in a chip. In addition, the peak value in Matrix is less sharper than that in the Music.

Table 1. Experimental parameters.

System parameters	Parameter values
Signal rate	10e+5
spreading code rate	3.1*10e+6
Spreading factor	31
Multipath number	3
Delay time	1/8, 4/8, 7/8 chips
sampling interval of MUSIC	The width of 0.5
Snapshots of MUSIC	2000

3.2 *RAKE receiver*

Simulation block diagram of RAKE receiver is shown in Figure 2 and Simulation of RAKE receiver is shown in Figure 3. According to the way of MRC and EGC, the delay estimations based on those three methods are merged. Figure 3 shows that the error rates based on the Music and the Matrix are lower than that based on matched filter, and the error rate of the Music is the lowest in the three methods. Estimation of the delay in a code chip helps to improve the reliability of system communication, and MRC does better than EGC when using the same algorithm methods.

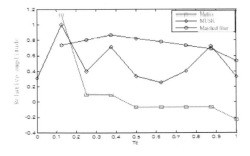

Figure 1.   Three ways to estimate the delay in a chip.

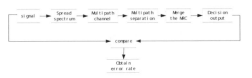

Figure 2.   The simulation block diagram of the RAKE receiver.

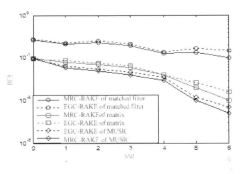

Figure 3.   RAKE receiver based on three methods.

4 CONCLUSIONS

The traditional RAKE receiver can identify and use the multipath information with more than a chip relative delay, while in real communication, especially in environments like the dense buildings, indoor, the coal mines and others, there are a large quantities of multipath with relative delay with a chip. As a result, it is of vital significance to research on the influence the relative delay has on multipath signals within a chip and effective detection methods. This paper analyzes and simulates the algorithms to estimate the delay within a chip—the Matrix and the MUSIC, and compares those two estimation methods with that using matched filter. Finally, multipath separation in CDMA code chip and RAKE receiver are realized by merging the MRC and EGC. The simulation result shows: the merging method of MRC does better than EGC, and both the Matrix algorithm and the MUSIC algorithm can estimate the delay in a code chip, whereas the matched filter fails in delay estimation in a chip. What's more, the system error rate is reduced by isolating the delay in a chip and the MUSIC has better performance than the Matrix.

This work is supported by Youth Science Foundation of Jiangsu Province under grant No. BK20130199, and National Natural science Foundation of China under grant No. 51274202.

REFERENCES

[1] S. SENMOTO, D.G. CHILDERS. Signal Resolution via Digital Inverse Filtering[J]. IEEE Trans, 1972, NO. 5:633~640.

[2] Ahmad Hatami.Application of Channel Modeling for Indoor Localization Using TOA and RSS[D].Worcester Polytechnic Institute, May 2006.

[3] VITERBI A J.CDMA:Principles of Spread Spectrum Communication[M].Addison Wesley Publishing Company, 1995.

[4] RAPPAPORTTS.Wireless Communications Principles& Practice [M].Prentice Hall:Inc, 1996.

[5] Jean-Jacques Fuchs.Multipath Time-Delay Detection and Estimation[J].IEEE, 1999, NO.1:237~243.

[6] Lijing, Caomaoyon, Pei liang.Resolution Analysis of the MUSIC-Based TOA Algorithm.IEEE, 2010:718~721.

Network Security and Communication Engineering – Chan (Ed.)
© 2015 Taylor & Francis Group, London, ISBN: 978-1-138-02821-0

A multi-mode jammer frequency synthesizer designed based on the DDS & PLL hybrid method

Y.K. Li & Y.C. Wang
Department of UAV Engineering, Ordnance Engineering College, Shijiazhuang, China

ABSTRACT: Frequency synthesizer is an important component in the modem electronic systems. This paper studied a kind of frequency synthesizer in a multi-mode jammer and put forward an improvement method for DDS (Direct Digital Synthesis) and PLL (Phase Locked Loop) hybrid frequency synthesizer. To further raise PLL performance in fast locking and noise immunity ability in the improvement method, a PLL with Analog Adaptive Bandwidth Control (AABC) has been designed in this paper. The new-type AABC has been presented based on conventional Charge pump PLL circuit. The PLL with AABC block can control the loop bandwidth according to the locking status and the amount of inputs' phase error adaptively; and consequently low-noise and fast-lock performance is achieved. In the end, the designed Frequency synthesizer satisfies an $800MHz$ to $1.8GHz$ output range and a $0.4ms$ maximum conversion time. Working at $900MHz$, the phase noises are $-93dBc/Hz$ at $200\ kHz$ offset, and the whole circuit draws $8.7mA$ current from$3.0V$ supply.

1 INTRODUCTION

Frequency synthesizer is an important component in the modem electronic systems. It has popular application in the field such as radar, communication, measurement instruments and electronic navigation, etc. As a well-developed frequency synthesis technology, PLL is good at spurious behavior but it has weakness on frequency switch speed and frequency resolution. The DDS technology has attracted popular attention for its excellent performance on frequency switch speed, frequency resolution and Phase noise. However, due to its digital structure limit, it just can work on a narrow bandwidth and its spurious level is not so good.

This paper studied a kind of frequency synthesizer in a multi-mode jammer by using a hybrid method that combines the DDS with the PLL technologies. A wide BW (Wide Bandwidth) frequency synthesizer is finished with the DDS & PLL hybrid frequency synthesis method. After a detailed debugging the synthesizer works well from $800MHz$ to $1.8GHz$, it can output arbitrary frequency with 1 kHz frequency step. The spurious level and phase noise Performance meet the requirements of the actual project.

2 INTEGRATED DESIGN

The DDS has very high frequency resolution and shorter conversion time in the frequency synthesizer, but it signals bandwidth and outputs the tallest frequency to be subjected to restriction; The PLL has

very high work frequency and wider BW, but the frequency convert time to be longer, and the frequency resolution be lower. As showed from this, if we put together DDS and PLL set, then it can acquire the frequency synthesizer of high performance. This paper presents a Nested-mixer DDS+PLL architecture that achieves very wide BW while maintaining the required frequency resolution and conversion time.

The proposed Nested-Mixer DDS+PLL architecture is show in Figure 1.

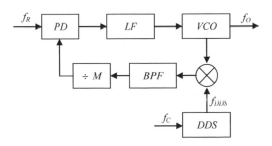

Figure 1. Nested-mixer DDS+PLL frequency synthesizer diagram.

The whole frequency resolution of frequency synthesizer is decided by the DDS, and then it can well develop the advantage of the DDS high resolution. When the PLL locks, its output frequency is an f_o.

$$f_o = Mf_R \pm f_{DDS} = Mf_R \pm \frac{Kf_C}{2^N} \qquad (1)$$

In this design project, the phase-noise performance of the frequency synthesizer mainly from the PLL and the DDS phase-noise decides. Because DDS output is not through PLL frequency double, so the DDS output phase-noise and reference spurs will not worsen further, so that project has lowly phase-noise and reference spurs. The electric circuit adoption PLL provides to take f_0 as the output frequency of partition and the DDS provides a fine frequency resolution to fill up a frequency cleft. Because the DDS has a very high frequency resolution, for promising that the frequency overlay, then request to have $BW_{DDS} > f_R$, in this project, BPF design and creation are more difficult, because the f_o is higher and the distance of $Mf_R + f_{DDS}$ and $Mf_R - f_{DDS}$ then is more near and this will beg BPF to have a very abrupt attenuation characteristic.

For this, the improvement project puts forward what Figure 2 shows.

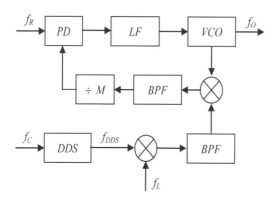

Figure 2. Improvement DDS+PLL frequency synthesizer diagram.

Mix f_C and local oscillator f_L First, move the DDS output to a higher frequency, then mix it and PLL output signal, the output frequency of this kind of circuit is f_0.

$$f_0 = Mf_R + f_L + f_{DDS} \qquad (2)$$

So, in the improvement of the project for promising that the frequency overlay, then request to have $BW_{DDS} + f_L > f_R$, as a result the f_R in this kind of project can obtain higher, the PLL division-ratio (M) then can be lower, contribute to improving phase-noise performance, and the higher basis frequency allows a wide BW, it can reduce conversion time, make time of frequency conversion shorten, but increase one mixer in this circuit.

According to this project, a frequency synthesizer of actual design is what Figure 3 shows. The circuit adopted the AD9858 DDS chip and the ADF4106 PLL chip of AD Company.

The AD9858 is a DDS featuring a 10-bit DAC operating up to 1GSPS. The AD9858 uses advanced DDS technology, coupled with an internal high speed, high performance D/A converter to form a digitally programmable, complete high frequency synthesizer capable of generating a frequency-agile analog output sine wave at up to 400MHz. The AD9858 is designed to provide fast frequency hopping and fine tuning resolution, (32-bit frequency tuning word). The frequency tuning and control words are loaded into the AD9858 via parallel (8-bit) or serial loading formats. The AD9858 contains an integrated charge pump (CP) and phase frequency detector (PFD) for synthesis applications requiring the combination of a high speed DDS along with PLL functions. An analog mixer is also provided on-chip for applications requiring the combination of a DDS, PLL, and mixer, such as frequency translation loops, tuners, and so on. The AD9858 also features a divide-by-2 on the clock input, allowing the external clock to be as high as 2 GHz.

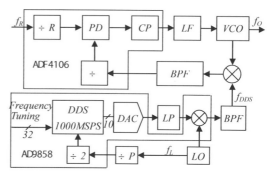

Figure 3. Actual of design DDS+PLL frequency synthesizer diagram.

The ADF4106 consists of a low-noise digital PFD, a precision charge pump, a programmable reference divider, programmable A and B counters and a dual-modulus pre-scaler (P/P + 1). The A (6-bit) and B (13-bit) counters, in conjunction with the dual modulus pre-scaler (P/P + 1), implements an N divider (N = BP + A). In addition, the 14-bit reference counter (R Counter), allows selectable REFIN frequencies at the PFD input. A complete PLL can be implemented if the synthesizer is used with an external loop filter and VCO (Voltage Controlled Oscillator). Its very high bandwidth means that frequency doublers can be eliminated in many high-frequency systems, simplifying system architecture and lowering cost.

3 CIRCUIT DESIGN AND SIMULATION

3.1 *General PLL circuit forms*

Figure 4 shows the functional diagram of the MB15E07 in an integer-N PLL system.

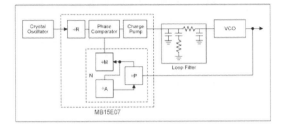

Figure 4. Simplified block diagram of a typical PLL system consisting of a PLL, crystal oscillator, loop filter, and VCO.

It consists of a phase detector (or comparator), an output charge-pump, a dual modulus pre-scalar, an N counter, and an R counter. The N counter consists of a main (M) counter and a swallow or auxiliary (A) counter.

$$N=P*M+A \qquad (3)$$

The N counter then works in conjunction with the dual modulus pre-scalar (P).

During power-up (assuming that the PLL was pre-programmed), the VCO would oscillate at the desired frequency plus some offset. This frequency is first divided by the integer N, and then compared to the reference crystal oscillator frequency, whose frequency has also been divided by an integer R. If there is a phase difference between the two frequencies, the voltage at the PLL output changes accordingly. For example, if the VCO frequency is lower than that of the reference, the charge-pump will charge the loop filter capacitors to increase the voltage. If the VCO frequency is higher than the reference, the charge-pump will discharge the loop filter capacitors to decrease the voltage. An increase in voltage results in an increase in frequency, and visa-versa. Hence, the PLL functions as a feedback loop that keeps the VCO output frequency locked at the desired frequency. The VCO frequency is a function of N, R ,and FREF and is calculated as follows:

$$FVCO= (N/R)* FREF \qquad (4)$$

Careful design and implementation of the clock circuit is required to ensure optimum performance. This can be achieved by choosing the proper components and providing a well-designed high-frequency PC board. These ensure that the VCO will oscillate and phase lock at the desired frequency, while providing the proper output power levels.

3.2 *Adaptive PLL circuit forms*

The adaptive method can be called variation BW again, it is the method which makes the PLL that has BW to regulate function at lock and unlock two kinds of state. The work principle is: make the PLL have bigger BW while being placed in to catch state in the beginning, can reduce catching time, realize fast locking; while being placed in lock state, make the BW automatically lessen, can get lower phase noise and jitter of output signal, and make the PLL have better noise immunity.

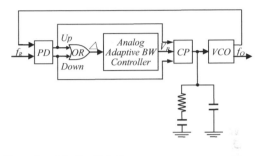

Figure 5. Adaptive PLL diagram.

It is shown as diagram, PFD exportation UP and DOWN connect to go into a logic or door and get the absolute value signal of phase shift $\triangle$e, then $\triangle$e will input AABC, Get the BW control voltage VBW that can control the charge pump current size.

3.3 *Simulation result*

Finally, the designed Frequency synthesizer satisfies a 0.8GHz to 1.8GHzoutput range and a 0.4ms maximum conversion time. Working at 0.9GHz, the phase noisesare-93dBc/Hz at 200 kHz offset, and the whole circuit draws 8.7mA current from 3.0V supply. Simulation wave forms such as in figure 6 to Figure 9.

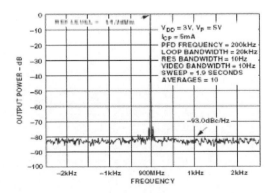

Figure 6. Phase noise (900 MHz, 200 kHz, and 20 kHz).

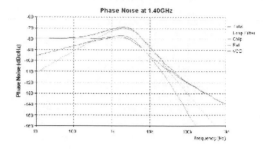

Figure 7.   Phase noise.

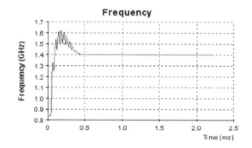

Figure 8.   Lock time.

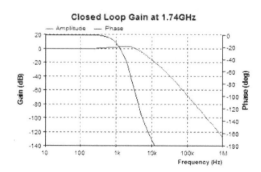

Figure 9.   Closed loop gain.

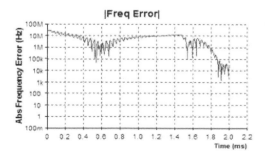

Figure 10.   Abs frequency error.

4   CONCLUSION

A PLL with AABC and an $800MHz$ to $1.8GHz$ output range have been designed in this paper. The Adaptive PLL can be widely used for Clock Data Recovery or Clock Generation in high-speed data communication systems since its special performance in fast locking and noise immunity. Finally, the designed Frequency synthesizer satisfies a $800MHz$ to $1.8GHz$ output range and a $0.4ms$ maximum conversion time. Working at $900MHz$, the phase noisesare-$93dBc/Hz$ at $200kHz$ offset, and the whole circuit draws $8.7mA$ current from$3.0V$ supply. The simulation results show that the designed PLL realizes the function of adaptive bandwidth control and meets the design specification.

REFERENCES

[1] AN535.Phase-locked Loop Design Fundamentals. Motorola Semiconductor Products, 1970.
[2] Analog DevieeIne, 1000MSPS14bit, SVCMOS Direet Digital Synthesizer, AD9858 Dtatsheet, 2005.
[3] Analog Devices Inc.Fractional-N Frequency Synthesizer ADF4106.Data Sheet, 2004.
[4] Chris O'Connor, "Develop Trimless Voltage-Controlled Oscillators," Microwaves & RF, July 1999 pp. 69–78, and January 2000, pp. 94–105.
[5] Guan-Chyun Hsieh, James C.Hung.Phase-Locked Loop Techniques-A Survey.IEEE.Transactions on industrial electronics, 1996, VOL.43, 609–615.
[6] Heung-GyoonRyu&Hyun-SeokLee, Analysis and minimization of Phase Noise of the Digital Hybrid PLL Frequency Synthesizer,IEEE Transactions Consumer EleetronieS, 2002.5,Vol.48(2),PP:304–312.
[7] Hittite Microwave Corporation. BT Digital Phase-Frequency Detector, 10-1300 MHz w/5-Bit Counter, 10-2800MHz HMC440QS16G Data Sheet, 2004.
[8] Ken Holladay, "Design Loop Filters For PLL Frequency Synthesizers," Microwaves & RF, September 1999, pp. 98–104.

Network Security and Communication Engineering – Chan (Ed.)
© *2015 Taylor & Francis Group, London, ISBN: 978-1-138-02821-0*

The symbolic method for time series based on mean and slope

Y. Wang & Y. Su
College of Computer and Communication, Lanzhou University of Technology, Lanzhou, China

ABSTRACT: Neither the algorithm of symbolic aggregate approximation (*sax*) nor the symbolic algorithm for time series data based on statistic feature (*sfvs*) will involve in the shape of the time series, so it cannot effectively represent the similarity of the time series. In this paper, a symbolic method for time series based on mean and slope is introduced to represent the similarity of the time series. It firstly, segments the time series based on key points, then symbolizes the mean and slope separately, records every symbol's occurrence times and position, finally uses every symbol's occurrence times and position as the metrics standard. The experiments show that this method can be used effectively for time series similarity matching, and also improve the correct rate.

KEYWORDS: Time series, Mean, Slope, Symbolization.

1 INTRODUCTION

Data mining is extracting or "mining" useful knowledge from large data. Since the thesis about the time series similarity search was first published by *Agrawal*[1], time series data mining has became a research focus of this technology.

How to extract time-related information and knowledge effectively is the key problem of time series data mining. High dimension is a characteristic of time series, so reducing its dimension is the major problem of the study. Symbolic representation is an effective method of discrete dimension reduction. In recent years, many scholars have proposed some effective methods, such as the algorithm of symbolic aggregate approximation[2] The extended algorithm of symbolic aggregate approximation is based on maximum and minimum values and mean[3] (*ext_sax*), and the symbolic algorithm for time series data is based on statistic feature[4] and so on. *sax* is the most popular and influential method in recent years. But dividing the time series averagely is the biggest flaw of the *sax*, as it just retains the sequence's mean information and does not consider the sequence's shape characteristic. It can't effectively represent the similarity of the sequences.

Aiming at these problems, a new method is introduced in this paper. It regards the mean and shape of a sequence as two important characteristics, and transforms them into symbols, using every symbol's occurrence times and position as the values to calculate the similarity distance.

2 THE BASIC IDEA OF SYMBOLIC ALGORITHM

The continuous numerical curve maps onto a limited symbol set by the discretization method is the basic idea of time series symbolization[5], then the time series is represented by the symbols that come from the symbol set *alphabet* = {*a*, *b*, *c*, ...}.

It is assumed that there are two time series *P* and *Q*, and the length of them are all n. They will be represented by the symbol set *alphabet* = {*a*, *b*, *c*, ...} of size k. According to the data used in this paper, it just needs to record the symbolic sequences when the mean and slope are symbolized. Slope is the angle tangent of two adjacent key points' connection and x-axis' positive direction. By using the method in this paper, it transforms these two sequences into the modes that is expressed by mean and slope like

$$\hat{X} = \{< mean_{symbol1}, slope_{symbol1} >, ..., < mean_{symbolN}, slope_{symbolN} >\}$$

3 RELATED THEORY AND DATA PREPROCESSING

3.1 *The data of standard normal distribution*

Time series follow the characteristic of normal distribution, which requires several intervals of equal area to calculate similarity distance in this paper, so the sequence must be transformed into standard normal distribution at first. The sequence $X = x_1, x_2, ..., x_n$ has the length of n, and it uses the following standardization equation:

$$x_i' = \frac{x_i - \mu}{\sigma} \qquad (1)$$

μ represents the mean of whole sequence, and σ represents the variance of whole sequence. The sequence will be represented by $X' = x_1', x_2', ..., x_n'$ after transforming X into standard normal distribution.

3.2   The selection of key points

When it transforms time series X into standard normal distribution X', traverses X' and finds all extreme value points, then it selects key points in these extreme value points. For a sequence or a segment of sequence, it will have many extreme value points when the curve fluctuation becomes bigger. Although the sequence is transformed into standard normal distribution, it will be still affected by outliers. So it must follow the following two rules when selecting the key points shown in Figure 1.

The first one: The sequence's first and last extreme value points are the key points, and the difference between two key points' values must not be less than the threshold value ε_1.

The second one: The time difference of pre and post key points must be not less than the interval T.

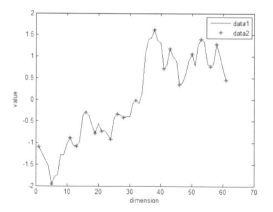

Figure 1.   The closing price of shanghai stock index.

As the example shown in Figure 1, it is the closing price of shanghai stock index in July 2013 to October 2013. In Figure 1, Data 1 represents 61 closing price

which has standardized, data 2 represents the selected key points. It is easy to know from Figure 1 that all the key points come from the extreme value points, but not all the extreme value points are the key points. The value of T is 2 in this case.

3.3   The equivalence of key points

The so-called equivalence is to compare the key points' time of two sequences, if one sequence's some time of key points does not appear in the other sequence's key points, it puts this time and the sequence's corresponding data of this time as a key point and adds it to the other sequence's key points, until the corresponding time of two sequences' key points are identical.

It is necessary to equal the key points' time of two sequences, because the method in this paper is based on every symbol's occurrence times and position after symbolizing every segment's mean and slope, and it will use deformation Euclidean equation to calculate the similarity distance. It may be a big difference between two unknown similarity sequences' key points. So it will be a big difference between the same symbol's occurrence times and position of two sequences after symbolizing every segment's mean and slope based on key points, and it may also be similar among parts of them, it will easily misjudge when making similar judgment.

3.4   The breakpoints and data discretization

The transformed time series follow the standard normal distribution, so the area is 1, which is surrounded by the standard normal distribution curve and the coordinate axis. It needs to set some breakpoints in order to divide the area into k equal parts. The breakpoints are the ordered list of value $Pt = \{pt_1, pt_2, ..., pt_{k-1}\}$, and it makes the area be $1/k$, which is surrounded by the standard normal distribution curve from pt_i to pt_{i+1} and the coordinate axis. So the core problems of time series symbolic are the selection of breakpoints and data discretization. The selection of breakpoints is based on k, which is the size of symbol set $alphabet = \{a, b, c, ...\}$. Then it finds the breakpoints in the standard normal distribution table after selecting value k.

The breakpoints are -0.67, 0, and 0.67 in the case of $k = 4$, these three breakpoints divide the area into 4 equal parts. So symbol a represents the mean (or slope) if $s_{mean} <= -0.67$, symbol b represents the mean (or slope) if $-0.67 < s_{mean} <= 0$, and the rest can be done in the same manner. An example is shown in Figure 2, 61, and the closing price of shanghai stock index is transformed into $aaaabaabbbddddcdddd$.

256

Table 1. The selection of breakpoints.

size	pt_1	pt_2	pt_3	pt_4	pt_5	pt_6	pt_7	pt_8	pt_9
2	[0]								
3	[-0.43]	[0.43]							
4	[-0.67]	[0]	[0.67]						
5	[-0.84]	[-0.25]	[0.25]	[0.84]					
6	[-0.97]	[-0.43]	[0]	[0.43]	[0.97]				
7	[-1.07]	[-0.57]	[-0.18]	[0.18]	[0.57]	[1.07]			
8	[-1.15]	[-0.67]	[0.32]	[0]	[-0.32]	[0.67]	[1.15]		
9	[-1.22]	[-0.76]	[0.43]	[-0.14]	[0.14]	[0.43]	[0.76]	[1.22]	
10	[-1.28]	[-0.84]	[-0.52]	[-0.25]	[0]	[0.25]	[0.52]	[0.84]	[1.28]

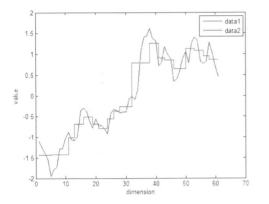

Figure 2. The symbolic about closing price of shanghai stock index.

It should be noted that the range of slope about the same segment may be different if T gets different values. The sequence is standardized, so the values of each point are mainly in $[-3,3]$. The mean and slope must be standardized in the same standard, therefore the range of slope also must be mainly in $[-3,3]$. For example, if T is 4, the range of slope will concentrate in $[-1.5,1.5]$, so it must transform the range into the interval $[-3,3]$, when transforming the range, it's just to multiply every slope by $T/2$.

3.5 The calculation about similarity distance

We calculate the similarity distance between these two sequences after transforming time series P and Q into the mode like $\hat{X}$. It is divided into two steps to calculate the similarity distance, the first step is to calculate the difference of occurrence times between the same symbol of two symbolized sequences, and the second step is to calculate the difference of position between the same symbols of two sequences. It is assumed that $\hat{P}$ and $\hat{Q}$ are the transformed sequences, and the calculation equation of times difference between the same symbol is shown as Equation 2.

$$d_1 = \sum_{i=1}^{k}[(N_1(i)-N_3(i))^2 + (N_2(i)-N_4(i))^2] \qquad (2)$$

k is the size of symbol set, $N_1(i)$ and $N_2(i)$ record the occurrence times of each symbol that represents mean and slope in the mode sequence $\hat{P}$, $N_3(i)$ and $N_4(i)$ record the occurrence times of each symbol that represents mean and slope in the mode sequence $\hat{Q}$.

If d_1 describes overall condition of each symbol's occurrence in the sequences, d_2 can describe the detail difference between the same symbol of two sequences. The calculation equation of position difference between the same symbols is shown as Equation 3.

$$d_2 = \sum_{i=1}^{k}[\sum_{j=1}^{l_1}(A[i,j]-C[i,j])^2 + \sum_{j=1}^{l_2}(B[i,j]-D[i,j])^2] \qquad (3)$$

k is still the size of symbol set, $A[i,j]$ and $B[i,j]$ records the occurrence position of each symbol that represents mean and slope in the mode sequence $\hat{P}$, $C[i,j]$, and $D[i,j]$ record the occurrence position of each symbol that represents mean and slope in the mode sequence $\hat{Q}$. It must select the minimum occurrence times of each corresponding symbol for l_1 and l_2 in Equation 4, which is shown as the following:

$$\begin{cases} l_1 = \min(N_1(i),N_3(i)) \\ l_2 = \min(N_2(i),N_4(i)) \end{cases} \qquad (4)$$

There are two reasons about the selection of the minimum occurrence: One reason is that the difference of occurrence times about the same symbol has been considered in d_1. The other reason is that it will have an error when it exceeds the less occurrence times if the maximum occurrence is selected. So the final similarity distance is shown as the following, N is the length of symbolized sequence in Equation 5.

$$dist(P,Q) = \frac{\sqrt{d_1 + d_2}}{\sqrt{2}N} \qquad (5)$$

257

Through the above theoretical introduction and analysis, the algorithm steps to calculate time series sequence similarity distance using this paper's method are shown as the following:

Step 1: Transforming the query sequence and the waiting detection sequence into the standard normal distribution according to Equation 1 and recording the standardized sequence.

Step 2: Traversing the standardized sequences, finding all extreme value points of them, selecting the key points from extreme value points according to the two rules, then equaling the key points' time of two sequences and letting the key points' time of two sequences identically.

Step 3: Segmenting the sequences according to the key points, calculating the mean and slope of each segment, and transforming the slope range into the same range of the mean, then symbolizing the mean and slope according to Table 1 and recording every symbol's occurrence times and position of each sequence.

Step 4: Calculating the similarity distance according to Equation 5, and recording the distance $dist$.

Step 5: Comparing with the similarity threshold ε_2 after getting the final distance $dist$, it is similar if the $dist$ is less than ε_2. On the other hand, it is not similar and going to the next cycle.

4 EXPERIMENT SIMULATION

The experimental simulation environment: *Windows 7, Inter(r) Core(tm)2 Duo cpu T5870 2.00ghz, Memory 2.00gb, MatlabR2010b.*

4.1 *The data of experiment*

This paper selects the dataset synthetic control chart time series of *uci kdd* achievement, this dataset is about 600 time series sequences, the total of categories is 6, and each category has 100 time series sequences, the length of each sequence is 60.

4.2 *The comparison of similarity measure*

In the experiment, it selects 10 sequences randomly in each of the 6 categories that come from *synthetic control chart time series*. Then it calculates the similarity distances from the first sequence with all the sequences respectively and counts the results about the correct rate and false reject rate. The correct rate is the ratio of the number of original similar sequences which are still judged to be similar by the similarity measure to the total number of original similar sequences. The false reject rate is the ratio of the number of original dissimilar sequences which are misjudged to be similar by the similarity measure to the total number of original dissimilar sequences. The experiment in this paper is repeated 5 times, and the final data is the mean of 5 times.

Experiment 1: This experiment is divided into two parts; first it compares the traditional method of *sax* with the method in this paper. Then it compares the improved method of *kp-sax* [6] with the method in this paper and the k value is 4. Table 2 and3 are the comparison of results.

It is easy to know from the results of Table 2 and 3 that the correct rate of the method in this paper is slightly higher than the traditional method of *sax* and the improved method of *kp-sax*, and the false reject rate is obviously lower than these two methods. Both the traditional method of *sax* and the improved method of *kp-sax* are not considering the shape of the time series. Although the *kp-sax* method is improved, it is based on average division of *sax* method and uses the key points that fall into every average segment to divide the segment again, and the expression of symbolic mean is closer to the original sequence at this time, but it still only uses mean. So their false reject rate has been high. This paper's

Table 2. The comparison of the traditional method of sax.

Dataset	The length of dataset	Correct rate(%)		False reject rate(%)	
		This paper	*sax*	This paper	*sax*
Synthetic control	[600]	[83.75]	[83.6]	[15.64]	[46.7]

Table 3. The comparison of the improved method of kp-sax.

k	Dataset	The length of dataset	Correct rate(%)		False reject rate(%)	
			This paper	*kp-sax*	This paper	*kp-sax*
4	Synthetic control	[600]	[77.17]	[69]	[18]	[32]

method not only considers the mean of sequence, but also adds the shape of sequence as the part of the similarity distance, so it greatly avoids mistaking the sequences which have same mean but different shape to be similar. The *sax* and *kp-sax* method only use the mean to judge, it may misjudge some original similar sequences to be dissimilar because of the mean deviation of a few outliers. This paper can ensure the correct rate and obviously reduce the false reject rate.

Experiment 2: This experiment only uses the method in this paper and compares the corresponding data under the different symbol set size k. The results are shown in Table 4.

This experiment just uses the method in this paper, and we can know the results of similarity search from the Table 4, when the symbol set size k is greater, the correct rate is higher and the false reject rate is lower. The breakpoints are taken more when the value k is greater, and the symbols that describe the mean and slope are more. Thus the symbolized mode sequences are more detailed and closer to the original sequences, so it has a smaller error when being compared similarly. However the value of k should not be too large, it is usually in range $k \in [2,10]$, and the range $k \in [3,6]$ is always used.

Table 4. The comparison of different k values.

k	Dataset	The length of dataset	Correct rate(%)	False reject rate(%)
3	Synthetic control	[600]	[76.7]	[19.33]
4	Synthetic control	[600]	[77.17]	[18]
5	Synthetic control	[600]	[82]	[12.7]
6	Synthetic control	[600]	[88.33]	[12.5]

5 SUMMARY

A symbolic method for time series based on mean and slope is introduced to represent the similarity of the time series in this paper, it mainly uses every symbol's occurrence times and position of symbolized mean and slope to measure similarity distance. The method starts from two features of the mean and slope of time series, it can reduce the dimension of sequence effectively and also can keep the main information and features when reducing dimension, and it also solves the problem that loses too much information in the traditional method of *sax*. It can be seen from the experiment results, that this paper provides a better way than *sax* method in the time series similarity analysis, and the experiments have proved the effectiveness of this method.

ACKNOWLEDGMENT

This research was financially funded by the National Natural Science Foundation of China (*No: 61263019*), the Natural Science Foundation of Gansu province (*No: 1112RJZA029*) and the Doctoral Foundation of *LUT*.

REFERENCES

[1] Agrawal R, Faloutsos C, Swami A. Efficient Similarity Search in Sequence Database[C]// Proc.Of the 4th International Conference on Foundations of Data Organization and Algorithms. Chicago, USA: [s.n.], 1993.

[2] Lin J, Keogh E, Lonardi S, et al. A Symbolic Representation of Time Series with Implications for Streaming Algorithms // Proc of the 8th ACM SIGMOD Workshop on Research Issues in Data Mining and Knowledge Discovery. San Diego, USA, 2003: 2–11.

[3] Battuguldur L, Yu S, Kyoji K. Extended SAX: Extension of Symbolic Aggregate Approximation for Financial Time Series Data Representation[EB/OL]. [2010-10-18]. http://citeseerx.ist.psu.edu /viewdoc/download? doi-10.1.1.149.9325&rep-repl&type=pdf.

[4] Zhong Qingliu, Cai Zixing. The Symbolic Algorithm for Time Series Data Based on Statistic Feature. Chinese Journal of Computers, 2008, 31(10): 1857–1864.(in Chinese)

[5] An Hongqiang, Niu Qiang. Symbolic Representation Algorithm for Time Series Based on Sliding Window and Local Features. Application Research of Computers, 2013, 30(3): 796–798.

[6] Yan Qiuyan, Meng Fanrong. A Key Point Based SAX Improving Algorithm. Journal of Computer Research and Development, 2009, 46(Suppl.): 483–490.

Network Security and Communication Engineering – Chan (Ed.)
© *2015 Taylor & Francis Group, London, ISBN: 978-1-138-02821-0*

Coefficient of variations on SDP matrix – a new feature for discriminating between explosion and earthquake

H.M. Huang, Y. Tian & C.J. Zhao
College of Computer Science and Information Engineering, Guangxi Normal University, Guilin, China

X.H. Shi
College of Physics Science and Technology, Guangxi Normal University, Guilin, China

ABSTRACT: This paper proposes an algorithm to calculate a Coefficient of Variations (CV) on SDP matrix. The CV of each seismic event can be derived from its SDP matrix, which is an array of polar coordinate graphs being clamped and projected from all effectual seismic signals measured in all available observatory stations of one event. The Coefficients of Variations (CVs) for 35 earthquake events and 27 explosion events have been calculated. Another discriminative feature – corner frequency is also calculated for the 62 seismic events. Experiments have been carried out by simply threshold dichotomy on the two features – CV and corner frequency. The results confirm that CV is an excellent discriminating feature for recognizing seismic source type.

KEYWORDS: Coefficient of Variations (CV), Seismic Source Type, Visualization, Symmetric Dot Pattern (SDP), SDP matrix.

1 INTRODUCTION

Seismic signals may be originated from natural earthquake events or non-natural earthquake events. Natural earthquakes are commonly tectonic earthquakes, and often occur in the depths of more than 10km underground[1], generally the depths of granite layers or basalt layers[2]. Non-natural earthquake events include chemical blasts or nuclear detonations, and also mine collapses, factory mechanical shakings and plane clashes etc. If seismic wave signals are appropriately processed or properly transformed, explosion events can be effectively distinguished from earthquake events with high probability [3].

This paper makes an attempt to calculate the coefficient of variations (CV) on SDP matrix for distinguishing earthquake and explosion events. SDP is a polar coordinate graph being proportionally projected from one single temporal seismic signal with some suitable algorithm. The CV value of each seismic event can be derived from its SDP matrix, which is an array of polar coordinate graphs being clamped and projected from all effectual seismic signals measured in all available observatory stations of one event.

2 SYMMETRIC DOT PATTERN (SDP)

By referring the visualizing principles of Cymascope[4], a visualization scheme – symmetric dot pattern (SDP) – is proposed here. The SDP algorithm is schematically listed in Fig.1.

1 Start - reading out seismic wave data. The seismic wave data is read out by the "Decision Support System[5] for Classification and Recognition of Earthquakes and Explosions". The read out seismic wave data can be expressed as:

$$S(i)|_{i=0,1,...,N-1} = | h_i | \qquad (1)$$

where N is total sample count, h_i is the recorded i-th sample value in the designated channel, $|\cdot|$ is a absolute value operator.

2 Wave normalizing - temporal wave normalization. For a digital seismic signal $S(i)|_{i=0,1,...,N-1}$, the corresponding normalized wave data is $S'(i)|_{i=0,1,...,N-1}$:

$$S'(i)|_{i=0,1,...,N-1} = \frac{(Max(S')-Min(S'))\times(S(i)-Min(S))}{(Max(S)-Min(S))} + Min(S') \quad (2)$$

where $Max(S)$ and $Min(S)$ are maximum and minimum of signal $S(i)|_{i=0,1,...,N-1}$; and $Max(S')$ and $Min(S')$ are maximum and minimum of normalized seismic signal $S'(i)|_{i=0,1,...,N-1}$, which are often designated as $Max(S') = 1$ and $Min(S') = 0$ or $Max(S') = 1$ and $Min(S') = -1$, and other values are also possible when it is necessary.

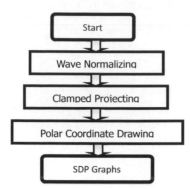

Figure 1.   Flow-chart of the SDP algorithm.

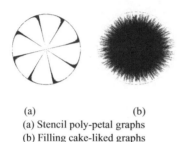

(a)   (b)
(a) Stencil poly-petal graphs
(b) Filling cake-liked graphs

Figure 2.   Two typical SDP graphs.

3 Clamped projecting. The normalized signal $S'(i)|_{i=0,1,...,N-1}$ are processed as the following with an adjustable clamped value *Amp*:

$$Amp = (Max(S') - Min(S')) \times A \ (0 \leq A \leq 1) \quad (3)$$

$$Y(i) = |Amp - S'(i)|, \ i = 1,2,...,N-1 \quad (4)$$

where $S'(i), \ i = 1,2,...,N-1$, are normalized wave data calculated as the above formula (2); *Max(S')* and *Min(S')* are maximum and minimum of $S(i)|_{i=0,1,...,N-1}$; and the *A* is an adjustable weight value between 0 and 1, here *A = 0.5*, by adjusting the value of *A*.

4 Polar coordinate drawing. The SDP graph–symmetrical dot pattern graph is a circularly integrated graph which is actually evenly distributed *K* copies of one polar coordinate graph.

$$\theta(i,k)|_{i=0,1,\cdots,N-1} = \hat{\theta} \times i + \frac{2\pi}{K} \times k, \qquad k=0,1,\cdots,K \quad (5)$$

$$\rho(i)|_{i=0,1,\cdots,N-1} = Y(i)|_{i=1,2,\cdots,N-1} \quad (6)$$

where $\hat{\theta}$ is angular difference for two successive sample points: $\hat{\theta} = 2\pi/(N \cdot K)$; *K* is the petal number, piece number or copies number of the configuration of one SDP graph, in here, the petal number *K = 8*; *N* is the total sample count.

Two typical SDP graphs are listed in Fig.2.

3   THE SDP MATRICES OF EARTHQUAKE AND EXPLOSION EVENTS

Seismic wave data are provided by China Earthquake Data Center (http://data.earthquake.cn/). The 35 earthquake events occurred in 2003~2007, with epicenters locating in neighbors of metropolitan Beijing (at about longitude 116°E, latitude 40°N) and 27 explosion events occurred in 2002~2008, with epicenters also locating in neighbors of metropolitan Beijing (at about longitude 115°E, latitude 40°N) which are selected. Three earthquakes and three explosions are exemplified in Table 1.

Table 1.   Fundamental information of some earthquake and explosion events.

Event Type	Date	Latitude	Longitude	Magnitude
Earthquake	2003-06-19	40°06.95′	116°18.83′	ML1.6
	2005-03-01	40°02.48′	115°57.93′	ML2.3
	2007-05-05	40°32.54′	115°78.53′	ML2.4
Explosion	2003-06-28	40°25.84′	115°28.89′	ML1.1
	2005-03-12	40°31.45′	115°29.61′	ML1.2
	2007-06-11	40°30.05′	115°25.83′	ML1.3

If discriminating earthquake and explosion by directly inspecting SDP matrices, when petal number (*n*) is 8 and weight value (*A*) is 0.5, the distinction of earthquake and explosion is intuitively quite obvious. For a pair of earthquake and explosion events in Table.2, 10 stations seismic wave data are randomly selected and their SDP graphs matrices are drawn in Fig. 3. By this Fig, it is very clear that earthquake events show much more diverseness than explosion events in SDP graphs matrices. Fig.4 shows more graphs – SDP matrices of 35 SPD graphs, whose seismic wave signals are randomly selected from 35 stations of earthquake and explosion events in Table 1.

(a) SDP matrix of earthquake (b) SDP matrix of explosion

(2005-03-01)   (2005-03-12)

Figure 3. Comparison of 10 stations SDP matrices of earthquake and explosion.

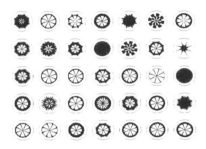

(a) SDP matrix of earthquake (2007-05-05)

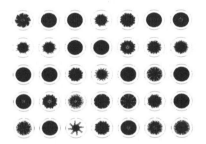

(b) SDP matrix of explosion (2007-06-11)

Figure 4. Comparison of 35 Stations SDP Matrices of Earthquake and Explosion.

It has been experimented for more stations seismic signal data of earthquake or explosion events. The diverseness property is consistent: the graph of SDP matrix of earthquake is more diverse and complex than that of explosion.

4 COEFFICIENT OF VARIATIONS (CV) BEING CALCULATED FROM THE GRAPHS OF SDP MATRIX

The coefficient of variations is a normalized measure of dispersion of a probability distribution or frequency distribution. It is defined as the ratio of the standard deviation σ to the mean μ[6].

The steps to calculate CV are listed as the following:

1 Calculating the set of degree of filling for an event from its SDP matrix. For an event with n graphs in its SDP matrix, the set of degree of filling is :

$$R = \left\{ R_k \big|_{k=1,2,\dots,n} \right\}, \qquad R_k = S_k / N (k=1,2,\dots\dots,n) \quad (7)$$

where S_k is the total number of filling pixels in the k-th 2D binary graph of SDP matrix: $S_k \big|_{k=1,2,\dots,n} = \sum_{i}^{L} \sum_{j=1}^{M} I(i,j)$, and $I(i, j)$ is the i-th row and j-th column pixel value ($(I(i,j) = 0 \text{ or } 1$) of the binary graph which being transformed from original SDP graph. All the graphs of SDP matrix for an event are the same size with L horizontal pixels and M vertical pixels. The total pixel number of the graphs is same: N ($N = L*M$).

2 Calculating CV value for an event from the set of degree of filling.

$$CV = \sigma / \mu \qquad (8)$$

where $\sigma = \sqrt{\dfrac{1}{n-1} \sum_{k=1}^{n} (R_k - \mu)^2}$, and $\mu = \dfrac{1}{n} \sum_{k=1}^{n} R_k$

The CV values of the 62 events (35 earthquakes and 27 explosions) have been calculated and are expressed as horizontal axis in Fig.5. The vertical axis is the corner frequency [7] for each of these events.

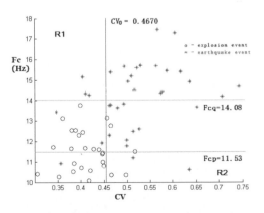

Figure 5. Distribution map of 62 Events by CV and corner frequency.
(horizontal axis: CV, coefficient of variations; vertical axis: Fc, corner frequency, unit: Hz)

5 CONCLUSION

It is easy to see that earthquake events are more diverse and have larger CV values than explosion events. If we discriminate the 35 earthquake events and 27 explosion events only by the CV value on threshold $CV_0 = 0.4670$, with the discriminating rule: when $CV < CV_0$, an event belongs to explosion, otherwise, belongs to earthquake, 6 earthquake events and 5 explosion events will be misclassified. The total correct recognition rate is $51/62 \approx 82.3\%$.

The averages of corner frequencies for 35 earthquakes and 27 explosions are $Fcq = 14.08Hz$ and $Fcp = 11.53Hz$ respectively. If the explosion region **R1** ($CV < CV_0$) and earthquake region **R2** ($CV \geq CV_0$) which are delimited by CV value are further refined by Fcq and Fcp, the discriminating rule can be updated as: when $((CV < CV_0) \cap (Fc < Fcq) \cup ((CV \geq CV_0) \cap (Fc < Fcp)$), an event belongs to explosion, otherwise, belongs to earthquake. Under the new discriminating rule, 5 earthquake events and 2 explosion events are misclassified, the total correct recognition rate is $55/62 \approx 88.7\%$.

The result suggests that the CV which is extracted from graphs of SDP matrix surely provides a novel and feasible feature to discriminate earthquake and explosion. CV can be regarded as a new feature for discriminating seismic source types and should be added to the arsenal of seismology for seismic source recognition. Further investigations with more seismic data are deserved for validating this new feature.

ACKNOWLEDGMENT

This research was financially supported by the National Science Foundation of China on contract No. 41264001.

REFERENCES

[1] P. Shen, Zh. Zh. Zheng. Application of Transient Spectrum to Discrimination of Nuclear explosions and Earthquakes[J](In Chinese). Chinese Journal of Geophysics, 42(2), 233–240. (1999).

[2] Sh. R. Li. Distinguishing Earthquakes and Explosions from Digital Seismic Recording Data[J] (In Chinese). Earthquake Research in Plateau, 12(3), 39–43.(2000).

[3] A. X. Wu. On Quantitative Identification of Explosion Earthquake Based on Cepstrum Computation of HHT and Statistical Simulation of Sub-Cluster[C]. Proceedings of the 31st Chinese Control Conference, July 25-27,2012,Hefei,China,5311–5316. (2012).

[4] Information on: http://www.cymascopc.com/

[5] H. M. Huang, H. J. Zhou, Y. J. Bian. The Development of Decision Support System for Identifying Explosion Events by Seismic Waves[J]. JDCTA: International Journal of Digital Content Technology and its Applications, 5(4), 107–112. (2011).

[6] Information on: http://en.wikipedia.org/wiki/Coefficient_of_variation.

[7] P. Zhang, F. Sh. Wei, K. Pan, Y. J. Bian and X. Q. Jiang. Comparison of Corner Frequency between Explosion and Earthquake[J]. Seismological and Geomagnetic Observation and Research, 30(3), 20–25. (2009).

[8] T. K. Hong. Seismic discrimination of the 2009 North Korean nuclear explosion based on regional source spectra[J]. J Seismol, 17,753–769. (2013).

[9] H. M. Huang, R. Li, Sh. J. Lu. Discrimination of Earthquakes and Explosions Using Chirp-Z TransformSpectrum Features[C]. 2009 World Congress on Computer Science and Information Engineering, 210–214. (2009).

Network Security and Communication Engineering – Chan (Ed.)
© 2015 Taylor & Francis Group, London, ISBN: 978-1-138-02821-0

Improved SURF algorithm based on proportional distance and angle for static sign language recognition

Y. Wang , M.Z. Yang & Y. Luo
Key Laboratory of Optical Fiber Communication Technology, Chongqing Education Commission, Chongqing University of Posts and Telecommunications, Chongqing, P.R. China

ABSTRACT: For the instabilities of SURF (Speeded-Up Robust Features) feature points and unconsidered matching cases of scale variations when using the method of judging interest points by distance characteristic, an improved eliminated false-matched SURF algorithm for static sign language recognition is proposed. A fusion gesture segmentation method of combining color image and depth image was used. Then the improved algorithm adopted proportional distance characteristic discrimination is combined with angle characteristic criterion to expurgate the false matched points synthetically. Experiments show that the fusion gesture segmentation method improves the accuracy of segmentation and its robustness against illumination and environment changes. Meanwhile, it solves influences on interference factors such as light and skin color objects. In addition, the improved SURF algorithm based on proportional distance and angle can solve the effects on factor of scale variations effectively and has relatively strong robustness against illumination, complex background, and angular variation.

KEYWORDS: Improved SURF, Distance discrimination, Angle discrimination, Proportional distance.

1   INTRODUCTION

In recent years, with the continuous development of machine vision technology, sign language recognition based on vision is becoming a research hotspot issue. Anant Agarwal, Manish K Thakur [1], and Quan Yang [2] all used Kinect sensor to access the depth image for sign language recognition. Recent years, Chongqing information accessibility and service robot engineering technology search center has applied SURF characteristics to identify the static alphabet and obtained good recognition results [3]. Zhiqiang Wei used the method which Euclidean distance is performed to get exact matching to avoid wrong matching [4].

However, a large amount of image matching researches based on SURF algorithm has exposed that the accuracy of image feature points matching is not high enough. Low feature points detection accuracy because of instability of SURF feature points lead to mistakes. In addition, false matched point pairs have been formed in early matching process. And gesture segmentation results based on depth image have unitary pixel value distribution, so it is much easier to detect incorrect interest points particularly in cases of angle and scale variations. By far, some methods are judging interest points by distance without considering matching of scale variations. In this paper an improved eliminated SURF algorithm combining depth image for static sign language recognition is proposed.

2   THE FUSION OF COLOR AND DEPTH GESTURE SEGMENTATION

Gesture segmentation results directly affect the subsequent feature extraction and recognition work. According to the distance between hand and the camera, the segmentation method based on depth image can recognize the gesture area from the background by using grayscale histogram with an appropriate threshold to carry on binary processing. Because the distance is insensitive to illumination and environment changes, this method has robustness against these cases. When only the depth image is processed, the wrist or the elbow will be mistakenly segmented for hand. However, when only color image is processed, hand gesture segmentation is susceptible to interference factors such as light and skin color objects. A fusion gesture segmentation method of the histogram and Gaussian skin color model, which respectively processes the depth images and color images from Kinect sensor output, improves the accuracy of segmentation. Figure 1 shows segmentation results using segmentation method based on Gaussian skin color model, depth image, and the fusion method respectively under different conditions.

(a) bright light conditions   (b) weak light conditions

(c) complicated background conditions

Figure 1.   Segmentation results based on different gesture segmentation methods.

3   IMPROVED SURF ALGORITHM BASED ON PROPORTIONAL DISTANCE AND ANGLE

SURF is a kind of fast robust features matching algorithm proposed by Herbert Bay in 2006 [5]. Improved SURF is shown in Figure 2. After the process of interest point detection and description, we can get the rough primary matched point pairs. Proportional distance characteristic which is adopted to improve the accuracy of eliminating false matched points is combined with angle characteristic to expurgate the false matched points synthetically. If these two characteristic parameters of the input image are equal to those of template image respectively, we conclude that the feature point pairs detected before are matched. Otherwise, the unmatched ones would not be taken into consideration.

4   FEATURE POINTS DETECTION AND DESCRIPTION

Given that the pixel of input image $I(\mathbf{X})$ is described as $\mathbf{X} = (x, y)^{\mathrm{T}}$, SURF improves the speed of interest point detection by image convolution filtering. A gradually increasing box-filter [6] which is approximate with Gaussian filtering was introduced to build scale space of integral image based on Hessian matrix. The determinant of the approximate of Hessian matrix can be obtained as follows:

$$\mathrm{D}_{SURF} = \det\left(\mathbf{Happrox}\right) = D_{xx}D_{yy} - \left(0.9D_{xy}\right)^2 \qquad (1)$$

To ensure the existence of centre pixel [7], the pixel difference value between two neighbor scale-spaces should be set as an even number. D_{SURF} is used to find extreme points in each scaled image. Then these points are compared with 26 neighboring ones in scale-space, and we'll obtain local maxima which is kept as an interest point. Then a Non-Maximum Suppression is carried on in 3-Dimensional neighborhood to acquire the location information x, y. and scale σ.

The Haar wavelet response of each feature point found in the x and y directions are calculated. Then the response values are weighted with Gaussian coefficients in order to make the contribution of the response close to the center point greater while the distal ones smaller. We determine the direction of the longest vector is equivalent to the direction of the feature point. At last, we can get a descriptor $\mathbf{V}$ which forms 64-dimensional vectors and it should be dealt with normalization so that it has robustness against light and scale.

5   FEATURE POINTS MATCH AND REFINEMENT

5.1   *Early match of feature points*

See Figure 2 for fast indexing during the matching stage. Through calculating and comparing the

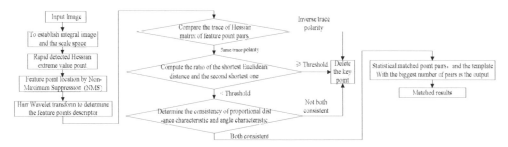

Figure 2.   Improved SURF algorithm based on proportional distance and angle.

Euclidean distances as well as the polarities of the traces in order to judge the similarity of two descriptors, we can determine whether the two descriptors match or not. Firstly, if they do not have the same type of contrast, this feature point is treated as a false one and it will not be taken into consideration. Conversely, if they have the same type of contrast, we go on calculating the Euclidean distance of them respectively according to:

$$D(i,j)=\left[\sum_{i=1}^{n}\left(D_{ik}-D_{jk}\right)^{2}\right]^{\frac{1}{2}} \qquad (2)$$

For each feature point, compute the ratio of the shortest Euclidean distance and the second shortest one. Next, the ratio is compared with the default threshold value (usually 0.7 in experiment). If the ratio is in the range of 0.7, we think that this point matches with the nearest point. Otherwise, it would be regarded as a wrong one. SURF algorithm can exclude some false feature points to some extent. However we still gain some false matched pairs. Therefore, other methods should be taken into consideration to improve the matching accuracy.

5.2 Refinement of feature point and match

In the previous chapter, we have got the early matched point pairs. Each feature point is put in small to big order according to the Euclidean distance of its descriptor. Then we will get two point assemblages $P=\{P_1,P_2,\cdots,P_n\}$ for input image I and $P'=\{P_1',P_2',\cdots,P_n'\}$ for template image I' where P_i and P_i' are regarded as the matched pairs. Since each feature point matches with the nearest interest point which has the shortest distance, we believe that the nearest descriptors are more matched and they cannot be the false matched pairs. Assume the first two point pairs named $P_1(P_1')$ and $P_2(P_2')$ are the reference point, and all the distances between P_1 and P_i as well as those between P_1' and P_i' are counted.

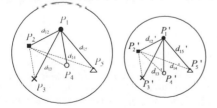

Figure 3. Distance and angle between corresponding feature points.

As Figure 3 shows, these distances are set as $d_1=\{d_{12},d_{13}\cdots d_{1n}\}$ and $d_1'=\{d_{12}',d_{13}',\cdots d_{1n}'\}$ severally. There is a problem with the method of

comparing the Euclidean distance value directly in the case of scale changing. Hence, a proportional distance method is applied. d_1 and d_1' are set as the standard. Meanwhile the ratios of d_1 and d_{1i} are recorded as R_{1i} according to Equation (3). In addition, do the same operation on d_1' and d_{1i}' and we can also get R_{1i}'. Assume the angle of P_iP_1 and P_iP_2 is α_i as well as the angle of $\overline{P_i'P_1'}$ and $\overline{P_i'P_2'}$ is α_i'. Then the difference value between R_{1i} and R_{1i}' and the difference between α_i and α_i' according to (4) are achieved. In case that the difference absolute value is in the range of θ and R_{1i} and R_{1i}' would be considered to be the same. It is the same with α_i and α_i'.

$$R_{1i}\left(R_{1i}'\right)=\frac{d_1\left(d_1'\right)}{d_i\left(d_i'\right)} \qquad (i=2,3\cdots n) \qquad (3)$$

$$\begin{cases} \left|R_{1i}-R_{1i}'\right|<\theta & (i=2,3\cdots n) \\ \left|\alpha_i\left(\angle P_1P_iP_2\right)-\alpha_i'\left(\angle P_1'P_i'P_2'\right)\right|<\varepsilon & (i=2,3\cdots n) \end{cases} \qquad (4)$$

There is a proposed criteria that if R_{1i} is equal to R_{1i}' and α_i is also equal to α_i', P_i is matched with P_i', or they are not. These unmatched point pairs should be eliminated. On basis of this, we have got precise matched feature point pairs and the number of matched pairs of input image with all the templates are counted. We set the template with the biggest number as the matched result. The proportional distances can eliminate effects on factor of scale variations, and corresponding proportional distances of input image and template image are compared with each other so as to judge the feature points.

6 EXPERIMENTAL RESULTS AND ANALYSIS

The experiments are realized on the base of Kinect sensor and a computer. The programming environment is Visual C++2008 includes OpenCV and OpenNI. The improved SURF can exclude the false feature point pairs effectively. Under normal lighting conditions, Figure 4 shows the results of interest point extracted under different scales (0.5 times and 0.7 times). The left picture is extraction result of template, the middle one is based on SURF, and the right one is based on improved SURF proposed by this paper. Comparing the following two images, we can find some false feature points detected under scale changes are excluded effectively using improved algorithm.

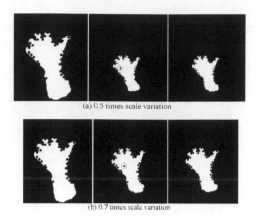

(a) 0.5 times scale variation

(b) 0.7 times scale variation

Figure 4. Detection results of the feature points extracted under different scales.

In order to verify the robustness of the improved algorithm against the changes of illumination, background, angle, and scale, letter "V" is taken as an example, and experiments are conducted in the cases of strong and weak illumination, complex background, angle, and scale variations. As shown in Figure 5, they are RGB image, depth image, and result image respectively in 5 cases above. The experiment results show that the method proposed can accurately acquire good recognition results in different environmental cases. As is written before, fusion method of

hand gesture segmentation is adopted to overcome the influences on the changes of illumination and complex background. In addition, feature extraction based on SURF has strong robustness against angle and scale variations.

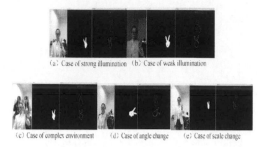

(a) Case of strong illumination   (b) Case of weak illumination

(c) Case of complex environment   (d) Case of angle change   (e) Case of scale change

Figure 5. Experimental results of recognition in 5 cases.

26 letters in manual alphabet are taken as test objects, and 85 experiments are conducted for each letter under the circumstance of normal illumination and scale. As shown in Table 1, the result 1 is based on SURF algorithm and result 2 is based on improved algorithm. We can see the average recognition rate based on SURF is only 94.7%, and the average recognition rate based on the improved algorithm can achieve 97.2%. Hence, there is a higher recognition rate using improved algorithm proposed by this paper.

Table 1. Statistics of recognition rate.

Case	Result 1	Result 2	Case	Result 1	Result 2	Case	Result 1	Result 2
A	95.5%	97.4%	B	94.0%	96.3%	C	93.5%	97.3%
D	94.3%	98.1%	E	96.1%	98.1%	F	96.0%	98.2%
G	96.5%	97.0%	H	91.0%	96.8%	I	94.3%	96.5%
J	93.6%	98.3%	K	93.7%	97.3%	L	93.4%	95.9%
M	88.9%	95.3%	N	94.6%	95.8%	O	94.4%	96.2%
P	96.9%	97.9%	Q	93.5%	95.5%	R	95.5%	98.6%
S	96.3%	97.3%	T	96.3%	96.4%	U	93.5%	97.7%
V	96.2%	98.2%	W	96.4%	98.4%	X	93.6%	95.1%
Y	97.5%	98.7%	Z	96.0%	98.5%			

7 CONCLUSIONS

This article uses fusion hand gesture segmentation based on the histogram to process depth image and Gaussian skin color model to process color image. The improved SURF algorithm proposed in this paper uses two matching characteristic parameters namely distance and angle between interest points to judge

the correctness of feature points synthetically. The proportional distances can eliminate effects on factor of scale variations, and solve the problem of match of false point pairs under the case of scale variation effectively. A large number of experiments show that this method can overcome the influence of illumination, complex background, angle and scale variations and improve the recognition rate.

ACKNOWLEDGMENT

This research was financially supported by Chongqing Municipal Education Commission of Science and Technology Research Project. (Item No.: KJ120519)

REFERENCES

[1] Agarwal A., Thakur M.K., Sign language recognition using Microsoft Kinect, in Contemporary Computing (IC3), 2013 Sixth International Conference on. IEEE, 2013, pp 181–185.

[2] YANG Quan PENG Jinye. Sign language recognition algorithm based on depth image information, J. Journal of Computer Applications. 33 (2013) 2882–2885.

[3] Hu Zhangfang, Yang Lin, Luo Yuan, et al. A novel static sign language letter recognition method based on improved SURF algorithm, J. Journal of Chongqing University of Posts and Telecommunications(Natural Science Edition)ISTIC. 25 (2013) 554–548.

[4] WEI Zhiqiang, HUANG Lei, JI Xiaopeng, Research on Sequence Image Matching Based on Point Feature, J. Journal of Image and Graphics, 3 (2009) 525–530.

[5] Baya H, Essa A, Tuytelaarsb T, et al. Speeded-up robust features (SURF), J. Computer Vision and Image Understanding. 110 (2008) 346–359.

[6] Pires B R, Singh K, Moura J M F. Approximating image filters with box filters, in Image Processing (ICIP), 2011 18th IEEE International Conference on. IEEE, 2011, pp 85–88.

[7] WANG Junben, LU xuanmin, HE Zhao. An Improved Algorithm of Image Registration Based on Fast Robust Features, J. COMPUTER ENGINEERING&SCIENCE, 33(2011) 112–117.

Network Security and Communication Engineering – Chan (Ed.)
© *2015 Taylor & Francis Group, London, ISBN: 978-1-138-02821-0*

A hybrid CPU-MIC parallel Gaussian elimination algorithm for solving Gröbner bases in binary field

M. Zhu, B. Tang, J. Zhao, H. Xia & J.C. Li
School of Computer Science, National University of Defense Technology, Changsha, China

ABSTRACT: Gröbner bases method is a classic method for solving polynomial system. Currently, F4 algorithm is one of the fastest algorithms for computing Gröbner bases, and the main computational cost in F4 is Gaussian elimination of F4 matrix. We first analyze the characteristics of Gaussian elimination algorithm proposed by Lachartre especially for computing F4 matrix in binary fields, then present a CPU+MIC heterogeneous parallel algorithm for pivot row reduction, which is one of the most time-consuming steps, and utilize a series of performance optimization techniques, such as OpenMP multi-thread organization, SIMD vectorization, memory access optimization, to improve the efficiency of the algorithm. The efficiency of the algorithm is demonstrated by the experiments about well-known HFE cryptosystem. For instance, for a medium size problem such as HFE80, the performance of optimized heterogeneous parallel program is about 1.732 times than that of multi-thread program implemented on CPU.

KEYWORDS: Gröbner Basis, F4 algorithm, Gaussian matrix elimination, CPU+MIC heterogeneous computing platform, Binary field.

1 INTRODUCTION

Solving polynomial system is one of the important research directions in computer algebra, which is widely used in robotics, cryptography, computational geometry and so on. The existing methods of solving polynomial system include Gröbner bases method, characteristic set method and resultant method. F4 algorithm [1] is one of the most efficient Gröbner bases methods [2], in which the matrix Gaussian elimination is the main computational cost. In the literature [3], an efficient serial algorithm is proposed for Gaussian elimination based on the sparse and triangular block of the matrix. It introduces the matrix multiplication to the elimination, and obtains the higher computational efficiency.

With the development of engineering and science, the polynomial system is increasingly complex; the scale of matrix constructed during solving process is also growing; the matrix Gaussian elimination needs more time, and all of these have hampered the application of the algorithm. Because of the increasing scale of application, the serial algorithm for solving polynomial system cannot meet the requirement. The parallel computing technology is a kind of effective mean to overcome the problem. Currently, many-core and heterogeneous processors gradually become the mainstream high-performance computing platform, such as Xeon Phi (MIC) coprocessor

which has 50 or more the processor cores and can get 1 TFlops/s or higher double-precision floating point performance by wide vectors designing. MIC can provide higher computing performance at the lower cost compared to CPU. This paper presents a new matrix Gaussian elimination algorithm based on the characteristics of matrix operations in binary field, and the algorithm is optimized from OpenMP multi-thread parallel, SIMD vectorization and memory access optimization. The experimental results about HFE35 and HFE80 case of HFE cryptography illustrate that the performance of the optimized heterogeneous parallel algorithm significantly improves with respect to that of multi-thread algorithm on CPU.

2 SERIAL MATRIX GAUSSIAN ELIMINATION

Supposing that the input matrix of algorithm is M_0, the serial matrix Gaussian elimination algorithm can be divided into five steps: (1) splitting of the matrix, (2) reduction of pivot rows (Trsm), (3) reduction of non-pivot rows (Axpy), (4) computation of new pivot rows, (5) reduction of matrix row. In these steps, the second step is the most time-consuming. The main work of the paper is to optimize the algorithm on MIC, and realize the parallel of the algorithm. For the completeness of the paper, the equivalent form of M_0 after step (1) and (2) is as follows:

$$M_0 \sim \begin{pmatrix} A & B \\ C & D \end{pmatrix} \sim \begin{pmatrix} Id & A^{-1}B \\ C & D \end{pmatrix}, \qquad (1)$$

where A is sparse upper triangular matrix, C is sparse matrix, B and D are dense matrixes, Id is unit matrix. A consists of pivot rows and pivot columns. B consists of pivot rows and non-pivot columns. C consists of non-pivot rows and pivot columns. D consists of non-pivot rows and non-pivot columns.

2.1 Serial Trsm algorithm

Because the Gröbner bases is solved in binary field in this paper, the elements of M_0 are only 0 or 1. The core idea of serial Trsm algorithm is that the column reduction of B is guided by the row reduction of A. That is, when the i-th row of A is reduced, if the j-th column of the row is 1, A_{ij} is set to 0, and the j-th row of B is added to its i-th row. Based on the idea, the reduction of A is from bottom to up by row. In order to reduce storage space, A is stored in the form of sparse. B is stored in the form of dense. The serial Trsm algorithm is shown in Algorithm 1.

Algorithm 1: The serial Trsm algorithm

```
for(i = N-2; i >=0;i–)
    pA = the i-th row of A;
    pBdata =the i-th row of B;
    for(k = 0; k<count;++k)
        bitrow = the column number of the k-th
                 element 1 in pA located in
                 matrix A;
        pBadd = the bitrow-th row of B;
        for(j=0;j<rowstride;++j)
            pBdata[j] = pBadd[j]^pBdata[j];
```

2.2 Serial Trsm algorithm analysis

According to Algorithm 1, we use Pthread model to realize the multi-thread program on CPU, and obtain a certain performance improvement. But the design and implementation of the algorithm does not take into account the features of MIC. In order to take full advantage of MIC, we use roofline mode [4] to analyze the serial Trsm algorithm for optimizing it.

The roofline model is as follows:

$$PDM = MB \times TF / TB, \qquad (2)$$

where PDM is the upper bound of performance limited to memory access bandwidth, TF is the total Flops of the algorithm, TB is the total DRAM bytes of the algorithm. For specific hardware platform, PD and MB is the peak performance and memory access bandwidth, respectively. Depending on the relation of PDM and PD to choose optimization strategy, there are three cases:

1 If $PDM \ll PI$, choose memory access optimization strategies;
2 If $PDM \gg PI$, choose computing optimization strategies;
3 If there is litter difference between PDM and PD, choose the combination of memory access optimization strategy and computing optimization strategy.

Based on formula (2) and the current peak performance of MIC, we know that PDM is much smaller than the peak performance of MIC. In practical applications, memory bandwidth is smaller than theoretical memory bandwidth. This makes PDM further smaller. According to previous analysis, Trsm is typical data access constrained application. So the optimization about Algorithm 1 emphasizes on data access.

3 THE MULTI-THREAD PARALLEL OF TRSM AND OPTIMIZATION

The optimization of Trsm on MIC includes three steps: (1) multi-core, multi-thread parallel optimization and migration to MIC, (2) memory access optimization, (3) the parallel optimization of Trsm algorithm on MIC platform. The following describes these in detail.

3.1 Multi-core, multi-thread parallel optimization and migration to MIC

3.1.1 Multi-core, multi-thread organization optimization

The multi-thread parallel of Trsm algorithm is realized based on OpenMP programming model. From Algorithm 1, we know the outmost loop has order correlation; the middle loop has written collision; there is not any correlation in the innermost loop, so it suits multi-thread parallel. In order to reduce the overhead of the repeated and dynamically create and destroy threads, we split the innermost loop into two loops that one locates in the outmost; the other in the innermost. The compiled guidance statement of thread parallel block is in the outmost loop after splitting to ensure the correctness of the program.

3.1.2 Binding between threads and processor cores

The dynamic migration of threads among cores leads to certain performance degradation on many-core computing platform. In order to take full advantage of computing performance of the platform, we set the KMP_AFFINITY [5] environment variable to bind between threads and processor cores.

3.1.3 SIMD optimization

The one of features of MIC is supporting 512 bits wide vector instruction. This is good at data-level parallelism of SIMD. To make use of MIC, the vector optimization of source program is necessary. Through the command-line option -vec-report3 [6], we know which part of the program has been vectored. In the case of ensuring the correctness, the loops are vectored by forced vectorization instructions. In this paper, we vector the innermost loop after splitting in multi-core, multi-thread organization optimization by the instructions.

3.1.4 Strategy of migration to MIC platform

There are usually native and offload modes to compile and run MIC source code. Native mode only uses MIC processors, however CPU is idle. Offload mode makes full use of CPU and MIC to compute, so we choose offload mode to realize CPU+MIC heterogeneous parallel computing.

3.2 Memory access optimization

3.2.1 Data continuity

The storage of B before optimization does not take into account the thread attribute of data, so the distance among data is relatively large which does not meet the data continuity and affects the performance of program. Let the number of thread in OpenMP is 4. The relation between the thread and the data in B is shown in I of Figure.1. The matrix B in original program is stored as it is shown in II of Figure.1. From Figure.1, the span of memory address is quite large in a thread, which hampers data accessing and prefetching.

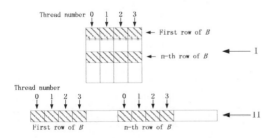

Figure 1. The thread data corresponding relation of matrix B and data storage structure before optimization.

In order to improve data continuity in threads, it's necessary to optimize the storage of data in B. The storage after optimization is shown in Figure.2.

Figure 2. The data storage structure of matrix B after optimized.

3.2.2 Partitioning strategy to increase cache hit rate.

Increasing the repeated utilization ration of data and raising cache hit rate can improve the performance of program. So we split the matrix A into blocks as it is shown in Figure.3.

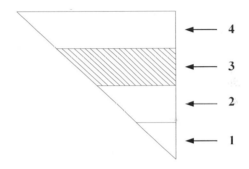

Figure 3. Partitioning strategy of matrix A.

From Figure.3, A is divided into 4 blocks. The numeric in Figure.3 is the serial number of block. Every block has the same number of rows (the bottom block may has fewer rows according to partitioning strategy). For original Trsm algorithm, the matrix A is computed from bottom to up by row (the outmost loop in Algorithm 1). However, the optimized Trsm is carried out from bottom to up by block in which the implementation is from right to left by column; the order of operation is from bottom to up inside column. Supposing that the 3-th block has N rows in which there are Y number of element 1 (the order of row is $n_{0,j}, n_{1,j}, \cdots, n_{Y-1,j}$) in j-th column. Thereby, when the column is executed, the j-th row of matrix B is reused Y times. For the k-th row in the block, there are X number of element 1 (the order of column is $m_{0,k}, m_{1,k}, \cdots, m_{X-1,k}$), thus the k-th row in matrix B is reused 2X times. In the partitioning strategy, if N is too small, the reusability of data is not enough; if N is too large, the cache hi rate reduces. In this paper, we choose N through the combination of experimental results and the size of cache. The strategy makes full use of the reusability of data, reduces the number of fetching data from the memory, and improves the performance of data access.

4 THE PARALLEL OPTIMIZATION OF TRSM ON MIC PLATFORM

In practical applications, the memory requirement is very large. Due to memory limitation, it's necessary that the matrix A and B are segmented by the memory of MIC to efficiently compute. Firstly, we allocate the M_A and M_B memory on MIC for the matrix A and B, respectively. Then, the matrix B is segmented according to M_B for reducing the date dependence so that the repeated data transfer between CPU and MIC decreases. Because there is not data correlation between the columns in matrix B, we segment the matrix B by column according to M_B as it is shown in Figure.4.

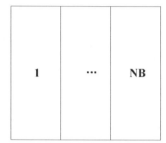

Figure 4.   Data segmentation strategy of matrix B.

The matrix B is divided into NB blocks by column, and the matrix A is split into NA blocks by the strategy. Thus, the parallel optimization algorithm of Trsm on CPU+MIC platform is shown in Algorithm 2.

Algorithm 2: The parallel optimization algorithm of Trsm

```
for(i = 0; i < NB; ++i )
    transfer_to_mic(B[i]);
    for(j = 0; j < NA; j += K)
        transfer_to_mic(A[j]);
        kernel_compute_mic(B[i], A[j]);
    transfer_to_cpu(B[i]);
```

---

5 THE EXPERIMENTS

5.1 *The experimental environment*

The experimental environment includes two 8-core Xeon E5-2670 CPU and one 57-core Xeon Phi 3120 (MIC) card, the memory is 128GB. The memory of a MIC is 6GB. The compiler is Intel icc 13.0, and the optimization option is -O3.

5.2 *The experimental results*

To compare the effect of different optimization strategies, we take the HFE35 and HFE80 of HFE

cryptography as cases. There are five versions of the program. For the sake of simplifying and discussion, PTHRAD denotes original multi-thread version; MIC_BASE is OpenMP multi-thread version based on original serial version; MIC_STO implements the strategy of data continuity based on MIC_BASE; MIC_CACHE is the version by increasing cache hit rate based on MIC_STO; MIC_FINAL is final version which is optimized on MIC platform based on MIC_CACHE.

We take HFE35 as example to test the contribution of strategies of memory access optimization. The corresponding versions are MIC_BASE, MIC_STO and MIC_CACHE. Especially, N is 200 in MIC_CACHE. The result is shown in Figure.5 (The time is only the computing time of MIC). From Figure.5, we know that the speedup of MIC_STO compared with MIC_BASE is 3.98; the speedups of MIC_CACHE compared with MIC_STO and MIC_BASE are 1.90 and 7.54, respectively. That is, the result illustrates that the optimization strategies is effective.

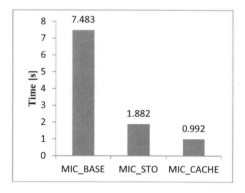

Figure 5.   Performance improvements contributed by the different optimization strategies.

We use PTHREAD and MIC_FINAL to compare final optimization Trsm on MIC and original multi-thread Trsm on CPU. The case is HFE80. The number of threads is 8 and 114 on CPU and MIC, respectively. N is 200. M_A and M_B are 0.3GB and 3.2GB, respectively. Table 1 is the experimental result. From table 1, the speedup of MIC_FINAL compared with PTHREAD is 1.732. The result illustrates the Trsm algorithm can obtain higher performance on CPU+MIC platform with large-scale computing requirement.

Table 1.   Speedup of MIC_FINAL relative to PTHREAD.

version	Total time [s]	speedup
PTHREAD	1331.695	1
MIC_FINAL	768.977	1.732

6 CONCLUSIONS

Solving large-scale polynomial system has increasing demands for high-performance computing technology. In this paper, we present a CPU+MIC heterogeneous parallel Trsm algorithm in F4 for computing Gröbner bases, and optimize the algorithm through OpenMP multi-thread organization, SIMD vectorization and memory access optimization based on the features of hardware architecture and application. The experimental results illustrate the proposed algorithm has the higher performance.

REFERENCES

[1] J.-C. Faugère, A New Efficient Algorithm for ComputingGroebner bases (F4), Journal of Pure and Applied Algebra. 139(1999) 61–88.

[2] B. Buchberger, An Algorithmical Criterion for the Solvability of Algebraic Systems, Aequationes Mathematica. 4(1970) 374–383.

[3] S. Lachartre, Algèbre linéaire dans la résolution de systemspolynomiaux Applications encryptologie. PhD thesis, Université Paris 6, 2008.

[4] Williams S, Waterman A and Patterson D, Roofline: An Insightful Visual Performance Model for Floating-Point Programs and Multicore Architectures, Communications of the ACM. 52(2009) 65–76.

[5] Intel, Intel Xeon Phi Coprocessor System Software Development Guide, Nov. 2012.

[6] Jim Jeffers, James Reinders, Intel Xeon Phi Coprocessors High-Performance Programming, Elsevier Inc., 2013.

Network Security and Communication Engineering – Chan (Ed.)
© 2015 Taylor & Francis Group, London, ISBN: 978-1-138-02821-0

Carrier frequency offset estimation of utilizing Subcarrier Residual Power for OFDM systems over time-varying multipath fading channel

F.C. Wu, T.C. Chen & W.J. Li
Changhua, Taiwan, R.O.C

T.T. Lin
Nan Shih Li, Miaoli City, Taiwan, R.O.C

ABSTRACT: In this paper, the method of utilizing Subcarrier Residual Power (SRP) is proposed to estimate Carrier Frequency Offset (CFO) for Orthogonal Frequency Division Multiplexing (OFDM) systems by using training sequence. The training sequence is composed of non-zero pilots and zero pilots. And SRP of training sequence is adopted to estimate CFO over time-varying multipath fading channel in frequency-domain. A low-complexity method utilized SRP is proposed, too. Simulation results show that the proposed methods are workable in existent systems and keep good performance.

KEYWORDS: Orthogonal frequency division multiplexing (OFDM), Carrier frequency offset (CFO), Pilot symbols, Subcarrier residual power (SRP).

1 INTRODUCTION

Orthogonal frequency division multiplexing (OFDM) technology is widely employed in wireless communication systems due to its high bandwidth efficiency. In addition, OFDM can effectively avoid inter-symbol interference (ISI) caused by multipath propagation. However, in wireless communication system, it will cause carrier frequency offset (CFO) by Doppler shift and oscillators mismatch of the transmitter and receiver. CFO will destroy the orthogonality between subcarriers and induce inter-carrier interference (ICI) which will decay performance. In order to solve the problem of CFO, many methods of CFO estimation were proposed [1–14]. Blind CFO estimation methods were proposed in [1–6]. CFO estimation methods with pilot symbols were proposed in [7–14]. In [7,8], a complexity method of channel residual energy (CRE) is calculated in time-domain.

In this paper, the method of utilizing subcarrier residual power (SRP) is proposed to estimate CFO in frequency-domain. The training sequence composed by non-zero pilots and zero pilots is considered. The rest of this paper is organized as follows. In Section 2, the signal model is introduced. Section 3, the proposed methods are described in details. Computer simulation results and analysis will be described in Section 4. Conclusion and reference are in the final.

2 SIGNAL MODEL

The architecture of an OFDM system is shown as Fig. 1. In the transmitter, the frequency-domain signal of j-th OFDM symbol can be expressed as

$$\mathbf{X}_j = [X_{0,j}, X_{1,j}, ..., X_{N-1,j}]^T, j = 0, 1, ..., N_{sym} - 1 , \quad (1)$$

Where N is the number of subcarriers and Nsym is the number of OFDM symbols in one transmission frame. The time-domain signal of j-th OFDM symbol can be expressed as

$$\mathbf{x}_j = [x_{0,j}, x_{1,j}, ..., x_{N-1,j}]^T = \mathbf{F}^H \mathbf{X}_j, j = 0, 1, ..., N_{sym} - 1 , \quad (2)$$

Where Γ is the $N \times N$ discrete Fourier transform (DFT) matrix with e–j(m–1)(n–1)/N to be its (m, n) element, and FH is the Hermitian transpose of F. The time-varying multipath fading channel with CFO and additive white Gaussian noise (AWGN) are considered. The discrete-time impulse response of fading channel with the maximum path delay L is denoted as

$$g(n,l) = [g(n,0)\ g(n,1) ... g(n,L-1)]. \quad (3)$$

Normalize CFO, ε, and AWGN with zero mean and variance $\sigma n2$ is involved. In the receiver, the time-domain signal of j-th OFDM symbol removed GI is given by

$$\mathbf{y}_j = \mathbf{E}(\varepsilon)\mathbf{G}\mathbf{x}_j + \mathbf{n}, \qquad (4)$$

Where

$$\mathbf{E}(\varepsilon) = diag\left[\; e^{\frac{j2\pi\varepsilon\cdot 0}{N}} \quad e^{\frac{j2\pi\varepsilon\cdot 1}{N}} \quad \cdots \quad e^{\frac{j2\pi\varepsilon\cdot(N-1)}{N}} \;\right], \qquad (5)$$

$$\mathbf{G} = \begin{bmatrix} g(0,0) & 0 & \cdots & 0 & g(0,L-1) & g(0,L-2) & \cdots & g(0,1) \\ g(1,1) & g(1,0) & 0 & & 0 & g(1,L-1) & \ddots & g(1,2) \\ \vdots & & \ddots & & & & \ddots & \vdots \\ 0 & \cdots & 0 & g(N-1,L-1) & g(N-1,L-2) & \cdots & & g(N-1,0) \end{bmatrix}. \qquad (6)$$

And

$$\mathbf{n} = [\,n(0) \quad n(1) \quad \cdots \quad n(N-1)\,]^T. \qquad (7)$$

And then the received frequency-domain signal is expressed as

$$\mathbf{Y}_j = \mathbf{F}\mathbf{E}(\varepsilon)\mathbf{G}\mathbf{F}^H\mathbf{X}_j + \mathbf{F}\mathbf{n}. \qquad (8)$$

In this paper, the training sequence placed in the first OFDM symbol of each frame is used to estimate CFO. The training sequence is composed of non-zero pilots and zero pilots like the short training sequence of transmission frame of 802.11 WLAN systems[15]. In this paper, the short training sequence of 802.11 WLAN systems is adopted to simulate the system performance. That is,

$$\begin{aligned} X_0 = \sqrt{13/6}[&0,0,0,0,-1-j,0,0,0,-1-j,0,0,0,1+j,0,0,0,1 \\ &+j,0,0,0,1+j,0,0,0,1+j,0,0,0,0,0,0,0, \\ &0,0,0,0,0,0,0,0,1+j,0,0,0,-1-j,0,0,0,1 \\ &+j,0,0,0,-1-j,0,0,0,1+j,0,0,0]^T. \end{aligned} \qquad (9)$$

In order to normalize the average power of each transmitted OFDM symbols, the factor $\sqrt{13/6}$ is multiplied because the non-zero pilot symbols utilizes 12 out of 52 subcarriers.

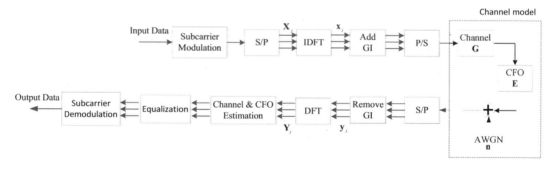

Figure 1. The architecture of OFDM systems.

3 CFO ESTIMATION OF UTILIZING SUBCARRIER RESIDUAL POWER

For an OFDM system, the CFO would induce the inter-carrier interference (ICI). Assume that ε is present, the frequency-domain signal of j-th OFDM symbol in the receiver is expressed as

$$\mathbf{Y}_j = [Y_{0,j},\; Y_{1,j},\; ...,\; Y_{N-1,j}]^T = \mathbf{F}\mathbf{E}(\varepsilon)\mathbf{F}^H\mathbf{X}_j, \qquad (10)$$

Where

$$\begin{aligned} Y_{i,j} = &\; X_{i,j} \cdot \frac{\sin(\pi\varepsilon)}{N\cdot\sin\left(\dfrac{\pi\varepsilon}{N}\right)} \cdot e^{j\pi\left(\frac{N-1}{N}\right)\varepsilon} \\ &+ \sum_{\substack{i'=0 \\ i'\neq i}}^{N-1} X_{i,j}\cdot e^{j\pi\left(\frac{N-1}{N}\right)(i'-i+\varepsilon)} \cdot \frac{\sin(\pi(i'-i+\varepsilon))}{N\cdot\sin\left(\dfrac{\pi(i'-i+\varepsilon)}{N}\right)}. \end{aligned} \qquad (11)$$

The first term of (11) represents the self-attenuation and phase rotation of the i-th subcarrier, and the second term is the ICI induced by the other subcarriers. Fig. 2(a) shows the power spectrum of the received training sequence with no CFO present. Obviously, the power converges on non-zero pilots, and the subcarriers of zero pilots are free of power. On the other hand, Fig. 2(b) shows the power spectrum of the received training sequence with CFO ($\varepsilon = 0.2$). The power is spread from non-zero pilots to zero pilots. Therefore, we define the total power of zero pilots as subcarrier residual power (SRP). The relation between SRP and ε is simulated and the result is shown as Fig. 3. As expected, the SRP increases as the CFO increases. Therefore, a CFO estimation method is proposed in this paper. We utilize SRP to estimate the CFO.

That is, a CFO estimate can be found from the candidate normalized frequency set,

$$S = \left\{ \varepsilon_k : \varepsilon_k = -0.5 + k \cdot \Delta\varepsilon \ , \ k = 0, 1, ..., Q, \Delta\varepsilon = 1/Q \right\} ,$$

$$(12)$$

Which can achieve the minimum SRP:

$$\hat{\varepsilon} = \min_{k} \ \text{SRP}(\mathbf{FE}(-\varepsilon_k)\mathbf{y}_0) , \qquad (13)$$

Where SRP(·) is the operation of total power of zero pilots and y0 is the received time-domain training signal. The choosing of Q (or $\Delta\varepsilon = 1/Q$) is trade-off between accuracy in CFO estimate and computational complexity. Fig. 4 shows the mean square error (MSE) for different values of Q. Obviously, the estimation error decreases as the value of Q increases. A large Q can achieve a better performance, but it induces high computational complexity.

In order to reduce complexity, another low-complexity method is proposed to estimate CFO with a coarse estimator concatenated with a fine estimator. The coarse estimator chooses a large value of Q, denoted as Q', to obtain the coarse CFO estimate:

$$\hat{\varepsilon}' = \min_{k} \ \text{SRP}(\mathbf{FE}(-\varepsilon'_k)\mathbf{y}_0) , \qquad (14)$$

Where $\varepsilon'_k = -0.5 + k/Q'$, $k = 0, 1, ..., Q'$. Then the fine estimator minimizes the SRP with the CFO estimate selected from the candidate frequency set, $\{\varepsilon''_k : \varepsilon''_k = \hat{\varepsilon}' - \Delta\varepsilon'/2 + k \cdot \Delta\varepsilon'/Q''$, $k = 0, 1, ..., Q''$, $\Delta\varepsilon' = 1/Q'\}$, to obtain the most accurate CFO estimate:

$$\hat{\varepsilon}'' = \min_{k} \ \text{SRP}(\mathbf{FE}(-\varepsilon''_k)\mathbf{y}_0) . \qquad (15)$$

OFDM systems with $N = 64$ are modulated over time-varying multipath fading channel. The 16-path exponentially decaying Rayleigh fading (EDRF) channel with root-mean-squared (RMS) delay spread $T_{rms} = 10T_s$, where T_s is sampling time. And normalize CFO, ε, assumed to be uniformly distributed over $[-0.5, 0.5]$ for each transmission frame. The performance of the proposed methods is shown in Fig. 5. The performance of $Q = 100$ is similar to $Q = 1000$. Consequently, the SRP method with $Q = 100$ is enough to estimate CFO accurately. Compared with the SRP method, the low-complexity method with coarse estimation ($Q' = 10$) and fine estimation ($Q'' = 10$) offers the same performance as the SRP method with $Q = 100$. The low-complexity SRP method with $Q' = 10$ and $Q'' = 10$ has the same computational complexity as the SRP method with $Q = 20$. However, the low-complexity SRP method can offer a better performance.

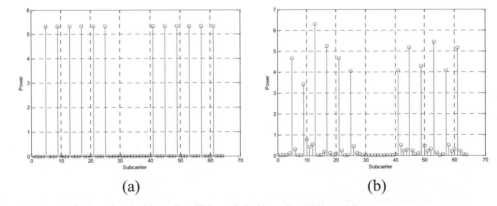

(a)	(b)

Figure 2. Power of subcarriers (a) Normalize CFO $\varepsilon = 0$ (b) Normalize CFO $\varepsilon = 0.2$.

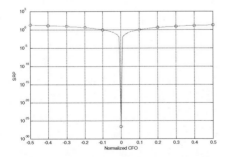

Figure 3. Normalize CFO versus SRP.

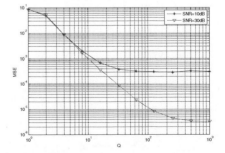

Figure 4. Q versus MSE.

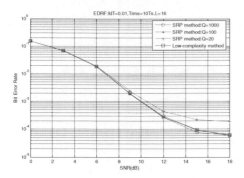

Figure 5. Performance of proposed methods over time-varying multipath fading channel.

4 CONCLUSION

In this paper, for OFDM systems with the training symbols composed of non-zero pilots and zero pilots, the CFO estimation of utilizing SRP was proposed. A low-complexity method was also proposed to simplify the computational complexity. The simulation results show that the proposed methods are workable, and keep good performance in an OFDM system over time-varying multipath fading channel.

ACKNOWLEDGMENT

This work was partially supported by the National Science Council, R.O.C, under the Contract NSC 102-2221-E-018-027.

REFERENCES

[1] W.J. Tai, Y.C. Pan, and S.M. Phoong, "A simple LS algorithm for improving ESPRIT-based blind CFO estimation in OFDM systems," *IEEE 23rd PIMRC*, pp.2303–2308, Sept. 2012.

[2] L. Wu, X.D. Zhang, *senior Member, IEEE*, P.S. Li, and Y.T. Su, "A closed-form blind CFO estimation based on frequency analysis for OFDM systems," *IEEE Trans. Comm.*, vol.57, no.6, pp.1634–1637, June 2009.

[3] L. Wu, X.D. Zhang, *senior Member, IEEE*, P.S. Li, and Y.T. Su, "A blind CFO estimator based on smoothing power spectrum for OFDM systems," *IEEE Trans. Comm.*, vol.57, no.7, pp.1924–1927, July 2009.

[4] H. Liu and U. Tureli, "A high-efficiency carrier estimator for OFDM communication," *IEEE Comm. Lett.*, vol.2, no.4, pp.104–106, April 1998.

[5] F. Gao and A. Nallanathan, "Blind maximum likelihood CFO estimation for OFDM systems via polynomial rooting," *IEEE Signal Processing Lett.*, vol.13 pp.73–76, Fed. 2006.

[6] T. Liu, and H. Li, "Blind carrier frequency offset estimation in OFDM systems with I/Q imbalance," *IEEE Signal Processing*, vol.89 no.11, pp.2286–2290, Nov. 2009.

[7] Y. H. Chung, Member and S. M. Phoong, "OFDM Channel Estimation in the Presence of Receiver I/Q Imbalance and CFO Using Pilot Symbols," *IEICE Trans. Comm.*, vol. E95-B, Feb. 2012.

[8] W.J. Li, "CFO estimation in time-varying multipath fading channel for OFDM system using pilot symbols," National Changhua University of Education Press, July 2012.

[9] J. Lei and T-S. Ng, "A consistent OFDM carrier frequency offset estimator based on distinctively spaced pilot tones," *IEEE Trans. Wireless Commun.*, Vol. 3, No. 2, pp.588–599, Mar. 2004.

[10] H. Minn, Y. Li, N. Al-Dhahir, and R. Calderbank, "Pilot designs for consistent frequency offset estimation in OFDM systems," in *Proc. ICC'06*, Istanbul, Turkey, vol. 10, pp.4566–4571, June 2006.

[11] M. Ghogho, A. Swami, "Semi-blind Frequency Offset Synchronization for OFDM," IEEE International Conf. on ICASSP, vol.3, pp.2333–2336, May 2002.

[12] C.W. Chang Y.H. Chung, S.M. Phoong and Y.P. Lin, "Joint estimation of CFO and receiver I/Q imbalance using virtual subcarriers for OFDM systems," *IEEE 23rd PIMRC*, pp.2297–2302, Sept. 2012.

[13] Y.F. Chen, "Frequency offset estimation based on virtual carrier in OFDM system," *IEEE Conf. IASP 2009*, pp.369–371, April 2009.

[14] T. Cui, F. Gao and A. Nallanathan, "Frequency offset tracking for OFDM systems via scattered pilots and virtual carriers," *IEEE Conf. ICC'07*, pp.5120–5125, June 2007.

[15] IEEE LAN/MAN Standards Committee, "Wireless LAN medium access control (MAC) and physical layer (PHY) specifications: High-speed physical layer in the 5 GHz band," *IEEE Standard 802.11a*, 1999.

Network Security and Communication Engineering – Chan (Ed.)
© *2015 Taylor & Francis Group, London, ISBN: 978-1-138-02821-0*

The feature extraction of the text based on the deep learning

X. Chen, S.F. Li & Y.F. Wang
College of Computer Science and Information Engineering, Tianjin University of Science & Technology, China

ABSTRACT: This thesis presents a program of the feature extraction of the text based on deep learning. Deep learning, a new learning algorithm, contains multilayer neural network. In this paper, the auto-encoder neural network is researched with the foundation of the approach of the deep learning. To utilize the MATLAB tool is to use the way of deep learning. The learning and training of the text samples are conducted. With the favorable network symmetry of the auto-encoder neural networks and the minimum number of the neuron in the center layer, the text feature can carry out the process of the dimensionality reduction. Through the test of the sample, the results show that there is consistence between the similarity of the feature and the similarity of the original text.

KEYWORDS: Deep learning, Auto-encoder neural network, Matlab, Text feature.

1 INTRODUCTION

The text is the unstructured data. The text should be translated into structured forms which are processible before the information processing. At present, the text feature selection uses the vector space model to describe the text vector. Generally speaking, by adopting the word segmentation and word frequency statistics method, we can get the features to represent every dimension of the text vector. Consequently, the dimension of this text vector is very large, which leads to the huge computational overhead in managing relevant processing, including the classification of the text content, clustering processing and the discovery of the interest patterns and knowledge discovery. Moreover, it will make the whole process have low efficiency, and damage the accuracy of classification and clustering algorithm. Therefore, it is significantly important to search for the text feature item, which is conducive to a more effective analysis.

The concept of deep learning originates from the artificial neural network[1]. The multilayer perceptron with multi hidden layer is a kind of deep learning structure. With the combination of low-level features, deep learning can form a more abstract high- level to represent the attribute, class or feature so as to find out the characteristics of the distributed data.

The concept of deep learning was proposed by Hinton et al in 2006. It puts forward the unsupervised greedy layer training algorithm based on the DBN, which makes the optimization of the related deep structure possible. Therewith, the deep structure of multi-layer auto encoder is presented. The convolutional neural network, proposed by Lecun et al, is the first real multilayer structure learning algorithm, which uses the space relation to reduce the number of parameters and to improve the training performance [2].

2 DEEP LEARNING

2.1 Deep learning theory

Deep learning is a new field of the research in the machine learning, and the motivation lies in the establishment and simulation of human brain for the analysis of neural network learning. It imitates human brain mechanism to explain the data, such as images, sound and text. Deep learning is an unsupervised learning. The concept of the deep learning originates from the artificial neural network. The multi-level perceptron of multi hidden layer is a kind of deep learning structure. With the combination of low-level features, deep learning forms a more abstract high-level to represent the attribute, class or feature. The essence of the deep learning is using the training data to construct the machine learning model with multi hidden layers and mass training data so as to learn more effective feature. Ultimately, it enhances the accuracy of the classification and the prediction. The characteristics of deep learning is as follows: 1) it emphasizes the depth of the model structure with 5 layer, 6 layer or more hidden layer nodes in general; 2) it clarifies the importance of the feature learning, that is to say, the feature of the sample in the original space is transformed into a new feature space so as to be convenient for the processing by the transformation of the feature layer by layer.

As a widely used method of deep learning, the auto-encoder neural network was proposed by Hinton et al in 2006. It is a kind of technique of the nonlinear dimensionality reduction in the neural network, which contains a number of hidden layers. The network is a kind of symmetric structure with the minimum number of neurons in the center layer. The network, after training, gets a group of weights which can express the input data in a low dimensional form. Sequentially, the feature, achieved by dimension reduction, can represent the input data.

2.2 Pre-training

2.2.1 Determination of the input parameters
The learning sample selects 50 groups of text feature with the same category, and each title is no more than 30 Chinese characters, which is shown in table 1.

2.2.2 Sentence pretreatment [5].
The learning sample sentences are preprocessed in pretreatment, that is to say, each Chinese character in the sentence is transformed into a digitized form which can be accepted by auto-encoder neural network. In order to accept external data by the neural network, firstly, the Chinese characters in the sentence will be encoded.

Each country has different coding in the aspect of information encoding, and unified use of the international code is the ASCII (USA Standard Code for information interchange), while our country uses GB code in Chinese) and BIG5 (in Chinese traditional)

The Spanish and digital symbol of GB2312-80 character set is 8×16 dot matrix. Since the text information is in western and digital symbols, and the training sample firstly transposes the dot matrix of each text information, such as Figure 1 and Figure 2. The transposed Chinese characters lattice is 16 x 16 and Spanish and digital one is 8 x 16. If the training samples have n Chinese characters and m Western

character or digital, transposed lattice code should be (2 * n+m) x 8 x 16, for example, the training sample "China has independently developed J55 heavy oil thermal recovery casing to successfully substitute the import", with total number of 18 Chinese characters, 1 Spanish, 2 numbers. After the transposition of each character, it composes 39 (2 x18+3) x 8=312 rows and 16 column of the two value image.

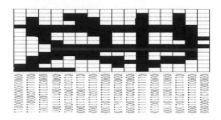

Figure 1.   "Wo" Chinese characters font dot matrix code transpose.

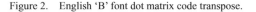

Figure 2.   English 'B' font dot matrix code transpose.

2.2.3 Pre-training
Pre-training partly is composed of "stacked" network structure constituted by several independent RBM. RBM is the special connection mode of BM

Table 1.   Learning samples.

1	Successfully developed with independent property rights 20G high pressure boiler tube in our country
2	Successfully developed with independent property rights K55 petroleum casing tube in our country
3	Successfully developed with independent property rights 4J36 precision alloy Marine pipe in our country
4	Successfully developed with independent property rights special high temperature resistant fire casing tube of the ship in our country
5	Successfully developed with independent property rights high heat resistant steel seamless steel tube in our country
6	Successfully developed with independent property rights steel chamfering machine in our country
7	Successfully developed with independent property rights crane booms seamless steel tube in our country
8	Successfully developed with independent property rights three kinds of limit specifications line successfully replace imported in our country
9	Successfully developed with independent property rights high strength and high toughness collapse casing in our country
10	Successfully developed with independent property rights the type 0460 steel in our country
...	......
50	Successfully developed with independent property rights medicine and chemical industry tube in our country

(Boltzmann machine). Figure 3 is the RBM network structure. It is a kind of hidden layer neurons with no connection, and only has two layers of neurons including the visible and hidden layers.

Figure 3.   Construction of restricted Boltzmann machine.

Weight adjustment formula for [6]

$$w_{ij}(t+1) = w_{ij}(t) + \Delta w_{ij} = w_{ij}(t) + \frac{\eta}{T}(<v_i h_j>^+ - <v_i h_j>^-) \quad (1)$$

In the formula, $w_{ij}(t)$ is the connection weights of neuron I, j in the tth step; η represents the learning rate; while T is the network temperature; $<u_i h_j>+$, $<u_i h_j>-$ respectively is the positive and negative average correlation.

In RBM, the neurons output of visible layer multiply by the output of hidden layer is the average correlation. Coefficient ε *is composed* of the unification and merge of *η* and T, while the iteration step is represented by the weight adjustment formula *ε*.

$$w_{ij}(t+1) = w_{ij}(t) + \Delta w_{ij} = w_{ij}(t) + \varepsilon(<v_i h_j>^+ - <v_i h_j>^-) \quad (2)$$

The weight training begins with t=0, with the completion of a RBM training, there will be a corresponding weight, and the weight adjustment formula is as follow

$$\Delta w_{ij} = \varepsilon(<v_i h_j>^0 - <v_i h_j>^1) \quad (3)$$

With multi trainings being conducted in this way, the completed output of the last positive training is regarded as the next RBM input for training.

2.2.4   *The realization of MATLAB*

The 50 groups of learning samples are no more than 30 characters, and each group is represented by16 column of two valued image with 30 * 16, and the lattice value less than 30 characters is 0.

The network hidden layer activation function is the logarithmic function of S type: Logsig. While the activation function in the output layer is the linear activation function: purelin, and the network training function is Adaptive lr momentum gradient descent algorithm for training the traingdx function.

The learning rate is set to be 0.01, the maximum training step is 300, and the error goal is 0.00001.

2.2.5   *The results of pre-training*

The structure of the established BP neural network model is:

[480, 250, 480], [250, 150, 250], [150, 80, 150], [80, 50 ,80], [50, 30, 50], [30,10,30] as shown in Figure 4.

Figure 4.   RBM network structure diagram.

The pre-training results as shown in table 2.

Table 2.   The results of pre training.

	Error	Learning rate	Step length
[480, 250, 480]	0.003048	0.01	170
[250, 150, 250]	0.0020809	0.01	150
[150, 80, 150]	0.0021866	0.01	120
[80, 50, 80]	0.0037575	0.01	100
[50, 30, 50]	0.014465	0.01	75
[30, 15, 30]	0.016332	0.01	50

2.2.6   *The deployment*

As shown in Figure 5, each RBM is connected to get the auto-encoder neural network. The weights of pre-training will serve as the initial weights of the auto-encoder neural networks to participate in the whole network of fine-tuning training.

Fine-tuning training is the further adjustments to the weights based on the initial weights in pre-training. By using the cross entropy algorithm for BP target function [5], the network of fine-tuning training is completed. Cross entropy is adopted to measure the difference between two probability distributions, which is a non-negative number, and the more similar the two distributions are, the smaller it will become. The original cross entropy is defined as

Figure 5. Development of RBM network structure fine-tuning training.

$$D(c \parallel d) = \sum_x c(x) \log \frac{c(x)}{d(x)} \tag{4}$$

In the formula, X is a random variable; d (x) is a known probability distribution; C(x) is the estimation of probability distribution.

As for the random variable x, when d (x) is used to estimate C(x), $c(x)$ will make adjustment of the affection of x to minimize cross entropy $D (c\|d)$. In the adjustment of weights in the auto-encoder neural network, the formula of the tuning algorithm of BP cross entropy is as following

$$H_m = -\sum_{i=1}^{m} [t_i \log y_i + (1 - t_i) \log(1 - y_i)] \tag{5}$$

In the formula, t_i is the target probability distribution, and y_i is the actual probability distribution.

The aim of network training is to adjust the weights to minimize the cross entropy function, and the weight adjustment formula is as following

$$\Delta w_{ij} = -\alpha \frac{\partial H_m}{\partial w_{ij}} \tag{6}$$

$$\frac{\partial net_i}{\partial w_{ij}} = O_j \tag{7}$$

Including

$$\frac{\partial H_m}{\partial net_i} = \frac{\partial H_m}{\partial O_i} \frac{\partial O_i}{\partial net_i} = \frac{\partial H_m}{\partial O_i} O_i (1 - O_i) \tag{8}$$

The output layer $O_i = y_i$

$$\Delta w_{ij} = -\alpha \frac{\partial H_m}{\partial w_{ij}} = \alpha (t_i - y_i) O_j \tag{9}$$

According to the weight adjustment formula above, the results of fine-tuning training network are shown in table 3.

Table 3. Results of the fine-tuning training.

Error	Learning rate	Step length
0.39286	0.01	300

2.2.7 Feature extraction

The depth of learning neural network is shown in Figure 7, which contains 6 hidden layers, and the number of neurons in central layer is 10 The center layer weight coefficient matrix W_6 has a coefficient of 300 (W_6=10 x 30), and the 310 parameter is the results of 7680 (30 x 16 row, 16 column) characters dot matrix transformed through auto-coder network layer. Moreover, it is 7680 Chinese lattice representation transformed into 310 new feature spaces.

3 TEST

If the text eigenvectors are similar to the similarity of the corresponding content, then the eigenvectors should be classified into same category. Vector is actually a line with direction in the multidimensional space. The angle cosine theorem is used to the calculation of vector. When the angle gets close to zero, the two vectors have similarity in the same direction.

Such as, the corresponding vector of text X and Y respectively is $x_1, x_2,...,x_{7680}$ and $y_1,y_2,...,y_{7680}$, therefore, the cosine of the angle is as following

$$\cos \theta = \frac{x_1 y_1 + x_2 y_2 + ... + x_{7680} y_{7680}}{\sqrt{x_1^2 + x_2^2 + ... + x_{7680}^2} \bullet \sqrt{y_1^2 | y_2^2 | ... | y_{7680}^2}} \tag{10}$$

A random sample of 10 groups of learning samples and their eigenvalues is used to test similarity of the cosine theorem. In samples, serial number 1 is regarded as the standard 1, the original text similarity and eigenvalue similarity of serial number 3 are 0.71 and 0.72 and the original text similarity and eigenvalue similarity of serial number 7 are 0.68 and 0.71. The test results show that the similarity is similar.

4 CONCLUSION

At present, deep learning is a new learning algorithm of the neural network. It requires enforcing a profound understanding and application to people. Nowadays, the application of the deep learning is still limited to the license plate recognition of the auto-encoder neural network. And the researches on the other aspects are still to be discovered. By using the research on the text feature of the auto-encoder neural network, this thesis converts text into two-dimensional lattice codes to get its feature. Moreover, we carry on the processing of the text sample by using the principle of reconstruction and the recognition of the auto-encoder neural network. Through the network learning and training, the dimensionality reduction is processed in the extraction of the text feature.

Results show that the similarity of the feature and the original text is consistent.

REFERENCES

[1] G.E.Hinton, R.R.Salakhutdinov. Reducing the Dimensionality of Data with Neural Networks,Science 313:504–507, 2006.
[2] Sun Zhijun,Xue Lei,Xu Yangming, Wang Zheng. Deep learning research were reviewed[J] Computer application research. 2012(08).
[3] HintonG E, Salakhutdinov R R.Reducing the dimensionality of data with neural networks[J]. Science, 2006,313:504–507.
[4] Wu Fenfen. Research on the algorithm of information extraction. The master degree thesis of Jilin University, 2006.
[5] Liu Gaoping, Zhao Dujuan, Huang Hua. The license plate number recognition using autoencoder neural network reconstruction [J]. Electron laser, 2011, 22(1): 144–148.LIU.
[6] ZHANG Jian, FAN Xiaoping, et al. Research on characters segmentation and characters recognition in intelligent LPR system[C]//Proceedings of the 25th Chinese Control Conference. Harbi: Beihang University Press, 2006: 7–11.
[7] Zhuo Jianfeng. Deep learning application in natural language processing[J] The China youth political college computer center.

Network Security and Communication Engineering – Chan (Ed.)
© 2015 Taylor & Francis Group, London, ISBN: 978-1-138-02821-0

The study on urban residential land allocation optimization model

W. Zhou & L.T. Jiao
Department of Science Management, Northwestern Polytechnical University, Xi'an, Shanxi, China

ABSTRACT: The purpose of the paper is to establish the land optimal allocation model based on genetic algorithm, which provides the decision makers with a variety of alternatives of the land configuration from multiple objective benefit of land use; and it can also be used to guide the process of decision-making; so it has the vital significance to promote the optimized allocation of land.

KEYWORDS: Optimal allocation model, Genetic algorithm, Land resource.

1 INTRODUCTION

Scientific, reasonable, and effective way of land allocation is an important aspect of the coordinated development of society, population, resources, and environment, so the optimal allocation of land resources becomes important. With the development of society and science, land resource optimal allocation research has made great progress, but there are few existing theories on land resource optimal allocation model, and the practicability is poor; it needs continuous development and improvement. So proper configuration of the urban land resource has become the hot topic on the study of land resources[1].

Land Resource Allocation Models are mostly concentrated on the hierarchy resource classification, resource evaluation, and regional development level, the scientific research on modern land resources with land system, the sustainable development strategy, the management, and the quantification are still in its infancy. It needs to consider quantity and the optimization of spatial structure instead of only focusing on the number of distribution; and it needs comprehensive consideration on many kinds of the optimization goal with social, economic, and ecological environment and so on, and makes continuously research and innovation to improve land allocation optimization system.

2 THE PROBLEMS AND MATHEMATICAL MODEL OF LAND CONFIGURATION

The land configuration optimization is the optimal allocation of land quantity and spatial structure of reasonable layout[2]. The optimal allocation of land use quantity is to make reasonable planning and allocation on land quantity to prevent the waste of land resources and guarantee the sustainable use of land resources;

reasonable layout of land spatial structure means that the space configuration is reasonable and the maximum benefits of land resources can be exerted, which involves the economic, social, ecological, and environmental aspects; it is a multi-objective optimization problem. There is certain limitation on traditional multi-objective solving method. So this paper puts forward the dynamic programming method of sustainable land use based on genetic algorithm. For the optimization problem, it can be described as a mathematical model with constraints, which can be seen in Formula 1.

$$\begin{cases} \max & f(x) \\ s.t. & X \in R \\ & R \subseteq U \end{cases} \tag{1}$$

where, $X = [x_1, x_2, \cdots, x_n]^T$ refers to decision variables, $f(x)$ is the objective function, U refers to the basic space, and R is a subset of U. The feasible solution is the solution which satisfies the constraint conditions, and R represents the set made up from all feasible solutions[3].

3 GENETIC ALGORITHMS

3.1 *The concept of genetic algorithms*

Genetic algorithm is a class of adaptive probabilistic randomized iterative search algorithm reference to the evolution rule of evolved biology [4]. It was first proposed by Professor JH Holland from United States in 1975. This algorithm can be used to solve optimization problems well [5, 6].

Genetic algorithm in genetic process has deterministic and random performance. Deterministic performance: when selecting operation, the law of biological evolution "survival of the fittest" is applied, that is, those individuals with higher fitness value are more likely to "genetic" to the next round of evolution iterative process. Randomness presents that the operation of selection, hybridization and mutation have a certain randomness[7, 8]. Its algorithms maintain multiple current solutions, which can increase the degree of the optimization on the approximate solution, and also can simplify the calculation process and obtain speedup. Genetic algorithms have many advantages compared to traditional optimization algorithms[9, 10].

3.2 The principle of genetic algorithms

Genetic algorithm is to simulate the biological evolution. First, it requires a coding method to express any potential and feasible solution of the research question as the "chromosomes". Then, it creates a group of random population made up of initial "chromosomes", and makes its evolution over a period of time; at the period when the "chromosomes" becomes changeable, it generates the new "chromosomes" which adapts to the environment. So after many generations of evolution, the optimal solution of the problem can be gotten.

4 LAND ALLOCATION OPTIMIZATION MODEL BASED ON GENETIC ALGORITHM

4.1 The thinking of the model

Problems of land allocation mainly involves two aspects, one is the configuration on the amount of land; the other is the configuration of the land spatial structure. These two aspects are the key of the research on land use, and the connotation of the two are unified[11]. This paper sets up a land optimal allocation model based on genetic algorithm, through genetic algorithm that there are varieties of cost-effective alternatives for land the configuration.

4.2 The elements of the model

The main elements of the model include: variables, constraints, and the optimization goal.

1 The set of variable
   Residential land according to routine can be divided into some aspects, such as residential building land, land for traffic, land for public services, landscaping and land for commercial services, set their variable X1 ~ X6.
2 Constraints

   1 The constraint of the total area
   There is contradictory between limited land resources and dynamic demand. The total amount of land must be considered in the process of the land allocation, and the number of land allocation must be strictly controlled to protect land resources.

2 The constraint of population pressure
Land is the carrier of the population. So as the population size changes, land allocation should be adjusted accordingly to adapt to social development and improvement of people's living standards.
3 The constraint of ecological balance
Public green land per capita refers to each resident average possession of public green space in urban area, which is calculated as: public green land per capita(m^2) = urban public green space area/city non-agricultural population.
4 The constraint of national policy
National policy plays a decisive guiding role in urban construction, and the purpose is to improve the utilization efficiency of land for construction.
3 The optimization goal
Optimization goal is to achieve multi-objective optimization for social, economic, and ecological benefits.

4.3 The optimization for number configuration

The optimization for land quantity configuration is a multi-objective optimization problem, and is the optimization process of the dual goals between social benefit and economic benefit.

1 The objective function

The objective function for maximized social benefits: $\max f_1 = \sum_{i=1}^{n} B_{1i} * x_i$, x_i refers to the x_i type of land allocation, B_{1i} refers to the Social benefit index of the i type of the land.
The objective function for maximized economic benefits: $\max f_2 = \sum_{i=1}^{n} B_{2i} * x_i$, x_i refers to the x_i type of land allocation, B_{2i} refers to the Social benefit index of the i type of the land.
2 The set of Constraints

   1 The constraints for planning total area: $\sum_{i=1}^{n} x_i = S_1$, where x_i indicates the type of land use, S_1 represents the total area of the sample planning;
   2 The constraints for population pressure: $p_t \cdot l_t = S_2$, where P_t represents the size of urban population growth in the target year; l_t represents the construction area per capita in the target year; S_2 represents the total area of construction with the increment of urban population.
   3 The constraints for land for traffic: $C_{min} \leq x_3 \leq C_{max}$, where C_{min}, C_{max} represent the traffic land scope according to Macro planning requirements;
   4 The constraints for green land: $T_{xt} \geq S_1 \bullet P$, where T_{xt} represents the area of green land according to Macro planning requirements in the specific urban specific target year, P indicates the proportion of green space in the planning total area.

Use the weight coefficient of transformation method to solve the problem, give weights $w_i (i = 1, 2, \cdots, n)$ to each function $f(x_i)(i = 1, 2, \cdots, n)$, in which w_i represents the important degree of multi-objective optimization problem corresponding to the function $f(x_i)$, set the linear weighted sum of each objective function $f(x_i)$: $u = \sum_{i=1}^{n} w_i f(x_i)$. The purpose is to transform the multi-objective optimization problem into single objective optimization so as to make use of genetic algorithm in single objective optimization to solve the problem.

4.4 The optimization for the spatial structure

The essence of spatial structure optimization is to make it rationality and coordination for regional space layout in land structure, and the goal is to achieve land use matching programs corresponding to the maximum space coordination coefficient sum.

First, the area of planning land should be divided into many independent physical block, then code it $k_i (i = 1, 2, \cdots, n)$. For example: dividing a residential land into 9 physical block, physical land coding is shown in Table 4.

Then, set the coordination coefficient of the land configuration space, the coefficient is the matrix with $m \times n$ corresponding to the land blocks m m and the land use types n.

Table 4. Physical land coding.

The land	k1	k2	k3	k4	k5	k6	k7	k8	k9
Code	1	2	3	4	5	6	7	8	9

Finally, seek the maximum benefits of the ecological environment, the objective function: $\max f = \sum_{i=1}^{n} C_{ik_i}$, k_i is use type of the number i for land, C_{ik_i} represents the coordination coefficient with configuration type k_i in the land of number i.

5 CASE ANALYSIS

In the reconstruction process of Xi'an city, there are a lot of land needed to be sorted, and renovated to make re-exploitation. This article selects the certain residential area in the western suburbs of xi'an as the object, and use optimization system to optimize the configuration of the land. The types of residential land can be divided into some aspects, such as residential building land, land for traffic in the area, landscaping, land for public services and land for commercial

services in the area, the types can be shown in Table 5, and the total area is 10 km².

5.1 The result for the number configuration optimization

Using the optimization module of the optimization system to optimize the planning condition of the residential area, the obtained results are shown in Table 5.

Table 5. The result for quantity configuration optimization.

The planning types for land use	Planning value(km²)	Optimal value(km²)
Residential building land	3.50	3.65
Traffic land in the area	0.72	0.73
Landscaping	2.40	2.24
Land for public service	2.66	2.78
Land for commericial services in the area	0.62	0.60

5.2 The optimization for spatial structure

The land could be divided into 9 physical blocks by the field measurement and detection, it is shown in Figure 1.

Land use types could be divided, such as: residential building land, land for traffic in the area, landscaping, land for public service, land for commercial

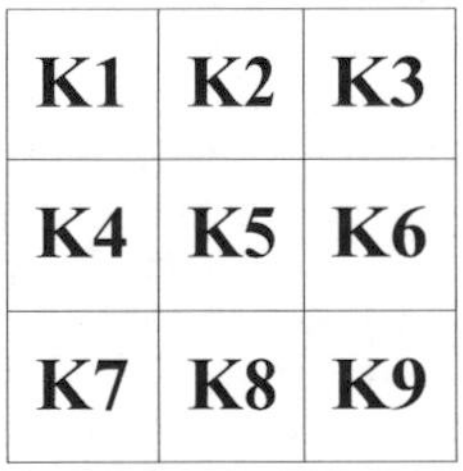

K1	K2	K3
K4	K5	K6
K7	K8	K9

Figure 1. The spatial structure identification of land use type.

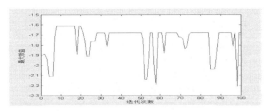

Figure 2. The process of space optimization.

services, the types can be respectively identified by an integer between 1~5.

Space coordinate coefficient is shown as below:

$$C_{ij} = \begin{bmatrix} 0.42 & 0.35 & 0.06 & 0.38 & 0.14 & 0.17 & 0.04 & 0.10 & 0.02 \\ 0.03 & 0.25 & 0.48 & 0.24 & 0.32 & 0.34 & 0.12 & 0.20 & 0.04 \\ 0.38 & 0.32 & 0.21 & 0.32 & 0.24 & 0.20 & 0.47 & 0.10 & 0.32 \\ 0.08 & 0.05 & 0.08 & 0.06 & 0.16 & 0.18 & 0.32 & 0.36 & 0.14 \\ 0.09 & 0.03 & 0.27 & 0.02 & 0.14 & 0.11 & 0.15 & 0.24 & 0.48 \end{bmatrix}$$

The result of the optimization: the optimization process is shown in Figure 2.

The alternative plans are shown in Figure 3:

In the 59th iteration, it produces the alternative plan (1), total benefit is 2.2100;

In the 98th iteration, it produces the alternative plan (2), total benefit is 2.2500;

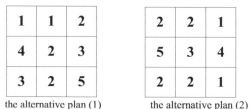

the alternative plan (1)　　　　the alternative plan (2)

Figure 3. The alternative plan.

Use optimization system to adjust the quantity configuration and spatial structure. The quantity optimization makes some adjustment for the land use in the area, which appropriately increases the construction land of the residential area, at the same time the land for public service also increases, while the quantity of landscaping and commercial service in the area correspondingly reduces; this makes it more reasonable. Spatial structure optimization, starting from the actual benefit, produces a variety of planning alternatives to provide decision-makers with reliable and selective decision services, in which rationality plays an important role for rationality of the planning.

6 SUMMARY

There exists a problem in land use planning that how to implement the land use structure to a specific geographical space to achieve the optimal allocation of land resources. The paper summarizes the current situation and deficiency of various land optimization studies, and then puts forward to the mathematical model to establish the optimization configuration model based on genetic algorithm; this provides the reference for the further research on optimized configuration of land resources and land resources management. The biggest advantage of the model is that it can fully reflect the desire of policy makers, which provides a variety of options for the best goals for the decision-makers.

REFERENCES

[1] WANG Jun-hong, HUANG Chen. The Appraisal Model of Optimized Allocation of Urban Construction Land Resource and Its Application[J]. Journal of Xinyang Normal University: Natural Science Edition, 2008.12(2):243–245.

[2] DONG Pin-jie, LAI Hong-song. A Method of Optimization Allocation for Land Use Spatial Structure Based on Multi-objective Genetic Algorithm[J]. Geography and Geo-Information Science, 2003,19(6):52–55.

[3] JIANG You_hua, WANG Xin_sheng. Using genetic algorithms to generate alternative schemes for urban planning[J]. Engineering Journal of Wuhan University, 2002.35(3):63–66.

[4] Han Chen, Wu Yong. Credibility research of the simulation of the integrative avionics electronic system which based on cloud focus judgement[J]. ELECTRONIC MEASUREMENT TECHNOLOGY, 2009,10:69.

[5] Marra M, Walcott B. Stability and optimality in genetic algorithm controllers [C]. Proceedings of the Intelligent Control, 1996, Proceedings of the 1996 IEEE International Symposium on, IEEE, 1996: 492–496.

[6] LIN Ya-hui, ZHONG Xiao-yao, ZHONG Cong-er, etc. Choice of Timber Logistics Center Location Based on Genetic Algorithm[J]. OPERATIONS RESEARCH AND MANAGEMENT SCIENCE, 2007,16(6):51–56.

[7] Lei Yingjie, Zhang Shanwen, Li Xuwu. Matlab genetic algorithm toolbox and application[M]. Xi'an: XIDIAN UNIVERSITY PRESS, 2005.

[8] Wang Xiaoping, Cao Liming. Genetic Algorithms: Theory, Applications and Software Implementation[M]. Xi'an: XI'AN JIAOTONG UNIVERSITY PRESS, 2002.

[9] Shi Zhongzhi, Wang Wenjie. Artificial Intelligence[M]. Beijing: National Defense Industry Press, 2007.

[10] Ma JinTao, L.L.Lai, Yang YiHan. Application of Genetic Algorithms in Reactive Power Optimization[J]. Proceedings of the CSEE, 1995,15(5):347–353.

[11] XI Yi-fan, YANG Mao-sheng, SHANG Yao-hua. Application of genetic algorithm in land function allocation planning[J]. J. Of NW Inst. of Arch. Eng. (Natural Science),2001,18(4):190–194.

Network Security and Communication Engineering – Chan (Ed.)
© 2015 Taylor & Francis Group, London, ISBN: 978-1-138-02821-0

Random walks for link prediction in networks with nodes attributes

Y.X. Chen & L. Chen
Department of Computer Science, Yangzhou University, Yangzhou, China

L. Chen
State Key Lab of Novel Software Tech, Nanjing University, Nanjing, China

ABSTRACT: The problem of link prediction has attracted considerable attention from various domains such as sociology, anthropology, information science, and computer sciences. A link prediction algorithm is proposed based on link similarity score propagation by a random walk in networks with nodes attributes. In the algorithm, each link in the network is assigned a transmission probability according to the similarity of the attributes on the nodes connected by the link. The link similarity score between the nodes are then propagated via the links according to their transmission probability. Our experimental results show that it can obtain higher quality results on the networks with node attributes than other algorithms.

KEYWORDS: Link prediction, Community, Modularity, Complex networks.

1 INTRODUCTION

Many social, biological, and information systems can be naturally described as networks, where vertices represent entities and links denote relations or interactions. Social network consists of individuals and their relations such as friendship and partnership. In the field of sociology, there has been a long history of social network analysis which investigates the relations among social entities. In recent years, social network analysis has attracted considerable attention from various business perspectives, such as marketing and business process modeling. In social networks, links among entities may vary dynamically. For example, email communications and cooperative interactions are changing over time.

Link prediction is an important task in social network analysis. It detects the hidden links from the observed part of the network, or predicts the future links. Link prediction has several applications including predicting relations among individuals such as friendship and partnership, and predicting their future behavior such as communications and collaborations. In social security network, link prediction is used to identify hidden groups of terrorists or criminals [1]. In the networks of human behavior, link prediction is used to detect and classify the behavior and motion of people [2]. In sensor networks, link prediction is used to explore the dynamic temporal properties [3], to ensure information transfer secrecy [4], and to make optimal routing [5].

A problem of increasing interest revolves around node attributes. In the attribute-inference problem, we aim to populate attribute information for network nodes with missing or incomplete attribute data. This scenario often arises in practice when users in online social networks set their profiles to be publicly invisible or create an account without providing any attribute information. The growing interest in this problem is highlighted by the privacy implications associated with attribute inference as well as the importance of attribute information for applications, including people search, collaborative filtering [6], and user identity resolution [7].

Several methods are proposed for link prediction in networks with node attributes, such as relational learning [8], matrix factorization, and alignment [9, 10] based on approaches, have been proposed to leverage attribute information for link prediction, but they suffer from scalability issues. More recently, Backstrom and Leskovec [11] presented a Supervised Random Walk (SRW) algorithm, but this approach does not fully leverage node-attribute information, as it only incorporates node information for neighboring.

In this paper, we proposed a link prediction algorithm based on probability propagation with nodes attributes. In the algorithm, each link is assigned a transmission probability according to the similarity of the attributes. The link similarity score between the nodes are then propagated by a random walk via the links according to their transmission probability. Our experimental results show that it can obtain higher quality results with node attributes than other algorithms.

2 PROBLEM DEFINITION

A network can be presented by $G = (V, E)$, where V is the set of nodes, and E is the set of edges. $A = [a_{ij}]_{n*n}$ is the adjacency matrix of the network. Each node in the network has its attributes reflecting the nature of the object corresponding to this node. For example, in the network of online community, the nodes represent the individuals in the community and the link of nodes shows the relations between friend, such as colleague, kinsfolk, classmates and so on. For each node, the attributes include the person's age, interest, address, profession etc. The relation of classmate of two nodes reflects that they may have similar attributes such as age, education and address. The relation of colleague of two nodes indicates that they may have similar attributes such as address and profession. In the social network, the attribute information of node is an important factor for link prediction. Using attribute information, we can get more accurate link prediction results than using only topological features of network. The more similar attributes two nodes share, the higher probability they have a relationship and are linked.

Let m be the number of attributes of the nodes, we use a vector $T_i=(t_{i1},t_{i2},...,t_{im})$ to represent the attributes of node v_i. Using vectors T_i $(i=1,2,...,n)$ as rows, the $n*m$ matrix $T = [t_{ij}]_{n*m}$ is formed as the attribute matrix of the network. Denote the attributes similarity between nodes v_i and v_j as $sim(i,j)$, the $n*n$ matrix $Sim = [sim(i,j)]_{n*n}$ is the attribute similarity matrix of the network.

Our goal is to predict potential links in the network using the topological information represented by adjacent matrix A, and the attribute information represented by attribute matrix T. The final result is represented by a link similarity score matrix $S=[s_{ij}]_{n*n}$, where s_{ij} indicates the probability of the existence of links between nodes v_i and v_j.

3 SIMRANK INDEX

Our algorithm is based on the SimRank [12] which is a link-based similarity measure and is built on the approach of previously existing link-based measures. SimRank approach is focused on "object-to-object relationships found in many domains of interest" [12]. It is according to the assumption that two nodes are similar if they are connected to similar nodes.

We use $\Gamma(x)=\{u|u \in V,(u,v) \in E\}$ to denote the set of all neighbors of node v, and define a link similarity matrix $S=[s_{ij}]_{n*n}$, and the initial value of element $s_{ij} = 1$ means a link exists between nodes v_i and v_j, and $s_{ij} = 0$ otherwise.

Then, the value of elements in S are updated by an iterative computation. Suppose notes x and y in Figure 1(a)

are similar, then a and b will be similar to some extent since they are connected to similar nodes. We can transfer the link similarity s_{xy} to the edge (a,b) via the edges (x,a) and (y,b).

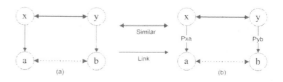

Figure 1. Link similarity $S(x,y)$ is transmitted to modify $S(a,b)$.

Therefore, the iterative formula for calculating s_{ab} is:

$$S_{ab}^{(k+1)} = \frac{c}{|\Gamma(a)| \cdot |\Gamma(b)|} \sum_{x \in \Gamma(a)} \sum_{y \in \Gamma(b)} S_{xy}^{(k)}. \tag{1}$$

$$S=cW^TSW+(1-c)I. \tag{2}$$

In formula (1), c is a constant called attenuation coefficient which shows that attenuation degree of link similarity in the transfer process. Formula (1) can be rewritten to Formula (2) in a matrix form. Here, W is the transformation matrix of adjacent matrix A.

The SimRank can also be interpreted by the random walk process, that is, s_{ij} measures how soon two random walkers, respectively starting from nodes i and j, are expected to meet at a certain node.

4 RANDOM WALKS MODEL FOR LINK PREDICTION IN NETWORKS WITH NODE ATTRIBUTES

A network with node attributes can be presented by adjacent matrix $A = [a_{ij}]_{n*n}$ and attribute matrix $T=[t_{ij}]_{n*m}$. For link prediction the network with node attributes, we should consider both the topological and the attribute information. Therefore, we first calculate the attribute similarity with cosine similarity measure, and we define it as the similarity between their attribute vectors. For two nodes v_i and v_j with attribute vectors T_i and T_j their attribute similarity is defined as:

$$sim(i, j) = \frac{\sum_{k=1}^{m} t_{ik} \cdot t_{jk}}{\sqrt{\sum_{k=1}^{m}(t_{ik})^2} \cdot \sqrt{\sum_{k=1}^{m}(t_{jk})^2}}. \tag{3}$$

We define a link similarity matrix $S=[s_{ij}]$, and the initial value of its element $s_{ij} = sim(i,j)$. Therefore, we assign a transmission probability $P_{ij}=sim(i,j)$ on each edge (i,j). In Figure 1(b), let the link similarity between notes x and y be s_{xy}. The transmission probabilities on edges (x,a) and (y,b) are P_{xa} and P_{yb}. We can transfer the similarity s_{xy} to the edge (a,b) via the edges (x,a) and (y,b).

If nodes x and y have high link similarity, and edges (x,a) and (y,b) have higher probabilities P_{xa} and P_{yb}, then a and b will be similar since they are closely connected to similar nodes.

Therefore, the iterative formula for calculating S_{ab} is:

$$S_{ab}^{(k+1)} = \frac{c}{P_x \cdot P_y} \sum_{x \in \Gamma(a)} \sum_{y \in \Gamma(b)} (P_{xa} + P_{yb}) \cdot S_{xy}^{(k)} / 2. \qquad (4)$$

Here, $P_x = \sum_{v \in \Gamma(x)} S_{xv}$ is the summation of similarities between node x and all its neighbors. Formula (4) also can be rewritten in a matrix form as (2).

The method can also be interpreted as the random walk process as follows. Suppose two walkers respectively start from nodes i and j, randomly walk in the network. They pass through edge (a,b) at the transmission probability P_{ab}. Then s_{ij} measures how soon the walkers are expected to meet at a certain node in the network.

5 EXPERIMENTAL RESULTS

To evaluate our proposed algorithm for link prediction in network with node attributes, we have tested it by a series of experiments on several data sets of networks. We use AUC (Area Under Curve) scores to evaluate the quality of the results by the algorithms tested.

In our experiments, we test our method on eight representative data sets of networks drawn from Digital Bibliography Library Project. We extract two matrixes from each network: the adjacency matrix and the attribute matrix. In each network, the nodes represent the authors, attributes present the key words in the title of papers. Then the datasets present the links between authors and their papers.

In our experiment, we have tested the accuracy of our algorithm RandWalk, and compared the AUC of the results with other ten link prediction algorithms. The test results are shown in Table 1.

Table 1. Comparison of algorithms' accuracy quantified by AUC.

	ACM	CEUS	ICICS	IJCGA	IJNS	JCMC	MSCS	NLDB
RandWalk	**0.8280**	0.9276	**0.9421**	**0.9201**	**0.9017**	**0.8937**	**0.8842**	**0.9298**
CN	0.8222	0.9319	0.9312	0.9116	0.8964	0.8781	0.8186	0.9091
Salton	0.8222	0.9321	0.9316	0.9119	0.8966	0.8782	0.8187	0.9093
Jaccard	0.4075	0.2894	0.3602	0.5551	0.3164	0.4737	0.4775	0.3398
Sorenson	0.8222	0.9321	0.9388	0.9115	0.8965	0.8782	0.8187	0.9093
HPI	0.8222	0.932	0.9388	0.9119	0.8966	0.8781	0.8186	0.9093
HDI	0.8222	0.9321	0.9387	0.9112	0.8965	0.8781	0.8186	0.9092
LHN-I	0.8222	0.932	0.9385	0.9106	0.8964	0.878	0.8186	0.9091
PA	0.5440	0.514	0.5481	0.6191	0.4676	0.6085	0.5611	0.4909
LP	0.8219	**0.9377**	0.9408	0.907	0.8994	0.8777	0.8175	0.916
Kaze	0.8219	**0.9377**	0.9388	0.8847	0.8992	0.8775	0.8139	0.9153

We can observe from Table 1 that our method has the best performance in all datasets except in CEUS. The reason for our algorithm getting higher quality results is that it integrates topological and attribute information. It precisely reflects the occurrence probabilities for the given network with node attributes, and eliminates the influence of bias data. This demonstrates that our proposed algorithm RandWalk can increase the quality of link prediction in networks with node attributes.

propagation in networks with nodes attributes. In the algorithm, each pair of nodes is assigned a transmission probability according to the similarity of the attributes on the pair of nodes. The link similarity score between the nodes are then propagated by the random walk through the links according to their transmission probability. Our experimental results show that algorithm RandWalk can obtain higher quality results on the networks with node attributes than other algorithms.

6 CONCLUSIONS

In this paper, we investigate the problem of link prediction in networks with nodes attributes. A link prediction algorithm RandWalk is proposed based on probability

ACKNOWLEDGMENT

This research was supported in part by the Chinese National Natural Science Foundation under grant Nos. 61379066, 61070047, 61379064, 61472344, Natural

Science Foundation of Jiangsu Province under contracts BK20130452, BK2012672, BK2012128, and Natural Science Foundation of Education Department of Jiangsu Province under contract 12KJB520019, 13KJB520026, 09KJB20013

REFERENCES

[1] L. Y. Lü, T. Zhou, "Link prediction in complex networks: A survey", *Physica A*, Vol.390, 2011, pp.1150–1170.

[2] H. S. Ahmed, B. M. Faouzi, J. Caelen, "Detection and classification of the behavior of people in an intelligent building by camera", *International Journal on Smart Sensing and Intelligent Systems,* Vol. 6, No. 4, September 2013, pp.1317–1342.

[3] L. T. Yang, S. Wang, H. Jiang, "Cyclic Temporal Network Density and its impact on Information Diffusion for Delay Tolerant Networks", *International Journal on Smart Sensing and Intelligent Systems,* Vol. 4, No. 1, March 2011, pp.35–52.

[4] Z. H. Liu, J. F. Ma, and Y. Zeng, "Secrecy transfer for sensor networks: from random graphs to secure random geometric graphs", *International Journal on Smart Sensing and Intelligent Systems,* Vol. 6, No. 1, February, 2013, pp.77–94.

[5] X. Jia, F. Xin, W. R. Chuan, "Adaptive spray routing for opportunistic networks", *International Journal on Smart Sensing and Intelligent Systems*, Vol. 6, No. 1, February, 2013, pp.95–119.

[6] P. Melville and V. Sindhwani. Recommender systems. In *Encyclopedia of Machine Learning*. Springer. 2010.

[7] S. Bartunov, A. Korshunov, S.-T. Park, W. Ryu, and H. Lee. Joint link-attribute user identity resolution in online social networks. In *Proceedings of the Workshop on Social Network Mining and Analysis* (SNA-KDD). 2012.

[8] A. P. Singh and G. J. Gordon. Relational learning via collective matrix factorization. In *Proceedings of the KDD*. 2008.

[9] A. K. Menon and C. Elkan. Link prediction via matrix factorization. In *Proceedings of the European Conference on Machine Learning and Principles and Practice of Knowledge Discovery in Databases (ECML/PKDD)*. 2011.

[10] J. Scripps, P.-N. Tan, F. Chen, and A.-H. Esfahanian. A matrix alignment approach for collective classification. In *Proceedings of the Intenational Conference on Advances in Social Networks Analysis and Mining (ASONAM)* , 2009.

[11] L. Backstrom and J. Leskovec. Supervised random walks: Predicting and recommending links in social networks. *In Proceedings of the WSDM Conference*. 2011.

[12] G. Jeh, J. Widom, SimRank: a measure of structural-context similarity, in: *Proceedings of the ACM SIGKDD International Conference on Knowledge Discovery and Data Mining*, ACM Press, New York, 2002, pp. 271–279.

Network Security and Communication Engineering – Chan (Ed.)
© 2015 Taylor & Francis Group, London, ISBN: 978-1-138-02821-0

Similarity propagation based link prediction in bipartite networks

F.Y. Yao
Department of Computer Science, Yangzhou University, Yangzhou, China

L. Chen
State Key Lab of Novel Software Tech, Nanjing University, Nanjing, China

ABSTRACT: The problem of link prediction has been studied extensively in literature. In this paper we present a new link prediction algorithm on bipartite networks. The algorithm uses a SimRank based on link similarity score. In the algorithm, each link in the network is assigned a transmission probability according to the similarity of the attributes on the nodes connected by the link. The link similarity score between different parts of the nodes are then propagated via the links according to their transmission probability. Our algorithm is empirically validated by extensive experimentation over real social networks of varying sizes.

KEYWORDS: Bipartite network, Link prediction, Random walk, Similarity.

1   INTRODUCTION

Both of the physical systems in nature and the engineered artifacts in human society could be modeled as complex networks, such as biology, economy, sociology, and other subjects. Complex systems could be modeled as networks or graphs, where nodes represent the objects and edges represent the interactions among these objects.

Link prediction is an important task in social network analysis. It detects hidden links from the observed part of the network, and predicts future links given the current structure of the network. Link prediction has several applications including predicting both relations among individuals such as friendship or partnership, and their future behaviors such as communications and collaborations. In a social security network, link prediction is used to identify hidden groups of terrorists or criminals [1], and to detect and classify the behavior and motion of people [2]. In sensor networks, link prediction is used to explore dynamic temporal properties [3], to ensure information transfer secrecy [4], and to realize optimal routing [5].

Bipartite network is an important category of complex networks (opposed to the general unipartite networks). Many real-world networks are naturally bipartite, such as scientists-paper cooperation network [6, 7], the actors-films network [1, 8], investors-company network [9, 10] disease-gene network [11], club member-activities network [12], audience-songs network [13], and computer terminals-data networks in P2P system [14] and so on. In this paper, we present a new link prediction algorithm on bipartite networks. The algorithm uses a SimRank based link similarity score. The link similarity score between different parts of the nodes are propagated via the links. Our algorithm is empirically validated by extensive experimentation over real social networks of varying sizes.

Given an undirected graph $G = (V^X, V, E)$ of a bipartite network, our goal is to predict these implicit links or the links that will appear in the future between nodes in V^X and V^Y. The basic idea of this paper is to use a similarity score to measure the probability of the occurrence of a link between each node pair (x, y), where $x \in V^X$, $y \in V^Y$. Node pair with higher similarity is more likely to be connected by a link. In our method, we use a SimRank based similarity score.

2   THE SIMRANK INDEX FOR BIPARTITE NETWORKS

SimRank[15] is a general effective similarity measure based on a simple graph-theoretic model. Let $s(a, b) \in [0, 1]$ denote the similarity between objects a and b, $s(a, b)$ can be defined as a recursive equation:

$$s(a,b) = \begin{cases} 1 & a = b \\ \dfrac{c}{|\Gamma(a)| \cdot |\Gamma(b)|} \displaystyle\sum_{x \in \Gamma(a)} \sum_{y \in \Gamma(b)} s(x, y) \end{cases} \quad (1)$$

Here, c is a constant between 0 and 1, $\Gamma(a)$ is the set of neighbors of node a.

In this paper, we extend the basic SimRank equation (1) to bipartite domains consisting of two types of objects. We use recommended systems as motivation, where nodes A and $B \epsilon V^X$ represent two customers, nodes x, y, w, and $z \epsilon V^Y$ represent the items. In such bipartite network, similarity of items and similarity of people are mutually-reinforcing notions. Namely, people are *similar* if they purchase *similar* items. Items are *similar* if they are purchased by *similar* people. The mutually-recursive equations that formalize these notions are analogous to equation (1). Let $S(A, B)$ be the similarity between customers A and B, and let $S(x, y)$ denote the similarity between items x and y. The initial values of $S(A, B)$ and $S(x, y)$ are:

$$S(A,B) = \begin{cases} 1, A=B \\ 0, A \neq B \end{cases}, \quad S(x,y) = \begin{cases} 1, x=y \\ 0, x \neq y \end{cases}. \quad (2)$$

Then, we can update the value of $S(A, B)$ by an iterative computation. The iterative formula for calculating $S(A, B)$ is:

$$S(A,B) = \frac{c_1}{|\Gamma(A)| \bullet |\Gamma(B)|} \sum_{x \in \Gamma(A)} \sum_{y \in \Gamma(B)} S(x,y). \quad (3)$$

Similarly, the iterative formula for calculating $S(x, y)$ is:

$$S(x,y) = \frac{c_2}{|\Gamma(x)| \bullet |\Gamma(y)|} \sum_{A \in \Gamma(x)} \sum_{B \in \Gamma(y)} S(A,B). \quad (4)$$

Here, $\Gamma(A) = \{x \mid x \epsilon V^Y, (A,x) \epsilon E\}$ stands for A's neighbors, and $\Gamma(A)$ stands for the degree of node A. Parameters c_1 and c_2 are both constants called attenuation coefficient which show that attenuation degree of link similarity in the transfer process. Such operations of link similarity transferring and modification will be performed repeatedly until the values of link similarity convergence. In fact, formula (3) stands for the average similarity of items bought by customer A and B and formula (4) stands for the average similarity of customers who bought x and y. Starting from the initial values initial values of $S(A, B)$ and $S(x, y)$ defined in (2), we can update the values of $S(A, B)$ and $S(x, y)$ alternatively using (3) and (4) until convergence. The output of $S(A, B)$ is the similarity of V^X nodes A and B, and $S(x, y)$ is the similarity of V^Y nodes x and y.

3 SIMILARITY PROPAGATION FOR LINK PREDICTION IN BIPARTITE NETWORKS

For link prediction in bipartite network, our goal is to estimate the similarity of the pair of nodes in different parts. Let nodes $A \in V^X$, $x \in V^Y$, and $Sim(A, x)$ be the similarity between nodes A and x. If there is a

link between node A and node x in the bipartite network, we set $Sim(A, x) = 1$, otherwise $Sim(A, x) = 0$. Therefore, the initial value of $Sim(A, x)$ is:

$$Sim(A,x) = \begin{cases} 1, (A,x) \in E \\ 0, (A,x) \notin E \end{cases}. \quad (5)$$

Then we interatively update the value of $Sim(A,x)$ for all the node pairs (A,x) which have no direct link.

We use the following iterative formula to calculate their similarity:

$$Sim(A,x) = \frac{c_3}{P_A . P_x} \sum_{y \in \Gamma(A)} \sum_{B \in \Gamma(x)} S(A,B).S(x,y).Sim(B,y). \quad (6)$$

Here, P_A stands for the summation of similarities of node A and nodes which have direct connection with node x in the bipartite network, and P_x stands for the summation of similarities of node x and neighbors of node A. Parameter $c3$ is a constant called attenuation coefficient which show that attenuation degree of link similarity in the transfer process. Based on the initial value of $Sim(A, x)$ defined in (5), the similarities are updated iteratively using (6) until convergence. The final value of $Sim(A, x)$ is just the occurrence probability of link (A,x).

The method can also be interpreted as the random walk process as follows. Suppose two walkers respectively starting from nodes A and x, randomly walk in the network. They pass though edge (B, y) at the transmission probability P_{By}. Then $Sim(A, x)$ measures how soon the walkers are expected to meet at a certain node in the bipartite network.

Our algorithm consists of two steps. First, we calculate the similarity of all node pairs within parts V^X and V^Y using formula (3) and (4) respectively. Second, we calculate the similarity of node pairs between parts V^X and V^Y. The final value of $Sim(A, x)$ is the element of the similarity matrix as the link prediction result.

4 EXPERIMENTAL RESULTS AND ANALYSIS

4.1 Experiments on southern women data set

Southern Women Data Set is a widely used data set for testing in bipartite networks. It was collected by Davis in 1930. This data set is made up of 18 women and 14 social activities, only when a women took part in some social activities, the corresponding node in the network can be connected, so this is a typical bipartite network. The structure of network for this data set is shown in Figure 1.

We randomly extract 10 edges from the data set as the testing set and the remaining 79 edges as training set.

The experiment was done 16 times in total, each time we randomly extract 10 edges from the data set

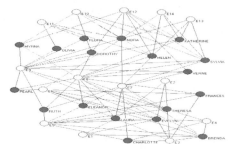

Figure 1. The structure of network for southern women data.

Figure 2. The structure of network for Scotland.

as the testing set. The detailed experimental results were shown in Table 1, where P represents the number of edges predicted by our proposed algorithm. We can observe from Table 1 that our proposed algorithm can predict most of the edges in the Southern Women Data Set. This shows that our proposed algorithm is feasible and efficient.

Table 1. The detailed experimental results of southern women data set.

No.	T	P	AUC	No.	T	P	AUC
1	10	9	0.8222	9	10	10	0.9153
2	10	10	0.9116	10	10	6	0.5483
3	10	10	0.9157	11	10	7	0.6191
4	10	9	0.8539	12	10	9	0.8139
5	10	8	0.7316	13	10	8	0.7158
6	10	9	0.8187	14	10	10	0.9377
7	10	7	0.6085	15	10	9	0.8032
8	10	9	0.8782	16	10	9	0.8833

4.2 Experiments on Scotland data set

In the second experiment, we adopted the data set of a chain enterprise in Scotland in the early 20th century for testing. The data set is made up of 136 shareholders and 108 companies in the early Scotland, each shareholder may work in different companies and each company may have different shareholders. Thus, it formed a bipartite network between the shareholders and the company. But different from the Southern Women Data Set, the graph of this data set is unconnected, there are many outliers. As is shown in Figure 2, the connected graph consisted of blue vertices is the largest connected sub-graph of this data set, the other colors of vertices are outliers.

We conducted two experiments on this data set. First, we tested them on the whole unconnected Scotland network. In the final similarity matrix between shareholders and the companies, we found

Figure 3. The biggest connected sub-graph of Scotland.

that some of the similarities of node pairs were zero, that's to say, our algorithm can detect the outliers. Then we focus on the connected bipartite network, so we extract one of the largest connected sub-graphs of the data set, which consists of 131 shareholders and 86 companies, as shown in Figure 3.

In Figure 3, 131 red vertices represent the shareholders; 86 blue vertices represent the companies. Then we test our algorithm on this dataset to predict the potential links. In the experiments, we also repeat the test for 16 times with different training sets of edges, and finally find that the average of AUC was about 0.8731. The test result shows that our algorithm is effective in link prediction, and can also be used to detect those outliers.

5 CONCLUSIONS

We present a new link prediction algorithm on bipartite networks. The algorithm uses a SimRank based link similarity score. In the algorithm, each link in the network is assigned a transmission probability according to the similarity of the nodes with each part. The link similarity score between different parts of the nodes are then propagated via the links according to their transmission probability. Our experimental results show that our algorithm our algorithm is effective in link prediction over real social networks of varying sizes, and can also be used to detect those outliers.

ACKNOWLEDGMENTS

This research was supported in part by the Chinese National Natural Science Foundation under grant Nos. 61379066, 61070047, 61379064, 61472344, Natural Science Foundation of Jiangsu Province under contracts BK20130452, BK2012672, BK2012128, and Natural Science Foundation of Education Department of Jiangsu Province under contract 12KJB520019, 13KJB520026, 09KJB20013.

REFERENCES

[1] L. Y. Lü, T. Zhou, "Link prediction in complex networks: A survey", Physica A, Vol.390, 2011, pp.1150–1170.

[2] H. S. Ahmed, B. M. Faouzi, J. Caelen, "Detection and classification of the behavior of people in an intelligent building by camera", International Journal on Smart Sensing and Intelligent Systems, Vol. 6, No. 4, September 2013, pp.1317–1342.

[3] L. T. Yang, S. Wang, H. Jiang, "Cyclic Temporal Network Density and its impact on Information Diffusion for Delay Tolerant Networks", International Journal on Smart Sensing and Intelligent Systems, Vol. 4, No. 1, March 2011, pp.35–52.

[4] Z. H. Liu, J. F. Ma, and Y. Zeng, "Secrecy transfer for sensor networks: from random graphs to secure random geometric graphs", International Journal on Smart Sensing and Intelligent Systems, Vol. 6, No. 1, February, 2013, pp.77–94.

[5] X. Jia, F. Xin, W. R. Chuan, "Adaptive spray routing for opportunistic networks", International Journal on Smart Sensing and Intelligent Systems, Vol. 6, No. 1, February, 2013, pp.95–119.

[6] Newman M E J. Scientific collaboration networks. I. network construction and fundamental results [J]. Physical Review E, 2001, 64: 016131.

[7] Newman M E J. Scientific collaboration networks. II. Shortest paths, weighted networks, and centrality [J]. Physical Review E, 2001, 64: 016132.

[8] LIU Ai-fen, FU Chun-hua, ZHANG Zeng-ping, CHANG Hui, HE Da-ren, An Empirical Statistical Investigation on Chinese Mainland Movie Network, Complex Systems and Complexity Science, 2007, 4(3): 10–16.

[9] Robins G, Alexander M. Small worlds among interlocking directors: network structure and distance in bipartite graphs [J]. Computational & Mathematical organization Theory, 2004, 10: 69–94.

[10] Battiston S, Catanzaro M. Statistical properties of corporate board and director networks [J]. European Physics Journal B, 2004, 38: 345–352.

[11] CHEN Wen-qin, LU Jun-an, LIANG Jia, Research in Disease-Gene Network Based on Bipartite Network Projection, 6(1), 2009, 13–19.

[12] Ergun G, Human sexual contact network as a bipartite graph[J]. Physica A, 2002, 308: 483–488.

[13] Lambiotte R, Ausloos M. Uncovering collective listening habits and music genres in bipartite networks [J]. Physical Review E, 2005, 72: 066107.

[14] Le Blond S, Guillaume J L, Latapy M. C lustering in P2P exchanges and consequences on performances[C]. Castro M, Renesse R. Peer- to-Peer Systems IV. Berlin: Heidelberg, 2005, 193–204.

[15] G. Jeh and J. Widom. SimRank: a measure of structural-context similarity. In KDD '02: Proceedings of the eighth ACM SIGKDD international conference on Knowledge discovery and data mining, pages 538–543. ACM Press, 2002.

Multimedia, Signal and Image Processing

Network Security and Communication Engineering – Chan (Ed.)
© *2015 Taylor & Francis Group, London, ISBN: 978-1-138-02821-0*

Electricity information collection system design based on WIMAX over 230 MHZ dedicated frequency band

Z.T. Lai

State Grid Electric Power Research Institute, Nanjing, China

ABSTRACT: Electricity information collection system has attracted great attention. However, existing research work is relatively insufficient. In this paper, electricity information collection system based on 230 MHz WiMAX system is designed. Moreover, networks topology and access scheme are presented. From the actual operation situation, the system function and performance can both satisfy the requirements.

1 INTRODUCTION

Nowadays, with the development of the application in electricity information collection system, the state grid corporation of China requires that the successful rate for metering reading should be greater than 99%, which has raised high demand for the coverage, bandwidth, reliability and security of the communication networks for electricity information collection system. However, the current wireless communication technology for electricity information collection system, such as optical fiber communication, power line narrow-band carrier communication, general packet radio service/code division multiple access (GPRS/CDMA) and wireless communication over 230MHz frequency band cannot meet the requirements of such a system.

Therefore, how to build an intelligent communication networks with supporting bidirectional communication, high real-time performance, high reliability and high security is a key problem when constructing a communication networks for an electricity information collection system. In particular, under the condition of limited communication resources, how to utilize the advanced communication technology to update and rebuild the means of communication is an urgent task to be solved. In this paper, we present a framework for the electricity information collection system based on worldwide interoperability for microwave access (WiMAX) wireless broadband networks using 230MHz dedicated frequency band for power communication and study the key issue related to information security in the system.

2 WIMAX WIRELESS BROADBAND NETWORKS OVER 230 MHZ ELECTRIC POWER DEDICATED FREQUENCY BAND

WiMAX is a kind of new broadband wireless access technology. It can provide high rate service and its Maximum coverage radius is 50 km. Moreover, WiMAX has some advantages, such as quality of service guarantee, high data rate, diverse services, and so on. A series of advanced technologies are used in WiMAX system, which includes orthogonal frequency division multiplexing (OFDM), adaptive antenna system (AAS), multi-input multi-output (MIMO), and so on. Mobility service can also be realized in WiMAX system. Aiming at the smart grid in China, WiMAX wireless broadband networks over 230 MHz dedicated frequency band is designed, which adopts several advanced technologies such as spectrum optimization, software radio, power optimization, and so on.

3 ELECTRICITY INFORMATION COLLECTION SYSTEM BASED ON 230 MHZ WIMAX

The networks topology of electricity information collection system based on 230 MHz WiMAX is shown in Fig. 1.

As is shown in Fig. 1, the system consists of collecting device, wireless terminal equipment, WiMAX wireless base station, and master station server. Collecting device includes switch monitoring device, camera, and electricity meter data collector, which can provide data source for the system.

Dedicated networks are composed of wireless terminal equipment and WiMAX wireless base station, which can provide data transmission platform for the system.

Master station server is in charge of receiving and processing data.

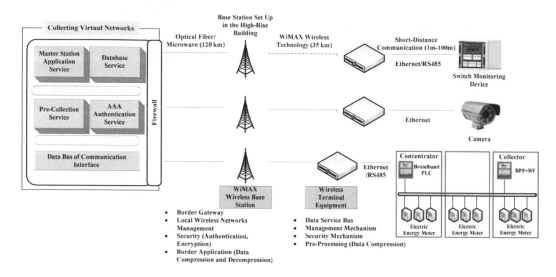

Figure 1.   The networks topology of electricity information collection system based on 230 MHz WiMAX.

The communication process between master station and collecting device under WiMAX networks is described as follows:

Collecting device is connected to wireless access equipment by Ethernet, RS485 or RS232 interface, which can build a virtual subnet.

Wireless access equipment is connected to WiMAX wireless base station by using WiMAX wireless communication technology. The maximum coverage radius of BS is 35 km.

WiMAX wireless base station can be accessed to collecting virtual networks by optical fiber or microwave, which can realize the end-to-end communication between collecting terminal and electricity information collection system.

Because the transmitted data is sensitive electricity information in electricity information collection system, it has a high requirement for information security.

4   ACCESS SCHEME IN WIMAX

Wireless wideband power system based on 230 MHz WiMAX is a kind of dedicated communication networks. Other user equipments cannot get access to the dedicated networks. Compared with other 3G networks, it can guarantee communication security

for the grid company. Access flow chart in WiMAX networks is shown in Fig. 2.

Subscriber station (SS) needs to synchronize with downlink channel in order to obtain DL-MAP and UL-MAP. DL-MAP and UL-MAP contain information of downlink and uplink sub-frames, respectively. The whole connection process includes ranging, capability negotiation, authentication, and registration.

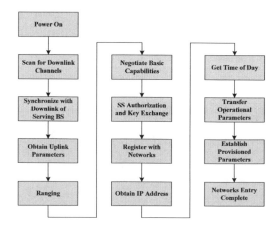

Figure 2.   Access flow chart in WiMAX networks.

Firstly, SS sends ranging request (RNG-REQ) to base station (BS) by the ranging competition slot. It will not send RNG-RSP at each frame until BS receives ranging response (RNG-RSP). Then, capability negotiation and authentication will be done. Finally, the way of request and response is also used in registration.

5 CONCLUSIONS

Installation and test of WiMAX BS, the replacement and file import of carrier communication module of electricity meter and dual mode communication module based on 230 MHz wireless cellular networks project have been finished on March 25, 2014. The average response time is 0.5-2 seconds. The success rates of broadband carrier and data receiving are more than 90% and 99.3%, respectively. The performance of data uploading and collecting can both satisfy the related requirements.

ACKNOWLEDGMENTS

This paper is supported by Science and Technology Project of State Grid Corporation (Grant No. 524600120043).

REFERENCES

[1] Pareit, D. & Lannoo, B. & Moerman, I. & Demeester, P. 2012. The History of WiMAX: A Complete Survey of the Evolution in Certification and Standardization for IEEE 802.16 and WiMAX. *IEEE Communications Surveys & Tutorials* 14(4): 1183–1121.

[2] Miao, K.X. 2010. Enterprise WiMAX building the next generation enterprise wireless infrastructure with WiMAX. Proceedings of the 2010 International Conference on Wireless Information Networks and Systems (WINSYS): 1–5.

[3] Abichar, Z. & Peng Y. & Chang, J.M. 2006. WiMax: The Emergence of Wireless Broadband. IT Professional 8(4): 44–48.

[4] More, S. & Mishra, D.K. 2012. 4G Revolution: WiMAX technology. Proceedings of the 2012 Third Asian Himalayas International Conference on Internet (AH-ICI): 1–4.

[5] Fourty, N. & Val, T. & Wei, A. 2009. A WiMAX physical layer use for emergency audio communications. Proceedings of the IEEE International Conference on Wireless and Mobile Computing, Networking and Communications: 435–440.

[6] Neves, P. & Fontes, F. & Monteiro, J. & Sargento, S. & Bohnert, T.M. 2008. Quality of service differentiation support in WiMAX networks. Proceedings of the International Conference on Telecommunications: 1–5.

[7] Lu, K. & Qian, Y. & Chen, H.h. & Fu, S. 2008. WiMAX networks: from access to service platform. IEEE Network 22(3): 38–45.

[8] Habib, M. & Ahmad, M. 2010. A review of some security aspects of WiMAX and converged network. Proceedings of the Second International Conference on Communication Software and Networks: 372–376.

[9] Lang W. & Wu R. & Wang J. 2008. A simple key management scheme based on WiMAX. Proceedings of the International Symposium on Computer Science and Computational Technology: 3–6.

[10] Wang, M. 2011. WiMAX physical layer: specifications overview and performance evaluation. Proceedings of the Consumer Communications and Networking Conference (CCNC): 10–12.

Network Security and Communication Engineering – Chan (Ed.)
© 2015 Taylor & Francis Group, London, ISBN: 978-1-138-02821-0

Video semantic classification based on ELM and multi-features fusion

P. Li & H. Wang

Key Laboratory of Fiber Optic Sensing Technology and Information Processing, Wuhan University of Technology, Ministry of Education, Wuhan, Hubei, China

ABSTRACT: Semantic video analysis plays an important role in video retrieval and computer vision. It bridges the semantic gap between video low–level features and high-level semantics. In this paper we proposed a method for semantic classification based on Extreme Learning Machine (ELM) and multi-features fusion. First we respectively extracted color, region, and texture characteristic vectors from the key frames in every video shot, then integrated the feature vectors and PCA was applied to get lower dimensional feature vectors which can be beneficial feature matrix for classifications. Finally, ELM and SVM were applied to classify semantics and we compared the effect of the single feature and the fusion feature. The experiment shows that the method promoted in this paper displays better classification accuracy for semantic video, and achieve performance at faster computing speeds than SVM.

1 INTRODUCTION

With the continuous development of Internet technology, Digital video has become an important way of network information recording and transmission. The video data is increasing at a tremendous speed in every day and the huge amounts of data bring great challenge for video classification and video retrieval. However, the traditional artificial way to analyze, manage and retrieve the video data, which not only consume large amounts of manpower, but also cannot adapt to the rapid growth of video data for classification and retrieval. In this case, there is an urgent need for a kind of content-base video retrieval system (CVBR) automatically. Because of the semantic gap between low-level features and high-level semantic concepts in video[5],how we bridge the semantic gap and extract semantic concept in video have attract more and more attention.

Currently, many scholars put forward a series method of content-base video semantic analysis, most of which apply SVM as the classifier for semantic classification. Niu [7] defines the color ratio feature and utilizes SVM as semantic video classifier, and the semantic is used for shot classification. Lu[6] presents a framework for semantic video event annotation that exploits three low-level features in video and uses SVM for classification. Ding[2] presents a method of video semantic annotation based on Bag-of-features which forms the classifiers under SVM for semantic annotation. However, the SVM classifier usually suffers from the high computational cost and the large number of parameters to be optimized.

To solve these problems, we apply Extreme Leaning Machine (ELM) as the classifier for semantic video classification. The ELM is proposed by Nanyang Technological University G.B. Huang[3][4][8], an associate professor, which is a kind of fast single-hidden layer feed forward neural networks. The parameters of the network are determined without any iteration, thus it greatly reduces the parameters adjusting time in the network, which has the advantages of fast learning speed and good robustness. This paper introduces a new method based on ELM and multi-features, first we extract visual features in video and then use the multi-features to train the ELM for semantic classification, which not only achieve performance at fast computing speed, but also has fewer optimization constraints.

2 THE FRAMEWORK OF SEMANTIC VIDEO ANALYSIS

In the process of acquiring of video semantic, the first step is video shot segmentation and then is key frame extraction in each shot. Key frame extraction is the basis for many video-retrieval related applications, being one of the well-investigated subjects that still gains research community's attention[1].In this paper we proposed a new way that applies the median center clustering point to get the key frame, choosing a frame which feature point has the minimum distance to other frames in every shot that includes almost all visual content of shot. As to semantic video classification, single visual feature usually couldn't get good performance at classification, so we apply the method of multi-feature serial fusion and PCA to get lower dimensional feature vectors which can be the most beneficial feature matrix for classifications and to improve the classification accuracy for semantic video. Figure 1 shows the structural model of semantic video classification

Figure 1. Structural model of semantic video classification.

The rest of this paper is organized as follows. Section 3 introduces the video low-level feature extraction and feature vectors standardization. In section 4, we describe ELM classification principle based on multi-semantics. Section 5 shows the experiment result. Finally, we conclude the paper in Section 6.

3 FEATURE EXTRACTION

The semantic video analysis is researching the relationship between the low-level features and high-level semantic concepts. In this paper three kinds of color, texture and region features are applied to improve classification accuracy, which effectively make up for the shortcoming of single feature after serial fusion.

3.1 Color feature extraction

Color feature is one of the most widely used features in the video and image processing, which is closely related to a large amount of information contained in video. We select HSV space for processing which conforms to the human eyes perception characteristics, and nonlinear quantitative is applied to the three components for non-uniform quantization. In order to get lower dimensional feature vectors, the Hue is divided into 16 grades, the Saturation and the Value is respectively divided into 4 grades. Given a image which pixel ratio is $M*N$, we need to convert the 3 dimensions color space into 1 dimension. The color value is calculated as:

$$f(x, y) = 16*h(i, j) + 4*s(i, j) + 4*v(i, j) \quad (1)$$

$i = 1, …, M, j = 1, …, N.$
The color histogram is given by:

$$H_d = \frac{1}{M \times N} \sum_{i=0}^{M-1} \sum_{j=0}^{N-1} \delta(f\, i, j, s), \forall d \in D \quad (2)$$

where D is the color set included in image, in this paper the set number is 256.

Figure 2 and Figure 3 shows the image and color histogram.

Figure 2. An image of NBA.

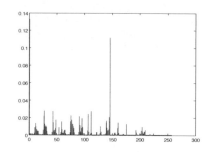

Figure 3. The color histogram of image.

3.2 Texture feature extraction

Different objects in image have different surface texture and organizational structure, which represent different semantics in video. The Gray-level co-occurrence matrix (GLCM) from an image is one of the most widely used methods for texture analysis, which can calculate the second-order statistical properties of the image. First we compress the grayscale of image from 256 to 16, calculate co-occurrence matrix for given direction 0°, 45°, 90°, 135° and distance between pixels 1, then we get four matrixes and calculate the texture parameters: Energy, Entropy, Moment of inertia and Correlation. Finally, the mean and standard deviation of the four texture parameters are calculated respectively and we get 8 dimensions texture feature. Table 1 shows the Texture feature vector of key frame in Figure 4.

3.3 Region feature extraction

In this paper we apply seven moment invariants of the image which are translation invariance, rotation invariance and scale invariance.

We standardized the feature vectors in order to adapt to the classifier's work. Given a feature vector

Table 1. Texture feature vector.

Semantic	Texture feature vector
Sports	0.1100, 0.0063, 4.8033, 0.1972, 1.9699, 0.6521, 0.9212, 0.0262
Forest	0.0796, 0.0044, 5.0819, 0.1950, 1.6705, 0.5239, 0.9323, 0.0213
Mountain	0.1086, 0.0038, 40.370, 0.1297, 0.5049, 0.1770, 0.9871, 0.0045
Desert	0.0919, 0.0045, 4.4490, 0.1858, 0.9413, 0.3737, 0.9799, 0.0080
Parade	0.1295, 0.0063, 4.3993, 0.1554, 2.0834, 0.7316, 0.8948, 0.0371

with n dimensions $V = \{v_1, v_2, \ldots, v_n\}$, the sample mean μ and the standard deviation σ are calculated as:

$$\mu = \frac{1}{n}\sum_{i=1}^{n} Vi, \quad \sigma = \sqrt{\frac{1}{n}\sum_{i=1}^{n}(Vi - \mu)^2} \qquad (3)$$

Then the standardization vector is obtained:

$$V' = \frac{Vi - \mu}{\sigma} \qquad (4)$$

4 THE PRINCIPLE OF ELM

In the traditional artificial neural network, the parameter of hidden layer nodes is obtained by certain iterative algorithm for optimization, which makes the parameters take up a lot time in training process. In order to enhance the overall performance of the building network, Nanyang Technological University G.B. Huang[5][6][7], an associate professor, put forward a new algorithm Extreme Leaning Machine (ELM). ELM is a kind of fast single-hidden layer feed forward neural networks. Figure 4 exhibits the structure of SLFN.

The main idea for the algorithm is that given N samples $X = [x_1, x_2, \ldots, x_n]^T$ and m semantic results $T = [t_1, t_2, \ldots, t_m]$, if the SLFN with Ñ hidden nodes, the input weight ω_i and the neurons threshold of the hidden layer b_i, then the output of SLFN is presented by:

$$\sum_{i=1}^{\tilde{N}} g(\omega_i \cdot x_j + b_i)\beta_i = t_j \quad j = 1, 2, \ldots, N. \qquad (5)$$

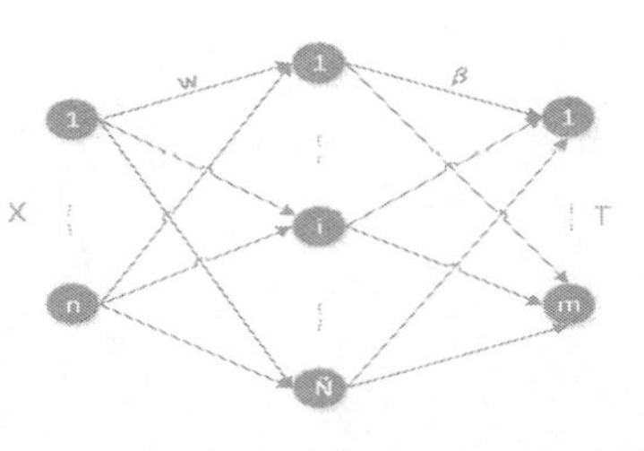

Figure 4. The structure of SLFN. SFLN includes three layers: input layer, hidden layer and output layer.

Where β_i is the output weight. In our method, we use sigmoid activation function of hidden nodes, which is formulated as:

$$g(\omega_i, x_j, b_i) = 1/(1 + \exp(-\omega_i \cdot x_j + b_i)) \qquad (6)$$

The Eq. (6) can be written as:

$$H\beta = T \qquad (7)$$

where

$$H = \begin{bmatrix} g(\omega_1, x_1, b_1) & \cdots & g(\omega_{\tilde{N}}, x_1, b_{\tilde{N}}) \\ \vdots & \ddots & \vdots \\ g(\omega_1, x_N, b_1) & \cdots & g(\omega_{\tilde{N}}, x_N, b_{\tilde{N}}) \end{bmatrix}_{N \times \tilde{N}} \qquad (8)$$

$$\beta = \begin{bmatrix} \beta_1 \\ \vdots \\ \beta_{\tilde{N}} \end{bmatrix}_{\tilde{N} \times m}, \quad T = \begin{bmatrix} t_1 \\ \vdots \\ t_N \end{bmatrix}_{N \times m} \qquad (9)$$

The output weight β is designed as:

$$\check{\beta} = H^+ T \qquad (10)$$

where H^+ is the Moore-Penrose generalized inverse of matrix H.

Algorithm ELM:

Input: Given a training set $\{(x_i, t_i)\}_{i=1}^{N} \subset R^n \times R^m$, activation function $g(\omega_i, x_j, b_i)$ and Ñ hidden nodes.

Step1: Randomly assign hidden node parameters $(\omega_i, b_i), i = 1, \cdots, \tilde{N}$.
Step2: Calculate the hidden layer output matrix H.
Step3: Calculate the output weight $\check{\beta} = H^+ T$.
Step4: Given a testing set $\{(x_i, t_i)\}_{i=1}^{N} \subset R^n \times R^m$.
Step5: Calculate the hidden layer output matrix H.
Output: The result of classification $T = H\beta$.

5 SIMULATION EXPERIMENTS

We select five semantics video in youku to evaluate the method proposed in this paper. Figure 4 shows part of the key frames. If a shot represents a kind of semantic, we express it as 1, or use 0 to express it, so that we form a semantic vector which is constructed by 0 and 1.

First we divide the video data into testing set and training set, then we apply ELM and SVM to classify the semantic video. The experiment result is shown in Table 2. Running environment is Win 7 of 32 bits, 3GHz Pentium (R) Dual-Core E5700, 2G internal storage, Matlab2010b.

Sports

Forest

Mountain

Desert

Parade

Figure 4. Part of the key frames.

Given the number of correct semantic a and error semantic b, then the Correct Rate Z is designed by:

$$Z = a / (a + b) \qquad (11)$$

We choose 110 key frames in this experiment, as shown in Table 2, ELM algorithm run 10 and 30 times faster and displays better classification accuracy rate than SVM, the accuracy of feature fusion is better than that of single feature as whole. The correct rate of fusion feature is highest in all features, and the performance of semantic Sports and Forests is generally better than Desert and Parade.

Table 2. The algorithm performance comparison results.

		ELM		SVM	
Feature	Semantic	Correct Rate	Time(s)	Correct Rate	Times(s)
Texture feature	Sports	0.909	0.01537	0.909	0.50572
	Forest	0.636		0.636	
	mountain	0.727		0.364	
	Desert	0.364		0.364	
	Parade	0.091		0.273	
Region feature	Sports	0.091	0.12679	0.091	0.57116
	Forest	0.364		0.091	
	mountain	0.273		0.091	
	Desert	0.273		0.273	
	Parade	0.182		0.091	
Color feature	Sports	0	0.03259	0.273	0.64732
	Forest	0.818		0.545	
	mountain	0.727		0.182	
	Desert	0.364		0.273	
	Parade	0.727		0.364	
Fusion feature	Sports	0.909	0.02878	0.727	0.32329
	Forest	0.909		0.636	
	mountain	0.909		0.909	
	Desert	0.545		0.455	
	Parade	0.455		0.364	

6 CONCLUSIONS

In this paper a classification algorithm of semantic video which is based on ELM and multi-features is proposed. First we combine three visual features and PCA is applied to get lower dimensional feature vectors which can be the most beneficial feature matrix for classifications. Then we compare the performance of ELM and SVM, the accuracy differences between fusion feature and single feature. The experiment shows that the algorithm ELM has advantages on computing speed and displays better classification accuracy rate than SVM, the accuracy of feature fusion is better than that of single feature as whole.

REFERENCES

[1] Chatzigiorgaki, M. & Skodras, A.N. 2009. Real-timekeyframe extraction towards video content identification, 2009 16th International Conference on Digital Signal Processing: 1–6.

[2] Ding Youdong, Zhang Hianfei, Li Jun. 2011. A Bag-of-Feature Model for Video Semantic Annotation. 2011 sixth International Conference on Image and Graphics: 696–701.

[3] Huang G.B.& Zhu Q.Y. 2006. Extreme learning machine: theory and applications. Neurocomputing70: 489–501.

[4] Huang G.B.& Wang Dian Hui, Lan Yuan. 2011. Extreme Learning Machines: A Survey, Inter. J. Mach. Lean. Cyber2(2):107–122.

[5] Li, B. & Sezan, I. 2003. Semantic sports video analysis: approaches and new applications. In IEEE Proceedings of 2003 international Conference on Image Processing: 17–20. Spain: Barcelona.

[6] Lu Jianjiang, Tian Yulong, Yang Li. 2009. A Framework for Video Event Detection Using Weighted SVM Classifiers. International Conference on Artificial Intelligence and Computational Intelligence4: 255–259.

[7] NiuZhenxing, Gao Xinbo, TaoDacheng, Li, Xuelong. 2009. Semantic Video Shot Segmentation Based on Color Ratio Feature and SVM, 2008 International Conference on Cyber worlds: 157–162.

[8] Zong W.W.&Huang G.B. 2011. Face Recognition Based on Extreme Learning Machine. Neurocomputing. In press.

Network Security and Communication Engineering – Chan (Ed.)
© *2015 Taylor & Francis Group, London, ISBN: 978-1-138-02821-0*

A study of the teaching methodology of digital signal processing

Z.H. Dong, Y. Li, D.Z. Mu & Y.J. Tang
Beijing Key Laboratory of Signal and Information Processing for High-end Printing Equipments, Beijing Institute of Graphic Communication, Beijing, China

ABSTRACT: Digital signal processing is a core curriculum of electronic and information study in tertiary education. In order to resolve some problems which exist in current curriculum teaching, this study examines the teaching content, teaching methods, assessment methods, and recommends some measures of teaching reforms, such as well-connected curriculum, well-organized teaching content, a variety of teaching and assessment methods etc. These enable students to understand and master the knowledge, and strengthen their problem solving capabilities.

KEYWORDS: digital signal processing, teaching content, teaching method, assessment method.

1 INTRODUCTION

Digital signal processing is an important core curriculum of electronic and information study[1]. The basic theory, basic analytical method and implementation method have been widely used in a number of related research areas, providing a solid foundation of knowledge for students engaging in theoretical research, application development and technology management. Being a highly theoretical curriculum, involving a wide range of knowledge, digital signal processing curriculum can be divided into three parts: digital signal analysis, digital filter design and digital signal processor. The first two parts are essential and they include time-domain analysis of discrete system, z-domain analysis, Fourier analysis of discrete signals, design and implementation of IIR and FIR filters[2]. These techniques are closely related to applied sciences such as communications engineering, industrial control, pattern recognition, speech and image processing, neural networks, and therefore widely used in modern life[3]. By studying digital signal processing, students shall master the basic concepts, principles and methods, and shall be equipped with the ability of self-learning, independent analysis, problem solving, engineering implementation and innovation.

2 PROBLEMS IN THE TEACHING PROCESS

Since the curriculum of digital Signal processing is highly theoretical, abstract has a wide coverage, there exist the following problems in the teaching process.

2.1 *Lack of connectivity in the curriculum*

Signal and system is the beginning course of digital Signal processing curriculum. It examines the continuous-time signal and the characteristics of the system, and analyses its time-domain, frequency-domain and complex frequency domain. It also introduces the discrete signal and its system and frequency-domain analysis. A focus of digital signal processing is to analyze the issues of discrete signal, its system and frequency-domain, as well as the design of digital filter. All these provide a theoretical foundation for the function implementation of the digital signal processing device, followed by the DSP technology which is the hardware part of digital signal processing[4]. Since the three above-mentioned parts are interrelated, it is very important to improve the transition of the relevant courses, especially when there is a repeat of teaching content, there is a need to maintain students' interest and at the same time enhance their understanding of topics.

2.2 *Lack of diverse teaching methods*

The theory of digital signal processing is based on a lot of mathematic work, including calculus, probability and statistics, as well as complex function, which requires numerous of algorithms, derivation and proof. If the classroom focus is on formula derivation and proof with the use of multimedia, it might fail to keep students' interest on learning, as they do not really understand the origin and application of the principles and methods. Therefore, it is essential to equip students with the basic concepts, principles and arithmetic skills; cultivate their learning interest; and improve their self-learning, independent analysis and problem solving capabilities [5].

2.3 *Inflexible assessment method*

Assessment is an important part of teaching process, and also an important tool to evaluate teaching effectiveness. As part of the teaching reform, assessment methods shall be flexible and diverse, transforming from the traditional "one examination judges everything" conclusive evaluation to both conclusive and progressive evaluations, which combine classroom teaching and independent learning evaluations. This makes the assessment more comprehensive, target-oriented, effective, progressive and developmental. Improving assessment methods would contribute to quality teaching, foster self-learning and innovation capabilities. The assessment score would then be a better indicator of the students' overall performance in the study [6].

3 MEASURES OF TEACHING REFORM

3.1 *Optimizing teaching content*

The curriculum of signal and system overlaps with the curriculum of digital signal processing. When the teacher teaches digital signal processing, the topics on time-domain of discrete-time and its system are not novel to the students. Therefore, syllabus adjustment is needed for both curriculums, avoiding too much overlapping between the two curriculums, so as to maintain students' focus in the topics, and hence minimize the impact on the learning of digital signal processing. Next, for the overlapping syllabus of the two curriculums, teaching method adjustment is needed. For example, students would feel tedious and have a lower interest at the beginning of digital signal processing curriculum, as the topics on time-domain of discrete-time and its system and frequency-domain analysis have been taught before. Therefore, when these topics are taught, a revision mode can be adopted, and how that discrete-time issue would affect digital signal processing will be focused on.

To enable students to master the basic knowledge and skills in a limited timeframe, it is essential to pick the right teaching content and streamline the syllabus. To achieve this goal, we need to integrate various teaching materials of digital signal processing organically, vetting for the best content, generating a number of knowledge focuses: (1) Time-domain analysis of discrete-time system: sequences, linear time-invariant system, difference equation, discrete convolution, causality and stability; (2) Z-domain analysis of discrete-time system: Z transform and

its ROC, inverse Z transform, Z-domain solution of differential equation, system function of discrete system; (3) Fourier analysis of discrete-time signal: discrete Fourier series, DFT, cyclic convolution, FFT, fast convolution, digital spectrum analysis based on FFT(aliasing, leakage effect, fence effect); (4) Digital filter: the essence of digital filter, design of IIR digital filter (impulse response invariant, the bilinear transformation, frequency band transformation), design of FIR digital filter (linear phase, window method, frequency sampling method), implementation structure of digital filter; (5) Application of digital signal processor.

In addition, after completing each chapter, we can highlight the knowledge focus, and help students understand what they have learnt by the use of "knowledge tree" or "knowledge chain". Knowing where they are standing in the knowledge web, students' logical thinking ability will be strengthened.

3.2 *Teaching methods reform*

1 Taking into account of learning pattern, flexible use of various teaching methods is necessary.

In the teaching process, taking into account of learning pattern and difference in learning level and knowledge focus, we can make use of various teaching methods in a flexible way, in order to cultivate students' interest in learning, and motivate their learning incentive. In the early stage of teaching process we can adopt a guidance approach, which means that using animated outcome to guide students through the understanding real-life application of theoretical knowledge, raising their learning interest. In the middle stage of teaching process we can adopt an inspiring and discussion-based approach. This is a student-oriented, teacher-led classroom, and discussion-based teaching, encouraging students to proactively participate in classroom learning, rather than a passive and one-direction manner. In the latter stage of teaching process we can adopt a research-based approach. Research-based approach is to introduce the scientific research method into classroom. With teacher's aid, motivation and guidance, students can identify, analyze and solve problems, as well as acquire knowledge, skills and innovation ability during the process.

2 Using modern technique to facilitate dynamic classroom teaching

Modern teaching technique contributes to teaching effectiveness and increase in information conveyed. Appropriate and reasonable use of multimedia can optimize teaching effectiveness. Multimedia teaching can convey large amount of information, transmit swiftly and enhance visual effectiveness. These characteristics can boost the quantity of knowledge that a student can learn within a timeframe. Nonetheless, teaching experience tells us that modern teaching technique must be used in a reasonable way in order to achieve desirable result, and it should be arranged in accordance to the characteristic of the curriculum. Digital signal processing curriculum has lots of conceptual content and formula derivation. Its signal and physical characteristics are not easy to understand. Therefore, it is better to combine the use of traditional and modern teaching techniques: for textual concept and theory, do not project all of them to the screen; for content about formula derivation, use blackboard; for difficult physical issues, slideshow presentation can be used. We can also use animated graphics to present the sampling theorem of continuous-time signal, according to changing parameters of signal bandwidth, sampling interval etc. It also helps to present the relationship between signal frequency extension period and sampling theorem, and how to select relevant parameters to avoid frequency aliasing. At the same time, teachers can provide detailed explanations, so that students will have a clear picture of the relationship between analog signal sampling and frequency spectrum, understanding and mastering the knowledge focus deeply.

3 Using mat lab simulation software to foster students' practical ability

Digital signal processing curriculum involves a lot of abstract mathematical concepts and algorithms, while mat lab software has powerful numerical calculation ability, chart analysis and simulation capabilities. Therefore, we can make use of mat lab's powerful simulation and visualization functions to present work processes and reveal their physical essence, giving students a clear picture and helping them to understand and master the abstract knowledge focus. At the same time, at the beginning of teaching, teachers can assign simulation tasks to students, requiring them to complete in a specific timeframe, in order to boost their problem analyzing and solving capabilities. An example of task is to conduct signal convolution and correlation operation by two different ways of calculation (time-domain and frequency-domain), and compare it with the signal's related calculation result. Through graphical comparison and analytical discussion, students can learn the topics from theory to essence, and better understand the concepts, formulas and principles. When introducing the design of digital filter, we can ask students to record a voice message, and then use different methods of programming to set up IIR or FIR filers to purify the voice message signal. By using mat lab graphical presentation, the characteristics, design processes and filtering results of different digital filters can be visualized. Students can also take part in choosing the topic of experiment, which would foster their practical skills and innovation ability.

3.3 *Diverse assessment methods*

To improve the assessment system, other than the final written examination, we shall also assess students' learning progress by including written and oral tests, as well as thesis assessment, so that students' comprehensive qualities and capabilities can be extensively evaluated. The overall score of the curriculum is determined by final examination (50%), thesis assessment (20%), laboratory performance (15%) and course work (15%). Final examination is mainly an evaluation of students' understanding of basic concepts and flexible utilization of knowledge. Excessive formula derivation and calculus shall be avoided. Thesis assessment is mainly an evaluation of students' mastering of the knowledge, their ability to conduct literature research, ability to implement, as well as thesis writing and presentation skills. Experience tells us that using multi-assessments can better find out students' de facto capabilities, and thus avoiding rigid assessment. It makes the assessment more recognizable from the students' point of view, enhancing their learning incentive and consciousness; and inspiring their potentials and personality development.

4 CONCLUSION

Based on the characteristics of digital signal processing curriculum, this paper analysed the problems existing in the teaching process, and proposed to tackle these problems by means such as optimizing teaching content, reforming teaching methods and assessment methods. It also discussed the plausible reform measures, including the use of various modern teaching methods, looking for the best combination to achieve teaching mission. The goal is to enhance students' understanding and mastering of digital signal processing, increase their incentive to learn, foster their innovation ability, and hence making digital signal processing a useful foundation curriculum.

ACKNOWLEDGMENTS

The research work was supported by BIGC projects of China under Grant No. 22150114002, No. 22150114041 and No. 22150114040.

REFERENCES

[1] Ye Xinrong, Zhang Aiqing. Research on application of comparative teaching method in digital signal processing course[J]. China Electric Power Education, 2013, 1:58–59.

[2] Chen Chun-kai, Guan Xue-mei, Tang Chun-ming. Teaching reform strategy of digital signal processing course[J]. Journal of Science of Teachers' College and University, 2014, 34(4):86–89.

[3] Wu Jie. Research on teaching methods of digital signal processing curriculum group for distinguished engineers. China Electric Power Education, 2014, 8:78–79.

[4] Deng Ji-yuan, Wu Jian-hui, Zhang Guo-yun. Discussing the teaching method of digital signal processing curriculum for electronic information engineering[J]. Journal of Hunan Institute of Science and Technology (Natural Sciences), 2012,25(1):86–88.

[5] You Jiale, Zhang Jianmin. Curriculum-driven teaching reform of digital signal processing. Journal of Electrical & Electronic Education, 2014, 36(4):85–86,118.

[6] Lan Lihui, Liao Fengyi, Wen Jiayan. Teaching reform and practice of digital signal processing course. China Electric Power Education, 2012, 3:86–87.

Network Security and Communication Engineering – Chan (Ed.)
© 2015 Taylor & Francis Group, London, ISBN: 978-1-138-02821-0

Detecting image authenticity based on Lambert illumination model and shadows

S.J. Fang & H.Y. Ge

College of Information Sciences and Technology, Engineering Research Center of Digitized Textile & Fashion Technology, Ministry of Education, Donghua University, Shanghai, China

ABSTRACT: We propose a method to detect image forgery based on the Lambert illumination model and shadows. This method describes a multitude of shading and shadow-based constraints by the wedge revealing the location of the distant light source. Simplify the parameters of the wedge and make it more accurate using Lambert model. There is linear programming problem and error function to explain the corresponding model. Detect the image tampering based on the inconsistency in light source direction of different areas and physically inconsistent of shading and shadows in an image. Experimental results show that our algorithm can calculate the light source direction and consistency in the shadows accurately. As a result, our developed algorithm can effectively identify the authenticity of an image.

1 INTRODUCTION

"Seeing is not necessarily believing, things are not always what." Photographs can no longer be trusted. In recent years, the images of forged and tampered with problems emerge endlessly and have plagued the people's production and life. Therefore, seeking effective image identification method is more and more urgent and important.

For an image, there must be some statistical violations to leave by an image processing operation, so many methods to detect image based on kinds of technology. One of them is based on the constraint method which makes the image forgery using some physical or geometric relations of stereotype. The consistency of shadows is a useful constraint to find suspicious shadow. In addition, for an image with the infinite light source, illumination and shadow based on the analysis of the physical constrains can detect inconsistencies.

E. Kee, J. O'Brien and H. Farid have proposed a physically based forensic method that simultaneously analyses shading and shadows in an image, especially in low resolution and quality images. This method is attractive because shading and shadows result from the 3-D interaction of lighting and geometry, and can therefore be difficult to accurately modify using 2-D photo editing software.

In this letter, by combining Lambert illumination model, we can estimate the location of the light. Meanwhile, by making a multitude of weak constraints, we can determine if the shading and shadows in an image are physically consistent.

2 RELATED WORK

2.1 Lambert illumination model

In real life, diffuse reflection phenomenon exists everywhere. Therefore, we calculate the direction of light source with the diffuse theory as the foundation in this paper. Lambert illumination model is one of the most simple diffuse reflection models, which can be used to calculate the direction of the light source.

Lambert illumination model is showed in Figure 1.

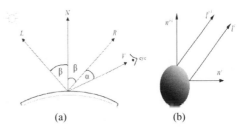

(a)   (b)

Figure 1. (a)Lambert illumination model describes the diffuse reflection phenomenon in the real-world.(b) Infinite light source illuminate the object, infinite light rays are nearly parallel to each other.

The image of the light intensity expression is:

$$I_p' = k_a + k_d (\| \bar{l} \| \times \| \bar{n}_p \| \times \cos\theta)$$
$$= k_a + k_d (\bar{n}_p \times \bar{l})$$

(1)

Among them, I_p' is the light intensity at the surface of the scenery at point p, k_a is light intensity from the environment light, k_d denotes parameter on the surface of the diffuse reflectance, $\vec{l}$ is a vector that expresses the direction of the point p points to the light source, $\bar{n}_p$ denotes the normal vector at point p in the surface.

2.2 *The model of the shadows consistency*

The location of the light source should lie within the intersection of these wedges. Note that the wedges are oriented from the shadow towards the corresponding object. If the light is behind the camera, these wedge constraints should be flipped 180 degrees. Meanwhile, due to perspective geometry there is a sign ambiguity as to the location of the projected light source.

For an authentic image there must be a location in the infinite two-dimensional surface to satisfy all cast shadow constraints. Therefore, the intersection of all the constraints should define a non-empty region.

In (a), General surface can be represented with curve, the curve of the (a) means the light-sensitive surface of the objects, the normal at any point on the curve can be confirmed using its tangent.

In (b), a wedge constraint defined by two solid lines and a half plane constraint defined by one curve. The projected location of the light source lies within the region (outline in black) formed by the intersection of these constraints. A sign inversion of these constraints is depicted in dashed line. The angel of the wedge be denoted as α

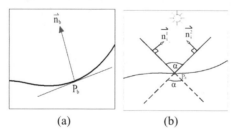

(a)   (b)

Figure 2. Caption is the interpretation with the model of the shadows consistency. Use this model establish a system of linear equations at any point on the curve.

3 METHODS

3.1 *Estimate direction of the light source*

Gray matrix of image is detected as f, the image grey value of any point (x, y) is denoted as $f(x, y)$ that is the point of the actual light intensity. The actual light intensity matrix:

$$I(x, y) = f(x, y) \qquad (2)$$

To be sure, if the image is detected as a gray image, the image data can be directly read. Otherwise, it is necessary to convert it to grayscale image, and read the image data to get the actual light intensity matrix I.

Image pixel gray level change in the most severe direction can be defined as the pixels on the edge of the normal direction. On this basis, neighborhood method calculates the surface normal direction. In the image, any pixel (x_h, y_h), $(1 \leq h \leq m \times n)$ calculate it with all points in the neighborhood gray level difference, choose has the largest gray difference of neighborhood points direction as the plane normal $(\bar{n}_h)$ direction of the pixel. Using this method is in order to calculate normal matrix at all the pixels in the image.

The Lambert lighting model shows that any pixel (x_h, y_h) of the light intensity in an image should be as follows:

$$I' = k_a + k_d \left(\bar{n}_h \times \vec{l} \right) \qquad (3)$$

There is an error function between its actual light intensity and calculated light intensity:

$$D(\vec{l}^1, \vec{l}^2 ... \vec{l}^k, k_a) = \| I' - I \|^2$$
$$= \| M\vec{v} - \vec{i} \|^2 \qquad (4)$$

In this function, M is the normal matrix on the plane, showed as follows:

$$M = \begin{bmatrix} M_1 & 0 & ... & 0 & 1 \\ 0 & M_2 & ... & 0 & 1 \\ : & : & : & : & : \\ 0 & 0 & ... & M_k & 1 \end{bmatrix} \qquad (5)$$

where $\vec{v}$ is made up of faceted light source direction vector of light intensity and environment (k_a), showed as follows:

$$\vec{v} = \left[L_x^1, L_y^1, ..., L_x^k, L_y^k, k_a \right]^T = \left[\vec{l}^1, ..., \vec{l}^k, k_a \right] \qquad (6)$$

where $\vec{l}^k$ is a vector that denotes the kth faceted light source direction:

$$\vec{l}^k = (L_x^k, L_y^k)^T \qquad (7)$$

where $\vec{i}$ indicates the actual light intensity matrix vector form of the I.

$$\vec{i} = \left[I(x_1^1, y_1^1), ..., I(x_q^1, y_q^1), ..., I(x_1^k, y_1^k), ..., I(x_q^k, y_q^k) \right]^T \qquad (8)$$

For the infinite light source images, convert problems into error function to obtain the minimum value. Therefore, the process of solving infinite light source direction of the light source images, can be summed up in equality constraint optimization process, the constraint problem is equivalent to solving the following equation:

$$\min \ D(\bar{v}) = \| M\bar{v} - \bar{i} \| \qquad (9)$$

We solve the global optimal solution of constraint conditions by calculating $\bar{v}$. The average k of faceted light source direction gets infinite light source direction of the light source, by converting them into the form of a rectangular coordinate system. Infinite light image of the light direction can be calculated with the rectangular coordinate system of the angle (α) between the x axis in this article.

3.2 Estimate consistency in shadows

Shadow constraints can be represented as linear inequalities in the plane. Shown in Figure 2 are two lines defined implicitly by their normal $\bar{n}_i^1, \bar{n}_i^2$ and the point p_a. The direction of the normal specifies the region in the plane in which the solution $\bar{x}$ must lie. The intersection of these two regions is the upward facing wedge. As shown in Figure 2 is a single line defined by its normal $\bar{n}_b$ and point p_b. In each case, a shadow constraint is specified by either a pair of lines (wedge cast shadow constraint) or a single line (half-plane attached shadow constraint).

Formally, a half-plane constraint is specified with a single linear inequality in the unknown $\bar{x}$:

$$\bar{n}_i \cdot \bar{x} - \bar{n}_i \cdot \bar{p}_i \geq 0 \qquad (10)$$

In this function, $\bar{n}_i$ is normal to the line and $\bar{p}_i$ is a point on the line. A wedge-shaped constraint is specified with two linear constraints with $\bar{n}_i^1$ and $\bar{n}_i^2$.

Given error-free constraints from a consistent scene, a solution should always exist in this system of inequalities. That is to say, we are looking for a solution with minimize slack variables, and satisfy all the factors and the shadow of the additional constraints.

If either the regular or inversion system has solutions in practice to solve linear programming, then we make the conclusion that constraints of shadows are consistent with each other. Otherwise, there is no the position of the light to meet all constraints, we conclude that some constraints is part of the generated image manipulation.

When an image generates inconsistent constraints, we hope to find which constraints are in conflict with others. These conflicting constraints provide the essential evidence that can be used to invalidate a forgery, and can be useful in determining what parts of an image that may have been manipulated.

3.3 Select the angle of the wedge

Infinite light irradiation in the image, such as Figure 3, on the rules of comparative object, facing the light source on one side of the higher brightness forms obvious radian marked with yellow solid line, we call it shading. The curve corresponds to the terminator – the contour at which the surface normal is oriented 90 degrees from the direction to the light source. Points to the right of the terminator are in an attached shadow. Blue arrows denote the direction of the light source based on the model of 3.1 section that calculate the direction, which determine the direction of the wedge area of a point on the boundary. At the same time, according to the radian of shading, we can calculate the angle of the wedge area (γ).

As showed in the Figure 3, we establish constraint equation:

$$\begin{cases} \gamma = f(\alpha, \beta) \\ f = x\alpha + y\beta + z \end{cases} \qquad (11)$$

Where β is the angle with the radian of shading. This function is the linear programming equation, which depends on the information of the shading in an image to obtain variables x, y, and z.

According to the shadow of consistency as evidence to judge the authenticity of the image, we select the several points on the wedge area to analysis if these areas have intersection. This suggests that the choice of the wedge area is very important:

Upper or lower of the direction of the wedge influences the region to make minimal set intersection. We use the direction of the light source to determine the direction of the wedge.

If the angle of the wedge area is much than normal to include error of point, making the error result; if angle is little to exclude the right point, the right result is t sentenced into wrong. The corresponding solution is:

1. Using the direction of the light source determines the direction of the wedge. We make the center line vector of the wedge as its direction. In the Figure 3(a), for example, the direction of this wedge is in line with the direction of the light source.
2. Using the object shading information and the direction of the light source chooses the right angle. In Figure 3 (b), for example, according to the formula (11), we employ the linear programming equations to establish constraints with angle of the wedge, and then we can calculate precisely the right range of the angle.

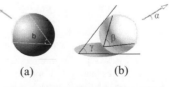

(a)   (b)

Figure 3. This model describes shadows, shadings and the direction of the light source on the rules of the objects. α is the angle between the direction of the light source and the x axis in this article.

4 EXPERIMENTAL RESULTS

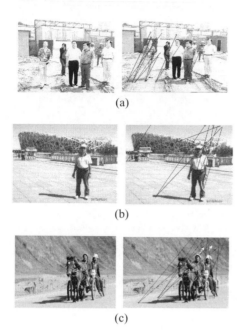

(a)

(b)

(c)

Figure 4. Captions are the results of the simulation using software. The figure (a) reveals this image to be a fake, and (b),(c) are real. All original images get from the network.

In the Figure 4, the left of (a) is the original with the naked eye observation without any disharmony. The results are shown in the picture on the right after the software MATLAB2013a simulation. On the right, the white arrows indicate the direction of the light source, yellow lines indicate information of the shading on the objects, blue lines show the wedge area. Red area is the intersection of wedge areas on the symbolic objects in the image. Obviously, these two areas have no intersection. So we think this image has been tampered.

In Figure 4 (b), we can acquire the information as you can seethe red area of the intersection in this right image, there is the mini set of the wedges, we choose a left experimental point to test the consistency and their wedge areas have intersection, so we determine the image is real.

As shown in Figure 4 (c), according to the above model calculates the direction of the light source and marks white arrow, red area and blue areas. The area of intersection illustrates the consistency of shadows in the image, in accordance with the yellow line denoted the shading information. We can judge the image is true.

We can summarize that the improved algorithm increases the image detection accuracy, and reduces the angle of wedge area according to shading information. Therefore our algorithm reduces the experimental error. Although most imaging conforms to

Lambert diffuse reflection model, but not all imaging surface are strictly comply with this model. Moreover, shadows have a negative impact for computing the image pixels on the image plane normal. These two aspects may be lead to a calculation error.

5 CONCLUSION

The proposed algorithm is suitable for the single pictures of lights, such as natural light. Consistency based on image shadow detection algorithm has a preliminary research, but there are a lot of short-ages. Thus, the improved algorithm is proposed based on previous studies. In order to obtain accurate test results, this paper made a simplified model, selecting evident photosensitive surface as the background of the object. At the same time, not all points on the object can be used to analysis. In the process of analysis, the interference of noise is not in conformity with the result of the constraint condition, we automatically ignore it. More clever counterfeiters modify the shadow on the picture information through the software based on projection methods, this algorithm will have a hard time to obtain convincing analysis results in this paper.

REFERENCES

[1] Popescu and H. Farid. 2005. Exposing digital forgeries by detecting traces of re-sampling.*IEEE Trans. Signal Process.*53(2):758–767.
[2] M. K. Johnson and H. Farid. 2007. Exposing digital forgeries in complex lighting environments. IEEE Trans. Inf. Forensics Secur.3(2): 450–461.
[3] H. Farid. 2009.Exposing digital forgeries from JPEG ghosts. IEEE Trans. Inf. Forensics Secur.4(1):154–160.
[4] M. C. Stamm and K. J. R. Liu. 2010.Forensic detection of image manipulation using statistical intrinsic ngerprints. IEEE Trans. Inf. Forensics Secur.5(3):492–506.
[5] W. Wei, S. Wang, X. Zhang, and Z. Tang.2010. Estimation of image rotation angle using interpolation-related spectral signatures with application to blind detection of image forgery. IEEE Trans. Inf. Forensics Secur.5(3): 507–517.
[6] H.P. Chen, X.J. Shen, Y.D. Lv, Y.S. Jin. 2011. Blind Identification for Image Authenticity Based on Lambert Illumination Model.Journal of Computer Research and Development,48(7).
[7] James F, O'Brien, H. Farid. 2012. Exposing photo manipulation with inconsistent reflections.ACM Trans. Graph,31.
[8] H. Yao, S.Z. Wang, Y. Zhao, X.P. Zhang. 2012. Detecting Image Forgery Using Perspective Constraints.IEEE Signal Process.Lett.19.
[9] Y. Li, Y.J. Zhou, K.G. Yuan, Y.C. Guo, X.X. Niu. 2014. Exposing photo manipulation with inconsistent perspective geometry. The Journal of China Universities of Posts and Telecommunications,21 (4).

Network Security and Communication Engineering – Chan (Ed.)
© 2015 Taylor & Francis Group, London, ISBN: 978-1-138-02821-0

Visualized system of location-based sandstone landforms

J. Ma & Z.Q. Yu

School of Software, Shanghai Jiao Tong University, Shanghai, China

ABSTRACT: Location-based service and data visualization are two widely used technologies. However, there are no mature information systems of sandstone landform that apply these technologies to improve efficiency of data collection, manipulation, and presentation. In this paper, we designed and built up a visualized system as an information platform for sandstone landform research. We implemented the mobile-based data collection service, data automatic classification, and data visualization services on the server. Through the experiments we prove the effectiveness of the system.

1 INTRODUCTION

In modern times, computer science and technology has been widely used and applied in different industries. With the support of information technology, users and scholars in a specific knowledge area can highly improve the speed of study and work. For example, Geographic Information System is a common computer system which is designed to collect, store, manage and analyze the geographic related data [1].

Danxia layer is initiated by Jinglan Feng in 1928. Now Danxia landform has developed as a new branch of geomorphology. The research of Danxia landform in China has experienced three stages in history since the establishment of the concept of Danxia Landform by Guoda Chen in 1939 [2]. It is considered as a kind of sandstone landform among international study, and we adopt the name of sandstone landform for the rest of this paper [8]. However, most of the study in this area use traditional methods (e.g., record the description of sandstone landforms with textbooks), and the lack of utility of modern technology has lowered the efficiency of the improvements. Especially there is no mature information system that can support data collection, data classification, and demonstration of the sandstone landform in China. International researchers are unable to communicate and share the achievements as well.

Location-Based Service or LBS, is the ability to find the geographical location of the mobile device and provide services based on this location information [3][9]. Generally speaking, it applies multiple technologies like wireless location, GIS, Internet, database, etc. It uses mobile equipment like cell phone or PDA to gather location data, and provide the location related services including vehicle navigation, shop store searching and so on [4].

Data visualization involves the study and applications of visual representation of data, so that information can be transferred and demonstrated more clearly and efficiently [5].

In this paper, we apply the LBS technology to get longitude and latitude data when users take photos of sandstone landforms. After users upload the photos to the server, it can automatically classify the photos into different sandstone landform folders according to the location data. We also use visualization methods to demonstrate the statistical data to make it easier to do analysis.

The organization of this paper is as follows: the structure and architecture of the system will be illustrated in section 2. Section 3 focuses on the implementation of the system and section 4 shows the experiments of the system. In section 5 we present the final conclusion.

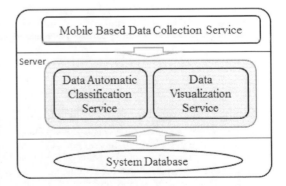

Figure 1. System architecture.

2 SYSTEM ARCHITECTURE

The system is organized with three parts as is shown in Figure 1. The first one is mobile-based data collection service which works on mobile phones. The second part is services on the server and the third part is the database.

2.1 Mobile-based data collection service

The service works on the mobile phone, and it is used to collect photos with the specific longitude and latitude of the position where photos are taken.

Table 1. Classes of sandstone landform data.

No.	Data Class	Data Field Samples
1	Basic Information	Landform name, Province
2	Geography and Geomorphology	Northwest Altitude/Latitude, Continuous Area
3	Natural Environment	Climate Type, Annual Avg. Temperature
4	Local Culture	Ethnic Group, Religion
5	Investigator Information	Investigator Name, Submission Time
6	Protective Class Name	International Class Name, National Class Name
7	Approver Information	Company Name, Telephone Number
8	General Picture	Geology Scene, Landmark Scene

The location data is acquired from GPS module on the mobile phone, and saved as file name of the photo. These photos as well as the location data will be uploaded into the server.

2.2 System database

The database stores more than 300 geographic areas of sandstone landforms in mainland of China. For each sandstone landform area, it stores 8 classes of data as is shown in Table 1.

2.3 Data automatic classification and data visualization

This part is deployed on the server and it has two functions. First, it automatically classifies photos into specific categories to indicate which sandstone landform they belong to. The file name of the photo incorporates location data, and geographic area location data of each sandstone landform are stored in the database. The service can then analyze the relationship between the photo and the sandstone landform. The second function is data visualization. Since the database has stored large quantity of data, it will generate much important statistical info and demonstrate corresponding visualized data on the web page.

3 IMPLEMENTATION

The whole system is made up of two parts. One part is mobile-based data collection service, which is implemented as an android-based application. The android system has already provided mature and robust libraries for functions of taking photos and accessing GPS modules.

The other part is a web-based application on the server. For the server side, we apply Struts, Spring, and Hibernate (SSH) to build up the web-based applications. Struts is an open source web application framework which adopts MVC (Model-View-Controller) design pattern. MVC is a design pattern that divides an application into three interconnected parts, so that it can separate representation of information from user interaction [6]. Spring is a lightweight J2EE framework and inversion of control container for Java platform. Hibernate is a data persistence framework which works for object relational database mapping [7]. The SSH is widely used in web applications since it is highly valuable for reducing coupling while enhancing maintenance of the system.

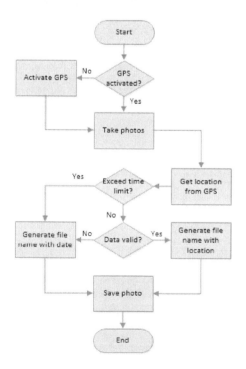

Figure 2. Flow diagram of mobile based service.

3.1 Mobile-based data collection service

For this part, we implemented an android-based application, which is used to take photos and record the corresponding location data as file name of photos. Once users take a photo, the program will try to get the longitude and latitude data from GPS module. If the valid location data is acquired successfully within a specific time limit, the program will generate the file name according to the location data and then save the photo. Otherwise, the program will generate the file name according to the date and save the photo. Thus location data is incorporated in the file name and program on the server can use these data to automatically classify the photos. For those photos that do not include location data, users have to classify them manually. Figure 2 shows the flow diagram.

3.2 Register and login

This function module is used to apply for and verify the authority. Currently the system is open only to experts and scholars in the industry. If users want to access the data, they need to register their user name, password, and email address. Once they are approved, a notification will be sent to their email address. Then the user can use the matched username and password to login into the system. This function is very important, since only authorized users have the right to manipulate data and publish news in the system.

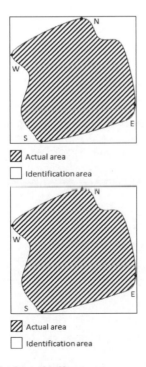

Figure 3.   Sandstone landform area.

3.3 Data automatic classification

The database stores information of more than 300 sandstone landform areas. For each sandstone landform area, it records longitude and latitude of four location points which mark the north, west, south, and east respectively, as is shown in Figure 3. The program first acquires the longitude and latitude value from the file name of photos which are taken from the mobile application. And then it compares the values to the four location points of each sandstone landform. Once the program successfully identifies the sandstone landform to which the photo belongs, it will copy the photo into the folder which belongs to that sandstone landform. Otherwise it will copy the photo to a backup folder and wait for users to classify manually.

For the comparing process, it will accept if the location of photo is either in actual area or identification area. Since no two sandstone landforms are geographically connected, they will be considered as a single one. We implemented an efficient algorithm, and pseudo code is as follows:

1. ***let*** *V represent all sandstone areas;*
2. ***let*** *p represent location point of the photo;*
3.   *area := NULL;*
4. ***for each*** *v ∈ V*
5.   ***if*** *(p.longitude>= v.west_longitude* ***and***
6.   *p.longitude<= v.east_longitude* ***and***
7.   *p.latitude>= v.south_latitude* ***and***
8.   *p.latitude<= v.north_latitude)*
9.   ***then*** *area := v;*
10. *break;*
11. ***end if***
12. ***end for***
13. ***return*** *area;*

Table 2.   Data collection testing results.

Venue	Success	Failure	Average Time (Seconds)
1	10	0	5.01
2	10	0	4.47
3	10	0	4.83
4	10	0	4.59
5	10	0	4.80

3.4 Data visualization

Statistical data is highly needed, and we provide 4 types of statistical data. Two bar charts include sandstone landform area distribution and average altitude distribution. Two pie charts include international protective class and national protective class.

The program reads data from the database, and calculates the corresponding result. After that, it demonstrates the visualized data by using bar charts and pie charts. We use Flot, which is a pure JavaScript plotting library for jQuery, to implement data visualization. It focuses on simple usage and attractive looks, and now works well on IE, Chrome, Firefox, and other common web browsers.

4 EXPERIMENT

4.1 Mobile-based data collection service

We have tested this function in 5 different outdoor venues, and for each venue we have taken 10 photos. The result is shown in Table 2. The second column indicates quantity of photos that successfully acquire the location data from GPS module. The third column indicates those that do not acquire location data within valid time limit. The last column indicates average time consuming for successfully acquiring longitude and latitude data.

4.2 Register and login

We send matched username and password to the system, and then login the system after we are successfully verified. Figure 4 shows the user interface after login.

4.3 Data automatic classification

We have implemented the web-based system and tested the important functions. Since traveling across all sandstone landforms is impractical, we prepared 30 pictures for data automatic classification function. The pictures are divided into three groups. File names of pictures in group one incorporate locations within 10 different sandstone landform areas. File names of pictures in group two incorporate locations that are outside of any sandstone landform areas. Pictures in group three are named by dates. According to the experiments, all pictures are uploaded into the system and automatically copied into expected folders.

4.4 Data visualization

Our experiment shows that bar charts and pie charts are normally presented on web pages. According to our manual calculation, all the statistical data generated by the program are correct.

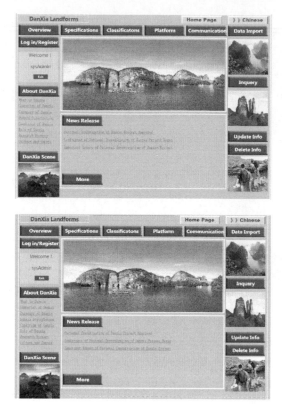

Figure 1. System user interface.

5 CONCLUSION

In this paper, we design and implement the visualized system which incorporates mobile-based data collection service, as well as data automatic classification and data visualization services on the server. This system should work as a highly efficient platform for experts of sandstone landforms to do research and data analysis.

ACKNOWLEDGMENT

This work is supported by National R&D Infrastructure and Facility Development Program (No. 2013FY111900).

REFERENCES

[1] K.C. Clarke, Advances in geographic information systems, Computers, Environment and Urban Systems 10 (1986) 175–184.

[2] H. Peng, 2001. Danxia Geomorphology of China: A Review. Chinese Science Bulletin, 46:38–44.

[3] Diep Dao, Chris Rizos, J.L. Wang. Location-based services: technical and business issues. GPS Solutions Journal, December 2002, Volume 6, Issue 3, pp 169–178.

[4] SubhankarDhar, San Jose. Challenges and business models for mobile location-based services and advertising. Communications of the ACM, Volume 54 Issue 5, May 2011 Pages 121–128.

[5] Vitaly Friedman. Data Visualization & Infographics, Graphics, Monday Inspiration, January 14th, 2008.

[6] T. Reenskaug, J. O. Coplien. The DCI Architecture: A New Vision of Object-Oriented Programming. http://www.artima.com/articlcs.

[7] F. Liu, H.X.Guo, B. Fu. The Research of Web Application Framework Based on SSH. ISBIM '08. International Seminar on Business and Information Management, vol.2, 2008, pp. 169–172.

[8] Robert W Young, Robert A L Wray, Ann R M Young. Sandstone Landforms.The Geographical Journal, Volume 176, Issue 1, pages 119–120, March 2010.

[9] B.Jiang, X.Yao. Location-based services and GIS in perspective. Computers, Environment and Urban Systems, 2006, 30(6): 712–725.

Network Security and Communication Engineering – Chan (Ed.)
© 2015 Taylor & Francis Group, London, ISBN: 978-1-138-02821-0

An imaging retrieval method based on block color texture and relevance feedback

B.P. Wang
Science and Technology on UAV Laboratory, Northwestern Polytechnical University, Xi'an, P.R. China

Y.L. Lan, X.T. Wang & H. Wang
School of Computer Science, Northwestern Polytechnical University, Xi'an, P.R. China

ABSTRACT: In this paper, we realize the image retrieve based on block color histogram and gray symbiotic matrix on the basis of the research and analysis on the color and texture feature extraction algorithm. On this basis, we use the relevance feedback technology to update the weights and to adjust the image color, texture feature coefficient and the weight corresponding to these two factors of each sub-block. And combined with feature matching technology, we realize the image retrieval algorithm based on color texture and relevance feedback. The algorithm uses different test libraries, and obtained experimental results are compared. From the experimental results, we can see the related feedback algorithm is based on the color and texture has better retrieval performance and gets perfect retrieval results.

KEYWORDS: Image retrieval, feature extraction, color, texture.

1 INTRODUCTION

Today, the digital image has become one of the main resources in the information society as its application increases. How to find the image that users are interested in from a large number of multimedia resources gradually becomes an urgent problem which needs to be solved, the research on image retrieval algorithms have emerged.

Currently, color features, including color histogram, color moment, color correlation vector, color clustering [1]and so on, and texture features, including statistical method and structural method[2], are often used for image retrieval. In this paper, we propose the method that describes the image color characteristics with the block color histogram and extracts the texture feature of the image by gray symbiotic matrix. Besides, the relevance feedback is used for content-based image retrieval methods. The relevance feedback method in this paper uses weights adjusted method [3]; the weights of each sub-block and the color and texture feature are adjusted to improve the retrieval accuracy of the image. Experiments show that retrieval method in this paper has higher accuracy and meets the users' cognitive requirements for image better.

2 THE IMAGE FEATURE RETRIEVAL

2.1 *The color feature retrieval*

Because color is the most intuitionistic feature and one of the most important characteristics of the image visual perception, the image retrieval methods based on the color feature are the most basic methods of content-based image retrieval method [4]. Among them, the color histogram is the most widely used color feature in image retrieval system, since it has the simple similarity calculation and is not sensitive to the image size change and the image rotation. However the traditional color histogram retrieval methods only record the global color statistics but miss the spatial distribution information of color [5]. In this case, an improved color histogram method is proposed.

In this paper, a fixed division to extract color histogram method [3] is used. First, convert the image from *RGB* color space to *HSV* space. Appropriate quantization of each component is necessary to reduce the dimensions of feature vectors to *HSV* space. Because the range of each component of *H, S, V* is large, and the colors distinguished by human eyes are very limited. The method to quantize the three components *H, S, V* will be unequally used in this paper. Divide the hue H space into eight parts; both saturation space and brightness V space are divided into three parts. Then compose the H, S, V quantized to an one-dimensional vector:

$$L = 9H + 3S + V \tag{1}$$

L is an integer and the its range is between 0 and 71 Then calculate L to get an one-dimensional histogram obtained 72 handle. Then divide the image space into 3*3 blocks, and calculate the L of each block. Each sub-block has the same initial weight. When calculating the similarity, we adjust the weights of each sub-block using the relevance feedback methods in this paper.

Suppose image I is the image to be retrieved in the image database, and image Q is the image submitted by users. The 72 handle color histogram vector L_{Ii} and L_{Qi} (where i = 1, 2... 9) of each sub-block can be calculated by the formula (1). Calculate the Euclidean distance between each corresponding sub-block by the formula (2), and the total distance is equal to the summation of weight distance of each sub-block, as shown in equation (3) below:

$$S_{ci}(Q,I) = \sqrt{(L_{Qi} - L_{Ii})^2} \tag{2}$$

$$S_c(Q,I) = \sum_{i=1}^{9} W_i S_{ci} \tag{3}$$

where W_i(i = 1, 2, ... 9) is the weights of each block.

2.2 The texture feature retrieval

In the study of computer vision, the texture can distinguish the different objects in images on the micro, since it reflects the visual features of homogeneity phenomenon in the image independent of color or brightness of the image. Hence, texture features is also one of the main features of content-based image retrieval system [8].

Haralick expressed the texture with gray symbiotic matrix in 1970s. The method to build the symbiotic matrix by the relationship between pixels' directions and distance is still used widely today [9]. First, set up a symbiotic matrix based on the direction and distance of pixels, and then extract meaningful statistics as texture features from the matrix. Commonly used are angular second moment, entropy, homogeny area and non-similarity and so on. The symbiotic matrix is calculated as follows: suppose $(f_{xy})_{M \times N}$ is a two-dimensional digital image, and the size of it is $M \times N$. Gray level co-occurrence

matrix is the number of the pixels which gray level is j and at a distance of $\delta = (dx^2 + dy^2)^{1/2}$ from pixel i. The following is abbreviated as P_{ij}. The symbiotic matrix is shown in Figure 1. And, it can be expressed as the equation (4):

$$p(i, j, \delta, \theta) = \{[(x, y), (x + dx, y + dy)] \mid f(x, y)$$
$$\tag{4}$$
$$= i, f(x + dx, y + dy) = j\}$$

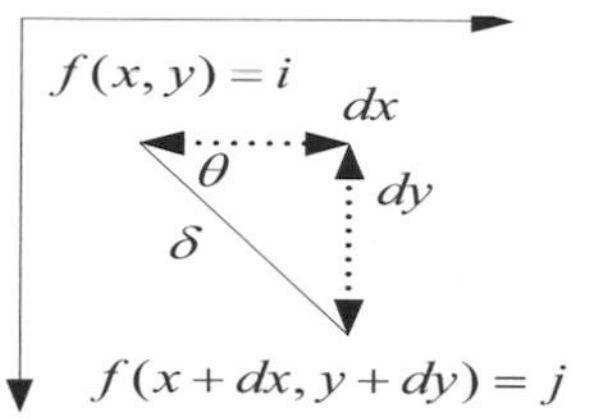

Figure 1. The gray symbiotic matrix.

Many parameters can be extracted from the symbiotic matrix as a measure of image texture features to quantize image texture features. Predecessors summarized 14 kinds of features to express the texture of an image. Commonly used are energy, entropy, the moment of inertia and interdependency.

Energy: It is the measure of the uniformity of gray level.

$$ASM = \sum (P_{ij})^2 \tag{5}$$

Entropy: Entropy is a measure of the amount of information of the image, and the texture information also belongs to image information.

$$ENT = \sum_{i,j} P_{ij} \bullet \log P_{i,j} \tag{6}$$

Moment of Inertia: In the image, the groove of texture deeper, the value of moment of inertia greater, the visual effect of the image sharper.

$$CON = \sum_{i,j} (i - j)^2 P_{ij} \tag{7}$$

Correlation: correlation is a measure of the gradation linear relationship, and is used to describe the degree of similarity between elements of matrix.

$$COR = \frac{\sum_{i,j} (i - \mu_x)(j - \mu_y) P_{ij}}{\sigma_x \sigma_y} \tag{8}$$

where $\mu_x = \sum_{i=0}^{L-1} i \sum_{j=0}^{L-1} P(i,j)$, $\mu_y = \sum_{j=0}^{L-1} j \sum_{i=0}^{L-1} P(i,j)$,

$$\sigma_x = \sum_{i=0}^{L-1}(i-\mu_x)^2 \sum_{j=0}^{L-1} P(i,j) , \quad \sigma_y = \sum_{j=0}^{L-1}(j-\mu_y)^2 \sum_{i=0}^{L-1} P(i,j)$$

The grey scale of an image is divided into 16 grades, so the scale of the symbiotic matrix is 16×16. Then, calculate each feature vector of each sub-block by the above four formulas, when θ is $0°$, $45°$, $90°$, $135°$. The symbiotic matrix of these four directions obtains energy, entropy, moment of inertia, and the correlation. Finally, the mean and the standard deviations of each parameter μ_{CON}, σ_{CON}, μ_{ASM}, σ_{ASM}, μ_{RNT}, σ_{RNT}, μ_{COR}, σ_{COR}, are regarded as the components of the texture feature vector. It is necessary to internally normalize the above eight components, since their physical meaning is different. The Gaussian normalization method is been used in this paper. A small amount of super or ultra-small value of elements has a little impact on the distribution of the value of elements normalized.

The partitioned method is used for the texture feature extraction in this paper. The two-dimensional space is divided into 3*3 blocks, and each block has the same initial weights. Calculate the feature vector of each block $F_i = \{f_{i1}, f_{i2}, f_{i3}, f_{i4}, f_{i5}, f_{i6}, f_{i7}, f_{i8}\}$, where $i \in [1,9]$. Calculate the distance of the corresponding block by the formula (3) and all blocks' weights distance is summarized by the formula (4). Where image I is the image to be retrieved, and image Q is the image submitted by users.

$$S_{ti}(Q,I) = \sqrt{\sum_{j=1}^{8}(f_{Qij} - f_{Iij})^2} \qquad (9)$$

$$S_t(Q,I) = \sum_{i=1}^{9} W_i S_{ti}(Q,I) \qquad (10)$$

where $W_i (i=1, 2, \ldots, 9)$ is the weights of each block.

2.3 The synthesis feature retrieval

The retrieval method based on color, texture feature has its own apply scope. So the retrieval method combining the color with texture feature is proposed in this paper. First, extract color and texture feature vector by the method described in the former two

parts, and then combine these two vectors to retrieve images. The main steps are as follows:

1 Calculate the similarity distance between the color feature vector of image I F_{IC} and the color feature vector of image J F_{JC};

$$S_{CI,J} = dis(F_I, F_J), I, J = 1, 2, \ldots, M, and I \neq J \quad (11)$$

2 Calculate the means m_c and the standard deviations σ_c of $M(M-1)/2$ distances calculated by the method in step1);
3 Calculate the color feature distance $S_{c1}, S_{c2}, S_{c3}, \ldots, S_{cm}$, between the image Q to be retrieved and the image I in the image database;
4 Normalized S_{ck} to the rang of [0, 1] by the above normalization formula;
5 The same with step 1), 2), 3), 4), and normalized the S_{tk} of texture to the rang of [0,1];
6 Calculate the total distance between the image to be retrieved and the target image, by the formula as follow:

$$S_k(Q,I) = W_c S_{ck}(Q,I) + W_t S_{tk}(Q,I) \qquad (12)$$

where W_c is the weight of color feature and W_t is the texture feature, and the default value of them is 0.5.

3 THE RELEVANCE FEEDBACK TECHNOLOGY

In this paper, we adjust the weights of images' color and texture feature and their corresponding blocks by adjusting the measure of distance. The main steps of the relevance feedback method in this paper are:

1 The color and texture feature vectors of each image in the image database are extracted and the corresponding feature database is established. We match the feature of the image Q to be retrieved with the images in database to get the S images most close to Q. Then, range the s images by ascending distance.
2 Calculate the variance summation σ_{ci}^2 of the color feature distance between sample image Q and relevant images for each image I_m of the n images returned by users. The weight of the image color feature is the multiplicative inverse of the variance summation of the sub-block. Likewise, calculate the variance and σ_{ti}^2 of the texture feature distance and the weights of texture feature of

the sub-block, and then normalize the weights. The formula is shown in (13) (14)(15),where $h \in [c,t]$

$$\sigma_{hi}^2 = \sum_{m=1}^{n} (dis(I_i(m), Q_i))^2 \qquad (13)$$

$$W_{hi} = \frac{1}{\sigma_{hi}^2} \qquad (14)$$

$$W_{hi} = \frac{W_{hi}}{\sum W_{hi}} \qquad (15)$$

3 Adjust the proportion of the color and texture features weights in the retrieval system by the formula (16), i.e. feature coefficient W_h, where $h \in [c,t]$. Where d_h is the mean color feature distance or the mean texture feature distance between the n images the system returned and the sample image. D_{max}, D_{min} is the maximum, minimum of the color feature distance and texture feature distance between the n images the system returned and the sample image. Normalize them by the formula (17).

$$W_h = \frac{1 - (d_h - D_{min})}{D_{max} - D_{min}} \qquad (16)$$

$$W_h = \frac{W_h}{\sum W_h} \qquad (17)$$

When the user retrieves the image again, the weights updated are substituted into the formula (3), (10), and (12) to calculate to improve the retrieval accuracy.

4 EXPERIMENT AND PERFORMACE ANALYSIS

4.1 *Experimental results*

The images used in the experiment belong to the class of animal, flower, trees in the "BaiWan Image Database" and the class of street, mountain in the "MIT Image Database". There are 400 images, and the resolution of them is 640*416 or 416*640.

The system extracts an image randomly as a retrieval example. First, the system calculates the feature vector of the image extracted; then the feature vector matches with the feature vectors of all images. Finally, return the images top 12 and range them from the top down, from the left right by the similarity. The experimental results is shown in Figure 2.

(a) The result of the color feature retrieval

(b) The result of the texture feature retrieval

(c) The result of the synthesis feature retrieval

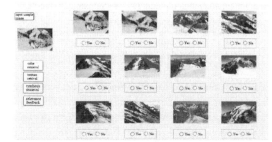

(d) The result of relevance feedback

Figure 2. The experimental results.

326

4.2 *Analysis of the results*

The precision ratio and recall ratio are used for testing the performance of the retrieval system. The precision ratio is defined as the proportion of relevant images in the search results. The recall ratio is defined as the proportion of the relevant images retrieved in all relevant images in the image database. The experiment chooses four image classes, and each of them includes 100 images. An image is extracted randomly and set its threshold value 0.33. Compare the result of single feature retrieval method and that of the relevance feedback retrieval method combining color and texture feature, the experimental results are shown in Figure 3. It can be seen from these two figures that the average precision and recall of the feature based on the color and texture are higher than those of the single feature. Besides, the retrieval accuracy is higher after the relevant feedback.

5 CONCLUSION

In this paper, we propose a new retrieval algorithm combining the sub-block histogram, texture symbiotic and relevant feedback. Compared with the retrieval algorithm based on the single feature, the performance of the algorithm based on the synthesis features is better. The color and texture feature coefficient of the image and the weights of each sub-block corresponding to these two features are adjusted by the relevant feedback method. Experiment shows that the performance of this relevant feedback method is good in image retrieval.

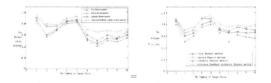

(a) The recall level of the image retrieval

(b) The precision of the image retrieval

Figure 3. The performance of these four methods.

ACKNOWLEDGMENT

This research is being supported by the National Natural Science Foundation of China under Grant 61472324 and the Defense Pre-research Fund (9140A25040214HK03401).

REFERENCES

[1] Swain M J, Ballard D H.Color indexing [J]. *International Journal of Computer Vision*, 1991, 7(1):11–32.

[2] He Yong Wang,Lei Li, Zheng An Yao.The Image Retrieval Method Based on Texture Feature[J],*Application Research of Computers*,2008(10):82–83.

[3] Xiao Juan Guo. The Research of Content-based Image Retrieval Technology [D]. Xi'an: Northwest University, 2007.

[4] Kokare M, Bisw as P K, Chatt erji B N. Texture Image Retrieval Using New Rotated Complex Wavelet Filters [J]. *IEEE Transactions* on 1168–1179.

[5] Mudigonda N R, Rangay R M, Desautels J E. Gradient and Texture Analysis for the Classification of Mammo Graphic Masses[J]. *IEEE Transactions on Medical Imaging*, 2000(10):1032 1043.

[6] John M, Sitharama S.Content-based image retrieval systems [C]. *IEEE Proceedings* of the 1999 *IEEE Symposium on Application-specific Systems and Software Engineering Technology*, 1999:136– 143.

[7] Sameer A, Rangachar K, Ramesh J. A Survey on the Use of Pattern Recognition Methods for Abstraction, Inde ing and Retrieval of Images and Video[J]. *Pattern Recognition*, 2002(35):945–965.

[8] Yu Li,Lei Li,The Research of Content-based Image Retrieval Method[J],*Computer Science*,2009, 9(26):6–11.

[9] Sobottka K, Pitas I. A Novel method for automatic face segmentation [J]. *Facial Feature Extraction and Tracking Signal Processing-Image Communication*, 2002, 9(7):221–257.

[10] Wan X, Jay C. Color distribution analysis and quantization for Image retrieval[J].*Proc.of SPIE Storage and Retrieval for Image and Video Databases*,2006,26(70):8–15.

Network Security and Communication Engineering – Chan (Ed.)
© *2015 Taylor & Francis Group, London, ISBN: 978-1-138-02821-0*

E-Learningization of computer science: From bytes to typeset

B. Barrett

American Public University, Charles Town, USA

ABSTRACT: While the English language has been evolving over the years, the impact of technology has helped to change the English language in many fields – especially in science and technology. As more businesses and organizations in all countries have seen a change in commonality of vocabulary – the acceptance therefore and implementation of new vocabulary may not be accepted – unless it is embraced by giants in the technology field. For example, while many people have been googling items, places, and peoples for years, some people may take offense by hearing that they have been "googled" by another person. However, while the main word form of "google" was embraced by leading American dictionary personnel, the term took off and at a binary rate that has surpassed all hopes and dreams that the main company, Google, could have ever imagined. The dream of one day having one's own projects and services become a regular household item has always been on many company's own "goals lists" for years or rather decades. However, for many the term google has opened up a new world of learning and breaking down of learning barriers. This particular group consists of adult learners with disabilities that have been previously limited in their educational opportunities due to the lack of virtual versus physical barriers. Namely, the use of today's use of technology in the field of computer science and learning opportunities, a new type of learning known as e-learning has been a guiding light for the possibility of yet another new term – e-learningization.

1   INTRODUCTION

1.1   *Changing computer jargon and vocabulary usage*

While the English language has been evolving over the years, the impact of technology has helped to change the English language in many fields – especially in science and technology. As more businesses and organizations in all countries have seen a change in commonality of vocabulary – the acceptance therefore and implementation of new vocabulary may not be accepted – unless it is embraced by giants in the technology field. For example, while many people have been googling items, places, and peoples for years, some people may take offense when hearing that they have been "googled" by another person. However, while the main word form of "google" was embraced by leading American dictionary personnel, the term took off and at a binary rate that has surpassed all hopes and dreams that the main company, Google, could have ever imagined. The dream of one day having one's own projects and services become a regular household item has always been on many company's own "goals lists" for years or rather decades. However, for many the term google has opened up a new world of learning and breaking down of learning barriers. This particular group consists of adult learners with disabilities that have been

previously limited in their educational opportunities due to the lack of virtual versus physical barriers. Namely, the use of today's use of technology in the field of computer science and learning opportunities, a new type of learning known as e-learning has been a guiding light for the possibility of yet another new term – e-learningization.

1.2   *Evolution of computer science enabling technology for all students*

Many companies today are realizing that earlier predictions of a shrinking workforce has become more of a reality. Whereas, other organizations have made strides in employing workers with disabilities as an act of social responsibility, other entities have started to realize the need and value of this untapped human resource (Thakker, 1997)[2]. A key point that needs to be kept in mind is that research has shown us that employees with disabilities have lower turnover rates, lower absenteeism, and higher motivation than some of their counterparts in the workplace (Fersh & Thomas, 1993)[3]. Thus, more employers are looking towards academic institutions for well-qualified graduates to fill the various positions growing as a result of this shrinking workforce. In particular, are today's educators able to prepare learners with the necessary technological skills and education to best prepare them for this new workforce outlook?

Currently, 54 million Americans with disabilities use information technology (IT) at colleges and universities, which accounts for 20 percent of the population (Oblinger & Ruby, 2004)[4].

Historically, most university instruction has been focused on content knowledge, rather than the learning styles of the students. Thus, this new type of learning, known as e-Learning, has started to help fill this void, especially for learners with disabilities. Meanwhile, there are additional barriers in education that do prevent these learners with disabilities from succeeding in both the face-to-face and online learning environments, especially in certain courses. As more educational institutions continue to create and implement newer and improved forms of learning technology, the field of e-learning will continue to evolve.

2 CHANGES AND CHALLENGES OF COMPUTER TECHNOLGY

2.1 As computer technology evolves more learning opportunities are available to all

Thus, we need to take a closer look at this newer computer technology and the growing members of this newer learning populations. In particular, we need to consider what is the relevance of disability statistics in terms online learning? Before we can go further into this topic, we need to explore current statistics that will help to provide an overview of why more and more people with disabilities are seeking further education. Second, once we realize the growing numbers of people in this segment of today's population, we can then formulate a baseline to understand why educators need to consider why these students are fighting not only barriers in the workplace, but also in the field of education, namely, higher education. According to the Disability Status 2000, Census 2000 Summary File #3; and Census 2000 Brief (March 2003)[5] (C2KBR-17), the following statistics about the employment of people of disabilities was noted: "The total number of people with disabilities aged 16-64 is 33,153,211, [of which], the total number employed is 18,525,862. The percent of people with disabilities aged 16-64 employed is 55.8" (Census, 2000, para. 4-6)[6]. Further, the reports went on to noted that "18.6 million people disabilities employed aged 16-64, 60.1% of men with disabilities are employed, and 51.4% of women with disabilities are employed" (Census, 2000, para. 7)[7]. As we can see from these statistics, there is still a disparity between the employment rates of people with disabilities versus others without disabilities. Nonetheless, one way that many Americans with disabilities are trying to overcome these figures is by seeking additional education. In particular, the field of e-learning courses

and programs of studies are increasing significantly as technology improves. Historically, many students with disabilities would have to avoid such programs or courses involving engineering, computer science or mathematics, depending on the type of disability they had. However, with this changing technology, more creative learning advances are pathing the way for individuals with psychomotor or mental challenges, as well as helping them to take more "control" over their learning abilities and skills.

We should note that many adult learners have different learning styles, as well as educators have different teaching styles. Also, we need to consider that if a student with a disability decides upon taking an e-learning course, the next question will be if the educator will be able to help with the student's learning accommodations (as required by U.S. federal law if federal monies are involve)? In the United States, the passage of the Americans with Disabilities Act of 1990 (ADA) help to change and correct many factors and circumstances affecting the lives of these individuals. Specifically, it provided a federal mandate compelling organizations, in particular in the field of education, to make changes in how it made accommodations for learners with disabilities. Consequently, while this federal mandate has been in effect for two decades, some universities and their instructors still have not brought their courses up to par. As a result, this federal mandate has caused many universities and their various departments to set up either to address the various learning needs of "all" students or perhaps perish as more students start to gravitate towards more popular and accepting universities.

2.2 From bits to bytes and online technological socialization

The rate of speed and responses provided by various technological companies has vastly increased the number of school being able to require the best students with top degrees, as well as develop a new social order of learning in which the learner has more control over the learning environment (live or virtual classrooms), as well as being online daily or lecturing at the local university on said days. As more schools move towards this option, there is the need to create, implement, and evaluate a number of different practices and procedures in order to prepare all students, especially learners with disabilities, to compete for quality and meaningful employment. And why more and more disciplines, such as computer science, need to update their courses with better technology and policies in order to better defend oneself, both in and out of various learning environments. Consequently, as more schools achieve and implement more technology, the image of the training student has started to evolve too. According to Preece (2000)[8], these online communities "consists of people who interact socially as they strive to satisfy their own needs or perform special roles; a shared purpose that provides a reason for the community; policies that guide people's interactions; and computer systems to support and mediate social interaction and facilitate a sense of togetherness" (p. 10). Finally, in the fields of business and management, this specific discipline has recognized the demographical changes of the student population and have been continuously brainstorming various methods of e-learning to obtain the best possible learning outcomes. As a result, the need for a greater of evolutionary change of events in the field of education is needed more than ever before.

2.3 *Evolution of what constitutes a learning environment*

As more and more instructors and administrators start to create, revise, design, develop and implement various instruction methods moving from a sheet of paper to the computer, one underlying historical fact remains. Despite the number of years that have gone by since the creation of ENIAC [9], the role and function of the technology has helped to increase many educational and work opportunities for many students – either with or without disabilities. Therefore, as more technological advancements start to appear in various learning environments, these colleges and universities will need to find more computers with larger storage capacity; less expense in operational costs; and helping to develop today's learner to become more confident in his or her approach to daily work life.

3 CONCLUSIONS

It should be emphasized that e-Learning is not for all students in each country or social setting. Nonetheless, for many students with disabilities, these many facets or approaches towards e-Learning have offered more opportunities than ever before. Let us consider some examples here. For example, visual learners were able to benefit from applications in PowerPoint and Flash Multi-Media technology. Auditory learners could benefit from online classrooms with auditory lectures, Podcasts for students, as well as live chats. From a blended-approach perspective, some online programs offer both auditory lectures, as well as PowerPoint slide presentations. Thus, live chats (both auditory and visual – i.e., Elluminate, Horizon Wimba, etc.) have offered more opportunities for a variety of learners. As we consider each of these various representatives of various parts of the e-Learning process, we also need to consider all of the factors for tomorrow' learning environments and populations, as well as the specific course matter.

Finally, we need to realize that many adult learners have different learning styles, and so do educators have different teaching styles. In addition, we need to consider why many students with disabilities take e-Learning courses over live, physical courses. This answer may vary from one adult learner to another. In the United States, the passage of the Americans with Disabilities Act of 1990 (ADA) has changed many factors affecting the lives of persons with disabilities. In particular, it mandated that in field of education, there were certain changes that needed to be made in order to meet the accommodations needs of learners with disabilities. Despite all of these changes, one key problem has been the view of one's peers if they are able and willing to work, as opposed to their counterparts with disabilities that may not be seen by others as having equal abilities, skills and/or education. However, as the various phases of e-Learning continue to evolve, more individuals are seeing changes as a vital part of the change management mission in organization.

REFERENCES

[1] Knowles, M.S. 1980. The Modern Practice of Adult Education. Andragogy Versus Pedagogy, Englewood Cliffs: Prentice Hall/Cambridge.
[2] Thakker, D. 1997. Employers and the Americans with Disabilities Act: Factors Influencing Manager adherence with the ADA, with special reference to individuals with psychiatric disabilities. Dissertations Abstracts International,. (University Microfilms No. 9727300).
[3] Fersh, D. & Thomas, P.W. 1993. Complying with the Americans with disabilities act: A guidebook for management with people with disabilities. Westport, CT: Quorum Books.
[4] Oblinger, D. & L. Ruby, L. 2004, January. Accessible technology; Opening doors for disabled students. Retrieved March 1, 2008 from http://www.nacubo.org/x2074.xml.
[5] Disability Status 2000, Census 2000 Summary File #3; and Census 2000 Brief (March 2003) (C2KBR-17) Retrieved on March 20, 2008 from http://www.dol.gov/dolfaq/go-dol-faq.asp?faqid=66&faqsub=Statistics&faqtop=People+with+Disabilities&topicid=11.
[6] Disability Status 2000, Census 2000 Summary File #3; and Census 2000 Brief (March 2003) (C2KBR-17) Retrieved on March 20, 2008 from http://www.dol.gov/dolfaq/go-dol-faq.asp?faqid=66&faqsub=Statistics&faqtop=People+with+Disabilities&topicid=11.
[7] Disability Status 2000, Census 2000 Summary File #3; and Census 2000 Brief (March 2003) (C2KBR-17) Retrieved on March 20, 2008 from http://www.dol.gov/dolfaq/go-dol-faq.asp?faqid=66&faqsub=Statistics&faqtop=People+with+Disabilities&topicid=11.
[8] Preece, J. 2000. Online communities: Designing usability, supporting sociability. Chichester: Wiley.
[9] Mainframe (2011). Retrieved from http://www.thocp.net/hardware/mainframe.htm.

Network Security and Communication Engineering – Chan (Ed.)
© *2015 Taylor & Francis Group, London, ISBN: 978-1-138-02821-0*

A measurement study of Google Hangouts video calls

Y.L. Wang, X. Zhang & W.D. He
Institute for Network Sciences and Cyberspace, Tsinghua University, Beijing, China

ABSTRACT: In this paper, we conduct several experiments on Google Hangouts to study its architecture and behavior under different network conditions. Through measurement results, we 1) give out the architecture of Hangouts and the constitution of its video/audio streams, and 2) give out the model of its rate control policy under different bandwidth and random packet loss rate. The results can be useful for research work afterwards and product designs for real-time multimedia applications.

1 INTRODUCTION

Communication has always been a basic demand for human beings, and the development of Internet offers us more and more choices. Having experienced the evolution from message chatting to VoIP, now we are in the age of video chatting or video conferencing. Video conferencing is not a new issue, and there have been many mature products based on H.323 protocol in use. But before its birth, problems of cost and scalability haven't been solved properly for decades. The central component MCU needs to decode each stream, mix them and then encode them, which requires more computation and more participants, so MCU needs to be powerful, and usually appears as expensive customized hardware. These expensive video conferencing solutions apparently can't satisfy our demands for communication, and in fact, mostly for business use. To meet the demands of end consumers, software-based video conferencing products utilizing new architecture have appeared in recent years, which offer services for free or at low costs. Undoubtedly, Skype and Google Hangouts are among the most popular ones on the Internet. Unlike Skype, which has been the leader of Internet communication tools since the age of VoIP, Google Hangouts is quite a young product. It offers free 720p HD video conferencing for at most 10 people, which is quite outstanding in this field, and with the support of SVC technology and powerful Google network, it can offer high quality video conferencing service to users all over the world, which is quite hard for other products. All these features make it be a strong competitor in this field and a good object for researchers to study.

In this paper, we conduct several experiments to study the following features of Hangouts: 1) *System Architecture*: There are two typical architectures for video conferencing: Peer-to-Peer and Server-Client. The choice of architecture can affect bandwidth consumption, packet loss recovery strategies and Quality-of-Experience like delay, so how to choose architecture for different scenarios is quite crucial for video conferencing applications. 2) *Rate Control Policy*: For end consumers, their network accesses can be quite variable and may change all the time. How to adapt to different network conditions to offer the best experience as possible is critical and challenging for video conferencing applications.

Packets of Hangouts are partially encrypted, so it's impossible for us to study every detail of it. We regard it as a black box, and through a set of experiments, we try to analyze how Hangouts solve the two problems mentioned above. Our findings are concluded as follows:

- Hangouts uses pure client-server architecture for data transfer, and clients never directly exchange data with each other. Every client exchanges video/audio streams with "Google Network", which consists of many servers.
- Hangouts uses RTP/RTCP (Schulzrinne et al. 2003) protocols to transfer multimedia data, and different parts of video/audio streams are distinguished by the payload type field defined in RTP protocol.
- Rate control policy of Hangouts under different bandwidth is quite complex due to its SVC technology. We give out a model to describe this policy which fits well with actual data.
- Rate control policy of Hangouts under different random packet loss rate is a two-phase policy with a boundary of 10% packet loss rate.

The rest of the paper is organized as follows. Related works is briefly discussed in Section 2. Section 3 introduces our measurement platform. We show our measurement results and analysis in Section 4. Section 5 concludes the whole paper.

2 RELATED WORKS

Most of previous work in real-time telecommunication is focused on Skype, especially its VoIP aspect (Baset & Schulzrinne 2006, Bonfiglio et al. 2008, Huang et al. 2009). As we know, video conferencing is quite different from VoIP, having far more bandwidth consuming and being much more sensitive to packet loss. There hasn't been much literature on software-based video conferencing or video chatting products. Cicco et al. (2008) investigated Skype video responsiveness to bandwidth variations. Zhang et al. (2012) conducted several experiments on Skype two-party video calls under different network conditions and proposed models for Skype video calls' rate control, FEC redundancy and video quality.

Google Hangouts has drawn researchers' interest in recent years. Xu et al. (2012) did measurement study on 3 popular video telephony systems for end-consumers: Google Hangouts, iChat and Skype, and presented their key design choices, including application architecture, video generation and adaptation schemes, loss recovery strategies, etc. Their work is quite instructive, but there're some mistakes in their experiment design and these applications have released new versions since then. Besides, they didn't offer a quantitative description or model for these products.

3 MEASUREMENT PLATFORM

Our measurement work started in May 2014, when Hangouts has published its HD version.

As is illustrated in Figure 1, our measurement platform consists of two computers, both connected to Internet and running Hangouts. Both computers send and receive video/audio at the same time, but for ease of description, we call them sender and receiver separately. The sender has a webcam supporting 720p/30fps at most. Besides Hangouts itself, we run two other assistant programs on the sender to do the experiments. One is NEWT (Network Emulator for Windows Toolkit), which is developed by Microsoft and can simulate different network conditions like bandwidth, packet loss rate, propagation delay and so on. It works as a filter and manipulates every packet sent into it according to the rules. With its help, we can find out how Hangouts behaves under different network conditions. The other one is Wireshark, which we use to capture packet level data for analysis. For the sender, video/audio data of Hangouts is first sent to NEWT, then captured by Wireshark and finally sent to the Internet. For the receiver, it just runs Hangouts and is the other part of the video conferencing session.

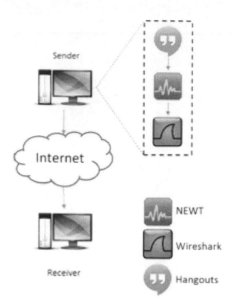

Figure 1. Our measurement platform.

4 MEASUREMENT RESULTS

In this section, we show the measurement results and our analysis. The results are demonstrated as four parts: system architecture, stream constitution, sending rate under different bandwidth, and sending rate under different random packet loss rate.

4.1 System architecture

By analyzing source and destination address of packets captured by Wireshark, we can find out how Hangouts clients interact with each other.

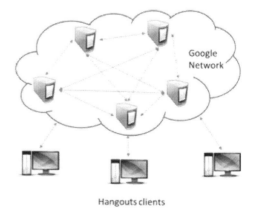

Figure 2. Google Hangouts architecture.

334

Surprisingly, both sender and receiver never exchange video/audio data directly, and they only interact with third-party servers from Google even when they are in the same subnet connecting to the same switch. Every server can transfer video and audio data at the same time, so Hangouts clients in fact interact with "Google Network", which consists of many servers belonging to Google allocated all over the world. This architecture is described in Figure 2.

It works well for scenarios where participants are far away from each other and the connection among them is not that ideal, so "Google Network" can play as a proxy to offer better connection. As we know, "Google Network" has pretty fast connections within it and access points that users can always find near their devices. But this architecture isn't always the best solution. For example, when clients are in the same subnet with propagation delay within 10ms, it's non-sense for all the data to go to the server and then come back, suffering from extra delay and packet loss.

4.2 *Stream constitution*

By combining captured packets and the information from the pop-up statistics window of Hangouts, we can see what protocols it uses. Except for exchange of some personal account information by using TCP, Hangouts uses UDP to transfer all media data. Like most network real-time multimedia applications, Hangouts uses RTP/RTCP protocol to pack video/audio streams. By decoding captured UDP packets according to RTP protocol, we can find more about media streams of Hangouts.

Hangouts uses random ports on client side and solid ports, 19305, on server side, so it's impossible to distinguish video or audio stream by ports. By analyzing decoded packets, we find that Hangouts uses payload_type field defined in RTP protocol to distinguish different types of streams. There're four types of payload_type value appearing in our trace:

- Payload_type is 100: These packets are video data packets, which consumes most of the bandwidth
- Payload_type is 103: These packets are audio data packets and appears when your mic captures sound. These packets form a stream of about 40kbps.
- Payload_type is 105: These packets are also audio data packets but they're only noise. For the use of CNG (comfortable noise generation), Hangouts still send out audio data even when you mute your mic. These packets have the same size of 81 bytes and form a stream of about 8kbps.
- Payload_type is 72 or 73: These packets are RTCP packets used to exchange statistical information between client and server.

Besides these four types of packets, STUN protocol packets are also found in UDP packets, which are typically used for NAT traversal and port binding information in Hangouts for random ports selection.

4.3 *Sending rate under different bandwidth*

As an application running on the Internet, how to adapt to variable network bandwidth is a critical consideration. In order to find out how Hangouts behaves under different bandwidth, we use NEWT to limit the available upload bandwidth for Hangouts and use captured packets to calculate its sending rate. For we are studying steady behavior, so for each bandwidth, we run the application for 1 minute and then start capturing packets. For each bandwidth, we run Hangouts for 5 minutes and calculate sending rate every 1 second, which we call a round, so for each round we get 300 data points and use mean value as the result for this round. To get rid of the impact of Internet's uncertainty, we run 10 rounds for each bandwidth and use the mean value of 10 rounds as our final result.

In the first step of our experiment, we try to find out the sending rate range of Hangouts. By trying different bandwidth, we find that 1) when bandwidth decreases below 70kbps, Hangouts will end the whole conversation, and 2) when bandwidth increases over 10Mbps, the sending rate will stay at about 2Mbps, which is quite a reasonable limit for 720p/30fps video stream using codec VP8.

With the upper and lower bound tested above, we test bandwidth of 3 intervals with different granularity: 70kbps to 100kbps with step of 10kbps, 100kbps to 2500bps with step of 100kbps and 3Mbps to 10Mbps with step of 1Mbps, and get the result illustrated in Figure 3, with x axis for bandwidth and y axis for sending rate. To see the result more clearly, we divide the result into two figures. The upper one is the result of bandwidth varying from 70kbps to 2Mbps, and the lower one is the result of bandwidth varying from 2Mbps to 10Mbps.

From Figure 3, we can see that Hangouts adjusts sending rate not simply linearly with the available bandwidth but uses different policy for different bandwidth intervals. In most intervals, the sending rate varies linearly to the available bandwidth and keeps close to it. Here're some points of interest:

- When bandwidth is 400kbps, the sending rate is quite low compared to its neighbors, so we test more bandwidth between 300kbps and 500kbps and find that when bandwidth is between 330kbps and 450kbps, the sending rate stays at about 300kbps. We infer this phenomenon is the result of SVC technology that Hangouts utilizes and 300kbps is the upper rate limit of layer 0 while 450kbps is the lower rate limit of layer 1.

- When bandwidth is between 1000kbps and 1700kbps, the sending rate increases more slowly compared to its neighbors. When checking calculated rate data points, we find that these points are distributed around 1000kbps or the bandwidth limits. We infer that the upper rate limit of layer 1 is 1000kbps and the lower rate limit of layer 2 is 1700kbps. The final results come from partially added layer 2 packets.

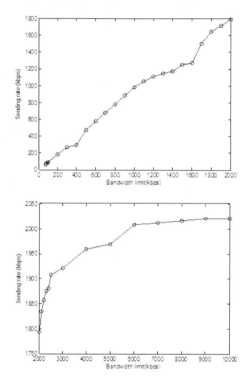

Figure 3. Steady sending rate under different bandwidth.

From data collected and analysis above, we can give out the rate control model of Hangouts for different bandwidth. Assuming b the bandwidth (in kbps) and $r(b)$ the sending rate (in kbps), we get:

$$r(b) = \begin{cases} 0, 0 \leq b < 70 \\ 0.8923b - 0.6494, 70 \leq b < 330 \\ 294, 330 \leq b < 450 \\ 0.996b - 15.22, 450 \leq b < 1053 \\ 0.4414b + 568.8, 1053 \leq b < 1700 \\ 0.9431b - 81.58, 1700 \leq b < 2019 \\ 0.1717b + 1476, 2019 \leq b < 2533 \\ 0.02723b + 1842, 2533 \leq b < 6180 \\ 2010, b \leq 6180 \end{cases} \quad (1)$$

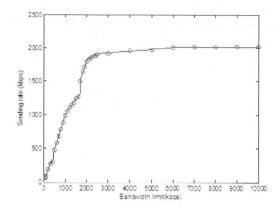

Figure 4. Steady sending rate model under different bandwidth.

Figure 4 demonstrates the curve of our model and the circles represent the data points we collect in experiments. We can see that our model fits the actual data quite well.

4.4 Sending rate under different packet loss rate

Besides bandwidth, packet loss is another consideration for Internet real-time telecommunication applications. Similar to experiments with bandwidth, we use NEWT to simulate different packet loss rate while keeping the bandwidth steady to find out how Hangouts behave under different packet loss rate. In this part, we only discuss random loss.

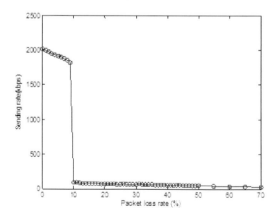

Figure 5. Steady sending rate under different packet loss rate.

Figure 5 demonstrates the result of our experiment, and here we limit the bandwidth to 10Mbps. Apparently there are two phases for sending rate adjustment policy. When packet loss rate is under

10%, the sending rate decreases linearly with packet loss rate. When packet loss rate goes over 10%, the sending rate will decrease to a very low level and keep going down linearly with packet loss rate. Finally when packet loss rate reaches 70%, the video call session will be torn down.

When packet loss rate is under 10%, there are two possible strategies to adjust sending rate: 1) calculating available bandwidth according to bandwidth limit and loss rate, and then adjusting sending rate according to this available bandwidth; or 2) calculating proper sending rate directly from current sending rate and loss rate. Here we limit the bandwidth to 10Mbps, so when loss rate is 9%, the available bandwidth is 9.1Mbps with sending rate of about 2Mbps according to our result in last part, which is apparently different from the actual data. So strategy 2) is more likely the strategy used by Hangouts.

By using data of loss rate under 10%, we get a line fitting in the following form, with r for sending rate in kbps and plr for packet loss rate.

$$r = -2095\,plr + 2017 \qquad (2)$$

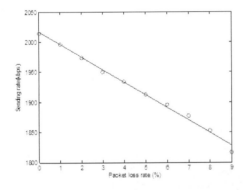

Figure 6. Sending rate under 10Mbps bandwidth constraint and packet loss rate <10%.

Figure 6 plots the curve, with circles for actual data points. We can see that they fit pretty well. Let's analyze the parameters in Equation 2. 2017 is sending rate with no loss and 2095 is close to 2017, which means when packet loss rate raises by 1%, the sending rate will decreases by 20.95kbps, about 1% of the no-loss sending rate.

When packet loss rate reaches 10% and more, Hangouts apparently adjusts its sending rate regardless of current bandwidth. From data, we get a line fitting in the following form, with r for sending rate in kbps and plr for packet loss rate.

$$r = -95.87\,plr + 91.85 \qquad (3)$$

Figure 7 plots the curve, with circles for actual data points.

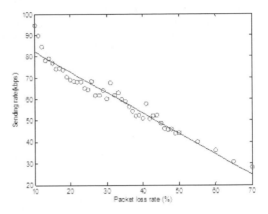

Figure 7. Sending rate under 10Mbps bandwidth constraint and packet loss rate >= 10%.

From the analysis above, we can get the sending rate model of Hangouts under different packet loss rate as Equation 4, with b for bandwidth in kbps, plr for packet loss rate, $r(b)$ in kbps as defined in Equation 1 and $r(b,plr)$ for sending rate in kbps under the condition of b and plr:

$$r(b,plr) = \begin{cases} r(b) - plr * r(b), & 0\% \leq plr < 10\% \\ -95.87\,plr + 91.85, & 10\% \leq plr < 70\% \\ 0, & 70\% \leq plr < 100\% \end{cases} \qquad (4)$$

5   CONCLUSION

In this paper, we conduct several measurement experiments on Google Hangouts to study its architecture and behavior under different network conditions. From experimental results, we find that: 1) Hangouts uses pure client-server architecture for data transfer; 2) Hangouts uses RTP/RTCP protocols to transfer multimedia data and distinguishes different streams by payload_type field in RTP; 3) rate control policy of Hangouts under different bandwidth is quite complex due to its SVC technology. We find that Hangouts video stream has 3 layers and give out a model to describe this policy which fits well with actual data; 4) rate control policy of Hangouts under different random packet loss rate is a two-phase policy. When packet loss rate is under 10%, the sending rate is adjusted according to no-loss sending rate and loss rate. When loss rate goes over 10%, the sending rate is drawn to a quite low level regardless of current bandwidth.

Hangouts is quite a outstanding product in software-based video conferencing field, drawing more and more attention from both academy and industry. We believe our measurement results are instructive for researchers afterwards and better product designs for real-time telecommunication applications over Internet.

REFERENCES

[1] Baset, S.A. & Schulzrinne, H.G. 2006. An analysis of the skype peer-to-peer internet telephony protocol. In Proceedings of IEEE INFOCOM, pages 1–11, April 2006.

[2] Bonfiglio, D. et al. 2008. Tracking down Skype traffic. In Proceedings of IEEE INFOCOM, pages 261–265, April 2008.

[3] Cicco, L.D. et al. 2008. Skype video re-sponsiveness to bandwidth variations. In Proceedings of ACM the 18th International Workshop on Network and Op-erating Systems Support for Digital Audio and Video, pages 81–86, May 2008.

[4] Huang, T. et al. 2009. Tuning Skype redundancy control algorithm for user satisfaction. In Proceedings of IEEE INFOCOM, pages 1179–1185, April 2009.

[5] Schulzrinne, H. et al. 2003. RTP: A Transport Protocol for Real-Time Applications, RFC 3550.

[6] Xu, Y. et al. 2012. Video telephony for end-consumers: measurement study of Google+, iChat and Skype. In Proceedings of Internet Measurement Conference, pages 371–384, November 2012.

[7] Zhang, X. et al. 2012. Profiling Skype video calls: rate control and video quality, In Proceedings of IEEE INFOCOM 2012, pages 621–629, March 2012.

Network Security and Communication Engineering – Chan (Ed.)
© 2015 Taylor & Francis Group, London, ISBN: 978-1-138-02821-0

Using SOA and Web 2.0 in web service applications

M.C. Huang
Department of Business Information Systems / Operation Management, University of North Carolina at Charlotte, North Carolina, USA

ABSTRACT: This paper proposes a three-tier architecture, which includes content network, social network and service network. It also presents a structure for web and load services in ad-hoc computer networks, which is a new system architecture using SOA (Service Oriented Architecture) and Web 2.0 concepts to implement functions for web and load services. It is a three-tier system structure based on Web service functions to implement services seeking and load distribution. Furthermore, this project would construct a knowledge sharing and learning platform based on the mentioned three-tier architecture. Different communities provide their services to each other by using our new knowledge platform and this forms a "virtual community". It leads to our desired accomplishments of "service reusability" and "service innovation". It also proposes the frameworks of new SOA and therefore provides a platform to enhance the interactions between academia and industry.

1 INTRODUCTION

Computer networks can provide parallel computation and services. It is important that hosts find services from other hosts, and send loads to other hosts for some certain function implementation through network transfer. With the increasing popularity of mobile communications and mobile computing, the demand for web and load services grows. When a computer is overloaded or it needs special services from other computers, it may send requests to other computers for web and load services. For example, a computer may need some jobs to be executed with higher quality of services or it needs some jobs to be done with a short period of time that its processor is too slow to perform the jobs; therefore, it may send part of those jobs to other computers with higher speeds of processors. Since wireless networks have been wildly used in recent years, how a host finds services it needs or how it transfers loads to other nodes has becomes a very important issue because not all wireless hosts have the ability to manipulate all their loads. For instance, a host with low battery power cannot finish all its jobs on time and should ask other nodes to provide services to finish the jobs, or it should transfer some of them to other hosts.

Before a wireless host transfers its loads to other hosts or asks for load services from other hosts, it has to find available hosts using resource allocation algorithms. There are several resource allocation protocols been developed, for example, IEFT Service Location Protocol (SLP) [8] and Jini[25] software package from Microsystems. However, these protocols address how to find the resources in wired networks, not in wireless networks. Maab[15] develops a location information server for location-aware applications based on the X.500 directory service and the lightweight directory access protocol LDAP[26]; while it does not cover some important issues about the movements of mobile hosts, for example, how to generate a new directory service and how a host gets the new services, when a directory agent moves away its original region. In an Ad-Hoc network, system structure is dynamic and hosts can join or leave any time. Therefore, how to provide load services and how to find available hosts providing load services become important issues in an Ad-Hoc network system. The goal of this paper is that users can easily find and share resources based on the concepts of "service reusability" and "service innovation"

Based on the population of Web Services techniques, this paper discusses a new architecture which uses SOA model with Web 2.0 [8, 11] for web services. By using Web 2.0 with SOA, the network resources should be easily found by the hosts which need services. Based on XML, the SOA load service system can be used in any computer system platform [1, 2, 3]. This is a very important characteristic for hosts to share or request services in different systems. With the help of Web 2.0, hosts can find the required services easily from the Internet [4].

Figure 1 shows the basic SOA structure [5, 6, 7], which is built by three major components – the Directory, Service consumer, and Service provider. SQA is operated by the following. The Directory provides a platform for information that a service provider can register in the Directory for providing services; a service consumer can find its desired service it needs in the Directory. Once the Directory finds services that a service consumer needs, it sends a query response back to the service consumer to notify the result. At this time, the service consumer has the information about the hosts which can provide services. Therefore the consumer contacts the service provider directory by sending requests. The service provider now will send responses back to the consumer for the services the consumer needs. This is also called the "invoke" process.

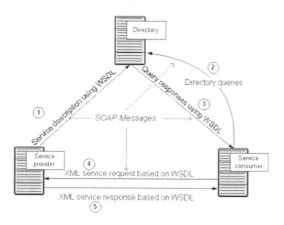

Figure 1.   SOA structure.

2   SYSTEM STRUCTURE

The system structure for the SOA model is illustrated in Figure 2. There are two layers in this structure – the Service Network layer, and the Service Logic layer.

The Service Network layer is the main network that is connected to the internet by using the regular network protocols. It receives requests from internet and forwards requests to the Service Logic layer in Web Service (WS) object forms. Each WS object is based on the SOA model which can communicate each other in social network way. The Service Logic layer is the main layer that uses WS objects to communicate each other in the sub-network under the Service Network layer. Different WS object has different objectives, for example, some WS objects are used for social network communication, while other WS objects are used for accessing contents in Content Networks. Since they are in SOA form, it is easy for them to find the resources they need for different purposes. Inside the Service Logic layer, there are sub-networks for different purposes and functions. For example, nodes can form social networks; storage devices can also form a content network for data accessing. All these operations are managed by the WS objects under the

SOA model. For the service reusability purpose, most WS objects are generated by the Service Network layer for data and object consistency.

In Figure 2, all the services and requests are in the forms of Web Service objects which are defined and implemented by XML. For users who need services, requests are sent by the users to the Service Network Layer. The Service Network is the gateway to accepting requests and sending back requesting results to the requesters.

All the requests are processed by the Service Logic Layer, which finds the required information and applications. The Content Network is a network which communicates databases. The Social Network contains the relations for social communities. The following procedure illustrates how it works.

a. Users send requests to a Service Network.
b. The Service Network forwards requests to Service Logic Layer via Web service functions. In this step, requests are transferred to objects that can communicate with the Service Logic layer.
c. When users send requests, Service Network has the ability to generate the desired WS objects according to the requests forwarded by the Service Network.
d. Service Logic Layer performs the required functions for the requests. It accesses data and information from Content Network using Web service functions. A User can also contact other users using WS objects under the Service Logic Layer.
e. After the Service Logic Layer has the results for the requests, it sends the results back to the Service Network using Web service functions.
f. Service Network then sends the results back to the users who sent initial requests.

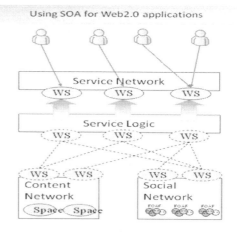

Figure 2.   Using SOA for Web 2.0 applications.

There are several advantages with the design structure.

a. With the system structure, users can join the desired networks anywhere once they are connected to the Internet.
b. With the characteristics of Web 2.0 with SOA model, users can join the desired networks they need to share or find resources easily they need.
c. Using WS objects for the communication makes it easy for service reusability and service innovation. Users do not need to construct special system or programs for data accessing and analysis.
d. Different Service Networks communicate with each other to find and share available resources.

3 IMPLEMENTATION AND SIMULATION

Based on the structure of SOA model with Web 2.0 application, the system can be built in a three-tier structure. The lowest level is the sub-networks including Content Network, Social Network, which provides resources for data sharing. The middle level is the Service Logic layer, which provides WS objects. The top level is the Service Network layer that accepts requests or sends request results back to the users.

This paper is going to construct a virtual community, and work on a simulation for data generation and analysis. Ten thousands nodes with a thousand Web services will be used in the simulation. The simulation will compare the performance for data sharing and load transferring to the system without using SOA structure.

4 CONCLUSION

Usually it is hard to find the required network resources in the Internet for load balance and load service purposes. Because of this, a new structure is proposed for hosts in the computer networks to find the resources for web services. This new structure provides new way that finds resources easily for web services. Especially, when a user needs services which are not very commonly provided in the Internet, with the help of Web 2.0 and SOA, users can find what they need because of the "long tail" property of Web 2.0 and the platform free property of SOA.

The system performance will be evaluated by using a simulation. Usually it is very hard to evaluate the performance for Web 2.0. Therefore this project should be useful in finding the performances for the usage of Web 2.0 and SOA.

REFERENCES

[1] Ross J. Anderson, "The eternity service," *Pragocrypt 1996*, 1996.
[2] StephanosAndroutsellis-Theotokis and DiomidisSpinellis, "A survey of peer-to-peer content distribution technologies," *ACM Computing Surveys*, 36(4):335–371, December 2004. doi:10.1145/1041680.1041681.
[3] Yvonne Balzer, "Improve your SOA project plans," *IBM*, 16 July 2004.
[4] Biddle, Peter, Paul England, Marcus Peinado, and Bryan Willman, "The Darknet and the Future of Content Distribution," *2002 ACM Workshop on Digital Rights Management*, 18 November 2002.
[5] Jason Bloomberg, "Mashups and SOBAs: Which is the Tail and Which is the Dog?," *Zapthink*.
[6] Channabasavaiah, Holley and Tuggle, "Migrating to a service-oriented architecture," *IBMDeveloperWorks*, 16 Dec 2003.
[7] Thomas Erl, "About the Principles," Serviceorientation. *org*, 2005–2006.
[8] E. Guttman, C. Perkins, J. Veizades and M. Day, "Service Location Protocol," Version 2, IEFT, RFC 2165, November 1998.
[9] Dion Hinchcliffe, "Is Web 2.0 The Global SOA?," *SOA Web Services Journal*, 28 October 2005.
[10] Huang, M., S. H. Hosseini, and K. Vairavan, *Load Balancing in Computer Networks*, Proceedings of ISCA 15th International Conference on Parallel and Distributed Computing Systems (PDCS-2002), Special session in Network Communication and Protocols. Held in the GALT HOUSE Hotel, Louisville, Kentucky, Sep. 19–21.
[11] Kim, A.J. (2000), "Community Building on the Web: Secret Strategies for Successful Online Communities," *London: Addison Wesley(ISBN 0-201-87484-9)*.
[12] Christopher Koch, "A New Blueprint For The Enterprise," *CIO Magazine*, Mar 1 2005.
[13] Peter Kollock (1999), "The Economies of Online Cooperation: Gifts and Public Goods in Cyberspace," *Communities in Cyberspace*, Marc Smith and Peter Kollock (editors). London: Routledge.
[14] Paul Krill, "Make way for SOA 2.0," *InfoWorld*, May 17, 2006.
[15] H. Maab, "Location-Aware Mobile Application Based on Directory Services," MOBICOM 97, pp23–33.
[16] Joe McKendrick, "Anti SOA 2.0 petition nears 100," *ZDNet.com*, June 29, 2006.
[17] YefimNatis& Roy Schulte, "Advanced SOA for Advanced Enterprise Projects," *Gartner*, July 13, 2006.
[18] Andy Oram et al., "Peer-to-Peer:Harnessing the Power of Disruptive Technologies," Oreilly 2001.
[19] Howard Rheingold (2000), "The Virtual Community: Homesteading on the Electronic Frontier," *London: MIT Press. (ISBN 0-262-68121-8)*.
[20] Ross, T. J., Fuzzy Logic with Engineering Applications, McGraw Hill, 1995.
[21] Antony Rowstron and Peter Druschel, "Pastry: Scalable, Decentralized Object Location, and Routing for Large-Scale Peer-to-Peer Systems," *Proceedings Middleware*

2001 : IFIP/ACM International Conference on Distributed Systems Platforms, Heidelberg, Germany, November 12–16, 2001. Lecture Notes in Computer Science, Volume 2218, Jan 2001, Page 329.

[22] Ralf Steinmetz, Klaus Wehrle (Eds), "Peer-to-Peer Systems and Applications," *ISBN: 3-540-29192-X,* Lecture Notes in Computer Science, Volume 3485, Sep 2005.

[23] Smith, M. "Voices from the WELL: The Logic of the Virtual Commons,"*UCLA Department of Sociology.*

[24] I. Stoica, R. Morris, D. Karger, M. F. Kaashoek, and H. Balakrishnan, "Chord: A scalable peer-to-peer lookup service for internet applications," *Proceedings of SIGCOMM 2001*, August 2001.

[25] J. Waldo, "The Jini Architecture for network-centric computing," Communication of the ACM, pp 76–82, July 1999.

[26] W. Yeong, T. Howes, and S. Kille, "Lightweight Directory Access Protocol," RFC 1777, March 1995.

[27] http://opengroup.org/projects/soa/doc.tpl?gdid=10632.

Network Security and Communication Engineering – Chan (Ed.)
© 2015 Taylor & Francis Group, London, ISBN: 978-1-138-02821-0

Current–mode signal processing based OTA design for multimedia data transmission

Jong-Un Kim, Sung-Dae Yeo, Dong-Ho Kim, Gye-Min Lee & Seong-Kweon Kim
Graduate School of NID Fusion Technology, Seoul National University of Science & Technology, Seoul, Korea

ABSTRACT: We implemented a low-power Operational Trans-conductance Amplifier (OTA) for a current-mode signal processing for a multimedia system. A current-mode signal-processing technique can be used for low power multimedia system including volumetric display and immersive acoustic system, because it has the potentiality of constant power consumption independent of operating frequency. The proposed OTA was implemented with 0.35 um CMOS process technology and the active chip size was 66.3 um × 11.35 um. According to the measurement, the power consumption was observed with less than 100 uW at $V_{dd} = 2.1$ V.

1 INTRODUCTION

Our multimedia environment has advanced to surround us with audio-visual information in 3-dimensinal space. To transmit and reproduce 3 dimensional space with all audio-visual information becomes the ultimate target of communication and broadcast service. However, Multimedia devices are required to realize the goal of low power consumption and high speed operation. Therefore, current-mode signal processing has been studied as a promising low power technique.

For example, in the current-mode circuits, the summation circuit is designed with wired-or structure because the input of current mirror is a relatively low impedance node and may be considered a signal ground. The summation is carried out with current quantities added for a small calculation time. Moreover, the current-mode circuit has a feature such as a constant power consumption that is independent of operating frequency. Due to a simple summation circuit structure and a small calculation time of current mode circuits, the power consumption of a current-mode multimedia device is drastically cut down. Current-mode circuits that rely on current mirrors are usually designed with sampled or switched current (SI) types, and generally implemented in complementary metal-oxide silicon (CMOS) technology which has compatibility of conventional digital signal processing. The current-mode analog finite-impulse-response (FIR) Filter with a variable tap circuit has been designed for the multimedia device with low power consumption and a high speed operation.

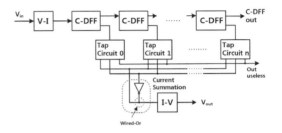

Figure 1. Current-mode FIR filter.

Figure 1 shows the block diagram of a programmable current-mode analog FIR Filter with a variable tap circuit. The input voltage signal is converted into the current signal with the operational trans-conductance amplifier (OTA). Then, the current signal is transferred to the current delay flip flop (CDFF) and weighted with the tap circuits. The output currents of the tap circuits are delivered to the summation circuit with a wired-or structure. Finally, the output current from the summation circuit is converted into the output voltage signal with the I-V converter.

In the design of the current-mode FIR Filter, the OTA is one of the key devices for a wireless mobile multimedia communication requiring low power consumption. The system performance is under the control of the OTA accuracy. For example, the operation speed of FIR computation is dependent on the operating frequency of OTA. For the current-mode FIR Filter, the condition of low power and small chip size for a single chip is required.

In this paper, OTA with low power consumption and small chip size is proposed.

2 CURRENT-MODE SIGNAL PROCESSING TECHNIQUE

Figure 2 shows the current memory (CM) circuit. CM circuit is the basic circuit of the SI circuit. CM has components of a current source, a memory MOSFET, an input switch, an output switch, a control switch and a dummy MOSFET. CM operation is performed with the charge in the gate-source parasitic capacitance of the memory MOSFET.

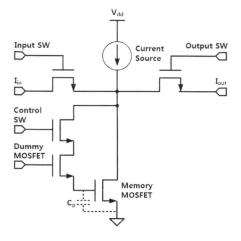

Figure 2.   Structure of CM.

Figure 3 shows the operation of CM. The operation is divided into three periods, "sample", "hold" and "output". In the "sample" period, the input switch and the control switch is connected, and the output switch is disconnected.

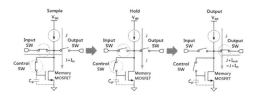

Figure 3.   Operation of CM.

The current through the control switch is memorized by the charge at the parasitic capacitance of a gate of the memory MOSFET. In the "hold" period, all switches are opened and the input current keeps being memorized. In the "output" period, the output switch is shorted and the input switch and the control switch keep in the disconnected condition. The output current can be obtained with the same value memorized at the gate of the memory MOSFET. The output current flows in the opposite direction to the input current. The CDFF is constructed using three CMs with one input node and two output nodes, as shown in Figure 4. The input current is delayed and mirrored to the tap circuit. For the operation of the FIR Filter, the tap circuit is designed with the programmable operation.

The programmable current-mode analog FIR Filter with a variable tap circuit can be designed for the multimedia device with a low power as shown in Figure 1.

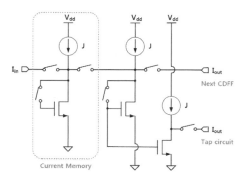

Figure 4.   Structure of CDFF.

3 DESIGN OF LOW POWER OTA

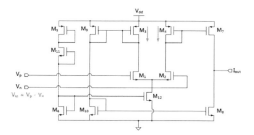

Figure 5.   Proposed OTA schematic.

Figure 5 shows the schematic of OTA. Basically the OTA was designed with simple current mirror structure because current mirror circuits are mainly used in the current-mode FIR Filter. The input stage of M_1 and M_2 is designed with MOS source-coupled pairs with source degeneration.

This source degeneration is used to robustly enhance the linearity of a differential pair at the cost of a reduced input voltage range. The output stage of M_5, M_6, M_7 and M_8 is designed with the same sized current mirror structure to realize the low offset current. Using these output stage structures, the dc offset current between OTA and the input of current mirror is drastically reduced. The input voltages, V_p and V_n are converted to the current in the input stage M_1 and M_2.

The converted current I_{d1} and I_{d2} are multiplied with gate-ratioed current mirror of the ratio of k. The output current is obtained with the difference of I_{d7} and I_{d8} multiplied with the ratio of k. The output current of I_{out} is approximately expressed as

$$I_{out} = k \times g_m \times \left(V_p - V_n \right) \qquad (1)$$

Here, $g_{m7} = g_{m8} = g_m$.

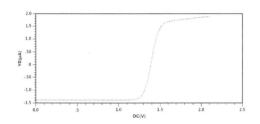

Figure 6. I_{out} as a function of input voltage V_p.

Figure 6 shows the simulation result of the linear output current characteristics. The output current is varied as a function of V_p. In the case of $V_n = 1.4$ V, the output current was linearly varied from -1.5 uA to 2 uA as function of the variation of input voltage V_p.

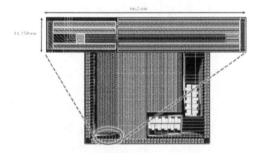

Figure 7. Chip photograph of OTA.

Figure 7 shows the chip photograph of the implemented OTA using 0.35 um CMOS technology. The active chip size of the circuit is 66.3 um × 11.35 um. According to the measurement, the power consumption was observed with less than 100 uW at $V_{dd} = 2.1$ V.

4 CONCLUSIONS

The OTA was designed and implemented with 0.35 um CMOS technology for current-mode multimedia system. The simulation result shows that the proposed OTA has the linearity for current mode signal processing. According to the measurement, the power consumption was observed with less than 100 uW at $V_{dd} = 2.1$ V. The proposed OTA promises a successful low power current-mode FIR Filter.

REFERENCES

[1] Fiez, T.S. et al. 1991. Switched-Current Circuit Design Issues. *IEEE Journal of Solid-State Circuits* 26(3): 192–202.
[2] Balachandran, G.K & Allen, P.E. 2002. Switched-Current Circuits in Digital CMOS Technology With Low Charge-Injection Errors. *IEEE Journal of Solid-State Circuits* 37(10): 1271–1281.
[3] Togura, K. et al. 2001. Low Power Current-Cut Switched-Current Matched Filter for CDMA. *IEICE Transactions on Electronics* E84-C (2): 212–219.
[4] Kim, S.K. et al. 2001. Novel FFT LSI for Orthogonal Frequency Division Multiplexing Using Current Mode Circuit. *Japanese Journal of Applied Physics* 40(4B): 2859–2865.
[5] Saigusa, S. et al. 2003. Switched-Current Analog Programmable Filter for Software-Defined Radio. *Japanese Journal of Applied Physics* 42(4B): 2185–2189.

Network Security and Communication Engineering – Chan (Ed.)
© 2015 Taylor & Francis Group, London, ISBN: 978-1-138-02821-0

Optimal switching intermittency correlations of information complexes DNA for fractal networks Shannon entropy inside living cells

N.E. Galich
Department of Experimental Physics, Saint-Petersburg State Polytechnic University, Saint-Petersburg, Russia

ABSTRACT: Experimental data on fluorescence of DNA complexes inside neutrophils and other cells in flow cytometry with nanometer spatial resolution in large populations of cells are analyzed. All DNA correlations inside living cells have super dense packing in complex fractal networks of the universal classes of 'exponentially small worlds', in twice-double (quadruple) logarithmic scale. Central moments for fluctuations of fluorescence intensity in networks of Shannon entropy have unstable behavior, i.e. strong exponential increasing moments of high orders, characterizing manifestation of intermittency. Instability and switching stability for even and odd central moments depend on the scale networks of information entropy in given cell. Distributions and switches central moments for DNA fluorescence depend on health status of given person in given time and can be used for sensitive diagnoses. DNA oxidation inside living cells reflects intermittency in information activity DNA and self-regulation of informational homeostasis for total Shannon entropy of full set DNA inside cells. Intermittency selects the optimal scale of correlations for large-scale DNA activity and reflects fundamental properties of packing, synchronization, and mixed topology varied correlations DNA in cells.

1 INTRODUCTION

We present results of novel nonlinear analysis of experiments flow cytometry on immunofluorescence with nanometer spatial resolution in the flow direction [1] for large populations~10^4 -10^5 of neutrophils in the peripheral blood and for other cells of human and chickens, for different, intercalating dyes and varied excitations of fluorescence [1-4]. Oxidative activity of DNA is visualized at fluorescence. In each experiment, it is observed that fluorescence of three-dimensional (3D) DNA nanostructures of all non-coding and coding parts of DNA in full set of chromosomes inside cells. Each cell makes a chaotic Brownian motion at chaotic rotations in the jet of blood, flowing through the laser beam, during measurements. Therefore, each fluorescence histogram defines a representative statistics for various two dimensional (2D) projections on the photomultiplier all possible detailed spatial images of fluorescing 3D DNA in large populations of cells.

Detailed descriptions of experimental procedures for physical measurements and dyeing DNA by ethidium bromide, *etc* are presented in [1-5]. This is a high sensitive method for diagnostics of many different and complex diseases, early diagnostics of illnesses, hidden diseases [1-4]. Main results are presented as the dependences for number of flashes N(I) on fluorescence intensity I. Usually normalized distributions (probability) of flashes $P(I)=N(I)/N_0$,

are used, where N_0 is a common number of flashes, I is dimensionless intensity, corresponding to given number of measuring channels for intensity. In our experiments maximal number of measuring channels in $I_{max}=256$. Figures 1 and 2 presented distributions for frequency of information $J(I)=-\ln(P(I))$ and information entropy $E(J(I))=-p(I)\ln(p(I))$, based on normalized distribution of information p(I) (see equations. (2),(3) and [1-5]) for fluorescing neutrophils in the blood different healthy and unhealthy men.

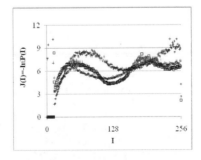

Figure 1. Dependence information $J(I)=-$ LnP(I) for frequency of flashes P(I) on their intensity I(r=256) for fluorescing DNA in neutrophils; cross points correspond to bronchial asthma, total number of flashes is N_0= 76 623; quadrate points correspond to the healthy donor, N_0= 40 109; circle points correspond to the oncology disease, N_0= 40 752.

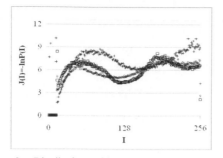

Figure 2. Distributions of information entropy E(J(I)) in the dependence on fluorescence intensity I(r=256).

At the present time we do not know how to extract, separate information and interpret immunofluorescence data. Main difficulties interconnect with rather strong irregularities and brokenness in fluorescence distributions. These irregularities reflect many unknown features DNA correlations for oxidative activity DNA in fractal networks inside living cells. These irregularities are illustrated in Figures.3.

We concentrate on the modern statistical analysis of these irregular distributions. Our analysis is based on invariance of total Shannon entropy for DNA information in all aerobic cells [2]. Therefore we analyze statistical peculiarities of self-regulation distributions on local fluctuations entropy near average level.

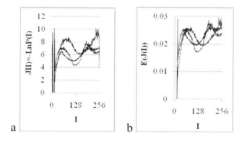

Figures 3. Continuous distributions of information J(I) (a) for Figure 1 and information entropy E(J(I)) (b) for Figure 2 in the dependence on fluorescence intensity I(r=256).

We focus on varied fractal properties non-Gaussian large-scale distributions information entropy of DNA inside all living cells.

2 DENSE FRACTAL PACKING AND SELF-REGULATION SHANNON ENTROPY DNA

Ranger or rank r of histogram fluorescence defines scale of multistage clusters in networks with structure of bronchial tree. Range r coincides with the number of columns in histograms or number measuring channels of fluorescence intensity at given maximal value

of dimensionless intensity, i.e. $r=I_{max}$. In the presented experiments, the maximal number of channels is r=256. Variations of range r, i.e. rank r of histograms or variations the scale r, provide varied changes in irregularity and brokenness for frequency distribution of fluorescence in histograms various ranks r [1,2,5] The quantitative measure of irregularity and brokenness for frequency distribution of flashes for any rank r, in all histograms, may serve as a Hurst index H. Hurst exponent H is determined by means of regression equation [6] and connected with fractal dimension D=2-H [6]. This definition of Hurst index H for initial fluorescence distributions in Figures 2,3b at varied ranks r =4, 8, 16. 32, 64, 126 and 256 leads to varied fractal dimensions D(r) for entropy E(I,r) in Figure 4.

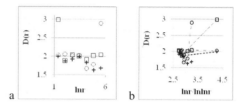

Figure 4. Dependence fractal dimensions D(r) of information entropy E(J(I,r)) on rank r (a) and on combination lnr/lnlnr (b).

These poorly organized and mixed distributions fractal dimensions can be presented in more accurate logical sequence if complex fractal networks of 'exponentially small worlds' [3,5,8,9] in twice-double logarithmic scale in Figure 5 is introduced.

At the present time networks of 'exponentially small worlds' are the best and denser networks for super dense packing information. Complex networks for information entropy DNA in all living cells are also characterized by invariance of total Shannon entropy E(J,r) for any rank r [2,8,9]. Dependence of total entropy E(J,r))=lnr on rank r is logarithmic

$$E(J,r)-=-<\ln p_i(r)>=-\sum p_i \ln p_i=\ln r \qquad (1)$$

Where the summation and index i are changed from i=1 up to i=r; r is rank (range); r=4, 8, 16, 32, 64, 128, 256; normalized probability density $p_i(r)$ is defined as probability distribution of information J_i for DNA oxidative activity $J(I,r)=J_i(r)=-\ln P(I,r)=-\ln P_i(r)$

$$p_i(r)=\left(\frac{J_i(r)}{\sum J_i(r)}\right), \quad J_i(r)=-LnP_i(r) \qquad (2)$$

Shannon entropy distribution for probability $p_i(r)$ is

$$E_i(r)=-p_i(r)\ln p_i(r) \qquad (3)$$

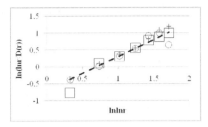

Figure 5. Dependence of ln(lnr/D(r))) on double logarithm of rank r in multi-scale networks of 'real worlds' for fluorescing DNA inside neutrophils: the dotted line corresponds to the ideal network of 'exponentially small worlds' (D=2), without fractals.

The empirical invariant E(J,r))=lnr defines informational homeostasis for oxidative activity 3D DNA for full set chromosomes inside all living cells of all aerobic beings [2,8,9]. Information transforms at DNA activity inside cells, at variations of inner and external conditions, for all cells, as in the all open systems, ensure stability, adaptability and vitality cell at self-regulation of homeostasis [2,8,9]. Self-regulation is defined as the local deflections information entropy from average homeostatic level to total information entropy E(J,r))=lnr [8,9]. Entropy's noises form the main signals for individual self-regulation of homeostasis for any cells in the body any aerobic beings in the case of any local disorders of any origin, such as inner diseases, infections, traumas, *etc*. This determines the defensive response and immunity during reconstruction of fractals and complex information networks DNA in all cells living inside us [1-3,5,8,9]. Thank self-regulation.

Let us consider local noises, scales and local redistributions entropy in complex networks entropy during self-regulation homeostasis E(J,r))=lnr. The stability of self-regulation of homeostasis ensures the exponential decrease central moments of high order for all distributions noise entropy in networks of any scale [2,9]. Consider a random variable $iE_i(r)$, the average value of this random variable $<iEi(r)>$ and the relative deviations from the mean value

$$c_i(r)= iE_i(r)\times(<iE_i(r)>)^{-1}-1,$$
$$<iE_i(r)>=(r^{-1})\times\sum iE_i(r) \quad (4)$$

Random variable $c_i(r)$ characterizes fluctuations of dimensionless intensity for distributions of information entropy, fluctuations in the number of measuring channels for a given rank r, i.e. scales of the changes and correlations of information entropy $E_i(r)$. Statistical peculiarities varied distributions $c_i(r)$ are characterized by the changes of central moments $<c(r)m>=M(c(r),m)=M_m(c(r))$ and power means or averages of Hölder $<c(m,r)>$ {5,9]

$$<c(r,m)>=[(r^{-1})\times((\sum c_i^m)]^{(1/m)} \quad (5)$$

Number m determines the order of moment $M_m(c(r))$ and the order of Hölder's averages $<c(r,m)>$. Distributions of M(c(r),m) and $<c(r,m)>$ are individual for given person in given time and have rather strong dependence on network's scale, i.e. on rank r. Typical illustrations are presented in Figures 6.

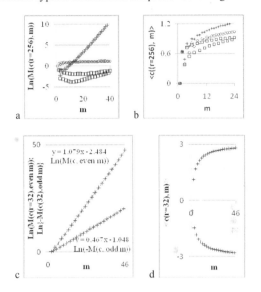

Figure 6. Two distributions of central moments M(c(r=256),m) (a) and M(c(r=32),m) (c) and corresponding distributions of Hölder averages $<c(r=256),m)$ (b) and $<c(r=32),m)$ (d) in the dependence on number m for moments M(c(r),m) and Hölder averages $<c(r,m)>$ for rank r=256 (a),(b) and three statuses of health various men and for rank r=32 (c),(d) at asthma disease. Identical points for various lines in Figures (a) - (d) belong to the even m (upper points) and odd m (lower points).

3 OPTIMAL SWITCHING INTERMITTENCY OF CORRELATION SCALES DNA IN NETWORKS INFORMATION ENTROPY

According to Figures 6c and 6d, dependences of central moments M(c(r),m) on their number m are an exponential and have two branches; positive branch corresponds to the even numbers m and negative branch corresponds to the odd numbers m. The same exponential and branching behaviors characterize all dependences of central moments M(c(r),m) on their number m for all ranks r≤128. Hölder's averages $<c(r,m)>$ in Figure 6d for asthma disease have the similar, but not exponential, branching for positive and negative branches, for even m and odd m, if rank r≤128. If r≥256, are observed more asymmetric, non-universal distributions M(c(r),m) and $<c(r,m)>$, as it is shown in Figures 6a, 6b.

The exponential growth of high-order correlations, i.e. the positive trends for the derivatives of the functions $(Ln(M(c(r),m))'_m \geq 0$ to the variable m for even m and $(Ln(-M(c(r),m))'_m \geq 0$ for odd m, as in Figures 6c,7a,7b, mean that multiple correlations (m>2) are dominant compared with pair correlations (m=2). It also means the participation (repeatedly) of all 46 chromosomes [5,9] in self-regulation of homeostasis E(J,r)=lnr. The exponential growth of <c(r,m)> for even and odd m=1....46, as in Figure 6c, is illustrated in Figure 7 for various ranks r.

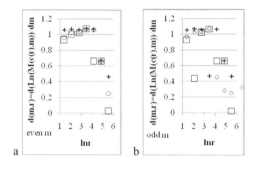

Figure 7. Dependence for derivatives d(m,r) of logarithm central moments $(Ln(M(c(r),m))'_m = d(m,r)$ to the variable m on lnr for even m (a) and $(Ln(-M(c(r),m))'_m = d(m,r)$ for odd m (b).

According to Figures 7 distributions of central moments M(c(r),m) have different exponential behavior in the dependence on number m: indexes of exponents $(Ln(M_m))'_m \geq 0$ and $(Ln(-M_m))'_m \geq 0$ for even m and odd m are positive and different for various states of various health people. Strong exponential growth of central moments M(c(r),m) by increasing their numbers m leads to the intermittency in the scales of correlations information entropy DNA, when the absolute value of each next moment |M(c(r),m+1)| is more than previous moment |M(c(r),m)|, etc. Fast exponential growth of higher statistical moments means that even a low level small noise for higher orders of correlation and autocorrelation of noise ensures intermittency of DNA activity, i.e. rare irregular bright flashes in time [3]. Thus, for all DNA inside cells exist chaotic mixes all scales of large-scale correlations <iE_i(r)>: all interconnected. These mixes ensured very quick and high effective self-regulation of information homeostasis E(J,r)=lnr for any local perturbation of iE_i(r). Here intermittency inevitably reflects tangled topology and complicated mixes of different scales for networks of 'small worlds' caused by irregular ambiguous switching, mixed up with ragged loops for fractals in networks of 'small worlds' with D~lnr/lnlnr [7], which are shown in Figure 4b and in [5]. Intermittency is fundamental property of DNA correlations [10]. According to Figures 7 most quick growth of moments M(c(r),m) and the optimal (most strong) intermittency occurs for even m at rank r=32. This conclusion, data in Figure 7 and [9] reflect more

effective changes for DNA activity inside cells in the large-scale networks of entropy, at rank r≤32 [9]. These values of rank r≤32 define the large scales of correlations, more than 1/32 part from cell size.

For odd m and rank r=32, at asthma disease in Figures 6c and 7b, the minimal intermittency is observed. In this case and in the similar cases in Figure 7 a strong difference between positive and negative growth for all corresponding moments M(c(r),m) with even and odd m of all perturbations E_i(r) for given rank r are observed. In the other cases, for other r, in Figures 7 very similar and identical values of |M(c(r),m)|, i.e. equilibrium for even and odd moments M(c(r),m) are observed. This reflects exponentially effective redistributions for any local perturbation of information entropy E_i(r) at self-regulation informational homeostasis inside cells. Self-regulation depends on health status of given person in given time.

REFERENCES

[1] Galich N.E. 2011. Complex Networks, Fractals and Topology Trends for Oxidative Activity of DNA in Cells for Populations of Fluorescing Neutrophils in Medical Diagnostics. *Physics Procedia*, v.22, pp. 177–185.

[2] Galich N.E. 2014a. Invariance and Noises of Shannon Entropy for Information on Oxidative Activity of DNA in All Living Cells for Medical Diagnostics. *American J. of Operations Research*, v.4, pp. 72–89.

[3] Galich N.E. 2013a. Fractal Networks of Real Worlds of Fluorescing DNA in Complete Set of Chromosomes inside blood cells for Medical Diagnostics. *Open J. of biophysics*, v.3, pp. 232–244.

[4] Filatov M.V., Varfolomeeva E.Y.,Ivanov E.A. 1995. Flow cytofluorometric detection of inflammatory processes by measuring respiratory burst reaction of peripheral blood neutrophils. *Biochem. Molec.medicine*.v.55, pp.116–121.

[5] Galich N.E. 2013b. Dense Fractal Networks, Trends, Noises and Switches in Homeostasis Regulation of Shannon Entropy for Chromosomes' Activity in Living Cells for Medical Diagnostics. *Appl. Math., Special Issue on Chaos and Fractal*, v.4, pp.30–41.

[6] Feder J. 1988. Fractals. New York: Plenum Press

[7] Newman M.E.J. 2003. The structure and function of complex networks. *SIAM Rev.*, v. 45, pp.167–256.

[8] N.E. Galich 2014b. Self-Regulation of Information Complexes DNA in Hierarchy Networks of Shannon Entropy inside Living Cells. *AASRI Procedia*, v.9, pp.123–130.

[9] N.E. Galich 2014c. Dense Patterns, Switching, Stability and Self-Regulation in Information Networks DNA inside Living Cells. *EIRI Procedia (2014 International Conference on Future Information Engineering)* v.10, pp.25–31.

[10] S. Rybalko, S. Larionov,. M. Poptsova and A. Loskutov. 2011. Intermittency as a universal characteristic of the complete chromosome DNA sequences of eukaryotes:- From protozoa to human genomes. *Phys,Rev.E.*, v.84, 042902-1-5.

Network Security and Communication Engineering – Chan (Ed.)
© 2015 Taylor & Francis Group, London, ISBN: 978-1-138-02821-0

A unified decoding method of system combination based on forest

Y.P. Liu
Harbin Engineering University, Harbin, China

C.G. Ma
Harbin University of Science and Technology, Harbin, China

ABSTRACT: The word level system combination, which is better than phrase level and sentence level, has emerged as a powerful post-processing method for Statistical Machine Translation (SMT). This paper introduces simple bracket transduction grammar for forest decoding. To optimize the more feature weights, we introduce Minimum Risk (MR) with/without Deterministic Annealing (DA) into the training criterion, and compare it with Minimum Error Training (MERT). The unified approaches of n-gram model based on forest are shown to be superior to conventional cube pruning in the setting of the Chinese-to-English track of the 2008 NIST Open MT evaluation. We get 1.84 BLEU score improvement in test set through two-pass decoding.

1 INTRODUCTION

System combination aims to find consensus translations among different statistical machine translation (*SMT*) systems. It has been proven that such consensus translations are usually better than the translation output of individual systems. Confusion network (*CN*) (Rosti et al. (2007)) for word-level combination is a widely adopted approach for combining *SMT* output, which is shown to outperform sentence re-ranking methods and phrase-level combination (Rosti, et. al. 2007). In order to construct confusion network, word alignment between a skeleton and a hypothesis is a key technic in this approach. The important alignment methods include Translation Edit Rate (*TER*, Snover et al. (2006)) based alignment, which is proposed by Sim et al. (2007) and often taken as the baseline, and a couple of other approaches, such as the Indirect Hidden Markov Model (*IHMM*, He et al.(2008)), the *ITG*-based alignment (Karakos et al. (2008)), incremental *IHMM* (Li et al.(2009)). Joint optimization (He et al.(2009)) integrates *CN* construction and *CN*-based decoding into a decoder without skeleton selection.

In this paper, we start from the view of decoding, and introduce a forest decoding algorithm of system combination which runs in a bottom-up manner. This algorithm parses the target words in the order of *CN* with reordered words and inserted null words. The forest decoding which is based on system combination is different from the conventional decoding algorithm, in particular:

- The formal grammar is used in forest-based system combination. Such a type of forest is generally used by parsing and machine translation. The algorithm produces the forest by simple bracket transduction grammar (*SBTG*) with lexical and non-terminal rules.
- Two pass decoding algorithm is adopted in the framework. The first pass uses a 5-gram language model, and the resulting parse forest is used in the second pass to guide search with the reestimated *n*-gram probability. We rescore the derivation through original viterbi model and reestimated *n*-gram model, using the product of inside and outside probability.
- *MR* with/without *DA*, which attempts the solution of increasingly difficult optimization problem, is introduced in two-pass forest decoding, because of fitting the curve of the probability distribute of log-linear model better and solving the large feature number.

This paper is structured as follows. After introducing the definition of the forest in Section 2, we will first, in Section 3, give the detail of the inside and outside algorithm and several *n*-gram probability computation methods for forest decoding; then in Section 4, we show training criterion, which is *MR* with *DA* for the smooth objective function to ensure that the decoder explores the search space as largely as possible to overcome overfitting and the limitation of feature number, which is different from *MERT*. In Section 5, experiment results and analysis are presented.

2 FOREST OF SYSTEM COMBINATION

Formally, a forest or hypergraph is a compact structure of all derivations for a given sentence under a synchronization context-free grammar (*SCFG*). There are many applications for forest structure such as model training (Tu et al. 2010) and decoding (Li et al. 2009, Kumar et al. 2009, Denero et al. 2009,) in machine translation. Forest gives more search space in decoding phase to avoid search error because of imprecision parameter estimation of the model or local feature.

Formally, a forest F is defined as a *4*-tuple $F=<V,E,G,R>$, where V is a finite set of hypernode, E is a finite set of hyperedge, $G \in V$ is the unique goal item in F, and R is the set of weights. For every input sentence $f_1^J = f_1,...,f_J$, each hypernode is in the form of X_i^j, which denotes the partial translation forest of source partial language $f_i,...,f_j$ spanning the substring from i-1 to j. Each hyperedge $e \in E$ is a triple tuple $e=<T(e),h(e),f(e)>$, where $h(e) \in V$ is its head, $T(e)$ is a vector of tail nodes, and $f(e)$ is a weight function from $R^{|T(e)|}$ to R.

Our forest-based system combination is represented by simplified bracket transduction grammar (*SBTG*). Formally, an *SBTG* can be defined as a tuple $G=<T_s, N_s, P>$, where T_s is the terminal word symbol in source language, N_s is the non-terminal symbol including three symbol $N_s = \{S, X_1, X_2\}$, P is a set of production rules including two type:

- Lexical rule: $X \rightarrow w$, $w \in D$
- Non-terminal rule: $S \rightarrow <X_1X_2>$, $X \rightarrow <X_1X_2>$, $X \rightarrow <X_1X_2>$

Where D is dictionary including ε(null word for normalization in system combination), start symbol, end symbol and regular word. Non-terminal rule is like monolingual glue rule in hierarchical phrase-based model or normal/inverse ordering in the model based on bracket transduction grammar.

To integrate forest into system combination, we model the process of the derivation generation using the probabilistic *SBTG*.

$$\hat{e}_1^I = \arg \max_d \sum_{d \in D} \prod_{r \in d} \Pr(r) \qquad (1)$$

where D is the set of all derivations generating the target language, d is the best derivation, r is the lexical or non-terminal rule. $\hat{e}_1^I$ is the best translation.

3 DECODING PROCEDURE

3.1 Inside and outside pruning

The inside-outside algorithm is a way of reestimating hyperedge probability in *PCFG/SCFG*. It was introduced as a generalization of the forward-backward algorithm for parameter estimation on *HMM*. It is used to compute expections, for example as part of the *EM* (Expectation Maximization) algorithm. Standard inside and outside recursion formulation is showed in Equation (2) and Equation (3), in which $I(v)$ and $O(v)$ is inside and outside score in hypernode, $f(e)$ and $w(e)$ are feature and weight vector ,respectively. $dim(f(e))$ is the dimension of feature vector. $IN(v)$ and $OUT(v)$ are incoming and outgoing hyperedge of v.

$$I(v) = \sum_{e \in IN(v)} \left[\sum_{dim(f(e))} f(e)w(e) \right] \cdot \left[\prod_{u \in T(e)} I(u) \right] \qquad (2)$$

$$O(v) = \sum_{e \in OUT(v)} \left[\sum_{dim(f(e))} f(e)w(e) \right] \cdot \left[O(h(e)) \prod_{\substack{u \in T(e) \\ u \neq v}} I(u) \right] \qquad (3)$$

Through Equation (4), we can get the posterior probability of hyperedge including language model feature score, which is computed after its hypernode is generated. After getting inside score and outside score, we can prune the score of hyperedges which is below the threshold. $P(e/F)$ is the posterior probability of specific hyperedge e, and p_e is the original weight of hyperedge e in forest F, and Z is normalization factor that equals to the inside probability of the root node in F.

$$p(e \mid F) = \frac{1}{Z} p_e . O(h(e)). \prod_{v \in T(e)} I(v) \qquad (4)$$

3.2 N-gram probability

We adopt three types of *n*-gram estimation model and compare these models which are proposed by Li et al.(2009), which describes an algorithm for computing it through *n*-gram model, by Kumar et al.(2009),which describes an efficient approximate algorithm through the highest hyperedge posterior probability relative to predecessors on all derivations in forest, and by Denero et al. (2009), which give the expectation count and exact algorithm.

In order to compute *n*-gram probability, we merge the hyperedge with the same left and right equivalent *n-1* gram state into the same hypernode, which already removeε.

3.3 Decoding algorithm

Different from conventional cube pruning, we use cube growing to exploit the idea of lazy computation, which get *n*-best from top to bottom in forest. Conceptually, complicate forest decoding incorporates the following procedure:

1 Generating the forest: generate all hypernodes and hyperedges in *F* bottom-up in topological order. Every hypernode have many hyperedges with *SBTG* rule for phrase structure. According to the generating order of a target sentence, the decoding category is bottom-up (Liu et al.2009). Finally, the forest has a distinguished goal item for convenient decoding.

2 Running inside recursion algorithm: for every hypernode, we compute the inside score from their hyperedges with tail nodes. The algorithm set the inside score of axiom item to zero, and run from bottom to top for the score through axiom item and inference rule.

3 Running outside recursion algorithm: for every hypernode, we compute the outside score of the hypernode, which is the left or right branch of the parent hypernode(two non-terminal in *PCFG/SCFG*).

4 Computing hyperedge posterior probability: according to inside and outside probability of hypernode and inside score of goal item and hyperedge weight, we have the posterior probability for forest.

5 Inside-outside pruning: to reduce the search space and improve the speed, pruning the hyperedges is important in the light of posterior probability. Our pruning strategy includes threshold and histogram.

6 Computing *n*-gram probability: the method assumes *n*-gram locality of the forest, the property that any *n*-gram introduced by a hyperedge appears in all derivations that include the hyperedge and thus we apply the rule of hyperedge *e* to *n*-grams on *T(e)* and propagate *n-1* gram prefixes or suffixes to *h(e)*.

7 Assigning scores to hyperedge: if the hyperedge introduce the *n*-gram into the derivation, the *n*-gram probability is assigned to hyperedge for search *k*-best translation from forest.

8 Reranking hyperedges of hypernode: we reestimate the *n*-gram model feature value by *n*-gram posterior probability and count expectation.

9 Finding the best path in the forest: cube growing compute the *k*-th best item in every cell lazily. But this algorithm still calculates a full *k*-th best item for every hypernode in the forest. We can therefore take laziness to an extreme by delaying the whole *k*-best calculation after generating the forest. The algorithm needs two phascs, which arc forward that is same as viterbi decoding, but stores the forest (keep many hyperedges in each hypernode) and backward phase that recursively ask what's your *k*-th best derivations from top to down.

The word posterior feature $f_s(arc)$ is the same as the one proposed by Rosti et. al. (2007). Other features used in our log-linear model include language model $f_{lm}=LM(e)$, real word count $f_{wc}=N_{word}(e)$ and ε word count $fε=N_{null}(e)$, where *CN* is confusion network. Equation (15) is decoding framework.

$$\log p(e \mid f) = \sum_{i=1}^{N_{arc}} \log(\sum_{j=1}^{N_s} \lambda_s f_s(arc)) + \alpha LM(e) \\ + \beta N_{null}(e) + \gamma N_{word}(e) \quad (5)$$

Where *f* denotes the source language, *e* denotes the consensus translation generated by system combination.λ_s, *α*, *β* and *γ* is weight of other feature. Cube pruning algorithm with beam search is employed to search for consensus translation.

4 TRAINING PROCEDURE

Although *MERT* performs quite well on models with small number of features, in general the algorithm severely limits the number of features that can be used since it doesn't use gradient-based update during optimization, instead of updating one feature at a time. This is almost certainly a side effect of *MERT* approach that was used to construct the models so as to maximize the performance of the model on its single best derivation, without regarding to the shape of the rest of the distribution.

While there have been at least several attempts to do *MR* training for *SMT*, Smith and Eisner et al.(2006) introduced the deterministic annealing into *MR* to attempt the solution of increasingly difficult optimization problems.

The risk, or expected loss, of probability translation model of every sentence, defined with respect to *BLEU*, where $\tilde{e}$ is the reference and *e* is the hypothesis translation, is given by:

$$L = \sum_{e \in N-best} [p(e \mid f)BLEU(\tilde{e}, e)] \quad (6)$$

Where *L* is continuous and differentiable with respect to parameter *θ*, so the objective function can be optimized by *BFGS* gradient descend. To overcome the overfitting, Smith et al.(2006) smooth *MR* by *DA* to ensure the search space is as large as possible before it attain the optimal weight set.

$$L = \sum_{e \in N-best} [p_{\lambda,\theta}(e \mid f)BLEU(\tilde{e}, e) \\ + T \cdot H(p_{\lambda,\theta}(e \mid f))] \quad (7)$$

Where $H(p_{\lambda,\theta}(e \mid f))$ is the entropy of probability distribution $p_{\lambda,\theta}(e \mid f)$, and λis the scale hyperparameter, when λ=0 give the uniform distribution; when λ=1 give the original model probability distribution;

and as $\lambda \to \infty$, the probability approach the winner-take-all viterbi function, and θ is feature weight vector.

In order to gradient descend, we need the gradient of two objective function. They are given as following:

$$\frac{\partial L}{\partial \theta_k} = \sum_{e \in N-best} \Big[BLEU(\tilde{e}, e)(\lambda h_k - E_{p_{\lambda,\theta}(e|f)}[\lambda h_k]) \\ p_{\lambda,\theta}(e \mid f) \Big] \quad (8)$$

$$\frac{\partial L}{\partial \theta_k} = \sum_{e \in N-best} \Big[(BLEU(\tilde{e}, e) - T(1 + \log p_{\lambda,\theta}(e \mid f))) \\ \times (\lambda h_k - E_{p_{\lambda,\theta}(e|f)}[\lambda h_k]) p_{\lambda,\theta}(e \mid f) \Big] \quad (9)$$

We perform optimization in two steps: first optimizing θ_k through Equation (8) and (9), whose difference between them is whether it has *DA* or not; the second optimizing λ through (10) and (11). So the optimizer could exactly compensate for increasing λ by decreasing the vector θ proportionately.

$$\frac{\partial L}{\partial \lambda} = \sum_{e \in N-best} \Big[BLEU(\tilde{e}, e)(h\theta - E_{p_{\lambda,\theta}(e|f)}[h\theta]) \\ p_{\lambda,\theta}(e \mid f) \Big] \quad (10)$$

$$\frac{\partial L}{\partial \lambda} = \sum_{e \in N-best} \Big[(BLEU(\tilde{e}, e) - T(1 + \log p_{\lambda,\theta}(e \mid f))) \\ (h\theta - E_{p_{\lambda,\theta}(e|f)}[h\theta]) p_{\lambda,\theta}(e \mid f) \Big] \quad (11)$$

5 EXPERIMENTS

In our Chinese-to-English translation experiments, the best test *BLEU* score of candidate systems participating in the system combination 27.33. The worst test *BLEU* score of candidate systems participating in the system combination is 21.45. The baseline is *n*-best decoding to *CN*, which is constructed by incremental *IHMM*.

5.1 *Forest decoding with first-pass*

We ran decoder on a 2.8 Ghz Xeon with 24GB of memory. By varying the beam size for first-pass, we can plot graphs of BLEU score versus various pruning setting (beam width set to 0.0001, stack size varying according to 10,100,200,300,400 and 500) under search time in test set as shown in Figure 1. Decoding quality is measured by the *BLEU* score. Decoding efficient is measured by search time. The decoding algorithm of *n*-best and forest uses cube pruning and cube growing, respectively. The mixed pattern mixes the *n*-best of both outputs.

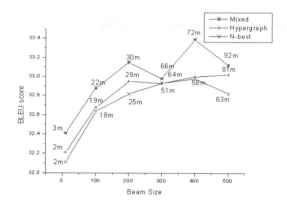

Figure 1. The quality and efficient of first-pass decoding on the test set.

5.2 *Effect of n-gram model in second-pass*

As it is shown in Table 1, decoding with a single *n*-gram model improves over the viterbi baseline with the mixture of forest and *n*-best decoding (beam size is 500) except the case of unigram model. Moreover, a bigram achieves the best performance among five different order of *n*-gram model. The results of the interpolation, which are showed in Table 2, between viterbi and *n*-gram model improve over the *n*-gram one. The best *n*-gram model shows an improvement of +1.03 and +1.35 *BLEU* point over decoding without forest.

Table 1. The quality and efficient of second-pass decoding on the test set.

n-gram model	Test(BLEU)
5-gram_1	33.54
5-gram_2	33.33
5-gram_3	33.27

Table 2. The quality and efficient of second-pass decoding on the test set.

viterbi+*n*-gram	Test(BLEU)
vi+5-gram_1	33.84
vi+5-gram_2	33.83
vi+5-gram_3	33.77

Compared with first-pass forest decoding, the performance of three types of *n*-gram probability can improve by at most 0.84 *BLEU* score in test set. If we compare baseline (incremental *IHMM*) with it, the performance of three types of *n*-gram probability can improve by at most 1.59 *BLEU* score in test set.

Different from Li et al.(2009) (denoted by ngram_1), Kumar et al.(2009)(denoted by ngram_2) and Denero et al.(2009) (denoted by ngram_3), we adopt the unified of two approach including the exact n-gram posterior probability.

5.3 Effect of minimum risk with deterministic annealing

The worst and best performance with first-pass forest decoding is better (up to +0.24% and 0.51% *BLEU* score in test set) than incremental *IHMM* with cube pruning.

We compare the three training schema: *MERT* vs. with/without *DA* with quenching scaling factorλ. With the entropy constrains, starting temperature T=1000; quenching temperature T=0.001. The temperature was cooled by half at each step; then we double λ at each step. Once the T is quite cool, it is common in practice to switch to rising λ directly and rapidly until some convergence condition. We locally optimize Equation (8), (9), (10) and (11) through *BFGS*.

Table 3. The MERT and MR with/without DA performance of the test set.

Training Criterion	Test(BLEU)
MERT	34.17
MR without DA&queching	34.19
MR&DA without quenching	34.28
MR&DA with quenching	34.43

The performance of *MR&DA* with quenching scale factor is better (up to +0.25%) than *MERT*.

6 CONCLUSIONS

This paper proposed forest decoding method based on word–level system combination of translation hypotheses from *SMT* systems of various paradigms, and then compared a set of n-gram feature on forest for two-pass decoding. The forest decoding includes three types of n-gram probabilities, which are n-gram calculating style. The two-pass forest decodings are shown to outperform cube pruning style decoding significantly. We get 1.59 *BLEU* score improvement in test set through two-pass decoding.

ACKNOWLEDGMENTS

The authors were supported by National Natural Science Foundation of China, Contracts 61300115, and science and technology research projects in Education Department of Heilongjiang Province, Project No. 12521073, and China Postdoctoral Science Foundation funded project No.2014M561331.

REFERENCES

[1] Antti-Veikko I. Rosti, Spyros Matsoukas, and Richard Schwartz 2007. ImprovedWord-level System Combination for Machine Translation. Proceedings of ACL 2007.

[2] M. Snover, B. Dorr, R. Schwartz, L. Micciulla, & J. Makhoul, 2006.A study of translation edit rate with targeted human annotation. Proc. Assoc. for Machine Trans. in the Americas.

[3] Khe Chai Sim, William J. Byrne, Mark J.F. Gales, Hichem Sahbi, and Phil C. Woodland. 2007. Consensus network decoding for statistical machine translation system combination. In Proc. of ICASSP, pages 105–108.

[4] Xiaodong He, Mei Yang, Jianfeng Gao, Patrick Nguyen, and Robert Moore. 2008. Indirect-HMMbased hypothesis alignment for combining outputs from machine translation systems. In Proc. Of EMNLP08.

[5] Chi-Ho Li,Xiaodong He,Yupeng Liu and Ning Xi.Incremental HMM Alignment for MT System Combination.In Proc. of ACL09.

[6] Damianos Karakos, Jason Eisner, Sanjeev Khudanpur, and Markus Dreyer 2008. Machine Translation System Combination using ITG-based Alignments.In Proceedings of ACL08.

[7] Xiaodong He, Kristina Toutanova. 2009. Joint Optimization for Machine Translation System Combination. In Proc. Of EMNLP09.

[8] Zhaopeng Tu, Yang Liu, Young-Sook Hwang, Qun Liu, and Shouxun Lin. 2010. Dependency Forest for Statistical Machine Translation. Proceedings of COLING 2010.

[9] Zhifei Li, Jason Eisner and Sanjeev Khudanpur. Variational Decoding for Statistical Machine Translation. In Proceedings of ACL 2009.

[10] Shankar Kumar,Wolfgang Macherey,Chris Dyer,Franz Och.Efficient Minimum Error Rate Training and Minimum Bayes-Risk Decoding for Translation Hypergraphs and Lattices. In Proceedings of ACL 2009.

[11] John DeNero, David Chiang, Kevin Knight .Fast Consensus Decoding over Translation Forest. In proceedings of ACL, 2009.

[12] David A. Smith and Jason Eisner. 2006. Minimum risk annealing for training log-linear models. In Proceedings of COLING-ACL, pages 787–794.

Network Security and Communication Engineering – Chan (Ed.)
© 2015 Taylor & Francis Group, London, ISBN: 978-1-138-02821-0

Adaptive anti-interference encoding strategy based on IMS

M. Zhou & C.Y. Zhou
Key Laboratory of Communication & Information Systems, Beijing Jiaotong University, Beijing, China

ABSTRACT: IMS (IP Multimedia Subsystem) has recently gotten wide application in the fusion of fixed network and wireless network. With the rapid development of network integration, Global networks are coming across more and more unprecedented challenges, and one of the most difficult challenges is the resistance to interference from various networks. So the anti-interference performance should also be brought into our attention. This article puts forward AACS—a kind of adaptive anti-interference strategy, which can adaptively change the number of check code words according to channel environment to make sure a high quality transmission at a low cost of bandwidth, ensuring good user experience. Finally, AACS is tested in the framework of IMS and a satisfactory result is obtained.

KEYWORDS: IMS, NGN, Error correction coding, AACS, Multimedia transmission.

1 INTRODUCTION

Large metropolitan cities are increasingly growing huge heterogeneous networks including co-existence of 3G/4G networks, wi-fi networks, and wimax networks and so on, which are able to provide multimedia service, in which streaming media is really significant. It can be divided into two types: live streaming and on-demand streaming. For live streaming, all users not only visit the same content, but also use the same broadcasting channels, so this kind of service is like traditional TV service. For on-demand streaming, any user can visit any content that he wants anytime, because each user is allotted an independent channel. So on-demand streaming service needs more network resources than live streaming service.

As lots of users access to networks via wireless networks, wireless networks have been becoming more and more important. However, wireless environment has more interference but less network resources than wired environment. So it is difficult to realize on-demand streaming service in wireless networks. We can compress content via source encoding to reduce the occupied network resources, and resist to interference via channel coding technology to solve the difficult problem. We will only discuss the channel coding technology in this article.

2 RELATED WORK

2.1 *Scalable layered coding*

Traditional multimedia stream adopts a single compression encoder, which compresses multimedia into a single video stream. We can get the original video content by decoding this single video stream. However, modern multimedia compression encoder adopts hierarchical coding strategy, which includes two types: Scalable Layered Coding SLC[1] and Multiple Description Coding MDC. In scalable layered coding strategy, the compression encoder compresses original data and outputs multilayer stream, which includes a base layer and some enhanced Layers, and different layers represent different time domain or frequency domain. Base layer carries the most important message of the video, and enhanced layers need to be decoded on the basis of base layer, besides that the more layers are decoded, the higher the resolution of the video is. When scalable layered coding is adopted, we can change the number of layers to adapt to the variation of network bandwidth[2].

2.2 *Periodic broadcast protocol*

As aforementioned, for on-demand streaming media, each user is allotted an independent channel, and many resources are occupied, which is particularly bad for wireless spectrum resources. To resolve this difficult problem, two methods are offered:1. On-demand services are limited to popular content, such as popular films or popular sports.2.Periodic broadcast protocol can be adopted to reduce the bandwidth of on-demand services, because only a few channels are needed to offer on-demand services for large number of users by using periodic broadcast protocol. Periodic broadcast protocol can be divided into two sections: a. First, divide video into fixed or variable length video segments. b. Then, the sender broadcasts periodically these video segments at different frequency. The receiver is thought to have enough bandwidth to receive broadcast video segments, and the protocol can set a proper broadcast cycle to make the receivers not feel the broadcast delay [3].

We can combine scalable layered coding with periodic broadcast protocol, because the former can satisfy different needs of different users, while the latter can save bandwidth(see Figure 1)[4].

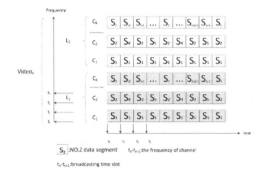

Figure 1. Scalable layered coding and periodic broadcast protocol.

2.3 IMS

IMS is short for IP Multimedia Subsystem, which is a network architecture of NGN(Next Generation Network), and its main advantage is supporting multimedia across the network architecture[5]. We can apply scalable layered coding and periodic broadcast protocol to transfer message across the network architecture on basis of IMS. Because IMS is not the emphasis of this article, so we only give the block diagram of IMS (see Figure 2)[6].

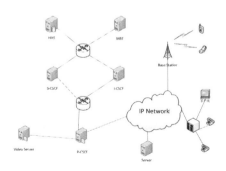

Figure 2. The block diagram of IMS.

3 ADAPTIVE ANTI-INTERFERENCE CODING STRATEGY (AACS)

3.1 Correlation theory

We first introduce the symbols used in the following equations. $V_n(n = 1, 2, ..., M)$ represents NO. n video in M videos. Because this system uses periodic broadcast protocol to transmit a video, we divide a video into D pieces of segments, each of which is the same in length, and we use $S_z(z = 1, 2, ..., D)$ to express NO. z segments, besides that scalable layered coding is used to transmit these segments in B layers, and $L_w(w = 1, 2, ..., B)$ represents NO. w layer. System distributes A channels in every layer, and uses to express NO. k channel. So the $C_k(k = 1, 2, ..., A)$ can be recorded as $S_{n,w,k,t}$.

3.2 AACS algorithm

When multimedia message is transferred in wireless environment, two problems should be cared for:

On one hand, it is the precious spectrum. Bandwidth occupied by check code should be smaller than bandwidth occupied by anti-interference code, and the equation is:

$$\sum_{n=1}^{M}\sum_{W=1}^{B_n}\sum_{K=1}^{A_n}P_{n,w,k,t} \times 2X_{n,w,k,t} \leq B_t^{RS} \times T \quad (1)$$

B_τ^{RS}—bandwidth used to resist to interference provided by the system

T—the length of the broadcast time

On the other hand, when the message is transmitted hierarchically, packet error rate of each layer should be lower than the maximum permissible, or it will lead to too many mistakes to provide enough information for decoding. The equation is:

$$e_{n,w,k,t}^{RS} \leq E_{n,w} \quad (2)$$

$(1 \leq n \leq M, 1 \leq W \leq B, 1 \leq K \leq A, 0 \leq t)$

$E_{n,w}$ - the highest packet error rate of video n that is transmitted in the layer w.

So we design AACS mainly from the above two aspects, and we divide AACS implementation process into two stages:

a. The distribution of the key check code word.

We want to use the least check code to achieve the highest packet error rate that the system can tolerate. Because we use the scalable layered coding, and the lower-layer media stream is more important than the higher, in which the higher-layer packet decoding is done on the basis of the low-level decoding, we distribute more check code to the lower layer. For example, the number of the base layer's check code is more than the first enhanced layer's, and the number of the first enhanced layer's check code is more than the second enhanced layer's. If there are some video packets belonging to the same layer, we will allot check code to these packets on basis of the number and the level of visitors. For example, the packet A that is visited by higher and more visitors will be distributed more check code than packet B that is visited by lower and fewer visitors. In this way, it is the strong anti-interference ability that the high level users have. For reflecting the number and levels of the visitors to distribute the check code reasonably, an impact factor named I is defined, which is used to describe the number and level of the visitors.

$$I_{n,w,k,t} = \alpha_n \times \frac{\beta_n}{f_{n,w,k,t}} \qquad (3)$$

Among (3) equation, a_n is access to the user level factor

$$a_n = \frac{\sum_{q=d}^{j} \psi_q}{\sum_{i=1}^{c} \psi_i} \ (j \geq q = d \geq 1, c \geq i = 1),$$

ψ_i represents the level of every visiting user, besides if the ψ_i is bigger, the level of user is higher. $\sum_{i=1}^{c} \psi_i$ represents the sum of the levels of all of the visitors, and $\sum_{q=d}^{j} \psi_q$ represents the sum of level of who visits the NO. n video. β_n represents the accessing rate of NO. n video, and because the access probability of video follows Zipf's law, we can give calculation formula:

$$\beta_n = \frac{n^{-(1-\alpha)}}{\sum_n n^{-(1-\alpha)}}, \text{ according to experience } \alpha=0.271.$$

$f_{n,w,k,t}$ is named as broadcast frequency, which represents the number of times that a piece of video is averagely broadcasted in a time slot whose length is t.

b. The distribution of the supplement check code word.

If the distribution of the key check code word is over, and the bandwidth that is given to check error is not used up. We can have the supplement check code word, which is distributed according to a factor named affecting weight that represents the impact factor of every video segments' proportion in the impact factor of all video segments.

$$w_{n,w,k,t} = \frac{I_{n,w,k,t}}{\sum_{n=1}^{M} \sum_{W=1}^{B_n} \sum_{K=1}^{A_n} I_{n,w,k,t}} \qquad (4)$$

B_r represents the remaining bandwidth that is used to check error, so $B_r \times \dfrac{w_{n,w,k,t}}{\sum_{n=1}^{M} \sum_{W=1}^{B_n} \sum_{K=1}^{A_n} w_{n,w,k,t}}$

is the remaining bandwidth that is used in $S_{n,w,k,t}$. $\dfrac{T}{h \times N_{n,w,k,t} \times P_{n,w,k,t}}$ represents the length of a bit in broadcasting time slot T. So the number of the supplement check code word is:

$$X_{n,w,k,t}^{sup} = B_r \times \frac{w_{n,w,k,t}}{\sum_{n=1}^{M} \sum_{W=1}^{B_n} \sum_{K=1}^{A_n} w_{n,w,k,t}} \times \frac{T}{h \times N_{n,w,k,t} \times P_{n,w,k,t}} \qquad (5)$$

4 SIMULATION AND PERFORMANCE COMPARISON

4.1 Simulation design

We use MATLAB to simulate AACS, and we have taken two other kinds of distribution strategy of check error code into consideration, including random distribution strategy and average distribution strategy to compare with AACS. And the simulating parameters are listed as follows:

Table 1. Simulation parameter setting.

Parameter name	Parameter value
The number of videos/n	6
The length of video/T_{video}	20min
Frame rate/R_{fram}	30/sec
The number of bits represented by a symbol/h	8 bit
The most packets in a video/P	50
The length of broadcasting/T	5/6s
Base layer's SNR/SNR^{base}	30dB
Enhanced layer's SNR/SNR^{enh}	5dB
Packet error rate in base layer/e^{base}	0.5%
Packet error rate in enhanced layer /e^{enh}	1.5%
Bandwidth of error correction/B^{error}	0%—15%
Bit mistake rate in wireless/e_b	10^{-3}—10^{-4}
The number of broadcasting channels/L_w	4
The show time of every packet/s	1/60
The number of byte in video packet/byte	191
The number of check error code in a packet/byte	[0,64]

4.2 Result and discussion

As mentioned above, the simulation have taken random distribution strategy and average distribution strategy into consideration. Although these two strategies are easy to realize, they cannot distribute code adaptively like AACS. So they can cause the waste of bandwidth, and cannot satisfy the need of transmitting multimedia video in dynamical channels. To compare the above three strategies better, we have defined average affecting weight:

$$G_{eva} = \sum_{n=1}^{M} \sum_{W=1}^{B_n} \sum_{K=1}^{A_n} w_{n,w,k,t} \times e_{n,w,k,t}^{RS}$$

$w_{n,w,k,t}$ is named as affecting weight. $e_{n,w,k,t}^{RS}$ is the packet error rate of every video packet. So in the same condition, the less the G_{eva} is, the better the strategy is, because the more important packets' packet error rate is lower.

According to the table 1, every video packet can contain 255byte, which contains 191byte data code words and 64byte check error code words. Downward transmission bandwidth is set to 2Mbps, and the bandwidth that check code words have occupied is 15% in downward transmission bandwidth(0.3Mbps).

According to $B_{UE} = X_{sum} \times \left(\dfrac{h}{T} \right) \times P$, X_{sum} represents the sum of the check codeword, and the number of h can be expressed in bits per symbol(1 byte=8 bit). In the average allocation strategy, 8.68055byte code words can be allotted to every packet at most, besides the average affecting weight $G_{eva} \in [2.525063 \times 10^{-4}, 4.461261 \times 10^{-5}]$, and the coding gain is 42.4444 (dB) at most. In random allocation strategy, G_{eva} decreases more slowly than the average allocation strategy's, and the convergence instability of G_{eva} is not suitable for dynamic allocation code word. At last, we adopt AACS to allot code words, and the data shows the allotted code words of every packet between 8 and 35. of AACS is 3.099506×10^{-5}, and the coding gain of AACS's average affecting weight coding gain is 44.0252 (dB). In terms of bandwidth usage, AACS uses less bandwidth to get a better correction effect, for example, in the simulation, AACS still has 173.280kbps bandwidth remaining compared to that random allocation strategy and average allocation strategy have used up the bandwidth.

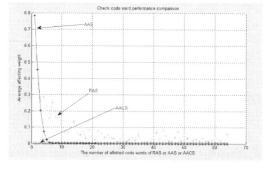

Figure 3. Performance comparison of distribution.

5 CONCLUSION

This article puts forward an adaptive allocation code words strategy AACS. Through simulating and comparing its performance with random allocation strategy(RAS) and average allocation strategy(AAS), we can draw the conclusion that AACS uses less bandwidth to get a better error-correcting effect. The core concept of AACS is adaptively allotting code words according to different degrees of importance of different video packets. The more important the video is, the more code words it can be allotted to. Thus, the transmitting system can ensure good anti-interference performance. Besides, the future research of AACS will mainly focus on the further improvement of operation efficiency and robustness of the algorithm to satisfy the need of multimedia service in the next generation Internet, and it will make it more targeted when combined with various industries.

ACKNOWLEDGMENTS

This paper is supported by the Technological Research and Development Programs of China Railway Corporation(2013X003-D).

REFERENCES

[1] Zhang D.Y. Jun W. & Lin S.Y.2011. A multi-path transmission algorithm for Layered scalable coding, 2011 International Conference on WirelessCommunications and Signal Processing (WCSP 2011):5 pp.

[2] Li C.H., Yuan C. &Zhong Y.Z.2009. LayeredEncryption for Scalable Video Coding, Proceedings of the 2009 2nd International Congress on Image and Signal Processing (CISP):4 pp.

[3] ZafalaoR.M. &da Fonseca N.L.S.2006. Periodicbroadcastingprotocolsfor clients with bandwidth limitation, IEEE Globecom 2006:6.

[4] Bagouet O., Hua K.A., Oger D.2003. Periodic broadcast protocol for heterogeneous receivers, Proceedings of the SPIE - The International Society for Optical Engineering:31–200.

[5] Zhang H.G, Hang N. & Eduardo M.G.2013. Scalable multimedia delivery with QoS managementin pervasive computing environment, Springer Science:317–355.

[6] Blum N., Jacak P. &Schreiner F.2008. Towards Standardized and Automated FaultManagement and Service Provisioning for NGNs, Springer Science:64–91.

Network Security and Communication Engineering – Chan (Ed.)
© 2015 Taylor & Francis Group, London, ISBN: 978-1-138-02821-0

The design of 474~858 MHz signal strip line directional coupler

Q.Y. Ma & Y.L. Guan
Department of Information Engineering, Communication University of China, Beijing

ABSTRACT: This paper introduced the principle of the strip line directional coupler. With the use of the even-odd mode analysis theory, the original parameters were gained through formulas. Then a 3 dB strip line directional coupler is analyzed and designed by the HFSS software. And based on simulation the initial curve of the directional coupler was gained. The model was optimized continually and an optimized coupler model corresponding to the design requirement was finally reached.

KEYWORDS: Analysis and design, Even-odd mode analysis theory, Optimization and simulation, Directional coupler.

1 INTRODUCTION

Directional couplers are the basic parts of many microwave systems [1,2]. Their importance comes from their many functions or applications. They are generally used as power splitters or multiplexers [3]. And its major application is in power combination or power division, especially in some microwave systems such as parts of antenna feeding networks, mixers or some balanced amplifiers [4,5] .

Nowadays, the major way of analyzing the strip line directional coupler is using the even-odd mode analysis theory[6,7,8] . Firstly the calculation of the odd and even mode impedances for the directional coupler is the important points, and it is the extremely significant in the design of the strip line directional coupler. According to the design principle of directional couplers, elements that influence coupling and coefficient are analyzed in this essay. Based on the principle of directional couplers and HFSS software [9, 10], I intend to design a strip line directional coupler with 474~858 MHz, 3±0.3dB coupling, directivity > 30dB, and the coupler has a center frequency of 666 MHz.

2 COUPLED LINE THEORY

When two unshielded and parallel transmission lines are close to each other, as the result of the electromagnetic interactions of each transmission line, there will be magnetic and electric energy coupled with each other. Such lines are regularly named as coupled transmission lines.

In general, there are three styles of coupled strip line transmission lines, parallel coupled strip line and offset coupled strip line posted by Shelton, and parallel broadside-coupled strip line posted by Cohn.

Usually couplers comprise two different or resembling transmission lines which are coupled with each other. Different styles of transmission lines such as micro strip, coaxial lines, strip line and waveguide are usually being used in the composition of the couples.

The traditional composition of the directional couples is shown in Figure 1.

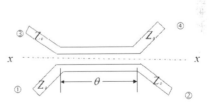

Figure 1.   Structure of a coaxial filter.

When the signal was input from port 1, then port 3 will be named as coupled port, port 4 is usually called isolated port and port 2 was output port.

As we all know, as one of the extremely important applications in microwave circuits, the function of the strip line directional coupler is to divide the input signal of the port 1 into two parts and export them to the port 2 and port 3 respectively, whose phase difference is 90°.

2.1 The odd-even mode theory

If we input a signal A from port 1 in Fig1, according to the even-odd mode analysis theory, it can be divided two steps. Firstly, we can input a signal A/2 from port 1 and a signal -A/2 from port 3. This time it is called the odd-mode excitation, and then the symmetry plane of the two coupling lines is equal to an electrical wall. Secondly, we can input a signal A/2 from port 1 and a signal A/2 from port 3. This time it is called the even-mode excitation, and then the symmetry plane of the two coupling lines is equal to a magnetic wall. In a word, from the even-odd mode analysis theory, the coupled lines are divided into two same parts by the electrical wall (odd-mode excitation) and magnetic wall (even-mode excitation). And we just only need to study half of them. It means the original four-port coupled strip line network is simplified to two two-port networks.

2.2 The design process of directional couplers

Before the odd and even mode theory, we must know Zoe is the impedance of the even mode signal, Zoo is the impedance of the odd mode signal, and θ is the electrical length of coupling lines at the working frequency. And coupling (C) and voltage coupling coefficient (Ko) is as follows:

$$C = 10 \lg |s_{21}|^2 = 10 \lg \left(\frac{k_0^2 \sin^2\theta}{1 - k_0^2 \cos^2\theta} \right) \quad (1)$$

$$k_0 = s_{21} = \frac{Z_{oe} - Z_{oo}}{Z_{oe} + Z_{oo}} \quad (2)$$

When the voltage of output port 2 is equal to that of coupled port 3, then we can know C = 3dB, ko=1.732 .Combined with the former formula, we can get

$$Z_{oe} = Z_0 \sqrt{\frac{1 + k_o}{1 - k_o}} = 120.9025\Omega \quad (3)$$

$$Z_{oo} = Z_0 \sqrt{\frac{1 - k_o}{1 + k_o}} = 20.6778\Omega \quad (4)$$

Then I intend to design the strip line directional coupler which dimension is 70mm*70mm. Thickness of the coupling lines is 2mm.

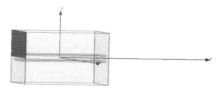

Figure 2. No port model.

Now we should design the model using the HFSS software which is shown in Figure 2. Firstly, according to the dimension of Zoe and Zoo, and then what we have to do is continue to optimize and simulate in the model, from it we can deduce that the width of the coupling lines is about 39mm,and the space of the two couplers is about 5.6mm.

And then we should apply the four ports to the former model which is shown in Figure 3 .

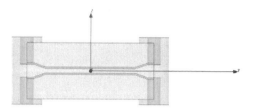

Figure 3. The original model.

As the former couplers is terminated in 50Ωports. And a coaxial line with a wave impedance of 50Ω should has an inner conductor with a diameter of about 16.9 mm and an outer conductor with a diameter of 38.8 mm. And the coupling lines of the directional couplers are about a quarter of the wave length at the center frequency of 666 MHz.

Based on the principle of directional couplers, we can know that the main factors that inflect the curve of S12 are the width of the coupling lines and the space of the two couplers. The main factors that inflect the curve of S11are the shape of the coupling lines and the connection between the ports and the coupling lines.

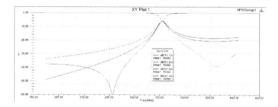

Figure 4. The original curve of S parameters of the designed coupler.

I'm surprised to see the simulation results which are shown in Figure 4. Then, what should I do is to analyze this circuit. After examining the circuit, I find the cause is that the width of the coupling lines is too wide. And it will be connected with the four ports, which will make short circuit. In order to improve the situation, I cut a corner in this position, which will make them detached which is shown in Figure 5.

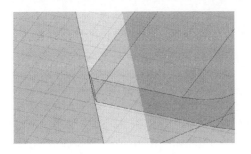

Figure 5. Cut a corner in this position.

According to the theory of directional couplers, in order to get better indicators, the coupling lines should be connected with the ports smoothly which is shown in Figure 6. By scanning each parameter of the directional couplers we can know the S parameter of the directional couplers is how to change when the parameter is changed. And then what we have to do is continue to optimize and simulate, and finally we get a good indicator which is shown in Figure 7.

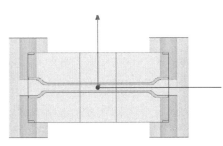

Figure 6. The final model.

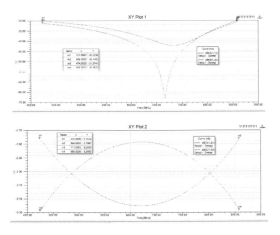

Figure 7. The final curve of S parameters of the designed coupler.

3 CONCLUSION

The article described the design process of a 3dB strip line directional coupler. In order to reflect the design strip line directional couplers with good return loss and directive characteristics, all the simulation and optimization are done using the HFSS software. And the simulation results declare that at working frequency (474~858 MHz), 3±0.3dB coupling, directivity >30dB, all of which reach the former requirements. And the odd and even mode theory is used to analyze and solve the strip line directional coupler.

REFERENCES

[1] Al-Taei, G.; Lane, L.; Passiopoulos, S .;"Design of high directivity directional couplers in multilayer ceramic technologies", Microwave Symposium Digest, IEEE MTT-S International, Vol.: 2 , 20–25 May 2001, Pages:61–64.

[2] D M Poza Andrews and Aitchison,"Microstrip lumped component quadrature couplers for use at the microwave frequencies," Proc. Elect Eng vol.147 pp.277–281, Aug. 2001.

[3] Juan de Dios Ruiz, Felix L."A stripline directional coupler adopting a nonhomogeneous dielectric method,"IEEE Trans. Microwave Theory Tech., vol. MTT-19, pp.716–722, Sept. 1969.

[4] Chul-Soo Kim; Chul-Sang Lim; Jun-Seok Park; DalAhn; Jae-Bong Lim; Seung-In Jack,"A design of the novel varactor tuned the directional coupler", Microwave Symposium Digest, IEEE MTT-S International, Volume: 5 , 14–19 June 1999, Pages:1715–1718 vol.4.

[5] Ling Kwon Kim, Kingsley N., Morton A., et al., "Fractal-shaped microstrip coupled-line filter for suppression of the second harmonic", IEEE Microwave Theory and Techniques, vol. 43, no. 8, pp. 2943–2946, Sep. 2005.

[6] Sundaram, M., Mobile Aspects India Pvt. Ltd, Maddela, A, et al., "Koch-fractal folded-slot antenna characteristics", IEEE Antenna and Wireless Propagation Letters, vol. 7, pp. 217–222, Apr. 2007.

[7] Unlu, H ; Topalli, K.; Sagkol, S.; Demir, S.; Civi, O.A.; Koc, S.S.; Akin, M."RF MEMS adjustable impedance matching network and adjustable power divider",International AP-S 2003. IEEE , Volume:3 Pages: 36–39 vol.2, June 2002.

[8] E. H. Cristal and L. Young, "Theory and Tables of Optimum Symmetrical TEM-Mode Coupled Transmission Line Directional Couplers," IEEE Trans. Microwave Theory & Tech., vol. 3, no. 5, pp. 554–559, September 1965.

[9] L. Young,"The analytical equivalence of tem-mode directional couplers and transmission-line stepped-impedance filters,"Proc. IEEE, vol.11, no. 2, pp.285–291,1963.

[10] Cristal, E.M.;"Coupled-Transmission-Line Directional Couplers with Coupled Lines of Unequal Characteristic Impedances", Microwave Theory and Techniques, IEEE Transactions on, Volume: 4 , Issue: 7, Jul 1966, Pages:347–350.

Network Security and Communication Engineering – Chan (Ed.)
© *2015 Taylor & Francis Group, London, ISBN: 978-1-138-02821-0*

Impact of partially coherent X-ray source on clinical phase contrast imaging

X.L. Zhang, J. Yang & S. Liu
School of Medical Instrument and Food Engineering, University of Shanghai for Science and Technology, Shanghai, China

X.L. Zhang & T. Xia
Shanghai Medical Instrumentation College, Shanghai, China

T. Xia
School of Optical-Electrical and Computer Engineering, University of Shanghai for Science and Technology Shanghai, China

ABSTRACT: In clinical applications, partially coherent light can actually be emitted by an X-ray source. Based on the phase imaging theory with the consideration of X-ray source coherence effects, the image intensity distribution formula of the Fourier transform is obtained. The impact of the actual X-ray source spot size on the phase effect and absorption effect is discussed with the elaboration of the relationships between the absorption and phase contrast and the distances from object plane to image plane. Furthermore, the optimal imaging position under actual clinical conditions can be obtained.

KEYWORDS: Phase imaging, Partial coherence, Phase contrast, Reduced complex degree of coherence.

1 INTRODUCTION

Traditional technology of X-ray imaging is based on different X-ray absorption of different biological tissues, also known as the absorption contrast imaging. This kind of technique can present clear images if the tested tissues are with significant difference in density, such as bones and muscles. But according to the theory, it can't give clearer result when densities are similar. Recently, the phase change of X-ray through biological tissues has been used to image, which is called phase imaging and can be used for biological tissues with similar destiny.

The potential use of X-ray phase imaging in clinical imaging is huge. Currently, this technique has become the focus that everybody cares about. We've made a lot of research about the theory of phase contrast imaging technology[1,2,3], but it has just been adopted in the clinical application. So we've conducted the relevant study on clinical application of phase imaging by taking partial coherence of X-ray into consideration.

Spatial coherence is the key aspect of image quality in phase imaging. The X-ray source used in clinical field is only partly coherent instead of being completely coherent or incoherent. Based on the theory of phase imaging and taking practical influence into account, this article attempts to analyze the reduced complex degree of coherence and radius of X-ray source with the elaboration of the relation between phase contrast and absorption contrast with the distance from the sample to detector. The optimal imaging position can be achieved with the consideration of the influence of being partly coherent to the actual X-ray source, thus we get a more general conclusion than the ideal X-ray source.

2 THE BASIC THEORY OF X-RAY PHASE IMAGE

During the interaction between X-ray and biological tissues, phase shift and attenuation of X ray can be expressed by T(x). To be convenient and general, only one-dimensional situation is considered:

$$T(x) = A(x)e^{i\varphi(x)} = \exp[i\varphi(x) - \mu(x)/2] \quad (1)$$

$$T(x) = 1 + i\varphi(x) - \mu(x)/2] \quad (2)$$

Based on axial Fresnel-Kirchhoff diffraction integral,

$$f(x;z) = (\frac{i}{\lambda z})^{1/2} \exp(-ikz) \int (\eta) \exp[\frac{-ik(x-\eta)^2}{2z}] d\eta \quad (3)$$

Regarding to the ideal point light source, light distribution on the detector can be Fourier transformed to[4]:

$$I(u;R_1,R_2) = \delta(Mu) - \psi(Mu)\cos(\pi\lambda R_2 Mu^2)$$
$$- 2\Phi(Mu)\sin(\pi\lambda R_2 Mu^2) \quad (4)$$

$\Psi(u)$ and $\Phi(u)$ are Fourier transformed from $\mu(x)$ and $\varphi(x)$. R_1 and R_2 represent the distances from the source to the object plane and from the object plane to the image plane. $M=(R_1+R_2)/R_1$ means the amplification factor.

3 PRACTICAL X-RAY PARTLY SOURCE COHERENCE EFFECTS

Considering part coherence of X-ray source used in clinical imaging, the formula (4) obtained from the ideal point source cannot be applied to clinical use any more, where the degree of coherence (γ) must be introduced. Based on Van Citter-Zernike theory[5], γ can be:

$$\gamma(s_1,s_2) = \frac{\int I(\zeta)\exp(i\bar{k}\,|s_1-s_2|\zeta/R_1)\mathrm{d}\zeta}{\int I(\zeta)\mathrm{d}\zeta}\cdot e^{i\varphi} \quad (5)$$

ζ refers to the coordinate on the light source plane.

As for the circular X-ray source with the radius of ρ and even light intensity distribution, the integral results of the formula above can be[6]:

$$\gamma = \frac{2J_1(v)}{v}e^{i\phi} \quad (6)$$

Where, $J_1(n)$ is the first order Bessel function, $v = \frac{2\pi}{\bar{\lambda}}\frac{\rho}{R_1}|\vec{s}_1-\vec{s}_2|$, $\phi = \frac{2\pi}{\bar{\lambda}}\frac{|\vec{s}_1-\vec{s}_2|^2}{2R_1}$, $|\vec{s}_1-\vec{s}_2|$ are the distance between two points on the object plane.

Complex degree of coherence shows the strength of X-ray source coherence. As the phase factor $e^{i\phi}$ has no effect on the module of complex degree of coherence, it makes no difference to the judge of the size of the contrast between the two points. The actual clinical imaging is in line with $\varphi=0$ and $\varphi<<\pi/2$, Therefore, we can introduce the reduced complex degree of coherence γ_{re}[7]. The reduced complex degree of coherence refers to the module value of the complex degree of coherence after the removal of the phase factor. As for the circular X-ray source with the radius of ρ and even light intensity distribution, reduced complex degree of coherence is equal to the optical system of optical transfer function OTF, the form of which is shown as the following:

$$\gamma_{re}(s_1,s_2) = \left|\frac{\int I(\zeta)\exp(i2\pi\zeta R_2 u/(R_1+R_2))\mathrm{d}\zeta}{\int I(\zeta)\mathrm{d}\zeta}\right|$$
$$= 2|J_1(v)/v| \quad (7)$$

The following formula is used[8]:

$$\vec{s}_1 - \vec{s}_2 = \frac{\bar{\lambda}R_2\vec{u}}{M} \quad (8)$$

Then (7) can be expressed as:

$$\gamma_{re}(u;R_1,R_2,\rho) = \left|\frac{2J_1(v)}{v}\right| = \left|\frac{2J_1(2\pi\rho R_2 u/R_1 M)}{2\pi\rho R_2 u/R_1 M}\right| \quad (9)$$

Then, the Fourier transform of light distribution on the image surface can be expressed as:

$$I(u;R_1,R_2) = \gamma_{re}(u;R_1,R_2,\rho)[\delta(Mu)$$
$$- \Psi(Mu)\cos(\pi\lambda R_2 Mu^2) - 2\Phi(Mu)\sin(\pi\lambda R_2 Mu^2)] \quad (10)$$

The formula above shows the intensity distribution formula of the frequency space obtained after considering the practical influence of partly coherence of X-ray.

Set $u=4\times10^4$Hz(20lp/mm)6, and make the X-ray wavelength range from 0.083nm to 0.124 nm. Based on formula (10), in the frequency space, the function of attenuation contrast is $\gamma_{re}\cos(\pi\lambda R_2 Mu^2)$, the phase contrast is $\gamma_{re}\sin(\pi\lambda R_2 Mu^2)$, and the items of attenuation contrast and phase contrast are:

$$\gamma_{re}\cos(\pi\lambda R_2 Mu^2) = \left|\frac{2J_1(4\pi\rho u R_2)}{4\pi\rho u R_2}\right|\cdot\cos(\pi\lambda R_2 Mu^2)$$

$$\gamma_{re}\sin(\pi\lambda R_2 Mu^2) = \left|\frac{2J_1(4\pi\rho u R_2)}{4\pi\rho u R_2}\right|\cdot\sin(\pi\lambda R_2 Mu^2) \quad (11)$$

4 DISCUSSION AND ANALYSIS

4.1 *Relationship among attenuation contrast, phase contrast and size of light spot*

Complex degree of coherence shows the strength of X-ray source coherence, the size of which should be consistent with the following formula:

$$0 \le \gamma_{re} \le 1 \quad (12)$$

When $\gamma_{re}=0$, X-ray source is incoherent source; $\gamma_{re}=1$, X-ray source is coherent source; $0<\gamma_{re}<1$, X-ray source is partly coherent source. Figure 1 is the variation curve of attenuation contrast and phase contrast under the situation that $\lambda=0.124$nm, setting the distance from source to object plane to be 1m, $0 \leq R_2 \leq 1$m, and the spot radius to be 0.5μm,5μm, 50μm and 500μm respectively, is conductive to find out the changing relation.

The conclusion can be made through the analysis of Figure 1 and 2: (1) Phase-effect will be reduced with the increase of the focus size, and the peak value of the phase curve is 0.98 when

ρ=0.5μm, which is 0.88 when ρ=5μm. In that case, the phase-effect weakens by 10% with the decupled focus size. When the focus size continues to increase to ρ=10μm, the peak value will be reduced to 0.70, and when ρ=15μm, it will be 0.40. This suggests that, phase-effect reduces slowly with the increasing of focus size. When focus size is less than 5mm, the declining speed of which will

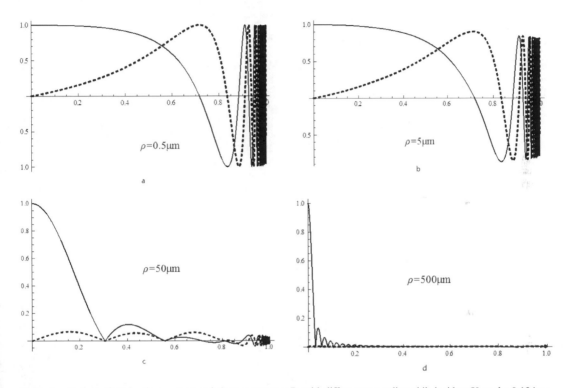

Figure 1.   Curves of attenuation contrast and phase contrast on R_2 with different spot radius while incident X-ray λ=0.124nm, in which ——attenuation contrast, ----- phase contrast.

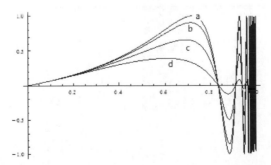

Figure 2.   Curves of phase contrast on R_2 with different spot radius while incident X-ray λ=0.124nm, a is ρ=0.5μm, b is ρ=5μm, c is ρ=10μm, d is ρ=15μm.

be very fast when the focus size is more than 5μm. (c) and (d) in Figure 1 suggest that when focus size increases to 50mm, phase-effect has become very small about 0.05, and when the focus size is 500mm, the phase-effect will disappear. The results above show that, the size of the focus size reflects the phase contrast by coherent coefficient, bigger focus results in lower coherence degree and none phase effect; larger focus results in higher coherence degree and strong phase effect. However, when it continues to decrease focus size, although it helps to improve the degree of coherence, its influence on the phase effects is insignificant. (2) When the focus size is less than 5μm, R_2 is about 0.7m at the first peak. And when $\rho=10\mu m$, R_2 is 0.69m, which suggests that focus size has less effect on the best position of imaging when it is less than 10mm, but when $\rho=15\mu m$, R_2 is about 0.64m at the first peak. Ideally, as for the point X-ray source, when M=2 ($R_1=R_2$), phase contract degree will reach the highest, and the attenuation contrast will reach the lowest. This conclusion is without considering the actual X-ray sources of the condition of the impact of partially coherent. As it is shown by Figure 2, the best imaging position and phase contrast equaling absorption contrast position differ

from the conclusion of the ideal point X-ray source with the effect of partially coherent light source[9].

4.2 Relations among phase contrast, absorption contrast and different data ranges of R_2

Phase contrast function and absorption contrast are both oscillation functions changing with R_2. In order to conduct further analysis of relationship between two effects of absorption and phase and R_2, function curves are shown with 4 kinds of different distances between source and image plane (R_2), which is shown in Figure 3.

From Figure 3 we can see that there is an interesting phenomenon that the best imaging position which accounts for the percentage of the total system can achieve maximum phase effect and minimum absorption effect simultaneously along with the change of the scope of R_2. As is shown by Figure 3 (a), the best imaging position is 0.42 m, accounting for 84% of the imaging system. And (b) shows that the best imaging position is 0.7 m, accounting for 70% of the imaging system. As it can be seen from figure 3 (c) and (d), that there are a variety of values for the best imaging position. The reason of this kind of condition is that system

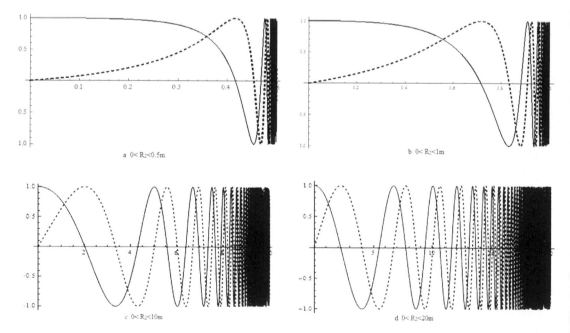

Figure 3. Curves of phase contrast and absorption contrast changing with R_2 while $\lambda=0.124$nm, $\rho=0.5$mm —attenuation contrast, ----- phase contrast.

length increases after increasing the distance from the light source to the image plane. In this way, the coherence of light source can be increased, and the change of coherence will get smaller. At this point, the best imaging position is mainly determined by sine and cosine factors of phase contrast function and absorption contrast function. And the two functions will change with R_2 oscillation. So there are multiple values. Although the increase of system length can make the best position imaging with a variety of choices, it should be noted that the increase of the system length will reduce the luminous flux through the object plane, resulting in the imaging fuzzy. In addition, when the system is short, the magnification will be larger because the best imaging position has a larger percentage in overall length according to the image magnification factor $M = (R_1 + R_2)/R_1$ in certain overall length, which is beneficial for imaging.

In conclusion, the influence of coherence degree on phase contrast is obvious. But there is no need to narrow the focus limitlessly. This effect can be ignored when focus reaches a certain value. The best imaging position depends not only on the size of focus but also on the system length. By calculating, we found that it's not necessary to improve the degree of coherence by increasing the system length, for smaller system length can also lead to satisfactory results.

REFERENCES

[1] Lewis R A. Medical phase contrast x-ray imaging: current status and future prospects[J], Phys. Med. Biol. 2004,49 :3573–3583.

[2] Xia T, Zhang X L, Zhang G Y. Analysis of the Absorption and the Phase-effect on Micro-focus X-PCI. Acta Photonica Sinica, 2009,38(10); 2516–2160.

[3] Xia T, Zhang X L, Ma J S. Analysis of the effect of spatial coherence and in incident X-ray photon energies on clinical X-ray in-line phase-contrast imaging[J]. Acta Photonica Sinica , 2011, 40(4): 627–635.

[4] Pogany A, Gao D, and Wilkins S. W.. Contrast and resolution in imaging with a microfocus x-ray source[J]. Rev. Sci. Instrum. 1997, 7: 2774~2782.

[5] Born M and Wolf E. Principles of Optics[M]. 6th ed., Pergamon, Oxford, 1980.

[6] Wu X Z, Liu H. A new theory of phase-contrast x-ray imaging based on Wigner distributions[J]. Med. Phys. 2004,9: 2378~2384.

[7] Wilkins S. W., Gureyev T. E , Gao D, et al. Phase-contrast imaging using polychromatic hard X-rays[J], Nature,1996,28:335~338.

[8] Wu X Z, Liu H, Yan A. Optimization of X-ray phase-contrast imaging based on in-line holography[J]. Nuclear Instrument and Methods in Physics Research B. 2005,234: 563~572.

[9] ZHANG X L , Liu S , Huang Y et al. Optimal Contrast for X-ray Phase-contrast imaging. Acta Photonica Sinica , 2008, 37(6): 1217–1220.

Network Security and Communication Engineering – Chan (Ed.)
© 2015 Taylor & Francis Group, London, ISBN: 978-1-138-02821-0

Azimuth multi-angle spaceborne SAR imaging mode for moving target indication

Y.M. Wang, J. Chen & W. Yang
School of Electronics and Information Engineering, BeiHang University, Beijing, China

ABSTRACT: Traditional multi-channel Synthetic Aperture Radar (SAR) has shortcomings like blind velocity and velocity ambiguity. To solve the problem, this paper introduces a novel approach for velocity measurement based on the azimuth multi-angle SAR imaging mode. First, the geometry model was built, and the functions to measure moving target's velocity were deduced as well. By analyzing the effects of the moving target on SAR image, the approach for measuring azimuth velocity with one satellite and range velocity with two satellites were proposed. Using this approach, we can not only estimate velocity along both azimuth and range direction with high accuracy, but also solve the problem of blind velocity and velocity ambiguity. The effective of this approach is validated by the computer simulation results.

KEYWORDS: SAR, Azimuth Multi-angle, Velocity Estimation, High Accuracy.

1 INTRODUCTION

Synthetic aperture radar (SAR) is attractive for global monitoring because of its flexible operability. From the 1970s, moving target monitoring develops gradually into the hot issue in spaceborne SAR. At present, Radarsat-2 and TerraSAR-X/TanDEM-X have carried out some experiments about moving target's velocity estimation. Besides, many systems have been proposed to estimate moving target [1][2]. Classic GMTI based on multi-channel mode exists blind velocity and velocity ambiguity problem and cannot estimate azimuth velocity [3]. 2012 German Bi-Directional SAR imaging mode (BiDi SAR) was proposed which improves the performance of velocity estimation [4]. However, it can just acquire less than three images with one satellite. Moreover, a Velocity Correlation Function (VCF) was proposed recently with considerable calculation load which was verified by real SAR data[5]. By azimuth multi-angle SAR, it can not only overcome these shortcomings, but also can be implemented easily.

The geometry and mathematical modes based on azimuth multi-angle SAR were presented in section 2. Besides, the reason for measuring azimuth velocity with one satellite and range velocity with two satellites is presented as well. In section 3, computer simulation results illustrate the effectiveness of this approach.

2 MOVING TARGET INDICATION AND VELOCITY ESTIMATION

Along the radar moving, we can receive many images of the target based on azimuth multi-angle SAR. This paper supposes the satellite illuminates two times of the moving target and defines the first look to the moving target as forward direction and the second look as backward direction. Figure 1 shows the azimuth multi-angle SAR acquisition geometry into the directions ϕ_{fore} and ϕ_{aft}, respectively. The same target area is acquired twice as the radar flies by with a short along track separation in-between the two acquisitions [6]. The images are separated in the azimuth spectrum. In this approach, the time lag between the two looks is an important parameter which can be used to measure target velocity.

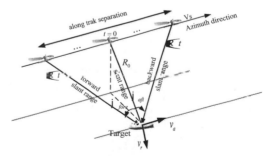

Figure 1. Azimuth multi-angle SAR acquisition geometry.

To focus our attention on the main subject, we first introduce the effect of the moving target. The range between the sensor and the target is as equation (1).

$$R(t) = \sqrt{\left(R_0 + v_r \cdot t\right)^2 + \left(\sqrt{V_g V_s} - v_a\right)^2 \cdot t^2} \qquad (1)$$

Where, t is the azimuth time and R_0 is center range at $t = 0$ i.e. the closest approach of the sensor to target. V_s and V_g are the satellite and the ground velocity, respectively, v_a and v_r are the target velocity in azimuth and range direction, respectively.

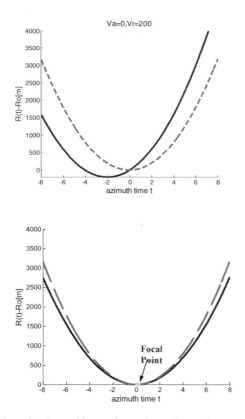

Figure 2. Range history change due to the moving target's velocity. The continuous black lines are with exaggerated target's velocity in range (left) and azimuth (right) directions of 200m/s respectively, and the dashed red lines are without velocity.

Figure 2 shows the effects of target's velocity on range history in azimuth time $t.v_a$ and v_r are exaggerated in the figure to enhance the visibility of the induced effects on the range history. From figure 2, we can know that azimuth velocity changes the curvature of range history. Moreover, the moving target's azimuth velocity does not shift the focal point, i.e. Doppler centroid is invariable. The slant range velocity results in azimuth time delay, which is known as Doppler shift.

In the next part, we will illustrate the above effects by means of theoretical derivation in detail.

2.1 Range velocity estimates by azimuth multi-angle observation

When $v_a = 0$, conduct first and second derivations of equation (1). We can find that

$$R'(t) = v_r, R''(t) = \frac{V_g V_s}{R_0} \qquad (2)$$

$$R(t) \approx R_0 + \frac{\lambda}{2}\left(f_d t + \frac{1}{2} f_r t^2\right) = R_0 + f'(0)t + \frac{1}{2} f''(0)t^2 \qquad (3)$$

If $v_a = 0$, i.e. a range velocity can cause changes as follows:

$$\begin{cases} \Delta f_d \approx \dfrac{2v_r}{\lambda} \\[2mm] \Delta t_{v_r} \approx -\dfrac{\Delta f_d}{f_r} = \dfrac{-2v_r}{\lambda} \bigg/ \dfrac{2V_g V_s}{\lambda R_0} = -\dfrac{R_0 \cdot v_r}{V^2} \\[2mm] \Delta r_{v_r} \approx \dfrac{\lambda}{2}\left(f_d \Delta t + \dfrac{1}{2} f_r \Delta t^2\right) = -\dfrac{\lambda}{4} \cdot \dfrac{\Delta f_d^2}{f_r} = -\dfrac{R_0 v_r^2}{2V^2} \end{cases} \qquad (4)$$

Where Δf_d is Doppler centroid shift, Δt_{v_r} is the shift of the closest approach in azimuth time, and Δr_{v_r} is the shift in slant range of the closest approach.

$$\Delta t = -\frac{\Delta f_d}{f_r}, V = \sqrt{V_g V_s}.$$

At the same time, we can find that the Δr_{v_r} is very small when the target with a small and medium range velocity. And it is identical in the fore and aft images. In comparison to a static target, the fore and aft range positions shift of the closest approach are $\Delta R_1, \Delta R_2$.

$$\Delta R_2 - \Delta R_1 = \frac{\lambda}{2}\left(f_{d2} - f_{d1}\right) + \left(R_2 - R_1\right) = -\frac{v_r R_0}{V}$$

$$\left(\varphi_{fore} + \varphi_{aft}\right) + \left(R_2 - R_1\right) \approx 0 \qquad (5)$$

Where $\left(R_2 - R_1\right) \approx v_r \cdot t_{lag}, t_{lag} = \dfrac{R_0 \cdot \left(\varphi_{fore} + \varphi_{aft}\right)}{V}$.

In conclusion, if the target just with range velocity, then the SAR images in the forward and backward are identical. Then consider two satellites flying with a short along track separation to measure the range velocity. The squint angles of the two satellites can be any angle.

$$\Delta n \cdot \frac{c}{2f_s} = v_r \cdot t_{lag} \Rightarrow v_r = \frac{\Delta n \cdot c}{2f_s \cdot t_{lag}} \qquad (6)$$

372

Where c is the light velocity, f_s is sampling rate and Δn is pixel shift in range direction. t_{lag} is the time lag between the two satellites.

2.2 *Azimuth velocity estimates by azimuth multi-angle observation*

If $v_r = 0$, i.e. azimuth velocity can cause changes as follows:

There is no closest approach azimuth time and range shift i.e. $\Delta t_{va} = \Delta r_{va} = 0$.

Doppler frequency modulated rate from

$$f_r \approx -\frac{2V^2}{\lambda \cdot R_0} \text{to} f_r' \approx -\frac{2(V-v_a)^2}{\lambda \cdot R_0}.$$

The mismatch Doppler frequency modulated rate due to v_a introduces not only defocusing, but also in the presence of squint a target azimuth shift. However, it is different in the fore and aft images. Using the azimuth shift Δt_{va} of the focused target between fore and aft images, v_a can be calculated.

$$\Delta t_{va,\,fore} = -\frac{R_0 \sin \varphi_{fore}}{V}\left(1-\left(\frac{V-v_a}{V}\right)^2\right)$$

$$\approx -\frac{2R_0 \sin \varphi_{fore} v_a}{V^2} = -\frac{t_{lag} v_a}{V} \qquad (7)$$

$$\Delta t_{va,\text{aft}} = \frac{R_0 \sin \varphi_{aft}}{V}\left(1-\left(\frac{V-v_a}{V}\right)^2\right)$$

$$\approx \frac{2R_0 \sin \varphi_{aft} v_a}{V^2} = \frac{t_{lag} v_a}{V} \qquad (8)$$

Then we can derive two equations to estimate v_a.

$$v_a = V - V \cdot \sqrt{1-\frac{2V\Delta t_{va}}{\left(f_{d,\text{aft}} - f_{d,\,fore}\right)\lambda R_0}} \qquad \text{or}$$

$$v_a = \frac{\Delta t_{va} \cdot V}{2 t_{lag}}. \qquad (9)$$

Where $\Delta t_{va} = \Delta t_{va,\text{aft}} - \Delta t_{va,\,fore}$, $f_{d,\,fore}$ and $f_{d,\text{aft}}$ are the forward and backward squint Doppler centroid, respectively. t_{lag} is the time lag between the two looks.

The different target position in azimuth multi-angle images provides the sign of v_a. Therefore, it is possible to measure azimuth velocity with one satellite.

3 SIMULATION:

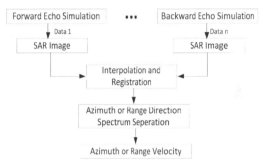

Figure 3. Moving target indication with azimuth multi-angle SAR.

Figure 3 shows the specific approach to carry out azimuth multi-angle simulation.

For the ship encircled in the left of Figure 4, v_r can be measured by two satellites. The squint angles of

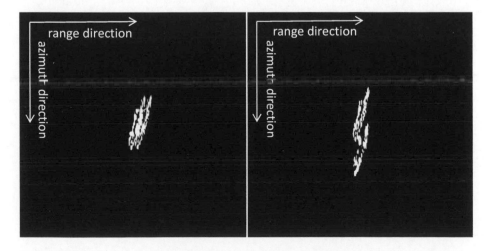

Figure 4. Left is for measuring range velocity. Right is for measuring azimuth velocity.

the two satellites are all 2.5°. The time lag between the two satellites is 6s. The result of range pixel displacement is 7 samples. From equation (6), the calculated result of v_r is 4.90m/s.

For the ship encircled in the right of Figure 4, v_a can be measured by one satellite. The squint angles of the two looks are ±2.5°. The result of azimuth displacement is 41 samples, from equation (9), the calculated result of v_a is 5.12m/s. One sample estimation error is 0.1m/s.

4 CONCLUSIONS

A new approach based on azimuth multi-angle for measuring target's azimuth velocity with one satellite and range velocity with two satellites has been discussed. This approach solves the problem of blind velocity and velocity ambiguity which is easy to implement. Moreover, it can achieve more than three times' observations to the target, that means more information about the target area. Furthermore, it can improve the moving target imaging quality with true motion parameter obtained by azimuth multi-angle SAR.

REFERENCES

[1] S.Barbarossa, Detection and Imaging of Moving Objects With Synthetic Aperture Radar, Part1: Optional Detection and Parameter Estimation Theory, IEE Proc. –f.1992, No1. PP79–88.
[2] J.Mittermayer, S. Wollstadt, Simultaneous Bi-directional SAR Acquisitions with TerraSAR-X, Proc. of EUSAR 2010, Aachen, Germany.
[3] DavideRizzato,Techniques for ground moving target detection and velocity estimation with multi-channel Synthetic Aperture Radar, April 2012.
[4] J. Mittermayer, P. Prats, S. Wollstadt, S. Baumgartner, P. Lopez-Dekker, Approach to Velocity and Acceleration Measuring in the Bi-directional SAR Imaging Mode, IEEE 2012.
[5] MotofumiArill,Efficient Motion Compensation of a Moving Object on SAR Imagery Based on Velocity Correlation Function, IEEE Trans. Geosci Remote Sens, vol.52.no.2.Feb.2014.
[6] J. Mittermayer, S. Wollstadt, P. Prats, P. Lopez-Dekker, G.Krieger, A. Moreira, Bi-directional SAR and Interferometric Short-term Time Series, Proc. of EUSAR 2012, 23–26 April,2012,Nuremberg,Germany.

Network Security and Communication Engineering – Chan (Ed.)
© 2015 Taylor & Francis Group, London, ISBN: 978-1-138-02821-0

Travel mode shares comparative analysis and policy suggestion

Y.B. Zhang, H.P. Lu & Y. Li
Institute of Transportation Engineering, Tsinghua University, Beijing, China

ABSTRACT: Travel mode share is a key part in building ecological city and sustainable transport system. This article studies the development trends and demand of urban traffic mode structure of China and advanced foreign experience, and then carries out the urban transportation development policy suggestions for China based on the analysis of current situations and characteristics of travel mode shares in China.

KEYWORDS: Travel mode shares, Sustainable transport mode share, Travel culture.

1 INTRODUCTION

Travel mode share means the proportion of different traffic modes used in resident trip generation which is under a comprehensive urban transport system. Travel mode share reflects the main characteristic of urban traffic demand and transportation system, the main function and role of various transport modes. As travel mode share affects the limited traffic resource allocation and the possibility for providing better choice for travelers, it has became a critical factor to decide the basic properties and overall efficiency of urban transportation system, and a key to solve urban traffic congestion, ease traffic pollution and reduce traffic energy consumption.

The main factors which influence travel mode share are traffic demand characteristics, natural geographical features, traffic culture concepts etc.

This article collects the resident travel data from 44 cities in 7 areas of China, and analyzed more than a dozen outstanding foreign travel mode shares, for trying to reveal the rules and proposing some policy suggestions for China.

2 THE MAIN CHARACTERISTICS OF TRAVEL MODE SHARE

2.1 *Sustainable transport mode share is relative stable*

Through the travel mode share analysis of 44 cities[1,2,3,4], it is clear that sustainable transport mode share (walking + cycling + public transport) is normally between 70% and 90%. There are 6 cities whose sustainable transport mode shares are less than 70%, accounting for 13% of the sample, which is as shown in figure 1.

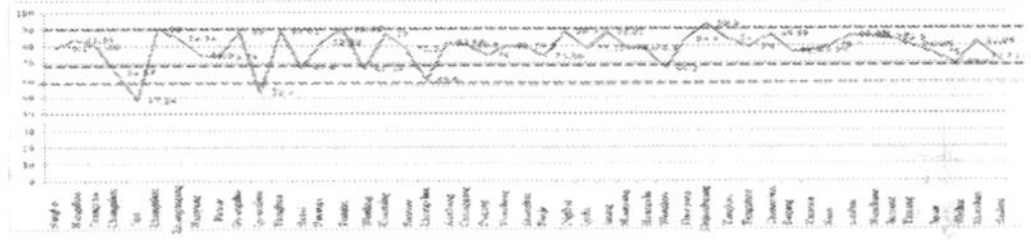

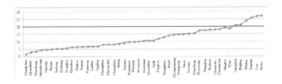

Figure 1. The different sustainable transport mode shares of city samples.

2.2 *The car mode share in most cities is still in the early stage of rapid growth*

There are 6 cities whose car mode shares are more than 20%, accounting for 13%; 23 cities are below 10%, taking 53% of the sample, which is as shown in figure 2.

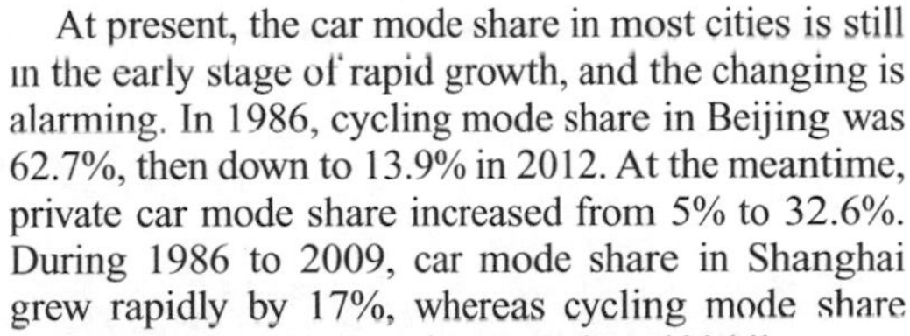

Figure 2. The different car mode shares of city samples.

At present, the car mode share in most cities is still in the early stage of rapid growth, and the changing is alarming. In 1986, cycling mode share in Beijing was 62.7%, then down to 13.9% in 2012. At the meantime, private car mode share increased from 5% to 32.6%. During 1986 to 2009, car mode share in Shanghai grew rapidly by 17%, whereas cycling mode share dropped by 17.5%. Despite more than 400 kilometers metro system had been built during 2004 to 2009, public transport mode share in Shanghai just grew 0.6%.

2.3 *Electric bicycle is booming in some cities*

There are 7 cities whose electric bicycle mode shares are greater than 20%, especially Quanzhou, which is up to 36.8%. Regarding ownership, there are already 4 million electric bicycles in Shanghai, the development of which cannot be ignored.

3 THE CHARACTERISTIC ANALYSIS OF TRAVEL MODE SHARE ON DIFFERENT CITY SCALES IN CHINA

In order to analyze the relationship between city scale and travel mode share, this article has regrouped the data depending on its city scale and analyzed its travel model share characteristics.

3.1 The characteristics of travel mode share on different population

The characteristic analysis of travel mode share on different city categories are shown as below.

3.1.1 The travel mode share in giant city
The travel mode share characteristics in giant cities are two "high":

The public transport stays at a relatively high level, and the mode shares are between 1/4 and 1/3;

The car mode shares are high, about 20%.

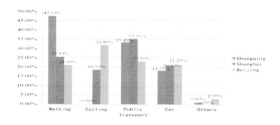

Figure 3. The comparison of travel mode shares between giant cities.

3.1.2 The travel mode share in super city
Regarding to public transport mode shares, that in super cities stay at the same level with that in giant cities, both around 30%; however, the car mode shares in super cities are significantly lower than that in giant cities; in contrast, the non-motorized traffic mode shares are slightly greater.

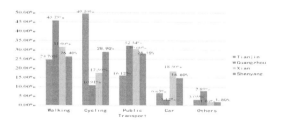

Figure 4. The comparison of travel mode shares between super cities.

3.1.3 The travel mode share in mega city
The non-motorized traffic mode shares are about 50%~70%, greater than the above categories;

The public transport shares are smaller than that in giant and super cities, only account for 15%~20%;

The car mode shares are also smaller, around 10%.

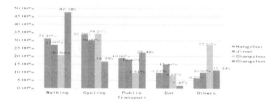

Figure 5. The comparison of travel mode shares between mega cities.

3.1.4 The travel mode share in big city
Most of the non-motorized traffic mode shares exceed 60%;

The public transport mode shares are between 15% and 20%;

Most of car mode shares are below 10%;

The electric bicycle mode shares are relatively high, more than 10% in half cities.

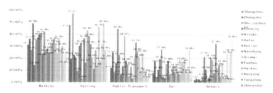

Figure 6. The comparison of travel mode shares between big cities.

3.1.5 The travel mode share in middle city
The non-motorized traffic mode shares are between 60% and 80%;

The public mode shares are below 10% in half cities, and around 20% in a few cities;

Most of car mode shares are below 10%.

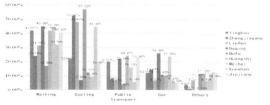

Figure 7. The comparison of travel mode shares between middle cities.

3.1.6 The travel mode share in small city

The public transport shares are generally low, less than 5%;

The car mode shares are around 5%;

The non-motorized traffic mode shares are relatively high, generally above 70%.

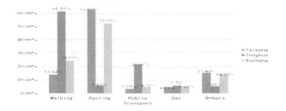

Figure 8. The comparison of travel mode shares between small cities.

3.1.7 There are three main characteristics of travel mode shares on different population

The public transport and car mode shares increase along with the increment of population;

The non-motorized traffic mode shares decrease along with the increment of population;

The sustainable transport mode shares (walking + cycling + public transport) are generally between 70% and 90%, regardless of the city scale.

4 REVIEWING OF THE TRAVEL MODAL SHARE IN SIMILAR CITIES AROUND THE WORLD

This article collects the data from 19 outstanding cities in the world, including Seoul, Tokyo, Paris, London, New York, Copenhagen, Stockholm, Munich etc. [5,6], and analyzed the travel mode share for each city. Then, some features are revealed as below:

4.1 The sustainable transport mode shares (walking + cycling + public transport) are between 60% and 90%

In this respect, most cities of China and foreign cities are at the same level, and a few cities have reached the advanced world level.

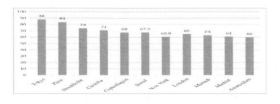

Figure 9. The sustainable transport mode shares in foreign cities.

4.2 The leading travel modes are different

The travel mode shares are different in foreign cities, because the difference of city scale, demand features, geographical conditions and history culture. For instance, the public transport share in Soule is 63%; the walking share and public transport share in Paris are 54% and 29% individually. In the perspective of sustainable transport, the travel mode shares in these cities are reasonable and dominated by sustainable transport.

Figure 10. The non-motorized traffic mode shares and public transport mode shares in foreign cities.

4.3 The different travel mode should be advocated by different travel distance, based on the travel demand and fuel economy

The different travel mode should be advocated by different travel distance. The public transport should be used for medium or long distance travel; the non-motorized traffic should be used for short distance travel, giving full play to their advantages.

5 MAIN CONCLUSIONS

At present, China is in the upsurge of old city transformation and new city construction, it is an opportunity to build eco-city and sustainable transport system. Moreover, a comprehensive transport system, which is dominated by sustainable transport, is the key to accomplish new-type urbanization.

Compared to foreign cities, the travel model share rationalities of cities in China are better in the perspective of sustainable transport share. However, there is a deteriorated danger of travel mode share in China, because the number of car grows rapidly, waling conditions are worsening, travel mode is shrinking, public transport mode share grows slowly and land is used unreasonably. Therefore, the main methods for building eco-city sustainable transport system are improving non-motorized traffic space, developing public transport and restraining car mode share from rapid growth.

Sustainable transport is the fundamental goal of urban transport development. Regardless of city type and scale, the fundamental requirements for building eco-city sustainable transport system are implementing sustainable transport priorities and assuring that sustainable transport mode share is greater than 85%.

Because the differences of city scale, travel demand, geographical conditions and historical culture, travel mode shares are also different. For evaluating the travel mode share rationality in a city, the fundamental basis is characteristics of travel demand, instead of subjective supposition or even an unreasonable mindless index for one travel mode. For instant, Copenhagen, Amsterdam, Seoul, Tokyo and Paris are all excellent cities in the world, dominated by sustainable transport. However, the dominating travel mode in Paris is walking, accounting for 55%; the dominating travel mode in Copenhagen and Amsterdam is cycling, taking 35% and 34% individually; the dominating travel mode in Seoul is public bus, with its proportion being 35%; but the dominating travel mode in Tokyo is rail transit, accounting for 58%. These cities are all dominated by sustainable transport, but dominating travel modes are significantly different. This is due to the different travel demand characteristics, travel culture and travel habit.

Prioritizing public transportation development should be unshakeable in big and mega cities. These cities should decide public transport mode share and provide distinctively diversified services. In addition, public transport should be putted at the head of building schedule.

There are two main factors for raising public transport mode share, one is improving public transport service level, and the other is controlling car ownership and usage. At present, China is still in the early stages of vehicle rapid growth. Restraining the speed of private car mode share growth in this time is the key to develop new-type urbanization and avoid worsening troubles of urban transport in the future.

The source management measures of travel demand, such as building compact city, using mixed land, promoting workplace and residence balance, improving ancillary facilities etc., could be helpful to reduce the total travel demand, shorten travel distance and improve travel mode share. These measures are first steps to solve urban traffic problems.

Driving the TOD mode, integrated and developing transport hub and surrounding land are necessary for building eco-city and sustainable transport system. Taking the chance of large scale transport hub construction, different travel modes could be connected seamlessly. In addition, transport hub and surrounding land should be developed integrated.

Electric bicycle is a new traffic tool, and it is not located at two extremes. Electric bicycle should be developed by considering travel demand, geographical features, resident travel behavior, and road infrastructure conditions.

REFERENCES

[1] Over the years summary for investigated reports of urban comprehensive transport, Institute of Transportation Engineering, 2013.
[2] H.P. Lu, Y.B. Zhang, The Comprehensive Transportation Planning of Qigihar, Institute of Transportation Engineering, 2012.
[3] Beijing Transport Annual Report, Beijng Transportation Research Center, 2013.
[4] H.P. Lu, Y.B. Zhang, The Comprehensive Transportation Planning of Liaoyang, Institute of Transportation Engineering, 2012.
[5] H.P. Lu, Eco-City and Green Transport: World Experience, China Architecture and Building Press, 2014.
[6] H.P. Lu, J. Wang, The Guiding Strategy of Relationship between Traffic Supply and Demand, based on Sustainable transport Theory, Comprehensive Transportation, 13(6), pp. 4–10.

Network Security and Communication Engineering – Chan (Ed.)
© *2015 Taylor & Francis Group, London, ISBN: 978-1-138-02821-0*

Ontology modeling for knowledge management in printing company

L. Wei, Y.B. Zhang, W.M. Zhang, X.H. Wang & M.J. Xu
School of Mechatronic Engineering, Beijing Institute of Graphic Communication, Beijing, China

ABSTRACT: Constructing ontology model for printing company is aimed at improving knowledge management efficiency among printing companies and realizing effective sharing of knowledge in printing related fields. Referring to experiences of ontology modeling in other domains and the investigation of the work flow in the printing factory of Beijing Institute of Graphic Communication (BIGC), the scope of the printing company ontology model was analyzed. New process and method for ontology modeling were proposed according to particular features of printing companies. Job Definition Format (JDF) helped to describe processes of the work flow and extract professional concepts and terminology. To illustrate how to define class hierarchy, examples of property, relations, axioms, examples on printing paper ontology and the ontology of printing process were presented. The establishment of printing company ontology model proved to improve knowledge management in printing company and provided a solution for further development in printing domain ontology modeling.

1 INTRODUCTION

1.1 *Research background*

Printing companies are in need of setting standard criteria on printing knowledge management in order to support researches in various fields related to printing. At present, the research on printing knowledge management mainly focuses on building the framework of knowledge management system, which cannot provide a practical solution for printing companies to manage and share knowledge together [1]. That is to say, no comprehensive and systematic methods on printing knowledge management were proposed up to now. Since printing has been a highly knowledge-intensive field, most researches such as printing materials, printing and packaging machinery, printed media design, are difficult to reach an agreement on classification and expression of concepts.

Successful applications of ontology in knowledge engineering have inspired researchers to introduce it into printing industry. After analyzing the demand in the printing factory of BIGC, this paper concluded concrete steps of printing company ontology modeling and gave some examples to illustrate them.

1.2 *Overview of ontology*

Ontology is originated from philosophy. Nowadays, it has been widely used in the field of information and science. As a common method of knowledge representation, ontology is quite different from Frame, Sematic Network and Agent, because it describes conceptual hierarchy relationships other than specific concepts. At the same time, it overcomes the problem of semantic heterogeneity and makes information exchange and sharing possible [2]~[3]. The main purpose of constructing an ontology model for printing company is, with collaborative participation of experts in printing, describing concepts in printing company by virtue of ontology. The printing company ontology will provide a comprehensive knowledge system and improve the efficiency of knowledge management [4].

2 ONTOLOGY MODELING

2.1 *The classification of ontology*

Although there is no unified standard on how to divide ontologies, they can be generally put into four categories as follows.

2.1.1 *Upper ontology*

Upper ontology is a model of common concepts that are generally applicable in a wide range of domains [5]. These concepts which are completely independent of the specific events and domains can also provide normative ontology modeling in sub-domains.

2.1.2 *Domain ontology*

Domain ontology is more specific than upper ones. It describes concepts and their relationships in one particular domain. In general, domain ontology is relatively in large scale. Ontology modularization was proposed to reduce the complexity of ontology, which benefited ontology maintaining and management [6].

2.1.3 Task ontology

Task ontology aims at describing one task which includes defining some problems or methods. It is not confined to a particular domain but focuses on dynamic knowledge [7], such as building automatic laminating machine malfunction ontology.

2.1.4 Application ontology

Building application ontology depends on a particular task in a particular domain. It describes more concrete concepts than upper ontology and domain ontology do.

2.2 Methods of ontology modeling for printing company

Ontology has been proved to be an effective tool for knowledge management in some fields. However, there is no systematic and standard method on ontology modeling [8]. Nowadays, there are some methods which are widely applied to ontology modeling, TOVE, METHONLOGY, IDEF5 and so on [9]. Since no existing ontologies of printing company can be reference samples, this paper introduced a new ontology modeling process which suits printing industry after studied the features of the main work flow in the printing factory of BIGC. The specific ontology building process is shown in figure 1.

Figure 1. Printing company ontology modeling process.

3 ONTOLOGY OF PRINTING COMPANY

3.1 Tools and methods

Printing industry is labor-intensive and has little experience in knowledge management. In fact, knowledge management can not only help companies improve production efficiency, but also motivate innovation among staff. Constructing printing company ontology can realize knowledge index, exchange and sharing among different companies. The main users of the ontology are experts, researchers and some practitioners in printing industry. Ontology engineers and printing experts are responsible for the maintenance of the ontology. Up to now, ontology languages mainly include OIL, DAML+OIL, OWL, etc. And OWL has perfect reasoning mechanism and semantic representation ability,

which is commonly accepted and applied. As research on ontology continues, the development of ontology tools also reaches a certain level. And Protégé is developed on the basis of OWL language and is becoming one of the main ontology developing tools.

3.2 Decide the scope of ontology

A comprehensive and complete ontology should involve interacting with both exterior and interior environment. However, this will make the ontology much larger and more complex. Since printing companies are all order-oriented and the work flow of printing company involves a large quantity of professional knowledge, this paper confined the scope of ontology to the interior environment of printing company to embody it.

3.3 Build the framework of printing company ontology

In order to ensure that the framework of printing company ontology is stable, this paper has referred to ontology modeling processes of other domains and consulted printing experts. Especially, to express the dynamical processes of completing an order, JDF helped to define the general work flow in most printing companies. As is shown in figure 2, printing company ontology can be divided into many branches.

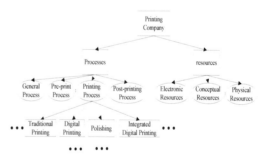

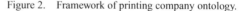

Figure 2. Framework of printing company ontology.

3.4 Divide and define prime concepts

Generally, ontology modeling consists of five meta languages, classes, relations, attributes, instances and axioms. As the knowledge system will expand with the development of printing industry and further studies, it is impossible to list all concepts. With regard to those concepts emerged later, we can add them into the ontology according to their attributes and relations and maintain the ontology.

JDF is a standard data exchange format which covered all the processes in printing. This paper identified the main work flow of most printing companies after studied JDF and investigated the printing factory in BIGC. Every task in the work flow can be presented with the format of "somebody—does something—on something". For example, binding can be described as "binders are binding books on the binding machine". We can extract important concepts such as binders, binding machine, binding and a connotative concept, prints. At the same time, the relations between different concepts are also apparent to clarify.

3.4.1 Classes

Ontology can be seen as a tree structure with many branches. Classes, the most basic elements in ontology modeling, are seen as nodes of the tree structure. In addition, there are numerous subclasses subordinated to classes. Defining the hierarchy of classes can avoid semantic repetitions and conflicts [12].

3.4.2 Relations

Classes have intricate relations with each other. As to defining the hierarchy and relations of classes, there are mainly three ways: top-down method, bottom-up method and synthetic method. Some of common relations are Hyponymy relations, Unity and parts relations, Property relations.

After analyzing printing paper classes and their relations and integrating the method proposed by paper [13], the hierarchy of relations was constructed from top to bottom as follows.

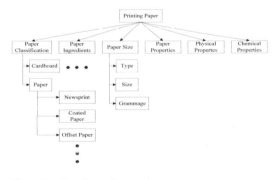

Figure 3. Relations of paper classes.

3.4.3 Attributes

Classes establish connections with each other through their relations and attributes. Therefore, defining attributes of classes can cover more detailed parts of professional knowledge and enrich the content of ontology. Attributes can be divided into two categories, conceptual attributes and relation attributes. The former ones embody the connotations of classes, while the later ones give further constraints to their relations. Since every class has one or more

attributes, we define attributes according to our research requirements.

3.4.4 Instances

Semantically speaking, instances are important ways that support inference and query. Constructing instances equals to establishing knowledge base, and we need to perfect all related attributes. There are many ways to store instances, and the particular one is stored in the database, thus we can come back to instances to have an accurate and thorough reference while we are retrieving something.

3.4.5 Axioms

Axioms represent eternal assertions which are mainly constraints to relations and attributes. The primary functions of axioms are reasoning through knowledge and checking consistency of different classes which refer to the same part of knowledge.

3.5 An instance of printing process ontology

The printing company is quite large that we take the printing process as an example to simply illustrate this specific process. After analyzing the printing process of an order named Xin Yin HuiShou(Vol. 4), prime classes, relations and attributes are listed in following tables.

Table 1. Some attributes of printing process.

Attributes	Value	Type	The Range of Attributes
Name	Xin Yin HuiShou (Vol. 4)	String	Conceptual Resources
Specification	210*285	Integer	Conceptual Resources
Number	500	Integer	Conceptual Resources
Positive Color+ Back Color	4+4	Integer	Physical Resources
Printing Method	Positive and Back	String	Physical Resources
Subdivision of Paper	6	Integer	Physical Resources
Puzzle Number	1	Integer	Physical Resources
Printed Sheet	500	Integer	Physical Resources
Press Spoilage	300	Integer	Physical Resources
Machine Number	Heidelberg CD 102	String	Physical Resources
Technological Requirements	None	String	Printing Process
Start Date	2014/3/26	String	Printing Process
Completion Date	2014/3/31	String	Printing Process

Table 2. Relations in printing process.

Class1	Class 2	Relations	Description
Cover	Content	Parallel Relation	Cover and content form the finished product.
Heidelberg CD 102 Printer	Cover, Content	Tool-Object Relation	Use the printer to print the book.
Printing Staff	Heidelberg CD 102 Printer	Subject-Tool Relation	Printing staff operate the printer.
Printing Staff	Printing	Subject-Activity Relation	Printing staff complete the printing task.
Heidelberg CD 102 Printer	Printing	Object-Activity Relation	The printer is used to print.

3.6 Evaluation of printing company ontology

After the printing company ontology is preliminarily established, it is of vital importance to examine and evaluate it. Generally, ontology development software can inspect itself to avoid grammatical mistakes and keep consistency between concepts. However, we need some ways to evaluate the reasonability of the ontology, such as listening to the advice of ontology experts and printing experts, assessing the quality of the ontology in practice, etc.

3.7 Maintaining and improving of printing company ontology

For any field, knowledge is infinite which always needs to be updated and amended. Thus, constructing the printing company ontology is an infinite loop which is called ontology circulation [14]. Constructing an ideal ontology needs researchers to keep studying again and again; sometimes we even need to come back to the first step to set the range of the ontology. Only through continually researching can we finally determine the printing company ontology.

4 CONCLUSIONS

From the perspective of ontology modeling, by taking other domains' ontologies as models and combining characteristics of printing industry, we determined the main technical route and constructed the printing company ontology. The specific method to define classes, relations and attributes was illustrated in detail by giving examples of printing paper. The knowledge system of printing company is presented clearly and reasonably by classifying and defining

important terminologies, which benefits knowledge sharing, reusing and retrieving in a printing company. Printing companies also benefit a lot in knowledge management by virtue of the ontology. Since the process of ontology modeling depends on studying many subjects related to printing, more work need to be done to improvecompleteness, accuracy and consistency of describing concepts and apply the ontology modeling to other aspects in printing industry.

ACKNOWLEDGMENT

Wei Li and Zhang Yongbin thanks Beijing Municipal Commission of Education for the finical support with grant number KM201310015008, the BIGC Key Project with grant number of E-a-2014-13 and the support with The Importation and Development of High-Caliber Talents Project of Beijing Municipal Institutions.

REFERENCES

[1] Zhang, H. & Deng, R.R. 2009. Research on Knowledge Management for Printing Enterprise Based on Business Processes. *Library and Information Service* (1):118–120.

[2] Deng, H.Z., Tang, S.W., Zhang, M., Yang, D.Q. & Chen, J. 2002. Overview of Ontology.*ActaScientiarum Naturalium Universitatis Pekinensis*38(5): 730–738.

[3] Li, S.P., Yin, Q., Hu, Y.J., Guo, M.& Fu, X.J. 2004. Overview of Researches on Ontology.*Journal of Computer Research and Development* 41(7):1041–1052.

[4] Feng, Z.Y., etc.2007. *Ontology Engineering and Application*. Beijing: Tsinghua University Press.

[5] Ontology. http://en.wikipedia.org/wiki/Ontology

[6] Fu, L. & Liu, Z. 2013.Review of Ontology Modularization. *Library and Information*(1):17–22.

[7] Tang, X.B., Wei, Z. & Xu, L. 2008. The Approach to Information System Modeling Based on Ontology. *InformationScience*26(3):391–395.

[8] Zhu, H.M., Ji, X.L., Huang, W.D. &Miao,M. 2008. Research on Ontology Modeling in Telecommunication Domain.*ModernInformation*1(1):184–186.

[9] Deng, X.Y. 2012. A Survey on the Research on Ontology Building.*ModernComputer*(5):19–22.

[10] Li, J., Meng, X.X. & Su, X.L. 2007. *Domain Ontology Modeling Methods(The 1st Edition)*.Beijing: Tsinghua University Press.

[11] JDF tutorials: http://www.cip4.org.

[12] Chen, L.F., Song, J.Y. & Shi, J. 2011. Specific Ontology Building and Analysis on Military Communication Domain.*Computer Technology and Development*21(7): 90–97.

[13] Xiong, D.H., Fang, K., Dai, X.P. & Huang, H. 2011. Study on the Approaches in Constructing Agricultural Ontology. *Journal of Agricultural Mechanization Research*33(11):48–51,55.

[14] Cheng, Y. 2012. The Research of Computer Information Modeling and Its Application Based on Ontology and RFID.*East China University Of Science and Technology*.

Speaker adaptive real-time Korean single vowel recognition for an animation producing

H.K. Yun & S.M. Hwang
School of Computer Science, Korea University of Technology and Education, Chonansi, Korea

B.H. Song
Department of Industrial Design Eng. Korea University of Technology and Education, Chonansi, Korea

ABSTRACT: Voice recognition technique has been developed and it has been actively applied to various information devices in Korea such as smart phones and car navigation systems. Since the basic research technique related to the speech recognition has been based on research results of other languages such as English and Japanese, it is possible to meet a sort of difficulties or some problems in point of view from the recognition. It should check once at least or check a margin for applying the Korean vocal sound system to improve the recognition of Korean speech, since 44 Korean phonemes always have their own phonetic values. However, the scope of this study is the recognition of single vowels for a digital contents producing, particularly lip sync animation, since the lip sync producing generally requires tedious hand work of animators and it seriously affects the animation producing cost and development period to get a high quality of lip animation. In this research, a real time processed automatic lip sync algorithm for virtual characters as the animation key in digital content is studied by considering Korean vocal sound system. The proposed algorithm is contributed to producing a natural condonable lip animation with the lower producing cost and the shorter development period. The system of real time vowel recognition for producing digital contents focused on formants frequencies was proposed. The recognition process consists of speech signal as the input, filtering, Fast Fourier Transform and identification. The algorithm based on the formant frequency using F1 and F2 was proposed, whose output was applied to the autonomic natural animating of the character's mouth shape for small and medium sized animation productions or e-learning contents productions. The result shows the proposed speaker dependent single vowel recognition system was able to distinguish Korean single vowels from dialogue of a dubbing artist with real-time. The average of the recognition ratio was 97.3% in the laboratory environment. It gives a possibility that more condonable lip sync can be produced automatically without any animator involved.

1 INTRODUCTION

Movement of the lips and tongue during speech is an important component of facial animation to increase the immersion of users in digital contents. Mouth movement during speech is ongoing and relatively rapid, and the movement encompasses a number of visually distinct positions. Also, the shape of mouth must be synchronized to the dialogue. Since the most of energy and time duration of speech signals are accumulated in vowels, it is required to recognize vowels from the dialogue of dubbing artists to produce animation key for the lip shape of virtual characters. Previous results of researches related to speech recognition, especially vowel recognitions, can be applied for synchronizing the mouth movement with a dialogue in digital contents such as animations and e-learning contents (Hwhang et el. 2013a). However, the principle techniques were mainly originated form the English-speaking world and the Japanese. Since the vocal sound system of Korean is different from other languages, it should be careful that the existing or previous technique is directly applied for recognizing Korean vowels. Furthermore, most of previous research results related to Korean language were derived using the digital signal processing technique which treats Korean-speaking as an input sound signal without considering the characteristic of the Korean vocal sound system. There is possibility to improve the recognition algorithm and rate though the Korean phonemes always have the same phonetic description or value.

Vowel recognition has been widely studied as a main topic in the speech signal processing area in order to find their effective speech recognition. Most of studies used MFCC (Mel-Frequency Cepstral Coefficients) and LPC (Linear Predictive Coding) algorithm. The most popular spectral based on parameter used in recognition approach is the MFCC. Due to its advantage of less complexity in implementation of feature extraction algorithm, only sixteen coefficients of MFCC corresponding to the Mel scale frequencies of speech cepstrum are extracted from spoken word samples in database. All extracted MFCC samples are then statistically analyzed for principal components, at least two dimensions minimally required in further recognition performance evaluation (Umesh et el 1997, 2002a, 2002b). A technique for the vowel classification using linear prediction coefficient with combination of statistical approach and ANN (Artificial Neural Network) was proposed (Paul et el.2009). A speech recognition system using fuzzy matching method was presented which was implemented on PC. MFCC and LPC with ANN are among the most usable techniques for spoken language identification. MFCC and LPC methods are used for extracting features of a speech signal and ANN is used as the recognition and identification method. However, the most of energy and time duration of speech signals is accumulated in vowels and the shape of mouth generally depends on the vowel of dialogue, which is well known from the result of previous researches related to speech recognition. Our concern is only for Korean single vowels in Korean dialogue to apply an automatic mouth animation or facial animation for a multimedia contents production, since most of lip sync in digital contents are manually animated by animators according to hearing dialogue. It is one of the extremely tedious and cost effective works to produce digital contents.

The proposed solution should meet the real time lip synchronization which the lip shapes mainly depend on vowels not consonants same as our previous studies. And general preprocessing speech analysis techniques are applied to recognize vowels without any specific instruments or tools to be utilized the solution at the small and medium sized animation producing studio or e-learning contents producing studio. To classify vowels, the result of FFT analysis is used to get the characteristic of each vowel. The incoming voice of a dubbing artist is analyzed the frequency component to extract vowels features to recognize vowels.

Since the mouth movement of virtual objects has to be synchronized with a dialogue exactly, lip-sync is one of tedious works for animators and a time consuming work. The mismatch or artificialness between mouth shape of characters and speaking reduces the immersion in the contents. This paper proposes a real time vowel recognition algorithm with the formant analysis which is a new technique to automatically perform lip synching for a computer generated character to match a real speech or a dialogue.

2 RELATED WORKS

2.1 Voice recognition

The studies related to voice recognitions were done by a lot of researchers or institutions (Shin et el. 2006, Kocharov 2004, Takao et el 2005). They mainly were concerned about the recognition words or speakers. Our concern is the real time vowels recognition to apply in producing a 3D animation. That should reduce the load of animators by minimizing simple monotonous hand working to synchronize with dialogues. Hence, the wanted recognition rate is not too high, which gives an affordable rate to be acceptable by producers at the current stage. But the lip sync solution should synchronize with the lip animation of character to the dialogues with real time. One of simple applications in Korea is the e-learning contents producing. Since the most of e-learning contents in Korea generally are taken recording of the lecturer's performance or the lecture in front of a blackboard. However, the proposed solution has a big additional advantage which is that the lecturer does not need to realize the camera, since some lectures repeatedly take pictures by their mistaking due to their performing in front of camera with trepidation. The proposed method will save the producing time and the cost. Furthermore, if the real human lecturer is replaced by a virtual lecturer in e-learning contents, the contents have more freedom to increase the scholastic achievement by adding some animation effects such as resizing objects with real time rendering techniques.

The proposed solution is applied to a general preprocessing for speech signals, a FFT analysis to get formants frequencies, comparing with reference data of the speaker to recognize vowels, converting the result of comparing with the reference table, and animating lip shape using the result as the key value for lip in the animation, as shown in Figure 1. The first stage is the feature extracting of vowels for the lecturer before starting the lecture. Lecturer can record his or her voice using the installed microphone in a notebook computer or a desktop computer. The lecturer pronouns 5 times of each single vowels such as ㅏ[a], ㅔ[e], ㅖ[i], ㅗ[o], and ㅜ[u]. Then vowels' features of lecturer are extracted using FFT algorithm as a reference data with real-time. The sampling rate was 8 KHz and 1024 FFT analysis was applied to get the formant frequencies as the references feature of each vowel since those are enough to show the mouth movement of the speaker if the Korean single vowels triangle is investigated. The next stage is to find the reference component in the dialogues, which is same as the test of similarity comparison in the image processing. The result is converted to one of indexes of vowels for lip shape animation for the virtual character of lecturer. The next step goes over the subject in this study, which is the animating lip shape using the index as an animation key. This stage should require a real time rendering technique to get more natural facial expressions.

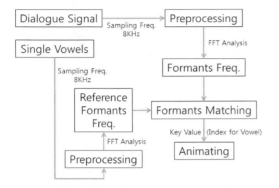

Figure 1. Block diagram for lip sync solution.

2.2 Characteristics of Korean vocalization

2.2.1 Vowel vocal sound system

The phonetic description of Korean vowels are al-ways invariable, which is of big difference from English. The vowels are grouped as single vowels (monophthongs) [Figure 2] and diphthongs [Table 2] according to the Korean vocal sound system.

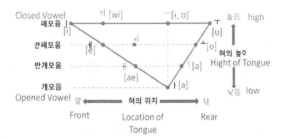

Figure 2. Korean single vowel triangle.

Table 1. Korean diphthongs.

group	Diphthongs
l[j]	ㅑ [ja], ㅕ [jə], ㅛ[jo], ㅠ[ju],
ㅗ/ㅜ[w]	ㅘ[wa], ㅝ[wə], ㅙ[wɛ], ㅞ[we],
-	ㅢ[ij or ʊj]

Korean single vowels are grouped into the rounded vowel and the unrounded vowel by the mouth shapes during pronouncing. The sound of vowels is determined by the lip shape, and tongue's location and height as shown in Table 2.

Table 2. Lip shape and tongue's position of single vowel.

max. point of tongue		Front vowel		Back vowel	
shape of lip		Unrounded vowel	Rounded vowel	Unrounded vowel	Rounded vowel
height of tongue	high	ㅣ [i]	ㅟ[wi]	ㅡ[ɨ, ʊ]	ㅜ[u]
	middle	ㅔ[e]	ㅚ[oi]	ㅓ [ə]	ㅗ [o]
	low	ㅐ[ae]		ㅏ [a]	

2.2.2 Vowel vocal sound system

The normal reading speed of Korean is 348 SPM (Syllables Per Minute) and the fast reading speed is 426 SPM. The average speed of spoken Korean is about 256 SPM and it varies from 118 to 409 SPM (Shin & Han 2006). The 256 SPM means that 4.4 syllables are pronounced in every second or a sylla-ble is pronounced for about 227msec. It is equal to about 6.8 frames for a 30 fps (frames per second) ani-mation. Since the maximum spoken speed is about 400 SPM, 6.7 or 6.8 syllables are pronounced per second. If the sampling frequency is 8KHz, the num-ber of samples is about 1200 per syllable. Hence, the real-time 1024 FFT could be handling the frequency analysis of Spoken Korean. Since the maximum spo-ken speed is about 400 SPM, 6.7 or 6.8 syllables are pronounced per second. If the sampling frequency is 8KHz, the number of samples is about 1200 per sylla-ble. Hence, the real-time 1024 FFT could be handling the frequency analysis of spoken Korean.

3 EXPERIMENT

The proposed single vowel recognition algorithm is for a natural and condonable lip animation which is automatically produced according to the dialogue of a dubbing artist. And the produced lip animation should minimize mis-matching dialogue with lip ani-mation and human errors of animators.

3.3 Experimental method and process

Our vowel recognition processing strategy in this paper is using the formant analysis as a result of FFT which is commonly used for speech recognition. FFT is a basic tool for digital signal processing applica-ble for spectrum analysis. Another transformation is Discrete Cosine Transform. DCT is a discrete trans-form whose kernel is defined by the cosine function. It is not popular to use in speech recognition, since DCT does not produce clear efficient third formant F3 in speech recognition. It has been known for many years that formant frequencies are important in determining the phonetic content of speech sounds. Several authors have therefore investigated formant frequencies as speech recognition features, using various methods. However, using formants for recognition can sometimes cause problems, and formant frequencies cannot discriminate between speech sounds for which the main differences are unrelated to formants. Thus they are unable to distinguish between speech and silence or between vowels and weak fricative. However, our previous study (Hwang et el. 2013a, 2013b) has shown that the recognition of single vowel is applicable for animation of the mouth shape more natural and faster than manual work for lip sync. The important thing to perform accurate lip synching to the real continuous speech such as dialogue, is the preprocessing for speech signal and is the extraction of feature from speech signal. The first step (left limb in Figure 1) is the feature extracting of vowels for speaker to prepare the reference of single vowels. Speaker phonate his or her voice using the microphone to process by a computer. Speaker pronouns each single vowel (e.g. 'a', 'e', 'e', 'o', 'u') 5times, then vowels' features of the speaker are extracted using FFT to get the frequency profile. The sampling rate is 8000 Hz and 1024 FFT analysis is applied to get the first and the second formant frequencies which are the reference feature of each vowel. The next step is to find the reference component from the dialogues, which is same as the similarity comparison test of the F1 and F2 for each vowel with the reference table of single vowels. The block diagram of the proposed system is shown in Figure 1. Here, F1 and F2 formant frequencies are defined to calculate as follows:

$$A_k = \sum_{i=0}^{bandwidth-1} a(f)_i \qquad (1)$$

Where A_K is defined as the sum of amplitudes for the k-th band and our experimental band width is 40 Hz, that means the frequency band of FFT analysis is divided into100 for helping the real time processing. Since the range of index k is $1 \leq k \leq 100$, the index k of the first maximum A_K is defined as F1 and the second value of A_K is also defined as F2. Then, each index and amplitude is stored for recognizing of vowels in dialogue as the reference which is compared with the incoming voice signal of the speaker (right limb in Figure 1). A typical result of FFT analysis is shown in Figure 3 which is short sentence, 'ga-ja ga a-ga' in Korean, same as 'Let's go kid' in English. The result shows that the proposed recognition algorithm is able to analyze frequency components of the incoming voice with real time. According to the output, the proposed FFT algorithm shows about 2 times per syllable, which means the velocity of speaking is about

0.25 sec per syllable. Therefore the speed of speaking should be normal since the average speaking velocity is 4.4 syllables per second or 0.23 sec per syllable.

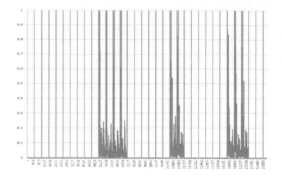
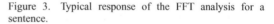

Figure 3. Typical response of the FFT analysis for a sentence.

3.4 *Experimental result*

The proposed algorithm was tested by 3 men subjects who were twenties on whether the algorithm was working or not. They attended the experiment 3 times for 2 weeks and each of them made 9 set of experimental data. Each data set is made up reference and tested data since the formant frequency is possible to change according to the physiological condition and the environment. Therefore the proposed method is a sort of limited speaker dependent recognition and takes reference data every time before speaking or dialogue to improve the recognition rate. As shown in Table 3, 'ㅗ[o]' and 'ㅜ[u]' are lower rate relatively. The reason is that they are the rounded back vowels and the only difference is the height of tongue. If the experimental data is analyzed, 'ㅗ[o]' is recognized as 'ㅜ[u]' frequently and vice versa.

Table 3. Lip shape and tongue's position of single vowel

Single vowel	Recognition rate[%]
'ㅏ' [a]	98.5
'ㅔ'[e]	98.6
'ㅣ'[i]	97.2
'ㅗ'[o]	96.3
'ㅜ'[u]	96.1
Average.	**97.3**

4 CONCLUSION

In this paper, a limited speaker dependent speech recognition solution for automatic producing lip-sync of objectives in digital contents is proposed. The conventional way to produce lip sync is to work manually by animators according to hearing of a dubbing artist's recording. It is one of tedious works for animator and also requires a plenty of time to get a good quality of lip sync. The experimental result shows that our proposed method is able to recognize single vowels without any problem in real-time. Furthermore, our method has an advantage since they have the same lip shape in spite of the misrecognition between the same group of vowels. Though the proposed solution is tested in the laboratory environment, the recognition of single vowels in the recording studio does not induce any trouble to analyze the voice in real time since the recording studio has better equipment and facilities and the proposed system is isolated from the dialogue recording equipment to keep the sound quality of digital contents. The next experiment is a field test and the proposed system is applied to a real dialogue recording in a recording studio. It has to be confirmed whether the system processes in real time at real situation, not in the laboratory environment.

Korean language, Hangul, has also several diphthongs which have the mouth shape varying characteristic during phonating. However lip sync is relatively faster than other event in digital contents and it is only about 4 or 5 frames per syllable. Therefore the approximate mouth shape is sufficient to implement lip sync of objectives in animation. Similarly, the proposed system could be easily applied for any other languages.

REFERENCES

[1] Cho Sung Moon, 2003, An Acoustic Study of Korean Vowel System, Korean Language Culture, vol. 20, pp. 427–441.

[2] Chung Hyun-yeol, Shozo Makino, & Keniti Kido, 1991, Analysis, Perception and Recognition of Korean Vowels, ICEIC,91, vol.2, pp. 195–198.

[3] Hwang S.M. & Yun H.K. 2013a. Extraction for Lip Shape using Real-time Vowel Recognition", Spring Conference, KIIECT 2013, vol. 6, no. 1, pp. 39–42.

[4] Hwang S.M., Yun H.K. & Song B.H, 2013b. Automatic Lip Sync Solution for Virtual Characters in 3D Animations, ICCT2013, vol. 2, no.1, pp. 432–433.

[5] Kocharov Daniil A., 2004, Automatic Vowel Recognition in Fluent Speech(on the Material of the Russian Language), 9th Conference Speech and Computer, SPECOM'2004, http://www.isca-speech. org /archive.

[6] Kodandaramaiah G.N., Giriprasad M.N. & Rao M. Mukunda, 2010, Independent Speaker Recognition for Native English Vowels, Int. j. of Electronic Eng. Research, Research India Publications, vol. 2, no. 4, pp. 377–381.

[7] Paul A.K., Das D., & Kamal M.M., 2009, Bangla Speech Recognition System Using LPC and ANN Advanced in Pattern Recognition, ICAPR '09, pp. 171–174.

[8] Shin Moonja & Han Sook-ja, 2006, A Study of Rate and Fluency in Normal Speaker, Speech vol.10, no.2, pp. 159–168.

[9] Takao Murakami, Kazutaka Maruyama, Nobuaki Minematsu, & Keikichi Hirose, 2005, Japanese Vowel Recognition Based on Structural Representation of Speech, INTERSPEECH, pp1261–1264.Umesh S. Cohen L. & Nelson D. 1997, Frequency warping and speaker-normalization, In ICASSP-97, IEEE International Conference on Acoustics, Speech, and Signal Processing, vol. 2, pp. 983–986, April.

[10] Umesh S. Cohen L. & Nelson D. 2002a, Frequency warping and the mel scale", Signal Processing Letters, vol. 9, no. 3, pp.104–107.

[11] Umesh S, Kumar S. B., Vinay M. K., Sharma R., & Sinha R., 2002b, A simple approach to non-uniform vowel normalization, ICASSP 2002, IEEE International Conference on Acoustics, Speech, and Signal Processing, vol. 1, pp. 517–520.

Data Mining

Network Security and Communication Engineering – Chan (Ed.)
© *2015 Taylor & Francis Group, London, ISBN: 978-1-138-02821-0*

Database-based online call graph tool applications

W.Z. Sun & D. Jia
College of Information Engineering, Capital Normal University, Beijing, China

Y. Xiang & R.D. Cao
Department of Computer Science and Technology, Tsinghua University, Beijing, China

ABSTRACT: Efficiency and scalability problems beset the existing file system access of the Call Graph based on Register Transfer Language (CG-RTL), a kernel function call graph tool based on file. Such problems have prompted the design and implementation of the Database-Based Call Graph on RTL (DBCG-RTL). The DBCG-RTL is a kernel function call graph tool based on database, and it can be used to analyze the function call relationship between user-specified modules. DBCG-RTL obtains a function's entry address, returns line number, and other relevant information by searching the kernel symbol table, thereby improving the accuracy of this analysis tool. DBCG-RTL also stores standardized format data tracked through a variety of dynamic analysis tools in the database. It provides perfect support to the virtual directory and extends the function of the dynamic function call analysis; so it is more efficient than CG-RTL.

1 INTRODUCTION

The call graph based on register transfer language (CG-RTL) and dynamic call graph based on RTL (DCG-RTL) are online function call graph tools with a function call relationship between user-specified modules. They can be used for static and dynamic analyses, but they are beset by certain shortcomings. Both online function call graph tools use many files to store data and need repeated traversal of the source code directory tree, particularly when processing virtual directory data. When generating a function call graph or a function list, the logical relationships and the process of traversing directories and files become progressively complex, thereby increasing the difficulty on the part of programming developers. In addition, the accuracy of some same-name functions and other information is insufficient.

The purpose of designing an online function call graph based on database (i.e. DBCG-RTL) is to use the advantages of the database. Such advantages include the following: high independence, easy retrieval, good scalability, storage of information of functions obtained by the kernel symbol table, dynamic analysis tools, and virtual directories. The design of the DBCG-RTL also aims to simplify the function call logic of CG-RTL and DCG-RTL in generating a function call graph or list. It also aims to generate a relatively more efficient and accurate function call logic for DBCG-RTL and to facilitate the expansion of the functions of dynamic function call and virtual directories.

2 DBCG-RTL: AN ONLINE FUNCTION CALL GRAPH TOOL BASED ON DATABASE

2.1 Design objectives and architecture of DBCG-RTL

This study designs a DBCG-RTL to achieve the following objectives: (1) retain the original capabilities of static analysis and improve the processing efficiency of the tool, (2) provide some support for dynamic function call relationship analysis and virtual directory analysis through standardized interfaces, and (3) improve the accuracy and correlation of function call information.

The architecture of DBCG-RTL includes the following modules: data preprocessing, function call graph generation, function call list generation, expansion interface, HTML caching, and other functional modules. The main functions of some of these modules are as follows:

Data preprocessing module: The data preprocessing module is the core module of DBCG-RTL. It extracts the function definition and calls information from the compiler intermediate results and stores them in the database. The module then uses the kernel symbol table and addr2line tool to search for functions and complements the database table of the function definition information. It also stores standardized format data tracked by dynamic analysis tools, such as S2E, in the database and generates a series of corresponding database tables of virtual directories based on database tables of real directories.

Function call graph or function call list generation module: According to the data in the database tables, this module generates the function call graph or list of a real or virtual directory.

Expansion interface module: This module uses the data in the database tables and provides data support for the vulnerability map, taint tracing, or other functions.

2.2 *Implementation of the milestones of DBCG-RTL*

The design and implementation process of DBCG-RTL includes the following milestones to perfect the corresponding functions of CG-RTL. First, the data required for the generation module of the function call graph or list are planned, and a number of database tables are designed. Second, the function-related information is complemented with the kernel symbol table and addr2line tool. Third, the normalized data obtained by the dynamic tracing tools, such as S2E, Vmxice, and Ftrace, are stored in the database to extend the dynamic function call data. Fourth, the virtual directory's information in the database is stored to establish a logical extension for the function call relationship with the original source code. Lastly, the data in the database is used to generate static and dynamic function call graph or list.

Considering the system described, DBCG-RTL needs, and most of the characteristics of the data, we designed 10 database tables. Among these dataset tables, five have covered the information on a real directory. (1) The FDLIST database table is used to save all the paths, names, and start and end locations in the source code and memory definition information. (2) The SLIST database table is used to hold a static position in the call and information on the calling and called functions. (3) The SOLIST database table is used to save the static relationship between the files and directories. The purpose of creating this table is to calculate the right of sides of a function call graph. (4) The DLIST database table is used to save the relevant information during the dynamic function calls. Unlike the SLIST database table, the DLIST adds time and process/thread information on the dynamic function calls. (5) The DOLIST database table is used to save the dynamic relationship between the files and directories.

A virtual directory is a concept relative to the real directory of the source code. The purpose of organizing a virtual directory is to provide users with different logical partitions for different uses of the same source code, thereby facilitating the analysis and modification of the kernel code. (6) The VLIST database table is used to save the virtual directory path and the corresponding real directory path. (7) Other database tables—VFDLIST, VSLIST, VSOLIST, and VDOLIST—contain the call information on virtual directories. Virtual and real directory database tables have similar design format and generation methods. However, before generating each virtual directory record, relevant information on the real directory in the VLIST database table must first be found.

2.2.1 *Function call graph generation*

Algorithm 1 explains what the DBCG-RTL obtains from a database table and how it obtains this information in generating a function call graph.

Algorithm 1: Call graph matrix generation algorithm

The nodes in a call graph are determined based on the modules specified by the user in a web page. This process is conducted by finding the function-defined path and function name from the FDLIST database table. Each node is then numbered.

The number of the static and dynamic calls between the nodes is calculated using the SOLIST and the DOLIST database tables.

The format required by Graphviz [8] according to the node number and number of calls between the nodes is used as the output.

2.2.2 *Function call list generation*

Algorithm 2 explains what the DBCG-RTL obtains from a database table and how it obtains this information in generating a static or dynamic function call list.

Algorithm 2: Call list generation algorithm

a. Obtaining relevant information on the static function call

The module paths are respectively specified as the calling function's path and the called function's path in the SOLIST database table, as the number of calls accumulates.

The module paths are respectively specified as the calling function's path and the called function's path in the SLIST database table, while the following are being searched and recorded: the calling function's unique identification, called function's unique identification, location where the function is called, and line number where the function is called.

The calling function and the called function's relevant information (i.e. function name, function definition path, function line number, and so on) are searched using their unique identification.

b. Obtaining relevant information on the dynamic function call

The module paths are respectively specified as the calling function's path and the called function's path in the DOLIST database table, as the number of calls accumulates.

The called function's relevant information (i.e. the function line number and so on) in the FDLIST database table is searched using the called function's path (i.e. the function name, function definition path, and so on) found in the DOLIST database table.

2.2.3 *Function call graph and list of a virtual directory*

The method of generating a function call graph or list of a virtual directory is similar to the method of generating a functional call graph or list of a real directory. However, the database tables read in generating a function call graph or list are changed from FDLIST, SLIST, SOLIST, and DOLIST to VFDLIST, VSLIST, VSOLIST, and DOLIST respectively. A comparison of the generation method of a real directory and that of a virtual directory leads to the conclusion that algorithms 1 and 2 in a virtual and real directory are exactly the same, and only the database tables used in the program need to be modified.

3 COMPARISON OF DBCG-RTL AND CG-RTL

DBCG-RTL retains all the functions of CG-RTL, but it improves and expands the capabilities of the latter and enhances the performance of the algorithm of the latter.

3.3 Improved and expanded features

3.3.1 Acquisition of relevant information on the functions

DBCG-RTL finds a means of obtaining relevant information on the functions using the kernel symbol table, thereby improving the recognition of the function with the same name. For example, given four-definition information on "abort_dma", a same-name function exists, in Linux-3.5.4 kernel and x86_32 platform, as shown in Figure 1.

```
+----------+--------------------------------------------+----------+
| f_name   | f_dfile                                    | f_dline  |
+----------+--------------------------------------------+----------+
| abort_dma | drivers/staging/comedi/drivers/cb_pcidas64.c | 3155 |
| abort_dma | drivers/staging/comedi/drivers/gsc_hpdi.c    | 1034 |
| abort_dma | drivers/usb/gadget/goku_udc.c                |  637 |
| abort_dma | drivers/usb/gadget/net2280.c                 | 1081 |
+----------+--------------------------------------------+----------+
```

Figure 1. Definition information on abort_dma in Linux-3.5.4 kernel and x86_32 platform.

CG-RTL always obtains the information on the function it first encounters, which is defined at location drivers/staging/comedi/drivers/cb_pcidas64.c, and whose function definition line number is 3155. By contrast, DBCG-RTL, which uses the kernel symbol table to search function information, obtains the function call information according to the actual situation, as shown in Figures 2-3.

```
+----------+--------------------------------------------+
| C_name   | C_dfile                                    |
+----------+--------------------------------------------+
| abort_dma | drivers/staging/comedi/drivers/cb_pcidas64.c |
| abort_dma | drivers/staging/comedi/drivers/cb_pcidas64.c |
| abort_dma | drivers/staging/comedi/drivers/cb_pcidas64.c |
| abort_dma | drivers/staging/comedi/drivers/cb_pcidas64.c |
+----------+--------------------------------------------+
```

Figure 2. Call information on abort_dma in CG-RTL in Linux-3.5.4 kernel and x86_32 platform.

```
+----------+--------------------------------------------+
| C_name   | C_dfile                                    |
+----------+--------------------------------------------+
| abort_dma | drivers/usb/gadget/net2280.c               |
| abort_dma | drivers/usb/gadget/goku_udc.c              |
| abort_dma | drivers/staging/comedi/drivers/cb_pcidas64.c |
| abort_dma | drivers/staging/comedi/drivers/gsc_hpdi.c  |
+----------+--------------------------------------------+
```

Figure 3. Static call information on abort_dma in DBCG-RTL in Linux-3.5.4 kernel and x86_32 platform.

3.4 Storage of information on the virtual directory

DBCG-RTL uses virtual directory database tables to generate function call graphs and lists (see 2.2.3. Function call graph and list of a virtual directory). In CG-RTL, handling the correspondence relationship between the real and virtual directories (see 1. INTRODUCTION) is difficult because the function of the virtual directory is not supported well.

3.5 Performance comparison

Table 1 shows the statistical results of the 19 175 test records on the top-level directory and subdirectories of fs and mm with CG-RTL and DBCG-RTL, respectively, in generating function call graphs.

Table 1. Comparative statistical results for CG-RTL and DBCG-RTL in function call graphs.

	Static function call	Dynamic function call
Average	seconds	seconds
CG-RTL	34.774373	13.64157438
DBCG-RTL	31.16096883	30.89251218

As Table 1 shows, for static function calls, the performance of DBCG-RTL is better than that of CG-RTL. However, for dynamic function calls, CG-RTL outperforms DBCG-RTL. An in-depth analysis of these results is provided below.

Table 2 shows the extreme value comparison between DBCG-RTL and CG-RTL for dynamic function calls.

Table 2. Extreme value comparison of DBCG-RTL and CG-RTL for dynamic function calls.

	Minimum	Maximum
	seconds	seconds
CG-RTL	9.702509	219.696264
DBCG-RTL	0.358781	210.269188

A comparative analysis of the respective original test data of CG-RTL and DBCG-RTL shows that the maximum and minimum times required by DBCG-RTL are

393

shorter than those required by CG-RTL. However, given that the calculation time of CG-RTL is directly generated by the pretreatment document, the test results are widely distributed on the average. By contrast, DBCG-RTL tends to reread data from the database tables and to recalculate according to different nodes every time it generates a function call graph. This tendency of DBCG-RTL increases the time change, along with the number of different nodes. (The higher the number of the nodes, the longer the computing time required). For example, 14 nodes exist in the arch's function call graph, which translates to a computing time of approximately only 0.358 781 s. By contrast, 354 nodes exist in the function call graph from fs/buffer.c to mm/mprotect.c, which translates to a computing time of approximately 210.269 188 s. A large number of nodes reduces the computational efficiency of the dynamic invocation. Furthermore, when the number of nodes is less than 100, the efficiency of DBCG-RTL is better than CG-RTL.

Table 3 shows the statistical results of 39 231 test records between the top-level directory and subdirectories of fs and mm with CG-RTL and DBCG-RTL, respectively, in generating function call lists.

Table 3. Comparative statistical results for CG-RTL and DBCG-RTL in function call lists.

	Static function call	Dynamic function call
Average	seconds	seconds
CG-RTL	0.156215133	0.018628627
DBCG-RTL	0.013235089	0.003751179

As Table 3 shows, the efficiency of DBCG-RTL is significantly better than that of CG-RTL for both static and dynamic function calls.

In summary, DBCG-RTL is more efficient than CG-RTL in generating static function call graphs (i.e. by an average improvement of 10.4%) and static and dynamic function call lists (i.e. by an average improvement of 90.3%). By contrast, CG-RTL presents certain advantages in generating dynamic function call graphs.

4 CONCLUSIONS AND FUTURE PROSPECTS

DBCG-RTL uses the kernel symbol table to look up function information, which increases the accuracy of data information. It also results in an expanded and more comprehensive dynamic function call. DBCG-RTL also introduces the information on virtual directories to a database, thereby expanding the original kernel code. To improve the efficiency of CG-RTL, DBCG-RTL uses data in the database to generate static and dynamic function call graphs and lists.

DBCG-RTL is currently operational on only one server. A follow-up to this article will attempt to improve the efficiency of DBCG-RTL through a data processing method in a distributed environment. In the near future, we hope to build an operating system kernel analysis test platform that runs in a distributed environment and will help developers test kernel analysis and find errors more efficiently.

ACKNOWLEDGMENTS

This work is supported by National Science and Technology Major Project of China under grant No.2012ZX01039-003 and 2012ZX01039-004-01-3.

REFERENCES

[1] Ge, H.S. Researches and Analyses on the Advantages of Database, *Journal of Shangqiu Vocational and Technical College, vol.6: 46-48, April 2007.*
[2] Sun, W.Z. & Du, X.Y. & Xiang, Y. & Tang, W.D. & Hou, H.R. CG-RTL: a RTL-based Function Call Graph Generator, *Journal of Chinese Computer Systems, vol.35: 555-559, March 2014.*
[3] Xiang, Y. & Tang, W.D. & Sun, W.Z. & Du, X.Y. Dynamic call graph base on OS kernel trace, *Application Research of Computers, 2015, unpublished.*
[4] Wu, J.B. & Xiang, Y. & Lu, H.M. & Li, L. Automatic Kernel Function's Dynamic Tracking on Real System, *The first session of the open-source operating system design and analysis Conference Proceedings. 2013.*
[5] S2E, https://sites.google.com/site/dslabepfl/proj/s2e
[6] Ftrace, http://elinux.org/Ftrace
[7] Meng, X.Z. Linux kernel debugger based on Inter VT, *Xian University of Posts & Telecommunications undergraduate thesis, 2012.*
[8] Graphviz, http://www.graphviz.org/
[9] Fattoriy, A. & Paleariy, R. & Martignoni, L. Dynamic and Transparent Analysis of Commodity Production Systems, *2010.*
[10] Zhao, W. Research and Design based on lookup method of SQL database table, *Technology Wind: 49-50, May 2012.*
[11] Zhang, G.R. Design and implementation of E-Form System based on Relational Database and Non-Relational Database, *Zhongshan University master's thesis, April 2012.*
[12] Gu, W. & Chen, L.J. Research on the Query Optimize Technology of MYSQL, *Microcomputer Applications, Vol.30: 48-50, July 2013.*
[13] Zhang, Z.Q. Query Method and Optimization Techniques Based on Relational Database, *Coal Technology, vol.31: 218-219, May 2012.*
[14] Shi, W.M. Data standardization of database design, *Modern Information: 80-82, August 2003.*
[15] Y.Tang, X.Tian, "Design and Implementation of Information Database based on B/S Structure," Modern Information, pp.73-74, August 2006.
[16] Wu, Z.B. Database Table Principal Linkage Design Method Discussion, *Computer Knowledge and Technology, 2006.*
[17] Ge, W.Y. Explore design methods of the Hydrologic relational database table structure, *Yangtze River, vol.22: 15-23, June 1991.*

Network Security and Communication Engineering – Chan (Ed.)
© 2015 Taylor & Francis Group, London, ISBN: 978-1-138-02821-0

Research on the correlation of Shanghai composite index and the S&P500 index based on M-Copula-EGARCH

C. Liu, X. Gao, H. Pan, J.H. Li & X.D. Liu*
Donlinks School of Economics and Management, University of Science and Technology Beijing, Beijing, China

ABSTRACT: As economic globalization is accelerating, financial products of different regions have become more closely linked. This paper analyzes closed price series of the Shanghai Composite Index and the S&P500 stock index. At first, it uses EGARCH function to portray the marginal distribution of stock index price series, and finds that the EGARCH function can obtain better results compared to traditional fitting functions; then it uses the M-Copula function to connect the portrayed model, and makes parameters estimation and model checking; finally, it applies Monte Carlo method to simulate the fitted M-Copula function model based on EGARCH function, and calculates the value at risk VaR. Through the effective combination of EGARCH, M-Copula, VaR and other technologies, we overcome the problems of non-linearity and non-normality among different assets in the portfolio, and provide a new idea for portfolio selection, risk measurement and control.

1 INTRODUCTION

In financial markets, investors seek the maximum return of the portfolio at a given risk level in order to spread the financial risk. In the risk management process, asset allocation problem is the foundation and core. The traditional Markowitz mean- variance model lays the foundation of modern portfolio theory. It reflects the uncertainty in investment as risk and the standard deviation or variance as a standard to measure risk. But with exacerbation of economic globalization, this model has been not sufficient to meet the risk management needs.

Since the 1990s, the value at risk (VaR) has become a new way to measure risk. In the process of risk measurement and portfolio strategy selection, the most important thing is to determine the distribution of assets, and we generally use the normal distribution to characterize it. However, when the kind of assets is increasing and the same kind of assets has different marginal distributions, traditional assumptions will generally fail. By introducing the Copula technology into the financial markets, you can have a deeper understanding of dependencies between financial assets and make better predictions of risk. If you want to predict the risk, it is the most appropriate to introduce the GARCH volatility model which is the most commonly used model to reflect characteristics of the market changes, and it can effectively capture the clustering and heteroscedasticity phenomenon of the volatility of return on assets. After the combination of them, we can not only analyze the VaR of portfolio through diversified distribution, but also effectively

capture the nonlinear relationship between financial markets.

Since Embrechts et al. [1] introduced Copula into the financial quantitative analysis, scholars have conducted related operations. Gabriel et al. [2] discussed the effects and limitations of elliptical Copula used to describe financial data; Davide and Walter [3] studied pricing and risk analysis of credit derivatives with some Copula and found that t-copula was more appropriate in the financial data analysis; Hull and White [4] established the pricing model of credit derivatives with multiple underlying assets by using Copula.

Engle [5] first proposed the ARCH model to describe the conditional heteroscedasticity that existed in the UK inflation. Zakoian [6] proposed TARCH, and Saidi and Zakoian [7] proposed NARCH. The former took into account that the variance had much to do with one of parameters, whereas the latter was an important nonlinear ARCH model.

When it was discovered that the ARCH model cannot express this information that "in some cases the auto-correlation coefficient subsides slowly", and that in the practical application, the estimation of completely free lag distribution often led to the destruction of non-negative constraints, GARCH model came into being. Ding et al. [8] proposed A-PARCH which was of high induction. GARCH models do not overcome all the shortcomings of ARCH models, and EGARCH models were proposed in order to solve this limitation.

In order to study dependencies and risks of financial variables dynamically, Wu et al. [9] established Copula-GARCH estimation of the joint distribution

---

*Corresponding author

of multi-asset yields, analyzing the investment risk of portfolios of multiple assets. Peng and Fu [10] proposed time-varying Copula-GARCH-Mt model. Xu et al. [11] conducted a comparative study of the impact that DCC-GARCH, M-Copula-GARCH and Copula-SV had on hedge ratios of copper and cotton future contracts.

This paper selects the most representative Chinese Shanghai Composite Index and the United States S&P500 as the research object, establishes the marginal distribution through EGARCH model, depicts the distribution of market's closing price indexes, describes the skew, spikes, and fat tail of market data and indicates the reason why it cannot establish appropriate model assuming the traditional normal distribution. And then we construct flexible multivariate distributions through Copula to adapt to the actual distributions of the portfolio investment in stock market, do a quantitative research about risk assessment, and make an analysis of VaR, volatility rate of portfolio risk under the premise that inter-dependence structure of assets are taken into consideration.

2 CONSTRUCTION OF RISK MEASUREMENT MODELS

The conditional distribution of financial time series usually presents skew, time-varying, volatility clustering, spikes, fat tail and other characteristics. The vector GARCH model (EGARCH) is suitable to portray the marginal distribution of each asset, so we use EGARCH model to describe the marginal distribution of financial time series in this section.

The correlation among financial markets is changing and the changes of correlation are usually asymmetric, so we introduce a more flexible mixed Copula function, namely M-Copula function to fit the joint distribution function and reflect market characteristics. The specific form is following:

$$(z_{st}, z_{ft}) \sim MC(T_{vs}(z_{st}), T_{vf}(z_{ft})) \quad (1)$$

$$MC(u, v) = \omega_1 C_{Gum}(u, v) + \omega_2 C_{Cl}(u, v) + \omega_3 C_{Fra}(u, v) \quad (2)$$

Where T_v whose mean is 0, variance is 1 and DOF is v obeys t distribution, that is $\sqrt{v/(v-2)}\xi_t \sim t(v)$. In (2), $C_{Gum}(u, v)$, $C_{Cl}(u, v)$, $C_{Fra}(u, v)$ represent Gumbel Copula, Clayton Copula, and Frank Copula respectively. $\omega_1, \omega_2, \omega_3$ are weights of the Copula, and $\omega_1 + \omega_2 + \omega_3 = 1$, $\omega_1, \omega_2, \omega_3 \geq 0$. Gumbel, Clayton and Frank all belong to the binary Archimedean Copula function, and they describe the tail correlation, low tail correlation and symmetry correlation among variables respectively. The mixed Copula function can also make use of weights to describe the asymmetric correlation pattern in which there are both up and low tail correlations among markets.

Finally, it provides measurement of portfolio's risk and choices for the optimal investment strategy by calculating the portfolio's VaR. We use Monte Carlo

simulation to calculate the VaR of a portfolio and compare this method with the traditional VaR model based on multivariate conditional normal distribution assumptions, finding that Copula model based on extreme value theory is dominant.

3 EMPIRICAL RESEARCH

3.1 Selection and pre-processing of sample data

This paper selects the daily closing price of Shanghai Composite Index (SH) and S&P500 as study sample. The time period is from January 4, 2000 to May 28, 2008, and there are a total of 2023 and 2111 set of valid data respectively. The SH data come from NetEase financial database, and S&P500 data come from Yahoo Finance.

After excluding non-trading days, the price series $\{P_{nt}\}$ is closing prices in market n at day t, and the yield sequence $\{R_{nt}\}$ is:

$$R_{nt} = 100(\ln P_{nt} - \ln P_{n,t-1}) \quad (3)$$

By using (3), we can make preprocessing to SH and S&P500 data through Eviews.

3.2 Basic statistical properties

Through the statistical analysis for above data, we find that the skewness of the two index yield series are not 0, that is, the yield distribution is not symmetric with deflection characteristics. Their kurtosis are also greater than 3, namely the tail of the yield distribution is greater than the tail of a normal distribution, indicating their yield series present "spikes and fat tail". Meanwhile, JB statistics are greater than the critical value of 5.9915 at 5% significant level, refusing the null hypothesis that the yield series obey normal distribution, that is, the two index yield series are not normally distributed, and we cannot use traditional mean - variance model to analog.

3.3 Stationary test

The EGRACH model requires that time series is stationary, so we need to do stationary test to the yield sequence, namely unit root test. We select the ADF test to validate stability of index yield of above financial time series.

In the ADF test, the P values of t statistic are less than 0.05, rejecting the null hypothesis that there exists unit root, and it indicates that the two index's yield sequences are both stationary.

3.4 Portraying marginal distributions

In order to overcome some shortcomings of the GARCH model in handling this kind of problem, this paper introduces the EGARCH model which has different assumptions for conditional variance.

Take EGARCH (1, 1) function to portray marginal distribution of each sequence, and the model structure is as follows:

$$y_t = \mu_t + \varepsilon_t,$$
$$\ln(h_t) = C + \alpha \left|\varepsilon_{t-1}\right|/\sqrt{h_{t-1}} + g\varepsilon_{t-1}/\sqrt{h_{t-1}} + \beta \ln(h_{t-1}). \quad (4)$$

Assuming standardized residuals obey t distribution, that means $\varepsilon_{t-1}/\sqrt{h_{t-1}}$ obeys t distribution.

By using preprocessed data, we can get the estimated results.

Table 1.　EGARCH (1, 1) model estimation by SH data.

Parameters	Parameters estimation	P value
C	−0.132454	0.0000
α	0.200559	0.0000
g	−0.042156	0.0022
β	0.978897	0.0000

The marginal distribution of SH data which is based on EGARCH (1, 1) model $F(\cdot)$ is:

$$\ln(h_t) = -0.13245 + 0.20056\left|\varepsilon_{t-1}\right|/\sqrt{h_{t-1}}$$
$$- 0.04216\varepsilon_{t-1}/\sqrt{h_{t-1}} + 0.97890\ln(h_{t-1}). \quad (5)$$

Table 2.　EGARCH (1, 1) model estimation by S&P500 data.

Parameters	Parameters estimation	P value
C	−0.056497	0.0000
α	0.068466	0.0000
g	−0.125256	0.0000
β	0.985738	0.0000

The marginal distribution of S&P500 data which is based on EGARCH (1, 1) model $G(\cdot)$ is:

$$\ln(h_t) = -0.05650 + 0.06847\left|\varepsilon_{t-1}\right|/h_{t-1}$$
$$- 0.12526\left|\varepsilon_{t-1}\right|/\sqrt{h_{t-1}} + 0.98574\ln(h_{t-1}). \quad (6)$$

In order to make the model portrayed more accurate, we should test if there is still ARCH effect, and GARCH effect after the model has been portrayed. This paper uses ARCH LM test.

The results suggest that the ARCH effect has been eliminated, that is, each model portrays marginal distributions successfully.

3.5　Connecting marginal distributions

According to parameters estimation results based on the EGARCH (1, 1) model, we use the two-stage maximum likelihood estimation method to make parameters estimation to M-Copula model combined with MATLAB. Parameters estimation results are in the table below:

Table 3.　M-Copula function parameters estimation results.

M-Copula function	Parameters	Parameters estimates	Likelihood value
$CM(\cdot,\cdot)$	α	0.3812	0.9328
	θ	0.0128	
	λ	0.0051	
	ω_1	0.1070	
	ω_2	0.8730	
	ω_3	0.0200	

Referring to the above data, in M-Copula models of the two stock market index yield sequence, Clayton Copula function accounts for the largest proportion, reaching 0.8730, which indicates that there is strong low fat correlation among data; the following is Gumbel Copula function whose weight is 0.1070, meaning the up tail correlation is weak among data; the smallest is Frank Copula whose weight is 0.0200, and there is almost no symmetry correlation among data.

3.6　Application of VaR model

In this paper, we use Monte Carlo simulation method to generate random numbers, and simulate to calculate VaR. In VaR of the portfolio, many assets obey a joint distribution, so the generation of random numbers should also be handed over to the joint distribution.

As for the Shanghai Composite Index and the S&P500 (they are defined as assets in this paper), the marginal distributions of their return of assets R_{SH}, R_{SP} are respectively $F(\cdot)$, $G(\cdot)$ (they have been portrayed by EGARCH function), but related structures are determined by M-Copula functions. So the random number obtained by Monte Carlo simulation $(R_{SH}, R_{SP}) \sim C(F(R_{SH}), G(R_{SP}))$.

In fact, $F(\cdot)$ and $G(\cdot)$ obey $(0,1)$ uniform distribution. As long as the random numbers $(u,v) \sim C(u,v)$ can be generated, we can obtain random numbers (R_{SH}, R_{SP}) required by Monte Carlo simulation through the inverse calculation of the marginal distributions, and $(R_{SH} = F^{-1}(u), R_{SP} = G^{-1}(v))$.

In summary, this paper combines with M-Copula function, EGARCH function and Monte Carlo simulation, and finds that calculation steps of the Shanghai Composite Index and the S&P500 portfolio's VaR under certain circumstance of fixed portfolio weights are as follows:

1 Generate two independent random numbers u, w, which obey $(0, 1)$ uniform distribution, and u is

the first pseudo-random number for simulation; then use the $v=C_u^{-1}(w)$ formula to generate another pseudo-random number;

2 Calculate the yield R_{SH}, R_{SP} corresponding, u, v according to the distribution functions, $F(\cdot)$, $G(\cdot)$ of the two stock markets, and $R_{SH} = F^{-1}(u)$, $R_{SP}=G^{-1}(v)$.

3 Given the weight θ of Shanghai stock in the portfolio, calculate the yield R of the portfolio, $R = \theta R_{SH} + (1-\theta)R_{SP}$.

4 Repeat step (1) ~ (3)n times, simulate n kinds of possible scenarios of the portfolio's future yield, and thus we can obtain empirical distribution of portfolio's future yield. As for a given confidence level $1-\alpha$ and $P[R<-VaR_\alpha] = \alpha$, we can obtain VaR of the investment portfolio under the confidence level $1-\alpha$, and VaR_α represents VaR value under confidence level $1-\alpha$.

Generate vectors (V_{SH}, V_{SP}) randomly which obey two-dimension M-Copula function $C_M(\cdot,\cdot)$, and calculate simulation yield sequence $R=(R_{SH}, R_{SP})$ of the two stock markets through the marginal distributions portrayed by EGARCH(1,1). Make 1000 times simulations at 95% confidence level, and the VaR empirical distribution of the two stock markets are obtained.

Combined with the above theory, the paper can calculate VaR of Shanghai Composite Index and the S&P500 are 1.68 and 1.11. Then, we calculate the empirical distribution of the portfolio's loss sequence consisting in the two markets, and then obtain that the portfolio's VaR is 1.03. By comparing the average VaR of a single asset with the VaR of investment portfolio, we can draw a general conclusion that the portfolio can effectively reduce investment risk.

4 CONCLUSIONS

This paper combines with EGARCH function and M-Copula function, does empirical analysis for return series of stock price index of the Shanghai Composite Index and the S&P500, calculates the portfolio's value at risk VaR, and forms the following conclusions.

First, through the statistical properties of the data of empirical analysis, we can find financial time series used in this paper do not follow normal distribution and volatility clustering, spikes, fat tail and other features, and the index return series of different stock markets have different marginal distribution. The EGARCH model can describe the leverage effect of the stock market, the asymmetry of positive and negative volatility, and there are no parameter limits, so it can describe the financial time series better.

Second, it mixes a variety of Copula functions into M-Copula function, builds M-Copula-EGARCH model to effectively capture the dynamic and static relationship among financial time series, and calculates the portfolio's value at risk VaR.

In short, the method described herein can overcome the limitations that the traditional mean—variance model cannot represent data of non-linearity or non-normality well, fit complex correlated time series effectively and provide a valuable reference for portfolio risk measurement.

ACKNOWLEDGMENTS

This work is supported by the Fundamental Research Funds for the Central Universities (FRF-TP-14-052A1).

REFERENCES

[1] P. Embrechts, A. McNeil, D. Straumann, "Correlation and dependence in risk management: Properties and pitfalls," *Risk management: value at risk and beyond*, pp. 176–223, 2002.

[2] G. Frahm, M. Junker, A. Szimayer, "Elliptical copulas: applicability and limitations," *J. Statistics & Probability Letters*, vol. 63, pp. 275–286, March 2003.

[3] D. Meneguzzo, W. Vecchiato, "Copula sensitivity in collateralized debt obligations and basket default swaps," *Journal of Futures Markets*, vol. 24, pp. 37–70, January 2004.

[4] J. C. Hull, A. D. White, "Valuation of a CDO and an n-th to default CDS without Monte Carlo simulation," *The Journal of Derivatives*, vol. 12, pp. 8–23, February 2004.

[5] R. F. Engle, "Autoregressive conditional heteroscedasticity with estimates of the variance of United Kingdom inflation," *Econometrica: Journal of the Econometric Society*, pp. 987–1007, 1982.

[6] J. Zakoian, "Threshold heteroskedastic models," *Journal of Economic Dynamics and control*, vol. 18, pp. 931–955, May 1994.

[7] Y. Saïdi, J. Zakoïan, "Stationarity and geometric ergodicity of a class of nonlinear ARCH models", *The Annals of Applied Probability*, vol.16, pp.2256–2271, April 2006.

[8] Z. Ding, C. W. Granger, R. F. Engle, "A long memory property of stock market returns and a new model," *Journal of empirical finance*, vol. 1, pp. 83–106, January 1993.

[9] Z. X. Wu, M. Chen, W.Y. Ye, B.Q. Miao, "Risk analysis for investment portfolio based on Copula-GARCH," *Systems Engineering-Theory & Practice*, vol. 3, pp. 45–52, June 2006.

[10] X. H. Peng, Q. Fu, "Time-varying Copula-GARCH-M-t model and portfolio's risk prediction based on the MCMC algorithm," *Journal of Applied Statistics and Management*, vol. 1, pp. 180–190, January 2013.

[11] J. Xu, X. N. Wang, Y.L. Liu, "Research on the application of DCC-GARCH, M-Copula-GARCH and Copula-SV in hedging commodity futures," *Review of Investment Studies*, vol. 4, pp. 50–64, April 2012.

Network Security and Communication Engineering – Chan (Ed.)
© *2015 Taylor & Francis Group, London, ISBN: 978-1-138-02821-0*

Big data for the Norwegian maritime industry

H. Wang & A. Karlsen
Aalesund University College, Aalesund, Norway

P. Engelseth
Molde University College, Molde, Norway

ABSTRACT: Big data is changing the business world in a profound way and successful adoption of big data analytics can increase productivity in a company. Lean production has been proven effective in reducing waste and improving productivity in a manufacturing or service process. In this paper, we briefly review the opportunities and challenges in the era of big data for the Møre maritime cluster of North-Western Norway, and propose that effective strategies combining BD and lean will enable the cluster to continue to be a global leader in the maritime industry.

1 INTRODUCTION

It is commonly agreed that we are in the era of *big data* (BD), and it is highlighted in a statement from the United Nations that "the world is experiencing a data revolution" (UN Global Pulse 2012). BD is characterized by *3Vs*: high *volume*, high *velocity*, and high *variety* (Laney 2001).

BD is changing all aspects of human society, esp. the business world. Through an extensive scientific business survey by MIT and IBM, it was detected that top performing companies are more likely to be sophisticated users of analytics than low performing companies and more likely to see their analytics are used as a competitive differentiator (Lavalle et al. 2010). Another study (Brynjolfsson et al. 2011) from MIT, after surveying 179 large companies, showed that companies that adopted "data-driven decision making" have output and productivity that is 5-6% higher than what would be expected given their other investments and IT usage. A 5% increase in output and productivity makes a significant difference of winning the fierce competition in most industries. This quantified study confirmed that the effective adoption of BD can enhance a company's decision making, insight discovery, and process optimization.

Lean production is a systemic method for the elimination of waste within a manufacturing process. Lean not only successfully challenged the mass production practices in the automotive industry, also led to a rethinking of a wide range of manufacturing and service operations (Holweg 2007). Lean principles have been widely studied and practiced in academia and many different sectors, the first book (Womack et al. 1990) that studied the lean principles has become one of the most widely cited references in operations management. Lean emphasizes on *just-in-time* (JIT) in that if production flows perfectly then there is no inventory; if customer valued features are the only ones produced, then product design is simplified.

The north-western Møre region of Norway is one of leading clusters in the global offshore supply vessel (OSV) shipbuilding and shipping (ship service) market. As pointed out in (Hammervoll et al. 2014), this cluster consists of companies representing all stages from upstream suppliers to downstream customers and this geography proximity plays an important role in the value creation process in the whole supply chain. However the globalization is counteracting the impact of this proximity. Lean is expected to increase the competitiveness, but some practices in reality can hinder its application in this particular cluster.

In this paper, we briefly review the opportunities and challenges in the era of BD for the Møre maritime cluster. Seeing the fact that JIT coincides with the high velocity of BD, we propose that effective strategies combining BD and lean can help a maritime company increase its output and productivity, and hence enable the cluster to con-tinue to be a leader in the global maritime industry.

2 LITERATURE REVIEW

2.1 *Big data*

A popular definition of BD is

Big data is high volume, high velocity, and/or high variety information assets that require new forms of processing to enable enhanced decision making, insight discovery and process optimization (Laney 2012).

The above definition is coherent with the 3Vs model for describing BD. IBM introduced a fourth 'V' – veracity, concerning about the quality of data, because biased data can only produce wrong answer.

Nowadays data are collected from all aspects in business, but the large volume of data would be of little use if it has not been effectively analysed so that insightful information can be extracted and applied in the decision making and business processes.

In the seminal paper of (Chen et al. 2012), big data analytics (BDA) is considered in the wider context of business intelligence and analytics (BI&A). With several decades' development of BI&A, we have many successful database-based technologies for structure data and web-based technologies for unstructured data. The Economist expected that the number of mobile devices, surpassing the number of laptops and PCs in 2011, will reach 10 billion in 2020. The overwhelming and fast-increasing number of mobile devices and sensors has inspired a lot of innovative technologies and created new businesses. At the same time, this new development brings a lot of value-adding opportunities to existing business sectors. E.g., the RFID technology with real-time data visibility brings instant location-awareness to commodities, vehicles, personnel and etc., which could fundamentally change how a supply chain is operated. Another example is that FedEx collects and processes data from airlines, connections hubs, location of assets, weather forecasts, and traffic information, which allows for real-time routing information to be pushed out to individual drivers and optimization of pickup/delivery and asset utilization (Løvoll & Kadal 2014). We can see that real or near-real time information delivery is one of the defining characteristics of BDA.

Another change BD brings to business is that it creates new possibilities for international development (Vital Wave Consulting 2012). With timely and correct BD analytics results, a company can have an accurate mapping of service needs so as to predict demand and supply changes globally.

A lot of BDA technologies are data-driven, but process-driven BDA like process mining (van der Aalst 2011) has emerged as a new area, highly relevant to business process improvement. Process mining makes use of event logs in e.g., health care or supply chain and enables new process discovery and conformance checking.

2.2 Lean production

Lean production originates from the *Toyota Production System* (TPS)(Ohno 1988). After the end of World War 2, Toyota could not copy American auto manufacturers' principles of mass production of autos to ensure profitability (Womack et al. 1990). There was simply not enough space in Japan to park defect new autos in an enormous lot while waiting for repair. In addition, Toyota had limited financial resources to substantiate large investments into the behemoth-type auto factories found in the USA at that time. Lean involves at core JIT manufacturing with focus on the timed technicalities of pull-based automobile production within a factory coordinated with JIT deliveries from suppliers at various stages of the assembly line. The 5 main principles of Lean are (Womack & Jones 2003):

1 *Value*: the principal starting point for lean thinking, which can be determined only by the final customer. Value is created by customer satisfaction providing the right product, for the right price at the right time. This is the essence of concept of "customer value"

2 *Value Stream*: a set of activities needed to design, produce and deliver a specific product, providing an optimal value to the customer through a full value creation process minimizing all possible waste. This is the essence of transformation purpose associated with customer value objectives.

3 *Flow*: smooth and unobstructed movement through value-creating stages. This implies how transformation should take place through production levelling.

4 *Pull*: actions taken solely to satisfy customer needs, instead of pushing often unwanted products onto the customer. This implies that production is ultimately directed by the customer.

5 *Perfection*: never ending process for improving value, value stream, flow and pull in different operations. This is in line with the statement of "zero defects" w.r.t. process output quality.

The priciple of JIT coincides with the real-time characteristics of BD, and we can expect that the emerging new BDA technologies will further enable JIT throughtout the value chain in the company level and the value network in the business cluster level.

3 OPPORTUNITIES AND CHALLENGES WITH BIG DATA AND LEAN

3.1 Møre maritime cluster of Norway

The North-Western Møre region of Norway is one of leading OSV clusters in the world, consisting of almost 200 companies with a total turnover in excess of NOK 50 billion. The cluster has a complete value chain with many leading international OSV companies like Rolls-Royce Marine, STX-OSV, Farstad and Bourbon, who all have their own development and design divisions in this region.

The core actors involved in contracting OSVs are ship operators, ship designers, and the shipyard.

Figure 1 shows the various functions found in the OSV maritime network.

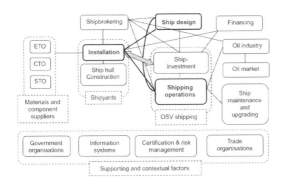

Fig 1. Maritime Network (Hammervoll et al. 2014) (The focal ship design – shipping operations – installation triad unit of analysis, indicated by bold interconnecting line; interconnectivity between ship investment and installation at a shipyard represents the functional basis for contracting an OSV ship, indicated by a double-headed arrow).

The North Sea and new oilfields in the Barents Sea, outside Brazil, and in the South China Sea present new *deep-sea* challenges like harsh weather conditions and deep sea levels.

Companies in the cluster have benefited from the close geographical proximity (Hammervoll et al. 2014). A relatively long history of previous collaboration within the cluster facilitates cooperation and effective information and knowledge exchanges, improves ship quality, and shortens lead times. However, globalization is counteracting the impact of this proximity. The turnover of contracts from shipping companies in the cluster for local design companies was reduced from 56% in 2006 to 25% in 2009. To an increasing degree, vessels now are being built at foreign locations such as Brazil, China, and Dubai. Advanced components produced in the cluster are assembled in a shipyard in Brazil or China. Shipyards constructing OSVs are becoming increasingly global, working with shipyards overseas. The largest Norwegian OSV operator now has 75 percent of its vessels in Asian and Brazilian waters. It is observed that international customers are less involved in the predesign process but more financially oriented, contrary to many local customers that are concerned about getting the best technological solutions.

3.2 Lean principles vs. reality in maritime

Facing the globalization and deep-sea challenges, cost and delivery speed are the key areas to seek improvement for the Møre cluster. Lean principles

are expected to help. However, observation of the current shipbuilding practices was presented in(Kristoffersen 2012), which seems discouraging to lean practitioners.

Here we briefly summarize some of the current practices. Local shipyards were described as having more of an artisan (instead of industrial) history and intuitive eye for shipbuilding, so local workers can understand the designers' intentions well, but this lack of industrial technical details will cause major problems if a design is to be built in a foreign yard. OSVs have long and complex production cycles, ship owners' common premature sales and constantly-changing situation of supplies often cause the specifications to be modified or even rewritten. In addition, the sales and marketing departments tend to accept orders with little consideration of their production capacity. These realities are difficult w.r.t. the "smooth flow" principle in lean. Unexpected needs to deliver very large and complex supplies are not uncommon. To fulfil customers in this manner, stocks are necessary in ship production. In addition, because a yard has to secure production capacity with suppliers, a ship order might need a slight "push" instead of the recommended "pull".

3.3 Big data for maritime

Recall the discussions in Section 2 on BDA and lean, BDA is characterised as real-time and lean emphasises on JIT manufacturing. In this light, we expect that BDA works well with the principles of lean, and hence increase productivity.

Deep seas pose challenges to both ship production and operation. New sensor technology enables new possibilities of remote ship monitoring in different situations, e.g., emerging solution like the Rolls Royce Hemos system can transmit sensor data from different components to land-based service centres. Moreover, we expect that by applying BDA on this detailed operation data, operation challenges for deep seas could be identified and prioritised so that new components and designs could be devised more quickly ("value", "pull" and "perfection").

Globalization drives OSV operators to have large number of vessels operating in different international locations. With BDA targeting to find the pattern between operation and maintenance, these companies can better predict and allocate resources for the vessels' maintenance cycle ("flow").

BDA on sensor-enabled operation data will also improve energy efficiency and environmental performance, safety verification and assessment, and monitoring of accidents and environment risks (Løvoll & Kadal 2014) ("value"). In addition, regulating bodies can use the opportunities presented by BD in the ship production and operation and introduce more quantified regulations for the administration of ships and seas ("value stream").

On top of sensor-enabled data, we should also look at the data generated from the production and operation processes. We can use process-driven BDA technologies like process mining to identify bottlenecks and challenges in the processes, and develop new (technological or managerial) methods to tackle them. By analysing external data like global maritime market, oil price, ocean environment, together with internal data like customers, employees, shipyards can better predict ship owners' needs and move towards "pull" in their ship-building. Similarly, ship owners, by analysing ship operation and customers' behaviour data, can have better idea of the requirements of their new ships, and hence less late costly modifications to ship designs; and they can better schedule their ship-purchasing activities ("smooth flow"). In addition, companies can use BDA to improve their positions in the overall value chain ("value stream").

Above discussions can be seen as that BDA enables the application of lean principles. The report of (Deloitte 2012) reminds us that it is also worth looking at the opposite direction – lean principles can help the deployment of BDA. Nowadays the volume and complexity of data and its sources are exploding, so is the provision of BDA technologies. In addition, wrong use of data or BDA technologies could generate wrong analytic results. This could easily overwhelm (or even scare away) a company, especially if this company is a newbie in the BD world. To make things even worse, the introduction of BDA incurs (normally) significant extra costs. In this situation, the lean principles can help. By narrowing down some most critical customers' needs, a maritime company can target specific data sources and choose only relevant analytic methods ("value" and "pull"). By quickly feeding back insights to the production and operation, the company can continually improve the processes ("perfection").

To sum up, we expect that strategies combining BDA and lean principles can greatly benefit a maritime company, and hence the Møre cluster.

4 CONCLUSIONS

In this paper, we look at the challenges for the north-western Møre maritime cluster of Norway, namely deep-sea and globalization. We also look at the difficulties in the reality practice w.r.t. lean principles, which is successful in many other business sectors. Seeing the fact that the JIT principle of lean coincides with the high velocity of BD, we propose that an effective strategy combining BD and lean can help a maritime company increase its output and productivity, and hence will enable the cluster to continue to be a global leader.

REFERENCES

[1] Van der Aalst, W., 2011. *Process Mining*, Springer.
[2] Brynjolfsson, E., Hitt, L.M. & Kim, H.H., 2011. Strength in Numbers: How Does Data-Driven Decisionmaking Affect Firm Performance? *SSRN Electronic Journal*.
[3] Chen, H., Chiang, R.H.L. & Storey, V.C., 2012. Business Intelligence and Analytics: From Big Data to Big Impact. *MIS Quarterly*, 36(4), pp.1165–1188.
[4] Deloitte, 2012. *Big data: Time for a lean approach in financial services.*
[5] Hammervoll, T., Lillebrygfjeld Halse, L. & Engelseth, P., 2014. The role of clusters in global maritime value networks. *International Journal of Physical Distribution & Logistics Management*, 44(1/2), pp.98–112.
[6] Holweg, M., 2007. The genealogy of lean production. *Journal of Operations Management*, 25(2), pp.420–437.
[7] Kristoffersen, S., 2012. Nextship - Lean Shipbuilding. *Molde University College*.
[8] Laney, D., 2001. 3D Data Management: Controlling Data Volume, Velocity and Variety. *Gartner Inc.*
[9] Laney, D., 2012. The Importance of "Big Data": A Definition. *Gartner Inc.*
[10] Lavalle, S. et al., 2010. Analytics : The New Path to Value. *MIT Sloan Management Review and IBM Institute for Business Value.*
[11] Løvoll, G. & Kadal, J.C., 2014. Big data - the new data reality and industry impact. *DNV GL.*
[12] Ohno, T., 1988. Toyota *Production System: Beyond Large-Scale Production*, Productivity Press.
[13] UN Global Pulse, 2012. Big Data for Development : Challenges & Opportunities. *United Nations.*
[14] Vital Wave Consulting, 2012. Big Data , Big Impact : New Possibilities for International Development. *World Ecnomic Forum.*
[15] Womack, J.P. & Jones, D.T., 2003. *Lean thinking: banish waste and create wealth in your corporation.*, Simon and Schuster.
[16] Womack, J.P., Jones, D.T. & Roos, D., 1990. *The machine that changed the world: The story of lean production*, Free Press.

Network Security and Communication Engineering – Chan (Ed.)
© 2015 Taylor & Francis Group, London, ISBN: 978-1-138-02821-0

A business metadata modeling approach for data integration based on SDMX

H. Ge & R.N. Rao
Distributed Computing Lab, Shanghai Jiaotong University, Shanghai, China

ABSTRACT: In data integration, business metadata can help technical metadata guide integration processes as well as provide business rule supports. SDMX provides a common model to describe data and it is suitable to construct business metadata. The paper is against the background of a provincial level transportation and logistics project, proposes a business metadata modeling approach based on SDMX information model, and presents a metadata management application to maintain mapping relations between technical and business metadata in data integration. Experiment results show that this metadata management application helps effectively manage transportation and logistics data and increases the efficiency of data integration.

1 INTRODUCTION

A massive growth of data is coming about, which leads to various information systems being used to manage data that is stored among enterprises with heterogeneous data structures, and therefore data integration is required. However, most of the data integration tools focus on the technical metadata which lacks consideration of business logic. To give technical metadata business meaning, so far some measures have been taken: using data dictionary to define universal data structures on logistics domain, using such data structures to rebuild enterprise databases. However, these approaches are not efficient since the cost of rebuilding databases cannot be ignored and is not scalable. To eliminate this problem, an effective management of business metadata as well as the mapping relations between technical metadata and business metadata is urgently needed.

Business metadata has been used to describe data and its lifecycle in business logic perspective to support various information systems, and it proved to be an effective way in managing data (Arun Sen, 1989). This paper is against the background of a provincial level transportation and logistics project, proposes a business metadata modeling approach based on SDMX information model which fully describes the structure of related business data; presents a metadata management application to manage this business metadata and maintain mapping relations between technical and business metadata in data integration; and finally it makes an evaluation of this application.

2 RELATED WORKS

2.1 Existing standards

At present, many different metadata schemes are being developed as standards in different fields, such as science, education, media, and finance (Baca, 2002).

CWM focuses on data warehousing (John Poole, Dan Chang, Douglas Tolbert & David Mellor, 2002) to enable easy interchange of warehouse and business intelligence metadata in distributed heterogeneous environment. RDF focuses on web resources (W3C, 2014), using a variety of syntax formats to describe conceptual description or modeling of information implemented in web resources. Dublin Core Metadata is used to describe web resources as well as physical resources (Dublin Core, 2013).

2.2 Existing researches

Research (Zhang Yu, Jiang Dongxing & Liu Qixin, 2009) proposed a flexible metadata standard to consolidate the utilities of heterogeneous datasets in distinct systems to achieve highly integrated information retrieval. Research (Zhonggui Ma, Chengyao Wang & Zongjie Wang, 2011) built a three-layered metadata model, including data source metadata, business metadata and topic metadata to realize heterogeneous data integration and improve efficiency and accuracy for data querying by metadata. Paper (Yagiz Kargin, Milena Ivanova, Ying Zhang, Stefan Manegold & Martin Kersten, 2013) proposed an ETL method called Lazy ETL, which extracts and loads metadata of a data source to give metadata views to users and for the coming query, only the required data items are extracted, transformed and loaded to lower costs of initial loading.

However, these researches are either made for general purposes, which are lack of pertinence, or designed as industry-specific tools, they may only work in particular fields like earth science. Transportation and logistics metadata management and data integration requirements quite differ from other fields, thus these researches may not be suitable among logistics enterprises.

2.3 *SDMX information model*

SDMX provides a common model to describe data and to structure the statistical content. It has provided a way to model statistical data, and has defined a set of metadata structures to support it as well. SDMX information model defines data structures by using concepts-including dimensions, attributes and measures- relationships, constraints and rules to describe data meaning and using code lists to enumerate the possible values of data (SDMX, 2011).

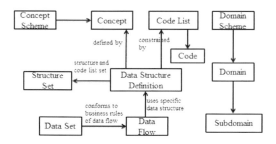

Figure 1. Information model elements.

As shown in Figure.1, a domain represents a specific theme like comprehensive law enforcement theme or credit theme in transportation and logistics filed, and it is composed of several subdomains such as enterprise information, vehicle information and cargo information. Business metadata in each domain is defined in data structures together with concepts and code lists. A data set contains business data and needs to conform to business rules of data flows while a data flow must use a specific data structure to organize data.

3 ANALYSES FOR BUSINESS METADATA

3.1 *Information model elements*

Transportation and logistics data involves numerous business domains from transportation infrastructure

construction in the social aspect to logistics order tracking in the commercial aspect, as well as logistics enterprise production information management in the management aspect. To model such data, we need a full view of transportation and logistics field before picking out key domains, and then we should find out concepts to describe the data ,so we can create data structure definition. Data set can be built using this data structure definition.

In transportation and logistics field, a domain represents for a specific theme, we have picked out 11 business domains:

- Transportation and logistics infrastructure construction theme
- Illegal transportation record theme
- Comprehensive law enforcement theme
- Transport administration service theme
- Decision support data service theme
- Order tracking theme
- Logistics enterprise production information management theme
- Vehicle maintenance and rescue theme
- Freight hub node information theme
- Transport information service theme
- Credit information service theme

Within these business domains 8 sharing subdomains are built, which provide sharing data among enterprises and organizations:

- Vehicle information
- Vehicle technical file
- Driver information
- Cargo information
- Cargo Type
- Companies
- Packaging information
- Order information

Figure.2 is a simple example of an order in order information subdomain. The trading took place in May 1, 2014 between company A and company B. The trading goods contain 10 televisions with $200 for each and 2 music CDs with $30 for each. The total price of the order is $2,300.

Order			
No 100000000001		Date	2014-05-01
Seller Company A		Buyer	Company B
Goods	Quantity		Unit Price
Television	10		$200
Music CD	2		$30
Total	$2,300		

Figure 2. An order example.

Concepts are used to describe this order, thus we can first pick out 8 concepts easily seen in the order form: 1) Order Number, 2) Trading Date, 3) Seller, 4) Buyer, 5) Goods, 6) Quantity, 7) Unit Price and 8) Total Price. Besides these, a formal order must have its customs rules of data, so we supplement the above concepts with 9) Date Format and 10) Price Format.

Dimensions are used to describe the data and form the identifier of the related data and the following dimensions can be detected: Order Number, Trading Date, Seller, Buyer, Goods, Quantity, Unit Price and Total Price, as we can observe an order through any of the above concepts and get different views such as "How many orders were created in May 1, 2014" or "How many orders were made in company A in the year of 2013". Meanwhile, attributes will contain additional information of the data and following concepts are identified as attributes in this example: Date Format and Price Format.

In order to be able to share and understand data, it needs to be declared, what the possible values for each concept are. Besides, the possibility of defining text formats for free text values is usually used for attributes, the commonly used approach is to attach lists of values.

Where possible, the values for code lists are taken from international standards, such as those proposed in ISO-code lists. We decide code lists for attribute Date Format and Price Format using ISO time code list and ISO currency code list, as shown in Figure.3.

Code List: Date Format		Code List: Price Format	
Code	**Description**	**Code**	**Description**
P1Y	Annual	CNY	China Yuan Renminbi
P1M	Monthly	HKD	Hong Kong Dollar
P7D	Weekly	TWD	Taiwan New Dollar
P1D	Daily	USD	United State Dollar

Figure 3. Code lists for attributes.

Generally, trading date is recorded in the format "YYYY-MM-DD" according to ISO standard, however, if the trading continuously takes place, such date should be a duration, which we use codes like P1Y or P1M to represent duration time. Besides, we may find that the trading only allows for 4 types of currency, thus we use currency code list to set price format value like CNY for China Yuan Renminbi.

3.2 The Common XML expression

Data structure definition provides structural metadata of a data set, using concepts identified above associated with code lists. This section discusses the common XML expression of data structure definition.

3.2.1 Expression of concepts
Part of the concepts will be defined as follow:

<Concept agencyID="DATACENTER" id="Order_Number">

<Name xml:lang="en">Order Number</Name>

</Concept>

<Concept agencyID="DATACENTER" id="Date">

<Name xml:lang="en">Trading Date</Name>

</Concept>

<Concept agencyID="DATACENTER" id="Seller">

<Name xml:lang="en">Seller in trading</Name>

</Concept>

<Concept agencyID="DATACENTER" id="Buyer">

<Name xml:lang="en">Buyer in trading </Name>

</Concept>

<Concept agencyID="DATACENTER" id="Goods">

<Name xml:lang="en">Name of goods </Name>

</Concept>

A data structure definition is defined by an enterprise or an organization, thus it must contain its own agencyID. We use "DATACENTER" to represent logistics enterprise data center.

Each concept is defined with the ID of agency responsible for the concept, the concept id, and a language-dependent description for the concept.

3.2.2 Expression of code lists
Part of the concepts will be defined as follow:

<CodeList agencyID="DATACENTER" id="CL_Price">

<Name xml:lang="en">Price code list</Name>

<Code value="CNY">

<Description xml:lang="en">China Yuan Renminbi</Description>

</Code>

<Code value="HKD">

<Description xml:lang="en">Hong Kong Dollar</Description>

</Code>

<Code value="TWD">

<Description xml:lang="en">Taiwan New Dollar</Description>

</Code>

<Code value="USD">

<Description xml:lang="en">United States Dollar</Description>

</Code>

</CodeList>

A code list contains the ID of agency, the code list id, a language-dependent description and a set of code values, which define the possible values taken by the dimensions and the coded attributes. Here, the code lists provide the value of attribute "Price_Format".

3.2.3 *Expression of data structure*

In a data structure definition, key family is used to combine concepts and code lists together and classify concepts into dimensions and attributes as well.

```
<KeyFamily          agencyID="DATACENTER"
id="EG1" >
    <Name xml:lang="en">OrderInformation</Name>
    <Components>
        <Dimension conceptRef="Order_Number" />
        <Dimension conceptRef="Date" />
        <Dimension conceptRef="Seller" />
    <Dimension conceptRef="Buyer" />
    <Dimension conceptRef="Goods" />
    <Dimension conceptRef="Quantity" />
    <Dimension conceptRef="Unit_Price" />
    <Dimension conceptRef="Total_Price" />
    <Group id="Group">
        <DimensionRef>Date</DimensionRef>
            <DimensionRef>Seller</DimensionRef>
    </Group>
    <Attribute    conceptRef="Date_Format"    codel-
ist="CL_Date" assignmentStatus=" Mandatory "/>
    <Attribute    conceptRef="Price_Format"    codel-
ist="CL_Price" assignmentStatus="Mandatory"/>
    </Components>
</KeyFamily>
```

The components of data structure definition are defined in key families starting with dimensions. Each dimension contains reference to a descriptor concept and the code list if needed, and some of these dimensions are grouped together to provide a specific perspective of data.

Attributes are listed after dimensions and groups. Each attribute should be declared where it is mandatory or conditional. As same as dimensions, attributes also contain reference to concepts and the code list if needed.

4 DESIGN OF METADATA MANAGEMENT APPLICATION

4.1 *System architecture*

The architecture of metadata management application includes four layers: data layer, logical layer, interface layer and web UI, which is shown in Figure.4. Data layer provides metadata repository to store business metadata, technical metadata and their mapping relations; Logical layer provides functional realizations to manage metadata stored in repository; Interface layer provides APIs of logical functions to web UI; and Web UI allows users to visually build business metadata, import technical metadata or configure their relations. The illustration of each layer is shown as follow:

Data layer: Business metadata together with technical metadata imported from data sources and their mapping relations are stored in MySQL to build metadata repository. Agencies are in charge of all other elements which they are responsible for. Domains and subdomains together provide a logical view of transportation and logistics themes. Concepts and code lists are separately defined and bound to the data structure definition in designing period.

Besides, data structure definition version information is stored directly and is used to build appropriate view of historical metadata, allowing users to trace the changes among different versions.

Logical layer: This layer mainly provides realizations of metadata management functions, including business metadata management, technical metadata management, mapping management and data source access functions.

Interface layer: This layer abstracts logical layer and provides interfaces to both web UI and external applications such as ETL systems and transportation systems. These interfaces include business metadata management interface, technical metadata management interface, mapping management interface and data source access interface.

Web UI: Web UI provides visual components and allows users to build business metadata, import technical metadata from external databases and map them to construct data structures with business meanings.

4.2 *Metadata management functions*

4.2.1 *Technical metadata import*

Technical metadata is always stored in local databases of logistics enterprises to describe their technical data of the information system. By importing technical metadata of these enterprises, we will be able to take charge of their data structures and provide it to external applications like ETL tools for further usage.

To import technical metadata from heterogeneous data sources, we need to build an adapter that can automatically load database connection driver and get metadata of the database or schema information whatever different techniques users choose like "Oracle" or "MySQL". Technical metadata is stored in repository by separating tables and columns with their own meta-info like column name or data type.

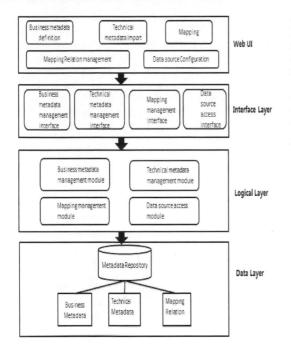

Web UI

Interface Layer

Logical Layer

Data Layer

Figure 4. Architecture of metadata management application.

To protect the privacy of data, transportation and logistics enterprises may deny accesses outside their local network, thus we design that technical metadata in our repository should be accessible through interfaces by external applications. One example is that ETL tools will read data structures of data sources from our metadata repository instead of directly reading them from local databases of logistics enterprises.

4.2.2 Business metadata modeling

As discussed above in Chapter 3, we separate modeling steps according to information model elements. Thus we need to manage all the elements used to describe transportation and logistics business metadata, and at the meantime we must maintain all relationships among these elements. For example, a codelist must contain one or more codes to describe an attribute and a code may be used in many codelists, so we use a many-to-many mapping to maintain such relationship.

Besides, like providing technical metadata to external applications, business metadata should also be accessible resources for data sharing and data communication. All data with heterogeneous structures will have their business meaning through business metadata, so business metadata should be the bridge to connect different enterprise data and provide communication approach to each other.

4.2.3 Mapping configuration

Mapping between technical metadata and business metadata builds the connection that data in physical storage like database will have high-level business meaning, thus it is more adaptive in business situations.

We implement mapping function by designing interfaces to relate database columns with concepts including dimensions and attributes. Usually, attributes are used to supplement dimensions. Figure.5 gives a mapping instance.

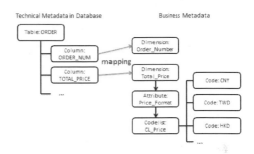

Figure 5. Mapping of columns in "OEDER" table in database between dimensions "Order_Number", "Total Price" and attribute "Price_Format".

4.2.4 Data integration based on business metadata

Traditionally, ETL tools provide name mapping between source data store and target data store in job design steps. However if the data stores are different in structure or naming among logistics enterprises, name mapping will not work at all. Therefore, we use the business metadata to solve this problem.

As shown in Figure.6, before an ETL job is designed, we first map technical metadata to business metadata to give data in heterogeneous sources a unique business meaning. When we start designing ETL jobs, such mapping works that data columns or items bind to what share the same business meaning.

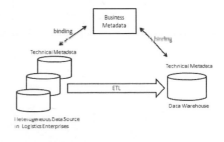

Figure 6. Steps of ETL processes.

Besides, since we use business metadata to bind technical metadata, data from various logistics enterprises can be traced wherever they flow to in ETL processes and it helps to take charge of the data quality.

For this purpose, we make some efforts in an open-source ETL tool as follow:

1 Adding metadata management supporting to ETL tool;
2 Modifying source and target data mapping using business metadata instead of name mapping;
3 Using transportation and logistics business instance to verify the business metadata.

5 EVALUATION

In this section, we evaluate the feasibility of our business metadata structure by using the metadata management application in data management and data integration.

Table 1. Usage of the application.

Index	Description
Running Time	more than 3 months
Service Status	all in good condition
Data Exchange	35,000 per day, more than 11 million in total
Integrated Data Item	1 million per day both from enterprises to data center and from data center to data analysis systems

The metadata management application we propose in this paper is being applied in a transportation and logistics cloud platform. It provides metadata management service interfaces to more than 1200 logistics enterprises for daily data processing, and supports data exchange and integration among cloud data center and these enterprises.

Table.1 gives the detail information about the usage of the metadata management application including application running time, the status of services providing metadata management functions to enterprises, data exchange amount among information systems and data item processed in data integration.

Figure.7 shows the efficiency improvement of data integration based on business metadata. The

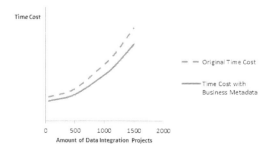

Figure 7. Efficiency improvement.

application has reduced the project configuration time by using business and technical metadata mapping to automatically configure the relations between source and target data.

Experiment results show that in real practices, the application is able to satisfy the daily requirements of data management and improve the efficiency of data integration among enterprises, and it proves to be well-worked. Therefore, our business metadata structure proposed in this paper is feasible and well-designed.

6 CONCLUSIONS

In this paper, we propose a business metadata modeling approach based on SDMX information model, and present a metadata management application to manage both technical and business metadata as well as their mapping relations. The information model defines the key parts of business metadata and provides a logical view of transportation and logistics business world,it proves to be well-designed through the daily use of metadata management application in real practices among logistics enterprises.

REFERENCES

[1] Arun Sen, Metadata management: past, present and future, Decision Support Systems 37 (2004) 151–173.
[2] Baca, M. (ed) (2002), Metadata Standards, Library Technology Reports, 38, (5), 19–41.
[3] Dublin Core Metadata Element Set, Version 1.1, Dublincore.org, 2013.
[4] John Poole, Dan Chang, Douglas Tolbert, and David Mellor (2002). The Common Warehouse Metamodel: An Introduction to the standard for Data Warehouse Integration. OMG Press (John Wiley & Sons), 2002 ISBN 0-471-20052-2.
[5] Resource Description Framework (RDF) Model and Syntax Specification: http://www.w3.org/TR/PR-rdf-syntax/.
[6] SDMX Technical Specification: http://sdmx.org/?page_id=10.
[7] Yagiz Kargin, Milena Ivanova, Ying Zhang, Stefan Manegold, Martin Kersten, Lazy ETL in Action ETL Technology Dates Scientific Data, Proceedings of the VLDB Endowment, Vol. 6, No.12, 2013.
[8] Zhang Yu, Jiang Dongxing, Liu Qixin, Metadata-based integration scheme for heterogeneous datasets. Journal of Tsinghua University(Science and Technology), VOL 49, NO 7, pp. 1037–1040, 2009.
[9] Zhonggui Ma, Chengyao Wang, Zongjie Wang, Research on three-layered metadata model for oil-gas data integration, proceedings of 2011 IEEE International Conference on Cloud Computing and Intelligence Systems (CCIS2011).

Network Security and Communication Engineering – Chan (Ed.)
© 2015 Taylor & Francis Group, London, ISBN: 978-1-138-02821-0

SQL injection attacks manual steps and defense

Y.L. Wang & B. Zhang
Xuchang Vocational Technical College, Xuchang, Henan, China

ABSTRACT: SQL injection attack is currently the most common method of attacking from hackers. This paper describes the definition of SQL injection attacks, analyzes the principle of injection attacks and SQL injection and makes a comprehensive analysis about SQL injection attacks steps. For most sites the SQL injection vulnerabilities and constraints of strict filtering problems are presented, from webmasters and Web application, two measures of defense SQL injection attacks are given.

KEYWORDS: Page, SQL injection, Attack step, Defense.

1 INTRODUCTION

With the development of B/S mode of application, a considerable number of programmers are writing code, with no input data to judge the legality of the user, so the application is under security risk. By submitting elaborated database query code and get a reply, the hacker can get the data they wanted. This is SQL Injection.SQL injection is a sensitive database information which may have been illegally viewed or deleted, and may even make the server been controlled by hackers and become "chicken".

2 SQL INJECTION PRINCIPLE

2.1 *Injection attacks principle*

The main reason for injection attacks is because the distinction between the program commands and user data is not strict [2]. An attacker has the opportunity to program commands entered by the user as data submitted to the Web page; the attack is after the illegal access to confidential information.

To launch injection attack, an attacker would need to be interpreted as a mixed command "data" in the conventional input. To make this work, three things must be done first:

1 Determine the technology used in Web pages

Injection attacks on the programming language or hardware closely. In order to determine the technology used, the attacker can inspect the footer of a Web page to view the error page, check the page source code or use a tool such as Nessus for spying.

2 Determine all possible input

Input ways used by users are very easy to find on the Internet, such as HTML forms. In addition, the attacker can hide the HTML form input, HTTP header, cookies, and even being invisible to the user's back-end AJAX requests to interact with the Web application.

3 Find a user input used for injection

To find out all user input in a proper way, it has to be screened for these input methods to identify those which can be injected into the command input mode. This task seems a little difficult, but if you look carefully at the error message on the attacker's Web page, you can often get what you want.

2.2 *SQL injection principle*

Because SQL injection is from the normal port access WWW pages [3], thus it enables an attacker to bypass the authentication mechanism by using the loophole which is caused by some pages of the target site's missing or uneffective control over the pass parameters. The attacker achieves his goal to acquire, modify, delete data and even control the database server and the destination Web server. In this attack, the attacker will insert some malicious code into an alphabetic string. Then the string is passed to an instance of SQL Server database through a variety of means for analysis and execution. As long as the malicious cold is in line with the rules of the SQL sentences, then, when the code is compiled and executed, the system will not be found. So that remote users can not only input data to a Web page, but also perform arbitrary commands on the database.

Most of Web application architecture has the MySQL database to store the relevant data. Because Web applications use SQL statements to manipulate MySQL database, and form user data in his browser, input, Web applications will use the $_POST variable to save. Hackers use the established pattern of this web design, adding special code in the form of data input of the client to change the original SQL statement. If the web application system directly submits SQL statement without checking the value of $_POST variables, the SQL injection attacks will start.

3 SQL STEP INJECTION ATTACKS

At present, in most of the web applications, users are provided with an interface to input data to verify permissions or search information and so on, such as the common online banking application, trading sites, etc [4][5]. Thus, there is a big defect injection vulnerability. There may have been injected into the hacker attack, so that the background data is completely exposed to the hacker. SQL injection steps are shown in Figure 1.

1 Search for possible SQL injection vulnerabilities link

General press systems, forum, guestbook and photo album have the similar links as list.php? Id= such links, the form can be judged from the calling SQL statement to query the database and display it, if this list. php page on the back of the parameter filter bad judgment on the legality or lack of user input, then it is possible SQL injection vulnerabilities. Hackers often try to find similar sites after the invasion by search engines and other tools.

2 Test whether the web sites have SQL injection vulnerabilities, determine the type of database

After finding available SQL injection vulnerabilities link by entering the special statement, you can return information based on the browser to determine whether there is a SQL injection vulnerability. Digital type, character fields, namely at the end of the link to join "and 1 = 1" and "and 1 = 2", " 'and' 1 '=' 1" and " 'and' 1 '=' 2", if the former normal return, which returns "no content of the column will be Back" message, indicating that the site must exist SQL injection vulnerability, there is a great chance you can get the administrator's user name and password.

After finding available SQL injection vulnerability page, by entering the special statement, you can return based on the browser information to determine the type of database, by building "and (select count (*) from msysobjects)> 0" statement, if returned to insufficient permissions information, you can determine that access databases; build "and (select count

(*) from sysobjects)> 0" statement, if it returns to normal, you can judge a MSSQL database. Further build "and db_name ()> 0" statement to return the database name.

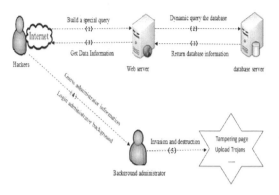

Figure 1.   SQL injection process.

3 The administrator account guesses the table name and the table and the length of the field

Many programmers in the design of the database will use some specific names as table names or field names. Table names, field names that stored in the database follow the laws and hackers' statement by building a special database administrator account in order to find a table names, table field names, user names, and passwords of length and content. This speculation process can be injected through a large number of online tools to quickly achieve and combine with other hackers to gather useful information. It is so easy to decipher the user's password which can also be added manually after SQL injection vulnerability in a special statement to crack the URL link. Hackers by constructing "and (select top 1 name from sysobjects where xtype = 'u')> 0", to get the first table name, such as user, on this basis, re-construction "and (select top 1 name from sysobjects where xtype = 'u' and name not in ('user'))> 0 ", then get a second table, the back and so on.

After Guessing the table name to the administrator account, you need to guess the table column names. Using the system comes with two functions namely col_name () and object_id (), to build such "and (select top 1 col_name (object_id ('table name'), 1) from sysobjects)> 0", to get the first table field column names, and so on, to get passwords and other column names.

Guessing the column length can build "and (select top 1 len (column name) from table name)> N" statement, where N is a number, Changing the value of N to guess this column length, when N is 6 it is correct while when N is 7 it is wrong, we guess the length is 7 . To guess the second record, we need to use the statement "select top 1 len (column name) from table where column name not in (select top 1 column name from the table name)".

4 Guess user names and passwords, to find Web Management Entrance

After successfully guessing the administrator account and password fields, cracking specific user name and password are followed. By constructing such as hackers and 1 = (select administrator id from (select * from table where an administrator id = 1) where asc (mid(column name,1,1)) <100) commands. A gradual guess solve a specific user name, in accordance with the above principle, (asc(mid (column name,1,1) column name in the previous statement replaced by PASSWORD, the corresponding password can be got. After obtaining a username and password, you can make use of some injection tools easily to manage and get backstage entrance into the site's backend management system for the invasion and destruction.

As a hacker, clear user is a built-in variable SQL Server. Its value is the user name of the current connection type is nvarchar. The hackers know clearly that user is a built-in variable SQL Server; its value is the current user's name. The type is nvarchar, so sometimes by constructing and user> 0 order, according to the error message SQ LServer, you can not waste Chuihuizhili in order to get the user name of the database.

5 The invasion and destruction

After the successful landing site background management system, then you can carry out any acts of vandalism, such as tampering with web pages, upload Trojans, leaving the back door, modifying data, user information leakage and further invasion of the database server.

4 SQL INJECTION PREVENTION

Because SQL injection attacks are from normal WWW port access, and superficially, it looks like the common web page, the loophole caused mainly by no strict filtered data and less restricted pass parameters submitted by the user. Therefore, with the development of webmasters and Web application, programmers must pay enough attention to the following steps in order to avoid the possible SQL injection attacks:

1 The data submitted by the user input parameters and strict filtering

When designing a Web page, numerical parameters, to judge whether it contains illegal characters for character parameters, such as the need for commas, single quotes, double quotes, semicolons, and other special symbols restrictions and filtering, if select, delete, from, *, string union appear like multiple simultaneous, they should pay attention to the length of the user submitting. It will be better if the length of the parameters can be judged. The program gives an error message whenever any unauthorized program appears.

2 Set permissions on the database server, try not to let the Web page identify the super administrator connection data, unless there are special needs, do not grant the permission to read the system tables and execute system stored procedures. As to user tables, some permission should be strictly considered, do not grant update, insert, and other privileges to only read operations.

3 Abandon dynamic SQL statements, Use user's stored procedures instead to access and manipulate data. After the establishment of a database, a careful analysis of the various operations of the Web page requires the database, and a careful analysis of the various operations of the Web page requiring on the database, one can build stored procedures, and then let the Web page calls the stored procedures to perform database operations. Thus, the data submitted by the user will not be used to generate dynamic SQL statements, but really and actually pass parameters to stored procedure as, effectively blocking the way to SQL injection.

4 Turn off or remove unnecessary interactive submission form page, the code layer to shield the script unsafe and dangerous character, so as to effectively prevent certain injection attacks.

5 The site administrator should patch and strengthen data timely and disable unnecessary services and functions. Meanwhile they should active database monitoring system and use tools or equipment to detect attacks on Web pages. Early prevention is very important.

5 CONCLUSION

With the popularity of the rapid development of network industries and Web applications, SQL injection techniques become the network's most popular hacking techniques. More and more hackers use the technology of the website or server attacking, causing leakage and confidential data loss, and the danger is very great. As long as webmasters and Web page designers understand SQL injection techniques, and pay enough attention to the security, we can maximally avoid being attacked.

REFERENCES

[1] Nan,Chen. 2005(01).SQL injection attacks and prevent the realization.Information Security and Communications Privacy:26–27.

[2] Yinhao, Xu.2009.Security Research and SQL Server, SQL injection. East China Normal University: 212–215.

[3] Guobiao, Wu. 2010(04).SQL injection attacks Principle and Prevention. Journal of Shaoxing University (Natural Science):98–99.

[4] Yun.wang.Waiping,Guo. 2010(05):Research on SQL injection attacks and guard method in web project. Computer Engineering and Design: 976–977.

[5] Jian,Huang. 2012(04).computer information security technology and pretection.Information Security and Technology:83–85.

412

Network Security and Communication Engineering – Chan (Ed.)
© *2015 Taylor & Francis Group, London, ISBN: 978-1-138-02821-0*

Real estate appraisal model based on Ann and its application

W.Q. Wang, W.H. Dong & X.D. Liu
Donlinks School of Economics and Management, University of Science and Technology Beijing, Beijing, China

ABSTRACT: With the establishment and development of real estate market, the demand for real estate appraisal is growing. The accurate appraisal of the fair value of real estate has a significant influence on the practice of real estate investment. As the inefficiency and inaccuracy of appraisal method of traditional estate appraisal theory, the real estate appraisal system based on BP neural network is established. Trade cases are trained and tested with the help of MATLAB software. The main aims of this paper are shown below: The enriching of the index system of the influence factors in the real estate appraisal; the collection data of several USA real estate investment project; evaluating above data using the traditional method and artificial neural network method; providing a new operation method for the assessment of real estate by the establishment of the real estate evaluation model based on artificial neural networks. The results show that the accuracy and objective of real estate appraisal are improved.

KEYWORDS: Real estate appraisal, traditional methods, bp artificial neural networks, MATLAB.

1 INTRODUCTION

The Real estate assessment is highly complex, and its accuracy can be influenced by data collection techniques and various types of errors. Traditional methods of appraisal include market method, cost approach and income method. Basic principle of market method is substitution principle, through which real estate price is estimated by selecting comparable real estate and modifying different index. Market method is widely used with high credibility, but in practice the various indicators of each real estate are rarely exactly the same, so they often need to be corrected, hence the market is affected by man-made factors greatly. The income approach is one of present value method, which is future income oriented. Its reliability depends on the reasonable assessment of the future earnings and capitalization rate whereas the uncertainty nature of future earnings method results in the inherent defects of the income approach. In cost method, real estate is valued by replacing cost, however the cost is not equal to the market value which must consider the market supply and demand, and therefore the cost value collectively refers to the intermediate results for complement.

For the inefficiency and inaccuracy of traditional appraisal method, the current real estate appraisal system needs to be improved. Therefore, this paper introduces the real estate assessments method through BP neural network and assesses the value of target real estate based on the principles of market comparison approach.

Artificial neural network is a nonlinear processing tool, without having to set the logical distribution of relationships between variables in advance. Nonlinear parallel system achieves the purpose of learning and memorizing by adjusting the input values and the desired output value of black-box process and then enters data to the trained neutral network and gets solutions. BP neural network can simulate the human brain information processing model, conduct nonlinear process to complex factors as well as parallel information, and form a comparable relationship. Artificial neural network collects a large number of training examples and memorizes it, which is similar to the appraiser's dealing instance memory.

Real estate prices are influenced by many factors, including market conditions, location factors, physical characteristics etc. Through detailed analysis of the above 3 factors, 26 sub-layer metrics are used in this paper. In practical analysis section, we use the above-mentioned indicators as benchmarks to quantify the comparable transaction instance data. In the practice section, we collect 24 real estate transaction instances for training and 6 for testing. After normalizing the variables, the real estate influence factors are set as input variables while the price is set as one of output variables. The BP neural network in MATLAB neural network toolbox provides BP function through which we can train and save these data and the real estate valuation can be conducted.

2 BP ARTIFICIAL NEURAL NETWORK STRUCTURE

2.1 Neural network model design

BP (Back-Propagation, BP) networks are multilayer feed forward neural networks, which consists of three layers: an input layer, a hidden layer and an output layer. Every node represents an artificial neuron. The basic principle of BP neural network is the steepest descent method, and the network is learned by setting the input value, output value and the range of allowable error. With a gradient search technique, the error passes backward and its weight can be constantly revised. Hence the differences between actual output and the expected output can be reduced gradually. The characteristic of BP neural network is information propagation and error back propagation, where information passes through the input layer into the hidden layer to the output layer after processing. If it does not conform to the desired output, the error will give feedback to the upper layer to adjust the weight repeatedly to achieve acceptable error. For example, a BP neural network structure in which the input layer has n nodes, hidden layer has j nodes and output layer has m nodes is shown as figure 1.

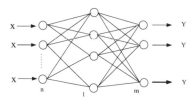

Figure 1.   3-layered BP network model.

2.2 Parameters design

2.2.1 The variable of input layer and output layer
The input variables must be non-linear between each other which have the largest effect on the output result. The different areas of real estate prices are not comparable, and the collected transaction example are almost rental property, so in this case, we choose 26 influencing factors as the input variables, and the real estate average rental price as the output variable.

2.2.2 The hidden layer design
The Kolmogorov theorem states that for any continuous functions in closed interval, single implicit layered BP network can be used to approximate, namely in the structure with appropriate weights, m dimension can be mapped with three layered BP network. So we will usually design hidden layer as 1 layer.

2.2.3 The hidden nodes determine
The number of nodes in hidden layer is essential to the model. The node numbers of hidden layer selection

are closely related to specific problems. And for different problems, there are different suitable numbers. To determine the specific values, the following formula can be referred to:

$$a = \sqrt{n + m} + b$$

where a stands for the number of hidden nodes; n stands for the number of input layer nodes; m stands for number of output layer nodes; b is a constant between 1 and 10.

2.2.4 The function selection
In order to ensure the system is nonlinear, in general using sigmoid function as the activation function is as follow:

$$f(u) = \frac{1}{1 + e^{-u}} .$$

The output layer transfer function is purelin, and the training function is the momentum gradient descent method traingdm.

2.2.5 The initial weight selection
Weight is of great influence on network. If an initial value is too large, the weighted input value will fall into saturated zone of activation function, which will lead to small derivative values. Due to the weight adjustment formula, derivative values tend to 0 which makes weighted input values to be close to 0. The adjustment process could stop. Small numbers are hence selected within (-1,1) interval by random to ensure that the neural weight adjusts in the most significant region of S activation function.

2.2.6 The learning rate selection
The learning rate determines the convergence rate of BP neural network. We generally select the values in (0.01, 0.9) interval.

2.2.7 The selection of expected error
The expected error of neural network is an appropriate value after training for many times. In practical part, we use several different expected errors to train the networks and then choose the most appropriate one as the expected value.

2.2.8 The data standardized processing
In order to eliminate the difference between the magnitude, data needs to be normalized to [0,1]. The prediction results fall in [0-1], which need to be anti-normalized to recover the initial unit. The prediction result is more intuitive.

For the positive index, the formula is as follows:

$$x_k = \frac{x - x_{min}}{x_{max} - x_{min}}$$

For the negative index, the formula is as follows:

$$x_k = \frac{x_{max} - x}{x_{max} - x_{min}}$$

where x_k is processed data, x is raw data, x_{min} is a minimum value for similar indicators, x_{max} is the maximum one.

2.3 Training process

The trained data is inputted into the MATLAB training, and training process are as follows:

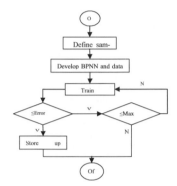

Figure 2. The BPNN training process.

3 EMPIRICAL ANALYSIS ON THE EVALUATION OF REAL ESTATE

3.1 Data preparation

This paper collects total 32 cases of real estate examples in Manhattan, Oakland, Washington for empirical analysis. Macro market factors and regional factors are collected from the Bureau of labor statistics and U.S. Department of Labor, and entity information are collected from the related real estate website. Excluding the impact factors of the overall real estate market, we retained 26 indicators and number the index using I_1 to I_{27} for convenience which is as following:

Table 1. Index set.

The index	No	The index	No
CPI	I_1	Educational facilities	I_{15}
House prices	I_2	Community safety	I_{16}
Rental price	I_3	Surrounding landscape	I_{17}
Energy price	I_4	Structure	I_{18}
Business condition	I_5	Elevator status	I_{19}
Housing expenditure	I_6	Intelligent degree	I_{20}
Income (hourly $)	I_7	Decoration	I_{21}

Unemployment rate	I_8	Parking lot	I_{22}
Safety factors	I_9	Green status	I_{23}
Average house No	I_{10}	Developer	I_{24}
House hold rate	I_{11}	Vacancy rate	I_{25}
Poverty population	I_{12}	The new situation	I_{26}
Public transportation	I_{13}	Rent level ($/SF)	I_{27}
Living facilities	I_{14}		

In the above table, CPI, house price, rental price and energy price are measured by statistic index; Business condition represents average company numbers per square kilometers while average house number are calculated by the same method; Housing expenditure is the proportion of total expenditure; Safety factor is the proportion of national criminal number; Unemployment rate, house hold rate, poverty population are all presented by percentage.

I_1—I_{12} and I_{27} are quantitative indicators, regional statistics data on the transaction example are shown as table 2 while the normalized macro factors are shown in table 3. I_{13}—I_{27} are qualitative index which are expert scored.

Table 2. The original data of the macro factors.

Index	I_1	I_2	I_3	I_4
Washington	154.6	163.9	191.9	185.2
Manhattan	260	276.1	344.8	209.8
Oakland	248.7	272.7	272.734	297.4
Index	I_5	I_6	I_7	I_8
Washington	315.8701	35.30%	31.1	5.10%
Manhattan	108.5882	39.70%	27.38	5.40%
Oakland	705.8971	35.20%	30.13	6.00%
Index	I_9	I_{10}	I_{11}	I_{12}
Washington	0.015514	1676.379	42.40%	18.50%
Manhattan	0.009126	236.7899	67.80%	2.90%
Oakland	0.007985	2764.958	41.00%	20.30%

Table 3. The normalized macro factors data.

Index	I_1	I_2	I_3	I_4	I_5	I_6
Washington	0.59	0.59	0.56	0.62	0.36	0.89
Manhattan	1.00	1.00	1.00	0.71	0.12	1.00
Oakland	0.96	0.99	0.79	1.00	0.80	0.89
Index	I_7	I_8	I_9	I_{10}	I_{11}	I_{12}
Washington	1.00	0.14	0.97	0.16	1.00	0.40
Manhattan	0.88	1.00	0.60	1.00	0.94	0.68
Oakland	0.97	0.09	1.00	0.14	0.85	0.77

3.2 Training result

After 33748 training steps, BP neural network achieve its set error. Specific training results are as follows:

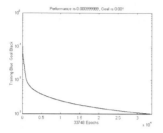

Figure 3.   The actual fitting effect.

Test results are as follows:

Table 4.   Real estate appraisal results.

Object	The output	Test rent $/SF	Actual rent $/SF	Relative error $/SF
1	0.8575	29.45	28.53	3.22%
2	0.9041	31.05	29.44	1.24%
3	0.8912	30.60	30.57	0.10%
4	0.8951	30.74	30.49	0.82%
5	0.8912	30.60	29.84	2.55%
6	0.8951	30.74	30.65	0.29%

4   CONCLUSION

The main results of this paper are: to establish the real estate appraisal model based on artificial neural network and MATLAB toolbox; to set up the index system of the influence factors of real estate price in the evaluation model; to use 24 transaction examples in three U.S. cities to train the neural network, and then to input 6 transaction instances for valuation. The results of BP neural network after training can relatively accurately simulate the relationship between the influence factors and real estate actual value.

The following problems are found in the research and need to be further perfected:

The scope problem. The evaluation of real estate price based on artificial neural network must have active trading market and a lot of comparable case. If the transaction is not active, the evaluation model can't realize the training and simulation process.

The indicators selecting problem. In the division of index system, the concrete data of market factor's collection is difficult. For different sections of the same city, due to too much refinement, we can't collect the macro economic data effectively.

The comparable cases selection. Comparable cases selection has great influence on the performance of the network. in the actual simulation process, while selection of training object still has certain subjectivity.

Although several problems exist, the real estate assessment based on artificial neural network effectively reduces the influence of subjective factors, and improves fairness and scientific to a certain extent. The established empirical model which has excellent analysis performance will have a positive effect on the promotion of real estate appraisal method.

ACKNOWLEDGMENTS

This work is supported by the Fundamental Re-search Funds for the Central Universities (FRF-TP-14-052A1).

REFERENCES

[1] Can, A. 1992. Specification and estimation of hedonic housing price models. *Regional science and urban economics* 22(3):453–474.
[2] Friesen, D. D. et al. 2011. A comparison of multiple regression and neural *networks* for forecasting real estate values. *Regional Business Review* 30:114–136.
[3] Guo, J. et al. 2013. An integrated cost-based approach for real estate appraisals. *Information Technology and Management*:1–9.
[4] Guo, P. 2010. Private real estate investment analysis within one-shot decision framework. *International Real Estate Review* 13(3):238–260.
[5] Liang, M. A., & Xiao-lin, Y. A. N. G. 2010. Study on the Residential Risk *Investment* Evaluation Based on the Ann-CVaR Model. *Journal of Engineering Management* 6:019.
[6] LIU, X. et al. 2011. Real estate appraisal system based on GIS and BP neural *network*. *Transactions of Nonferrous Metals Society of China* 21:s626–s630.
[7] Lizieri, C. et al. 2012. Pricing inefficiencies in private real estate markets using total return swaps. *The Journal of Real Estate Finance and Economics* 45(3):774–803.
[8] Lützkendorf, T., & Lorenz, D. 2005. Sustainable property investment: valuing *sustainable* buildings through property performance assessment. *Building Research & Information* 33(3):212–234.
[9] McCluskey, W. et al. 1997. Interactive application of computer assisted mass appraisal and geographic information systems. *Journal of Property Valuation and Investment* 15(5):448–465.
[10] Pagourtzi, E. et al. 2003. Real estate appraisal: a review of valuation methods. *Journal of Property Investment & Finance* 21(4): 383–401.
[11] WANG, D., & YU, S. 2012. Analysis into the Effects of Urban Subway or the House Price Nearby: A Ease Study on the No. 4 Line of Beijing. *Urban Studies* 4:018.
[12] Xuhui, W., & Feihua, H. 2007. The Model *of* Teaching Quality Evaluation Based on BP Neural Networks and Its Application. *Research in Higher Education of Engineering* 5: 78–81.
[13] Zhaoyu, P. et al. 2012. The application of the PSO based BP network in short-term load *forecasting*. *Physics Procedia* 24: 626–632.

Network Security and Communication Engineering – Chan (Ed.)
© 2015 Taylor & Francis Group, London, ISBN: 978-1-138-02821-0

Manufacturing information model for design of manufacturing

H.J. Liu, Q.M. Fan & B. Chen
Science and Technology on UAV Laboratory, Northwestern Polytechnical University, Xi'an, P.R. China

Q.M. Fan
School of Computer Science and Technology, Xi'an Technology University, Xi'an, P.R. China

B. Chen
Key Laboratory of Contemporary Design & Integrated Manufacturing Technology, Ministry of Education, Northwestern Polytechnical University, Xi'an, P.R. China

ABSTRACT: In the light of growing global competition, organizations around the world today are constantly under pressure to produce high-quality products at an economical price. The integration of design and manufacturing (DFM) activities into one common engineering effort has been regarded as a key strategy for their survival and growth. In this paper, the importance of DFM and parts information model are introduced in today's manufacturing enterprises. It describes the classification of parts information model; namely: product models, process models, and cost models, the main content and research status that the model contains. The author has proposed a semantic network description method of the manufacture parts information and details of its main content, which is the core of this paper.

KEYWORDS: Manufacturing information, Design for manufacturing, Semantic network, Express.

1 INTRODUCTION

In the 1960s, manufacturing workshop courses disappeared in design students' curricula in the United States. As a result, manufacturability analysis of the design has been neglected over the years. Substantial consideration has been given to the design of products for performance (functionality, quality, aesthetics and ergonomics, etc.). However, since the designers ignore the manufacturability of the design, sometimes it is not possible to manufacture the part or the design which justifies high manufacturing cost and long delivery time.

With the increase of global competition, the pressure to put quality products in market in time at a competitive price also increases as well. At present the manufacturing industry is featured with more emphasis being paid on improving productivity, product quality and reducing costs and time-to-market. These market pressures are transformed into optimized designs and manufacturing processes. The most challenging and influential part of the design process is the conceptual design stage. It has well proved that poor design at the conceptual stage can never be transformed into a successful product. It is estimated that around 80% of the cost of commercializing a product is determined during the early design stage [1-2], although the early design stage itself only accounts for a fraction of the total development costs. Concurrent engineering and more specifically DFM function during the product realization process. Design for manufacture (DFM) is the practice of designing products with manufacturing in mind. Its goal is to reduce costs required to manufacture a product and improve the ease with which that product can be made. Performing DFM analysis needs to choose DFM metric and method, specify DFM tasks and the sequence to perform them, define manufacturing information model and choose or develop DFM tools [3]. The objective of DFM is that considering manufacturing early in the design process, the design can be favorably influenced to improve quality, reduce cost and decrease time-to-market. DFM is the process whereby design teams evaluate the manufacturability of the product based on manufacturing information model, and then provide feedback so that the product design can be improved with respect to manufacturability.

2 THE MANUFACTURING INFORMATION MODEL

Informational model for manufacturing systems is based on general informational model, which is the fundament of the manufacturing branch. The matter of our interest is not a general system, but the particular manufacturing (production) system. We can apply the results from informational model of manufacturing systems to many useful things.

- Documentation of manufacturing system
- For design of manufacturing system
- For reengineering of manufacturing system
- For understanding of manufacturing system
- For simulation of manufacturing system
- For implementation of informational system of manufacturing system

In general, three types of models may be needed to support manufacturing evaluation: product models, process models, and cost models. A product model needs to model geometry, manufacturing features, and some non-geometry information such as tolerance and surface finish at different abstraction level of manufacturability analysis. A process data model describes a process activity, its sub-activities, and the associated data. A manufacturing process model encodes the capabilities of a manufacturing process including shape producing capabilities, dimensions, tolerance and surface quality capabilities, geometric and technological constraints and manufacturing cost. Traditionally, there are two methods for process shape producing capability modeling: process-based methods and part-based methods. In a process-based method, machine tools, fixture devices, cutting tools and kinematics motions as well as operation precedence in manufacturing processes are utilized to capture the capability of the process. In a part-based method, feature types, attributes and numbers in a machining process are adopted to define its capability. The capability of a manufacturing process can be also expressed in the form of constraints. Constraints can be classified into three levels: universal level constraints, shop level constraints, and machine level constraints [4]. Manufacturing resources are defined as the equipment which enables industry to turn raw materials into marketable products [5]. Different representations of manufacturing resources have been employed by a variety of software tools, which perform various tasks. Resource model should include: tooling/materials, resource descriptions, equipment/labor, materials knowledge and so forth [6]. Several manufacturing resource models are developed in MO system, and NIST rapid response manufacturing (RRM) project. Proposed the requirements specification for the manufacturing resource information modeling. An information model for the manufacturing resource is available. The model is in EXPRESS and developed for the NIST Rapid Response Manufacturing Intramural Project.

Cost models can be classified into activity-based cost models, scaling cost models and statistical cost models. Activity-based cost model decomposes the cost into elementary cost items and then respective cost of these items is estimated. It gives relatively precise results but detailed product and manufacturing process information which is needed. The task of gathering all the required date is time consuming. Scaling cost model estimates the cost by interpolation or extrapolation of historical data for closely related product. It assumes that there is a simple relationship between the considered parameter and the final cost. Statistical cost model is constructed based on statistical relationships and operates as a black box. Cost estimation's formula plays an important role in the cost models. The formula regards the cost as a dependent variant to one or more independent cost drivers.

Industry and academia long for a standardized data format that is independent platform and application. The recent effort on AP240 reflects the trend that the product model, manufacturing process and resource model will become standardized in the near future.

3 MANUFACTURING INFORMATION MODEL EXPRESS

A semantic network is a labeled directed graph representing objects/entities, their properties and their relationships. The structure of a semantic net is shown graphically in the terms of nodes and the arcs connecting them. Nodes are often referred to as objects/entities and the arcs as links or relationship between two nodes. Figure 1 shows an example of

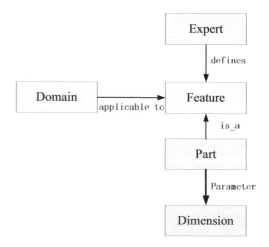

Figure 1.   A semantic network example.

application of semantic network to model the objects/entities and their relationships in DFM analysis.

One problem with using semantic network is that there is no standard definition of arc names. However, two commonly used links can be identified as "HAS-A" and "IS-A". For example:

feature HAS-A name, feature HAS-A parameter, machine IS-A manufacturing resource.

The three items of objects/entities, attribute, and value occur so frequently in modeling DFM information that it is possible to build a simplified semantic net just by using them. An entity-attribute-value (EAV) or triplet could be used to characterize the part model, material library and manufacturing resource. The EAV triple representation is convenient for listing knowledge in the form of a table and thus translating the table into computer. Some examples of an EAV triple table are shown in Table 1.

Table 1. Entity-attribute-value triplet example of parameters.

Entity	Attribute	Value
Parameter	Diameter	8
Parameter	Length	10

The first row in Table 1 represents that diameter is a parameter and has a value of 8. EAV triples are especially useful for representing facts, and the patterns to match the facts in the antecedent of a rule. Another example is given in Table 2.

Table 2. Entity-attribute-value triplet example of a hole.

Entity	Attribute	Value
feature	name	Hole
feature	ID	1
feature	Parameter-name	diameter
feature	Parameter-value	6

The content in Table 2 can be easily translated into PROLOG predicts, For example

is_a (hole, feature)
has_ID (hole, 1)
has_parameter_value (hole, diameter, 6)
Or CLIPS:
(deftemplate Parameter-of-Feature (slot Feature-ID)(slot Feature-Name) (slot Feature-Parameter-Name) (slot Feature-Parameter-Value));
(Parameter-of-Feature (ID 1)(Feature "HOLE") (Parameter "diameter") (Value 6))
However, that modeling the part model, material information, operation information, resource information and so on makes the design of EAV triple table

more concise and unambiguous. Entity-Relationship-Diagram (ERD) has been chosen as the information modeling method to this work. Information modeling methods include Entity-Relationship met hod (ER), Function Modeling method and object-oriented method (OO). Function modeling approach focuses on decomposing system functionality and the information flow between different objects; O-O approach defines the object as the basic element which contains both data and functions, thus it is easy to model complex objects and drivers.

Industry and academia long for a standardized data format that is independent platform and application. The recent effort on AP240 reflects the trend that the product model, manufacturing process and resource model will become standardized in the near future and provides good extensibility. ER approach emphasizes on identifying the entities, their attributes, and the relationships among the entities. As discussed above, each type of manufacturing information has entities, attributes, and relations, thus ER is appropriate to model the manufacturing information.

Another desired characteristic is to model the hierarchy and inheritance of the manufacturing information, as Figure 2 shows:

Such a representation can be achieved by adding parent item to the current item, which is used as a pointer to high-level entity.

By specifying the parent item of each feature, the feature hierarchy can be established flexibly according to different manufacturing processes and different experts. Figure 3 shows that some parameter examples relate to certain features. These parameters can be added, modified, and deleted dynamically through standard database operation.

In a similar way, the material, operation, resource, cost/time, process structures can be modeled.

4 CONCLUSION

This paper concentrates on the manufacturing information model for design for manufacturing activity. Manufacturing information model becomes an integral part of communication in enterprise. Today DFM is performed at various abstraction levels by comparing design requirements, specifications and features versus manufacturing process capability models. The manufacturing information model proposed here in addressing the issues and requirements raised concerning management of DFM information. The semantic net express creates a neutral information model for easy exchange of manufacturing capability information. It is envisioned that DFM applications among other applications would access the manufacturing information model just as CAD systems are being built by semantic net structure format.

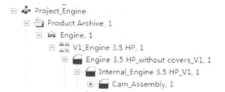

(a) Product structure hierarchy.

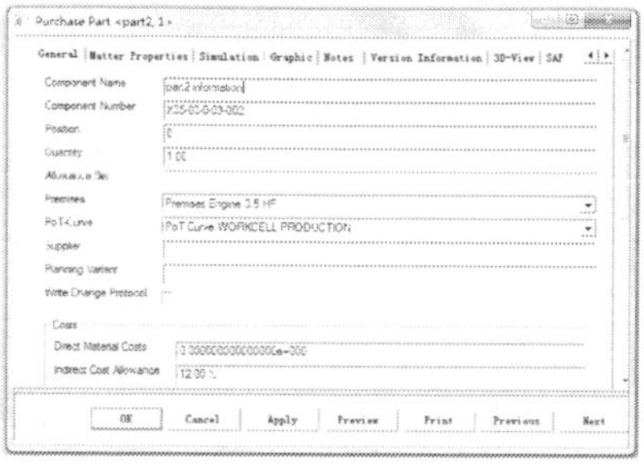

(b) Product information.

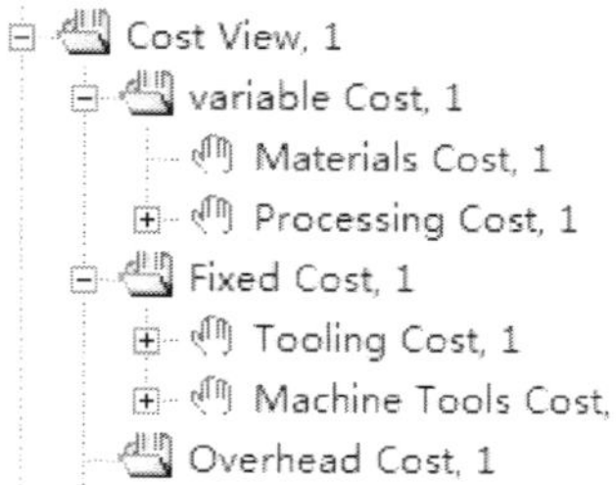

Cost View, 1
  variable Cost, 1
    Materials Cost, 1
    Processing Cost, 1
  Fixed Cost, 1
    Tooling Cost, 1
    Machine Tools Cost,
  Overhead Cost, 1

(c) Cost structure hierarchy.

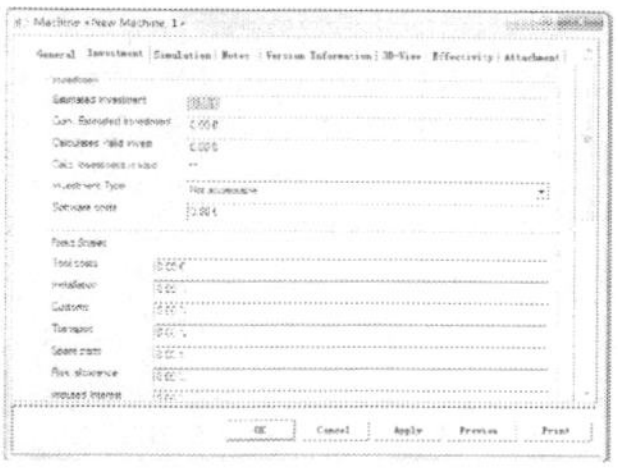

Figure 2. Hierarchal nature of manufacturing information.

(d) Cost information.

Process Planning Information, 1
  Process Plan Engine WORKCELL PRODUCTION, , 1
    Main trial Engine WORKCELL PRODUCTION, 1
      Position engine block on tote board, 1

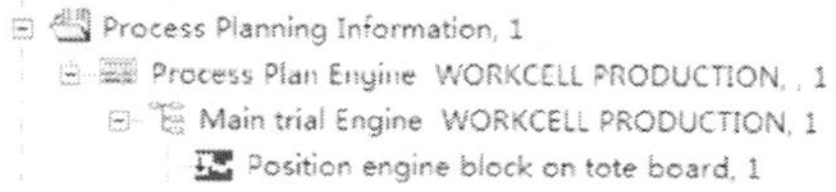

(e) Process structure hierarchy.

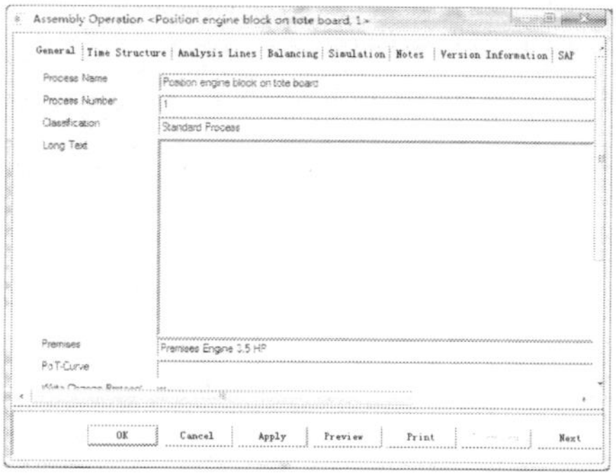

(f) Process information.

Feature	Parameter	Practical Limit?	Upper Limit	Lower Limit	Description
Part	Batch Size	--	--	--	The batch size of...
Part	Tolerance	V	0.13	0.05	tolerance
Part	Surface Finish	V	3.2	0.8	Surface finish
Part	Thickness	V	5	0.1	There is a range f...

(a) Sample part parameters.

Feature	Parameter	Practical Limit?	Upper Limit	Lower Limit	Description
HOLE	diameter	V	6.35	1.58	There is a range for...
HOLE	x	--	--	--	X
HOLE	y	--	--	--	Y
HOLE	thickness	V	2	0.1	There are limits for t...
HOLE	z	--	--	--	Z

(b) Sample hole parameters.

Figure 3. Part features and parameters.

ACKNOWLEDGMENTS

This work is supported by the Science and technology plan projects of xi'an (CY1313, CXY1338(5)), the National Science and technology support program (2012BAB15B01),The Defence Pre-research Fund(9140A250402).

REFERENCES

[1] Boothroyd G, J. *Comput Aided Des.* 26,7 (1994).
[2] Rehman S, Guenov MD. *Comput Ind Eng.* 35, 3 (1998).
[3] Anderson, D. M., Design for Manufacturability & Concurrent Engineering; *How to Design for Low Cost, Design in High Quality, Design for Lean Manufacture, and Design Quickly for Fast Production, Publisher*: C I M Pr(2003).
[4] C.X. Feng, A. Kusiak, J. CAD 5, 27 (1995).
[5] Jurrens K, Fowler J, Algeo MB, Editor, National Institute of Standards and Technology (1995).
[6] Y.T. Lee, Editor, *National Institute of Standards and Technology* (1999).

Network Security and Communication Engineering – Chan (Ed.)
© *2015 Taylor & Francis Group, London, ISBN: 978-1-138-02821-0*

Design of Momordica grosvenori E-commerce platform

Y. Liu & Y. Deng
College of Information Sci & Engin, Guilin University of Technology, Guilin, China

Z.Z. Jiang & J.L. Jiang
Guilin Layn Natural Ingredients Corp, Guilin, China

ABSTRACT: The design of the Momordica grosvenori e-commerce system mainly releases Siraitia grosvenori, planting technology, information processing technology, industry dynamics, policy information, research dynamic, researchers, and other relevant information. Platform using Web Services technology is based on the B/S model, with the development of the PC website and mobile client. The PC site uses JavaScript and Struct framework, using a database called MySQL. The mobile client is developed under the J2ME platform, which provides the iOS and Android versions. Through the platform we are able to master the latest domestic Momordica grosvenori industry. We can also carry out online trading platform about the Momordica grosvenori commodity. The platform has played a positive role in promoting the development of Guangxi district, as well as expands the market of Momordica grosvenori momordica industry both at home and abroad.

KEYWORDS: Momordica grosvenori electronic commerce, Web Services, JavaScript, mobile client, Guangxi.

1 INTRODUCTION

Mobile commerce can truly make everyone get the information of the entire network and service at any time, any place, and the exchange of information whenever and wherever possible means increasing demand of diversification, it also brings more business opportunities to enterprises. Momordica grosvenori Guangxi local characterized as crops, has a high edible value, and it is also an important medical raw material. Due to geographical conditions, the agricultural products should be promoted in the national reputation. With the help of the Guangxi science and technology promotion projects, Momordica grosvenori e-commerce system is designed, in order to build market publicity and promote the Momordica grosvenori industry in our region.

Momordica grosvenori e-commerce platform is designed based on Web Services technology on the basis of the B/S model, and the PC website and mobile client are developed. The PC site uses JavaScript and Struct framework, using a database called MySQL. The mobile client is developed under the J2ME platform, which provides the iOS and Android versions. The platform not only provides information publication and inquiry service, but also makes Siraitia grosvenori goods trade online.

2 PLATFORM ARCHITECTURE

The reason of construction of the platform is mainly to promote the Momordica grosvenori industry in Guangxi region, and to expand the influence of the Momordica grosvenori in domestic and foreign market. Therefore, the platform columns are very abundant, including: industry dynamics, activity of the exhibition, the latest announcement, product display, industrial alliance, plant introduction, planting, processing technology, introduction of the scientific research personnel, research articles, research results, policies and regulations, economic and trade policies, data download, online transactions and website column. From the columns we can see, the platform is based on a full range of Momordica grosvenori and Momordica industry is introduced, and we strive to create the first e-commerce platform of Momordica grosvenori.

3 DETAIL DESIGN OF PLATFORM

3.1 *Display page*

The presentation layer of electronic commerce platform uses JavaScript and JQuery technology, with the struct framework and the logic control layer interaction.

To achieve the following functions:

Industry dynamics: introduce the latest information of the domestic and foreign Momordica grosvenori industry, with a large amount of information, real-time advantages;

Industry Alliance-the Guangxi Momordica grosvenori industry alliance organization information, including the members of the unit information introduction and product introduction;

Activities show-introduce the latest activities of Guangxi Momordica grosvenori industry alliance, including activities related to the member units;

The latest announcement-Guangxi Momordica grosvenori Industry Alliance release the latest notice information, including the web related information bulletin;

Product display-in the form of pictures showing Guangxi Momordica grosvenori finished and Guangxi Momordica grosvenori industry alliance of the member units of product, click the link to open, you can view detailed product information and purchase information, the function to search similar products also provide;

Plant introduction-Introduce Guangxi characteristic agricultural products of Momordica grosvenori plant growth environment information and other relevant information;

Plant introduction-Explanation of Siraitia grosvenori plant related technology;

The processing technology of Momordica grosvenori-Introduction to information processing technology, storage technology and other production processes;

Scientific research personnel-Introduce Siraitia grosvenori research field of experts and scholars research information, promoting academic exchanges of Momordica grosvenori;

Research results of relevant research achievements information-an introduction of Siraitia grosvenori, convenient for the user to know the latest progress in the research of Momordica grosvenori;

Policies and regulations: promoting national, autonomous region, city and county four government agencies of Siraitia grosvenorii policy information;

Preferential investment policies and trade policies-each link of propaganda government planting, processing and sales;

Download channel image download-provide website content data sets, video, documents and other various types of data;

Provide online trading in each member units of Momordica grosvenori Industry Alliance products, agricultural products, such as mangos teen Momordica grosvenori germchit merchandise to browse, order and purchase payment and other electronic trading function, promote Guangxi District of Siraitia grosvenori network transactions;

The background system provides all front desk display content of online management and transaction information management, user management, website state management, background database management, website maintenance management function;

Website-information introduces the site overall information, site map navigation, website production, website contact information;

In website homepage, we provide processing information browsing, user login, data download, and guestbook, submit the information and other functions. More specific settings for the web site columns. To facilitate the development and maintenance at the same time, the website uses the CSS and J query. For user login and transaction information submitted by customers, in the foreground of network transmission before using bank level 128 bits encryption technology encryption, which ensure the safety and integrity of user information and transaction information. Background pages provide platform for management personnel use, complete the information release and maintenance and release of trading systems and maintenance, user management, database maintenance and other functions.

In order to adapt to the development of the situation, a mobile client platform has been developed. It can not only meet the needs of different mobile devices (mobile phone, tablet, PDA) access, but also develops the corresponding versions of the mainstream mobile operating system, greatly expands the application platform in the new situation. The mobile client contains 3 modules:

3.2 Logic control module

A variety of foreground requests were submitted by the Struct framework, transferred to the corresponding logic processing module. These modules using Java function package, involving module unity require user information authentication before execution. Specifically related to the module contains:

Information browsing-according to the user the front click column and the title content retrieval database, the retrieval results for the front display;

Information inquiry-according to the user the front where the columns and search for content retrieval database, the retrieval results for the front display;

Information release-according to the background administrator chooses to release the columns and released into the database and returns the results of the operation for the confirmation;

User login-according to the user login information for authentication, the password comparison operation, complete user identity verification;

User management: according to the actual needs of work, to the Guangxi Momordica grosvenori industry alliance members unit managers, transaction user management, and password management operation;

Transaction modules-for confirmation, delivery, payment operation according to the foreground single Momordica grosvenori orders; for consultation, evaluation, response, and other operations on the transaction;

The database module of database backup and recovery, log-operating functions such as operation;

This part is the platform logic control center, and it is the link of the front display and database operation.

3.3 *Database*

All data stored in the database of the whole platform. These data include:

All the information for the user to each column of information-all front desk release columns browsing;

User information-Guangxi Momordica grosvenori industry alliance of the member units to the administrator information, electronic transaction registration administrator information of user authority design, website operation of user information system data;

The related transaction data transactions data-Guangxi Momordica grosvenori e-commerce platform trading system;

Platform state information-Guangxi Momordica grosvenori e-commerce platform running state data;

All the information is encrypted algorithm running after storage, and the realization of disaster backup and recovery function, greatly improve the platform security and disaster recovery capability.

4 CONCLUSION

This article has carried out on the design of Siraitia grosvenori e-commerce platform. The utilization of the network and the mobile terminal test can be safe and stable to access to the platform. With rich and perfect content of this platform, as the characteristic agricultural products of Guangxi area, Momordica grosvenori and Momordica grosvenori industry will gradually open the market, for the vast number of consumers and the people familiar. We will promote the process of internationalization of the platform in the next step by the launch of the important markets in Southeast Asia and Europe etc.

ACKNOWLEDGMENTS

Guangxi science and technology development project (1355001-2-5); Guangxi Natural Science Foundation (2013GXNSFBA019277).

REFERENCES

[1] Wu Zhongtang, Feng Jiuchao. 2012 J2ME and chaotic encryption mobile commerce security scheme of [J]. Computer science, 34 (11): 110–112.
[2] Chen Zhen, Fan Huafeng. 2011 based on the Web Services data exchange platform design [J]. Fujian computer 5:118–119.
[3] Li Yonggang, Zhou Xiaofei, Xu Ming. 2013 based on the application of mobile commerce security design of [J]. Computer J2ME and XML technology, 11:105–108.
[4] Zhao Wentao, he Peizhen. 2013 research university application integration of Web services and implementation of [J]. Micro computer information.26 (9): 155–156.
[5] Zhang Xin. 2010 smart card in a single sign on Model in SAML Application Research Based on [D]. Southwest Jiao Tong University, 83–85.
[6] Zhang Zongping, Zhang Yong, Qin Hui. 2011 Web Services data exchange platform based on the application of [J]. Computer today, 07:60–68.

Network Security and Communication Engineering – Chan (Ed.)
© 2015 Taylor & Francis Group, London, ISBN: 978-1-138-02821-0

The impact of oil price shocks on Vietnam's stock market

D.T. Le
School of Economic and Trade, Hunan University, Changsha, PR China
Faculty of Banking and Finance, Industrial University of Hochiminh City, Vietnam

Q. Zhang
School of Economic and Trade, Hunan University, Changsha, PR China

ABSTRACT: This paper examines the impact of oil price shocks on Vietnam's stock market by using the AR-JI(h_t)-EGARCH model. To better understand how the price shocks in the oil market are transmitted to the stock market, we considered expected, unexpected and negatively unexpected oil price volatilities, incorporating these into the model of stock returns. Empirical results indicate that world oil prices have a positive effect on Vietnam's stock returns and the expected events of modifications in oil spot prices have significant positive impacts on the stock market. Inversely, no strong evidence is found of a relation between unexpected volatility of oil price and stock market returns. The results also show that the volatility in the Vietnam's stock market during 2006-2008 was led by the fads/crowded trend and foreign investors' trading led the domestic investors to follow and make a crowded psychology.

KEYWORDS: Vietnam, Stock returns, Volatility, Oil prices.

1 INTRODUCTION

Oil is an important source of energy worldwide, and it has played a prominent role in the economy. In the seminal work of (Hamilton 1983), he showed that oil price increase was responsible for almost every post-World War II US recession, and a large number of researchers have focused on the relationships between oil prices and economic activity. (Jones and Kaul 1996) examined the impact of oil prices on stock returns, and they found that the change of oil prices did impact aggregate stock returns for Canada, the United States (US), the United Kingdom (UK) and Japan. (Huang, Masulis and Stoll 1996) adopt a vector autoregressive (VAR) model to identify the relationship between oil futures prices and aggregate stock returns of US.

With the US stock market, for example, (Sadorsky 1999) and (Papapetrou 2001) found a negative relationship between oil price shocks and aggregate stock returns for the USA and for Greece respectively. (Kilian and Park 2009) concluded that the propagation of oil price shocks on stock prices was primarily through effects on final demands for goods and services.

(Chiou and Lee 2009) described the relationship to be asymmetrical, with oil price increase having a more significant effect on stock market returns than oil price decrease. They used the ARJI-GARCH model to analyze the relationship between oil prices and stock returns.

The structural modified characteristic has been approved to examine the time-varying effects of oil price shocks on stock returns. (Aloui and Jammazi 2009) used the Markov switching specification to supersede the time-invariant specification. (Park and Ratti 2008) appreciated the effects of oil price shocks on stock returns for the US and 13 European countries by using first-differenced VAR. (Lee, Yang and Huang 2012) utilized the methodology of regression with asymmetric effects to re-examine the issue of the oil price-stock markets relationship. (Arouri and Rault 2012) examined the relationships between oil prices and stock market returns for Gulf Cooperation Council (GCC) countries. In addition to the multi-economy examinations, there are also a small number of works studied on single country. Some of these studies used data from emerging economics. For example, (Papapetrou 2001) studied this relations on Greece market whereas China was considered in (Cong, Wei, Jiao and Fan 2008) and (Zhang and Chen 2011), and the studies on Vietnam were done by (Narayan and Narayan 2010).

In order to contribute the literature for developing oil importing countries, this paper examines the impact of oil price shocks on Vietnam's stock returns, by using daily data for the period 2002-2013. The contributions of this paper are as follows.

First, this paper explores the relation between oil price shock and stock returns by considering expected, unexpected and negatively unexpected constituent of oil price volatilities. Furthermore, the paper applies the Autoregressive Conditional Jump Intensity (ARJI) method, combined with the EGARCH process to Vietnam's stock returns.

Second, although (Narayan and Narayan 2010) had researched the impact of oil prices on Vietnam's stock prices, yet their analysis was based on data for the period 2000-2008. In this period, the Vietnam's stock market is still imperfect. Therefore, it is necessary to further examine whether the rise of oil price shocks raises Vietnam's stock market when it has become one of the frontier emerging markets in the region.

Third, Vietnam is a net importer of oil and is an emerging economy. This paper contributes to the literature that investigates the relation between oil price shocks and stock returns of emerging oil importing countries.

The rests of the paper are organized as follows: we begin with a description of the methodology adopted for this study in Section 2, followed in Section 3 by the data and the empirical results. In the Section 4, we provide some concluding remarks and possible future works.

2 METHODOLOGY

2.1 The ARMA-GARCH model

It is to examine whether expected and unexpected volatility of oil price can expound the volatility behavior for stock market returns. By following (Chan and Maheu 2002), (Chiou and Lee 2009) and (Zhang and Chen 2011), the ARMA-GARCH model used to depict the modifications of oil prices is as follows:

$$r_t = \mu + \sum_{i=1}^{p} \varnothing_i r_{t-i} + \sum_{j=1}^{q} \theta_j a_{t-j} + a_t \quad (1)$$

$$a_t = \sqrt{h_t} \varepsilon \quad (2)$$

$$h_t = \omega + \sum_{i=1}^{k} \beta_i h_{t-i} + \sum_{j=1}^{l} \alpha_j a_{t-j}^2 \quad (3)$$

where a_t denotes the jump innovation assigned to be a conditionally mean zero, which is equal to 1 and 2, respectively, in our empirical analysis.

Turning to disassembling r_t, the expected changes (e_t) can be taken as the difference between the changes of oil price and the estimated residual:

$$e_t = r_t - a_t \quad (4)$$

2.2 The ARJI(-ht)-EGARCH model

(Chan and Maheu 2002) and (Zhang and Chen 2011) modified the model of ARJI method with the exponential generalized conditional heteroscedasticity

(EGARCH) framework, and the jump model of stock returns can be written as:

$$R_t = \mu + \theta_1 R_{t-1} + \theta_2 \varepsilon_{1,t-1} + k_1 e_t + k_2 up_t + k_3 up_t + \varepsilon_{1,t} + \varepsilon_{2,t} \quad (5)$$

$$\varepsilon_{1,t} = \sqrt{h_t} z_t, \quad z_t \sim NID(0,1) \quad (6)$$

Define the information set at time t to be the history of returns, $\varphi t = \{Rt,...,R1\}$. The conditional variance of $\varepsilon 1,t$ equals h_t, is measurable with respect to φt. Hence, EGARCH (1, 1) process is defined by:

$$h_t = \exp\left(\omega + \beta lnh_{t-1} + d\varepsilon_{1,t-1}/\sqrt{h_{t-1}} + \alpha\left(\left|\varepsilon_{1,t-1}/\sqrt{h_{t-1}}\right| - \sqrt{(2/\pi)}\right)\right) \quad (7)$$

The return series include the normal stochastic process and the stochastic jump process that is assumed to have the Poisson distribution with a time-varying conditional intensity parameter λt and have a density function given by:

$$P\left(n_t = j/\boxed{}_{t-1}\right) = \exp(-\lambda_t)\lambda_t^j / j!, \quad j = 0,1,2,... \quad (8)$$

$$\lambda_t = \lambda_0 + \rho\lambda_{t-1} + \gamma\xi_{t-1} \quad (9)$$

The conditional jump intensity at time t is related to one past lag of the conditional jump intensity plus one lag of ξ_t. The ex-ante error ξ_{t-1}, captures the unexpected number of jumps in the previous period, and it is calculated as follows:

$$\xi_{t-1} \equiv E\left[n_{t-1}/\varnothing_{t-1}\right] - \lambda_{t-1} = \sum_{j=0}^{\infty}(n_{t-1} = j/\varnothing_{t-1}) - \lambda_{t-1} \quad (10)$$

$$\varepsilon_{2,t} = J_t - E\left[J_t \mid \varnothing_{t-1}\right] = \sum_{k=1}^{n_t}\pi_{t,k} - \theta_t\lambda_t \quad (11)$$

2.3 An extension of ARJI: the ARJI(-ht) model

To this point, we have postulated the distribution of the jump size to be Gaussian; however, it may also change and display conditional dynamics. We could then consider an extension of the model that allows the conditional mean of the jump size to be a function of lagged returns, and consider the variance to be changed with h_t, which we defined to as ARJI-h_t:

$$\theta_t = \eta_0 + \eta_1 R_{t-1} D(R_{t-1}) + \eta_2 R_{t-1}\left(1 - D(R_{t-1})\right) \quad (12)$$

$$\delta_t^2 = \xi_0^2 + \xi_1 h_t \quad (13)$$

where $D(x)$ is an indicator function which equals 1 if x>0 and 0 otherwise, and $\eta0$, $\eta1$, $\eta2$, $\xi0$ and $\xi1$

are parameters to be estimated. This specification of the conditional mean of the jump size allows some flexibility regarding where jumps are centered. This formulation may capture the rally after a stock market crash through a change in jump direction, when $\eta2$ has a negative value. H_t is a prediction of the EGARCH volatility component of our model in time t, and if the variance of the jump size is sensitive to contemporaneous market volatility, this extended model may capture a better effect.

3 DATA AND EXPERIMENTAL RESULTS

The data employed in this study comprises 3,000 observations of daily closing stock price index, obtained from the website of Hochiminh Stock Exchange (www.hsx.vn). The sample period is from 1 March 2002 to 31 December 2013 which is the most updated data to the time of this study. Europe Brent Spot Price is obtained from the Wind Database and U.S Energy Information Administration respectively.

3.1 Descriptive statistics

Table 1 shows the descriptive statistics for daily stock market returns with daily returns computed as below:

$$R_t = \ln(P_t) - \ln(P_{t-1}) = \ln(P_t / P_{t-1}) * 100 \qquad (14)$$

where P_t is the daily price at time t and P_{t-1} is the daily price at t-1.

Table 1 summarizes the returns of oil prices along with the returns of the VN-Index statistics. The statistics point out that the daily average returns are positive and very small compared with the standard deviation. The returns series of the VN-Index and oil price are normally distributed. Skewness is not close to zero and kurtosis is much higher than two for the return series. The positive skewness coefficient of -0.037471 and -0.018560 indicates that series are skewed towards the left. The kurtosis of the Vietnam stock market's return is 4.3511, implying that leptokurtic or the series have distribution with fatter tail and are more peaked at the mean than a normal distribution. All return series are negatively skewed.

Figure 1 provides an illustration of the oil prices and returns for the oil spot contracts and stock index. According to Figure 1, volatility clustering is obvious, that is, high volatility in one period tends to be followed by high volatility in the next period, suggesting the sensibleness of the GARCH structure.

Table 2 shows the results of the application of the (Dickey and Fuller 1979) test and the tests to the prices and the first-order differences in the VN-Index and the oil spot prices. We also show the results of the

Table 1. Descriptive statistics, 2002-2013.

	VN-Index	Oil prices
Mean	0.032872	0.055615
Standard error	1.529531	2.157541
Maximum	7.741532	18.12974
Minimum	-4.971353	-16.83201
Skewness	-0.018560	-0.037471
Kurtosis	4.459315	8.060562
Jarque-Bera	262.2878***	3201.863***
Ljung-Box Q2(24)	297.59***	221.084***

***Significance at the 1% level; **Significance at the 5% level; and *Significance at the 10% level

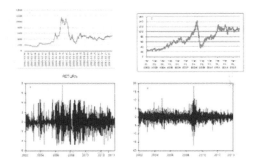

Figure 1. (a) Vietnam stock index (VN-Index); (b) returns of Hochiminh stock exchange index; (c) Europe Brent Spot Price; (d) returns of Europe Brent Spot Price.

Table 2. Unit root and stationarity tests for stock price returns and oil price returns.

	VN-Index		Oil prices	
Test	Level	Difference	Level	Difference
Panel A: ADF test				
Model C	-3.432376	-21.08986***	-2.432341	-27.55842***
Model C/T	-3.961153	-21.08634***	-2.961102	-27.55384***
Panel B: PP test				
Model C	-3.432374	-18.05261***	-1.943608	-25.74235***
Model C/T	-3.961150	-18.06323***	-3.216020	-25.73891***
Panel C: KPSS test				
Model C	2.934284	-17.99825***	-1.940213	-25.23825***
Model C/T	-3.905621	-17.99806***	-3.160152	-25.14962***

***Significance at the 1% level.

stationarity test based on the null hypothesis that a time series is trend stationary.

The results on non-stationarity for the level and first-order difference, which are recognized by using the above tests, reveal that both of the series are non-stationary in level, but are stationary in first-order differences, suggesting that all of the series are integrated, with an order of one, I(1).

3.2 *Application of the ARMA (1, 1)-GARCH (1, 2) model to changes of oil spot prices*

In the following part, we used the maximum likelihood estimation method to estimate the parameters of the GARCH model. Table 3 shows the statistics of oil price volatility. It indicates that the coefficients of ARMA(1,1), $BETA_1$, $GARMA_1$ measurements of GARCH (1,2) effect, W_0, B_0, A_0 and A_{00} are all significant, and the statistics of Ljung-Box Q^2 are relatively smaller, suggesting that the model is suitably simulated.

Table 3. ARMA (1, 1)-GARCH (1, 2) model on the change of oil prices.

Variable	Coefficients	Standard error
MU1	0.081622360***	0.076625519
BETA1	0.630796281***	0.187133512
GARMA1	-0.514623817***	0.125008471
W0	0.087649351**	0.024796187
B0	0.926528304**	0.013549962
A0	0.051668713*	0.016750018
A00	0.089183232***	0.011050368
Ljung-Box Q2(15)	18.7264	
Ljung-Box Q2(25)	25.3817	
Function value	-5246.67035124	

*,**,*** indicate significance at the 10%, 5% and 1% levels, respectively.

3.3 *Application of the ARJI(-ht)-EGARCH (1, 1) model to returns of VN-Index*

The results of the ARJI(-h_t)- EGARCH model with consideration of the changes of oil spot prices into the model of stock returns, including expected (e_t), unexpected (up_t) and negatively unexpected., are reported in Table 4.

In conformity with the statistics reported in Table 4, the EGARCH (ω, α, β, d) effect for VN-Index is discovered to be significant in both models. We can see that, from the results shown in Table 4, the value of d is significant and negative (-0.065 and -0.079), which shows that the stock market responds are more sharply to negative innovations than positive one of equal size.

Regarding the impact of oil spot price market to the stock returns, the empirical evidence is found of a relation between expected volatility of oil spot price and stock market returns. The results in Table 4 show that the statistical significance of both k_1 values (0.042 and 0.055), interprets that the expected

Table 4. ARJI(-ht)-EGARCH estimation by BFGS.

Parameter	ARJI		ARJI-ht	
	Coefficients	Std error	Coefficients	Std error
μ	0.02961	0.03120	0.04119	0.03774
Φ1	-0.49761	0.36190	-0.32080	0.41651
Φ2	0.43659	0.35126	0.38422	0.40115
k1	0.04194**	0.04614	0.05491**	0.04179
k2	0.01496	0.08494	0.02449	0.08168
k3	0.03716	0.04670	0.04036	0.04159
ω	-0.07626***	0.02186	-0.02692***	0.01654
α	0.09896***	0.02936	0.08542***	0.01148
β	0.98551***	0.02395	0.99118***	0.01126
d	-0.06524***	0.02185	-0.07951***	0.01976
η0	-0.01206	0.09633	-0.17554**	0.05416
η1			0.12633***	0.01947
η2			-0.19763***	0.02868
ξ0	1.53279***	0.23553	1.38769**	0.18053
ξ1			0.67125***	0.31841
λ0	0.00558**	0.00349	0.03419***	0.00451
ρ	0.94733***	0.00841	0.95815***	0.00468
γ	0.38147**	0.09413747	0.42516***	0.06252763
Q2(15)	8.169[0.581]		7.152[0.861]	
Q2(25)	11.943[0.631496]		10.854[0.821413]	
Log-likelihood	-6070.58513679		-6064.18753126	
LR test	20.0511***			

*,**,*** indicate significance at the 10%, 5% and 1% levels, respectively. Q2(15) and Q2(25) denote Ljung-Box test for serial correlation in the squared standardized residuals with 15 and 20 lags. P values are in square brackets. The likelihood ratio (LR) test is used to compare the fit of the two models.

events of modifications in oil spot prices have significant positive impacts on the VN-Index. Inversely, the results in Table 4 also demonstrate that the t-statistics of k_2 and k_3 are all smaller than the critical values. It means that no evidence is found of a relation between unexpected volatility of oil price and stock market returns, which proposes that investors in the stock market react much less to wonders of volatility than volatility they can warn in the world oil market.

As to the jump size ($\pi_{t,k}$) distribution, all the coefficients are statistically significant in both of the models. Particularly, the value of η2 in the ARJI-h_t model is significantly negative (-0.198), which indicates that after a stock market downward trends, the direction of a jump in the next period is more likely to be positive than negative. As shown in the Table 4, the value (η0) of jump size is not significant, and all the coefficients depicting the jump from the ARJI-h_t-EGARCH model indicate considerable significance.

Regarding the jump intensity (λ_t), all the coefficients are also statistically significant, which can explain the presence of time-varying jumps on the arrival of news events. The positive value of γ parameter (0.381 and 0.425) which estimates the sensitivity of λ_t to the most recent intensity residual (ξ_{t-1}) is statistically significant, indicating that a unit increases in ξ_{t-1} results in a increased effect on the next period's jump intensity. The positive value of ρ parameter (0.947 and 0.958) estimates the persistence in the conditional jump sensitivity of λ_t, which indicates that a high probability of many (few) jumps today tends to be followed by a high probability of many (few) jumps tomorrow.

The likelihood ratio test was used to investigate whether oil price shock plays an important role in the dynamics of stock returns. Table 4 shows that the log-likelihood for the ARJI-h_t- EGARCH model displays an increase of 6.3976, indicating that the null hypothesis of no difference between $\eta1$ and $\eta2$ is rejected at the 1% significance level.

4 CONCLUSIONS

This paper examined the impact of oil price shocks on Vietnam's stock market, by using the returns series of VN-Index and Europe Brent Spot Price from March 1, 2002 to December 31, 2013. The primary purpose of this paper was to investigate the time series features of the stock market in connection with jump phenomena by employing an autoregressive conditional jump intensity (ARJI) and its extended model (ARJI-ht), incorporated with the exponential generalized conditional heteroscedasticity (EGARCH) method. This model can analyze the possible impacts of oil price shocks on stock returns and recognize the oil prices effect on the stock returns.

To better understand how the price shocks in the oil market are transmitted to the stock market, we considered expected, unexpected and negatively unexpected oil price volatilities, incorporating these into the model of stock returns.

We found that the volatility in the Vietnam's stock market during 2006-2008 was led by the fads/crowded trend, and foreign investors' trading led the domestic investors to follow and make a crowded psychology. Regarding the impact of oil spot price market to the stock returns, the empirical evidence is found of a relation between expected volatility of oil spot price and stock market returns. The results obtained from this study are of potential significant interest to investors and financial market participants. Future studies can investigate the mechanisms via the changes in oil price that affect firm behaviors and stock prices in order to create an economic model relating oil prices to firms' dividends and performance.

REFERENCES

[1] Aloui, C., and R. Jammazi, 2009. The effects of crude oil shocks on stock market shifts behaviour: A regime switching approach, Energy Economics 31, 789–799.

[2] Apergis, N., and S. M. Miller, 2009. Do structural oil-market shocks affect stock prices?, Energy Economics 31, 569–575.

[3] Arouri, M. E. H., 2011. Does crude oil move stock markets in Europe? A sector investigation, Economic Modelling 28, 1716–1725.

[4] Arouri, M. E. H., and C. Rault, 2012. OIL PRICES AND STOCK MARKETS IN GCC COUNTRIES: EMPIRICAL EVIDENCE FROM PANEL ANALYSIS, International Journal of Finance & Economics 17, 242–253.

[5] Chan, W. H., and J. M. Maheu, 2002. Conditional Jump Dynamics in Stock Market Returns, Journal of Business & Economic Statistics 20, 377–389.

[6] Chiou, J.-S., and Y.-H. Lee, 2009. Jump dynamics and volatility: Oil and the stock markets, Energy 34, 788–796.

[7] Cong, R.-G., Y.-M. Wei, J.-L. Jiao, and Y. Fan, 2008. Relationships between oil price shocks and stock market: An empirical analysis from China, Energy Policy 36, 3544–3553.

[8] Hamilton, 1983. Oil and the macroeconomy since World War II, Journal of Political Economy 92, 228–248.

[9] Huang, B.-N., M. J. Hwang, and H.-P. Peng, 2005. The asymmetry of the impact of oil price shocks on economic activities: An application of the multivariate threshold model, Energy Economics 27, 455–476.

[10] Huang, R. D., R. W. Masulis, and H. R. Stoll, 1996. Energy shocks and financial markets, Journal of Futures Markets 16, 1–27.

[11] Huang, W., Q. Liu, S. G. Rhee, and L. Zhang, 2010. Return Reversals, Idiosyncratic Risk, and Expected Returns, Review of Financial Studies 23, 147–168.

[12] Johansen, S., 1991. Estimation and Hypothesis Testing of Cointegration Vectors in Gaussian Vector Autoregressive Models, Econometrica 59, 1551–1580.

[13] Jones, C. M., and G. Kaul, 1996. Oil and the Stock Markets, The Journal of Finance 51, 463–491.

[14] Kilian, L., and C. Park, 2009. THE IMPACT OF OIL PRICE SHOCKS ON THE U.S. STOCK MARKET*, International Economic Review 50, 1267–1287.

[15] Lee, B.-J., C. W. Yang, and B.-N. Huang, 2012. Oil price movements and stock markets revisited: A case of sector stock price indexes in the G-7 countries, Energy Economics 34, 1284–1300.

[16] Lee, C.-C., and J.-H. Zeng, 2011. The impact of oil price shocks on stock market activities: Asymmetric effect with quantile regression, Mathematics and Computers in Simulation 81, 1910–1920.

[17] Narayan, P. K., and S. Narayan, 2010. Modelling the impact of oil prices on Vietnam's stock prices, Applied Energy 87, 356–361.

[18] Nguyen, C. C., and M. I. Bhatti, 2012. Copula model dependency between oil prices and stock markets: Evidence from China and Vietnam, Journal of International Financial Markets, Institutions and Money 22, 758–773.

[19] Papapetrou, E., 2001. Oil price shocks, stock market, economic activity and employment in Greece, Energy Economics 23, 511–532.

[20] Park, J., and R. A. Ratti, 2008. Oil price shocks and stock markets in the U.S. and 13 European countries, Energy Economics 30, 2587–2608.

[21] Sadorsky, P., 1999. Oil price shocks and stock market activity, Energy Economics 21, 449–469.

[22] Scholtens, B., and A. Boersen, 2011. Stocks and energy shocks: The impact of energy accidents on stock market value, Energy 36, 1698–1702.

[23] Zhang, C., and X. Chen, 2011. The impact of global oil price shocks on China's stock returns: Evidence from the ARJI(-ht)-EGARCH model, Energy 36, 6627–6633.

Network Security and Communication Engineering – Chan (Ed.)
© *2015 Taylor & Francis Group, London, ISBN: 978-1-138-02821-0*

Formal verification of concurrency errors in JavaScript web applications

N. Bhooanusas & D. Pradubsuwun

Department of Computer Science, Thammasat University, Patumthani, Thailand

ABSTRACT: This paper proposes a timed trace theoretic verification to detect concurrency errors in Java Script web applications. They are caused because the parsing operation stops parsing unexpectedly and event handling operations are executed in an unexpected order. Experimented with the eight benchmark applications, the proposed approach shows its effectiveness.

1 INTRODUCTION

Developing web applications by JavaScript is becoming much more popular because it supports a lot of components on client-side and executes independently on the web browser (Swiech & Dinda, 2013).Therefore, several frameworks for implementing JavaScript such as JQuery, Dojo, node.js etc. have been proposed. It is likely that each of them works asynchronously e.g. Ajax (Lin 2008). This causes concurrency errors in JavaScript web applications because of unexpected execution orders of it. Presently testing has been introduced to detect concurrency errors in JavaScript web applications(Hong et al. 2014). A number of test cases for JavaScript web applications must be generated and simulated with the system. Checking the correctness of the system is not only a test generation and simulation, but also formal verifications.

One of them is the timed trace theory. It is used to verify a concurrent system i.e. a synchronous circuit (Pradubsuwun et al. 2005) and web service compositions (Kongburan & Pradubsuwun 2014, Pradubsuwun & Kongburan 2014, Pradubsuwun 2014). The failures are detected by conformance checking between implementation and its specification. Both are represented by time Petri nets.

In this paper, we proposed a timed trace theoretic verification to detect concurrency errors in Java Script web applications. The JavaScript web application is modeled by the time Petri net. Then, it is verified by a verification tool (Pradubsuwun et al. 2005). Similarly, the proposed method is demonstrated with the eight benchmark applications (Hong et al. 2014).

This paper is organized as follows. Concurrency errors in JavaScript web applications are described in the section 2 and Timed trace theoretical verification is explained in the section 3. The proposed method and experimental results with the benchmarks are shown in the section 4, and section 5, respectively. Finally, the conclusion is given in section 6.

2 CONCURRENCY ERRORS IN JAVASCRIPT WEB APPLICATIONS

A web application can be implemented by JavaScript. The JavaScript provides several components for developing the web application. The model of execution of a web application can be classified into two operations i.e. parsing operations and event handling operations (Hong 2014).

A parsing operation is used to construct a Document Object Model (DOM) tree and identify the event handler function in JavaScript. Generally, multiple parsing operations can be contained in an execution of the web application. However, it must be paused when a waiting point is reached. The waiting point consists of a script element for an external file, an HTML element whose content is longer than a buffer size, and a JavaScript instruction alert().

An event handling operation is defined as an interaction between the components in JavaScript. It includes network event operations e.g. img tag or AJAX, user input event operations e.g. button tag and timed event operations e.g. set Timeout function.

Normally, all of them are atomically and sequentially executed through a web browser. Otherwise, they satisfy order violations and atomicity violations (Zheng et al. 2011). Hence, concurrency errors in JavaScript web applications will occur.

```
1: <html>
2: <head>
3: </head>
4: <body>
5:      ..
6:      <script src="jquery-1.9.1.js"></script>
7:      <script>
8:          function button1Clicked() {
9:              list = null;
10:             jQuery.getJSON("case8.txt",{},function(result) {
11:                 list = result;
12:                 ...
13:             });
14:             ...
15:         }
16:         function button2Clicked() {
17:             ...
18:                 avg += list[i];
19:                 ..
20:         }
21:     </script>
22:     ...
23:     <button onclick="button1Clicked()">Button1</button>
24:     ...
25:     <button onclick="button2Clicked()">Button2</button>
26:     ...
27: </body>
28: </html>
```

(a) The shortened source code of B8.

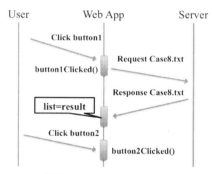

(b) Correct execution scenario.

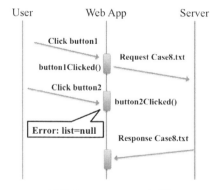

(c) Concurrency errors in B8.

Figure 1. Concurrency errors in a JavaScript web application.

We now give you an example of a concurrency error in a JavaScript web application derived from the eight benchmark applications. It is a benchmark 8 or B8.The shortened source code of B8 is shown in Figure 1(a). This is an interaction between user input event operations and network event operations. The whole task is explained as follows. Firstly, the user clicks the button1 (user input event operation) in order to get a text file (network event operation) i.e. "*case8.txt*" from a server. The result from this text file is then stored into a *list* variable. After that the user clicks the button 2 (user input event operation).This is a correct execution scenario shown in Figure 1(b). In Figure 1(c), it shows a situation when the concurrency error in the B8 occurs. This is because the last user input event operations happen before the network event operation is completed. It means that the user clicks the button 2 before the result from the text file finish storing into a *list* variable.

3 TIMED TRACE THEORETIC VERIFICATION

In this paper, the timed trace theory (Pradubsuwun et al. 2005) is used as a formal method for detecting concurrency errors. Here, we briefly described its concept. Initially, the implementation and specification are modeled by modules and a semi-module, respectively. A module M consists of three-tuple i.e. I is a set of input transition, O is a set of output transition and N is a time Petri net. The time Petri net N is composed of six-tuple i.e. P is a finite non-empty set of places, T is a finite set of transitions ($P \cap T = \varnothing$), F is a set of flow relations ($F \subseteq (P \times T) \cup (T \times P)$), $lb : T \rightarrow R^+$ and $ub : T \rightarrow R^+ \cup \infty$ are functions for the earliest and latest firing time of transitions, satisfying $lb(t) \leq ub(t)$ for all $t \in T$, and μ_0 is the initial marking of the net. Each transition t must be enabled and fired within time bound $lb(t)$ and $ub(t)$. A timed trace structure of a module M or $T(M)$ is a four-tuple (I, O, S, F) where S is a success trace set and F is a failure trace set. A semi-module has the same tuple as the module, but its timed trace structure, $T_s(M)$, is defined differently. The semi-module of a module M is a semimirror of M denoted by $M^{sm} = (O, I, N)$. Thus, the implementation modules conforms to its specification module if the composition of $T(M_1)$ and $T_s(M_2^{sm})$, written by $T(M_1) \parallel T_s(M_2^{sm})$, is failure-free. Intuitively, the implementation modules behave the same as the specification module with any environment without failures. Conformation checking can be done by Theorem1, which is shown as follows.

Theorem1: $\{M_1,...M_{k-1},M_{k1},...,M_{km},M_{k+1},...,M_n\}$ conforms to M_s, if $\{M_{k1},...,M_{km}\}$ conforms to M_k, and $\{M_1,...,M_{k-1},M_k,M_{k+1},...,M_n\}$ conforms to M_s.

Note: that implementation of system may consist of several modules including $M_1,...M_{k-1},M_k,M_{k+1},...,$ and M_n such as $M_k = \{M_{k-1},...,M_{km}\}$ and its specification is represented by module M_s.

4 PROPOSED METHOD

An overview of a proposed method is illustrated in Figure 2. Firstly, a JavaScript web application source code is modeled as the time Petri net by an Algorithm 1. The output of the Algorithm1 is the time Petri net of JavaScript web applications. We call that the implementation. The specification is also modeled by the time Petri net. Then, the implementation is verified with its specification by a verification tool (Pradubsuwun et al. 2005). If concurrency errors happen, the counterexample will be produced. It helps us to locate the position of errors.

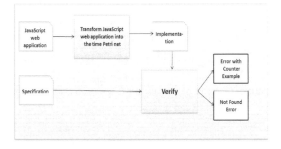

Figure 2.   The overview of a proposed method.

The Algorithm 1 is used to transform HTML files or JavaScript files representing a web application into the time Petri net. The input of this algorithm is the JavaScript web application. It is modeled as the time Petri net corresponding to the parsing operations, the event handling operation and the waiting point contained in the file. We demonstrate this algorithm with the eight benchmark applications i.e. the benchmark B1 - B8. The result is shown in Figure 3(a) – 3(h), respectively.

Algorithm 1: Transforms the JavaScript web application into the time Petri net.

Input: The JavaScript web application source code
Output: The time Petri net $N(P, T, F, lb, ub, \mu_0)$
1:    **begin**
2:       *nested browsing ← iframe tag*
3:       *network event operation ← img tag ∪ AJAX*
4:       *user input event operation ← button tag*
5:       *timed event operation ← setTimeout*
6:       *waiting point ← script tag with src attribute*
           ∪ html whose content is longer
               than a buffer size
                  ∪ alert() function
7:    Extract *nested browsing, network _ event _ operation,*
         user _ input _ event _ operation,
         timed _ event _ operation, waiting point
         from ZHTML file.
8:    Initial a time Petri Net N.
9:       **for** each *tag ∈ waiting point* **do**
10:       identify tag as a start point and an end point for parsing
11:    **for** start point tag **to** end point tag **do**
12:       Read the HTML file line by line.
13:       **if** *tag ∈ nested browsing, network _ operation*
, *user _ input _ operation, time _ operation* **then**
14:          subnet = call model the time Petri net (tag)
15:       **end if**
16:       *append*(the time Petri net N, subnet)
17:    **end for**
18:    **end for**
19:    **end**

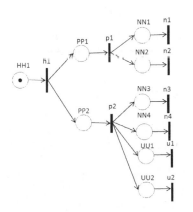

(a) The implementation of the time Petri net of B1

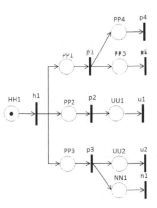

(b) The implementation of the time Petri net of B2.

433

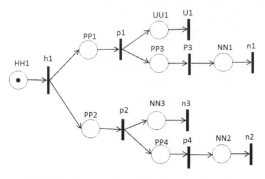

(c) The implementation of the time Petri net of B3.

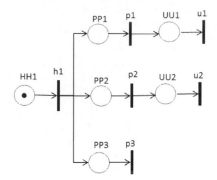

(f) The implementation of the time Petri net of B6.

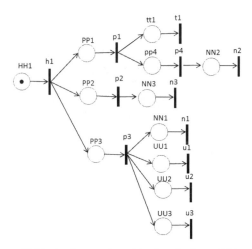

(d) The implementation of the time Petri net of B4.

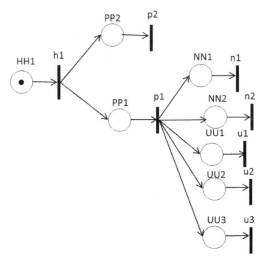

(g) The implementation of the time Petri net of B7.

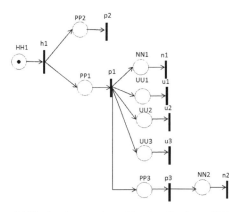

(e) The implementation of the time Petri net of B5.

Figure 3.   The time Petri net of the eightbenchmarksB1-B8.

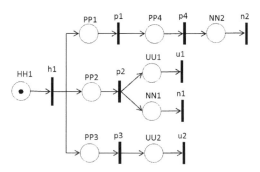

(h) The implementation of the time Petri net of B8.

The specification is obtained from the design phase. Then, it is represented by the time Petri net. For example, the specification of the benchmark B8 expects that the execution of the network event operation (i.e. jQuery.getJSON()) must complete before the user input event operation (i.e. click on the button 2) occurs. Hence, the time Petri net of the specification of the benchmark B8 is shown in Figure 4. In Figure 4, jQuery.getJSON() and click on the button 2 operation are represented by the transition n1 and u2, respectively. We then verified the JavaScript web application. For example, the implementation of the benchmark B8 in Figure 3(h) and its specification in Figure 4 are verified to detect concurrency errors.

5   EXPERIMENTAL RESULTS

In this section, we demonstrated our proposed method with the eight benchmark applications. The benchmark B1, B2, B3, B4, and B8 are concurrency errors because of unexpected executing order in the JavaScript web application. Also, the benchmark B5, B6, and B7 are concurrency because the parsing operation stops parsing unexpectedly. Here, the experiment has been done on a 2.2 GHz Intel corei7 with 8 gigabytes of memory. Table 1 shows the concurrency error traces when verifying the benchmark B1-B8. The experimental results show that concurrency errors can be detected by the proposed method.

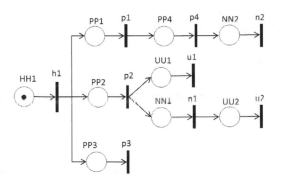

Figure 4.   The specification of the time Petri net of B8.

Table 1.   Concurrency error traces of the B1-B8.

Benchmark	Concurrency error traces
B1	h1, p1, n2, n1
B2	h1,p1,p4,u2,p5
B3	h1,p1,p2,n3,p4,p3
B4	h1,p1,p2,t1,n3
B5	h1,p1,u3,p2
B6	h1,p1,u2,p3
B7	h1,p1,p2,u1,u2,n1,u3,n2
B8	h1,p1,p2,u2,n1

6   CONCLUSION

In this paper, we have proposed a timed trace theoretic verification to detect concurrency errors in Java Script web applications. Demonstrating the proposed method with the eight benchmark applications, the concurrency errors can be detected effectively. The error traces are also expressed for locating the position of a bug. In the future, we plan to verify the real-world JavaScript web application.

REFERENCES

[1] Hong, Shin & Park, Yonbae & Kin, Moonzoo 2014. *Detecting Concurrency Error in Client-side JavaScript Web Application*.Cleveland:IEEE.
[2] Kong buran, Wutthipong & Pradubsuwan, Denduang .2014. Formal *Verification* of WS-BPEL using Timed Trace Theory. *Advance Materials Research*931–932:1452–1456.
[3] Lin, Zhijie & Wu, *Jiyi*& ZHANG, Qifei & ZHOU, Hong 2008. *Research on Web Applications Using Ajax New Technologies*.Three Gorges: IEEE.
[4] Pradubsuwun, Denduang & Yoneda, Tomohiro& Myers, Chris 2005, Partial Order Reduction for Detecting Safety and Timimg Failure of Timed Circuit. *IEICE* E88-D(7): 1–15.
[5] Pradubsuwun, Denduang 2014, *Hierarchical Verification of WS-BPEL*.Thailand.
[6] Pradubsuwun, Denduang & Konghuran, Wutthipong 2014. Patial order *reduction* for verification of WS-BPEL, *International journal of Modeling and Optimization* 4(3):239–245.
[7] Swiech, Maciej & Dinda, Peter 2013.*Make JavaScript Better By Making It Even Slower*.San Francisco:IEEE.
[8] Zhen, Yunhui & Bao, Tao & Zhang, Xiangyu 2011. *Statically Locating Web Application Bugs Caused by Asynchronous Calls*.Hyderabad:ACM.

Network Security and Communication Engineering – Chan (Ed.)
© *2015 Taylor & Francis Group, London, ISBN: 978-1-138-02821-0*

The impact of import on economic growth: Evidence from China

X.D. Liu & J. Zhang
Donlinks School of Economics and Management, University of Science and Technology Beijing, Beijing, China

ABSTRACT: In exploring the effect of foreign trade on economic growth, the major concern is always about export while import is totally overlooked. In fact, import plays a key role in national economic growth. In this paper we set a VEC model to analyze the influence of import on the economic growth of China, by using data on GDP, total imports, R&D investment, the investment in fixed assets, and exchange rate. The database covers the period from 1978 to 2013. The empirical results show that import has negative effects on GDP growth in the short term, but it also has positive effects in the long term. One explanation is that the increase of imports was not coordinated with the national economic growth in the past. That means the structure of imports is unreasonable. Specifically, China lacks the importation of high technology, which can drive the economic growth.

1 INTRODUCTION

After the reform and opening up, international trade has become one of the most important engines for economic development in China. Although the free-trade policy is dominant, in general, trade theorists and policy makers are inclined to encourage export and restrict import (Tan, 2012). China also pursues export-oriented trade policy. Import has played a relatively passive role in foreign trade for years (Wang, 2008). After the 2008 global financial crisis, which was triggered by the U.S. sub-prime mortgage crisis, international trade protectionism has spread over the world. The collapse in exports highlights the defect of China's economic growth pattern which overly depends on export and investment.

At present, there are internal and external conditions for China to expand imports and import growth rate (Sun, 2007). The Twelfth Five-year Plan of China adjusted the foreign trade policy. China began to value the import and export equally. Therefore, to explore the relationship between import and economic growth is quite meaningful.

Import is one of vital determinants for economic development (Awokuse, 2008). As Krugman (1993) said: "The import, rather than export, is the purpose of trade. Export is not the purpose of trade, however, it's the burden when imports cannot meet the demand." In the long run, the ultimate goal of export is import. For a country, the aim of export is to exert its own advantages, while the purpose of import is to learn and to use other country's strengths (Xiong & Yang, 2005).

Technical progress is a key factor for long-term economic growth. New technologies are usually embodied in intermediate goods and capital goods. However, in the process of industrialization, developing countries cannot produce them. Import helps developing countries to acquire foreign technology from R&D intensive countries (Busse & Groizard, 2008). When they have the access to variety of higher quality intermediate inputs in foreign markets, the efficiency of capital will be improved (Amiti & Konings, 2007).

In addition to being intermediaries for technology transfer, import may also motivate domestic innovation through competition, thus promoting the rise of labor productivity (Awokuse & Yin, 2010). Countries involved in international trade are more likely to survive and thrive, because when participating in international trade, enterprises are forced to raise the production efficiency.

Throughout research about the relation between the import trade and economic growth of China, we found that several points need to be improved as follow. Firstly, many of the literature modeled without dealing with non-stationary data, resulting in a "spurious regression" problem. Secondly, differences before and after the reform and opening up were not taken into consideration. Thus, although the sample expansion was expanded, the data quality was significantly decreased (Sun et al., 2009). Finally, the effect of inflation was not eliminated. To this end, we have made various improvements about the defects to analyze the role of import in national economic development more comprehensively.

2 MODEL SUMMARY

2.1 VAR model

VAR model is one of the most user-friendly models to achieve the analysis and forecast of multiple correlated economic indicators. VAR model is built based on the statistical nature of the data. We construct the model by regarding each endogenous variable as a function of all endogenous variables' lagged values in the system, and then extend univariate auto-regressive model to vector auto-regressive model built by multivariate variables of time series. The VAR model of single variable is shown as Equation 1 below:

$$\mathbf{y}_t = \Phi_1 \mathbf{y}_{t-1} + \cdots + \Phi_p \mathbf{y}_{t-p} + \mathbf{H}\mathbf{x}_t + \varepsilon_t, t = 1,2,\cdots,T \quad (1)$$

where $\mathbf{y}_t$ = k-dimensional column vector of endogenous variables; $\mathbf{x}_t$ = d-dimensional column vector of exogenous variables; p = lag order; T = the number of samples; ε_t = k-dimensional column perturbation vector; and $\phi_1,\ldots\ldots,\phi_p$ and H are matrices of coefficients need to be estimated. ε_t is the white noise sequence.

VAR model can also be extended to be multivariate, and the formula is as Equation 2.

$$\begin{pmatrix} y_{1t} \\ y_{2t} \\ \vdots \\ y_{kt} \end{pmatrix} = \Phi_1 \begin{pmatrix} y_{1\,t-1} \\ y_{2\,t-1} \\ \vdots \\ y_{kt-1} \end{pmatrix} + \cdots + \Phi_p \begin{pmatrix} y_{1\,t-p} \\ y_{2\,t-p} \\ \vdots \\ y_{kt-p} \end{pmatrix} + \mathbf{H} \begin{pmatrix} x_{1t} \\ x_{2t} \\ \vdots \\ x_{dt} \end{pmatrix} + \begin{pmatrix} \varepsilon_{1t} \\ \varepsilon_{2t} \\ \vdots \\ \varepsilon_{kt} \end{pmatrix} \quad (2)$$

2.2 VEC model

In the VAR model, all endogenous variables y_{kt} should be stationary. On condition that all endogenous variables contain unit roots and that there is co-integration relationship among variables, we can get the VEC mode as Equation 3.

$$\Delta \mathbf{y}_t = \alpha\beta' \mathbf{y}_{t-1} + \sum_{i=1}^{p-1} \Gamma_i \Delta \mathbf{y}_{t-i} + \mathbf{H}\mathbf{x}_t + \varepsilon_t \quad (3)$$

Each equation in the VAR model is an auto-regressive lag model; therefore, we can regard the VEC model as a VAR model with co-integration constraints.

3 EMPIRICAL ANALYSIS

3.1 Data collection and ADF test

We collect data on Chinese GDP, total imports, R&D investment, the investment in fixed assets, exchange rate, CPI based on previous year and average annual exchange rate of the RMB Yuan against US Dollar. Due to the effect of inflation, they don't reflect the real situation of the national economy. Therefore, we decide to use the CPI to eliminate the effect of inflation. Selecting the base year as 1978, through Equation 4, we can get the CPI based on the year of 1978,

$$\text{CPI}_N = \text{CPI}_{N-1} \times \text{cpi}_N, \text{CPI}_{1978} = 1 \quad (4)$$

where the base year of CPI_N is 1978, and the base year of cpi_N is the previous year.

By dividing related variables by CPI respectively, we can get the standardized data, whose base year is 1978.

To a further analysis, we take the logarithm of all data (except exchange rate) and it will not be explained again.

To test whether each variable is stationary, we make trends drawing of GDP (LNGDP), total imports (LNIM), R&D investment (LNRD), the investment in fixed assets (LNRD), and exchange rate (ER) as Figure 1.

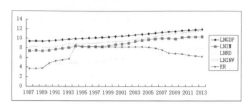

Figure 1. Trend chart of variables.

In Figure 1, it can be seen that all series are not stationary. Therefore, we conduct a first-order difference of all data, and then test whether they are stationary through the ADF test. As is shown in Table 1, in confidence level of 10%, all series are stationary after the first-order difference.

Table 1. Result of ADF text.

Statistic	T-Statistic	Prob.*
D(LNGDP)	-2.679691	0.0915
D(LNIM)	-3.702545	0.0105
D(LNRD)	-3.040723	0.0447
D(LNINV)	-2.954041	0.0534
D(ER)	-4.359078	0.0023

3.2 Lag order and co-integration test

Set VAR model and conduct the lag order selection test. The result is shown in Table 2. According to the AIC criterion and the SC criterion, the lag order is 3. Because the making co-integration test and setting VEC model will lose a degree of freedom, the corresponding lag order is 2.

Table 2. Result of lag order selection test.

Lag	LogL	LR	FPE	AIC	SC	HQ
0	3.54	NA	7.77e-07	0.12	0.37	0.19
1	164.54	241.49	9.82e-12	-11.21	-9.74	-10.82
2	212.04	51.46*	2.06e-12	-13.09	-10.39	-12.37
3	253.74	27.80	1.41e-12*	-14.48*	-10.55*	-13.44*

There are commonly two methods of co-integration test, namely Engle-Granger two-step test and Johansen test. As effective samples are relatively few, we adopt Johansen co-integration test and use Max test as a reference.

438

As is shown in Table 3 and 4, the result of the two tests suggests that there are 4 co-integration equations. The corresponding co-integration equation containing all the variables could be draw as Equation 5:

Table 3. Result of Johansen test.

Hypothesized No. of CE(s)	Eigenvalue	Trace statistic	0.01 critical value	Prob.**
None *	0.966049	174.8891	77.81884	0.0000
At most 1 *	0.827553	93.70067	54.68150	0.0000
At most 2 *	0.690162	51.51664	35.45817	0.0000
At most 3 *	0.547662	23.39570	19.93711	0.0026
At most 4	0.165977	4.355870	6.634897	0.0369

Table 4. Result of Max test.

Hypothesized No. of CE(s)	Eigenvalue	Max-Eigen statistic	0.05 critical value	Prob.**
None *	0.966049	81.18843	39.37013	0.0000
At most 1 *	0.827553	42.18403	32.71527	0.0004
At most 2 *	0.690162	28.12094	25.86121	0.0044
At most 3 *	0.547662	19.03983	18.52001	0.0081
At most 4	0.165977	4.355870	6.634897	0.0369

$$LNGDP=0.089LNIM+0.2766LNINV \\ +0.4477LNRD+0.0350ER \quad (5)$$

According to the Equation 5, the coefficient of LNIM is negative. It means that in the past the development of import was not in harmony with and well-adapted to the development of national economy. The main reasons are that structure of imports is not very reasonable and import is not in line with the optimization and upgrading of national economic industrial structure.

Besides that, the coefficients of LNRD, LNINV and ER are positive. It means that increasing R&D investment, building fixed assets and adjusting the RMB exchange rate regime would promote economic development in import system. Among them, the effect of R&D is the most obvious. On the one hand, the R&D investment will lead to innovation, and promote the assimilation of innovation (Cameron et al. 2005).On the other hand, the complementary nature of the capital accumulation and the innovation activities caused by R&D investment plays a strong role in promoting economic growth.

3.3 Setting VEC model and impulse response

According to the above test result, the VEC model can be established. Subsequently, Granger Causality test should be conducted. The P-value of the joint statistics (ALL) is 0.0013, less than 5%. Therefore, there exists the causal relationship among the variables. Hence, it's reasonable to set the VEC model by using GDP as dependent variable and LNIM, LNRD, LNIM, and ER as independent variables. The VEC equation can be shown as Equation 6:

$$D(LNGDP) =-0.5704LNGDP(-1) \\ +0.0377LNIM(-1)+0.1019LNINV(-1) \\ + 0.2415LNRD(-1) + 0.0180ER(-1) \\ + 0.4878D(LNGDP(-1))-0.2649D(LNGDP(-2)) \\ -0.0990D(LNIM(-1))-0.0816D(LNIM(-2)) \quad (6) \\ + 0.0035D(LNINV(-1))+ 0.0078D(LNINV(-2)) \\ + 0.035D(LNRD(-1)) + 0.0383D(LNRD(-2)) \\ -0.0115D(ER(-1)) + 0.0101D(ER(-2)) \\ +3.2309$$

As can be seen from the VEC equation, the growth of import hinders the growth of GDP, while the change of exchange rate, making R&D investment, and the investment in fixed assets have promoting effects on GDP growth. Compared with R&D investment, the influence of investment in fixed assets is relatively small, while the effect of exchange rate is the smallest.

In order to describe the process of Granger Causality and the relationship between LNGDP and LNIM, LNRD, LNINV, and ER more vividly, we use the impulse response to analyze their dynamic characteristics.

As is shown in Figure 2, in the first 2 years after importing goods, import had a negative impact on GDP, but it was not very obvious. After that, import began to have gradually positive effects. In the first 3 years, the investment in fixed asset had a promoting effect on economy. Then it began to present a negative effect that increased gradually. R&D investment had a positive influence on GDP throughout the sample period. The influence of exchange rate was relatively small in the first five years, and after that period the positive influence was obvious gradually.

The impulse response indicates that over the past 35 years although import had promoting effect on economic growth, but in the short term, there were negative points. It is mainly due to the fact that China had practiced the export-oriented strategy and paid less attention to import than export since the reform and opening up in 1978. At the beginning of the reform, owing to the shortage of foreign exchange, China implemented strict management policies of import. Although the capacity of foreign exchange-earning had been enhanced greatly and the foreign exchange reserves had increased rapidly, this conception still existed. Due to few overall and long-term strategies to stimulate import, import in foreign trade was always in a relatively dependent and passive position. Therefore, the function of import was not exerted fully and timely.

The impulse response indicates that over the past 35 years although import had promoting effect on economic growth, but in the short term, there were negative points. It is mainly due to the fact that China had practiced the export-oriented strategy and paid less attention to import than export since the reform and opening up in 1978. As a result, the development of import and economic growth is not very coordinated and the amount of imports fluctuated tempestuously. At the beginning of the reform, owing to the shortage of foreign exchange, China implemented strict management policies of import. Although the capacity of foreign exchange-earning had been enhanced greatly and the foreign exchange reserves had increased rapidly, this conception still existed. Due to few overall and long-term strategies to stimulate import, import in foreign trade was always in a relatively dependent and passive position. Therefore, the function of import was not exerted fully and timely.

4 CONCLUSIONS AND SUGGESTIONS

Based on the China's data of GDP, total imports, R&D investment, the investment in fixed assets and exchange rate from 1978 to 2013, we build a VEC model, which describes the equilibrium relationship of them. The empirical study shows that import has negative effect on GDP growth in the short term. But the increase of imports could contribute to economic growth in the long term. The conclusion affirms the positive effect of import on economic growth. However, it was not formed in the right year when goods were imported. Instead, import had an inhibitory influence on economic then. Through accumulating, import finally could promote economic growth. The result shows that

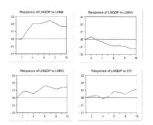

Figure 2. Pulse graphs of each variable to LNGDP.

in the past 35 years, the increase of imports was not coordinated with the national economic growth. The import and export trade policies were unsuitable and the import structure of China was unreasonable.

Therefore, in order to achieve the aim of import better and faster, we propose several recommendations. Firstly, governments at all levels need to change perceptions, pay attention to the importance of import, and adjust the structure of imports actively. Secondly, Chinese government should encourage technical imports, strengthen the introduction of high-tech talents and achievements, and promote the improvement of technology level. Thirdly, it must expand the import of energy resources and raw materials that China lacks and energy-guzzling, highpolluting, resource-intensive products, and limit the import which may hinder the industrial development and crowd out employment opportunities. Fourthly, import consumer goods appropriately, which aims at enhancing the quality of domestic consumer products, enriching consumption levels and optimizing the residents' consumption structure and consumption habits.

ACKNOWLEDGMENTS

This work was supported by the Fundamental Research Funds for the Central Universities (FRF-TP-14-052A1).

REFERENCES

[1] Amiti, M. & Konings, J. 2007. Trade liberalization, intermediate inputs and productivity: evidence from Indonesia. *American Economic Review* 97(5): 1611–1638.
[2] Awokuse, T.O. *2008*. Trade openness and economic growth: is growth export-led or import-led? *Applied Economic* 40(2): 161–173.
[3] Awokuse, T.O. & Yin, H. 2010. Does stronger intellectual property rights protection induce more bilateral trade? Evidence from China's imports. *World Development* 38(8): 1094–1104.
[4] Busse, M.J. & Groizard, L. 2008. Technology trade in economic development. *World Economy* 31(4): 569–592.
[5] Cameron, G., Proudman, J. & Redding, S. 2005. Technological convergence, R&D, *trade* and productivity Growth. *European Economic Review* 9(3): 775–807.
[6] Krugman, P.R. 1993. *What* do Undergrads Need to Know about Trade. *American Economics Review* 83(2): 23–26.
[7] Sun, F.Q., Shao, J.C. & Li, X. 2009. The Import and Export Trade and *Economic* Growth of China: Based on the Research of VAR Model. *Statistics and Decision* 279(3): 99–101.
[8] Sun, J.S. 2007. Studies of the Import-based Trade Contribution to China's Economic Growth. *International Economics And Trade Research* 23(1): 13–18.
[9] Tan, Z.Y. 2012. The Economic-Growth Effect of China's Investment-Oriented Importation. *International Business Research* 33(5): 16–22.
[10] Wang, J. 2008. The Growth Effect of China's Import— The Analysis Based on the Co-integration Test Presenting Structural Breaks. *Contemporary Finance & Economics* 278(1): 90–95.
[11] Xiong, Q.Q. & Yang, S.E. 2005. Reconsidering the Role of Import Played in Economic Growth–Based on an Empirical Research of China. *Journal of International Trade* (2): 5–10.

Network Security and Communication Engineering – Chan (Ed.)
© 2015 Taylor & Francis Group, London, ISBN: 978-1-138-02821-0

A log ontology learning approach to discover non-taxonomic relationships between events

D.Z. Wang, M. Sun & B. Chen
School of Computer Science and Engineering, University of Electronic Science and Technology of China, Chengdu, China

ABSTRACT: Ontology learning has become a hot topic in Semantic Web. To discover the efficient relationship between users' access behaviors, the paper proposes an approach to learning Event domain hierarchical relations. In this approach, Event is used to represent the users' behavior and log ontology is defined to express web usage knowledge. After generating the Event pair based on hierarchical association rules, we refine the results for reducing redundant rules by direct-parent relations. The experimental results show that our approach is efficient and can be used to generate log ontologies subsequently.

1 INTRODUCTION

Web usage mining is one type of Web mining[1] that aims to capture, model, and analyze patterns of user behavior that is recorded in the logs of Web servers as users interact with Web sites. However, information contained in log files is fragmented and lack of the semantic expression which was difficult for sophisticated analyses we intended to perform. The semantic web relying on formal ontologies helps to structure data for comprehensive and transportable machine understanding. Primarily, domain specialists and knowledge engineers obtained the ontology with assistance of the ontology learning[2,3]. While most of researches focus on constructing the web content ontology[4,5], how to construct web usage ontology on semantic web becomes a mandatory requirement[6].

User behavior varies in accordance to the background knowledge of a certain field when user interacts with web sites. The definition of log ontology in ONTOLOGER[7] advances the user behavior to the semantic level, which gives analyzers easy access to the abundant ontology language for the research of web using knowledge. However, this method is inflexible and limitary, because the commonness and relationship are taken into account by the system abstraction. It cannot meet the requirement of data mining for complex semantic web. This paper proposes a new hierarchical relationship learning method for Complex *Event* of log ontology, which is based on web document, user request and user access information.

2 EVENT AND LOG ONTOLOGY

The paper analyzes the user behavior and applies their commonness to a specified field. Log ontology is defined based on the core concept.

Analysis focused on user behavior and strategy is abstracted by semantic information. The concept *Event* proposed in this paper describes the semantic information of user access[8].

2.1 *Event*

Event is the formal description of all the visiting tasks to the business model of a web site

Event contains the commonness and characteristics of web usage information. Properties of *Event* collected from visiting information of page content and log files are employed to describe the features of user visiting from different levels. *Event* can be divided into two categories: Atom *Event* and Complex *Event* in terms of the purpose of visiting the site, interest and kinds of semantic strategies. Atom *Event* describes the user behavior of one-time request to the site. Complex *Event* based on the users' visiting mode of holistic information, which is propitious to acknowledge users' semantic behavior for analyzers.

Atom *Event* (*EA*): describes the user behavior of one-time request to the site. The request includes content Event and service Event. Atom Event can be divided into two categories

Content *Event* (*EC*): $EC \subset EA$, EC is the semantic behavior of users' visit to the page content of sites. It can express the users' interest to the content of a certain web page.

Service *Event* (*ES*): $ES \subset EA$, ES is the semantic behavior of users' access of web service. It means users utilize the web service provided by sites to satisfy the specified request.

Complex *Event* (*EX*): EX is an ordered sequence of EC and ES. $EX = <e_1, e_2, ..., e_n>, e_i \in EC \cup ES, i = 1, 2, ..., n.$

The results of web usage mining are more reasonable under such classification. It is akin to the definition of application *Event* by Stumme[9] classifying *Events* according to the types of users' request. Nevertheless, there is a significant distinction that definition proposed in this paper can perform a better task in describing users' semantic behavior, together with the analysis of commonness and applications in a certain field.

Further steps in web usage mining and inferring users' behavior in the semantic level can be achieved with inference rules under the definition of the concept *Event*. As a domain ontology of web usage, log ontology lies the foundation of the core concept *Event* describing the formal structure of web usage information in the semantic level.

2.2 Log ontology

Definition 1. Given a Description language$L$, Log Ontology (LO) is a sextuple, $LO := \{E, \leq E, R, \leq R, F, A\}$, where: E is a set of *Events*; $\leq E$ is a taxonomy of E; $R = \{hasPart, previouOf, domainR\}$; $\leq R$ is a hierarchy of R; F is a set of relationship function, $F = \{E \times E \rightarrow hasPart, E \times E \rightarrow previous\ Of, E \times E \rightarrow domainR\}$; A is a set of axioms based on L..

Figure 1. The basic structure of LO.

3 THE HIERARCHICAL RELATIONSHIP LEARNING OF EVENT

3.1 The granularity of events

Complex *Event* is defined as the regular expressions of Atom *Event*. Obviously, the former one is concerned with the latter one in *hasPart* relation. Under such restriction, we describe Complex *Event* as a sequence

of Atom *Event-EX* $=< EA_1, EA_2, ..., EA_n >$ and the granularity of *Event* in the following approach based on association rules can be determined. Namely, Complex *Event* corresponds to a specified Atom *Event* transaction and Complex *Events* construct the domain relationship learning space

3.2 Overview of our approach

Based on the algorithm for discovering generalized association rules proposed by Srikant and Agrawal [10], this section discusses the learning approach for the domain hierarchical relationship. The discussion, after extending the algorithm to ontology learning, states that Atom *Event* EA_k and EA_i are related to each other if instances of EA_k and EA_i always exist simultaneously in a Complex *Event*.

The learning method is provided with a set of Complex *Event*:$\mathcal{E}_{EX}' = \{EX_1, EX_2, ..., EX_r\}$ where $EX_i =< EA_{i1}, EA_{i2}, EA_{i3}, ...EA_{ir} >$ and association rules $EA_k \Rightarrow EA_i(EA_k, EA_i \subset EC \cup ES, EA_k \cap EA_i = \varnothing)$ describe the probable relationship between the two Atom *Event*. Thereby, support of a rule $EA_k \Rightarrow EA_i$ is the percentage of Complex *Event* that contains $EA_k \cup EA_i$ as a subset, and confidence is defined as the percentage of all the Complex *Event* that EA_i is seen when EA_k appears in a Complex *Event*, viz

$$support(EA_k \Rightarrow EA_i) = \frac{\left|\{EX_i \mid EA_k \cup EA_i \subseteq EX_i\}\right|}{n}$$

$$confidence(EA_k \Rightarrow EA_i) = \frac{\left|\{EX_i \mid EA_k \cup EA_i \subseteq EX_i\}\right|}{\left|\{EX_i \mid EA_k \subseteq EX_i\}\right|}$$

The paper mainly focused on domain relationships between two Atom *Events*, So we have $| EA_k |=| EA_i |= 1$. For the consideration that hierarchical relationship has influences on the users' access behavior, learning method of association rules will be applied after the expansion to Complex *Event* space based on *Event* hierarchical structure and the re-determination to the granularity of transactions. Expansion means to make $\mathcal{E}'_{EX} = \{EX'_1, EX'_2, ... EX'_r\}$ where $EX'_i = EX_i \cup EA_{il} \mid (EA_{il}, EA_{ij}) \in He, EA_{ij} \in EX_i\}$.

Moreover, *minsup* and *minconf* are defined to represent the given minimal threshold of support and confidence respectively, and the algorithm for discovering the *Event* domain relationship by association rules is given below:

Input: Complex *Event* space $\mathcal{E}_{EX}$, Atom *Event* hierarchical relationship tree T.

expand $\mathcal{E}_{EX}$ to $\mathcal{E}'_{EX}$
initialize *Event* Pair set $EP = \varnothing$,
Domain Relationship Library $R = \varnothing$
for each $< EA_k, EA_i >\in LO$ core Library
if $EA_k \neq EA_i \&\& EA_k \neq He(EA_i) \&\& EA_i \neq He(EA_k)$

$EP = EP \cup \{< EA_k, EA_l >\}$;

for each $< EA_k, EA_l >\in EP$

    if $support(EA_k \Rightarrow EA_l) \geq minsup$

        put $EA_k \Rightarrow EA$ into R

    if $confidence(EA_k \Rightarrow EA_l) < minconf$

        delete $EA_k \Rightarrow EA_l$ from R

Output: Domain Relationship Library R

3.3 Refinement of domain relationship

A lot of unimportant rules will be produced by comparing with 2 threshold because of the existence of present hierarchical relationship. Therefore refinement must be done to the generated rules.

Definition 2. Given $EA_k' = Hd(EA_k)$(Hd means EA_k' is the parent node of EA_k) and $EA_l' = Hd(EA_l)$, we call relations $EA_k' \Rightarrow EA_l$, $EA_k \Rightarrow EA_l'$ and $EA_k' \Rightarrow EA_l'$ are direct-parent relations of $EA_k \Rightarrow EA_l$.

Therefore, the confidence *Econfidence* and support *Esupport* are defined as follows:

$$Econfidence_{EA_k' \Rightarrow EA_l'}(EA_k \Rightarrow EA_l) = \frac{confidence(EA_k \Rightarrow EA_l)}{confidence(EA_k' \Rightarrow EA_l')}$$

$$Esupport_{EA_k' \Rightarrow EA_l'}(EA_k \Rightarrow EA_l) = \frac{support(EA_k \Rightarrow EA_l)}{support(EA_k' \Rightarrow EA_l')}$$

Refinement: k is a given threshold of interest to the rule $EA_k \Rightarrow EA_l$ and rules will be discarded because either measure *Esupport* or *Econfidence* fall below the k.

4 STIMULATION EXPERIMENTS

To test the proposed method, the paper develop the prototype system with Java language and Text2Onto, an open source software for ontology learning. It works well with concepts discovery, hierarchical and non-hierarchical relationship learning between concepts. Meanwhile, the accuracy of Text2Onto varies within a range between 70% and 90% according to different corpuses.

The experiments were conducted on a personal computer with a 2.4GHz Intel Core i5, 4G RAM, windows 2012 SQL Server, J2SDK1.7. The test data came from the website http://www.aifb.uni-karls-ruhe.de/Projekte and CO stands for the corpus after pre-processing[11], as is shown in the table.

Table 3 shows coefficient *support* and *confidence* of some rules. Lined ones will be deleted since either of their coefficient was not satisfied and also, because of much higher support and thought to be ordinary, we will ignore these highlighted relations.

Figure 2 shows the performance testing of the algorithm. The following function curves stands for

Table 1. Corpus CO.

Corpus	Data set	Data Structures	Number of Component	Component dimensions
CO	Page theme set	set	1013	1
	transaction file	sequence	24533	1/27/8.879 (min/max/average)
	Request Set	tipple	1143	2
	Page eigenvector space	VSM	1116	1013

Table 2. Support and confidence.

Atom *Event* Pair	Support	Confidence
(ViewSamsung,ViewIphone)	0.38	0.72
(SearchSamsung,SearchIphone)	0.34	0.71
(ViewSamsung, ScarchIphonc)	0.02	0.25
(ViewSamsung, ViewSE)	0.14	0.38
(ViewSamsungMobile,View Samsung Accessory)	0.19	0.68
(ViewSamsungMobile, ViewIphone Accessory)	0.03	0.67
(ViewSamsungMobile,View SamsungDC)	0.04	0.54
(HelpContact,Help)	0.16	0.71
(HelpContact,HelpDetails2)	0.14	0.63
(SearchProduct,ViewSonyDC)	0.13	0.13
(SearchProduct,View SamsungDC)	0.01	0.09
(ViewSonyDC,HelpContact)	0.14	0.49
(ViewSamsungDC, HelpContact)	0.13	0.47
(ViewSamsungMobile,Payment)	0.12	0.36
(ViewSonyDC,Payment)	0.04	0.36
(ViewSEDC,Payment)	0.03	0.27
(AddToCart, Payment)	0.36	0.99
(SearchProduct,Payment)	0.10	0.69

the influences that coefficient *minsup* have on the efficiency of the algorithm in the case of 5000, 7500 and 10000 transactions.

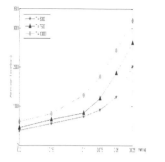

Figure 2.   Performance testing.

As is shown in the figure, algorithm time complexity is linear when *minsup* ranges between 0.2 and 0.075. However, when *minsup* goes below 0.075, time consuming of the algorithm grows extraordinarily fast, that is, the setting of *minsup* matters the efficiency a lot. Therefore, to avoid producing more redundant rules and lowering the efficiency, *minsup* is set to be 0.1 in the stimulation. It's worth mentioning that the interest threshold k could also affect the extraction of relations. So *minsup* and k are determined dynamically for the optimum in the learning process in terms of applications

5   CONCLUSION

In this paper, we propose a learning approach to domain hierarchical relationship of log ontology with the combination of conventional mining methods and semantic web usage mining. This approach based on association rules compute the support and confidence for every *Event* pair, extract those *Event* pairs satisfying the requirement as rules and finally refine the rules with definition of threshold k. Results exhibited in the stimulation expressed the semantic features of the site usage data. Namely, they can play an important role in Semantic Web usage mining as well as guidance to generate log ontology. Above all, this approach is thought to be effective.

ACKNOWLEDGEMENTS

This work is supported by "the Fundamental Research Funds for the Central Universities" (No. ZYGX2012J076) and Jiangsu Provincial Key Laboratory of E-business, Nanjing University of Finance and Economics.

REFERENCES

[1] Stephen G. Matthews, Mario A. Gongora, Adrian A. Hopgood, Samad Ahmadi, Web usage mining with evolutionary extraction of temporal fuzzy ssociation rules, Knowledge-Based Systems, 2013,66–72.

[2] V. Gorodetsky, V. Samoylov, S. Serebryakov, Ontology based context dependent personalization technologyl, International Conference on Web Intelligence and Intelligent Agent Technology, 2010, 278–283.

[3] D. Vuljani, L. Rovan, Mirta Baranovi, Semantically enhanced web personalization approaches and techniquesl, 32nd IEEE International Conference on Information Technology Interfaces (ITA'10), 2010. 217–222.

[4] L. Stojanovic, N. Stojanovic, J. Gonzalez, et al. OntoManager-a system for the usage-based ontology management. In: On The Move to Meaningful Internet Systems 2003: CoopIS, DOA, and ODBASE. Heidelberg: Springer Berlin,2003, 858–875.

[5] Wang, zhe, Semantic web mining based on ontology learning In: WIT Transactions on Information and Communication Technologies, 54 VOLUME 1, p 1103–1110, 2014.

[6] M. Thakur, Y. Kumar, G. Silakari. Query based Personalization in Semantic Web Mining. International Journal of Advanced Computer Science and Applications, (2)2011, 177–123.

[7] N. Stojanovic, J. Gonzalez, L. Stojanovic. Ontologer-a system for usage-driven management of ontology-based information portals. In: International Conference On Knowledge Capture. New York: ACM, 2003, 172–179.

[8] Sun Ming, Shang Qinghong, Chen Bo, A Taxonomic relationship Learning Approach for Log Ontology Content Event, Journal of Digital Information Management, 2012.4.

[9] B. Berendt, A. Hotho, G. Stumme. Towards Semantic Web mining. In: Proc. of the 1st International Semantic Web Conference on The Semantic Web (ISWC). London, UK: Springer, 2002, 264–278.

[10] A. Maedche, S. Staab. Discovering conceptual relations from text. In: ECAI 2000. Proc. of the 14th European Conference on Artificial Intelligence. Amsterdam, Berlin: IOS Press, 2000, 321–325.

[11] W. Wenshan, L. Haihua, Base on rough set of clustering algorithmin network education application, IEEE International Conference onComputer Application and System Modeling (ICCASM 2010), 2010, 481–483.

Network Security and Communication Engineering – Chan (Ed.)
© 2015 Taylor & Francis Group, London, ISBN: 978-1-138-02821-0

A formal study of event ontology assertion elements

P.F. Xia, Z.T. Liu & F.J. Liu
School of Computer Engineering and Science, Shanghai University, Shanghai, China

ABSTRACT: More and more people had been attracted by construction of the event ontology system, now its formal description is very important. In this paper, Z language combining with other formal methods, formally describe the assertion elements of the event ontology. Through a series of formal description, we give a set of formal methods suitable to the event ontology.

KEYWORDS: Event ontology, Assertion elements, Formal methods, Z language.

1 INTRODUCTION

With the rapid development of computer information processing technology, the role of the ontology has been widely recognized by scholars. Professor Liu Zongtian research team proposed an event ontology model in 2009[1]. The event ontology is composed of six elements of the event. Different from the traditional ontology, the event ontology emphasizes the dynamic characteristics of the event.

According to the definition of the event ontology, the assertion that one of the six elements fully reflects the dynamic nature of the event changes, pre-event status changes to the state of event occurring over time, and then into the post-event status. In order to achieve the formal description of the event ontology better, the formal description of the assertion elements is particularly important.

In this paper, through the analysis of corpus emergencies (CEC) texts, we try to adopt a formal method based on Z language, to show the feasibility and effectiveness of the formal description of the event ontology assertion elements.

2 RELATED CONCEPTS

2.1 *Event ontology assertion elements*

Early in the research process, we proposed an event ontology representation model. The model consists of six elements representation of events and event class. Event is defined as occurring at a specific time and circumstance, participation by a number of roles, and it exhibits some action characteristics of one thing. Six-tuple representation of the event, $e = (o,a,t,v,p,l)$, represent object, action, time, environment, assertions, and language performance.

This paper studies the assertion elements, including pre-condition, post-condition and intermediate assertions. Pre-condition denotes the set of states that has to be satisfied for triggering the event. Post-condition is a result set of states after event happens. Intermediate assertions denote intermediate states and the process of state change during the event over time. Through the early experimental study, it shows that the event ontology model has viable solution and research value, but there are still many problems and difficulties for us to solve.

2.2 *Uncertainty*

Uncertainty in the event ontology is mainly reflected in the probability of an event occurring or not. We must obtain a large number of similar events to get the statistical data. In real life, some events may certainly happen, and there also exists an event which will not occur. Random events are existing between them. Each of random events occurs with a certain probability. To get the probability of the events state change or events which are happening, we should require a large amount of experimental data by statistics or knowledge of experts' recommendations [2, 3].

According to the analysis of traffic accidents texts in CEC (have been marked 170 texts).

Table 1. Accidents trigger words statistics.

trigger words	Occurrences	proportion
collision	121	71.18%
traffic accidents	82	48.24%
rear-end	32	18.82%
rollover	27	15.88%
out of control	6	3.53%

From Table1, we can see that, in traffic accidents texts, there is a 71.18% probability of an event triggered by the collision; of course we omitted some less frequently triggered words. However, because news text corpus is different from the real life, news is just a small part of life. The statistics results are only one-sided, not very objective. In this research, we combine formal description and the corresponding probabilities, and unity is expressed by μ.

2.3 Formal methods

Obviously, the use of natural language description of the non-formal specification has the advantage of easy understanding, but it usually has fuzziness and ambiguity. In order to overcome this defect of natural language, scholars have proposed a new regulatory model through formal and standardized theory to describe the system requirements. There are many formal methods, such as description logic, scenario logic, Z language, etc.

Description logic is an object-based knowledge representation formal language, suitable for describing knowledge about concepts and concept hierarchy. Description logic is a first-order predicate logic that can determine the subset, and has a rich symbolic representation system [4]. A unified approach is used to describe the state and action. Description logic generally describes the concept which contains a clear definition. But in the real life, there are many uncertainties in the concept description, so we usually combine description logic and fuzzy logic. Because the concept is static and event is dynamic, description logic models can not represent the framework of the event well.

Scenario logic is used to describe an event related with a certain place, time, occasion and object. At the time of application, scenarios logic is often combined with temporal logic. Space and time are combined to make an issue described in greater image, like scenario reproduction. Situational logic is described by the corresponding symbol rules, in favor of reasoning calculus between scenarios, and is also suitable reasoning applied between events. But it is not entirely suitable for the process of change before and after the event to formally describe [5,6].

Z language is based on the form of a first-order predicate logic and sets theory specification language. The operation of the pre-and post-conditions is described in thinking [7]. Z language has a special graphic mode (Figure 1). At the top, identifier for example "Sname" is called schema name. Middle horizontal line is read as "such that". Above the horizontal part is called the declaration section, the following is the predicate part.

Figure 1. Z language model framework.

Z language formal mode integrates the features of description logic, scenario logic and the modeling framework. Because pre-and post-event status changes characteristic elements assertion, Z language has a unique advantage of formal representation. This article will adopt Z language formal method, with accidents domain knowledge framework as background, doing formal study of event ontology assertion elements.

3 THE STUDIES OF ASSERTION ELEMENTS FORMAL DESCRIPTION

In the six elements of the event ontology, assertion is irreplaceable. Formal is expressed as $P=(Ppre,Ppro,Ppost)$[1]. In this equation, Ppre means the pre-condition of assertion, Ppost means the post-condition of assertion and Ppro means the intermediate condition of assertion.

3.1 The pre-condition of assertion

The pre-condition such as pre-condition, pre-condition denote the states that have to be satisfied for triggering the event. In the field of traffic accident, conditions before the event are as follows: state of each object, state of the environment, state of time, etc. Summing up from our CEC texts, descriptive words about the object are usually used to describe the pre-condition of assertion, such as "out of control", which is expressed as pre-condition of assertion of a traffic accident. Assertions in the texts are not necessarily front, middle, and rear co-occurrence, it may be a single and omitting other state.

3.2 The pro-condition of assertion

The pro-condition such as the state is caused by an event that happens. In the field of traffic accident, some state after the events mainly reflect in the state of the object transfer. Analysis of traffic accidents texts in CEC, we are finding that after traffic accident may cause a variety of situations, such as "broken", expressed as post-assertion of a traffic accident.

3.3 Instances of Z language formal descript the traffic accident events

Global entity type definitions, {person, vehicle, location, time}, person is the set of all possible human, it may contains the drivers, walkers, perpetrators, injured, fire and rescue personnel, medical and rescue personnel, police, passengers, etc. Vehicle is the set of all possible modes of transport, it may include cars, trucks, electric cars, motorcycles, bicycles, etc. Location is the set of all possible locations. Time is the set of all possible time [8].

Entities concept description, firstly, the definition of the concept of human entities, mode "person" defines the object person's information (see Figure 2). Secondly, the definition of the concept of transport entity, mode "vehicle" defines the object transport information (see Figure 3).

```
----------------------------person--------------------------
ID: 00000001
Name: person
Category: walkers| drivers| passengers|*
Description: a human being regarded as an individual.
State: ok| broken| dead
Ok: can drive |can walk| move freely |*
Broken: injured| bleed | fractures|*
Dead: lose any functionality
```

Figure 2.   Mode description about person.

```
-----------------------vehicle-----------------------
ID: 00000002
Name: vehicle
Category: cars| trucks| bicycles|*
Description: a thing used for transporting people or goods, especially on land
State: ok| broken| dead
Ok: normal driving | appearance of harmony |*
Broken: some parts failure| damaged|*
Dead: lose any functionality
```

Figure 3.   Mode description about vehicle.

Now we begin to analyze the pre-condition formal model of traffic accident between people and cars (see Figure 4). The pre-condition model of traffic accident between cars are as follows (see Figure 5).

```
------pre_ traffic accident between people and cars --------
ID: 00001001
P: {walkers, drivers, *}
V: {cars, trucks,*}
p.in(ok)
#p≥2
v.in(ok)
p.in{ crossing the road | running a red light | drunk |*}
v.in{ crossing the road | speeding | out of control |*}
dom where = available
dom time = available
```

Figure 4.   Pre-traffic accident between people and cars.

```
----pre_ traffic accident between cars and cars -------
ID: 00002001
P: {drivers, passengers, *}
V: {cars, trucks,  *}
p.in(ok)
#p≥2
v.in(ok)
#v≥2
p.in{fatigue | drunk | running a red light | *}
v.in{running a red light| speeding| out of control| *}
dom where = available
dom time = available
```

Figure 5.   Pre-traffic accident between cars and cars.

```
----- traffic accident between people and cars ----
ID: 00001000
P: {walkers, drivers, *}
v: {cars, trucks,*}
p.in(ok)
v.in(ok)
p₀'=p↦ p.in(ok)    μ(p₀')
p_b'=p↦ p.in(broken)   μ(p_b')
p_d'=p↦ p.in(dead)   μ(p_d')
v₀'=v↦ v.in(ok)   μ(v₀')
v_b'=v↦ v.in(broken)   μ(v_b')
v_d'=v↦ v.in(dead)   μ(v_d')
dom where = available
dom time = available
```

Figure 6.   Traffic accident between people and cars.

447

Then we study the formal model of traffic accident which is happening between people and cars (see Figure 6) and the formal model of traffic accident is happening between cars (see Figure 7).

----- traffic accident between cars and cars ---

ID: 00002000

p: {drivers, passengers, *}

v: {cars, trucks, *}

---

p.in(ok)

v.in(ok)

$\#v \geqslant 2$

$p_o' = p \mapsto p.in(ok) \quad \mu(p_o')$

$p_b' = p \mapsto p.in(broken) \quad \mu(p_b')$

$p_d' = p \mapsto p.in(dead) \quad \mu(p_d')$

$v_o' = v \mapsto v.in(ok) \quad \mu(v_o')$

$v_b' = v \mapsto v.in(broken) \quad \mu(v_b')$

$v_d' = v \mapsto v.in(dead) \quad \mu(v_d')$

dom where = available

dom time = available

Figure 7.   Traffic accident between cars and cars.

Finally, we describe the post-condition formal model of traffic accident between people and cars (see Figure 8). The post-condition formal model of traffic accident between cars are as follows (see Figure 9).

------post_ traffic accident between people and cars ---

ID: 00001002

$\triangle$pre_ traffic accident between people and cars

---

p.in(ok) $\mu(p_o')$

p.in(broken) $\mu(p_b')$

p.in(dead) $\mu(p_d')$

v.in(ok) $\mu(v_o')$

v.in(broken) $\mu(v_b')$

v.in(dead) $\mu(v_d')$

location : unchanged

time : longer

Figure 8.   Post-traffic accident between people and cars.

------post_ traffic accident between cars and cars ---

ID: 00002002

$\triangle$pre_ traffic accident between cars and cars

---

p.in(ok) $\mu(p_o')$

p.in(broken) $\mu(p_b')$

p.in(dead) $\mu(p_d')$

v.in(ok) $\mu(v_o')$

v.in(broken) $\mu(v_b')$

v.in(dead) $\mu(v_d')$

location : unchanged

time : longer

Figure 9.   Post-traffic accident between cars and cars.

After the accident occurred, because people were injured, a large probability would conduct follow-up rescue operations. Formal description of the assertion of the rescue operations are shown below (seeFigure 10, 11, and 12).

-------pre_ rescue operations -------

ID: 12012001

P: { the wounded, doctors, *}

$\#p.in(broken) \geqslant 1$

dom where = available

dom time = available

Figure 10.   Pre-rescue operations.

-------- rescue operations -------------

ID: 12012000

P: { the wounded, doctors, *}

$\#p.in(broken) \geqslant 1$

$p_1' = p1 \mapsto p1.in(ok) \quad \mu(p_o')$

$p_1' = p1 \mapsto p1.in(broken) \quad \mu(p_b')$

$p_1' = p1 \mapsto p1.in(dead) \quad \mu(p_d')$

dom where = available

dom time = available

Figure 11.   Rescue operations.

----post_ rescue operations -----

ID: 12012001

△pre_ rescue operations

$p.in(ok)\quad \mu(p_o')$

$p.in(broken)\quad \mu(p_b')$

$p.in(dead)\quad \mu(p_d')$

dom where = available

dom time = available

Figure 12. Post-rescue operations.

In the figures above, "p.in (S)" represents the people who are in state S. "#p.in (S) ≥ 1" represents at least one person who is in state S. "$\longmapsto$" represents the change of state from front to the back, accordingly, "$\mu(p)$" represents the probability of state change. "dom W" represents the domain of W. Certainly the definition of symbols can refer to the corresponding literature[9,10].

4 REASONING

Hoare presented Hoare logic in 1969[11], it is the inference rules of assertions. Its basic form is {P} Q{R}. Condition P establishment, will lead to the implementation of the event Q, and after event Q termination, it will cause the condition R. Hoare logic for expansion, we can get that, there is another one event S, its basic form is {R'}S{T}. If the condition R and condition R' are the same or closely related, and from this we can infer that an event Q may lead to the event S.

With the formal description of the pre- and post-assert state, making appropriate reasoning becomes clearer and more effective. It fully reflects the Z language formalized model validity and superiority.

5 SUMMARY

This paper attempts to combine the advantages of the formal description of the description logic and scenario logic, and the Z formal language is used to solve the problems of event ontology assertion elements. Through the above analysis and formal description of the accidents class assertion elements, the unique advantages of Z formal language model framework are demonstrated, and it is conducive for events reasoning. The next step is to study the formal description of the other elements of the events and count the probability of uncertainty in the events. Ensure that it reflects the dynamic nature of the formal description of events. Formal description realizes the sharing of domain knowledge. Achieving formal description of the event ontology helps to build the system.

ACKNOWLEDGMENT

This research was financially supported by the National Natural Science Foundation of China under Grant No.61273328.

REFERENCES

[1] Liu Z T, Huang M L, Zhou W, et al. Research on event-oriented ontology model [J]. Computer Science, 2009, 36(11): 189–192.

[2] Tversky A, Kahneman D. Utility, Probability, and Human Decision Making[M]. Springer Netherlands, 1975:141–162.

[3] Cross V. Uncertainty in the automation of ontology matching[C]. Uncertainty Modeling and Analysis, 2003. ISUMA 2003. Fourth International Symposium on. IEEE, 2003:135–140.

[4] Franz Baader, Ian Horrocks, and Ulrike Sattler. Description Logics. In Frank van Harmelen, Vladimir Lifschitz, and Bruce Porter, editors, Handbook of Knowledge Representation. Elsevier, 2007.

[5] Kugler H, Harel D, Pnueli A, et al. Temporal logic for scenario-based specifications[M]. Tools and Algorithms for the Construction and Analysis of Systems. Springer Berlin Heidelberg, 2005: 445–460.

[6] G. P. Zarri, Semantic Web and knowledge representation, in the 13th International Workshop on Database and Expert Systems Applications. DEXA 2002, pp. 75–79.

[7] Potter B, Till D, Sinclair J. An introduction to formal specification and Z [M]. Prentice Hall PTR, 1996.

[8] Liu L, Miao H K. Z specification construction method [J] Computer Engineering, 2000, 26 (2): 39–41.

[9] Whittle J, Schumann J. Generating statechart designs from scenarios[C]. Software Engineering, 2000. Proceedings of the 2000 International Conference on. IEEE, 2000: 314–323.

[10] Harel D, Pnueli A, Schmidt J, et al. On the formal semantics of statecharts[J]. 1987.

[11] Pratt V R. Semantical consideration on floyo-hoare logic[C]. Foundations of Computer Science, 1976., 17th Annual Symposium on. IEEE, 1976:109–121.

Network Security and Communication Engineering – Chan (Ed.)
© 2015 Taylor & Francis Group, London, ISBN: 978-1-138-02821-0

New similarity measure for Intuitionistic Fuzzy Sets

J.H. Park
Department of Applied Mathematics, Pukyong National University, Busan, Korea

Y.C. Kwun
Department of Mathematics, Dong-A University, Busan, Korea

J.H. Koo
Department of Fiber Plastic Design, Dong-A University, Busan, Korea

ABSTRACT: This paper presents a new method for measuring similarity between Intuitionistic Fuzzy Sets (IFSs) and its application to pattern recognition. The geometrical interpretation of IFSs is utilized to generate a new method for measuring similarity in order to calculate the degree of similarity between IFSs. We attempt to prove some properties of the proposed similarity measures. Numerical example is provided to compare the proposed similarity measures with existing measures.

KEYWORDS: Intuitionistic fuzzy sets, Similarity measures, Pattern recognition.

1   INTRODUCTION

The theory of fuzzy set, proposed by Zadeh [15], has been a realistic and practical means to describe the objective world in which we live, and thus has successfully been applied in various fields. Therefore, over the last decades, several higher order fuzzy sets have been introduced in the literature.

As one of the higher order fuzzy sets, intuitionistic fuzzy set (IFS) was proposed by Atanassov [1] to deal with vagueness. The main advantage of the IFS is its property to cope with the uncertainty that may exist due to information impression. Because it assigns to each element a membership degree, a nonmembership degree, and a hesitation degree, IFS constitutes an extension of Zadeh's fuzzy set which only assign to each element a membership degree. So IFS is regarded as a more effective way to deal with vagueness than fuzzy set. Although Gau and Buehrer [4] later presented vague set, it was pointed out by Bustince and Burillo [3] that the notion of vague sets was the same as that of IFSs.

The definition of similarity measure between two IFSs is one of the most interesting topics in IFSs theory. A similarity measure is defined to compare the information carried by IFSs. Measures of similarity between IFSs, as an important tool for decision making, pattern recognition, machine learning, and image processing, have received much attention in recent years. Many similarity measures have been proposed. A few of them come from the well-known distance measures. The first study was carried out by Szmidt and Kacprzyk [11] extending the well-known distances measures, such as Hamming distance and Euclidean distance, to IFS environment and comparing them with the approaches used for ordinary fuzzy sets. Grzegorzewski [5] also extended the Hamming distance, the Euclidean distance, and their normalized counterparts to IFS environment. Hung and Yang [6] extended the Hausdorff distance to IFSs and proposed three similarity measures. On the other hand, instead of extending the well-known measures, some studies defined new similarity for IFSs. Li and Cheng [7] suggested a new similarity measure for IFSs based on the membership degree and nonmembership degree. Mitchell [10] showed that the similarity measure of Li and Cheng had some counterintuitive and had modified the similarity measure based on statistical point of view. Moreover, Liang and Shi [9] presented some examples to show that the similarity measure of Li and Cheng was not reasonable for some conditions and therefore proposed several new similarity measure for IFSs. Li et al. [8] and Ye [14], respectively, conducted comparative study of the existing similarity measures between IFSs. Xu [13] introduced a series of similarity measures for IFSs and applied them to multiple attribute decision making problem based on intuitionistic fuzzy information.

Among similarity measures for IFSs proposed by other authors, some of those, however, cannot satisfy the axioms of similarity or provide counterintuitive cases. Therefore, we propose new similarity measures based on the geometrical representation for IFS. The proposed similarity measure depends on the triplet, membership degree, nonmembership degree, and hesitation margin. This paper proves that the proposed measures satisfy the properties of axiomatic definition for similarity measures. In addition, numerical example is provided to compare the proposed measure with a number of existing measures.

2   BASIC NOTIONS OF IFSS

In the following, we firstly recall basic notions and definitions of IFSs which can be found in [1, 2].

Let X be the universe of discourse. An IFS A in X is an object having the form

$$A = \left\{ \left(x, \mu_A(x), v_A(x) \right) \mid x \in X \right\}. \tag{1}$$

where $\mu_A, v_A : X \to [0, 1]$ denote, respectively, membership and non-membership functions of A with the condition $0 \le \mu_A(x) + v_A(x) \le 1$ for any $x \in X$. Let IFS(X) denote the set of all IFSs in X.

For each $A \in \mathrm{IFS}(X)$, we call $\pi_A(x) = 1 - \mu_A(x) - v_A(x)$ the hesitancy degree of the element $x \in X$ to the set A, and $0 \le \pi_A(x) \le 1$ for all $x \in X$ [1, 2]. $\pi_A(x)$ is also called the intuitionistic index of x to A. It is obviously that $0 \le \pi_A(x) \le 1$ for all $x \in X$. For example, let A be an IFS with membership function $\mu_A(x)$ and nonmembership function $v_A(x)$, respectively. If $\mu_A(x) = 0.7$ and $v_A(x) = 0.1$, then we have $\pi_A(x) = 0.2$. It can be interpreted as the degree that the object x belongs to the IFS A is 0.7, the degree that the object x does not belong to the IFS A is 0.1, and the degree of hesitation is 0.2. By employing the geometrical interpretation, an IFS A can be expressed as

$$A = \left\{ \left(\mu_A(x), v_A(x), \pi_A(x) \right) \mid x \in X \right\} \tag{2}$$

If A is an ordinary fuzzy set, then $\pi_A(x) = 0$ for each $x \in X$. It means that third parameter $\pi_A(x)$ cannot be casually omitted if A is a general IFS, not an ordinary fuzzy set. Therefore, this representation of IFSs will be a point of departure when considering a new method to calculate the degree of similarity between IFSs.

For $A, B \in \mathrm{IFS}(X)$, Atanassov [1] defined the notion of containment as follows: $A \subseteq B \iff \mu_A(x) \le \mu_B(x)$ and $v_A(x) \ge v_B(x)$ for all $x \in X$.

As above-mentioned, we cannot omit the third parameter (hesitancy degree) in the representation of IFSs and then redefine the notion of containment as

follows: $A \subseteq B \iff \mu_A(x) \le \mu_B(x)$, $v_A(x) \ge v_B(x)$ and $\pi_A(x) \ge \pi_B(x)$ for all $x \in X$.

Definition 1. ([10]) Let $S : \mathrm{IFS}(X) \times \mathrm{IFS}(X) \to [0, 1]$ be a mapping. $S(A, B)$ is said to be the degree of similarity between $A \in \mathrm{IFS}(X)$ and $B \in \mathrm{IFS}(X)$ if $S(A, B)$ satisfies the properties (SP1)-(SP4):

- (SP1) $0 \le S(A, B) \le 1$;
- (SP2) $S(A, B) = 1$ if and only if $A = B$;
- (SP3) $S(A, B) = S(B, A)$;
- (SP4) $S(A, C) \le S(A, B)$ and $S(A, C) \le S(B, C)$ if $A \subseteq B \subseteq C$, $A, B, C \in \mathrm{IFS}(X)$.

Definition 2. ([10]) Let $D : \mathrm{IFS}(X) \times \mathrm{IFS}(X) \to [0, 1]$ be a mapping. $D(A, B)$ is called a distance $A \in \mathrm{IFS}(X)$ and $B \in \mathrm{IFS}(X)$ if $D(A, B)$ satisfies the properties (DP1)-(DP4):

- (DP1) $0 \le D(A, B) \le 1$;
- (DP2) $D(A, B) = 1$ if and only if $A = R$;
- (DP3) $D(A, B) = D(B, A)$;
- (DP4) $D(A, B) \le D(A, C)$ and $D(B, C) \le D(A, C)$ if $A \subseteq B \subseteq C$, $A, B, C \in \mathrm{IFS}(X)$.

Because distance and similarity measures are complementary concepts, similarity measures can be used to define distance measures and vice verses.

3   SOME NEW SIMILARITY MEASURES BETWEEN IFSS

Let $X = \left\{ x_1, x_2, \cdots, x_n \right\}$ be a universe of discourse. $A \in \mathrm{IFS}(X)$ and $B \in \mathrm{IFS}(X)$ are two IFSs in X, denoted by $A = \left\{ \left(x, \mu_A(x), v_A(x) \right) \mid x \in X \right\}$ and $B = \left\{ \left(x, \mu_B(x), v_B(x) \right) \mid x \in X \right\}$, respectively.

Li and Cheng [7] defined similarity measure for IFSs as follows:

$$S_d^p(A, B) = 1 - \sqrt[p]{\frac{\sum_{i=1}^{n} \left(\varphi_A(i) - \varphi_B(i) \right)^p}{n}}, \quad 1 \le p < \infty \tag{3}$$

where $\varphi_A(i) = \left(\mu_A(x_i) + 1 - v_A(x_i) \right)/2$, $\varphi_B(i) = \left(\mu_B(x_i) + 1 - v_B(x_i) \right)/2$.

Mitchell [10] found that the similarity measure $S_d^p(A, B)$ would characterize two different IFSs as identical. To overcome this drawback, he proposed a more realistic strong similarity measure in the following form:

$$S_m^p(A, B) = \frac{1}{2} \left(\rho_\mu(A, B) + \rho_v(A, B) \right), 1 \le p < \infty, \tag{4}$$

where $\rho_\mu(A, B) = 1 - \sqrt[p]{\sum_{i=1}^{n} |\mu_A(x_i) - \mu_B(x_i)|^p / n}$, $\rho_v(A, B) = 1 - \sqrt[p]{\sum_{i=1}^{n} |v_A(x_i) - v_B(x_i)|^p / n}$.

Liang and Shi [9] proposed the following similarity measures:

$$S_e^p(A, B) = 1 - \sqrt[p]{\frac{\sum_{i=1}^n \left(\phi_\mu(x_i) - \phi_v(x_i)\right)^P}{n}}, \qquad (5)$$

where $\phi_\mu(x_i) = \left| \mu_A(x_i) - \mu_B(x_i) \right|/2$, $\phi_v(x_i) = \left| (1 - v_A(x_i)) - (1 - v_B(x_i)) \right|/2$ and $1 \leq p < \infty$.

Wang and Xin [12] proposed the following similarity measures:

$$S_f(A, B) = 1 - \frac{1}{n}\sum_{i=1}^n \left(\frac{|\mu_A(x_i) - \mu_B(x_i)| + |v_A(x_i) - v_B(x_i)|}{4} + \frac{\max\left(|\mu_A(x_i) - \mu_B(x_i)|, |v_A(x_i) - v_B(x_i)|\right)}{2} \right)$$
(6)

Example 1. Assume that there are two patterns denoted with IFSs in $X = \{x_1, x_2, x_3\}$. Two patterns A_1 and A_2 are denoted as follows:

$$A_1 = \{ (x_1,\ 0.2,\ 0.6),\ (x_2,\ 0.2,\ 0.6),\ (x_3,\ 0.2,\ 0.5) \},$$
$$A_2 = \{ (x_1,\ 0.4,\ 0.6),\ (x_2,\ 0.2,\ 0.6),\ (x_3,\ 0,\ 0.3) \}.$$

Assume that a sample $B = \{ (x_1, 0.3, 0.7),\ (x_2, 0.3, 0.5),\ (x_3, 0.1, 0.4) \}$ is given. It is obvious that $S_e^p(A_1, B) = S_e^p(A_2, B)$ and $S_f(A_1, B) = S_f(A_2, B)$. The patterns cannot be differentiated using (5) and (6). In other words, we cannot obtain correct recognition results.

Now, to overcome the drawbacks of Liang and Shi's and Wang and Xin's similarity measures, we take into account three parameters describing IFSs to propose a new similarity measure between IFSs based on the geometrical representation of IFSs.

Let $A = \{ (x, \mu_A(x), v_A(x)) \mid x \in X \}$ and $B = \{ (x, \mu_B(x), v_B(x)) \mid x \in X \}$ be two IFSs in X.

We propose a new similarity measure:

$$S_g(A, B) = 1 - \frac{1}{n}\sum_{i=1}^n \left(\frac{|\mu_A(x_i) - \mu_B(x_i)| + |v_A(x_i) - v_B(x_i)| + |\pi_A(x_i) - \pi_B(x_i)|}{4} + \frac{\max\left(|\mu_A(x_i) - \mu_B(x_i)|, |v_A(x_i) - v_B(x_i)|, |\pi_A(x_i) - \pi_B(x_i)|\right)}{2} \right),$$
(7)

where $\pi_A(x_i)$ and $\pi_B(x_i)$ are, respectively, the hesitancy degree of the element $x_i \in X$ to the sets A and B.

Theorem 1. $S_g(A, B)$ is the similarity measure between two IFSs A and B.

Proof. For the sake of simplicity, IFSs A and B are denoted by $A = \{ (\mu_A(x_i), v_A(x_i), \pi_A(x_i)) \mid x_i \in X \}$ and $B = \{ (\mu_B(x_i), v_B(x_i), \pi_B(x_i)) \mid x_i \in X \}$ respectively. Obviously, $S_g(A, B)$ satisfies (SP1) and (SP3) of Definition 1. We only need to prove $S_g(A, B)$ satisfies (SP2) and (SP4).

(SP2): From (5), we have $S_g(A, B) = 1 \Leftrightarrow \mu_A(x_i) = \mu_B(x_i),\ v_A(x_i) = v_B(x_i),\ \pi_A(x_i) = \pi_B(x_i)$, for any $x_i \in X \Leftrightarrow A = B$.

(SP4): For any IFS $C = \{ (\mu_C(x_i), v_C(x_i), \pi_C(x_i)) \mid x_i \in X \}$, if $A \subset B \subset C$, then we have

$$S_g(A, C) = 1 - \frac{1}{n}\sum_{i=1}^n \left(\frac{|\mu_A(x_i) - \mu_C(x_i)| + |v_A(x_i) - v_C(x_i)| + |\pi_A(x_i) - \pi_C(x_i)|}{4} + \frac{\max\left(|\mu_A(x_i) - \mu_C(x_i)|, |v_A(x_i) - v_C(x_i)|, |\pi_A(x_i) - \pi_C(x_i)|\right)}{2} \right).$$

It is easy to see that
$$|\mu_A(x_i) - \mu_C(x_i)| \geq |\mu_A(x_i) - \mu_B(x_i)|,$$
$$|v_A(x_i) - v_C(x_i)| \geq |v_A(x_i) - v_B(x_i)|,$$
$$|\pi_A(x_i) - \pi_C(x_i)| \geq |\pi_A(x_i) - \pi_B(x_i)|.$$
So we have

$$\frac{|\mu_A(x_i) - \mu_C(x_i)| + |v_A(x_i) - v_C(x_i)| + |\pi_A(x_i) - \pi_C(x_i)|}{4}$$
$$+ \frac{\max\left(|\mu_A(x_i) - \mu_C(x_i)|, |v_A(x_i) - v_C(x_i)|, |\pi_A(x_i) - \pi_C(x_i)|\right)}{2}$$
$$\geq \frac{|\mu_A(x_i) - \mu_B(x_i)| + |v_A(x_i) - v_B(x_i)| + |\pi_A(x_i) - \pi_B(x_i)|}{4}$$
$$+ \frac{\max\left(|\mu_A(x_i) - \mu_B(x_i)|, |v_A(x_i) - v_B(x_i)|, |\pi_A(x_i) - \pi_B(x_i)|\right)}{2}$$

and thus we get $S_g(A, C) \leq S_g(A, B)$. By the same reason, we can get $S_g(A, C) \leq S_g(B, C)$.

Example 2. We consider the two patterns A_1 and A_2 and the sample B discussed in Example 1. Then, using (7), we find $S_g(A_1, B) = 0.833$ and $S_g(A_2, B) = 0.866$ and thus obtain the reasonable result $S_g(A_1, B) \neq S_g(A_2, B)$. Hence we obtain the correct recognition result.

However, the elements in the universe may have different importance in pattern recognition. We should consider the weight of the elements so that we can obtain more reasonable results in pattern recognition.

Assume that the weight of x_i in X is w_i, where $w_i \in [0, 1]$, $i = 1, 2, \cdots, n$. and $\sum_{i=1}^n w_i = 1$. The similarity measure between IFSs A and B can be obtained by the following form:

$$S_{gw}(A, B) = 1 - \sum_{i=1}^n w_i \left(\frac{|\mu_A(x_i) - \mu_B(x_i)| + |v_A(x_i) - v_B(x_i)| + |\pi_A(x_i) - \pi_B(x_i)|}{4} + \frac{\max\left(|\mu_A(x_i) - \mu_B(x_i)|, |v_A(x_i) - v_B(x_i)|, |\pi_A(x_i) - \pi_B(x_i)|\right)}{2} \right).$$
(8)

Likewise, for $S_{gw}(A, B)$, the following theorem holds.

Theorem 2. $S_{gw}(A, B)$ is the similarity measure between two IFSs A and B.

Proof. The proof is similar to that of Theorem 1.

Remark 1. Obviously, if $w_i = 1/n$ $(i = 1, 2, \cdots, n)$, (8) becomes (7). So, (7) is a special case of (8).

4 APPLICATIONS TO PATTERN RECOGNITION PROBLEM

Assume that a question related to pattern recognition is given using IFSs. Li and Cheng [7] used the principle of the maximum degree of similarity between IFSs to solve the problem of pattern recognition. We also apply the principle of maximum degree of similarity measures between IFSs to solve the problem.

Assume that there exist m patterns which are represented by IFSs $A_i = \{ \left(x_i,\ \mu_A(x_i),\ \nu_A(x_i) \right) : x_i \in X \} \in \mathrm{IFS}(X)$ $(i = 1, 2, \cdots, n)$, where $X = \{ x_1, x_2, \cdots, x_n \}$. Suppose that there is a sample to be recognized which is represented by $B = \{ \left(x_i,\ \mu_B(x_i),\ \nu_B(x_i) \right) : x_i \in X \} \in \mathrm{IFS}(X)$. Set

$$S_{g\,w}\left(A_{i_0},\ B \right) = \max_{1 \le i \le n} \left\{ S_{g\,w}\left(A_i,\ B \right) \right\}. \tag{9}$$

According to the principle of the maximum degree of similarity between IFSs, it can be decided that the sample B belongs to some pattern A_{i_0}.

In the following, two examples are given to show that the proposed similarity measure can overcome the drawbacks of existing similarity measures and can deal with the problems more effectively and reasonably than those similarity measures.

Example 3. Assume that there are three patterns denoted with IFSs in $X = \{ x_1, x_2, x_3 \}$. Three patterns A_1, A_2 and A_3 are denoted as follows:

$$A_1 = \left\{ \left(x_1,\ 0.2,\ 0.3 \right),\ \left(x_2,\ 0.6,\ 0.1 \right),\ \left(x_3,\ 0.3,\ 0.3 \right) \right\},$$

$$A_2 = \left\{ \left(x_1,\ 0.5,\ 0.2 \right),\ \left(x_2,\ 0.5,\ 0.4 \right),\ \left(x_3,\ 0.1,\ 0.3 \right) \right\},$$

$$A_3 = \left\{ \left(x_1,\ 0.5,\ 0.2 \right),\ \left(x_2,\ 0.4,\ 0.3 \right),\ \left(x_3,\ 0.6,\ 0.3 \right) \right\}.$$

Assume that a sample $B = \{ (x_1,\ 0.4,\ 0.3),\ (x_2,\ 0.5,\ 0.2),\ (x_3,\ 0.2,\ 0.4) \}$ is given. To interpret the notions of these patterns, we borrow the idea of Wang and Xin [12]. Given three kinds of mineral fields, each is featured by the content of three minerals and contains one kind of typical hybrid minerals. The three kinds of typical hybrid minerals are represented by IFSs A_1, A_2 and A_3 in X, respectively. Given another kind of hybrid mineral B, to which field does this kind of mineral B most probably belong to?

For convenience, assume that the weight w_i of x_i in X are equal and $p = 1$. By (5), (6) and (7), we have

$S_e^1(A_1,\ B) = 0.900$, $S_e^1(A_2,\ B) = 0.900$, $S_e^1(A_3,\ B) = 0.850$,

$S_f(A_1,\ B) = 0.883$, $S_f(A_2,\ B) = 0.883$, $S_f(A_3,\ B) = 0.825$,

$S_g(A_1,\ B) = 0.867$, $S_g(A_2,\ B) = 0.833$, $S_g(A_3,\ B) = 0.800$.

From this data, the proposed similarity measure S_g shows the correct classification according to the principle of the maximum degree of similarity between IFSs. That is, the sample B belongs to the pattern A_1. However, both Liang and Shi's similarity measure S_e^p and Wang and Xin's similarity measure S_f cannot classify this sample.

5 CONCLUSIONS

In this paper, we attempt to propose a new method for similarity measures between IFSs and discuss its application of the similarity to pattern recognition. Based on the geometrical interpretation of IFSs, taking into account three parameters describing IFSs, the new similarity measures are illustrated to calculate the degree of similarity between IFSs. The core properties of the proposed similarity measures are proved by a series of steps and several examples are illustrated for comparisons between the proposed similarity measures and the existing methods. The results allow us to make the conclusion that the proposed similarity measures are not effective and reasonable than both Liang and Shi's and Wang and Xin's measures.

REFERENCES

[1] K. Atanassov, Intuitionistic fuzzy sets, Fuzzy Sets and Systems 20 (1986) 87–96.

[2] K. Atanassov, More on intuitionistic fuzzy sets, Fuzzy Sets and Systems 33 (1986) 37–46.

[3] H. Bustince and P. Burillo, Vague sets are intuitionistic fuzzy sets, Fuzzy Sets and Systems 79 (1996) 403–405.

[4] W.L. Gau and D.J. Buehere, Vague sets, IEEE Trans Systems Man Cybernt. 23 (1994) 610–614.

[5] P. Grzegorzewski, Distances between intuitionistic fuzzy sets and/or interval-valued fuzzy sets on Hausdorff metric, Fuzzy Sets and Systems 148 (2004) 319–328.

[6] W.L. Hung and M.S. Yang, Similarity measures of intuitionistic fuzzy sets based on Hausdorff distance, Pattern Recognition Lett. 25 (2004) 1603–1611.

[7] D. Li and C. Cheng, New similarity measures of intuitionistic fuzzy fuzzy sets and applications to pattern recognitions, Pattern Recognition Lett. 23 (2002) 221–225.

[8] Y. Li, D.L. Olson and Z. Qin, Similarity measures between intuitinistic fuzzy (vague) sets: A comparative analysis, Pattern Recognition Lett. 28 (2007) 278–285.

[9] Z. Liang and P. Shi, Similarity measures on intuitionistic fuzzy fuzzy sets, Pattern Recognition Lett. 24 (2003) 2687–2693.

[10] H.B. Mitchell, On the Dengfeng-Chuitian similarity measure and its application to pattern recognition, Pattern Recognition Lett. 24 (2003) 3101–3104.

[11] E. Szmidt and J. Kacprzyk, Distances between intuitionistic fuzzy sets, Fuzzy Sets and Systems 114 (2000) 505–518.

[12] W. Wang and X. Xin, Distance measure between intuitionistic fuzzy sets, Pattern Recognition Lett. 26 (2005) 2063–2069.

[13] Z.S. Xu, Some similarity measures of intuitionistic fuzzy sets and their application to multiple attribute decision making, Fuzzy Optimization and Decision Making 6 (2007) 109–121.

[14] J. Ye, Cosine similarity measures for intuitionistic fuzzy sets and their applications, Mathematical and Computer Modelling 53 (2011) 91–97.

[15] L.A. Zadeh, Fuzzy sets. Inform and Control 8 (1965) 338–353.

Network Security and Communication Engineering – Chan (Ed.)
© 2015 Taylor & Francis Group, London, ISBN: 978-1-138-02821-0

Experiments and numerical analysis of flow rate in unstably stratified field

Motoo Fumizawa, Yoshiharu Saito & Naoto Uchiyama
Shonan Institute of Technology, Tsujido-Nishikaigan, Fujisawa, Japan

ABSTRACT: In the flow mechanism of unstably stratified field, the exchange flow occurs after Rayleigh-Taylor instability. Buoyancy-driven exchange flows were investigated the helium-air flow in the vertical narrow pathway between upper air chamber and lower helium chamber. Exchange flows may occur following the opening of a window for ventilation, when fire breaks out in a room, as well as when a pipe ruptures in a high temperature gas-cooled nuclear reactor. The numerical analysis and experiment in this paper was carried out in a test chamber filled with helium and the flow was visualized using the smoke wire method. The flow behavior was recorded by a high-speed camera combined with a computer system. The image of the flow was transferred to digital data, and the flow velocity was measured by PTV and PIV software. The mass fraction in the test chamber was measured using electronic balance. The detected data was arranged by the densimetric Floude number of the exchange flow rate derived from the dimensional analysis. The volumetric exchange flow rate was evaluated from the mass increment data. In the case of inclined openings, the results of both methods were compared. The inclination angle for the maximum densimetric Floude number decreased with the increase of the length-to-diameter ratio of the opening. For a horizontal opening, the results from the method of mass increment agreed with those obtained by other authors for a water-brine system.

KEYWORDS: Buoyancy, Exchange flow, Helium, moving particle method, Clockwise flow.

1 INTRODUCTION

In the flow mechanism of unstably stratified field, the exchange flow occurs after Rayleigh-Taylor instability. Buoyancy-driven exchange flows were investigated the helium-air flow in the vertical narrow pathway between upper air chamber and lower helium chamber. Exchange flows may occur following the opening of a window for ventilation, the outbreak of fire in a room or over an escalator in an underground shopping center, as well as when a pipe ruptures in a modular high temperature gas-cooled nuclear reactor. The fuel loading pipe is located in an inclined position in a pebble bed reactor such as the Modular reactor [1,2] andAVR[3,4].

In safety studies of High Temperature Gas-Cooled Reactor (HTGR), the failure of a standpipe at the top of the reactor vessel or a fuel loading pipe may be one of the most critical design-based accidents. Once the pipe ruptures, helium immediately rushes up through the breach. Once the pressure between the inside and outside of the pressure vessel has balanced, helium flows upward and air flows downward through the breach into the pressure vessel. This means that buoyancy-driven exchange flow occurs through the breach, caused by the density difference of the gases in the unstably stratified field. Since an air stream corrodes graphite structures in the reactor, it is important to evaluate and reduce the air ingress flow rate when a standpipe rupture occurs.

Studies have been performed on the exchange flow of two fluids with different densities through vertical short tubes. Epstein[5] studied the exchange flow of water and brine through various vertical tubes, experimentally and theoretically. Mercer et al. [6] studied an exchange flow through inclined tubes with water and brine experimentally. The latter experiments were carried out in the range of 3.5 <L/D < 18 and 0 deg < θ < 90 deg, and indicated that the length-to-diameter ratio L/D, and the inclination angle θ of the tube are the important parameters for the exchange flow rate. Most of these studies were performed on the exchange flow using a relatively small difference in the densities of the two fluids (up to 10 per cent). However, in the case of a HTGR standpipe rupture, the density of the outside gas is at least three times larger than that of the gas inside the pressure vessel. Few studies have been performed so far using such a large density difference. Kang et al. [7] studied experimentally the exchange flow through a round tube with a partition plate. Although one may assume that the partitioned plate, a kind of obstacle in the tube, would reduce the exchange flow rate, Kang found that the exchange flow rate was increased by the partition plate because of separation of the upward and downward flows.

The main objectives of the present study are to investigate the behavior of the exchange flow in the vertical narrow pathway then to evaluate the exchange flow rate using mass increment with the helium-air system. The following methods are applied in the present study.

1 Optical system of the Mach–Zehnder interferometer.
2 Numerical analysis.
3 Mass increment method.

2 OPTICAL SYSTEM OF MACH-ZEHNDER INTERFEROMETER

2.1 *Experimental apparatus and procedure*

The optical system of the Mach–Zehnder interferometer, MZC-60S to visualize the exchange flow is shown in Figure 1. After being rejoined behind the splitter, the test and reference laser beams interfere, and the pattern of interference fringe appears on the screen. If the density of the test section is homogeneous, the interference fringes are parallel and equidistant [8]. If the density is not homogeneous, the interference fringes are curved. Inhomogeneity in the test section produces a certain disturbance of the non-flow fringe pattern. The digital camera and high-speed camera using a D-file can be attached to the interferometer.

2.2 *Results*

To investigate the flow pattern, the exchange flows are visualized by the Mach–Zehnder interferometer. Figure1 shows the typical interferogram of the fringes for the vertical narrow tube. The upward helium plume and the downward air plume break intermittently through the passage and swing from left to right in the lateral direction. The period of the swing is~2s. It is clearly observed that the up flow of helium and the down flow of air do not take place smoothlybecause they interact strongly with each other. The flow is fluctuating and unstable.

3 NUMERICAL ANALYSIS

3.1 *Moving particle method*

Figure 2 shows sectional view and bird-eye view of numerical boundary conditions of moving particle method of 3D coordinate, respectively. Table 1 shows the calculation condition of moving particle method. Lagrange method is adopted to calculate program of

moving particle method. Therefore, the numbers of particles are around two million. The calculation program code is adopted Particleworks3.01 and possessor is composed of TESLA C2070.

Figure 3 shows unstably and asymmetric flow patterns became of itsdominance in the vertical narrow pipe. It is interesting that the horizontal cut view, in the figure,it reveals that the clockwise upward flow was observed, which means spiral upward flow pattern. Figure 4 shows that the upward plume behaves like the snake movement in the upper chamber. Numerical calculation of moving particle method predicts that the clockwise movement of the plume and snake behavior of plume and the behavior is consistent with the result that Mach–Zehnder method shows random behavior of the plume from left to right.

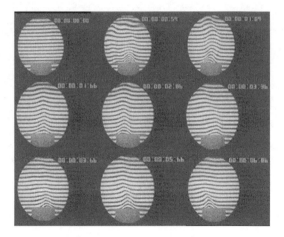

Figure 1. Visualization result of Mach–Zehnder method upper part of vertical narrow tube.

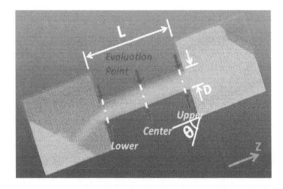

Figure 2. Cross sectional view of numerical boundary conditions of moving particle method.

Table 1. Calculation condition of moving particle method.

Number of particle N	1~2million
Inclination angle θ	0°,15°,30°,45°,90°
Particle diameter a	0.815~1.0mm
Flow path length L	100mm
Diameter flow path D	20mm
Container sizes of height(h) and diameter(d)	80mm ($h=d$)

3.2 HSMAC method

Two dimensional unsteady code of HSMAC method is adopted to the buoyancy-driven exchange flow system[1]. In the code, FORTRAN program is described as basic equation of mass, momentum and energy. Analysis coordinate is shown in Figure 5. The left part is the test chamber filled with helium gas and the right part is outside region filled with air. Typical calculation result is shown in Figure 6, where it is the narrow channel, between the left and right. The exchange flow occurs with vortex in the narrow channel. Therefore, the center flow rate Q1 is larger than the right edge flow rate Q2, as shown in Figure 7.

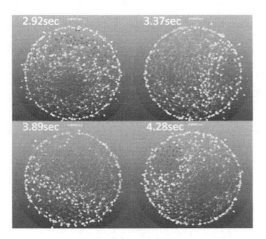

Figure 3. Horizontal view of exchange flow in the narrow tube.

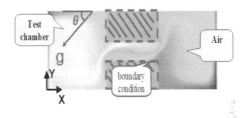

Figure 5. Analytical coordinate and conditions.

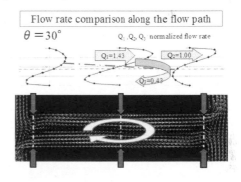

Figure 6. Analytical result of narrow flow path of exchange flow.

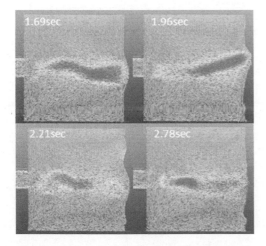

Figure 4. Vertical view of light plume behavior in the narrow tubeupper chamber on the narrow tube.

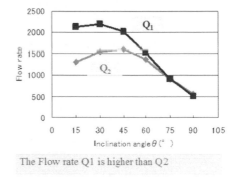

Figure 7. Relation of exchange flow rate andinclination angle of analytical results.

4   METHOD OF MASS INCREMENT

4.1   *Comparison between experimental and numerical analysis*

It is already known that the densimetric Froude number is regarded as constant within a time duration when the gas in the upward flow is assumed to be helium [9]. Figure 8 shows the relationship between Fr and inclination angle θ with L/D as a parameter. For inclined tubes, Fr is larger than that for vertical tubes. (i.e. L/D = 10). The densimetric Froude number reaches the maximum at 30 deg for the long tube. It is found that the angle for the maximum Fr decreases with the increase of L/D in the helium-air system.

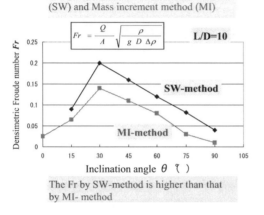

Comparison between Smoke wire method (SW) and Mass increment method (MI)

$$Fr = \frac{Q}{A}\sqrt{\frac{\rho}{g\,D\,\Delta\rho}}$$

The Fr by SW-method is higher than that by MI- method

Figure 8.   Relation of densimetric Froude number and inclination angle in both experiments.

5   SUMMARY

Numerical calculation of moving particle method predicts the clockwise movement of the plume behavior flow occurs in the narrow flow path.

Numerical calculation of HSMAC method predicts the circulation flow of the tube length direction occurs in the narrow flow path.

Numerical calculation of moving particle method predicts snake behavior of plume and the behavior is consistent with the result that Mach–Zehnder method shows random behavior of the plume from left to right.

Densimetric Froude number Fr by SW-method is higher than that by MI- method.

ACKNOWLEDGMENT

The authors are deeply indebted to Mr. Shuhei Ohkawa, working Sanwa-koki Co. Ltd., for hisunfailing interest and helpful input to this study.

REFERENCES

[1] Fumizawa,M. ; Proc. HT2005 ASME Summer Heat Transfer Conference, HT2005 -72131, Track 1-7–1, pp.1–7 (2005).
[2] Kiso et.al.;JSME Annual MTG, pp.339–340 (1999).
[3] M.M.El-Wakil, Nuclear Energy Conversion, Thomas Y. Crowell Company Inc., USA (1982).
[4] ArbeitsgemeinschaftVersuchs-Reactor   AVR, "Sicherheitsberichtfuer das Atom-Versuchskraftwerk, Juelich" (1965).
[5] Epstein,M., Trans. ASME J. Heat Transfer, pp.885–893 (1988).
[6] Mercer, A. and Thompson.H., J. Br. Nucl. Energy Soc., 14, pp.327–340 (1975).
[7] Kang,T. et al., NURETH-5, pp.541–546 (1992).
[8] Merzkirch.W., "Flow Visualization", Academic Press (1974).
[9] Fumizawa,M. et. al., J. At. Energy Soc. Japan, Vol.31, pp.1127–1128 (1989).

Network Security and Communication Engineering – Chan (Ed.)
© 2015 Taylor & Francis Group, London, ISBN: 978-1-138-02821-0

Configurable formula calculation tool design based on tagged data structure

G.Q. Wang, S.Y. Long & P. Shao
EPRI Nanjing Branch, Nanjing, Jiangsu, China

W.Z. Zhang
State Grid Chongqing Electric Power Company, Chongqing, China

Y.T. Zhao
State University of New York at Buffalo, New York, U.S

ABSTRACT: Nowadays, as information systems are developed and used in power grid, the procession and calculation of the large amounts of data in these systems plays important role in the operation and decision of power companies. In most information systems, as the different data structure, the data calculation and procession can only be realized by developing specific function module. In fact a great deal of data of each information system usually can be transformed into structure with its body, application and time tag. Based on this tagged data structure, this paper proposed a configurable computation formula tool to do the data calculation, by setting the components, calculating logic, the starting time and cycle of calculation and deploying a scheduling task in the background. This tool can realize the configurable and automatic data procession in the power information system.

KEYWORDS: Formula calculation, Tagged data structure, Power system, Configurable.

1 INTRODUCTION

There are more and more production and management data collected and produced in power information systems, their procession and calculation play an important role in the operation and decision of power companies[1-2]. As data structure in the information system is different from each other, it always needs to develop specific g function module to realize data procession which is difficult to extend or reuse [3], the solution is to design a configurable tool to do the data procession and make sure it isn't limited to the data structure[4].

In fact, normal data in power system can be determined by its body, application and time which is called as data tag and a great deal of data in complex structures usually can be transformed into it. So in this paper we designed and developed a configurable formula calculating tool based on this tagged data structure. If data structure in the system is different, the first thing needs to do is to change the data structure to the tagged type. By setting the components, calculating logic, the starting time and cycle of calculation and scheduling task in the background, this tool can realize the configurable data procession in the power information system. When the demand of data procession changes, it only needs to modify the settings.

2 DESIGN PRINCIPLE

The functional architecture of the configurable formula calculating tool is shown in figure1 which mainly includes functions as follows.

2.1 Formula configuration

In this module, the formulas are registered and the components, calculating logic, result of the formulas as well as start time, cycle of auto calculating are set;

2.2 Calculation schedule server

It runs backward and calls the formula calculation cyclically on the set time according to the configuration of formula;

2.3 Formula level divide

This module iterates through the list of formulas and sets the calculating level of each formula in accordance with whether it has formula components and the level of formula components, the formula's level decides the order its calculation be called;

2.4 Formula calculation

This module calculates the formulas in the order of its calculating level, during the calculating process, it reads the configuration of formula, searches and determines the value of each component, calculates the formula according to the configured logic and saves the result into the database;

2.5 Status supervisor

This module supervises the calculating status of each formula, records and displays process information and error logs during the calculating process in given time. When the auto calculate schedule has something wrong, this module can offer the users to start formula calculation manually.

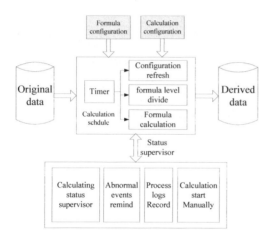

Figure 1.   Function distribution of compute tool.

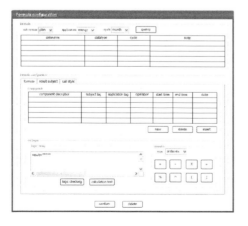

Figure 2.   View of formula configuration.

3 REALIZATION

The formula calculating tool realizes configurable and automatic data calculation of power system based on the configuration of components, calculating logic, auto calculation cycle and the calculation schedule. It can offer the function of data processing, transformation and statistics.

3.1 Formula configuration

The formula result is the same as the original data in the structure, so it can be registered as a tagged data with derived property and formula definition which include the components, calculating logic and calculation schedule.

The definition of formula components means selecting the data used in the formula calculation and seting the conditions it be retrieved, it can be original data or derived. Because of data tag, the data can be uniformly retrieved by set tags and conditions.

Formula calculating logic is a string connected by components and operators, which usually includes four arithmetic operations, logical operations, string operations as well as extreme, average, total value of whole grid or plant and so on.

The definition of formula calculation schedule sets start-up cycle and starting time of formula, the time shift between components' time and calculating time. The graphical interface is shown in figure2.

3.2 Calculation schedule server

Calculation schedule server plays the schedule role in the formula calculation tool, which runs backward and calls the formula calculation process on the set time according to the configuration and displays the output information and log during the processes.

3.3 Formula level divide

Formula calculation needs to select the components and get the value of each component for the calculation, but in fact there are some components which are other formulas' calculating result themselves. So it is necessary to determine the calculating order before the formulas are calculated, it is the formula level divide, which set the formula's level according to whether it has the components with formula and the level of component's formula, the higher level means the formula is calculated later. Formula level divide can make sure the components' formula calculates before the result formula.

Formula level divide is a cyclical and iterative process, if the formula has no components with formulas; its level is set 1, if the formula has components with formula, then it needs to determine the component formula's level first, and if the component formula has formula component itself, just do the same till the sub-formula has no formula component, set the sub-formula's level 1, then set the parent formula's level in the opposite order, which can be set as its sub-formulas' max level +1. The flow chart of Formula level divide is shown in figure 3.

5 Get the values of components according to cycle of components and the calculating time by the uniform method of data retrieving with tags;
6 Do the calculation according to the formula logic which may includes mathematical, logical and other special operations;
7 Get the date-time tag of the result data from the cycle of formula and the calculating time and save the result data with the date-time, physical and application tag.

The flow chart of formula calculation is shown in figure 4.

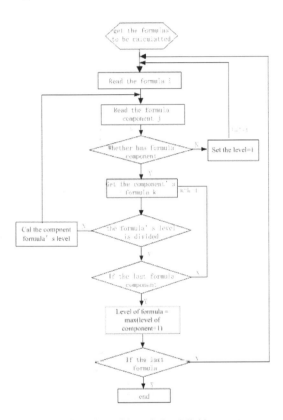

Figure 3.   Flow chart of formula level divide.

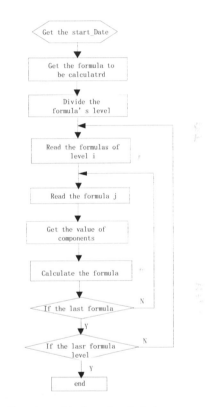

Figure 4.   Flow chart of the formula calculation process.

3.4   *Formula calculation*

The formula level is a step of the formula calculation and the complete calculation process has steps as follows.

1 Judge if it is the time to do the formula calculation, if yes, set the system time as the calculating time;
2 Get the configuration of all the formulas to be calculated;
3 Do the formula level divide, set calculating level for each formula to be calculated;
4 Iterate over the formulas to be calculated in the order of calculating level, do the step 4-7;

3.5   *Status supervisor of formula calculation*

This module does supervise the formula calculating time and calculating status and the error information emergence in the calculation process, it can promote users to know the progress status of the data calculation. Besides, users can select specific formulas and time periods and call the formula calculation manually, which is an import addition to the automatic schedule of the formula calculation. The graphical interface of status supervisor is shown in figure 5.

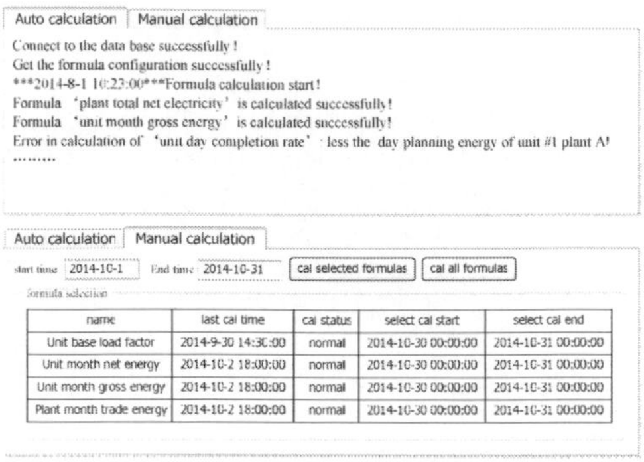

Figure 5.   Status supervisor of formula calculation.

If the demand of data process changes, users just need to register new formula or modify existing formula configuration that is relevant to the data process. The formula calculation schedule server will cyclically refresh the formula configuration (the cycle can be set and usually be 1 hour), users also can restart the formula calculation tool to refresh the formula configuration immediately, then the formula calculation process will calculate the new formula list in the next calculation job.

4 CONCLUSION

The configurable formula calculation tool proposed in this paper can realise the configurable data process in the power information system, it is especially fit for the changeable demand of data statistics and analysis in power companies. In the formula calculation tool, the data process is independent from the business of power information system, when there is new demand of data process, it just needs to add or modify relevant formula configuration by engineering work, the calculation schedule server running backward can do the formula calculation automatically according to the formula configuration. In normal condition users just need to supervise the calculation at specific time by the module of calculation supervisor. When something is wrong in some special circumstances, they can also do the formula calculation manually.

ACKNOWLEDGMENTS

The work described in this paper is financial supported by the project of Research on Key Techniques of United Power Market Operation (DZN17201300045) under State Grid Corporation, China and the project of Research on Key Techniques of Data Warehouse in Electric Power Transaction Operation (DZ83-13-002) under Electric Power Research Institute, China.

REFERENCES

[1] Zhao-hui LI, Gui-shi DENG, Qiang FENG, Feng CHEN, Jun HOU. The Design and Implementation of a Configurable Platform for Information Management. Computer Engineering and Applications, 2002,11:9–11,37.
[2] Xiao-dong BIAN, Mu HU. Application of intergrated data center in power system dispatching [J].Jiangsu Electrical Engineering. 2005,24(3):55–57.
[3] Song YU, Xiao-ping FAN, Zhi-fang LIAO, Xi YANG. Design and implemention of common data calculating based on SQL[J]. Computer Engineering and Applications, 2007,43(21):93–95,111.
[4] Ying-hua LIU, Jian CAO, Shen-sheng ZHANG, Xiao-guang ZHANG. Research on configurable comprehensive evaluation system[J]. Computer Integrated Manufacturing Systems, 2004,10(10):1291–1295.

Network Security and Communication Engineering – Chan (Ed.)
© 2015 Taylor & Francis Group, London, ISBN: 978-1-138-02821-0

Study on the efficiency of data sharing among windows multi-process

Z. Wu & Q.Z. Jia

School of Aerospace Engineering, Beijing Institute of Technology, China

ABSTRACT: Data sharing is the basis of mutual cooperation among processes. We studied the feature of data sharing among Windows' processes. Based on atomic access, we also proposed a new exchange mode of data among processes and described the principle and implementation of this mechanism in detail. The performance of our data sharing method was compared with other data synchronization method in Windows.

KEYWORDS: Inter-process Communication, Shared memory, Process synchronization.

1 INTRODUCTION

With people's higher demands on application programs, the single process cannot meet their expectations in many occasions. Thus, it becomes a trend to write programs by multi-process or multi-thread. During the design of a multi-process or multi-thread program, inter-process communication is necessary and there are many ways to execute it, like, Pipe, Mailslots, Clipped Board, Sockets, Message Queue, Memory-Mapped File, Shared Memory and so on [1]. Among the above methods, the most direct and fundamental one is to use Shared Memory which is, in fact, the basic of all other methods. Inter-process communication exists restriction when multi-process shares data. Program produces the desired effect by harmonizing the restriction.

This paper analyzes the features of the communication method using Shared Memory and regular process synchronization, and then designs a high efficiency and safety communication method among multi-process whose performance is tested by programming.

2 SHARED MEMORY COMMUNICATION METHOD

Every process believes it has the whole machine resources under the operation of Windows. So it's impossible for one process to get the address of others. Shared memory means to create a memory area that can be visited by all processes simultaneously. It is the most efficient way to exchange data by mapping the memory area to the virtual address of each process, thus making each process access the data simultaneously [2]. The detail procedure is as follows:

1 Set a certain area of the memory: create a core object of file-mapping and set the size of the shared memory by calling CreateFileMapping function then map the shared memory to the address space of the process by calling MapViewOfFile function.
2 Find out the shared memory: If the shared memory is used as a point-to-point format, every process must create an area for shared memory and initialize it. However, if the shared memory is used in the Client/Server framework, then only the Server process has to create the area for shared memory and initialize it. All the Client processes call OpenFileMapping function to get a handle pointing to the file-mapping object directly created by Server processes and get a pointer of the shared memory.
3 Clear after being done: Once the operation on shared memory is finished, call UnmapViewOfFile function will release the pointer of the shared memory and close the handle of the file-map object created before by calling CloseHandle function.

The synchronization and exclusion of the data must be taken into consideration when using the shared memory because the read and write operation happen in different process or thread. A global sign can be used to mark that the shared memory has been created and initialized successfully. The shared memory can also be controlled by other data synchronization method of Windows.

3 DATA SYNCHRONIZATION METHOD IN WINDOWS

3.1 Thread synchronization in user mode

3.1.1 Atomic access

Atomic access refers to a thread's ability to access a resource with the guarantee that no other threads will access the same resource at the same time. Windows offer interlocked functions to implement atomic access. A call to the interlocked function usually causes just a few CPU cycles (usually less than 50) to execute, and there is no transition from user mode to kernel mode (which usually requires more than 1000 cycles to execute), so they execute extremely quickly [3]. But interlocked functions has many limitations, they can only manipulate integer, Boolean values and singly linked list.

3.1.2 *Critical section*
A critical section is a small section of code that requires exclusive access to some shared resources before the code can execute. This is a way to have several lines of code "atomically" manipulate a resource. The system will not dispatch any other threads that want to access the same resource until your thread leaves the critical section [4]. You can use critical section as follows: you should create a CRITICAL_SECTION structure, the code that touches a shared resource must be wrapped inside EnterCriticalSection and LeaveCriticalSection functions. The critical sections are easy to use and they use interlocked functions internally, so they execute quickly. The disadvantage of critical sections is that you cannot use them to synchronize threads in multiple processes.

3.1.3 *Slim reader-writer locks*
An SRWLock has the same purposes as a simple critical section: to protect a single resource against access by different threads. However, unlike a critical section, an SRWLock allows you to distinguish between threads that simply want to read the value of the resource(the readers) and other threads which are trying to update these values(the writers). It allows all the reader threads simultaneously access a shared resource [5]. When a writer thread wants to update the resource, the access should be exclusive: no other thread, whether it is writer or reader threads.

3.1.4 *Condition variables*
Condition variables is a mechanism for synchronizing by using shared global variables between threads, including two actions: a thread hangs himself until a condition is true; another thread notifies waiting threads to continue when the condition is met [6].

3.2 *Thread synchronization with kernel objects*

3.2.1 *Event kernel objects*
Event kernel objects contain a usage count, a Boolean value indicates whether the event is an auto-reset or manual-reset event, and another Boolean value

indicates whether the event is signaled or nonsignaled [7]. There are two different types of event objects: manual-reset events and auto-reset events. When a manual-reset event is signaled, all threads which wait on the event become schedulable. When an auto-reset event is signaled, only one of the threads waiting on the event becomes schedulable. Events are most commonly used when one thread performs initialization work and then signals another thread to carry out the remaining work.

3.2.2 *Semaphore kernel objects*
Semaphore kernel objects are used for resource counting. They contain a usage count, but they also contain a maximum resource count and a current resource count. The rules for a semaphore are as follows: if the current resource count is greater than zero, the semaphore is signaled; if the current count is zero, the semaphore is nonsignaled; the current resource count cannot be negative and greater than the maximum resource count. The thread gains access to the protected resource by calling a wait function, passing the handle of the semaphore. Internally, the wait function checks the semaphore's current resource count and if its value is greater than zero (the semaphore is signaled), the counter is decremented by one and the calling thread remains schedulable [8]. If the wait function determines that the semaphore's current resource count is zero (the semaphore is nonsignaled), the system places the calling thread in a wait state.

3.2.3 *Mutex kernel objects*
Mutex kernel objects ensure that a thread has mutual exclusive access to a single resource. A mutex object contains a usage count, thread ID, and recursion counter [3]. Mutexes behave identically to critical sections. However, mutexes are kernel objects, while critical sections are user-mode synchronization objects. This means that threads in different processes can access a single mutex and mutexes are slower than critical sections. If the thread ID is zero (an invalid thread ID), then the mutex is not owned by any thread and is signaled, and if thread ID is a nonzero value, a thread owns the mutex and the mutex is nonsignaled. As it always happens, these checks and changes to the mutex kernel object are performed atomically.

3.3 *Software algorithm*

3.3.1 *Peterson algorithm*
Peterson's algorithm is a concurrent programming algorithm for mutual exclusion that allows two processes to share a single-use resource without conflict, using the only shared memory for communication [9]. The algorithm uses two

variables, flag and turn. A flag[i] value of true indicates that the process i wants to enter the critical section. The turn value indicates the process ID has access to the shared resources.

```
1 flag[i] = true;
2 turn = 1 - i;
3 while (flag[turn] == true && turn == 1 - i)
4 {
5     ;// busy wait
6 }
7 // critical section
8 flag[i] = false;
9 // end of critical section
```

Figure 1.   The Peterson algorithm.

3.3.2 Bakery algorithm

Bakery's algorithm maintains the first-come-first-served property by using a distributed version of the number-dispensing machines which can often be found in bakeries: each thread takes a number in the doorway, and then waits until no thread with an earlier number is trying to enter it [10]. In the Bakery algorithm, flag[i] is a Boolean flag indicating whether i wants to enter the critical section, and label [i] is an integer that indicates the thread's relative order when entering the bakery, for each thread i.

```
1 flag[i] = true;
2 label[i] = max(label[0],...,label[n - 1]) + 1;
3 while(flag[k] && (label[k],k) < (label[i],i) )
4 {
5 ;//busy wait
6 }
7 //critical section
8 flag[i] = false;
9 //end of critical section
```

Figure 2.   The Bakery algorithm.

The Peterson algorithm can only achieve mutual exclusion between two processes, Bakery algorithm can achieve mutual exclusive among multiple processes. The principal of Bakery algorithm is the need to read and write n distinct location, where n is the number of concurrent processes. Bakery algorithm has a relatively large overhead when they obtain the order of the process entering the critical section and especially the number of processes is large.

4 THE REALIZATION OF SHARED MEMORY WITH HIGH EFFICIENCY

4.1 Improved atomic access

Critical section can protect the shared data in multi-thread, but data cannot be synchronized in multi-process. Most of the time is wasted on the kernel object like event, mutex and semaphore which are used to implement the synchronization and exclusion of the processes. Yet, the expense of time and space can be a problem even if we implement process synchronization by software algorithm, especially when the number of process which needs to be synchronized is larger. There is no need for atomic access to transfer from user mode to kernel mode. However, processes cannot communicate with complex data structure in atomic access. So it is possible to make the communication among processes safer and more efficient by designing reasonable data structure in the shared memory by calling InterlockedCompareExchange function. This function compares the destination value (pointed by the first parameter) with the compared value (the third parameter), if they are equal, exchange the destination value with the exchange value (the second parameter) and the function returns initial value of the destination [11]. The whole procedure is locked in the memory and other processes cannot visit the memory at this time. The function prevents more than one thread from using the same variable simultaneously. The data structure in the shared memory is as follows:

```
1 struct MemoryBuff
2 {
3     volatile LONG iFlag ;    // the sign of data operation
4     vlatile struct Data ;    // data that are shared
5 };
```

Figure 3.   The data structure of shared memory.

The idea of improving atomic access: To create copy of the shared memory in the process following the data structure above and operate on the copy after which the copy is replicated and saved in the shared memory. Then, to compare iFlag in the data structure and the process ID by calling InterlockedCompareExchange function. If they are not equal, then it continues the replication procedure until it succeeds which means that the process finishes the operation of shared memory.

```
1 {
2     m_pBuff->iFlag = processID ;
3     MemoryBuff data ;
4     data.iFlag = ProcessID ;
5
6     //operation on the copy of the memory
7     do
8     {
9         memcpy(m_pBuff , &data , sizeof(MemoryBuff));
10    }while(!(InterlockedCompareExchange(&m_pBuff->iFlag, -1,ProcessID) == ProcessID));
11 }
```

Figure 4.   The improved atomic access.

4.2 Accuracy analysis

Improved atomic access can avoid the deadlock problem by operating on the shared memory. It also meets the demand for mutual exclusion. For example, suggesting that A and B are two processes that are working at the same time, if A could not complete the task on the shared memory and B correct the mark bit at the beginning in the process, at this time, the correction would be detected by A and the replication of the shared memory would continue.

5 PERFORMANCE TEST

A laptop with 32 bit Windows8 (Inter core i3 CPU 2.1GHz, 4.00G RAM (3.41G available)) is used for this test. The benchmark compared improves atomic access, critical section and mutex, and the elapsed time whichare recorded. Figure 5 records times of different times of protecting shared memory and Figure 6 records time of different number of concurrent processes that execute the same task using the improved atomic access, the critical section and the mutex.

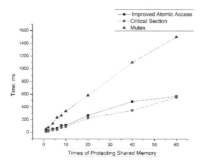

Figure 5. Result of different times of protecting shared memory.

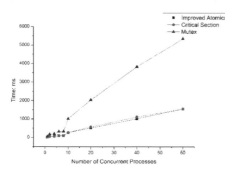

Figure 6. Result of different number of processes.

From the analysis of test results, the improved atomic access performs as well as the critical section both in times of protecting shared memory and in the number of concurrent processes, significantly better than the mutex.

6 CONCLUSION

Based on the atomic access of the Windows, this paper discusses a communication method among multi-process in the shared memory and calling the *InterlockedCompareExchange* function to synchronize processes. The method performs as well as the critical section by reducing the times of transferring from user mode to kernel mode. At the same time, the problem that critical section cannot be used across process boundaries is solved.

REFERENCES

[1] Zhou Huiqin, The Exploration and Research of Process Synchronization and Communication, Natural Science Journal of Harbin Normal University, 2. (2014).
[2] Jim Beveridge, Robert Wiener, Multithreading Applications in Win32: The Complete Guide to Threads, First ed., Addison-Wesley Professional, Boston, 1996. Chapter 13.
[3] Jeffrey Richter, Christophe Nasarre, Windows via C/C++, Fifth ed., Microsoft Press, Washington, 2008. Chapter 8.
[4] Li Penghui, Yan Jun, Research on Data Communication Methods among processes, Computer Knowledge and Technology, 8. (2012).
[5] SUN Jian-jie, CHEN Jia-pin, Implementation of Reader-writer Lock by Critical Section, Computer and Modernization, 9. (2011).
[6] Tie Jun, Jiang Tianfa, Application of Condition Variables in Thread Synchronization Mechanism, Journal of WuHan University of Technology, 3. (2007).
[7] Zheng Shangzhi, Chen Zujue, Han Yun,Lu Jun, Research on Mechanism of Signal Quantity in Linux, Computer Technology and Development, 17. (2007).
[8] Liu Hui, Liu Hong, Zhou Mingjun, Process of Synchronization and Mutual Exculsion, Network Security Technology & Application, 5. (2014).
[9] Maurice Herlihy, Nir Shavit, The Art of Multiprocessor Programming, First ed., Morgan Kaufmann Publishers, Burlington, 2008. Chapter 2.
[10] Information on http://en.wikipedia.org/wiki/Lamport%27s_bakery_algorithm.
[11] Information on http://msdn.microsoft.com/en-us/library/windows/desktop/ms683560(v=vs.85).aspx.

Network Security and Communication Engineering – Chan (Ed.)
© 2015 Taylor & Francis Group, London, ISBN: 978-1-138-02821-0

Distribution of wages and industrial agglomeration: Evidence from China

Y.M. Song
Economics School of Central University of Finance and Economics, China

ABSTRACT: This paper presents the analysis of the population flow from the perspective of new economic geography, which verified the relationship between wage distribution and population flow and market accessibility in China. Considering the reality of the situation Chinese population, with the introduction of the agricultural sector wage equation into the new economic geography, relax in the previous assumption on agriculture, we build a new wage equations in line with China's facts. The new wage equation contains the demographic variables, reflected the relation between the population migration and regional wage. According to empirical analysis, with the data of 28 provincial Chinese administrative regions, we draw a conclusion that the relationship between the agglomeration in China and the regional demand basically meet the new economic geography theory. The analysis of current spatial agglomeration phenomenon has important significance to formulate relevant policy.

1   INTRODUCTION

One of the important features of China's dual economic structure is the existence of a large number of surplus rural labor. When some unexpected factors causes the manufacturers to migrate to an area, they will need a lot of industry labors. The rise in demand will increase the level of real wages in the region, which will lead to workers migrating from other areas into the area, the influx of workers will further increase the demand for industrial products in the area, at the same time, with the increasing number of manufacturers, the price of the product will decline. With further expansion the size of the market and rising real wages, a new round of agglomeration and workers migration will be formed. When agglomeration achieves to a certain degree, the crowding effect will appear, manufacturers' trend to shift outwards; and too much workers in the area will produce a competitive effect, which will reduce regional wage level. Krugman (1991) established an effective model to explain why economic activity will gather in one place, what time will agglomeration appear, and the mechanism of the agglomeration process evolution. Krugman describes the relationship between wages and working flow, taking the spatial join into consideration.

Harris (1954) defines a regional market potential for other parts of industrial production in the area of gross purchasing power, which depicts the relationship between the demand for industrial products in the area of a regional market potential and other areas. The market potential depends on the level of income in the regions and regional transport cost. He just put

forward the concept; the function did not give. By new economic geography model, Fujita (1999) redefined the market potential function as the forms of wages, also is the formal spatial wage equation. Closer to the agglomeration center, the wage level is higher, the market potential is greater. The spatial distribution of wages reflects the market potential. An empirical study on the spatial wage equation is mainly applied to the new economic geography, foreign theory and empirical study is applied more widely, while there is less domestic application. Hanson (2004) through an improvement in the Fujita wage equation based on wage equation show spatial wage structure of USA meets the wage equation of new economic geography. With the same method, Redding and Venables (2004) verified core-periphery (CP) model with the international trade data. At home, according to the Fujita's wage equation, Shi Minjun (2007) simulated Chinese regional market potential by data, the index of market potential of China decreases from the coastal area to the inland area gradually, formed in coastal areas as the center, the inland area of the outer CP pattern, which meets the actual situation in China. Based on the method proposed by Shi Minjun, Zhao Zhao (2008) calculated and simulated the market potential by 329 regional administrative units in China, with different value of the transportation cost and the market scale, got the process wage change with the transportation cost and the size of the market changes. By Hanson (2004) method, using 28 provincial administrative regions' Chinese data, Zhang Xiaoxu (2008) verified that spatial structure of Chinese workers' wages were consistent with the new economic geography theory.

Market accessibility is the market scale to measure the distance of each area after total, regional wage is a function of market accessibility, represents an area of the market potential, the function relationship between them will be discussed in the later part of the paper. In this paper, the research mentality and the analysis method of Hanson is based on Venables, the theoretical and Empirical Analysis on the results of previous studies. One is to change the Department hypothesis. Hanson (2004) model is based on the Helpman (1998) to Krugman (1991) extended CP model. Helpman ard Krugman are two departments that follow agriculture into the real estate industry. According to Helpman, to improve sector hypothesis, combined with wage equation Fujita, proposed an improved wage equation. In this paper, reducing the agricultural sector hypothesis, which is more in line with actual China. In Europe and the United States and other developed countries, real estate prices are indeed in a certain extent can reflect the degree of congestion area, the real estate industry and manufacturing industries' strong association. However, in Europe and the United States this has been completed based on the premise of agricultural industrialization, and the share of its GDP proportion is relatively low. In China, agriculture accounts for the major proportion of GDP, and there are problems of rural surplus labor force to transfer to the city. Agriculture in China is also the foundation of industry. Agriculture and industry are closely linked. The two is to relax the assumption of agricultural sector. Krugman' assumption on agriculture is very strict; there is a great difference between it and the reality. This paper assumes that agriculture production returns to scale invariant; agricultural output unchanged; the price of agricultural products with increasing population rise; residents of the price elasticity of demand for agricultural products of the infinitesimal; the number of agricultural labor unchanged; farmers income elasticity of consumption is large. The third is to discuss the impact of agriculture on agglomeration. Krugman (1991) did not discuss the impact of agriculture on industry, the main reason is that if the interaction with the agricultural and industrial sector, the final model is too complicated. He discussed the stability of agglomeration, mainly using the replacement cost of manufacturers. Joining the agricultural sector, is difficult to describe the dynamic process. This paper mainly discusses the problem of spatial wage, is considered a foreign attraction, need is to measure the market appeal of the standard, to the influence factors into the wage equation as possible, does not need the complex process of dynamic derivative.

With regard to market potentials have different measurement methods, in addition to Harris (1954) proposed the "market potential function", Redding and Venables (2004) constructed Ma and Sa

indicators using bilateral trade data streams. The latter uses Ma and Sa indicators although more precise, there are support microscopic theory, but because it is difficult to get around the stage area between China's internal trade data, coupled with a situation in the region among countries around the level of labor can flow under Ma and Sa no distinction between the two effects even necessary (Ottaviano and Pineli, 2006)

2 THEORETICAL MODEL

2.1 Industrial sectors

Assuming the existence of multiple regions, each region can be represented by the j or K, each area only two departments of industry and agriculture. If industrial production increasing returns to scale; industry needed labor can flow freely; each firm produces a product, and can form a monopoly in the market, but there are alternatives, and manufacturers to enter the market without hindrance. Using the Cobb–Douglas production function definition of consumer preference:

$$U = C_M^\mu C_A^{1-\mu} \tag{1}$$

Cm is the consumption of residents of industrial products, CA is the consumption of residents to agricultural products, said the residents for industrial goods expenditure share, $0 < <1$, for industrial goods expenditure Cm uses the same alternative definitions of elastic production function:

$$c_m = \left[\sum_{i=1}^{n} c_i^{\sigma-1/\sigma} \right]^{\sigma/\sigma-1} \tag{2}$$

σ is the elasticity of substitution of arbitrary two kinds of industrial products, while $\sigma > 1$, n is the number of types of industrial products, i means the index of industrial products. The increasing return to scale industrial products can be expressed as:

$$L_{Mi} = \alpha + \beta x_i \tag{3}$$

α is the fixed cost of industrial production, β is constant marginal cost in the industrial production, both of which are a constant, Xi is the number of the goods i, Lmi is necessary labor force for the product i. With the expansion of production, cost of labor force for unit product is declined, which reflects the scale economy.

Under the condition of the market clearing, according to the consumer's budget constraint, namely to industrial goods expenditure is equal to the industrial sector of the income, we could get the maximization utility function about consumption of industrial products:

$$c_j = p_j^{-\sigma} \mu Y_j I^{\sigma-1} \qquad (4)$$

while, $c_j = p_j^{-\sigma} \mu Y_j I^{\sigma-1}$ (5)
is price index of industrial product.

Assuming no transport costs of agricultural products, industrial products from one place to another, the transportation cost, the iceberg cost terms, is between the two prices for:

$$p_{ijk} = p_{ij} e^{\tau\, d_{jk}} \qquad (6)$$

Suppose P_{ij} is the the FOB price (FOB) of commodity I which was produced in J, τ is the unit transportation cost, ijk is the distance between k and j. This is satisfied with the economic geography theory mentioned above, because of transportation costs, manufacturers hope to set up a factory in the high population density areas, it can achieve economies of scale, but also can save transport costs.

2.2 Department of agriculture

Agricultural production is assumed constant returns to scale; agricultural output is constant; the price of agricultural products rise with higher population density; the price elasticity of demand for agricultural products is infinitesimal; the number of agricultural labor is constant; farmers' income elasticity of consumption is large. Population variation is mainly caused by the migration of workers; this number of worker becomes the main factors affecting the prices of agricultural products. Therefore, we can simply assume that the agricultural products price index is a linear function about worker:

$$A_j = \gamma L_{jm} \qquad (7)$$

With the joining of the agricultural products price index, the model is more in line with the actual situation in China. In 2009 the Engel coefficient is 46% in China, if we only adjust the nominal wage for the real wage, the price index of industrial products do not meet the present situation of the Chinese.

The impact of agriculture on industry mainly depends on industrial goods consumption by farmers, while the agricultural product price index is worked. Because of the increase of population, most of them are workers; the number of areas required an increase of agricultural products. However, agricultural output restricted by the land resources and technical conditions, the supply of agricultural output is not adequate to the demand, will inevitably lead to rising prices of agricultural products. Rising prices of agricultural products can be expressed with the increase of agricultural products price index, impact of agricultural product price index will affect the overall price index, and eventually making workers' real wages fall. At the same time, the rising prices of agricultural products will bring more income to farmers, increase in income will bring increased demand for industrial products, which will promote a new round of industrial production, to produce further industrial agglomeration. Rising prices of agricultural products, is the area of congestion characteristics, this crowding effect can produce dispersion force impede agglomeration in economic activities. On the other hand, with much more income of farmers, accelerating industrial agglomeration would appear due to more large-scale industrial production is needed. This agglomeration is the farmers' income effect. Industrial agglomeration or industrial diffusion is deepened on the relationship between the two opposing forces.

2.3 Wage equations

The market is opening; firms can freely enter and exit. Because the market is monopolistic in competition, each manufacturer's products are unique, the manufacturer can forestall production. With the two hypothetical model of monopolistic competition proposed by Dixit-Stiglitz (1977): each manufacturer could make price while other manufacturers keep unchanged; manufacturers do not take into account the effect of price changes on the industrial product price index. These two hypotheses are set up in the number of products to many situations. Because of the existence of product alternatives on the market, according to the principle of profit maximization, the pricing method of manufacturers is addition cost pricing:

$$p_{ij} = \frac{\sigma}{\sigma-1} \beta w_j \qquad (8)$$

We can see that the price is based on marginal cost multiplied by a constant by (8), the constant values depend on the size of the elastic coefficient σ of products; price and yield nothing, this is the performance of the market products in quantities large enough. According to the pricing formula, the total scale we can come to a region of industrial products:

$$\sum_k \sum_i p_{ijk} c_{ijk} = n_j \sum_k \mu Y_k \left[\frac{\sigma}{\sigma-1} \beta w_j e^{\tau\, d_{jk}} \right]^{1-\sigma} I_k^{\sigma-1} \qquad (9)$$

I_k is price index of k, Y_k is income of k. Because of the industrial market structure of monopoly competition, the firm's profit will become zero with the addition of more other manufacturer. Therefore, the total income of the industrial enterprises is equal to the total income of wages paid to workers (no excess profits). When profits are zero, representative of the manufacturer's production scale is

$$x = \frac{\alpha(\sigma - 1)}{\beta}$$ (The detailed derivation can refer

to Monopolistic Competition and Optimum, Dixit & Stiglitz 1977 Product Diversity"). At this time, total wages of area is $w_j n_j \alpha \sigma$ (The total wage of j is $W_j L_{imj} n_j, L_{imj}$ is the number of workers hired by the manufacture which produce I in region j, when the profit is zero, for $x = \alpha(\sigma - 1)/\beta$, under the assumption of the same technique used by every manufacture. $L_{imj} = \alpha + \beta x_{ij} = \alpha \sigma$) Combining with (9), we can draw a wage equation without the agricultural sector (The wage equation was derived by Fujit in 1999.):

$$w_j = \theta \left[\sum_k Y_k e^{-\tau(\sigma - 1)} d_{jk} I_k^{\sigma - 1} \right]^{1/\sigma}$$ (10)

$$\theta = \left[\frac{\mu \beta^{1-\sigma}}{\alpha \sigma} \left(\frac{\sigma}{\sigma - 1} \right)^{1-\sigma} \right]^{1/\sigma}, \text{ it can be considered}$$

as a constant. From (10) we could find that wages rise with income rising, and decreased with the increasing transportation costs; increases with the increase in industrial price index. Regional Wage depends on regional income level Y, area distance d, product substitution elasticity σ, transportation cost τ, and industrial price index I. By adjusting the overall price index over the real wage must be equal, otherwise, do not consider the impact of other factors; the real workers will migrate to the areas of higher wage. The real wage will not change in the short run, can be considered as a constant, so, real wages for relationship among different area:

$$F = \frac{w_j}{A_j^{1-\mu} I_j^{1-\mu}} = \frac{w_k}{A_k^{1-\mu} I_k^{\mu}} \qquad A_k^{1-\mu} I_k^{1-\mu}$$ (11)

is price of unit utility

Put the expression of agricultural products price index (7) into (11) type, (12) can be obtained:

$$I_k = \left(\frac{w_k}{F L_{km}^{1-\mu}} \right)^{1/\mu}$$ (12)

In consideration of the agricultural sector of the influence of industrial sectors, the (13) type substitution (10) type, is different from Hanson (2004), more in line with the actual Chinese wage equation:

$$w_j = \lambda \left[\sum_k Y_k L_{km}^{\frac{(1-\sigma)(1-\mu)}{\mu}} w_k^{\frac{\sigma-1}{\mu}} e^{-\tau (\sigma-1) d_{jk}} \right]^{1/\sigma}$$ (13)

while $\lambda = \theta F^{1-\sigma}/\mu\sigma$ could be regarded as a constant.

The new wage equations describe the relationship between wages in the area and consumer purchasing power, reflects the relationship between the needs in the region and spatial agglomeration: the first is that the demand for products J area will increase, if the income level improved in surrounding area of J; the second is that increase in the number of workers in each area leads to change in wage in J area, the results depend on the value of μ and σ; the third is the rise in wage levels in the vicinity, and will raise the relative price of those areas, thereby increasing the needs of the product from J area; increased, with much more production of J area, namely σ increases, it will improve the demand for workers in the region j, thus wages in the area of J will increase demand for products made in J; the fourth is that the number of area J of agricultural products is limited, the migrate of workers will reduce real wages, thereby reduce the demand for products made by area j. It can be used to analysis distribution of labor in China in the view of new economic geography, when we put the population factors into the new wage equation by means of the agricultural sector.

3 INSPECTION BASED ON CHINESE PROVINCIAL ADMINISTRATIVE DATA

3.1 An econometric model

The new wage equation is a complex nonlinear equations, direct examination of its process is quite complex, and the result is not significant because of the complexity of the relationship between the independent variable. Hanson (2004) used the original model as the econometric model directly, takes the Y, H, L, d as independent variables, and takes μ, τ, σ as the parameters to be estimated, by controlling the parameters, has estimated the impact of the various independent variables on the dependent variable estimates, which show that wages and transportation costs as well as regional market size are related. Hanson had to verify the relationship between wages and consumer purchasing power through the U.S. 3075 county data, to test the basic theory of the new economic geography. Redding and Venables (2004) transformed the wage equation consisting of accessibility of market and accessibility of the supply functions, to verify the wage equation with the view of accessibility from both demand and supply, and then verify the basic theory about the new economic geography spatial agglomeration. We build the wage equation, through Hanson's research ideas in empirical tests, and with the analytical methods proposed by Redding and Venables, to verify the relations among regional wage, market accessibility and the number of workers in the region. Furthermore, we will verify whether relationship of the regional demand and spatial agglomeration consistent with the basic theory of the new economic geography in China.

According to Fujita's wage equation, Venables (2004) defined the size of the market (market capacity) as: $MC_k = Y_k I_k^{\sigma-1}$

Accessibility is defined as the market size of the market in each region plus the total distance beyond measure, namely

$$MA_j = \sum_k Y_k I_k^{\sigma-1} e^{-\tau(\sigma-1)} d_{jk}$$

Regional wages are a function of market accessibility. Market with high accessibility means a manufacturer could get larger external effects if the manufacturer set up factories there, lower market accessibility represents that manufacturer cannot reap the benefits of economies of scale if the manufacturers set up factories there. The decision of industrial location depends on the market accessibility, and market accessibility depends on market size and transportation costs in turn. Under condition of higher transportation costs, manufacturers set up factories in the region to generate economies of scale, but high transport costs could lead to a reduction of supply to the region. Manufacturers would set up factories in different regions, to supply the market in different regions. Economies of scale will appear if a regional transport costs are lower, firms will choose the region to save fixed costs and relatively low transport costs paid to service multiple areas of the market.

By definition given by Venables, the new wage equation in market size and accessibility are:

$$MC_k = Y_k L_{km}^{\frac{(1-\sigma)(1-\mu)}{\mu}} w_k^{\frac{\sigma-1}{\mu}}$$

$$MA_j = \sum_k Y_k L_{km}^{\frac{(1-\sigma)(1-\mu)}{\mu}} w_k^{\frac{\sigma-1}{\mu}} e^{-\tau(\sigma-1)d_{jk}} \quad (14)$$

The population determines the size of the regional market, but also determines the price index of agricultural products. Through the "overcrowding effect" and "farmers income effect" in the agricultural sector,

mobility of labor would influence agglomeration level, combined with the impact of the local market effects and price effects and competitive effects of the industrial sector, the interaction of these effects, the ultimate agglomeration effects result is hard to be determined. The number of workers L_m is taken into the model, which is more in line with China's practical situation. Migrant workers had a huge impact on local industry, whether production or consumption. The influx of migrant workers will have a huge impact on local wage level. In this sense, the analysis of spatial distribution of wages, we must consider this important factor.

The wage equation can be obtained in the form of wages and logarithmic equation with the introduction of the accessibility:

$$\log(w_j) = \frac{\lambda}{\sigma} \log(MA_j) \quad (15)$$

The new wage equation we built and the Hanson & Venables is different in considering the large number of Chinese population, with introduction of the number of workers In addition to verifying the relationship between market accessibility and wages, we will also verify the relationship between wages and movement of workers in the process of agglomeration. With considering L, we can form an econometric model as (16):

$$\log(w_j) = \xi + \varphi_1 \log(MA_j) + \varphi_2 \log(L_{jm}) + \eta_j \quad (16)$$

3.2 Data and variables

According to "China Statistical Yearbook 2008", we can get constitution of the average urban disposable household spending and consumer spending, and the average rural household discretionary spending and consumer spending constitutes. Calculated by weighing, we can find that the proportion of expenditure for industrial production is about 0.5. For computing convenience considerations, μ is taken as 0.5. Minjun Shi (2007) formed a coastal center, inland areas of the periphery to the center-periphery pattern, which is in line with China's reality. Using his method, the product substitution parameter is set as 0.7, the product elasticity of substitution σ is 3. Due to the complexity of the transaction cost structure, a more reasonable value is difficult to estimate, therefore, the estimated transaction cost could be calculated by adding transport costs, insurance costs, and a certain percentage of transaction cost. The average railway freight is 0.0861 CNY / t • km, the average Yangtze River waterway freight is 0.05 CNY / t • km, shipping prices of accommodation changes with the seasons and the number of orders, but compared with the railways and inland waterways, shipping has a significant price advantage, so the average freight shipping is 0.03 CNY / t • km. In the transport of goods, rail transport has the largest proportion, the weighed unit transportation cost τ is taken as 0.9.

This paper took mainly China's 28 provincial-level administrative data for analysis, which takes into account the special nature of the Tibetan agriculture, transportation, population, etc., which are not included in the analysis, while Chongqing, Hainan, and Taiwan are not taken into account either, for the traffic is hard to be measured. Total income Y provinces to nominal GNP approximate representation of the provinces, the nominal GDP can be obtained from the National Bureau of Statistics Yearbook that was published. Wages in different provinces and cities are the average wage of workers, which can also be the value of China Statistical Yearbook. Statistical Yearbook has the dates only by industry or by the number of workers in urban employment by industry in each region, there is no direct approximate value of a reasonable indicator, and therefore no employment in the provinces (municipalities) secondary industry and tertiary industry representatives and approximately the number of workers in the region, L_m. The distance between the provinces (municipalities) and the railway distance between the capital cities of provinces and cities through an approximate measurement method, although imprecise, but can be accepted as an indicator to measure the distance between the regions before the new standard was released. We get railway distance between provinces according to the odometer calculated by the Ministry of Railways. According to the data and selected fixed parameters above, we can calculate the market accessibility (MA) of different provinces.

3.3 Estimates

Due to changes in wages it is faster in richer provinces than the poorer provinces, the provinces which develop quickly have much amplitude changes in wages than the slowly development of the province, so the richer provinces and fast development of the provinces have greater variance in wages. Also, because there is a difference in various regions of the labor force, this difference is not only reflected in the differences between regions, but also reflects individual differences in the internal region, the difference will be reflected in the labor wage differentials, it is even more increased than the variance of wages. Therefore, taking the weighted least squares estimation method to estimate the model, they may be overcome possibly by heteroscedasticity which appears to have significant influence on the estimated results.

Many factors affect wages, economic activity substantially affects spatial agglomeration factors such as knowledge spillovers and social welfare. These are all very important factors, because of the difficulty in measuring it, we all take it as a random disturbance term in empirical analysis. However, MA is the function of the income level of the region Y, the number of workers between regions L_m, regional wage level W, the distance d and the three regional fixed coefficients consisting of each of these factors and the independent variables are closely related, and they would eventually affect the MA and Lm. In order to exclude the influence of random noise this may occur between the items and the argument of the correlation estimation results using the GMM procedure once again on the estimated model. Hanson (2004) has used population growth rate as a tool variable, on the grounds that the change reflects the population of long-term trends of economic activity, rapid population growth and rapid economic development in the region, reflecting the higher wages. Demographic factors have been introduced in the wage equation in this article, we can take the reverse approach considering Hanson's model of the real estate sector, the gross value added taxes real estate regions and the total profit tax of each regional real estate as an instrumental variable. Real estate growth rate, reflecting the region's economic trends, the total amount of real estate profit taxes is a manifestation of a regional real estate development, but also reflects the state of economic development. In Hanson's wage equation, if the surrounding areas with high housing stock level, which means that house prices are high, then a lot of employment is in those areas, the demand and industrial agglomeration contact link up with the real estate industry. Therefore, growth in real estate value and the total amount of profit taxes is an ideal tool for variables.

Estimation results can be seen from Table 1, in China, regional market accessibility and spatial region and the number of workers and wage distributions are highly correlated; this proves the situation in China in line with the basic theory of the new economic geography. Column (1) in Table 1 is the number of workers L and regional market accessibility MA is set to direct the independent variable, create a simple binary linear model and estimated results can be seen from the regression coefficients and t test volume, since variables and the dependent variable have significant relationship, however, R^2 means that the lower value goodness fit test, which may be prevalent caused by heteroscedasticity when we use cross-sectional data, after using the weighted least squares estimation, we got a result in column (2). The goodness of fit at this time there is a significant improvement, rising from 0.517 to 0.921, and the independent variables in the model did not significantly reduce

Table 1. Market access and lab to wage.

log(w)	(1)	(2)	(3)	(4)	(5)	(6)
Observation	28	28	28	28	28	28
Year	2007	2007	2007	2007	2007	2007
Wage	15556.53	13768.25	8.765	8.125	8.892	6.377
	(5.975)	(4.213)	(20.729)	(16.23)	(21.223)	(6.192)
market access	56.592	56.592	0.451	0.619	0.368	1.313
	(5.167)	(7.536)	(4.957)	(6.698)	(3.952)	(3.236)
Lab	−5.069	−5.069	−0.183	−0.221	−0.147	−0.551
	(−3.023)	(−4.135)	(−3.53)	(−4.075)	(−2.841)	(−2.573)
Estimation	OLS	TSLS	OLS	TSLS	OLS	GMM
σ	0.517	0.921	0.500	0.996	0.471	−1.36
D.W	1.7252	2.112	1.614	1.918	1.881	2.056
F-statistic	13.365	30.236	12.523	23.00	6.832	2.369

When MA value, μ is 0.5, σ is 3, μ is 0.9.

Source: "China Statistical Yearbook 2008", the length of railways between the capital cities published according to the Ministry of Railways Railway odometer been finishing.

the degree, DW has greatly improved. Column (3) is based on econometric models (16); the results of the basic construction of the least squares estimation after a significant degree of each independent variable are high, however, and (1) the same column, the overall model fit is not high, since DW is only 1.614, model serial correlation may exist. After its use weighted least squares estimation, the (4) column shows the results of all the statistical indicators are improved, but did not exceed the threshold DW 2, and so there may exist a serial correlation model. Column (5) is the use of Generalized Difference Method for serial correlation results which were corrected, the results did not improve DW value, but reduces the goodness of fit of the model. Re-use of LM test of (4) to test the resulting column to obtain a low F statistics, accompanied by a large probability, then, cannot confirm the presence of serial correlation model. Taking into account the random variables may exist multicollinearity, the GMM estimates can be obtained using (6) columns. After the introduction of the real estate value added and gross profits tax of these two instrumental variables, the model saliency respective variables is still relatively high, which fully proved the model to better describe the relationship between market accessibility and the number of workers and regional wages.

According to each symbol of regression coefficients in Table 1, the wage is high in the region with high market accessibility, which is consistent with the basic theory of the new economic geography. Manufacturer's location choice depends on market accessibility, regional market accessibility is bound to attract manufacturers of high concentration to the region, will expand the local demand for workers; the increase in demand will lead to rising wages. According to the regression coefficients, we can find that with the increasing number of workers, the number of workers in the region increases, the lower the wage level, which is also basically consistent with the same relationship as depicted in the new economic geography and population movements in wage levels. According to the CP model (Krugman, 1991), when the industries began to gather in a place, the increase in the demand for workers caused the rise in the level of wages. Higher wages attract workers to migrate to the area, and then expand the size of the market, driven by demand for industrial products, the formation of further industrial agglomeration would appear, which is called a process of cause and effect cycle. Only when the scale in the region reached a certain degree, resulting in a crowding effect of excessive competition, and the workers' wages will be reduced, which basically occurred in late industrialization.

473

4 CONCLUSIONS

Based on the dual economic structure and population number of the actual situation in China, a new wage equation was built in the paper to verify that the spatial wage equation by China's data. According to the empirical analysis, we found that China's spatial distribution of wages is in line with the new economic geography. It contains a wealth of policy implications. Firstly, manufacturer's location choice depends on MA, while MA is a function which is a composition of market size and the transportation costs. According to the function, one of the regional important tasks is to improve the MA of the region's market. There are many ways to improve the MA, it can reduce transport costs between regions; it can be an expansion of market size in the region. Meanwhile, there are many ways to expand the size of the market, and it can expand consumer demand; it also can attract industry gathering in the region. As to the number of workers, there exists a large number of rural labor force, and the process of solving the employment problem will be related to the industrial transfer in China, and a series of questions about urbanization. With MA of the region with good traffic conditions, it is best to allow the region with high MA to achieve the employment of rural surplus labor locally, so it is more conducive to local development. The model is defined as a new wage equation which is in line with China's practical conditions, through modeling some of the related previous theory, making quantitative analysis more feasible.

REFERENCES

[1] Hanson, G. H. Market Potential, Increasing Returns, and Geographic Concentration, 2004, mimeo.

[2] Masahisa Fujita, Jacques-François Thisse New Economic Geography: An appraisal on the occasion of Paul Krugman's 2008 Nobel Prize in Economic Sciences. Regional Science and Urban Economics 39, 2009, 109–119.

[3] Ottaviano, G. , and D. Pinelli, "Market Potential and Productivity: Evidence from Finnish Regions", Regional science and Urban Economics, 2006, 36(5), 636–657.

[4] Ottaviano, G. , and F, Thisse, "On Economic Geography in Economic Theory: Increasing Return and Pecuniary Externalities", Journal of Economic Geography, 2001, 1(2), 153–179.

[5] Paul Krugman. Scale Economies, Product Differentiation, and the Pattern of Trade The American Economic Review, Dec., 1980, 950–959.

[6] Paul. Krugman, Increasing returns and economic geography, Journal of political economy, 1991, 99, 483–499.

[7] Paul Krugman. Space: The Final Frontier The Journal of Economic Perspectives, Spring, 1998, 161–174.

[8] Reding, S., "Spatial Income Ineuqality", Swedish Economic Policy Review, 2005, 12(1), 29–55.

[9] Reding, S. and A. J. Venables. Economic Geography and International Inequality, Journal of International Economics, 2004, 62.

[10] Minjun Shi, Zhao Zhao, Jin Fengjun. China regional market potential evaluation [J]. Acta geographica Sinica. 2007.10.

[11] Zhao Zhao. The market potential and resource exploitation in the western area [J]. economic geography, 2008, 11.

[12] Xiaoxu Zhang. The spatial structure of wages, wage equation and the new economic geography Chinese – An Empirical Analysis Based on Chinese regional data [J]. statistics and decision making. 2008.1.

Network Security and Communication Engineering – Chan (Ed.)
© 2015 Taylor & Francis Group, London, ISBN: 978-1-138-02821-0

Mining user behavior patterns in massive open online courses

X.Y. Wu & P. Wang

School of Electronic and Information Engineering, Xi'an Jiaotong University, Xi'an, Shaanxi, China

ABSTRACT: Massive open online courses have attracted a large number of participants worldwide. We have studied more than 40,000 user's time-stamped log data which describes user's interactions with online courses from a university in China to address user's different behavior patterns in different qualitative ways. We have examined the use of course components, such as lecture videos, in terms of user's time allocation. We have processed the logs into separate time-series for each user and parsed user-level statistics on course component usage, including number of unique courses accessed, total time spent per course etc. The research results revealed a remarkably different user patterns of access and time allocation among different course components. The results also reveals that the improvement of education contents and the delivery methods in online courses as well as in traditional on-campus courses are necessary.

KEYWORDS: MOOCs, Data mining, Log data, Behavior patterns.

1   INTRODUCTION

Due to the information explosion nowadays, to provide the information that users are interested in is a high-priority issue involved in the development of Internet. The visiting data of users include the interest pattern of users [1], so users' interest can be intelligently mined by analyzing the visiting data.

Interest is a positive tendency in human's recognition of objects. It is only when this tendency of recognition is stable that human interest can be formed. Learning interest is a kind of interest that carries positive emotion and promotes students to learn more. Learning interest plays an important role in the maintaining of students' motivation, the improving of students' learning proactivity and the enhancing of their learning efficiency.

The modeling of users' learning interest is the basis for the recognition of users' learning interest. There are two solutions to build user interest model. explicit approach and implicit approach. However, the explicit approach involves the users and takes much time and effort, but user interest may change over time.

A behavior-based approach observes a user's action on the internet, such as clicking, visiting retention time, visiting times, navigation path, action of saving, editing, revising and downloading, and the input keywords in search engine, etc.

A content-based approach analyzes the contents on webpages that a user has visited. Kim et al. proposed a divisive hierarchical clustering approach to build the user interest hierarchy model that can be learned

from the contents on webpages bookmarked by a user [2]. This approach can only recommend the resources similar to those which have been visited by users, whereas some researchers have found that users consider the unexpected more valuable.

A hybrid approach observes user's behavior and the contents on the webpages visited by a user. Trajkova and Gauch have built a user profile based on concepts from a predefined ontology [3]. Tan Qiong etc. used the local autonomous agents to percept user's action and adopted a learning algorithm to get user interest profile [4]. These techniques show that the usefulness and accuracy of the resulting recommendations have been increased.

This article uses the approach of invisible tracking which does not require users to provide information and the system recognizes users' interest automatically by inspecting user actions in massive open online courses.

2   MINING USER BEHAVIOR PATTERNS IN MASSIVE OPEN ONLINE COURSES

The framework of our personalized web search approach is as shown figure 1. Users send queries at the entrance of the tool. First, we get the search results referring to the user's query. Then we compute the interest value of each searching item based on the user's interest model and the recommendation value of each item based on collaborative filtering. Then we combine these two kinds of value according to some strategies and reorder the search results as the second step. Third, the reordered search results are presented to the user and the user's clicking actions would be gathered as logs. Finally, the interest model and collaborative filtering model are updated according to the analysis of the user's logs.

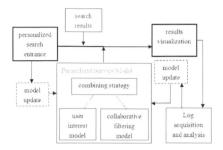

Figure 1. Framework of personalized web search.

User interest model is the formalized description of the user's interest information. There are three typical kinds of models, namely static model, implicit dynamic model and explicit dynamic model. In this paper, we adopted one of the explicit dynamic models-the weighted keyword vector model.

The weighted keyword vector model is described as following:

$$Interest_i = \{(k_1, w_1),\ldots,(k_n, w_n)\} \quad (1)$$

where, $Interest_i$ represents the interest model of user u_i. k_i is the $i\text{-}th$ keyword which is either extracted from the user's logs, queries or input by the user in advance. w_i is the weight of keyword k_i which represents how interested the user is in k_i. The weight is also called interest value.

For each search result r_j, we have obtained the interest value I_{ij} and recommendation value R_{ij} of user u_i to r_j. Then we compute the corresponding reordering value $Score_{ij}$ as following:

$$Score_{ij} = \alpha \times I_{ij} + (1 - \alpha) \times R_{ij} \quad (2)$$

3 EXPERIMENT

In the experiment, users launched many queries. The queries with high search frequency are system-structure, topic-map, entropy, personalization, neural-network, memory, visualization, machine-learning (in Chinese), SVM, LDA, KMP and Dijkstra etc. in massive open online courses.

For each user and each query, we compute the NDCG@n values when α in formula (2) varies from 0 to 1 with the step 0.01. When α changes to a new value, we can get a new order of the search results and

obtain a new NDCG@n value of the new sequence. So, while α changes, we get 101 NDCG@n values. Then we can compute the highest NDCG@n (which means the best reordering), the lowest NDCG@n (which means the worst reordering) and the average NDCG@n. Besides, the NDCG@n value of the original order provided by Google API can also be obtained. Then we compare these NDCG@n values with that of the original Google order to evaluate the effect of our approach.

For some queries, our approach has a better reordering effect than the original order provided by Google. Here, we take "topic-map"(in Chinese) as an example. In table 1, we randomly choose 8 users' experimental results and the row Avg is the average NDCG@n value of the corresponding column. The column best shows the highest NDCG@n value we can obtain when α changes from 0 to 1 while the column worst is the lowest one. The column average shows the average of all the NDCG@n values referring to different α (0-1). And the column original represents the effect of the order provided by Google API. The row Avg is the average value of each column (all users).

Table 1. Experimental results.

Query	best	worst	average	original	improving
topic-map	0.7770	0.4691	0.7156	0.5010	55.10%
entropy	0.7061	0.3788	0.6277	0.8915	-20.79%
memory	0.5947	0.2501	0.5222	0.7138	-16.69%
SVM	0.7773	0.4989	0.7170	0.7882	-1.38%
LDA	0.8229	0.3197	0.7314	0.6799	21.03%
KMP	0.6302	0.2137	0.5545	0.4779	31.87%
Dijkstra	0.7891	0.3545	0.7101	0.7425	6.28%
visualization	0.5111	0.1958	0.4319	0.7246	-29.46%
neural-networks	0.9438	0.3339	0.7922	0.9400	0.40%
machine-learning	0.9680	0.3394	0.8537	0.8951	8.15%
Avg	0.7788	0.3509	0.6919	0.7366	2.26%
standard deviation	0.1456	0.0993	0.1299	0.1562	0.2435

4 CONCLUSION

We can conclude from the experiment that our approach can obtain a better effect when a) the query is ambiguous, b) the search results contain too much spam information and c) the search results can be divided into several categories. But the approach doesn't work well when the query is explicit and the search result is precise. On the whole, our approach can make some contribution in personalized web search in massive open online courses.

ACKNOWLEDGMENTS

The research was financially supported in part by China National Key Technology R&D Program under Grant No. 2012BAI34B01; Ministry of Education of

China Humanities and Social Sciences Project under Grant No. 12YJC880117.

REFERENCES

[1] Yan Guo, Shuo Bai, Zhifeng Yang, Kai Zhang. (2005). Analyzing scale of web logs and mining users' interests. Chinese journal of computers, 28(9), 1483–1496, 2005.

[2] Hyoung-Rae Kim, Philip K. Chan. (2008). Learning implicit user interest hierarchy for context in personalization. Applied intelligence, 28(2008), 153–166.

[3] Joana Trajkova, Susan Gauch. (2004). Improving ontology-based user profiles. Proceedings of the RIAO, Vaucluse, France, 380–389.

[4] Tan Qiong, Li Xiaoli, Shi Zhongzhi. (2002). A method of providing individual service in search engine. Computer Science, 29(2002), 23–25.

The impact of XBRL classification criteria on accounting conservatism——an example of the wholesale and retail trade

R.Y. Hu & F.Y. Hu

Accounting Informatization And Finance Decision Support Center, East China University of Science and Technology, Shanghai, China

ABSTRACT: This research took the wholesale and retail trade industry listed on the Shanghai Stock Exchange as an example. The author analyzed the impact of the use of XBRL classification criteria on accounting information by taking the data from 58 listed companies in this industry from 2005-2010 as a sample for statistical and regression analysis. In this paper, by taking Shanghai wholesale and retail trade data as a sample, the establishment of XBRL, asset-liability ratio, the proportion of the largest shareholder, whether the general manager holdings, changes in accounting standards and other relevant independent variables, including a dummy variable XBRL, investment opportunities, cash flow, asset size and other control variables and the dependent variable of accounting conservatism were measured by building a model. Through the establishment of regression models, the author took regression analysis and used empirical research methods to explore whether there was a significant impact of XBRL classification criteria on Accounting conservatism and put forward relevant suggestions. Consequently, the author finds, accounting conservatism from 2008- 2010 was more significant, compared to the conservatism from 2005-2007, where the use of XBRL had a significant effect. According to relevant conclusions, the author proposed some suggestions to the implementation of XBRL.

KEYWORDS: Accounting conservatism, XBRL classification criteria, Listed companies.

1 INTRODUCTION

XBRL is an XML-based markup language for defining business and financial information and exchange. XBRL technical framework is mainly composed of three parts: XBRL specification, XBRL taxonomy and XBRL instance documents.

AICPA Chahes Hoffman proposed XBRL in April 1998. Since 1998when XBRL was proposed, its advantages have been widely applied in many aspects, including: reducing the cost of exchanging information exchange, improving the relevance of the information, and lessening the limitation in automatic exchange and excerption of financial information without restriction from individual software companies and information systems, etc. And it is also obvious that XBRL has developed rapidly.

But not every country has reached an agreement for the XBRL standard yet. The resolution that the U.S. listed companies and public funds could use interactive data proposed by SEC has led to a lot of controversies among investors (SEC, 2008).

The development of XBRL in China started relatively late, but the speed of its advancing is so high and shows many obvious potential advantages. In May 2002, China Securities Regulatory Commission set standard for electronic disclosure of listed companies and XBRL was selected. In early 2004, the Shanghai Stock Exchange began implementing XBRL pilot projects. On December 3, 2007, XBRL China- Chinese regional organization has been formally established. And in October 2010, the Ministry of Finance released the Common Enterprise Accounting Standards classification criteria at the 21st International XBRL Conference.

Meanwhile, XBRL taxonomy in China was also confronted. Although the Shanghai Stock Exchange and Shenzhen Stock Exchange joined the XBRL implementation plan in 2003 and 2004 respectively, they have not yet reached the harmonization of standards.

2 LITERATURE REVIEW

2.1 *Foreign research literature review*

XBRL, as a new form of XBRL financial reporting and disclosure of financial information, shows its advantage mainly by reducing the cost of searching for information, enhancing the transparency of accounting information and improving earnings quality.

Bonso'n et al (2001) found that XBRL-based financial reports could provide matched reports according to their preferences and their needs for a particular user. Kernan's (2004) study found that the main advantage of XBRL applications was to improve the transparency of corporate information and to accelerate the availability of information.

Diane J. Janvrin (2012) and Vicky Arnold (2012) believe that the implementation of XBRL has affected the investment decisions of investors. The former comes to the conclusion that the use of XBRL would provide accounting information and confirm the existing implementation plans and opportunities for future research foundation, which is researched through the establishment of four themes. The latter uses the experimental method and leads to the result that the implementation of XBRL has significant effects on non-professional investors.

But some scholars have come up with the idea that there are difficulties in XBRL implementation. Because each organization owns its own XBRL tags, it leads to the difficulty of the unified implementation of XBRL.

There are many scholars involved in the research about Accounting Conservation, but few researches are about the impact of XBRL classification criteria on Accounting Conservatism. Among them, the measure models proposed by Basu, Ball, Givoly and Ryan are most famous.

Basu first proposed his model in 1997, where he used positive and negative stock returns to represent the effective "Good News" and "Bad news" in capital markets. Ball (2005) proposed to use positive and negative cash flows to represent good news and bad news. Givoly (2000) used the extent and speed of negative cumulative accrued to reflect the Accounting Conservatism of the financial reports. Ryan (2000) used the carrying amount of the asset-the market price ratio as a measure to Accounting Conservatism.

2.2 Research literature review in China

The research about XBRL in China is very few and it began late. Besides, most of the researches lack empirical data to support the conclusions.

Shen Yingling (2005) elaborated the advantage of extensible Business Reporting Language (XBRL) through analyzing the limitations of existing network technologies- HTML. Lei Lei (2011) also believed that the use of XBRL-based financial reports could improve the transparency of financial information. However, there was not significant effect on the timeliness of financial information.

Zhao Mingxian and Zhang Tianxi (2010) used event analysis and regression analysis to test the information content of annual reports, and concluded that, although the market did not produce abnormal fluctuations overall in the annual report, the disclosure of XBRL had the impact on ERC on 10% significance

level for the Shanghai Stock Exchange listed companies. That means the disclosure of the annual report of XBRL has worked.

Li Fengui (2012) carried out qualitative research and empirical research to test the effects of XBRL on the quality of accounting information. Studies from the theoretical analysis and empirical research have shown that XBRL-based information can significantly affect CAR and enhance the quality of accounting information which could help investors make more effective decisions.

Meanwhile, scholars in China have built more suitable models for the Chinese market according to the four foreign classics models used to measure accounting conservatism. The basic principles were been changed, but the models are much targeted for Chinese listed companies.

According to the research by Givoly and Hayn (2000), Xu Xinhua, Sun Zheng(2008) and Quan Hongjie (2012) established a more appropriate method to measure accounting conservatism in China as follows:

Non–operating accruals
  = total accruals
  – Operating accruals
Total accurals = (net income + depreciation) – Cash flows from operating activities

$$\text{Item} = \text{increase in operating receivables accrued (– decrease) Increase in inventories (– decrease) + prepaid expenses increase (– decrease) – an increase of accounts payable (+ decrease) – payable increase (+ reduction)} \quad (1)$$

This model adds CI value ratio as a measure to the degree of robustness of the measurement values according to the measure of Ball and Shivakumar (2005). CI is the contrast ratio of non-operating accruals and number of asset value at the beginning. The greater the value is, the higher the degree of Accounting Conservatism is. It is easier to understand.

$$CI = -\frac{\text{Non–operating accruals}}{\text{The value of assets at the beginning}} \quad (2)$$

3 EMPIRICAL RESEARCHES

3.1 Research hypothesis

According to previous research about XBRL classification criteria, the author proposed a research hypothesis as follows:

H1: Accounting conservatism of 2008-2010 data with XBRL classification criteria standards has been improved greatly, as contrast to the conservatism level of 2005-2007 data without XBRL classification criteria.

H2: The implementation of XBRL classification criteria has a great impact on the enhancement of accounting conservatism.

H3: The implementation of XBRL classification criteria helps enhance accounting conservatism and keeps it stable.

3.2 *Empirical research sample*

This paper selects 71 Shanghai Stock Exchange-listed companies in wholesale and retail trade as a sample. The data from 2005-2007 is acquired from the annual report of listed companies (annual gathering sources Reuters - Stock page), and the data from 2008- 2010 from the Shanghai stock Exchange XBRL-on line. All the data in this paper are collected, organized and calculated by the author manually,

Samples are removed in the following order:

1 Excluding the companies after 2004 IPO (a total of eight);
2 Excluding ST companies since 2004 (a total of five);

Therefore, the actual number of samples obtained was 58 per year.

3.3 *Descriptive analysis and correlation test*

In 2008, the China Shanghai Stock Exchange and Shenzhen Stock Exchange required all listed companies to disclose XBRL reports. So the data in this paper is separated in 2008. The data from 2005-2007 is acquired from the annual report of listed companies (annual gathering sources Reuters - Stock page), and the data from 2008-2010 from the Shanghai stock Exchange XBRL-on line.

According to the method of measuring accounting conservatism, the descriptive analysis of CI is as shown in Table 1.

Table 1. Descriptive statistics of CI.

	2005	2006	2007	2008	2009	2010
Ave.	2.376%	1.596%	−0.422%	5.023%	4.894%	4.168%
Max.	33.497%	54.537%	22.797%	45.862%	77.681%	102.461%
Min.	−40.851%	−58.026%	−58.970%	−23.768%	−115.708%	−74.504%
Median.	2.953%	3.341%	3.255%	4.667%	4.614%	4.316%
Standard dev	10.448%	15.642%	14.280%	10.296%	24.051%	20.595%
Variance	1.092%	2.447%	2.039%	1.060%	5.785%	4.242%

From Table1, we can see that the mean and median of CI from 2008 to 2010 is significantly greater than that from 2005-2007. It shows that accounting conservatism has been significantly improved after XBRL implementation and the average value has been improved to some extent, increased from -0.5% - -2.5% in 2005- 2007 to 4% -5.2% in 2008-2010. That is 3% or so. The median has been increased by about 1.5 percent generally. However, whether the key reason of improvement of accounting conservatism is the implementation of XBRL needs further studies. However, we can see from the maximum, minimum and variance in Table 1 that although accounting conservatism has been improved, but its stability is not assured. The maximum, minimum and variance of CI in 2008-2010 have changed greatly during the period studied.

3.4 *Regression analysis*

In this paper, the regression model of regression analysis is as follows:

$$CI = \alpha_0 + \alpha_1 Lev + \alpha_2 First + \alpha_3 XBRL + \alpha_4 CF + \alpha_5 Accsted + \alpha_6 TQ + \alpha_7 Size + \alpha_8 Holder + \varepsilon$$

This paper uses time series data. CI is the dependent variable, and XBRL, Lev, First, Holder, Accsted are independent variables, and TQ, CF, Size are control variables.

Table 2. Variables.

CI	Accounting conservatism indicator
XBRL	Dummy variable, value is 1 if the company adopted XBRL in that year, otherwise 0
Lev	Asset-liability ratio
First	The company's largest shareholder stake
Holder	If the manager holds the stock from the company, 1;otherwise,0
Accsted	The change of accounting principles,1 if changed once; 2 if changed twice
TQ	TQ = (non-tradable shares * net assets per share + tradable shares * price + book value of liabilities) / total assets
CF	Operating cash flow, the ratio of cash flow from operations divided by the end use of the assets
Size	The natural logarithm of assets

There are a total of 348 samples from 2005-2010. And the author makes multiple linear regression analysis in SPSS 19 of Chinese version.

After correlation test and mean test, the author makes regression analysis. The data is as follows:

Table 3. Coefficient.

Model	Non-standardized coefficient		Standard coefficient			95.0% confidence interval of B	
	B	Standard err	Trial	t	Sig.	lower limit	Upper limit
1 (Constant)	−0.209	0.194		−1.074	0.284	−0.591	0.174
Lev	0.119	0.05	0.123	2.38	0.018	0.021	0.217
First	0.104	0.058	0.096	1.795	0.074	−0.01	0.218
XBRL	0.039	0.023	0.117	1.66	0.098	−0.007	0.085
CF	0.501	0.074	0.412	6.772	0.000	0.356	0.647
Accsted	0.039	0.027	0.11	1.428	0.154	−0.015	0.093
TQ	−0.036	0.008	−0.318	−4.707	0.000	−0.051	−0.021
Size	0.004	0.01	0.025	0.409	0.683	−0.015	0.023
Holder	−0.006	0.019	−0.019	−0.347	0.729	−0.043	0.03

a. Dependent variable: CI

During the time of multiple regression analysis, the ENTER was used to prevent part of the dummy variables and numerical variables getting into the model. The results are shown above.

According to Table 3, the result is carried out separately for each factor. The significance level of XBRL and First is lower than 0.1, although higher than 0.05, indicating that a great impact on CI. The significance level of XBRL reaches up to 0.098, from which we can conclude that XBRL does have a significant impact on the accounting conservatism, and significance levels of First can reach 0.074. This is also the key point of this study.

4 CONCLUSIONS AND SUGGESTIONS

This paper mainly focuses on the impact of the implementation of XBRL classification criteria on accounting conservatism.

According to the research carried out, the accounting conservatism has been enhanced. We can conclude that the implementation of XBRL classification criteria has a significant impact on accounting conservatism of listed companies based on the regression analysis. But it has little impact on the stability of accounting conservatism. The author believes that there will be a significant improvement of stability in the long term.

And there are some suggestions for the implementation of XBRL.

1 The implementation of XBRL classification criteria should focus on the difference of defferent companies.

2 The processed XBRL application should enhance the spread of XBRL standard.

3 In the process of promoting XBRL classification criteria, the government and the relevant regulatory authorities should force companies to disclose relevant information and at a certain time.

REFERENCES

[1] Huang Changyin, Wu Zizhong. Empirical research of factors on voluntary disclosure - based on XBRL classification criteria perspective.[J].Economic and Management Research,2011,(8).

[2] Jang Yihong. An Empirical Study of the quality of accounting information in listed companies.[M]. Shanghai: Shanghai University of Finance and Economics Press,2008,56–202.

[3] Lei Lei. Impact of XBRL on the quality of accounting information:[Master's degree thesis on Jinan University].[D].Jinan: Jinan University,2011.

[4] Li Fengui. Empirical Research of the impact of XBRRL on Accounting information quality. [Master's degree thesis on Changsha University of Science and Technology].[D]Hunan:Changsha University of Science and Technology,2012.

[5] Li Zengquan, Lu Wenbing. Accounting Conservatism of accounting earnings: discovery and enlightenment. [J].Accounting Research, 2003, (2).

[6] Quan Hongjie. Factors on accounting conservatism and its economic consequences:[Master's degree thesis on Hunan University].[D].Hunan:Business Administration College in Hunan University,2009.

[7] Shen Yingling .XBRL:Innovation on Access Network Financial Reporting [J]:Collected Essays on Finance and Economics,2002,(8):24–28.

[8] Tangwen. Research of XBRL taxonomy on small businesses:[Master's degree thesis on East China University of Science and Technology].[J].Shanghai: East China University of Science and Technology,2011.

[9] Xu Quanhua. Economic consequences and influence factors of accounting conservatism. [J]. Guangxi Social Sciences,2010(4).

[10] Xu Xinhua, Sunzheng. An Empirical Analysis of China's stock market cycle and accounting conservatism.[J]. Finance Research,2008,(12):53–57.

[11] Zhao Mingxian,Zhang Tianxi. Annual Information Content of XBRL standards-based [J]. Research on Economics and Management,2010,(2):26–30.

[12] Diane J. Janvrin,Won Gyun No.XBRL Implementation: A Field Investigationto Identify Research Opportunities. [J] Inf. Sys: 2012:169–197.

[13] Kamran Ahmeda, Darren Henry. Accounting conservatism and voluntary corporate governance mechanisms by Australian firms. [J]Accounting and Finance 52 (2012) 631–662.

[14] Vicky Arnold a, Jean C. Bedard, Jillian R. Phillips, Steve G. Sutton. The impact of tagging qualitative financial information on investor decision making: Implications for XBRL. International Journal of Accounting Information Systems 13 (2012) 2–20.

Network Security and Communication Engineering – Chan (Ed.)
© *2015 Taylor & Francis Group, London, ISBN: 978-1-138-02821-0*

Quantitative factor analysis in evaluating the satisfaction degree of teaching effects

Z.Y. Tang & Q.F. Yang
Department of Computer Science, Hunan Radio & TV University, Changsha, China

ABSTRACT: The most important step of teaching evaluation is to evaluate the teaching result, while the result through data processing can demonstrate the effectiveness of evaluation. The original data of teaching evaluation can be processed by many calculation methods, such as the factor analysis and fuzzy evaluation. This paper mainly performs the factor analysis, and interprets the way to conduct quantitative evaluation and quantitative analysis, which is intended for the satisfaction degree of teaching effects.

KEYWORDS: Satisfaction degree, Quantitative factor analysis, Teaching effects.

1 INTRODUCTION

The continuous development of advanced education calls for an increasingly improved teaching quality, so the top priority for teachers is to improve the teaching quality. The long-term school development cannot be realized without the support of the improved teaching quality. In this sense, it is important to promote the teaching reform and improve the teaching quality. This is to be done by proactively exploring the unbiased and effective method to evaluate the teaching quality, and building up a scientific and reinforced system to evaluate and monitor the teaching quality.

In last few years, most colleges have adopted the student-centered method for evaluating teachers. The significant edges of student-centered evaluation can be described as follows:

1 It is students that experience the teaching quality and result, while the teacher's word and deed would influence the student;
2 Students have understood all respects of a teacher, for example, their teaching quality, discipline, and capability, and they are more familiarized with the teaching.

To that regard, the teaching quality of each teacher can be fully analyzed by students' evaluation of the teacher's capability and teaching quality.

2 BRIEFING THE FACTOR ANALYSIS

The factor analysis is a kind of statistical method. It can be performed to probe into the internal structure of sample coefficient matrix, research the internal dependence among variables, discover the basic structure in the observed data, and infer some "potential variables" that are hardly observed from a huge amount of data. One major purpose of performing the factor analysis is to infer the observed facts to the maximum while using the minimum factors. By doing so, it cansimplify the data structure better, and profoundly uncover the essential relations among varied things.

Through the factor analysis, a group of variables are divided into the combinations of special factors and common factors according to the sample data, which have the potential domination. In this sense, the factor analysis highlights the special and public nature of one variable. Wherein, the special nature means that one variable carries the unique information, while other variables cannot possibly have it; the public nature refers to the case that there is correlation between every two variables, as they are under the influence of common factors. Its mathematical model can be displayed as follows:

$$X_i = a_{i1}F_1 + a_{i2}F_2 + \cdots a_{im}F_m + \varepsilon_i \qquad (i = 1, 2, \cdots, p) \qquad (m \le p)$$

wherein, X_i is a group of original variables, $F_1, F_2, \cdots, F_m$ are referred to as the common factors, ε_i is referred to as the special factor of X_i, which can be expressed by the matrix form:

$$X_i = AF + \varepsilon$$

wherein, $A = (a_{ij})_{p \times m}, a_{ij}$ is the factor load, which can better reflect the correlation between ith variable and jth common factor, $a_{ij} = r_{X_i}F_i$. The quadratic sum of all row elements contained in A $h_i^2 = \sum_{j=1}^{m} a_{ij}^2$ $(i = 1, 2, \cdots p)$ can be referred to as the variance contribution of the common factor

F_j, therefore S_j is the scale of measuring the relative importance of the common factor F_j. The quadratic sum of all line elements contained in A

$$S_j = \sum_{i=1}^{p} a_{ij}^2 \quad (i = 1, 2, \cdots p)$$ can be referred to as the

commonality of the variable X_i, which depicts the contribution made by all common factors to the total variance of the variable X_i. Therefore, the higher commonality of X_i would mean the better explanation available in this factor analysis model.

The factor analysis can be applied only by satisfying the condition: if the majority of relevant coefficients in factor matrix are bigger than 0.3 through the statistical test, the variable is suitable for factor analysis, vice versa.

Generally, there are two prevailing statistical methods for carrying out the factor analysis.

2.1 *KMO test*

This method is used to test if the partial correlation coefficient between variables is small enough (The partial correlation coefficient calculates the net correlation coefficient on the premise of controlling the influence of other variables on the bivariate, if there is strong influence of overlapping pass, or if the variable can be equivalent to the common factor, the partial correlation coefficient after controlling such influence is definitely small enough). KMO statistical value is between 0 and 1, and KMO value is close to 1. The closer correlation between variables would mean a higher accuracy by adopting such method. But if the final value<0.5, KMO Test is not recommended to use.

2.2 *Bartlett sphericity test*

It examines if the relevant coefficient matrix is unit matrix, so as to judge if the factor analysis is applicable. The hypothesis of this test is that the relevant coefficient matrix of original variable is unit matrix. If the Bartlett statistical value is bigger, then the corresponding concomitant probability is smaller when compared with the defined significance level, so the factor analysis is applicable; if not, just accept the hypothesis, while all variables are independent upon one another, so the factor analysis is not recommended to use.

3 EVALUATION RESULT ANALYSIS OF TEACHING EFFECT AND SATISFACTION DEGREE

3.1 *Selection of sample and variable indicator*

In terms of sample selection, this paper conducts a questionnaire survey of Open Remote students in one phase about the teaching result satisfaction in the second 2012~2013 semester. To make the survey results more representative, this paper chooses 25 teachers of international trade as the evaluation object. In these classes, a certain percentage of students are selected from each section of scores, and they are told to fill in the questionnaire. The total number of delivered questionnaires is 180, and there are 150 valid pieces.

In respect of variable indicator, this paper mainly chooses 10 indicators, which are listed in the table below:

Table 1. Indicators for evaluating the satisfaction degree of teaching effect.

Variables	Content
X_1	Demonstrate noble teacher's ethics, stringent teaching style, modeling effect, discipline observance and punctuality
X_2	Individualized quality education, clear teaching objective, and emphasis on the development of student's capability and quality
X_3	Flexibility in understanding the content of teaching materials, and carefully prepared lessons
X_4	Rich teaching materials and a wide range of knowledge
X_5	The suitability of teaching method for course feature, flexible and diversified teaching methods.
X_6	Concise language and good logic
X_7	Demonstrate proficiency in open thinking, proactive interaction among students and interesting classroom
X_8	On-time and careful homework correction, and demonstrate carefulness and patience in giving instruction and answering question
X_9	Clear and well-organized concept, emphasis on the difficult points
X_{10}	Link theory with reality, and put the acquired knowledge into use

As shown in Table.1, 10 indicators are seen as the variables of one data analysis, "10 scores" is the indicator mark for each one, while the data of each indicator is the average of all original data.

3.2 *The result of factor analysis process*

1 Data standardization. Input the original data into SPSS13.0, which automatically standardize the data.
2 Correlation analysis. During the process of correlation analysis, accurately calculate the correlation coefficient matrix through SPSS13.0, and

the final result indicates the majority of correlation coefficient is bigger than 0.3. In addition, all of these correlation coefficients conform to the significance test. Bartlett test value is 283.197, which is relatively big, P<0.0001. KMO statistical value is 0.678>0.5; therefore, 10 indicators of evaluating the teaching satisfaction can be aided by the factor analysis.

3 Solution to the factor load. Aided by SPSS13.0, the eigenvalue and variance contribution of correlation coefficient matrix concerning all indicators can be obtained.
4 Calculate the factor scores, and make a comprehensive evaluation of the satisfaction degree of teaching effects.

The score function of common factor can be obtained through the foregoing examples:

$$F_1 = 0.023X_1 + 0.055X_2 - 0.058X_3 - 0.014X_4 - 0.035X_5 - 0.100X_6 + 0.256X_7 + 0.280X_8 + 0.264X_9 + 0.258X_{10}$$

$$F_2 = -0.059X_1 - 0.172X_2 + 0.117X_3 + 0.350X_4 + 0.364X_5 + 0.408X_6 - 0.019X_7 - 0.095X_8 - 0.041X_9 - 0.041X_{10}$$

$$F_3 = 0.409X_1 + 0.410X_2 + 0.280X_3 - 0.140X_4 - 0.021X_5 - 0.023X_6 + 0.007X_7 + 0.065X_8 - 0.008X_9 - 0.012X_{10}$$

Therefore, the equation for constructing the evaluation function can be defined as:

$$F = a_1F_1 + a_2F_2 + a_3F_3$$

wherein, a_1, a_2, a_3 is the variance contribution of each common factor, or the corresponding weight as directly given by expert. This paper defines the evaluation function through the former one.

Table 2. The comprehensive scores and ranking of the satisfaction degree of teaching effects.

Teacher Numbering	F_1	F_2	F_3	Comprehensive Scores	Ranking
1	−0.610 86	0.391 77	−0.454 11	−0.230 00	18
2	0.668 37	−0.349 13	1.138 05	0.374 40	9
3	−0.296 39	−0.273 56	0.243 98	−0.170 36	17
4	0.537 22	0.457 71	0.180 13	0.395 27	8
5	1.924 53	−0.820 43	−0.739 36	0.502 65	5
6	−0.186 28	0.617 4	1.589	0.341 04	10
7	−1.210 06	0.630 92	−0.877 99	−0.493 35	20
8	0.237 77	0.800 72	0.955 25	0.481 25	7
9	0.201 64	1.247 44	−1.279 34	0.242 54	12
10	0.036 16	1.108 65	1.346 62	0.540 54	3
11	0.306 81	−1.392 53	1.263 86	−0.061 33	13
12	1.429 44	0.225 65	−0.282 77	0.651 54	2
13	−0.757 67	−0.680 54	−0.224 53	−0.562 78	21
14	−0.907 77	−1.964 61	−0.649 62	−1.059 33	24
15	2.102 28	−0.887 55	−0.263 38	−1.223 14	25
16	0.721 4	1.536 42	−1.704	0.487 76	6
17	−0.596 72	−1.623 26	−0.10. 62	−0.739 80	23
18	−0.184 03	−1.192 15	−1.026 38	−0.579 50	22
19	−0.286 6	0.508 67	−1.110 16	−0.156 11	16
20	2.220 53	−1.175 99	−0.927 63	0.503 33	4
21	−1.217 11	0.272 37	1.395 13	−0.242 73	19
22	1.060 45	0.363 32	1.709 08	0.838 79	1
23	−0.951 83	1.720 64	−0.878 32	−0.070 73	14
24	−0.239 26	−0.008 84	0.214 87	−0.074 67	15
25	0.205 57	0.487 51	0.484 63	0.304 72	11

$$F = 44.200\%F_1 + 28.329\%F_2 + 15.631\%F_3$$

According to the above equation, the comprehensive factor scores of 25 teachers can be summarized, which outputs the most satisfying teacher. In this paper, "0" is the basis for comprehensive score, if the score is much bigger than "0", which indicates a higher satisfaction degree, vice versa.

4 SUMMARY

From the above, the factor analysis can effectively evaluate the teaching effects. Through the final score of common factors, teacher could read the student's evaluation about teaching effects, and find out the strength and weak point, thereby allowing him or her to improve accordingly. The education administration may have an accurate judgment about the teaching effect according to the result displayed, and ultimately give incentives to teachers accordingly. By doing so, the education administration can positively improve the teaching quality.

REFERENCES

[1] Ding Xingfu. Remote Education[M].Beijing: Beijing Normal University Press, 2009.
[2] Yang Qiufen. Computer-aided Teaching Research[M]. Xi'an Jiaotong University Press, 2014.
[3] Zhang Xiuwei. The Application of Factor Analysis Model into the Teaching Evaluation for New Course Classroom[J].Journal of University of Science and Technology of Suzhou (Natural Science), 2008, 9:12–17.
[4] Lu Lihong. The knowledge of college English teacher about the evaluation criteria for course teaching[J]. Open Education Research, 2011,8:79–83.
[5] Cai Guochun, Xi Fei, Hu Rendong.The characteristics of college study during the development in Europe - The comparative study with American universities and colleges[J].Journal of Higher Education, 2011, 2:56–60.

Wireless Communications and Sensor Networks

Network Security and Communication Engineering – Chan (Ed.)
© 2015 Taylor & Francis Group, London, ISBN: 978-1-138-02821-0

Controlling algorithm for energy consumption of radio bandwidth to manage coverage area problem

K.A. Marsal & I. Abdullah
Faculty of Science and Technology, Universiti Sains Islam Malaysia, Bandar Baru Nilai, Negeri Sembilan, Malaysia

ABSTRACT: There is a wide range of application in wireless ad-hoc networks which are related to crucial operations of the government and security, private sectors, health services as well as environmental assessment. Based on the importance of these networks, the coverage area problem has remained an area of focus among researchers and is implemented as one of the most important design criteria of mobile devices. Efficiency consumption of energy is one of the key areas of wireless ad-hoc networks due to the limitation of batteries. Therefore, the protocol that uses low energy consumption is an important factor in these networks due to their impact on the network's coverage. In this study, the main objective is to determine the energy-efficient algorithms for coverage problems from the literature. It has been determined that to accommodate a large wireless ad-hoc network, distributed algorithms can be used to obtain a higher energy efficiency through coverage regulating algorithms.

KEYWORDS: Energy-efficient algorithms, wireless ad-hoc networks, coverage area problem, coverage regulating algorithms.

1 INTRODUCTION

Currently, almost all offices and households have adapted to using wireless LAN communication standards namely WiFi 802.11b for the unrestraint Internet connection and access to remote data (Molisch, 2010). Indeed, various engineering complexities have been encountered in the construction of mobile networking systems especially the energy efficiency aspect of mobile devices like 802.11-activated computers, tablets and smart phones. Therefore, low energy consumption is a crucial factor in the hardware design of these devices in order to optimize and overcome the limited energy source of batteries (Molisch, 2010). The minimization of energy consumptions and the improvement of energy efficiency pose a big challenge in the industry of mobile networks (Goldsmith & Wicker, 2002). Over the years, the advancement in the mobile devices industry has resulted in the invention of hardware designs with low energy consumption as the device's mobile network interface requires the highest source of energy input. Energy saving can indeed be enforced by deactivating the device's receptor, however in the practical sense, this may not be a straightforward step as a node must be arranged for turning on the receptor to receive, acknowledge as well as to take part in high levels of control and routing protocols (Öström, 2010). Therefore, the alignment between energy efficiency and routing protocols are especially important in multiple hops and ad-hoc mobile networks where packers are being forwarded to each receptor (Sohrabi et. al., 2000).

Procedures that are commonly used to achieve energy-efficient communication include the reduction of communicative time by deactivating the receptor, compression of data size during transmission and the reduction of transmission power. Controlling the transmission power complements the procedure of receptor deactivation and data compression for a simultaneous energy-efficient communication (Öström, 2010). The control of transmission power during active transmission of radio bandwidth is simultaneous with distribution of transmission power across various wireless networks such as IR-based wireless sensor systems and Bluetooth RF. The transmission power can be ideally reduced to the lowest stage without affecting its capability of receiving the accurate packets despite the interventions of path loss and weak signals (Sohrabi et.al., 2000). Controlling algorithm for energy consumption of radio bandwidth enables the adaptation of mobility and the fluctuation of noisy multipath in received signal strength (RSS) as well as the adaptability to asymmetric networks (Sheth & Han, 2003). In the early study of energy control of wireless ad hoc networks, the focus was restricted to stimulating environments via theoretical assessments such as the presence of two different networks for regulation and data that enables the implementation of 802.11b ad hoc networks. A second example includes the selection of identical lowest transmission energy in accordance to all nodes with the common lowest energy as displayed by the network with the identical connection as shown at the highest transmission energy (Sheth & Han, 2003).

Nevertheless, an adjustment to the MAC packet header as well as the maintenance of network connections routing table prior to each transmission power are required in this procedure that restrict the additional categorization (Mainwaring et.al., 2002). Another example of energy regulation procedure inspects the differences of the carrier-sense regions where transmissions are identified from the transmission block in which packets are interpreted. RTS or CTS are transmitted by using the highest energy while data packet is transmitted at an interval period at the same time also with the highest energy to avoid crash (Stankovic et al., 2003). In addition, these procedures are conducted by assuming that transmission losses are equal between nodes and do not acknowledge the issue of portability. The adjustment of transmission energy affects the connections of wireless ad hoc routing networks in complicated manners that are not well comprehended. Lowering and adding the transmission energy between two nodes result in networks graph to gauge the lowest path routings and in fluctuating through dispersed and compact interconnectivities. Node transmissions occur at a stable default energy level without an energy transmission control and enable the authorization of link energy costs to every links. Additionally, the existing energy level of batteries can be monitored by every node (Mukherjee et al., 2003).

By focusing on the cost factor of node energy and link energy, numerous control algorithms for energy consumption of radio bandwidth have been established in accordance to different measurements such as selecting paths with the highest amount of battery capacity, choosing routes which results in the lowest amount of energy consumed as well as in optimizing the usage of minimum power consumption (Agarwal et al., 2001). During the optimization of minimum power, the duration to network partitions is maximized and results in a higher level of complexity during the introduction of transmission energy. Given this, the main focus in controlling algorithm for energy consumption of radio bandwidth is often on the decision to employ energy control rather than the procedures of applying them. The categorizations of the topology control algorithms include multi-stage algorithms and energy-control algorithms (Agarwal et al., 2001). Further categorizations include heterogeneous and homogeneous networks. In heterogeneous networks, different maximum transmission power present in nodes while only identical power ranges are present in homogeneous networks (Polastre et al., 2004).

2 CONTROLLING ALGORITHM FOR ENERGY CONSUMPTION

In the recent years, there has been an increase in the number of studies in the development of control algorithms in wireless sensor networks. Wireless sensor networks or mobile ad-hoc networks are autonomous collections of portable nodes that enable communications across constricted mobile networks (Murthy & Manoj, 2004). Ad-hoc networks operate by using batteries with limited power input. Given this, the control of energy consumption is a crucial factor in the design of mobile ad-hoc networks (Murthy & Manoj, 2004). Given this, the present study focuses on different aspects of energy-control algorithms such as the efficiency of energy control algorithms, control schemes in distributed energy and energy regulating, the impact of layers and hardware design on energy requirements, the protocols of transmission energy and energy control via energy saving.

2.1 Energy control algorithms

Energy employed by scheduled users in executing their communication is determined by algorithms as transmission energy in networks could not be minimized by users. Various algorithms have been developed that pack the highest frequencies of transmission and in scheduling node power levels in minimizing transmission energy (Mainwaring et al., 2002). The study (ElBatt & Ephremides, 2004) have contributed to the development of an algorithm that packs the highest frequency of transmission of the existing slots without causing any disruption in transmissions. Subsequently, the access to system data within the network limit can be maximized (ElBatt & Ephremides, 2004). Another set of algorithms have been developed in scheduling node energy levels in accordance to the node energy levels within nodes. Signal to noise ratio from any existing connectivity is used to maintain the energy level by assuming a fitting arrangement of node energy levels. It has also been reported that energy level of nodes could not be effectively aligned with fitting encoded schemes (Liang & Dandekar, 2007).

Generic algorithms are developed to provide solutions to issues pertaining to lowest energy levels regarding nodes in wireless sensor networks. Permutation encoded schemes are more efficient in securing data transfer by using networks. By employing the theory of convex optimization, researchers have developed a collective energy regulation, channel scheduling and scale-regulating algorithms (Zheng & Kravets, 2003). It is used to direct the energy regulating issues in the channel scheduling procedures based on interferences and addresses the high traffic issues on the transport layer. Given this, algorithms are not efficient to be applied in larger networks (Zheng & Kravets, 2003). This problem is addressed by other researchers that came up with heuristic algorithms to handle large-size networks. The algorithms are directed in data transmission by using lowest transmission power across all scheduled users within a network. However, the weakness of this algorithm includes inefficiency in maintaining node energy levels in large networks based on interferences (Singh & Kumar, 2011).

Based on the algorithms developed from literature survey, it can be concluded that algorithms determine the energy required for schedule users in minimizing the transmission energy in networks. However, it cannot be decided that individual node power in networks is efficient in determining separating node energy. It can be summarized that the algorithms are not effective when used in encoded schemes of networks, and the issue of energy regulation cannot be addressed via a collective energy regulation and channel scheduling because of the obstructions by network size and channel scheduling based on interferences. Lastly, it can also be summarized that heuristic algorithms are not efficient in maintaining node energy levels for channel scheduling in large networks based on interferences.

2.2 Control schemes in distributed energy and energy regulating

An exceptional efficiency in increasing the network' holding power through lower energy requirement of nodes is provided by the protocols of energy regulating medium access when transmission occurs in wireless sensor networks (Martincic & Schwiebert, 2005). A distributed energy control scheme has been developed that results in a significant lower transmission energy requirement per bit, thus saving power and extending the mobile nodes. However, the protocols of energy regulating medium access are not efficient because of lower transmission energy per bit during higher rates of transmissions (Shih et al., 2001). An efficient energy allotment between sources and every prospective relays have been developed to determine the higher energy effectiveness capacity at particular transmission rates. In addition, networks have lower reliability in general because of issues associated with traffics and congestions (Shih et al., 2001).

Joint opportunistic energy scheduling and end to end rate-controlling schemes in wireless ad-hoc networks have been assessed in increasing network's efficiency through energy allotment control of every links and data rate of every node within the network. In general, more energy is consumed by networks in joint opportunistic energy scheduling and end to end rate-controlling schemes (Martincic & Schwiebert, 2005). A method has been developed to explain the higher consumption of energy by nodes as compared with the compelling choosing of nodes. This selection requires less energy consumption and does not result in network disruption. However, the performance of networks will be decreased at higher rates of transmissions (Akyildiz on et al., 2006). Based on the literature survey on control schemes in distributed energy and energy regulation, persistence

of energy regulation is caused by congestions and large data transmission in wireless ad-hoc networks.

2.3 Energy-efficient algorithms for coverage problems

In the literature, energy-efficient algorithms for coverage problems proposed can be categorized into centralized and distributed algorithms (Cardei & Wu, 2006). Decision processes are decentralized in distributed algorithms. Distributed algorithms are referred to as a set of distributed decision procedures of every node that utilizes localized data across a fixed amount of hops. Algorithms and procedures have to be designed in a distributed and localized manner to acknowledge a higher scale as a dynamic topography is present in wireless ad-hoc networks which requires a high amount of sensors (Younis et al., 2006). From the literature, the problems that are given the most attention pertaining to coverage issues can be categorized into region, points as well as barriers of coverage. Specific designs decisions are used when formulating solutions to address coverage problems include the features of algorithms, sensor distribution procedures, range of sensing and communicating, additional important demands.

3 CONCLUSION

The study have shed light on issues pertaining to energy regulation that needed to be acknowledge in order to improve the energy regulation of wireless ad-hoc networks. In wireless ad-hoc networks, the existing mobile nodes do not contain fixed infrastructures and are self-organized by convention. The main challenges in the field of wireless ad-hoc networks include a restricted energy supply, route disruptions, synchronization as well as security problems. Based on the energy restriction factor, wireless ad-hoc network nodes must be designed in a particular approach that minimizes the consumption of energy. Wireless ad-hoc network nodes function as routers and there fore demand a higher energy for the routers in the forwarding and relaying of packets. To address this challenge, the study has determined the energy control algorithms from the recent studies by briefly proposing its formulations and theorizations. In wireless ad-hoc network, the range of coverage is one of the most important characteristics in the system's quality of service. In general, coverage is correlated to power efficiency and the connections of networks and is one of the most crucial elements in mobile networks. From the study, it has been determined that in accommodating a large wireless ad-hoc network, a better performance can be obtained in coverage regulating algorithms and procedures via distributed algorithms.

REFERENCES

[1] Agarwal, S., Katz, R. H., Krishnamurthy, S. V., & Dao, S. K. 2001, September. Distributed power control in ad-hoc wireless networks. In *Personal, Indoor and Mobile Radio Communications, 2001 12th IEEE International Symposium on* (Vol. 2, pp. F–59). IEEE.

[2] Akyildiz, I. F., Lee, W. Y., Vuran, M. C., & Mohanty, S. 2006. NeXt generation/dynamic spectrum access/cognitive radio wireless networks: a survey. *Computer Networks*, 50(13), pp. 2127–2159.

[3] Cardei, M., & Wu, J. 2006. Energy-efficient coverage problems in wireless ad-hoc sensor networks. *Computer communications*, 29(4), pp. 413–420.

[4] ElBatt, T., & Ephremides, A. 2004. Joint scheduling and power control for wireless ad-hoc networks. *IEEE Transactions on Wireless Communications*, 3, pp.74–85.

[5] Liang, C., & Dandekar, K. (2007). Power management in MIMO ad-hoc networks: A game-theoretic approach. *IEEE Transactions on Wireless Communications*, 6, pp. 1164–1170.

[6] Goldsmith, A. J., & Wicker, S. B. (2002). Design challenges for energy-constrained ad hoc wireless networks. *Wireless Communications, IEEE*, 9(4), pp. 8–27.

[7] Mainwaring, A., Culler, D., Polastre, J., Szewczyk, R., & Anderson, J. (2002, September). Wireless sensor networks for habitat monitoring. In *Proceedings of the 1st ACM international workshop on Wireless sensor networks and applications* (pp. 88–97). ACM.

[8] Martincic, F., & Schwiebert, L. (2005). *Introduction to wireless sensor networking* (pp. 1–40). John Wiley & Sons: New York, NY, USA.

[9] Molisch, A. F. (2010). *Wireless communications* (Vol. 15). John Wiley & Sons.

[10] Mukherjee, A., Bandyopadhyay, S., & Saha, D. (2003). *Location management and routing in mobile wireless networks*. Artech House.

[11] Murthy, C. S. R., & Manoj, B. S. (2004). *Ad hoc wireless networks: Architectures and protocols*. Pearson education.

[12] Öström, E. (2010). Building and experimentally evaluating a smart antenna for low power wireless communication.

[13] Polastre, J., Hill, J., & Culler, D. (2004, November). Versatile low power media access for wireless sensor networks. In *Proceedings of the 2nd international conference on Embedded networked sensor systems* (pp. 95–107). ACM.

[14] Sheth, A., & Han, R. (2003, May). Adaptive power control and selective radio activation for low-power infrastructure-mode 802.11 lans. In *Distributed Computing Systems Workshops, 2003. Proceedings. 23rd International Conference on* (pp. 812–818). IEEE.

[15] Shih, E., Cho, S. H., Ickes, N., Min, R., Sinha, A., Wang, A., & Chandrakasan, A. (2001, July). Physical layer driven protocol and algorithm design for energy-efficient wireless sensor networks. In *Proceedings of the 7th annual international conference on Mobile computing and networking* (pp. 272–287). ACM.

[16] Singh, V. P., & Kumar, K. (2011). Literature survey on power control algorithms for mobile ad-hoc network. *Wireless Personal Communications*, 60(4), pp. 679–685.

[17] Sohrabi, K., Gao, J., Ailawadhi, V., & Pottie, G. J. (2000). Protocols for self-organization of a wireless sensor network. *IEEE personal communications*, 7(5), pp. 16–27.

[18] Stankovic, J. A., Abdelzaher, T. F., Lu, C., Sha, L., & Hou, J. C. (2003). Real-time communication and coordination in embedded sensor networks. *Proceedings of the IEEE*, 91(7), pp. 1002–1022.

[19] Younis, O., Krunz, M., & Ramasubramanian, S. (2006). Node clustering in wireless sensor networks: Recent developments and deployment challenges. *Network, IEEE*, 20(3), pp. 20–25.

[20] Zheng, R., & Kravets, R. (2003). On-demand power management for ad-hoc networks. In *Proceedings of IEEE Infocom, year*.

Network Security and Communication Engineering – Chan (Ed.)
© 2015 Taylor & Francis Group, London, ISBN: 978-1-138-02821-0

A workflow hierarchical system modeling approach for dust concentration network sensor

X.Z. Cheng, P. Wang, X.Z. Bai, Y.J. Yu, J.H. Liu & L.W. Cui
College of Electrical Engineering and Automation Shandong University of Science and Technology, Qingdao, China
State Key Laboratory of Mining Disaster Prevention and Control Co-founded by Shandong Province and the Ministry of Science and Technology, Shandong University of Science and Technology, Qingdao, China

ABSTRACT: In view of the problems that dust concentration sensor information (lower) based on field bus cannot swap and output (upper) network's incompatibilities, in this paper, a workflow hierarchical modeling method with the combination of Unified Modeling Language (UML) based on object-oriented and Petri net is proposed. Based on IEEE1451 standard, the method has built a deployment diagram of the static UML use case diagram, sequence diagram, and system; by converting UML to Petri net use case diagram, the system dynamic model is obtained; based on Petri net advantage in dynamic modeling, the method has analyzed, optimized and simplified the dynamic behavior of the system, and has confirmed the fact that the model structure has accessibility and boundedness instead of an existing deadlock. Therefore, we get the dust concentration network sensor system model which lays the foundation for hardware system design based on the model.

KEYWORDS: Dust concentration, network sensor, UML, petri net, IEEE1451, workflow modeling.

1 INTRODUCTION

Workflow modeling techniques have been developed for many years, and have achieved a series of results[1,2,3], such as in network sensor system modeling, Kang Lee and others[4] have established object-oriented sewage sensor system based on workflow, SORRIBAS J and others[5] have developed distributed marine measurement and control sensor system. The literature[6] proposes a workflow dynamic testing model on BNF (Backus-Naur form) as the main body, which can be used for a given workflow definition language. To the best of our knowledge, literatures using the ideas of workflow modeling in dust detection field are very rare. If we can use this kind of thoughts on the dust concentration sensor network modeling[7,8], it is of great significance to solve the problem that dust concentration detection system (lower) based on field bus sensing does not swap and the problem of the output (upper) network's incompatibility. This paper intends to use workflow modeling approach to combine UML and Petri and build a universal model of dust concentration sensor network system based on IEEE1451 standards, which is designed to address the problem of the industrial on-site dust concentration to achieve the sensor swap of the collection terminal (lower) and the network compatibility of the output terminal (upper).

2 DUST CONCENTRATION NETWORK USING SYSTEM MODELING

Standardized sensor is the trend of the development of network sensor, and IEEE1451 standard is the representative of the development direction. It is an ideal solution to design the network sensor on this standard. In order to realize a standard network sensor system, the sensor system formal model must be constructed firstly, which can help the fast implementation of network sensor system.

Currently in the mainstream modeling languages, unified modeling language UML is a well-defined, easy-to-express, powerful and widely-used visual modeling language[9]. Its strong ability of static structure description and object-oriented design thought is suitable for application in model construction based on IEEE1451 network sensor. But UML lacks effective analysis and verification method[10], Petri net can just make up for the shortcoming. Compared with UML, Petri net can provide a variety of ways to verify the analysis[11] because of the outstanding modeling advantage of dynamic characteristic. But the modeling method of Petri net is not intuitionistic, feeble in demand and interaction[12]. Combining UML and Petri net modeling method and making full use of their advantages can effectively capture requirements, analyze design, also strictly verify the model. Therefore, system modeling of dust concentration sensor network is introduced in this paper, based on the workflow hierarchical modeling method with combination of UML and Petri net. Figure 1 is the diagram of modeling method combining UML with the characteristics of IEEE1451 network sensor workflow and Petri net.

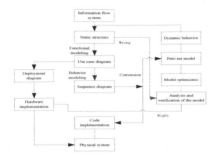

Figure 1. Block diagram modeling method based on UML and Petri net.

3 THE STATIC MODE CONSTRUCTION BASED ON UML

3.1 UML using case diagram model building

3.1.1 System function requirement analysis

The function of dust concentration sensing system is to complete data acquisition, processing and transmission. Participants are the local operators and remote operators. The on-site dust sensors realize the collection of dust information, actuators perform control command sent by a remote operator. Serving for the local and remote operators, the system makes the sensor have functions of self-test, self-calibration, self-diagnosis and network communication through integrating the dust sensors, signal processing unit and network interfaces to realize the standardization of information acquisition, processing and transmission. Therefore, the specific functions of the system to be implemented are analyzed respectively from the local operators, remote operators, the dust sensors and actuators.

3.1.2 Confirmed cases

By the system demand, the local operators need initialization, data collecting, data control, signal processing, and TEDS configuration. Dust sensors require to have data acquisition and data control. Execution units need to have output control, and data processing. So the system use case is concluded. Because use case got by the demand is not completely suitable in the use case diagram, there are always various relations in use case, such as generalization, contain, and extension. Through case analysis, the relationships between use cases are found out and are suitable for use in a use case diagram as shown in Figure 2.

Figure 2. Dust concentration sensing system use case diagram.

3.1.3 Determination of the relationship between the use case

As shown in Figure 2, it can be found that signal conditioning and data processing cases belong to the same module through analysis of the relationship between use cases, and they can merge. Based on the definition of IEEE1451 standard, use cases that STIM and NCAP can contain are shown in Figure 3 and Figure 4.

Figure 3. Inclusion relation among NCPA use case.

Figure 4. Inclusion relation among STIM use case.

3.1.4 Use case diagram of network sensor system

After the relationship between use cases is determined, it is time to see the relationship between the final use case and the participants. The local operators, sensors and actuators need to manipulate STIM module; Remote operators need to manipulate NACP module. Combined with Figure 3 and Figure 4, the dust concentration sensor network system use case diagram is obtained as shown in Figure 5.

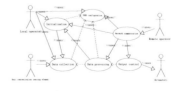

Figure 5. Network sensor system use case diagram.

3.2 UML sequence diagram model building

UML interaction diagrams describe the actual operation of the system. After determining the system use case diagram, it is necessary to use interaction diagrams to describe the actual operation and interaction of the system object. There are three kinds of interaction diagrams: sequence diagram, communication diagram, and time diagram, among which sequence diagram is the most widely used. This article uses the sequence diagram to describe the actual operation of the system.

The first is the interaction of remote operators with the system. Use case of interaction with the remote operators includes: network communication, initialization, data acquisition, data processing, TEDS configuration, output control. Second is the interaction of local operators with the system. Use cases of interaction with the local operators include: initialization, data acquisition, data processing, TEDS, and configuration.

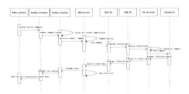

Figure 6.   Remote operators sequence diagram.

3.3 UML deployment diagram model building

Network sensor sensing system based on IEEE1451 standard in model construction, there are two modeling ways of object-oriented system in physics: component diagram and deployment diagram. This paper adopts the deployment diagram to build object-oriented modeling of system.

Deployment diagram is used for the physical deployment of modeling system, which is mainly related to the physical structure and the relationship between them. Users of network sensor system are remote and local operators, and the physical structure of the sensor network system includes the NCAP module and STIM module. Figure 7 is the deployment diagram of the system.

Figure 7.   Network sensor deployment diagram based on IEEE1451.

So far, the network sensor system static model has been established. First, the function of the system is analyzed to determine the role of the model and use cases, as well as the relationship between the various roles and use cases. After that, the networked intelligent sensor system sequence diagram is drawn, it has a detailed statement of the information flow when the system is designed and a precise statement of the signal flow of the internal system is made. Finally, the modeling of the system in physics established through the deployment diagram determines the system modules and the relationship between the internal modules components. The static representation of the system model is further improved.

4   PETRI NET DYNAMIC MODEL BUILDING

When designing the system, first of all we need to identify the system elements, especially the identification of passive elements and active elements. Petri net provides a strong support for this kind of duality. An object in the system as active elements or passive elements to build modeling is determined by the context of the system. Through the mapping relationship between the two, UML use case diagram of the system is transformed into Petri net model, the transformed graphics are shown in Figure 8.

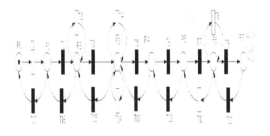

Figure 8.   Petri net model of network sensor system.

5   OPTIMIZATION AND SIMPLIFICATION OF THE SYSTEM MODEL

Through building the dynamic model, the dynamic behavior of the system can be analyzed. Through the analysis, the expression of the dynamic behavior of signal processing in the internal system is not very clear, optimization and simplification of the system model will be needed further. The Petri net model is obtained by simplifying as shown in Figure 9.

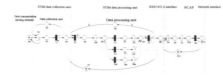

Figure 9. The optimized network sensor system Petri net model.

Using CPN Tools to verify and analyze this system model, it is confirmed that a deadlock does not exist in the model structure, and the system model has good accessibility and boundedness. The hardware system can be constructed according to the model.

6 CONCLUSIONS

For the problems that the dust concentration sensor cannot achieve the sensor swap of the collection terminal (lower) and the network compatibility of the output terminal (upper) in reality, on the basis of the IEEE1451 standards, we adopt the workflow hierarchical modeling method by combining the object-oriented unified modeling UML idea and Petri net and building the static UML use case diagram, sequence diagram, and the deployment diagram of the system. By using the transition relation from UML to Petri net, we get the dynamic model of the system. By taking the advantages of Petri net in the dynamic modeling to analyze, optimize and simplify the dynamic behavior of the system, we confirm that there does not exist deadlock in the model structure which has accessibility and boundedness, and finally we obtain the model of dust concentration network sensor system, which lays foundation for the later development of hardware systems based on the model.

ACKNOWLEDGMENT

The project was supported by the Major Program of the National Natural Science Foundation of China (Grant No. U1261205), Shandong Province Large Scientific Instruments and Equipment Upgrade Program (No.2013SJGZ05), the Promotive research fund for young and middle-aged scientists of Shandong Province (No.BS2013DX012) and the Postdoctoral Fund of China (No.2014M551934) and Science and technology innovation fund for Graduate students Shandong University of Science and Technology (YC140339).

REFERENCES

[1] KARNIEL A, REICH Y. Formalizing a workflow-net implementation of design-structure-matrix-based process planning for new product development [J]. IEEE Transactions on Systems, Man, and Cybernetics-Part A: Systems and Humans, 2011, 47(3): 476–491.

[2] ABRISHAMI S, NAGHIBZADEH M, EPEMA D H J. Cost-driven scheduling of grid workflows using partial critical paths [J]. IEEE Transactions on Parallel and Distributed Systems, 2012, 23(8): 1400–1414.

[3] ZHENG Chang-you, LIU Xiao-ming, YAO Yi, and REN Zheng-ping. Modeling Approach for Workflow Testing Based on Petri Nets [J]. Journal of University of Electronic Science and Technology of China, 2014, 43(1):119–124.

[4] LEE K, SONG E Y. UML model for the IEEE 1451.1 Standard [C]. Proceedings of the 20th IEEE Information and Measurement Technology, United States, 2003: 1587–1592.

[5] SORRIBAS J, DEL RIO J, TRULLOLS E, et al. A smart sensor architecture for marine sensor networks [C]. International Conference on Networking and Services, Piscataway, New Jersey, United States. 2006:1–8.

[6] HWANG G, LIN C, TSAO L T, et al. A framework and language support for automatic dynamic testing of workflow management systems [C]//Third IEEE International Symposium on Theoretical Aspects of Software Engineering. Tian jin, China: IEEE, 2009: 139–146.

[7] CHEN Ju-shi, JIANG Zhong-an, JIANG Lan. Experimental research on dust distribution and its influencing factors in belt conveyer roadway [J]. JOURNAL OF CHINA COAL SOCIETY, 2014,39(1):135–140.

[8] Hao Jiang, Jiannan Zhai, Hall strom, J.O., Flash Track: A Fast, In-network Tracking System for Sensor Networks[C], DCOSS IEEE International Conference, 2014,51(5):26–28.

[9] CHAI Yu-mei, FENG Qiu-yan, WANG Li-ming. Research on Methods for Generating Test Cases of Inter-Classes Interaction Based on UML Models and OCL Constraints [J]. ACIA ELECTRONICA SINICA, 2013, 41(6):1242–1248.

[10] QIN Huai-bin, LIANG Bin, SHAO Ming-wen, GUO Li, DAI Jian-guo. Modeling for System Services Gained Flow Based on UML and Petri Nets [J]. Journal of Anhui Normal University (Natural Science), 2011,34(1):38–41.

[11] YOON B J,VAIDYANATHAN P P. Context-sensitive hidden Markov models for modeling long-range dependencies in symbol sequences [J].IEEE Transactions on Signal Processing, 2006,54(11):4169–4184.

[12] Alireza Khoshkbarforoushha, Pooyan Jamshidi, Ali Nikravesh,Fereidoon Shams. Metrics for BPEL process context independency analysis [J].Service Oriented Computing and Applicadons, 2011, 5(3):139–157.

Network Security and Communication Engineering – Chan (Ed.)
© 2015 Taylor & Francis Group, London, ISBN: 978-1-138-02821-0

Network communication based on 6LoWPAN and ZigBee

F. Di
Sanjiang University, Nanjing, Jiangsu, China

B.Q. Luo & Z.X. Sun
Nanjing University of Posts and Telecommunications, Nanjing, Jiangsu, China

ABSTRACT: This paper realizes the interoperability between 6LoWPAN and ZigBee network based on some standard protocols. A mixed network architecture is put forward for interoperability between ZigBee nodes and 6LoWPAN nodes in a mixed wireless sensor network. We analyze the key issues of the protocol interoperability, design the border gateway function to implement the protocol transition, and detail data transmission flows among 6LoWPAN nodes, gateway and ZigBee nodes. This paper puts forward a new communication model and functional framework. In our design, the basic protocol stack remains unchanged; we achieve the basic intercommunication between ZigBee and 6LoWPAN nodes through a powerful gateway. The comprehensive scheme lays a good foundation for more in-depth studies on heterogeneous network integration.

1 INTRODUCTION

This paper realizes the interoperability between 6LoWPAN and ZigBee based on standard protocols as follows.

IEEE 802.15.4 [1] defines the protocol and enables the interconnection of data communication devices in a low-data-rate, low-power, and low-complexity short-range radio wireless personal area network (WPAN). As the basis of 6LoWPAN and ZigBee, it specifies the PHY and MAC of WPAN.

ZigBee [2] specifies the network layer and application layer on top of IEEE Std 802.15.4 MAC so as to provide a standards-based protocol for interoperability of sensor networks. It offers abundant application framework and security management system, and defines different specifications for different applications.

6LoWPAN [3] (IPv6 over Low-power and Lossy Networks) is a set of standards defined by IETF 6LoWPAN WG. 6LoWPAN standards enable the efficient use of IPv6 over low-power, low-rate wireless networks (include IEEE802.15.4) of simple embedded devices through an adaptation layer and the optimization of related protocols. At present, the 6LoWPAN WG has released three specifications, which specify the underlying requirements and goals of the initial standardization[3], the 6LoWPAN format and functionality [4] and the stateful header compression.

CoAP [5](Constrained Application Protocol) is a specialized web transfer protocol designed by IETF CoRE (Constrained Restful Environments) WG for using with constrained networks like 6LoWPAN which usually locates over UDP and provides a method/response interaction model between application end-points, supports built-in resource discovery, and includes key web concepts such as URIs and content-types.

UPnP [6] technology defines architecture for pervasive peer-to-peer network connectivity of intelligent appliances, wireless devices, and PCs of all form factors. It provides a distributed, open networking architecture that leverages TCP/IP and Web technologies to enable seamless proximity networking in addition to control and data transfer among networked devices. Through this technology, a device can dynamically join a network, obtain an IP address, convey its capabilities, and learn about the presence and capabilities of other devices. Finally, a device can leave a network smoothly and automatically. SSDP (Simple Service Discovery Protocol) is a multicast service discovery mechanism that uses a multicast variant of HTTP over UDP. It defines how devices send their own service and device information and how the control point searches its interesting service and device. UPnP networking process includes following six steps: addressing, discovery, description, control, eventing and presentation.

The network communication architecture of this paper is shown in Figure 1 [7]. The 6LoWPAN nodes (both routers and hosts) and ZigBee (both FFDs and RFDs) nodes are all located in the same mixed network. Through a complex and powerful border gateway B, the heterogeneous nodes can communicate with each other, and they can also access to the Internet to communicate with outbound nodes (IPv6, 6LoWPAN or ZigBee).

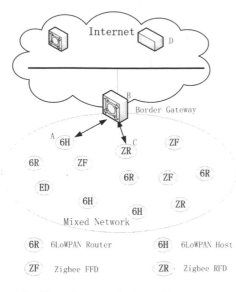

Figure 1.　Network communication architecture.

2　RELATED WORKS

With an adaptation layer, 6LoWPAN protocol stacks can easily offer interoperability between 6LoWPAN nodes and IPv6 nodes, while ZigBee nodes still require a more complex application layer gateway. There have been more communication techniques between ZigBee and IPv6 networks than those between ZigBee and 6LoWPAN.

2.1　6LoWPAN-ZigBee

At present, there are few solutions to the internetworking between ZigBee and 6LoWPAN, which are summarized into two approaches of dual stack node and dual stack gateway. Dual stack node can realize both ZigBee's and 6LoWPAN's functions, and only one of them can be used at the same time [8]. Thus, it is not an internetworking solution. Furthermore, it requires a more complex and powerful device which is not fit for the low-power and low-complexity WPAN. A dual stack gateway can efficiently simplify the functions of nodes and reduce resource and energy consumption.

Both solutions only stay in frame level, and the critical problems like addressing and service discovery remain unsolved and how to deal with the 6LoWPAN header compression and packet fragmentation is one of the most important problems for ZigBee to be resolved. Due to the close relationship between 6LoWPAN and IPv6, we can find some ZigBee-IPv6 solutions to most problems above.

2.2　ZigBee-IPv6

The straight approach to realizing the internetworking between ZigBee and IPv6 is to add IPv6/UDP layer between ZigBee NWK and APL layer, which is shown in figure 2. Each ZigBee node is assigned with an IPv6 address. When an IP packet is received from the IPv6 network at the gateway, it will be encapsulated into ZigBee NWK header and forwarded to the ZigBee network. On the other hand, if the gateway receives a ZigBee packet, it will de-capsulate the packet and continues the transmission with the IPv6 payload inside. This method changes the original ZigBee stack and cannot solve the packet size problem between 802.15.4 and IPv6.

ZigBee APL	Application		ZigBee APL
	UDP/TCP		
Zigbee NWK	IPv6		IPv6/UDP
	6LoWPAN adaptation		ZigBee NWK
IEEE 802.15.4 MAC	IEEE 802.15.4 MAC		IEEE 802.15.4 MAC
IEEE 802.15.4 PHY	IEEE 802.15.4 PHY		IEEE 802.15.4 PHY
ZigBee	6LoWPAN		IPv6 over ZigBee

Figure 2.　IPv6 over ZigBee approach.

An address translation mechanism [9] which is realized in the dual stack gateway is designed by Sakane S, etc. Take Figure 1 as an example. When the network initiates, IPv6 host D must register its IPv6 address (IPd) to pre-assigned Gateway B (IPv6 address: IPb, ZigBee address: Zb). B will help D to get its ZigBee address (Zd). ZigBee node C must register its ZigBee address (Zc) to B, too. If C wants to communicate with D, it sends out the packet to Zd. B will translate the packet into IPv6 format with "destination IP address = IPd" and "source IP address = IPc". In the reverse path, for communication from D to C, C will send packet to IPb with a data payload which contains "destination ZigBee address" and "source ZigBee address=Zc". After B receiving the packet, it de-capsulates the packet, and translates the payload into 802.15.4 format. This framework works. But it doesn't solve the service discovery, one of the most important functions of ZigBee.

With the improvement of the address translation solution, a novel integrated mechanism of interconnecting between ZigBee and IPv6 nodes is put forward by Reen-Cheng Wang [10], etc. The dual stack gateway realizes not only the translation between ZigBee address and IPv6 address, but the transition between ZigBee service discovery and SSDP service discovery of IP network. In this design, each ZigBee device is assigned with a global IPv6 address so that every IPv6 node can communicate with it directly. On the other hand, each IPv6 node is also assigned with a ZigBee short address, which is different from the solution above. The IPv6 Multicast Group is also established in all correlated IPv6 nodes for relaying broadcast messages from ZigBee network.

2.3 *Summary*

As 6LoWPAN is a protocol aiming to connect the low-power WPAN with the standard IPv6 network, the communication between 6LoWPAN and ZigBee cannot realize without IPv6. And the intercommunication of the heterogeneous network protocol commonly requires layer 3 forwarding. The ZigBee-IPv6 solutions above have a great reference for the intercommunication between 6LoWPAN and ZigBee, especially in the address and service discovery transition mechanism. Based upon the unchanged protocol stack structure, this paper offers a complex and powerful gateway equipped with all the basic protocol stacks to realize the address translation and protocol transition. Finally, an integrated data flow diagram is put forward to achieve the intercommunication between 6LoWPAN and ZigBee.

3 KEY ISSUES

3.1 *Addressing*

IEEE 802.15.4 defines two kinds of link-layer address: the global unique 64-bit EUI-64 address set by device manufacturer, and the 16-bit short address dynamically assigned in the PAN. However, the IPv6 address formed from EUI-64 address is the only one for the outbound transmission.

IP addressing with 6LoWPAN works just like in nomal IPv6 network, IPv6 addresses are typically formed automatically from the prefix of the LoWPAN and the interfaces identifier (IID). The difference in a 6LoWPAN addressing is that a direct mapping between the link-layer address and the IPv6 address is used for packet compression. ZigBee doesn't offer an adaptation layer like 6LoWPAN, so the border gateway needs a module for transition between link-layer address and IPv6 address for ZigBee nodes using the transition method 6LoWPAN defines. 6LoWPAN and ZigBee nodes in the same mixed network share the same prefix configured by the border gateway.

6LoWPAN and ZigBee networks both generally use the 16-bit short address for inbound communication respectively. 6LoWPAN supplies a Duplicate

Address Detection (DAD) mechanism in the gateway to assign a short address for nodes while ZigBee uses the CSkips algorithm. Therefore, the border gateway in this paper configures both 16-bit ZigBee address and 6LoWPAN address for nodes, which need heterogeneous communication in the mixed network.

In general, the border gateway sets an address transition module between link-layer address and IPv6 address for ZigBee nodes and an address mapping table for the mapping of ZigBee short address, 6LoWPAN short address and EUI-64 of each node (ZigBee, 6LoWPAN and IPv6 node) which needs to communicate with the heterogeneous nodes. Take Figure 1 for example, if 6LoWPAN node A, ZigBee node C and IPv6 host D communicate with each other through gateway B. The mapping table in B will be set as table 1.

Table 1. Address mapping table in gateway B.

EUI-64	ZigBee short address	6LoWPAN short address
E64(A)	Z16(A)	L16(A)
E64(C)	Z16(C)	L16(A)
E64(D)	Z16(D)	L16(D)
...	...	...

3.2 *Protocol transition and packet size consideration*

The protocol transition between ZigBee and 6LoWPAN is based on the header transition between ZigBee and IPv6. The address translation module above is also included in the header transition module in the border gateway. After the MAC header of ZigBee packet is the 8-byte NWK frame header and then the 2-10 bytes APS frame header. Besides the basic source and destination addresses and ports' information, the routing and security information is also to be consideration. As the transition is based on the specific application, how to realize the header transition is out of this paper.

For the packet size matching problem between IPv6 and IEEE802.15.4, 6LoWPAN put forwards solutions from two aspects, one is the header compression and decompression; the other is the packet fragmentation and reassembly. However, ZigBee stacks don't support the fragmentation or the reassembly mechanism. It is difficult to realize this mechanism in every ZigBee node even the relevant module is set up in the border gateway. Therefore, the communication between 6LoWPAN and ZigBee nodes in the scenery of this paper requires small packet size data as far as possible.

3.3 Service discovery

After the network initiates, an important process before data communication is service and device discovery. Nodes search for the interesting service and device in the network to communicate with and broadcast their own device information and service function information for other nodes to inquiry.

ZigBee alliance designs a set of service and device discovery mechanism, which is defined SD in the ZDO of APL. This paper adopts the UPnP architecture and its SSDP service discovery protocol. UPnP is built upon the HTTP and UDP, and 6LoWPAN supports the UDP and the simplified HTTP protocol (CoAP protocol of CoRE working group). Therefore, the UPnP architecture is also fit for the 6LoWPAN application layer, and uses the lightweight CoAP protocol and the RD (resource discovery) to realize the service and device protocol transition between the 6LoWPAN, ZigBee or IPv6 nodes. The service discovery protocol stacks comparison is shown in Figure 3. In conclusion, for service discovery, the border gateway in this paper realizes the protocol transition among ZDO-SD based on ZigBee, HTTPU-SSDP based on IP and the CoAP-RD based on 6LoWPAN.

UPnP	UPnP	APL
CoAP (RD)	HTTPU/MU (SSDP)	ZDO (SD)
UDP	UDP	
IPv6	IPv6	NWK
6LoWPAN		
802. 15. 4 MAC/PHY	802. 3 MAC/PHY	802. 15. 4 MAC/PHY

| 6LoWPAN | IPv6 | Zigbee |

Figure 3. The protocol comparison among 6LoWPAN, IPv6 and ZigBee.

4 THE SCHEME OVERVIEW

The scheme of this paper aims at the intercommunication between heterogeneous nodes using the ZigBee and 6LoWPAN basic protocol unchanged. This section gives the scheme overview including the protocol stack transition model in a border gateway.

4.1 Protocol stacks transition model

As shown in Figure 4, the intercommunication between ZigBee nodes and 6LoWPAN nodes of the same mixed network relies on the network layer forwarding between ZigBee stack and 6LoWPAN stack. Besides, the IPv6 stack is also configured in the gateway for outbound communication. To handle the multicast service discovery message and convert the different service discovery protocols, the gateway needs to analyze the upper layer protocol.

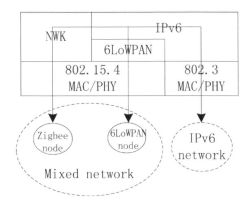

Figure 4. The protocol stacks transition model.

4.2 Gateway functional discription

The border gateway integrates the ZigBee coordinator with 6LoWPAN edge router function. Besides, the router function is indispensable. The functions in the border gateway including:

The basic network protocol stacks: including ZigBee protocol stack, 6LoWPAN protocol stacks and IPv6 protocol stacks.

Network layer protocol transition: it offers the header transition module to convert the NWK header of ZigBee packet and the 6LoWPAN header, also the APS header of ZigBee and UDP header when necessary.

The function of the address translation: the border gateway sets an address transition module between link-layer address and IPv6 address for ZigBee nodes and an address mapping table for the mapping of ZigBee short address, 6LoWPAN short address and EUI-64 of each node (ZigBee, 6LoWPAN and IPv6 node) which needs to communicate with the heterogeneous nodes.

Service discovery transition: the transition between ZDO-SD and CoAP-RD through the application layer data format transition between ZigBee-APL and CoAP-HTTP.

4.3 Overall data traffic flows

The intercommunication between nodes requires three basic phases: network initialization, service discovery and data transmission. Figure 5 mainly depicts the data transmission flows in the three phases when 6LoWPAN node initiates the ZigBee join request. The intercommunication initiated by ZigBee nodes is the reverse path.

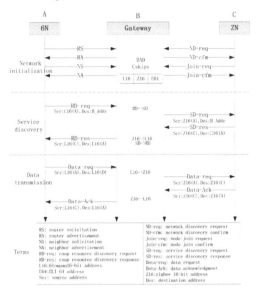

Figure 5. The overall data transmission flows.

1 Network initialization

The 6LoWPAN and ZigBee nodes share the address prefix and the multicast group ID of the border gateway and conduct the address registration to the gateway through 6LoWPAN neighbor discovery protocol and the ZigBee network discovery protocol respectively. Then, the border gateway assigns thea 16-bit 6LoWPAN address and ZigBee address through DAD mechanism and CSkips algorithm for each node and fills the address information into the address mapping table.

2 Service discovery

6LoWPAN node an initiates the service discovery request to B through multicast; B conducts the service discovery transition from RD to SD and forwards the request to the ZigBee area. C fulfills the service request from A and replies to B the unicast SD-response. After the service discovery protocol and address transition, B forwards the RD-response to A.

3 Data transmission

When the peering is set up, the data transmission is quite simple. A sends out a packet to L16(C) directly.

B will transform the packet to Z16(C) with header transition. The acknowledge (Ack) message from C will reply to Z16(A), which is also simply transformed by B to A.

5 CONCLUSION

In this paper, we fully consider the 6LoWPAN and ZigBee protocol characteristic and put forward a new communication model and functional framework. In our design, the basic protocol stack remain unchanged; we achieve the basic intercommunication between ZigBee and 6LoWPAN nodes through the complex and powerful gateway. The comprehensive scheme lays a good foundation for more in-depth studies of heterogeneous network integration.

This paper offers the frame level functional module, the primary interactive process and solutions to the key issues without some implementation details such as the network-layer header transition. Besides, this scheme still remains theory stage requiring more simulation and practice verification process.

ACKNOWLEDGMENT

This work is supported by the National Natural Science Foundationof China (60973140, 61170276, 61373135), The research project of Jiangsu Province (BY2013011), The Jiangsu provincial science and technology enterprises innovation fund project (BC2013027), the High-level Personnel Project Funding of Jiangsu Province Six Talents Peak and Jiangsu Province Blue Engineering Project. The major project of Jiangsu Province University Natural Science (12KJA520003)

REFERENCES

[1] COMMITTEE, L. M. S. 2003. Part 15.4: wireless medium access control (MAC) and physical layer (PHY) specifications for low-rate wireless personal area networks (LR-WPANs). *IEEE Computer Society*.

[2] SPECIFICATION, Z. 2004. ZigBee Document 053474r06 Version 1.0. *ZigBee Alliance*. December.

[3] KUSHALNAGAR, N., MONTENEGRO, G. & SCHUMACHER, C. 2007. IPv6 over low-power wireless personal area networks (6LoWPANs): overview, assumptions, problem statement, and goals. *RFC4919, August,* 10.

[4] MONTENEGRO, G., KUSHALNAGAR, N., HUI, J. & CULLER, D. 2007. Transmission of IPv6 packets over IEEE 802.15. 4 networks. *Internet proposed standard RFC,* 4944.

[5] SHELBY, Z., HARTKE, K., BORMANN, C. & FRANK, B. 2012. Constrained Application Protocol

(CoAP), draft-ietf-core-coap-13. *Orlando: The Internet Engineering Task Force–IETF, Dec.*

[6] PRESSER, A., FARRELL, L., KEMP, D. & LUPTON, W. Upnp device architecture 1.1. UPnP Forum, 2008.

[7] SHELBY, Z. & BORMANN, C. 2011. *6LoWPAN: The wireless embedded Internet*, John Wiley & Sons.

[8] HONGKE, W. D. Z. 2006. Research of ZigBee and 6LowPan Based on IEEE 802.15. 4. *Modern Science & Technology* of *Telecommunications*, 4, 014.

[9] SAKANE, S., ISHII, Y., TOBA, K., KAMADA, K. & OKABE, N. A translation method between 802.15. 4 nodes and IPv6 nodes. Applications and the Internet Workshops, 2006. SAINT Workshops 2006. International Symposium on, 2006. IEEE, 4 pp.-37.

[10] WANG, R.-C., CHANG, R.-S. & CHAO, H.-C. 2007. Internetworking between ZigBee/802.15. 4 and IPv6/802.3 network. *SIGCOMM Data Communication Festival.*

Network Security and Communication Engineering – Chan (Ed.)
© 2015 Taylor & Francis Group, London, ISBN: 978-1-138-02821-0

Feature reconstruction based on compressed information in cognitive radio networks

S.Y. Liu, Y. Zhao, Z. Sun, S.S. Wang & X.T. Chen
Key Laboratory of Universal Wireless Communications, Ministry of Education
Beijing University of Posts and Telecommunications, Beijing, China

ABSTRACT: Cognitive Radio (CR) networks have attracted more and more attention in recent years for the ability of dynamic spectrum management and intellectual learning. To meet the challenge of big data and ultra wide band signal processing, Compressed Sensing (CS) theory is introduced to break the bound of Nyquist-Sampling rate. In this paper, we proposed an algorithm of indirect feature reconstruction with good performance and low complexity of both time and space. Considering that most communication signals own cyclostationarity, single recognition and detection can be achieved by the use of the Cyclic Autocorrelation Function (CAF). In our study, we evaluated the algorithm in comparison of methods which directly recover CAF. By theoretical analyses and simulations, we demonstrated that the method proposed in the paper performed well and attracted future usage.

1 INTRODUCTION

In modern communication systems, spectrum is quite a valuable resource. To improve the spectral efficiency, cognitive radio technology is used in the network, helping do dynamic spectrum management. A main goal of the technology is to sense the spectrum and detect the Primer User (PU) over a wide band [1]. In traditional cognitive radio networks, Nyquist sampling is conducted at the receiver. However, when it comes to ultra-wide band signals, it is hard to do the high-rate sampling and the system requires enormous data capacity. Considering the sparsity that most communication signals own, compressed sensing (CS) is introduced in the process of spectrum sensing and signal detection, and it not only successfully solves the problem but also lowers the sampling rate. Compressed sensing is a technology proposed by Candies and Donoho in 2004 that uses fewer samples than Nyquist theory and can perfectly recover the signal [2].

In applications of spectrum detection and signal recognition, most studies are based on feature detection considering the cyclostationarity that communication signals generally have. So taking the sparsity of signal's cyclic spectrum and cyclic autocorrelation into consideration, instead of completely recovering the signal, we do incomplete reconstruction in some sparse feature domain with lower sub-Nyquist sampling rate to reduce the computational complexity in compressed sensing [3]. The superiority is especially highlighted in muti-carrier signal systems like 4G Orthogonal Frequency Division Multiplexing (OFDM) systems [4], which is the object our study mainly focuses on.

Studies based on the above idea include the following two core issues: the reconstruction of signal feature and the strategy of signal recognition and detection. A method proposed by Zhi Tian and others is widely accepted by recent fellows. In the framework, signal's 2-D cyclic spectrum was reconstructed based on CS theory and spectrum occupancy estimation was done on the frequency sets of 2-D cyclic spectrum according to binary hypothesis test [5]. However, the reconstruction formulation they used is quite complex with multiple Kronecker product and pseudo inverse operations, which makes the reconstruction hard to be implemented in practice. Besides, when it is used in OFDM systems, the vector length dramatically increases as the number of subcarriers gets larger and matrixes in the formulations are of tremendous scale which is hard to implement in practice.

In our study, by taking signal's cyclostationarity into consideration, signal detection and recognition are accomplished by use of cyclic-autocorrelation function [6]. Inspired by the process of 2-D cyclic spectrum reconstruction, we would like to reconstruct the signal's sparse CAF according to CS theory. However, during the study, we find that the computational complexity is still quite high and the requirement of memory is large. So in this paper we propose an approach based on the indirect reconstruction of autocorrelation function with sub-Nyquist samples [7][8], which gets good performance in the least time with the least space complexity. In the method, CAF is obtained by the reconstructed autocorrelation function, and then signal detection is triggered in searching of the type of signals, multi-carrier signal's number of subcarriers and the length of cyclic prefix (CP). In future studies we will also focus on the recognition of signal's modulation type. Another important part of this paper is the theoretically performance analyses of the proposed indirect recovery method including reconstruction performance and detection performance. At the same time, we make the comparison with the method of directly recovering CAF. Simulations are conducted to show and verify our statements and conclusions.

The remaining parts of this paper are organized as follows. Section 2 describes the feature reconstruction of signal autocorrelation. Simulation and analysis are present in Section 3. Finally, we present the conclusion in Section 4.

2 FEATURE RECONSTRUCTION OF AUTOCORRELATION

According to the theory of compressed sensing, to reconstruct the target signal with high probability, we need to confirm that: the signal is sparse in some transformation domain; the measurement matrix we use is unrelated to the transformation matrix and the reconstruction algorithm we choose can solve the reconstruction problem by obtaining the optimal or the suboptimal solution.

2.1 *Sparsity of the autocorrelation domain*

As is referred to in Section I, signal recognition and detection are conducted in the 2-D (two dimensional) CAF domain. However to achieve well estimation performance and lower reconstruction complexity, we would like to set the 2-D autocorrelation function as the sparse transformation domain to be recovered. $\forall n$ and τ are integers, and the autocorrelation function of the signal is

$$
\begin{aligned}
r_x(n,\tau) &= E\{x(nT_s)x^*(nT_s + \tau T_s)\} \\
&= E\{x[n]x^*[n+\tau]\}
\end{aligned}
\tag{1}
$$

So, the autocorrelation matrix R_x generated according to the definition is symmetric. As the method given in [4], we formulate the autocorrelation vector r_x and the 2-D autocorrelation matrix as follows with a relationship of $vec\{R\} = Br_x$, where r_x is of size $N(N+1)/2 \times 1$ and $vec\{\cdot\}$ stacks all columns of a matrix into a vector.

$$
\begin{aligned}
r_x = [&r_x(0,0), r_x(1,0), ..., r_x(N-1,0), \\
&r_x(0,1), ..., r_x(N-2,1)......, r_x(0,N-1)]^T
\end{aligned}
\tag{2}
$$

$$
R = \begin{bmatrix}
r_x(0,0) & r_x(0,1) & \cdots & r_x(0,N-1) \\
r_x(1,0) & r_x(1,1) & \cdots & 0 \\
\vdots & \vdots & & \vdots \\
r_x(N-1,0) & 0 & \cdots & 0
\end{bmatrix}
\tag{3}
$$

Under the system model, OFDM signal is cyclostationary, so its autocorrelation function is periodical with period P: $r_x(n,\tau) = r_x(n+kP,\tau)$. Considering a finite signal block length N, only elements of the first column and $\tau_0 + 1$ column are non-zero in the matrix R. τ_0 is shorter than the length of a OFDM symbol and is determined by the length of CP. The length of CP is obviously shorter than half the OFDM symbol. So the number of nonzero elements K in matrix R is bound to satisfy: $K \le 3N/2$. The total non-zero element ratio satisfying $R_k \le (3N/2)/N^2 = 3/2N$. When the signal block N is large enough, the sparsity is greater. Although the sparsity is a little worse than that of its 2D-CAF, the method can save quite a lot of calculation time in the reconstruction. We'll present the demonstration in Section III B. And next, to minimize the data size and make the result convincing, we set $N = 2(N_{FFT} + N_{cp})$. Where in cases of SCLD signal, the value of P is zero.

2.2 *Linear relationships*

The sampling can be modeled as $z_t = Ax_t$, where A is the measurement matrix of size $M \times N$ that achieves sub-Nyquist sampling. Therefore we obtain the following equation:

$$
R_z = AR_x A^H
\tag{4}
$$

where R_x and R_z are the autocorrelation matrix of the original signal and the sampled signal. We denote the vector of original signal AF as $vec\{R_x\} = P_N r_x$. When plugging (4) into the sampled autocorrelation vector $r_z = Q_M vec\{R_z\}$, we can express the measurement vector r_x as a linear function of the vector-form autocorrelation R_x using the property $vec\{UXV\} = (V^T \otimes U)vec\{X\}$ as:

$$
\begin{aligned}
r_z &= Q_M vec\{AR_x A^H\} = Q_M (A \otimes A)vec\{R_x\} \\
&= Q_M (A \otimes A)P_N r_x = \Psi' r_x
\end{aligned}
\tag{5}
$$

where P_N and Q_M are some specific mapping matrices, $\otimes$ denotes the Kronecker product and $\Psi' = Q_M (A \otimes A)P_N$ is the sensing matrix of size $M(M+1)/2 \times N(N+1)/2$.

2.3 *Autocorrelation reconstruction*

As is analyzed in section III.A, the autocorrelation of a signal is sparse enough. According to the compressed sensing theory, the recovery of sparse object vector r_z comes down to solving the NP-hard puzzle by using the sampled autocorrelation vector r_z as follows:

$$\left.\begin{array}{l} \hat{r}_x = \arg\min\|r_x\| \\ s.t. \qquad r_z = \Psi' r_x \end{array}\right\} \qquad (6)$$

According to the sparsity reconstruction theory in CS, (6) can be transformed to a $l1$ norm least square programming problem bellow and it is proved to be convex that exists a unique optimum solution.

$$\min_{r_x}\|r_z - \Psi' r_x\|^2_{1/2} + i\|r_x\| \qquad (7)$$

The sampling matrix we choose in the front of the receiver is of Gaussian random distribution with a sub-sampling rate of M/N. Accordingly, we choose OMP algorithm when doing the reconstruction [9]. In the method of OMP, the estimated result for each iteration is the optimal and gets a rapid convergence. Thus it shows well performance on incomplete reconstruction.

3 SIMULATION AND ANALYSIS

3.1 Simulation method

In our simulation, each sub-carrier of OFDM is modulated with QPSK. The whole signal bandwidth is 480 kHz, the number of sub-carrier is 32 and the symbol period is set as 0.8us with CP of 0.2us. We simulated 1000 OFDM symbols as the signal length and transmitted it under the AWGN channel.

At the receiver, we deal with two symbols as a unit to conduct the sub-Nyquist sampling and the sample autocorrelation is sent as reconstruction input. To confirm the compliment of AF reconstruction, we set the iterations as 120. In the tests we choose the probability of correct detection as the evaluation criterion

3.2 Results

When testing the performance of the algorithm, there are generally three factors that may affect the accuracy of the detection: the signal to noise ratio (SNR), sub-Nyquist sampling rate M/N and the number of OFDM symbols L.

To study the performance of the proposed indirect recovery method of CAF, we firstly do simulations on the comparison with the directly method. From Fig.1, we can clearly find the differences.

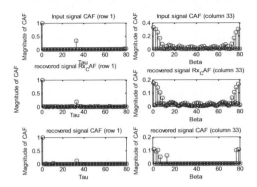

Figure 1. The comparision of CAF by the two methods.

During the test we set the sampling rate as M/N =0.3 which is proved to be sufficient for both AF and CAF reconstruction with SNR of 0dB. From the figures we can find that indirect recovery can get a more similar trace as the original signal, which also proves that the indirect reconstruction can perform well.

In the next case, the normalized mean square error (NMSE) of recovered AF, indirectly recovered CAF based on AF and directly recovered CAF has been compared with different value of system SNRs, which is presented in Fig.2.From that, we can see that the performance on the recovery of CAF by the two methods show almost the same, and the indirect method is even a little bit better. However the performance of AF recovery is unsatisfactory, and the fact also demonstrates the analyses in IV. B. What's more, both the methods perform a gentle deterioration with SNR, and we can conclude that the method can perfectly tolerate the noise.

Unless the above hopeful results, from Fig. 3 we find that, when doing signal detection, method

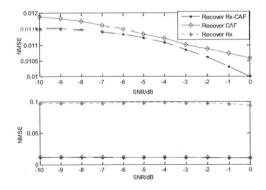

Figure 2. The NMSE of recovered AF, CAF based on AF and recovered CAF.

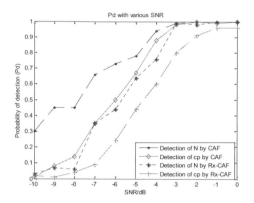

Figure 3.   The correct probability of detection with various SNR.

we prefer is not that ideal as we supposed. As the SNR decreases(especially when it is below -3dB), the points with higher value over cyclic frequency domain get a dramatic decrease. As a result, the distribution of the cyclic frequency domain is just like the white noise that stop the reliability.

However, we realize that a bit lower detection performance is tolerable and acceptable. By the proposed method, we get a dramatic decrease of implementation complexity.

3.3   Complexity of the algorithm

Considering that the time complexity can be evaluated by the calculation complexity. In the process of directly reconstruction of CAF, the used sensing matrix is $\Psi = Q_M (A \otimes A) P_N \left(\sum_{\tau=0}^{N-1} (D_\tau^T \otimes G_\tau) B \right)^\dagger$,

while $\Psi' = Q_M (A \otimes A) P_N$ for the AF. It is obvious that is quite complex. The Kronecker procedure and the pseudo-inverse matrix calculation take a long time.

In the case when sampling rate is 0.3 and symbol block length is 80, the simulated time consumption Ψ' is 2.3s while Ψ is 742.2s.

When considering the space complexity, in calculation of Ψ we need at least 7 big matrices of size $N^2 \times N^2$ while only two are needed in the reconstruction of AF. Accordingly, in the simulation we got space consumption of 27M for Ψ'. However, for Ψ we used 340M more RAM space.

4   CONCLUSIONS

We have proposed a method of CAF recovery by AF reconstruction with sub-Nyquist samples that dramatically lower the time and space complexity of most recovery puzzles. By Theoretical analyses and simulations, we have successfully proved that the proposed methods perform well on both the accuracy of the autocorrelation reconstruction and the signal detection. In the future work, under the idea of indirectly reconstruction, there may be a rapider and more convenient approach in detection of signal's high-order cumulants such as the modulation type.

ACKNOWLEDGMENT

This work was supported by the National Science and Technology Major Project of China under grant 2013ZX03001003-003.

REFERENCES

[1] G. Ganesan and Y. Li, "Cooperative spectrum sensing in cognitive radio, Part I: Two user networks," IEEE Trans.Wireless Commun., vol.6, pp. 2204–2213, 2007.

[2] Donoho, David L. "Compressed sensing." Information Theory, IEEE Transactions on 52.4 (2006): 1289-1306.

[3] Yingming, T., C. Yen, and Xiaodong Wang. "Wideband spectrum sensing based on sub-Nyquist sampling." (2013): 1–1.

[4] Ariananda, Dyonisius Dony, and Geert Leus. "Compressive wideband power spectrum estimation." Signal Processing, IEEE Transactions on 60.9 (2012): 4775–4789.

[5] Zhi Tian, Yohannes Tafesse, and Brian M. Sadler, "Cyclic Feature Detection With Sub-Nyquist Sampling for Wideband Spectrum Sensing" IEEE Journal Sel. Signal Processing, Vol. 6,Feb.2012 pp.58–69.

[6] O. A. Dobre, Q. Zhang, S. Rajan, and R. Inkol, "Second-order cyclostationarity of cyclically prefixed single carrier linear digital modulations with applications to signal recognition," in Proc. IEEE GLOBECOM, 2008, pp. 1–5.

[7] Bien, Jacob, and Robert J. Tibshirani. "Sparse estimation of a covariance matrix." Biometrika 98.4 (2011): 807–820.

[8] Nasif, Ahmed O, Zhi Tian, and Qing Ling. "High-dimensional sparse covariance estimation for random signals." Acoustics, Speech and Signal Processing (ICASSP), 2013 IEEE International Conference on. IEEE.

[9] Tropp, Joel A., and Anna C. Gilbert. "Signal recovery from random measurements via orthogonal matching pursuit." Information Theory, IEEE Transactions on 53.12 (2007): 4655–4666.

Network Security and Communication Engineering – Chan (Ed.)
© 2015 Taylor & Francis Group, London, ISBN: 978-1-138-02821-0

Electric field intensity calculation method of vertical antenna using wireless ad hoc network

X.Z. Hou & H.L. Sun
State Grid Chongqing Electric Power Research Institute, Chongqing, China

Y.G. Li, C.X. Yang, L. Tang & J.D. Zhou
Beijing Techrefine Technology Co.,Ltd, Beijing, China

ABSTRACT: Vertical antenna is used in wireless ad hoc network of smart power grids. The attenuation of ground surface can be generated due to electromagnetic wave along ground surface propagation, so electric field intensity must be calculated accurately for communication quality and robustness of the network. According to Sommerfeld dipole electromagnetic field theory, analysis of expression of electric field intensity is derived and simplified and the calculating method is provided in this paper.

KEYWORDS: Vertical antenna, Electric field intensity, Earth attenuation.

1 INTRODUCTION

Wireless Ad Hoc Networks (including special ad-hoc network, WSN wireless sensor network) is a hot area of research and application in recent years. It is considered to be the second-largest network after the Internet, and it is seen as a IOT of peripheral nerves. It is very important in all sectors of the national economy and national defense military.

Wireless ad hoc communication, will be widely used in smart grids. It is one of the downlink communication modes of energy meter remote automatic meter reading system [3] in intelligent electrical systems (last mile). Each meter (or collector connected to several meters) is equipped with a wireless communication module, and it becomes a node of the network topology, with hair, collection, and relay function. Since the group network has a multi hop, rounded, path optimization, and other functions, it has good robustness.

Measurement for wireless communication transmission power of National Radio Management Committee is 50mW (micro power), and the frequency is 470 ~ 510MHz. Due to the short distance between the meters (Collector) or meter and concentrator, the inter antenna propagation of electromagnetic waves are surface waves in the visible range, and vertical antenna (vertical polarization) can increase the field strength, but the electromagnetic wave will produce the earth attenuation in travel along the ground. For diffraction attenuation of obstacles on the ground wave, it can be solved by Wireless ad hoc network node relay function (or with repeaters). So, land on ground wave attenuation for the micro power wireless ad hoc network is the objective of existence. In the calculation of field intensity, the earth attenuation should be taken into consideration to ensure that the minimum signal strength of the receiving antenna can meet the requirement of signal to noise ratio (S/N), to guarantee the quality of communication. This paper will discuss this in detail.

2 THE INTENSITY OF THE ELECTRIC FIELD OF A VERTICAL ELECTRIC DIPOLE

According to the Sommerfeld dipole electromagnetic field theory between the air and the earth, if the vertical dipole and the receiving point are in the air (see Figure), the current I stands for the vertical electric dipole. The vertical component of the electric field intensity in arbitrary receiving point in considering the geodetic effect is[4][5].

$$E_{0z} = -\frac{j\omega\mu_0 I}{4\pi}[\frac{e^{jk_0 R_1}}{R_1} + \frac{e^{jk_0 R_2}}{R_2} -$$

$$2k_0^2 \int_0^\infty \frac{\beta_1 u}{k_0^2\beta_1 + k_1^2\beta_0}\frac{e^{-\beta_0(h+z)}}{\beta_0}J_0(ru)du](V/M) \quad (1)$$

In the formula

h-vertical dipole height from the ground (m);

Z -receiver point height from the ground (m);

x—Horizontal distance along the X axis direction from the origin of coordinates to the receiving point (m);

y—Horizontal distance along the Y axis direction from the origin of coordinates to the receiving point (m);

$r = \sqrt{x^2 + y^2}$—The horizontal distance from the origin of the coordinate to the receiving point (m);

$R_1 = \sqrt{r^2 + (h-z)^2}$—The distance between the vertical dipole and the receiver point (m);

$R_2 = \sqrt{r^2 + (h+z)^2}$—The distance between the vertical dipole image and receiving point (m);

$\beta_0 = \sqrt{u^2 - k_0^2}$;

$\beta_1 = \sqrt{u^2 - k_1^2}$;

$k_0 = j\sqrt{j\omega\mu_0(\sigma_0 + j\omega\varepsilon_0)}$ -Air propagation constant (1/m),

When $\sigma_0 = 0$, $k_0 \approx -\omega\sqrt{\mu_0\varepsilon_0} = -\dfrac{2\pi}{\lambda}$;

$k_1 = j\sqrt{j\omega\mu_1(\sigma_1 + j\omega\varepsilon_1)}$ —The earth propagation constant (1/m),

$$k_1 = j\sqrt{j\omega\mu_0(\sigma_1 + j\omega\varepsilon_0\varepsilon_r)} = -\omega\sqrt{\mu_0\varepsilon_0(\dfrac{\sigma_1}{j\varepsilon_0\omega} + \varepsilon_r)}$$

$$= -\omega\sqrt{\mu_0\varepsilon_0}\sqrt{(\dfrac{\sigma_1}{j\varepsilon_0\omega} + \varepsilon_r)} = k_0\sqrt{\varepsilon_r'}$$

ε_r' -The earth complex permittivityis

$$\varepsilon_r' = \dfrac{\sigma_1}{j\varepsilon_0\omega} + \varepsilon_r = -j\dfrac{\sigma_1}{2\pi f} \times 36\pi \times 10^9 + \varepsilon_r$$

$$= -j\dfrac{\sigma_1}{3 \times 10^8} \times 18\lambda \times 10^9 + \varepsilon_r$$

$$= \varepsilon_r - j60\lambda\sigma_1$$

μ_0—Air magnetic permeability ($\mu_0 = 4\pi \times 10^{-7}$ H/m);
μ_1—The earth's magnetic permeability ($\mu_1 \approx \mu_0$);
ε_0—The dielectric constant of the air
($\varepsilon_0 = 8.854 \times 10^{-12} \approx \dfrac{1}{36\pi} \times 10^{-9}$ F/m);
ε_1—The dielectric constant of the earth (F/m);
$\sigma_0 \approx 0$—Air conductivity (S/m);
σ_1—Earth conductivity (S/m);
$\omega = 2\pi f$-Angular frequency (rad);
f-Frequency (Hz), λ-Wavelength (m);
$J_0(ru)$- the first class of Bessel functions of order 0;
u—Eigen value.

In Formula (1) the first term is the direct wave field strength of the antenna, the second term is the antenna field emission in a perfectly conducting ground mirror, and the third term is finitely conducting ground attenuation.)

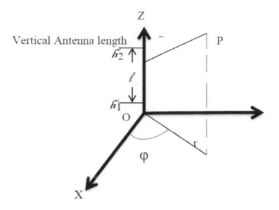

Figure 1. The mutual position between the vertical antenna and receiver.

In the third infinite integrals

$$f(u) = \dfrac{\beta_1}{k_0^2\beta_1 + k_1^2\beta_0} \tag{2}$$

With the method of Ward [6] to apply the formula (2) is expanded as a power series

$$f(u) = \sum_{m=0}^{\infty} a_m u^m \tag{3}$$

In the formula

$$a_m = \dfrac{1}{m!}\left[\dfrac{d^m}{du^m} f(u)\right]_{u=0} \tag{4}$$

To get the formula (3) into the formula (1)

$$E_{0z} = -\dfrac{j\omega\mu_0 I}{4\pi}\left[\dfrac{e^{jk_0R_1}}{R_1} + \dfrac{e^{jk_0R_2}}{R_2}\right.$$

$$\left. -2k_0^2 \sum_{m=0}^{\infty} a_m \int_0^{\infty} u^m \dfrac{u}{\beta_0} e^{-\beta_0(h+z)} J_0(ru)du\right] \tag{5}$$

The coefficient am: When m=0,

$a_0 =$

$$\left[\dfrac{\beta_1}{k_0^2\beta_1 + k_1^2\beta_0}\right]_{u=0} = \dfrac{k_1}{k_0^2 k_1 + k_1^2 k_0}\dfrac{1}{k_0^2 + k_1 k_0}$$

After the a1 is zero
To get a0 into the formula (5) is

$$E_{0z} = -\dfrac{j\omega\mu_0 I}{4\pi}$$

$$[\frac{e^{jk_0R_1}}{R_1}+\frac{e^{jk_0R_2}}{R_2}-$$

$$2\frac{k_0}{k_0+k_1}\int_0^\infty \frac{u}{\beta_0}e^{-\beta_0(h+z)}J_0(ru)du] \qquad (6)$$

In Formula (6) infinite integrals are for the third term of Sommerfeld integral, and the above formula evolved into

$$E_{0z}=$$

$$-\frac{j\omega\mu_0 I}{4\pi}[\frac{e^{jk_0R_1}}{R_1}+\frac{e^{jk_0R_2}}{R_2}-2\frac{k_0}{k_0+k_1}\frac{e^{jk_0R_2}}{R_2}] \qquad (7)$$

In another words, it is

$$E_{0z}=-\frac{j\omega\mu_0 I}{4\pi}\left(\frac{e^{jk_0R_1}}{R_1}+\frac{k_1-k_0}{k_1+k_0}\frac{e^{jk_0R_2}}{R_2}\right) \qquad (8)$$

This is the analytic expressions of vertical dipole electric field intensity vertical component.

3 THE ELECTRIC FIELD INTENSITY OF VERTICAL ANTENNA

If the current on the antenna is I and the length is l. The coordinate of the two points are h1, h2, and the integral of formula (8) is

$$E_v=\int_{h_1}^{h_2}E_{0z}\,dh$$

$$=-\frac{j\omega\mu_0 I}{4\pi}\int_{h_1}^{h_2}\left[\frac{e^{jk_0R_1}}{R_1}+\frac{k_1-k_0}{k_1+k_0}\frac{e^{jk_0R_2}}{R_2}\right]dh \qquad (9)$$

The two terms in the formula (9) can be expressed as a Sommerfeld integral, is

$$\frac{e^{jk_0R_1}}{R_1}=\int_0^\infty\frac{u}{\sqrt{u^2-k_0^2}}e^{-\sqrt{u^2-k_0^2}(h-z)}J_0(ru)du \qquad (10)$$

$$\frac{e^{jk_0R_2}}{R_2}=\int_0^\infty\frac{u}{\sqrt{u^2-k_0^2}}e^{-\sqrt{u^2-k_0^2}(h+z)}J_0(ru)du \qquad (11)$$

The integral of the formula (9), is the computation of integrals of exponential functions related to height of the two formulas, is

$$\int_{h_1}^{h_2}e^{-\sqrt{u^2-k_0^2}(h-z)}=e^{\sqrt{u^2-k_0^2}z}\int_{h_1}^{h_2}e^{-\sqrt{u^2-k_0^2}h}\,dh$$

$$=\frac{1}{\sqrt{u^2-k_0^2}}\left(e^{-\sqrt{u^2-k_0^2}(h_1-z)}-e^{-\sqrt{u^2-k_0^2}(h_2-z)}\right) \qquad (12)$$

As same

$$\int_{h_1}^{h_2}e^{-\sqrt{u^2-k_0^2}(h+z)}$$

$$=\frac{1}{\sqrt{u^2-k_0^2}}\left(e^{-\sqrt{u^2-k_0^2}(h_1+z)}-e^{-\sqrt{u^2-k_0^2}(h_2+z)}\right) \qquad (13)$$

In order to calculate simply, the exponential functions of formula (12) and formula (13) are expanded in Maclaurin series, and the first two items is

$$\int_{h_1}^{h_2}e^{-\sqrt{u^2-k_0^2}(h-z)}\approx\frac{1}{\sqrt{u^2-k_0^2}}\left[1-\frac{\sqrt{u^2-k_0^2}(h_1-z)}{1!}-1+\frac{\sqrt{u^2-k_0^2}(h_2-z)}{1!}\right]$$

$$=(h_2-h_1)=l \qquad (14)$$

So

$$\int_{h_1}^{h_2}e^{-\sqrt{u^2-k_0^2}(h+z)}\,dh\approx l \qquad (15)$$

It is

$$\int_{h_1}^{h_2}\frac{e^{jk_0R_1}}{R_1}\,dh\approx l\int_0^\infty\frac{u}{\sqrt{u^2-k_0^2}}J_0(ru)\,du=l\frac{e^{jk_0r}}{r} \qquad (16)$$

$$\int_{h_1}^{h_2}\frac{e^{jk_0R_2}}{R_1}\,dh\approx l\int_0^\infty\frac{u}{\sqrt{u^2-k_0^2}}J_0(ru)\,du=l\frac{e^{jk_0r}}{r} \qquad (17)$$

To get the two formulas into the formula (9), and it is

$$E_v=-\frac{j\omega\mu_0 Il}{4\pi}\left(\frac{2k_1}{k_1+k_0}\right)\frac{e^{jk_0r}}{r} \qquad (18)$$

If the antenna input current is I0, no loss of current on the antenna is[9]

$$I=I_0\sin\beta(l-h)\ (A)$$

I0- antenna input the current in formula:

$$I_0=\sqrt{\frac{P_r}{Z_i}}(A)$$

P_r-The power of antenna feeding termination (W);

Z_i-The input impedance of the antenna feeding termination (Ω)[9].

The phase shift constant

$$\beta = \frac{2\pi}{\lambda}.$$

The average current in the antenna is

$$I = \frac{I_0}{l}\int_0^l \sin\beta(l-h)\,dh = -\frac{I_0}{\beta l}[1-\cos\beta l] \qquad (19)$$

To get the formula (18) into the formula (17), it is

$$E_v = \frac{j\omega\mu_0 I_0}{2\pi\beta}(1-\cos\beta l)\left(\frac{k_1}{k_1+k_0}\right)\frac{e^{jk_0 r}}{r} \qquad (20)$$

The formula (20):

$$\frac{\omega\mu_0}{2\pi\beta} =$$

$$\frac{2\pi f \times 4\pi \times 10^{-7}\lambda}{2\pi \times 2\pi} = \frac{2f \times 10^{-7}}{f} \times 3 \times 10^8 = 60$$

To get it into the formula (20), it is

$$E_v = j\frac{60 I_0}{r}(1-\cos\beta l)\left(\frac{k_1}{k_1+k_0}\right)e^{jk_0 r}\ \text{(V/m)} \qquad (21)$$

The formula (21) is the vertical component of the electric field intensity of the simplified calculation formula without loss of vertical transmitting antenna in air medium any receiving point.

Do not calculate ground attenuation, when k1=k1, the space medium is only for air, and formula (20) [8] is

$$E_v = j\frac{30 I_0}{r}(1-\cos\beta l)e^{jk_0 r} \qquad (22)$$

The formula (22) is free space field intensity.

If the earth is an ideal conductor, then the formula (1) is no third term, and the image of the ground effect is instead of[9], then the formula (21) is

$$E_v = j\frac{60 I_0}{r}(1-\cos\beta l)e^{jk_0 r} \qquad (23)$$

The electric field intensity is represented by relative unit, and it is

$$E_v = 20\log\frac{E_v\,(\mu V)}{1\mu V}(dB_\mu) \qquad (24)$$

In the formula (21)

$$t = \frac{k_1}{k_1+k_0} \qquad (25)$$

Is coefficient of the earth attenuation. The earth propagation constant is

$$k_1 = j\sqrt{j\omega\mu_0(\sigma_1+j\omega\varepsilon_0\varepsilon_r)} = k_0\sqrt{\varepsilon_r}$$

To get into the formula (25), it is

$$t = \frac{\sqrt{\varepsilon_r'}}{1+\sqrt{\varepsilon_r'}} \qquad (26)$$

In the formula, ε_r' -'The complex relative permittivity of the earth is

$$\varepsilon_r' = \varepsilon_r - j60\lambda\sigma_1$$

If the electrical conductivity or the wavelength is very small, the formula (26) is

$$t = \frac{\sqrt{\varepsilon_r}}{1+\sqrt{\varepsilon_r}} \qquad (27)$$

To calculates the attenuation coefficient:
City: $\varepsilon_r = 4$, is

$$t = \frac{\sqrt{\varepsilon_r}}{1+\sqrt{\varepsilon_r}} = \frac{\sqrt{4}}{1+\sqrt{4}} = 0.667$$

$$t = 20\log 0.667 = -3.517dB$$

Wetland: $\varepsilon_r = 20$, is

$$t = \frac{\sqrt{\varepsilon_r}}{1+\sqrt{\varepsilon_r}} = \frac{\sqrt{20}}{1+\sqrt{20}} = 0.817$$

$$t = 20\log 0.817 = -1.756dB$$

We can see, the earth attenuation of city is about 3.5dB, and the earth attenuation of wetland is about 1.8dB. The transmit power of micro power ad hoc network antenna is 17dBm. The earth attenuation should be taken into consideration in the conductivity and permittivity of the smaller area.

4 CONCLUSION

1 In this paper, based on the Sommerfeld dipole theory, we derived analytical expressions of the vertical component of the vertical dipole electric field intensity and the simplified calculation formula of the vertical component of the electric field intensity vertical antenna, and provided a practical calculation method for the strength of electric field.

2 According to the theoretical analysis, the earth attenuation should be calculated in the calculation of electric field strength, for cities, towns or hilly and mountainous areas. Because of the complexity of the field[10], field measurements are necessary to determine the wireless circuit in the field.

3 In wireless ad hoc network planning, if we do not know the specific conditions of the site, we can estimate the circuit attenuation or the field by the above method [11][12].

REFERENCES

[1] The technologyof smart grid.Beijing: China Electric Power Press,2010.Write by Liu Zhenya.

[2] Intelligent Residential District.Beijing: China Electric Power Press,2012.Write by He Jianjun.

[3] The intelligent electric energy meter.Beijing: China Electric Power Press,2010.Write by ZongJianhua, Yan Hua etc.

[4] SommerfeldA.Partial differential equations in physic. New York.

[5] Academic Press Inc.1949.

[6] The electromagnetic coupling of AC transmission lines and telecommunication lines.Beijing: China Electric Power Press,2013.WritebyZhangWenliang, Cui Dingxin etc.

[7] Ward S.H.,Electromagnetic theory for geophysical application,Mining Geophisics,vol.II,1971.

[8] Calculation of vertical wire alternating current electric field intensity. Smart Grid 2010,36(6): 19~22.Write by Cui Dingxin, QuXuedi, Yang Yong.

[9] Kraus J.D.,MarhefkaR.J.,Antennas: For All Applications,Third Edition.(China Press, 2006).

[10] Kraus J.D.,MarhefkaR.J.,Antennas: For All Applications,Third.

[11] Edition.(translate by Zhang Wenxun, the antenna (Third Edition) Beijing:Electronic Industry Press,2006).

[12] Microwave technique and antenna.Beijing:Electronic Industry Press,2004. Write by Yin Jijie.

[13] The principle and application of the antenna of radio wave propagation. Beijing: people's posts & telecommunications publishing house the people's posts and telecommunications press.2008.7.Write by XieYixi.

[14] ITU-R .Propagation data and prediction methods require for the design of terrestrial line-of-sight systems.ITU-Rec.P.530-7,1997.GB/T 14617.3-93.Land mobile service and the fixed service propagation characteristics.1994.

Network Security and Communication Engineering – Chan (Ed.)
© 2015 Taylor & Francis Group, London, ISBN: 978-1-138-02821-0

Time-Interleaved Analog-to-Digital converter used for high-speed communication

Y.H. Gao, D.B. Fu, G.B. Chen & Z.P. Zhang
The NO.24th institute of CETC, Chongqing, China

ABSTRACT: A Time-Interleaved Analog-to-Digital Converter (TI ADC) used for high-speed communication is presented in this paper. To realize the timing of interleaving architecture, a 4-phase low jitter clock is designed to control four 1.25 G sample/s sub-ADCs. The sub-ADC adopts folding and interpolating architecture which includes sample and hold circuit, preamplifiers, folding and interpolating stage, coding and LVDS. Channel mismatch is the key problem of this kind of ADC. Thus digital calibration is used to adjust the offset error and to gain error of sub-ADCs. Besides, serial peripheral interface (SPI) is adopted to adjust mismatch of gain and sample time between sub-ADCs. The whole TI ADC is realized with a 0.18μm BiCMOS process. The result of the testament shows the whole ADC has a SNR of about 45dB at an input frequency of 495MHZ which is very suitable for high-speed communication system.

1 INTRODUCTION

With the explosion of service, it starves for the increment of the communication bandwidth which seems more and more difficult for the backbone network. These years, 10Gb/s, 40Gb/s, and 100Gb/s data transmission systems have been the focus for the next generation optical fiber network which requires supper high-speed analog-to-digital converter(ADC).

Usually, sample-rate beyond 1 GHz can be realized by the full flash architecture[1,2], the time-interleaved architecture[3]. Due to the large number of parallel comparators in a flash converter, the input capacitance of flash converter becomes tremendously large for resolutions above 6 bits, limiting the input bandwidth. Time-interleaved parallel channels can increase the maximum sampling rate of analog-to-digital converters (ADCs). TI AD takes a multi-phase clock to control sub-ADCs to sample and convert the same analog input signal to digital signal in several successive time phases. Then the output of those sub-ADCs is processed by the digital signal processor (DSP), those sub-ADCs are clocked such that the combination appears as a single fast ADC. The performance is sensitive to offset mismatches, gain mismatches, and sample-time errors between the interleaved channels, thus calibration is needed.

The arrangement of this article is as follows: Part 2 introduces the architecture of the presented four-channel TI ADC; Part 3 shows the design details, including circuit and design consideration; Part 4 is the design verification and Part 5 is the conclusion.

2 TIME-INTERLEAVED ADC

Time interleaved sampling technique is one of the most important methods to implement super high-speed analog-to-digital converter. As shown in

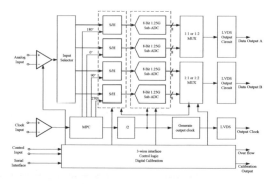

Figure 1. The architecture of the presented TI ADC.

Figure 1, a typical TI ADC contains a multi-phase clock (MPC), common analog input, 4 sub-ADCs and digital multiplexer. It works as follows: The MPC controls 4 channel ADCs to sample the common analog input signal at a fixed time interval, then each channel of ADCi (i=1,…,4) converts the corresponding sampled signal to digital codes and puts these codes to a following digital multiplexer to obtain a whole digital output as the whole ADC's output. Equivalently, the sampling rate of the whole ADC is multiplied by a factor of 4. The total sampling rate of the whole ADC is up to 5GHz.

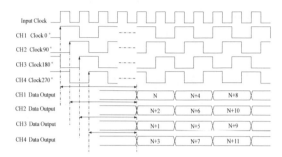

Figure 2. Timing diagrams of the presented TI ADC.

3 DESIGN DETAILS

3.1 *Low jitter multi-phase clock*

A Current Mode Logic (CML) clock is used for the T&H circuit and a CMOS logic clock is used for the sub-ADCs and the MUX. At high frequency, the impedance mismatch of input PIN would result reflection and timing error, so we usually use a sine wave signal instead of a square wave signal as the input clock. The schematic of the clock buffer as shown in Figure 3, contains two CML buffers and each CML buffer followed by an emitter follower. The input common voltage is biased at 1/2VDD.

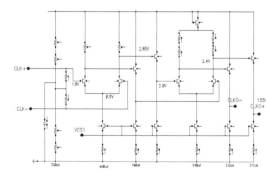

Figure 3. The front clock buffer.

The errors resulted from sampling clock jitter of analog-to-digital converter are shown in Figure 4. If the sampling clock is ideal, the sampling moment is shown as the vertical solid line, but the fact is not the situation. The sampling point is not fixed because of the existence of timing jitter.

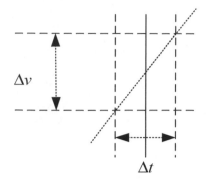

Figure 4. Voltage error resulted from sampling clock jitter.

Given a sine wave with an amplitude of A and frequency of f. We can estimate the effect of timing jitter to SNR as follows:

Given the input clock is

$$v(t) = A\sin(2\pi f t) \tag{1}$$

The slope of the input clock is

$$dv/dt = 2\pi f A\cos(2\pi f t) \tag{2}$$

Given the rms value of timing jitter is δ_{jitter}, then the total voltage error is

$$\Delta v_{rms} = \delta_{jitter} \times dv/dt \,|_{rms} = 2\pi f A\delta_{jitter}/\sqrt{2} \tag{3}$$

The effective value of the input clock is $A/\sqrt{2}$, the total SNR is

$$
\begin{aligned}
SNR &= 20\log_{10}\left[\frac{A/\sqrt{2}}{\Delta v_{rms}}\right] \\
&= 20\log_{10}\left[\frac{A/\sqrt{2}}{2\pi f A\delta_{jitter}/\sqrt{2}}\right] \\
&= 20\log_{10}\left[\frac{1}{2\pi f\delta_{jitter}}\right]
\end{aligned}
\tag{4}
$$

As we can see, the error from timing jitter is one of the determining factors of ADC's SNR. Ideally, N-bit ADC can reach SNR=6.02N+1.76, where N denotes the number of ADC bit. Thus, for our 8-bit ADC, the best SNR=6.02*8+1.76 ≈50 dB, given an input signal with the frequency f_{in}=495MHz, if we assume the ADC has a SNR of 45 dB, then the calculated $\sigma_{jitter} \approx 1.8 ps$,if the frequency of input signal increases, the requirement for low jitter is more strict.

3.2 S/H circuit

Sample/hold (S/H) circuit is one of the most important part of an ADC, which converts analog input signal to discrete amplitude signal. Thus the performance of S/H circuit determines the performance of the whole ADC, especially the dynamic performance. For our 1.25Gsample/s sub-ADC, an open-loop architecture is adopted, as shown in Figure 5(a). PMOS transistor instead of bipolar transistor is used as an input transistor in order to eliminate the effect produced possibly by input current. As we know, the errors both in gain and offset of open-loop S/H circuit are unavoidable, D/A controlled trimming technique is adopted, an 8-bit inverse R-2R network D/A control circuit is added to the base of the output transistors Q17 and Q18.

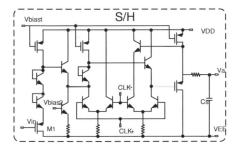

(a)

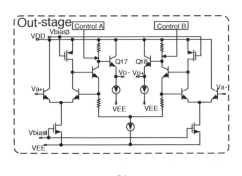

(b)

Figure 5. S/H circuit (a) sample and hold; (b) output stage.

3.3 Folding and interpolating ADC

The block diagram of the presented F&I sub-ADC is shown in Figure 6. The signals out from S/H are compared with the voltage across reference tap and pre-amplified further. Then the amplified signals are folded by the first stage folding amplifier by a factor of 3 and interpolated by the first 4-times interpolating resistor network. After finishing the first stage processing, there may be an averaging circuit to average the interpolated signals. After this processing, the averaged interpolating signals are feed into the second folding circuit where all the input signals are folded by a factor of 4 again, then a 4-times interpolating resistor network followed. Finishing the above two stage folding and interpolating, there gets 32 comparison signals. The 32 signals are encoded to the lower 5 bits of the sub-ADC. The higher 3 bits are acquired by the coarse quantizer. The higher 3 bits and lower 5 bits are feed into coder to get 8 bits as the output data of sub-ADC.

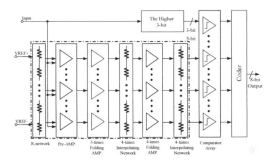

Figure 6. Architecture of folding and interpolating sub-ADC.

As depicted above, the pre-amplified signals out from pre-amplifiers are folded by 3 reference voltages using 3 differential pairs, as shown in Figure 7. The frequency of folding signal is equal to the product of input frequency and folding degree, which is 3 in the case, thus it is required that bandwidth of folder should be 4GHz[4].

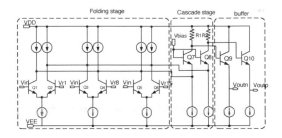

Figure 7. Folding by 3 circuits.

Usually the uniform resistive interpolation is chosen to implement the interpolating network in CMOS technology because of the process corner variations. Actually the resistive interpolation is done using a resistor voltage divider that interpolates more residue signals using only two residues produced by folders, as shown in Figure 8. Resistors R1', R2', ...R6' act as the averaging networks which can make zero-crossings shift towards their ideal positions, thus improve INL and DNL.

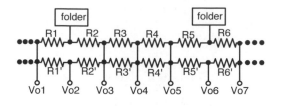

Figure 8.   Interpolation by 4 and averaging circuit.

3.4   Coding

Comparing to flash ADC, the encoder of folding and interpolating is more complicated because the output of F&I ADC`s comparator array is constituted by two parts: One is cyclic thermometer code from the fine quantizer which will be converted to form the lower LSBs and the other is common thermometer code from the flash ADC which will be converted to form the higher MSBs. Actually, these two parts operate separately,

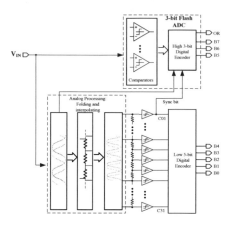

Figure 9.   Encoder realization of F&I ADC.

thus synchronization must be considered. Figure 9 shows the encoder realization blocks, one path of signal VIN flows into a 3-bit flash ADC and the output of which combined with the part output of the first stage folder, the synchronization bit C01 to be encoded into the higher 3-bit "B7B6B5" and out of range (OR for short)signal; the other path of VIN flows into the fine quantizer, including the first stage folders , the first stage interpolators, the second stage folders, the second stage interpolators, after those operation the output is cyclic thermometer code. These 25-1 codes are converted by the low 5-bit digital encoder to the lower 5-bit"B4B3B2B1B0".

That is to say, both output of the coarse and fine quantizer are first converted to Gray code on the consideration of bubble error reduction.

3.5   Calibration and SPI adjusting

Calibration of the whole ADC includes self-calibration of DC offset error and gain error of sub-ADC, gain mismatch between 4 sub-ADCs and manual SPI adjusting sampling clock between 4 sub-ADCs. Figure 10 shows the self-calibration of sub-ADCs. After powering on, calibration circuit gets ready for a new calibration, if CAL. signal equals 1, calibration starts, first DC offset is calibrated then shut off DC offset calibration mode ,entering gain error calibration, then all the calibration results are stored in registers for usage.

Figure 10.   The state transition of self-calibration.

The gain mismatch between sub-ADCs is also calibrated automatically. First, the digital output of current sub-ADC is detected and calculated by the digital calibration circuit where then output an 8 bits calibration signals. These signals are used to control a high precision digital-to-analog converter (DAC) which supply a variable biasing current for sample and hold amplifier, adjusting the gain of sub-ADC under calibration.

Moreover, sampling point error of each sub-ADC will results serious degradation of performance. A serial peripheral interface is designed to manually adjust the sampling point of each clock. As shown in Figure 11, the serial input data include the first 3 bitsv address (A2A1A0)

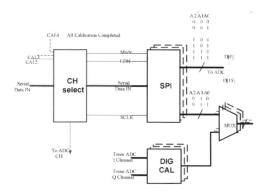

Figure 11.   The block of SPI and digital calibration.

and the following 16 bit data (D[15]…D[0]). A channel selector selects different sub-ADC under control signals CAL2, CAL3, and CAL4 and generates Mode and LDN control signal for each channel under expected timing.

4 DESIGN VERIFICATION

The whole TI ADC is designed with a 0.18μm SiGe BiCMOS process. As shown in Figure 10, the whole TI ADC has a SNR of 45dB with the input frequency of 495MHZ at the sample rate of 5GHz, owing an ENOB of 7.2 bits equivalently. The main concern parameters are summarized in Table 1.

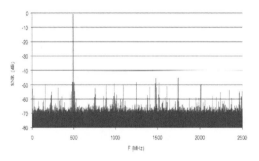

Figure 12. The SNR of the whole TI ADC.

Table 1. The main parameters of the whole TI ADC.

Parameters	Typical value
VDDA(V)	3.3
VDDD(V)	3.3
Power(W)	3.8
SNR(dB)	45
ENOB(bits)	7.2

5 CONCLUSION

In this paper, an 8 bit 5Gsample/s time-interleaved ADC is presented. The whole ADC has a SNR of about 45 dB at the input frequency of 495 MHZ and an equivalent ENOB of 7.2 bits. Some ideas of this work may be valuable for similar projects.

REFERENCES

[1] A.Barna,D.I.Porat, Integrated Circuits in Digital Electronics (Wiley, New York, 1973).
[2] A.G.F. Dingwall, Monolithic expandable 6 bit 20 MHz CMOS/SOS A/D converter. IEEE J.Solid-State Circuits 14(6), 926–932 (1979).
[3] W.C. Black, D.A. Hodges, Time interleaved converter arrays. IEEE J. Solid-State Circuits15(6), 1022–1029 (1980).
[4] Zhang Zhengping.etc. An ultra high-speed 8-bit timing interleave folding & interpolating analog-to-digital converter with digital foreground calibration technology. Journal of Semiconductors, Vol. 32, No. 9,2011.

Network Security and Communication Engineering – Chan (Ed.)
© 2015 Taylor & Francis Group, London, ISBN: 978-1-138-02821-0

Chaotic communication with new boss map for the ultra-high reliability in wireless communications

J.H. Lee, C.Y. An, B.J. Kim & H.G. Ryu
Department of Electronic Engineering, Chungbuk National University, Cheongju, Korea

ABSTRACT: It is possible that Chaos communication system can be used to improve the system security by using chaos signal. What is more, it is possible to reduce the chance of eavesdropping, and the system has strong characteristics for interfering and jamming signal. However, BER performance of chaos system is worse than digital communication system. For this reason, researches on BER performance improvement are being done actively. In previous studies, we proposed a novel chaos map for BER performance improvement, which is called 'Boss map'. Also, we proposed a novel chaos transceiver for BER performance improvement. However, BER performance is evaluated differently according to the delay time of chaos transceiver. Therefore, as for effective using of Boss map, we should find the optimal delay time in CDSK system and propose chaos transceiver. In this paper, when Boss map is used, we evaluate BER performance of CDSK system and propose chaos transceiver according to delay time. Besides, we find a delay time that is possible to have best BER performance in CDSK system and novel chaos transceiver after evaluation of BER performance according to delay time.

1 INTRODUCTION

Previous digital communication technology continually used a linear system theory and linearity in order to improve its performance, and nonlinear system was used to reach specific level of linear system. However, as this technology reaches basic limit, people have started to improve performance of nonlinear communication systems applying chaos communication systems to nonlinear system[1]. Chaos communication system is nonlinear system, and has the characteristics such as non-periodic, wide-band, non-predictability and easy implementation. Also, chaos communication system is decided by initial conditions of equation, and it has sensitive characteristic according to initial condition, because chaos signal is changed into different signal when initial condition is changed [2]. By combining characteristics of chaos signals with a digital communication technology, it becomes difficult to detect other transmitted signals, interference waves, and it has strong features in the multi-path interference. As a result, research can improve security to reduce the probability of eavesdropping [3]. However, generally speaking, security of chaos communication system is superior to digital communication system, but BER performance is not good. In case of CDSK system, chaos signal is transmitted with reference signal. So, it has many self-interference signals. As a result, BER performance is bad according to evaluation[4]. For BER performance improvement of chaos communication system, many researches are being

done. If we do the research on the existing chaos system, many chaos map and modulation methods will be used to evaluate BER performance. We found chaos map and modulation methods. By using it, a BER performance is the best. Moreover, the encoded system used the chaos signal or research applying the security system was accomplished.

In conventional research, we analyzed PDF trends of chaos map to improve the BER performance. Based upon this, we proposed the new chaos map[5]. And, we named the proposed chaos map 'Boss map'. In CDSK modulation method, we know that BER performance of Boss map is better than Tent map BER performance with 5dB[5]. Also, we proposed new chaos transmitter and receiver to enhance BER performance by reducing self-interference signal[6]. When proposed chaos transceiver is used, improved BER performance is similar to the BER performance of digital communication systems[6]. But when we use the boss map, BER performance of CDSK method and new chaos transmitter and receiver are changed according to time delay. Therefore, for effective using of Boss map, we should find the optimal delay time of CDSK and propose chaos transceiver.

In this paper, when Boss map is used, we evaluate BER performance of CDSK system and novel chaos transceiver according to delay time. Also, we find a delay time that is possible to have best BER performance in CDSK system and novel chaos transceiver after evaluation of BER performance according to delay time.

2 BOSS MAP

In previous studies, we showed the PDF of chaos signal, and analyzed PDF trends by changing the initial value, equation and parameter. And, based on the results of this trend, we proposed a novel chaos map.

There are two PDF trends that we can improve the BER performance. First, it should have low signal power, because distance between symbols is further than case of high signal power when low signal power goes through the normalization. Second, it should have the low probability of value near to 0. CDSK modulation method which distinguishes data based on 0. Namely, BER performance is good when probability of value near to 0 is low[5].However, two PDF trends for the BER performance improvement get opposite results. Therefore, two PDF trends require the proper trade-off in order to improve the BER performance. We proposed a new chaos map through the proper trade-off, and we named the new chaos map 'Boss map'. And, Boss map has BER performance which is about 5 dB better than the tent map in the CDSK method.

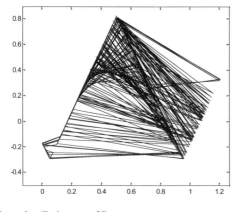

Figure 1.   PDF of Boss map.

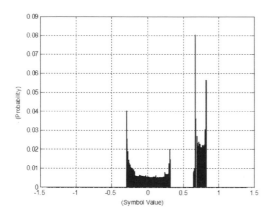

Figure 2.   Trajectory of Boss map.

Figure 1 shows PDF of Boss map that was proposed in order to improve the BER performance. We produced the PDF shown in figure 1 through the control of two factors having an effect on BER performance which we have explained in [5].

$$x_{n+1} = \alpha \left| 0.45 - \left| 0.503 - x_n \right| \right|$$
$$y_n = x_n - 0.3 \tag{1}$$

Equation (1) indicates the chaos equation of Boss map. The Equation (1) was changed from Tent map.

If the initial value is 0.1 and the parameter alpha is 2.5, trajectory of this equation is expressed as Figure 2, and PDF of this equation is expressed as Figure 1.The x-axis and y-axis means each x_n, y_n in Figure 2, Boss map is displayed in the form of a pyramid that is different from Tent map[5].

3 PROPOSED CHAOS TRANSCEIVER

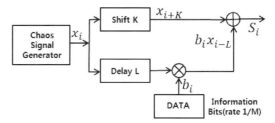

Figure 3.   Proposed chaos transmitter.

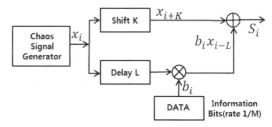

Figure 4.   Proposed chaos receiver.

In previous studies, we proposed a novel chaos transceiver in order to improve the BER performance. CDSK method has bad BER performance because it has so many self-interference signals. But, proposed chaos transmitter and receiver innovatively minimize self-interference. So, it can enhance BER performance [6].

Figure 3 shows a novel chaos transmitter that is possible to recover information signal by using only proposed chaos receiver. Proposed receiver is very similar to structure of the existing transmitter, but there is a process that shifts the chaos signal according to particular value in proposed receiver. Figure 4 shows proposed chaos receiver in order to improve the BER performance. Proposed chaos receiver needs to know the information of chaos signal.

$$S_i = b_i x_{i-L} + x_{i+K} \quad (2)$$

Equation (2) indicates a transmitted signal of proposed receiver. Transmitted signal of proposed receiver indicates a sum that includes delay chaos signal multiplied information bit and shifted chaos signal according to particular value[6].

$$S = \sum_{i=1}^{M} (x_i + b_i x_{i-L} + \xi_i) x_{i-L} \quad (3)$$

$$= b_i \sum_{i=1}^{M} x_{i-L}^2 + \sum_{i=1}^{M} n_i$$

$$n_i = x_i x_{i-L} + \xi_i x_{i-L} \quad (4)$$

Equation (3) and (4) are expresses the corelator output of proposed chaos receiver. Looking at the equation(3), first term means the desired signal, and second term means self-interference signal. Looking at the equation(4), the received signal is multiplied by chaos signal, and self-interference signal is evaluated much smaller than CDSK receiver[6].

4 SIMULATION

BER performance is significantly improved when we use Boss map and proposed chaos transceiver. But, when Boss map is used, BER performance is changed according to delay time of CDSK system and proposed chaos transceiver. Therefore we have to find the delay time which has the best BER performance.

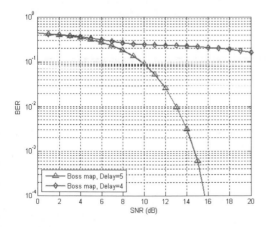

Figure 5. BER performance of Boss map according to delay in CDSK system.

In CDSK system, Boss map has absolutely better BER performance than Tent map, Logistic map, and Henon map[5]. CDSK system is added chaos signal and delayed chaos signal which is multiplied information signal and transmits [7]. Figure 5 shows BER performance of Boss map according to the delay time. In case a delay time is 5, figure shows improved BER performance, but when a delay time is 4, figure shows deteriorated BER performance.

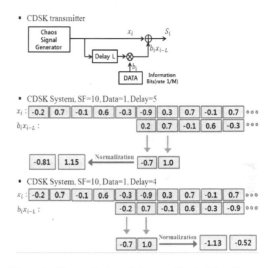

Figure 6. Transmitted symbols according to delay time.

In the CDSK system, before transmitting a signal, the signals go through the Normalization process which does the average signal a scaling with 1. In the figure 6, value of the transmission symbol is changed according to a delay time. Also, normalization changes the transmission symbol too. The time giving an example of two symbols, BER performance is better in the case where the delay of two symbols is 5 after normalization. Figure 6 takes two symbols as an example. In fact, BER performance of Boss map is changed according to a delay as the reason shown in Figure 6.

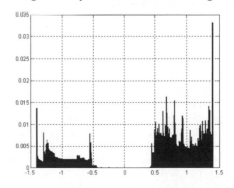

Figure 7. PDF after normalization. (Delay time = 5).

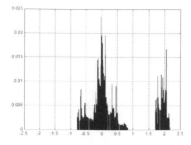

Figure 8.   PDF after normalization. (Delay time = 4).

Figure 7 and 8 show PDF after normalization according to delay time. In case of delay time 5, transmission signal near 0 is almost none, so probability of bit error rate is so low. But in case of delay time 4, transmission signal near 0 exists. So, noise or interference signal increases probability of bit error rate. So far a delay time is evaluated in a simulation only in case of 4 and 5, but in fact in case of odd numbered delay time, BER performance is improved, and in case of even numbered delay time, BER performance is deteriorated. Therefore, when Boss map is used in CDSK system, odd numbered delay time makes best BER performance.

In this paper, we evaluate BER performance of Boss map applying proposed chaos transmitter and receiver including CDSK system.BER performance of proposed chaos transceiver is changed by values of two delay times.

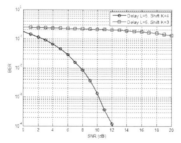

Figure 9.   BER performance of Boss map according to two delay time in proposed chaos transceiver.

In the figure 4, we can know that proposed chaos transmitter and receiver have delay time and shift value. In the figure 9, if the difference of delay time and shift value is odd number, BER performance is good. And, if it is the even number,I t means that BER performance deteriorates. The reason is the same, which has been explained in figure 6 after normalization symbol distance is far away in case of odd numbered delay time. Therefore, when Boss map is used, difference of delay time and shift value has odd number for effective using of Boss map.

5   CONCLUSIONS

Boss map has improved BER performance more than existing chaos map. Therefore, for effective using of Boss map, we should find the optimal delay time of CDSK system and shift time of proposed chaos transceiver. In this paper, when Boss map is used, we evaluate the BER performance of CDSK system and novel chaos transceiver according to delay time and shift time. Simulation result, when Boss map is used in CDSK system, delay time is necessary to be set to an odd number for the best BER performance. Also, when Boss map is used in proposed chaos transceiver, difference of delay time and shift value must be set to an odd number.

ACKNOWLEDGMENT

This research was supported by Basic Science Research Program through the National Research Foundation of Korea (NRF) funded by the Ministry of Education, Science and Technology (No. 2013R1A2A2A01005849) and this work was supported by the ICT R&D program of MSIP/IITP.(14-000-04-001, Development of compact MIMO antennas).

REFERENCES

[1] N. F. Rulkov, and M. M. Sushchik, "Digital Communication Using Chaotic Pulse Position Modulation", *IEEE Trans. Circuits Syst.* Vol. 48, pp. 1436–1444, 2001.

[2] W. M. Tam, F. C. M. Lau, and C. K. Tse, "Generalized Correlation-delay-shift-keying Scheme for Noncoherent Chaos-based Communication Systems," *IEEE Transactions on, Circuits and Systems I: Regular Papers*, vol. 53, no. 3, pp. 712–721, March 2006.

[3] Xiaofeng Li, Wei Pan, Bin Luo, and Dong Ma, "Mismatch robustness and security of chaotic optical communications based on injection- locking Chaos Synchronization," *IEEE JOURNAL OF QUANTUM ELECTRONICS*, Vol. 42, No. 9, Sep. 2006.

[4] M. Sushchik, L. S. Tsimring, and A. R. Volkovskii, "Performance analysis of correlation- based communication schemes utilizing chaos,"*IEEE Trans. Circuits Syst. I, Fundam. Theory Appl.*, vol. 47, no. 12, pp. 1684–1691, Dec. 2000.

[5] Jun-Hyun Lee, and Heung-GyoonRyu, "New chaos map for BER performance improvement in chaos communication system using the CDSK system," *Korea Institute of Communication Sciences on*, vol. 38, no. 8, pp. 629–637, Aug. 2013.

[6] Jun-Hyun Lee, and Heung-GyoonRyu, "A novel transmitter and receiver design of CDSK-based chaos communication system," *The Korean Institute of Electromagnetic Engineering and Sciences on*, vol. 24, no. 10, pp. 987–993, Oct. 2013.

[7] S. I. Hong and E. Y. Jang, "FPGA implementation of digital transceiver using chaotic signal,"*Korea Inst. Inform. Technol. Review*, vol. 8, no. 8, pp. 9–15, Aug. 2010.

Network Security and Communication Engineering – Chan (Ed.)
© 2015 Taylor & Francis Group, London, ISBN: 978-1-138-02821-0

A routing protocol for two-level clustering in Wireless Sensor Networks

Alghanmi Ali Omar, ChunGun Yu, Woon Sagong, Hyunjin Park & ChongGun Kim
Department of Computer Engineering, Yeungnam University, South Korea

Mary Wu
Department of Computer Culture, Yongnam Theological University and Seminary, Korea

ABSTRACT: The increasing demand for sensor networks inspires researchers towards efficient design and management of Wireless Sensor Networks (WSN). Energy efficiency, lifetime aware, and uniform energy dissipation are the key factors to design green wireless sensor networks. Considering those key factors, this paper proposes two-level hierarchical clustering based on routing protocols for data dissemination in WSN. In addition to cluster head of conventional clustering approach, this paper introduces Subordinates of the Cluster Head (SCH) as the local clustering to ensure energy and lifetime aware intra-cluster communication. The dominating set based on novel SCH selection procedure, considering the energy level, degree of node, and distance, makes the proposal unique in this sensor network study. We compare the performance of the proposal with existing cluster based approach through extensive simulation and find superiority to the previous approaches in terms of energy efficiency, lifetime awareness, and balanced energy dissipation. Furthermore, the proposed routing approach combines the facilities of proactive and reactive data dissemination and routing and thus it is suitable for both time-sensitive and time-insensitive applications.

1 INTRODUCTION

The application of micro- and nano-sensors [1] are numerous e.g. monitoring environment, battlefield, underwater world [2], space shuttle, patients' physiology [3] and psychology [4], natural disasters, industrial productions, mining and target tracking. Sensors are also embedded in consumer products and home appliances e.g. automobiles, electric woven, refrigerators, washing machines, mobile phones, and smart meters. The organized collection of sensor nodes, which is called sensor wireless sensor network (WSN), communicates wirelessly with one another to perform a definite application. The modern theories of internet of things (IoT) [5], device to device communication (D2D), machine to machine communication (M2M), smart grid (SG) [6], smart home for ambient assisted livings [7], location-aware systems [8] are basically developed on top of wireless sensor networks. Thus wireless sensor network remains in one of the most significant research areas throughout the last two decades.

The newfangled dimensions of sensor applications introduce diverse research challenges in wireless sensor network research. The up-to-date research challenges include miniaturization of sensor nodes from micro to nano-scale, antenna design for tiny sensors, embedding capability to access cognitive radios, embedding energy scavenging capability, energy efficient routing protocol design, energy aware MAC, lifetime aware and uniform energy dissipation algorithm designing, and development of green sensor networks, and low power communication protocol design [9]. The nodes of any wireless sensor networks [10] generally work as a unit of a system to complete certain obligations. Expiring of any sensor nodes from the network creates data deficiency, and as a result the whole sensor network produces erroneous results, incorrect and imperfect vision of environment and network becomes paralyzed. Therefore, the purpose of the research is to propose two-level cluster based power efficient routing protocol, which ensures balanced consumption of energy among the sensor nodes of the WSNs.

The proposed routing protocol follows hierarchical clustering architecture, where we are focusing on intra-cluster communication to gain energy efficiency. In addition to cluster head (CH) [11] or general cluster head (GCH), we introduce subordinate cluster head (SCH) for efficient and optimized cluster management. We propose a dominating set based SCH selection method to find out the subordinate nodes and their members to announce local administrations inside a cluster. To the best of our knowledge, this is the first proposal, which introduces local administrations using dominating set for intra cluster communication including near field communication (NFC) [12].

2   PROPOSED METHOD

In this paper, we propose additional sub-clusters in a cluster namely subordinate cluster head (SCH) for effective energy management of a cluster. We select the SCH nodes by considering residual energy and degree of nodes of any sensor node. Unlike relay node of a cluster, the SCH creates the opportunity of reactive and proactive routing of any clusters while maintaining data relays.

The system model is shown in Figure 1 and routing procedure of our proposed power efficient and energy balanced routing for wireless sensor network is considered. Wireless sensors are deployed in various patterns based on application requirements. In this research, we consider that the deployment of sensor nodes is followed by two-dimension (2D) hierarchical network topology and general cluster heads (GCH) are selected randomly following the same model as LEACH [13]. Here, in this paper, the neighbor nodes are defined on the basis of 1-hop communication model.

2.1   Description of the system model

The cluster of sensor nodes is formed on the basis of nodes within the radio distance, multi-hop communication is accepted. To shorten the radio communication, the cluster head which we call here general cluster head (GCH) appoints some subordinate nodes namely subordinate cluster head (SCH) to collect data from other general sensor nodes (SN). The SCH node always remains active to receive data from general sensor nodes (SNs) and send those data immediately to general cluster head (GCH). To save energy the general sensor nodes follow TDMA schedule to be active from sleep mode and sends collected data to SCH within its time slot, if and only if the sensed observation is significantly changed with respect to the hard and soft threshold. GCH sends the collected data towards the base station (BS) directly or through first level GCH. Figure 1 shows the two level clustering.

2.2   Proposed routing procedure

The dynamic cluster is formed firstly based on the general cluster head (GCH) considering the GCH as the center node of the cluster. Soon after the first phase, the GCH selects some of its member nodes as the subordinate node (SCH) using dominating set approach presented in algorithm 3.1, where dominating node is determined by considering the residual energy and degree of the node i.e. the number of eligible neighbors of the subordinate nodes. Here eligible neighbors are the set of neighbor nodes which can act as the relay or forwarder of the subject node.

$$EligibleScore(N) = \beta*(1/N_e)*EnergyLevel +$$
$$(1-\beta)*(1/N_d)*DegreeOfNode + \lambda*$$
$$(1/N_s)*Distance(1) \qquad (1)$$

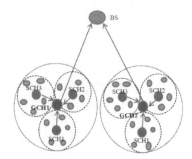

Figure 1.   The concept of two level clustering.

where, β=priority constant; N_e = normalization factor of energy level; and N_d=normalization factor of degree of node; N_s = normalization factor of distance from the GCH node; here the '*DegreeOfNode*'=Number of 1 hop successor neighbors of that node.

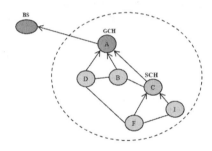

Figure 2.   Routing in a cluster.

After receiving data from plain sensor nodes, the SCH sends data to GCH along with the self-sensed data. So, the nodes, which follow the schedule to send data from multi-hop communication range, functions the same as the proactive nodes. Figure 2 shows routing in a cluster.

Algorithm 3.1:   Selection dominating_Set_SCH_( )

---

1. Initially mark GCH node as gray, mark eligible 1 hop neighbor nodes of GCH node as green and all other nodes of a cluster marks as white.
2. Determine eligibility score of each of the white nodes using function (1).
3. Select a node among the white adjacent nodes (also neighbors of green nodes) which has maximum eligibility score. Color it reddish.
4. Mark all of its eligible neighbor nodes as green.
5. Repeat step 2 to 5 until all nodes become green or reddish.
6. GCH defines all the reddish nodes as the SCH node. (The set of all reddish nodes is also called dominating set shows in Figure 2).

Algorithm 3.2: Hierarchical_Clustered_Routing_( )

1. Base Station (BS) selects general cluster heads (GCH) randomly to form clusters.
2. GCH Broadcast membership advertisement to the sensor network.
3. Sensor nodes (SN) transmit membership request to GCH node.
4. GCH select SCH nodes using Dominating_Set_SCH_Selection() algorithm, and broad cast the ID of SCH node to its member SNs. It also broadcast hard threshold and soft threshold for the current round of data transmission.
5. SN nodes senses environment and transmit data to SCH by following TDMA protocol and following threshold policy.
6. SCH nodes senses environment continuously and transmit data to GCH.
7. GCH compress data and send to base station (BS).
8. New round begins from step 1 again after base stations redistribution of new set of GCH nodes.

The SCH nodes also act as the proactive node because they also send data to the GCH continuously, and remain active all the time. Receiving data from the environment, each of the sensor nodes compare the data to its hard threshold value, if the observation exceeds the hard threshold, it sends the inspected observation to SCHs. Then the sensor node begins to monitor the environment constantly and inspects its observations, if it finds the significant change in its inspection with respect to the soft threshold, then it sends the observation results again to the SCHs. This is the reactive nature of the proposed routing procedure. If the sensed data is below the hard threshold value then sensor node will not send anything to SCH in its time slot. But, if the number nothing to send slot time exceeds the time threshold/slot threshold value then the sensor node only sends the beacon to inform that the node is still alive but has nothing to send at this moment.

3 SIMULATION RESULTS

To evaluate the performance of the proposed multi-level cluster based on energy efficiency routing protocol (MERP), the MATLAB R2010a simulation tools are used to analyze the energy efficiency and lifetime awareness. For simulation purpose, we simulate our proposed algorithm firstly without considering the distance value i.e. it considers only the energy level and degree of node to determine the eligibility score of node to be an SCH. Secondly; we consider the distance value with energy level and degree of node to determine eligibility score of a node to be an SCH. We compare our proposed algorithm with the benchmarked LEACH routing protocol. The simulation scenario is presented in Table 1. There are total 100 nodes deployed in 100x100 square meters of area. Total 10% of the sensor nodes are selected as the GCH for each round.

Table 1. Simulation scenario for performance study of MERP algorithm.

Parameters	Values
Topology	Random number of neighbors within 1 hop
Number of nodes	100
Simulation Area	100x100 m^2
Packet size	512 bits
Transmitter circuitry energy	50 nJ/bit
Transmitter amplification energy	100 pJ/bit/m^2
Receiver circuitry energy	50 nJ/bit
Initial energy of each node	0.5 Joules

Figure 3 shows the average energy dissipation of LEACH, TEEN and the proposed MERP protocols. The proposed protocol always dissipates less energy than the LEACH and TEEN protocol.

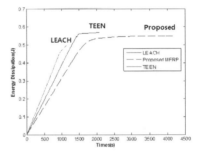

Figure 3. Comparison of average energy dissipation.

The energy dissipation also studied considering the base station (BS) outside the sensor area and 100 meters from the nearest sensor position i.e. in position (0,–100). In our simulation study we found that energy dissipation was becoming higher with the increment of distance of base stations as shown in Figure 5. Following the routing protocols LEACH, TEEN and proposed MERP, the last node dies at 912s, 2114s and 3472s respectively.

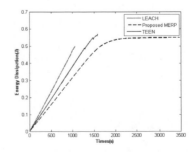

Figure 4. Comparison of average energy dissipation considering BS.

527

In Figure 5, we studied the number of packet received by the base station (BS) using LEACH, TEEN and MERP. In our study we observed that the number of packet of LEACH is larger than any other protocol because in this protocol all the generated packets have been sent towards the BS through CH, thus consuming huge energy. The packet reception in TEEN is the lowest among these three protocols because it uses hard and soft threshold barrier to transmit packets towards BS. The proposed MERP has a larger packet reception rate than TEEN because it uses time threshold to inform the network regarding its activeness (i.e. livelihood) in addition to hard and soft threshold. As the proposed MERP allows some nodes in sleep mode when some other neighboring sensor nodes remain active, this phenomenon helps the protocol to become energy efficient while enhancing a more successful packet reception rate than TEEN.

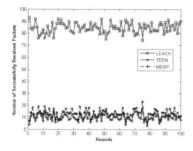

Figure 5. Number of packets up to 100th round.

4 CONCLUSIONS AND FUTURE WORK

The proposed two-level clustering of WSN is studied to outperform in energy efficiency and network longevity. The proposed method is a routing protocol with the combined proactive and reactive functional mode. There is less power consumption of the WSNs using the proposed routing protocol and the existing legendary routing protocol for WSNs is observed. Furthermore, the proposal has effective method to select the subordinates of the cluster head for dynamic cluster management. The SCH selection policy ensures the fairness in workload distribution of any clusters and it also certifies balanced energy consumption of the nodes of the cluster as well as sensor networks.

ACKNOWLEDGMENTS

This work has been funded by the BK21+ program of the National Research Foundation (NRF) of Korea.

REFERENCES

[1] Faezeh Arab Hassani, Yoshishige Tsuchiya and Hiroshi Mizuta, "In-Plane Resonant Nano-Electro-Mechanical Sensors: A Comprehensive Study on Design, Fabrication and Characterization Challenges", Sensors 2013, 13(7), 9364–9387.

[2] Jaime Lloret, "Underwater Sensor Nodes and Networks", Sensors 2013, 13(9), 11782–11796.

[3] Subrata Ghosh, Krishna Aswani, Surabhi Singh, Satyajit Sahu, Daisuke Fujita and Anirban Bandyopadhyay, "Design and Construction of a Brain-Like Computer: A New Class of Frequency-Fractal Computing Using Wireless Communication in a Supramolecular Organic, Inorganic System", Information 2014, 5(1).

[4] Alam, M. G. R., Cho, E. J., Huh, E. N., & Hong, C. S. (2014, January). Cloud based mental state monitoring system for suicide risk reconnaissance using wearable bio-sensors. In Proceedings of the 8th International Conference on Ubiquitous Information Management and Communication (p. 56). ACM.

[5] Fang, Shifeng, Li Da Xu, Yunqiang Zhu, Jiaerheng Ahati, Huan Pei, Jianwu Yan, and Zhihui Liu. "An Integrated System for Regional Environmental Monitoring and Management Based on Internet of Things." IEEE Trans. Industrial Informatics 10, no. 2 (2014): 1596–1605.

[6] Ahmed, Mohammad Helal Uddin, Md Golam Rabiul Alam, Rossi Kamal, Choong Seon Hong, and Sungwon Lee. "Smart grid cooperative communication with smart relay." Journal of Communications and Networks 14, no. 6 (2012): 640–652.

[7] James McNaull, Juan Carlos Augusto, Maurice Mulvenna and Paul McCullagh", Flexible context aware interface for ambient assisted living", Human-centric Computing and Information Sciences 2014, 4:1.

[8] Haeryong Cho, Min Choi, "Personal Mobile Album/ Diary Application Development", Journal of Convergence, Vol. 5, No. 1, Mar. 30, 2014, pp 32–37.

[9] Do-keun Kwon, Ki hyun Chung and Kyunghee Choi, "A Dynamic Zigbee Protocol for Reducing Power Consumption," Journal of Information Processing Systems, vol. 9, no. 1, pp. 41~52, 2013.

[10] Tarun Dubey and O. P. Sahu, "Self-Localized Packet Forwarding in Wireless Sensor Networks," Journal of Information Processing Systems, vol. 9, no. 3, pp. 477~488, 2013.

[11] Min Yoon, Yong-Ki Kim, Jae-Woo Chang, "An Energy-efficient Routing Protocol using Message Success Rate in Wireless Sensor Networks" Journal of Convergence, Vol. 4, No. 1, Mar. 30, 2013, pp. 15–22.

[12] Jae-Suhp Oh, Chan-Uk Park, Sin-Bok Lee, "NFC-based Mobile Payment Service Adoption and Diffusion", Journal of Convergence, Vol. 5, No. 2, Jun. 30, 2014, pp. 8–14.

[13] Heinzelman, Wendi Rabiner, Anantha Chandrakasan, and Hari Balakrishnan. "Energy-efficient communication protocol for wireless microsensor networks (LEACH) "System Sciences, 2000. Proceedings of the 33rd Annual Hawaii International Conference on. IEEE, 2000.

Network Security and Communication Engineering – Chan (Ed.)
© 2015 Taylor & Francis Group, London, ISBN: 978-1-138-02821-0

A compact 13-port circularly polarized antennas for MIMO communication systems

D.J. Lee, S.J. Lee, S.T. Khang & J.W. Yu
Department of Electrical Engineering Korea Advanced Institute of Science and Technology (KAIST), Daejeon, Republic of Korea

ABSTRACT: Antenna diversity is a well-known technique to enhance the performance of wireless communication systems and quality of services by increasing channel capacity. To overcome the physical limitation of diversity antenna due to strong mutual coupling between adjacent antennas, there are still much work in progress. The purpose of this paper is to design a compact integrated broadband and circularly polarized diversity antenna that has 13 feed ports with more than 15 dB isolation between them. Several methods were used for that purpose. More specifically, printed quadrifilar spiral antenna which has good axial ratio characteristics against size of ground plane was selected and optimum simulation for structure of MIMO antenna was conducted. It finds that antenna has an impedance bandwidth of 700MHz ranging from 2.2GHz to 2.9GHz (S11 < 10dB) and 13 ports antenna is integrated in half wavelength volume ($0.5\lambda * 0.5\lambda * 0.5\lambda$). The proposed antenna shows 74.54 b/s/Hz estimated channel capacity at an SNR of 20 dB.

1 INTRODUCTION

Multi-input multi-output (MIMO) wireless communication has the potential to significantly increase capacity without increasing transmitting power or spectrum. If M transmit and M receive antennas are used in Rayleigh flat-fading communication environment, a theoretical linear M scale increase in channel capacity can be achieved. However, to construct enough antennas within a small volume is always a challenge [1].

In a rich scattering environment, the MIMO cube takes advantage of spatial and polarization diversities in a compact volume. Further increases in the number of antennas on a cube can further increase the theoretical channel capacity, but on the other hand, it can also produce high mutual coupling between individual antennas, leading to increasing correlation and decreasing capacity [2~3].

In this paper, 13-port circularly polarized (CP) MIMO antennas are proposed and tested. For the 13-port antennas, printed quadrifilar spiral antennas are used and geometry of antennas are optimized to reduce mutual coupling between each antennas. All 13-antennas are built on 100 X 100 X 95 mm3 cube with an operating frequency of 2.4GHz ISM band. Channel capacity is a key parameter to use proposed antennas in MIMO wireless communications. Simulated capacities with antenna radiation patterns are presented in this paper.

2 ANTENNA DESIGN

The antenna design procedure starts with the design of printed quadrifilar antenna. The quadrifilar spiral antenna with four port feeding structure is selected to make uniform CP radiation. Under the quadrifilar radiators, four port feeding network is integrated and it is composed of a Wilkinson power divider, 90° delay line, T-junction power divider and 180° delay line. The reflection coefficient of one radiator and the mutual coupling between the radiator and the opposite radiator are designed to be equal for the high radiation efficiency. Fig. 1 shows final fundamental antenna element, the printed quadrifilar antenna (W_a = 25 mm, W_g = 50 mm, H_a = 10 mm, l_1 = 18 mm, l_2 = 2 mm, l_3 = 2 mm and l_4 = 7 mm). The short stubs are used for the impedance matching of radiator and the feeding networks.

To take advantage of spatial and polarization diversity in a rich scattering environment with low mutual coupling, printed quadrifilar antennas are integrated. Five quadrifilar antennas which have the 120 ° 3dB beam width are arranged, as shown in Fig. 1, to cover hemisphere area. 13-port quadrifilar antennas are consequently organized to cover whole sphere as shown in Fig. 2. (Wx = 100 mm, Wy = 100 mm and Wz = 95 mm). A prototype antenna was fabricated of FR4 PCB with dielectric constant 4.6 and thickness of 0.6mm for operating around 2.4GHz.

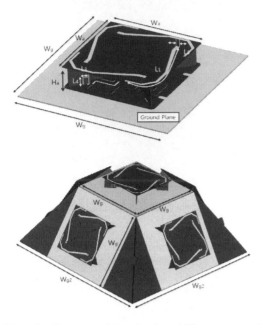

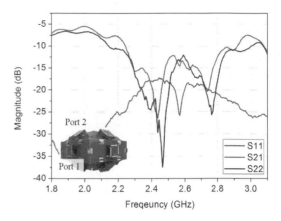

Figure 3. Measured S-parameters of the proposed 13-port MIMO antennas.

Figure 1. Geometry of the printed quadrifilar antenna and its integration.

3 MEASURED AND SIMULATED ANTENNA PERFORMANCE

In order to verify the proposed antenna, the proto-type antenna is tested. Fig. 3 illustrates the measured S-parameters of the adjacent antennas. It is difficult to present all combination of S-parameters between ports, and two antennas are selected which have the worst mutual coupling. The antennas resonate at 2.4GHz with around 29% impedance bandwidths for S11 < –10dB. The worst mutual coupling maintains the values better than –15dB. All other combinations of antenna ports have better than –20dB, including those results not shown in this paper.

The measured radiation patterns of the proposed antenna for x-z plane and y-z plane are shown in Fig. 4. Each antenna well radiates circular polarization pattern at 2.4GHz. The achieved peak gain at broadside is about 2.5dBic and small cross-polarization shows the purity of the synthesized circular polarization.

In order to verify the capacity of 13 port antennas, 3D polarized channel modeling simulation of 13-port antenna capacity includes the effect of the antenna, patterns, antenna location and channel [4]. The radiation patterns for all antennas obtained by CST and antenna locations are also used for the capacity estimation. For reflecting the realities of propagation environment, we control the scatter distribution in the channel. Fig. 5 shows the estimated channel capacity of 13-port MIMO antennas in a rich scattering environment. The ideal 13-port MIMO antennas obtain 86.56 b/s/Hz with independent identically distributed fading channel and the proposed antennas show 74.54 b/s/Hz at an SNR of 20 dB. Even though it is slightly lower than ideal case, it would be the good candidate for a MIMO communication antenna in a rich scattering environment.

4 CONCLUSION

Novel compact 13-port CP MIMO antennas using the printed quadrifilar antennas are presented. With compact and simple feeding network below the quadrifilar radiator, the proposed antenna can generate CP polarization and they are integrated within small volume to utilize spatial and polarization in a rich scattering environment. They exhibit better than –15dB isolation and it is decreasing correlation coefficient of antennas. At last, the capacity of the whole 13 port antennas is shown with simulation which includes the

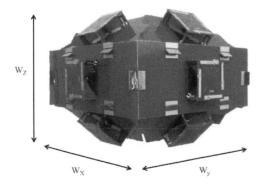

Figure 2. Prototype of the 13-port MIMO CP antennas.

effect of antenna, patterns, mutual coupling and channel and measured antenna properties and it shows the good MIMO potential.

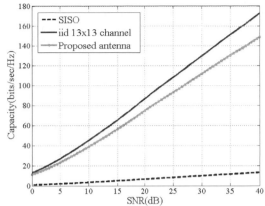

Figure 5. Estimated channel capacity of the proposed antennas.

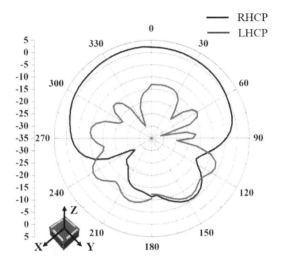

(a)

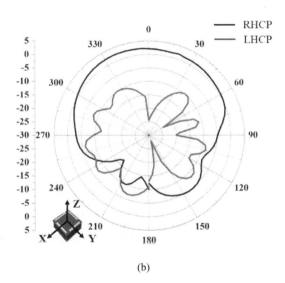

(b)

Figure 4. Measured radiation pattern of the proposed antenna: (a) x-z plane (b) y-z plane.

REFERENCES

[1] M. A. Jensen and J. W. Wallace, "A review of antennas and propagation for MIMO wireless communications," *IEEE Trans. Antennas Propag.*, vol. 52, pp. 2810–2824, Nov. 2004.

[2] C. Y. Chiu, J. B. Yan, and R. Murch, "24-port and 36-port antenna cubes suitable for MIMO wireless communications," *IEEE Trans. Antennas Propag.*, vol. 56, no. 4, pp. 1170–1176, Apr. 2008.

[3] J. Zheng, X. Gao, Z. Zhang, and Z. Feng, "A compact eighteen- antenna cube for MIMO systems," *IEEE Trans. Antennas Propag.*, vol. 60, no. 2, pp. 445–455, Feb. 2012.

[4] Manh-Tuan Dao, Vict-Anh Nguyen, Yun-Taek Im, Seong-Ook Park and Yoon Giwan, "3D Polarized Channel Modeling and Performance Comparison of MIMO Antenna Configurations With Different Polarizations," *IEEE Trans. Antennas and Propag.*, vol. 59, no.7, pp. 2672-2682, July. 2011.

WebRTC based remote rehabilitation guidance system

Q.H. Shi, J. Zheng, W. Wang & Y. Wu
Zhejiang Provincial Key Laboratory of Service Robot, College of Computer Science, Zhejiang University, Hangzhou, China

Z.J. Yan
Zhejiang Yuying College of Vocational Technology, Hangzhou, China

ABSTRACT: Compared with the large number of disabled population, medical resource is scarce and partially distributed. With the development of Internet and information technology, more and more services can be provided through the network, not restricted by distance. In this context, this paper developed a P2P instant communication system based on the WebRTC technology and realized a remote rehabilitation guidance system. The goal of this system is to make sure that a user can receive remote rehabilitation guidance scheme through computer or mobile phone. The system has no need to install additional software and plug-ins. The open web rehabilitation guidance can be realized through a browser only.

1 INTRODUCTION

The appearance of WebRTC (Web Real-Time Communication) has an impact on real-time communication. WebRTC is becoming a standard that allows users to quickly and easily perform voice and video interaction.

We build a system based on WebRTC, add support to information accessibility, and focus on system compatibility, low latency, and less traffic. In this paper, we improve WebRTC from two aspects. On the one hand, message will be serialized by a code algorithm named value base encoding (VBE) before being sent. This will reduce the time cost when pushing messages and improve network throughput. On the other hand, we proposed NAT information analysis algorithms. The improvement is intended to reduce the delay of establishing communication.

Based on the above work, this paper designed and implemented a P2P instant messaging system based on WebRTC to offer remote rehabilitation guidance for the disabled. This paper also explained the implement detail both of the server and client. This paper tested the system and received good experimental results.

2 RELATED TECHNOLOGIES

2.1 *WebRTC framework*

WebRTC (Lozano A. 2013) is an open source project aiming to enable the web with Real Time Communication (RTC) capabilities. It provides a complete set of media communications solutions, with audio engine, video engine and network transmission being the core part.

2.2 *Information accessibility technology*

Information Accessibility technology refers to the technology which enables people with disabilities conveniently access to and to make use of information in all cases.

2.3 *NAT traversal technology*

NAT(Network Address Translation) (Tsirtsis G. 2000) brings instant communication system a problem of losing the support of P2P (Ohzahata S& Kawashima K. 2011) which limits P2P technology.

The NAT equipment applied methods are not applicable to our system. Here we mainly introduce the methods applied to the client.

STUN (Simple Traversal of UDP NAT) (Rosenberg J et al. 2003) is a simple C-S protocol. Host client can detect IP and port through the exchange of UDP packets with the public STUN server.

TURN (Traversal Using Relay NAT) is a way to relay. It is also a C-S protocol which realizes the NAT travers all through a relay server.

ICE (Interactive Connectivity Establishment) (Rosenberg J. 2010)is actually a combination of STUN and TURN. ICE makes full use of the environment in the internal private network. WebRTC uses it to realize NAT traversal because of its comprehensive applicability.

3 MESSAGE SERIALIZATION (VALUE BASED ENCODING)

3.1 *Comparison of serialization*

Message serialization can make it convenient for the message storage and transmission. To do network data serialization, there are a lot of methods.

XML (Bray T et al. 1997), which is based on keywords and convenient to expand and makes visualization of data easier, may leads to large storage space used and encoding (decoding) speed very slow.

JSON (Crockford D. Json. 2006), in spite of having the convenience of visualization and slightly superior to XML in space utilization and rate of encoding (decoding), still do not have good performance.

Other serialization methods, e.g. Thrift (Slee M et al. 2007), with internal defined format, needs to introduce a set of complex framework, which is not conducive to the customers.

In this paper, we proposed an improved binary serialization scheme named Value Based Encoding (VBE).

3.2 *VBE algorithm*

VBE is based on binary. It changes the number of bytes according to the value of integer. In other words, the smaller the value of the integer is, the fewer number of bytes will be used.

Integer encoding process has two steps. Firstly, we represent the integer with binary, taking every seven bits as a group, each group as a byte (8 bits, first bit as flag), and arranging low byte at the top. Secondly, for a single integer, the flag of last byte is set to be 0, others to be 1(for convenience to identifying the border).

Through compression of bytes, VBE tries best to make use of the space. Also it is easy to put into practice.

4 IMPROVEMENT OF ICE

4.1 *NAT information analysis (improved ICE)*

Categories of candidate address (IP and port)include local address, server reflexive address and relay address(Rosenberg J. 2010). The function of ICE is to find out the couple of address that can be connected.

The process of ICE includes address information collection, address priority definition, connecting test and address couple decision.

4.2 *Improved address information collection*

In addition to collecting local candidate address, server reflection address and relay address,our improved ICE also needs to collect the location of the host (public/private network), NAT types, NAT Hairpin information (whether to support).

4.3 *Improved address priority definition*

Based on the collection of information, improvedaddress priority is

$$priority = \left(2^{24}\right) \times address\ type +$$
$$\left(2^{8}\right) \times modified\ value + \left(2^{0}\right) \times \left(256\text{-}Component\ ID\right) \tag{1}$$

The value of address type has the relationship that local address > reflexive address = predicted port address > relay address.Address type represents different types of address.The modified value changes with different situations.ComponentID is serial number of the inside parts of streaming media (such as RTP and RTCP),with size between 1and 256.

4.4 *Improved connecting test*

The two groups of address information of both sides of communication have a variety of combinations. We need to define priority to make connecting test easier. Priority definition is shown as(2).

$$pair\ priority = priority1 + priority2 \tag{2}$$

Traditional ICE will test all candidate address to make sure that a pair of address can be connected. Useful information listed in chapter 4.2 is analyzed so as to find out corresponding special connecting test orders. After information analysis, we can redefine the priority of candidate address.

5 EXPERIMENTS

We use Node.js (Tilkov S & Vinoski S. 2010) as our server and chrome as our client.

5.1 *Performance of VBE based serialization*

We use serialization time, deserialization time and the size of storage space to compare the serialization performance. All methods are applied to 100000 pieces of message. Results are shown in Figure 1.

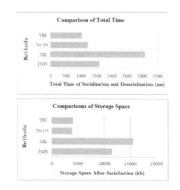

Figure 1. Comparisons of serialization and deserialization.

Obviously, VBE is far superior to XML, JSON, and as good as Thrift. However, our VBE is very concise with no need to adding additional framework and can be easily customized under the framework of our system.

5.2 Performance of improved ICE

5.2.1 Experiment design

To evaluate the performance of improved ICE, we measure average delay of connection, success rate of direct traversal, average times of address test.

Network structure is shown in Figure 2. And considering different situations, we have designed three scenarios.

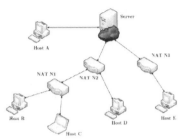

Figure 2. Experimental network structure.

5.2.2 Scenario 1

One end of the session is in the public network while the other end is in the private network. As shown in Figure 2,

host A is in the public network, and host B, C, D, E are in the private network. Results are shown in Table 1.

As the experimental results shown, if one end of the session is in the public network, the success rate of direct traversal will be the same. Since ordinary ICE needs to test local address and need to be in the private network, our improved ICE has smaller number of address test and smaller delay of connection.

5.2.3 Scenario 2

Both sides of the session are under the same private network. As shown in Figure 2, host B, C, D are in the same private network. Results are shown in Table 2.

As the experimental results shown, within the same LAN, if Hairpin is supported, ordinary ICE connection is faster than the improved ICE for improved ICE has collecting and analyzing cost. If Hairpin is not supported, the rate of traversal is 0. In other words, every message is sent through transmit server. Since improved ICE test transits server address directly, it has shorter time of address test.

5.2.4 Scenario 3

The two sides of the session are under different private network. As shown in Figure 2, host B, C, D are in the same private network while host E is in another private network. Results are shown in Table 3.

As experimental results shown, our improved ICE has better performance than ordinary ICE for all these three indicators.

Table 1. Results of scenario 1.

NAT type & ICE type	Average delay*	Average times of address test	Traversal times	Transmit times	Rate of traversal
NAT & ICE	945	2	50	0	100%
NAT & Improved ICE	870	1	50	0	100%
S_NAT** & ICE	1200	3	49	1	98%
S_NAT** & Improved ICE	1034	1.2	49	1	98%

* Average delay is measured in milliseconds (ms).
** S_NAT means symmetrical NAT.

Table 2. Results of scenario 2.

Hairpin & ICE type	Average delay*	Average times of address test	Traversal times	Transmit times	Rate of traversal
hairpin_S** & ICE	608	2	50	0	100%
Hairpin_S** & Improved ICE	788	1.5	50	0	100%
Hairpin_NS** & ICE	3420	4	0	50	0
Hairpin_NS** & Improved ICE	1240	2.5	0	50	0

* Average delay is measured in milliseconds (ms).
**Hairpin_S means Hairpin supported and Hairpin_NS means Hairpin not supported.

Table 3. Results of scenario 3.

NAT type & ICE type	Average delay*	Average times of address test	Traversal times	Transmit times	Rate of traversal
TS_NAT** & ICE	4320	9	0	50	0
TS_NAT** & Improved ICE	3640	5	9	41	18%
OS_NAT** & ICE	3930	9	32	18	64%
OS_NAT** & Improved ICE	2980	4.5	41	9	82%

* Average delay is measured in milliseconds (ms).
** TS_NAT means Two Symmetrical NAT and OS_NAT means One Symmetrical NAT.

6 SYSTEM DESIGN AND IMPLEMENTATION

6.1 *Server side design*

Core components of server side consist of signaling server, ICE server and web application server. The three kinds of servers are both independent and have mutual cooperation. Server side uses cluster architecture to ensure that services have nice concurrency and efficiency. Construction of the server side is shown in Figure 4.

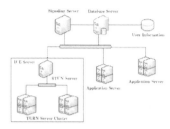

Figure 3.   Construction of server side.

6.2 *Client side design*

The main components include the audio engine, video engine, network interactive transmission, information accessibility and business application modules. In this paper, we designed browser client and WebRTC client.

6.2.1 *Browser client design*

For the browser client, WebRTC mainly provides 2 classes of API. One is named as MediaStream, which is used to get the media data. The other one is named as RTCPeer Connection, which is used to make calls and establish connections between peers, for media stream transmission.

6.2.2 *WebRTC client design*

Our WebRTC Client is derived from the source code of WebRTC released byGoogle. Our work is to streamline the module and, in some cases, to improve the module.

6.3 *System performance*

The lost rate and delay rate are measured to verify the performance of message transmission. We capture data sent and received between two clients directly. With the server locating in the public, and the load value of concurrency being 100, the two sides are set to send 10000 messages respectively. The result of lost rate is 0. And the amount of data being sent is 65% of the data being sent based on the traditional binary.

To estimate the performance of media interaction, we measure the connecting time and delay time. We take the arithmetic mean of 50 experimental results.

Thereby, we get the average connecting time of 3130ms and average delay time of 980ms respectively.

Obviously, we achieved nice performance.

7 CONCLUSION

In this paper, we study and design the system for remote rehabilitation guidance. The improvement of our system has an effect on instant message transmission and network traffic. In the process of P2P communication, our system can realize NAT traversal applied to a variety of different network architecture. However, there is still space for improvement in some aspects.

For signaling server, we adopt the WebSocket as signaling channel which leads the large number of users to becoming a bottleneck. For transmission of message, more secure encryption needs to be added in future business.

Generally speaking, remote rehabilitation system is helpful for people with disabilities. And it is an issue worthy of further study.

ACKNOWLEDGMENT

This work is supported by National Key Technology R&D Program (Grant No. 2012BAI34B01).

REFERENCES

[1] Bray T, Paoli J, Sperberg-Mcqueen C M, et al. Extensible markup language (XML)[J]. World Wide Web Journal, 1997, 2(4): 27–66.

[2] Crockford D. Json (javascript object notation). Technical report, json.org, 2006.

[3] Fette I, Melnikov A. The websocketprotocol[J], 2011.

[4] Lozano A A. Performance analysis of topologies for Web-based Real-Time Communication(WebRTC)[D]. Department of Communication and Networking, aalto university school of electrical engineering, 2013.

[5] Ohzahata S, Kawashima K. An experimental study of peer behavior in a pure P2P network[J]. Journal of Systems and Software, 2011, 84(1): 21–28.

[6] Rosenberg J, Weinberger J, Huitema C, et al. STUN-simple traversal of user datagram protocol (UDP) through network address translators (NATs)[R]. RFC 3489, IETF, Mar, 2003.

[7] Rosenberg J. Interactive connectivity establishment (ICE): A protocol for network address translator (NAT) traversal for offer[J]. Answer Protocols, 2010.

[8] Slee M, Agarwal A, Kwiatkowski M. Thrift: Scalable cross-language services implementation[J]. Facebook White Paper, 2007, 5.

[9] Tilkov S, Vinoski S. Node. js: Using JavaScript to build high-performance network programs[J]. Internet Computing, IEEE, 2010, 14(6): 80–83.

[10] Tsirtsis G. Network address translation-protocol translation (NAT-PT)[J]. Network, 2000.

Network Security and Communication Engineering – Chan (Ed.)
© 2015 Taylor & Francis Group, London, ISBN: 978-1-138-02821-0

Fast sparse compressive sensing based on the wavelet package hierarchical trees

J.X. Chen, T.M. Wang, W.L. Niu & L.J. Du
College of Applied Science and Technology, Beijing Union University, Beijing, China

ABSTRACT: For efficient and consistent signal compressive sampling and reconstruction, with the analysis of the sparsity of mutation smooth signal and its structural in wavelet space, a new Fast Sparse Compressive Sensing (FSCS) method, which decomposes signal with the rule of fast sparse based on its structural characteristics, was developed based on the multi-level tree model of wavelet packet. Combined with the established reconstruction algorithm, the effectiveness of the proposed Compressive Sensing method was evaluated with typical industrial mutation smooth signals, e.g., the pipeline leakage signal. With the acceptable root-mean-square error and signal to noise ratio of reconstructed signal, the compression ratio increased significantly, which was helpful to reduce the burden of communication bandwidth of industrial monitoring networks. With the compression ratio of 30:1, the parameter of ERP of the developed method is less than $1 \times 10-3$, which ensures the accuracy of reconstructed signal for pipeline leakage detection and positioning.

KEYWORDS: Compressed sensing, wavelet packet, data compression, pipeline leakage.

1 INTRODUCTION

As a convenient and efficient format, pipeline network based on transportation enables petroleum or natural gas to be transported to any places where it is needed. However, the pipeline leakage resulting from nature aging or intentional stealing can destroy the safety of network and its transportation. A couple of intelligent pipeline leakage detection methods based on different principles, including negative pressure wave, sound wave, and distributed optical fiber, have been developed and widely used. For all of them, a state monitoring network has to be built to keep collecting signal data from different measurement points and analyzing them for leakage positioning[8-14], which may overload especially when excess quantity of measurement data needs transmission and analysis. Therefore, it is necessary to develop new monitoring strategies, especially the intelligent algorithm with the concept of Compressive Sampling to achieve optimized performance with acceptable system cost.

In this paper, a new Fast Sparse Compressive Sensing (FSCS) method was proposed and applied to pipeline leakage positioning successfully. Different from existing methods, the new method was developed with respect to the sparse characteristics of signals in wavelet domain. Motivated by the tree structure of sparse signal, the Fast Sparse representation model was built based on the multi-level tree model of wavelet packet. The effectiveness of the proposed Compressive Sensing method was evaluated with data of pipeline monitoring network for leakage positioning. It was demonstrated that, compared with the existed ones, the statistical accuracy of signal reconstruction didnot decrease with fewer measurement data, which is beneficial to decrease system complexity with lower sampling rate.With the acceptable root-mean-square error and signal to noise ratio of reconstruction signal, the compression ratio increased significantly, which washelpful to reduce the burden of communication bandwidth of industrial monitoring networks. The developed CS method can be applied not only topipeline leakage positioning in industrial monitoring, but also toother places where excessive measurement data need transmission and analysis.

2 EFFICIENT COMPRESSIVE SENSING BASED ON FAST SPARSE REPRESENTATION MODEL

With the Fast Sparse (FS)representation model,the mutation smooth signal X can be decomposed as follows:

$$X = \psi^{FS} S \qquad (2)$$

where, $\psi^{FS} = [\psi_1, \psi_3, \psi_5, \ldots, \psi_L]$, $S_i = \langle x_i, \psi_i^{FS} \rangle$, $L < N$

According to the definition of signal sparsity, the sparse coefficient vector **S** satisfy the condition that $\|\mathbf{S}\|_p \equiv (\sum |S_i|^p)^{1/p} \le R$, $0 < p < 2$, $R > 0$, and $\|\mathbf{S}\|_0 = K, K \le N$, K is the sparseness of **S**.

Then, the compressive sensing measurement can be expressed in matrix form:

$$y = [y_1, \ldots y_m]^T = \Phi X = \Phi \Psi^{FS} \mathbf{S} \qquad (3)$$

Here, the size of measurement matrix Φ is $M \times N$, and each row a basis vector ϕ_i. y is a $M \times 1$ column vector.

The measurement matrix Φ should satisfy the restricted isometric property(RIP), that

$$1 - \delta \le \frac{\|\Phi X\|_2^2}{\|X\|_2^2} \le 1 + \delta \qquad (4)$$

Here, $0 < \delta < 1$.

While the Gaussian measure matrix was commonly used in Compressive Sensing with good universality and RIP property, we optimized the Gaussian measurement matrix with QR decomposition to decrease the value of δ, and recovered the signal X from the M compressive measurements faster, and time improved its measurement performance further at the same.

Let, $\Phi' = [\phi_{ij}]_{M \times N}$, $\phi_{ij} \in N(0, 1/M)$, and then do QR decomposition with Φ'.

The optimized the measurement matrix was shown in(5),

$$\Phi'' = (R')^T (Q)^T \qquad (5)$$

Further, if we assume the CS measurements are noisy, with zero-mean Gaussian noise n, we have(6),

$$y = \Phi'' X + n = \Phi'' \Psi \mathbf{S} + n = DS + n \qquad (6)$$

Finally, we can achieve perfect reconstruction by solving problem of linear regression, as shown in (8).

$$\hat{\mathbf{S}} = \arg \min \| y - DS \|_2^2 \qquad (7)$$

By the combination of measurement matrix Φ, measurement values y and the CS or the normal bases ψ. Under given prior distribution, the estimate of S in (7) can be archived through maximum a posteriori (MAP) estimate. According to Bayesian framework and Relevance Vector Machine, we set a zero-mean Gaussian prior distribution over S,

$$p(\mathbf{S}|\alpha) = \prod_{i=0}^N N(s_i|0, \alpha_i^{-1}) \qquad (8)$$

With α a vector has Gamma prior distributions. Its likelihood functions are shown in (9):

$$p(\alpha) = \prod_{i=0}^N Gamma(\alpha_i|a, b) = \Gamma(a)^{-1} b^a \alpha_i^{a-1} e^{-b\alpha_i} \qquad (9)$$

Where, $\Gamma(a) = \int_0^\infty y^{a-1} e^{-y} dy$, the 'Gamma function', a, b are its parameters.

As for the measurement values y, its likelihood functions satisfy (10),

$$p(y|s, \alpha, \sigma^2) = (2\pi\sigma^2)^{-M/2} e^{\left\{ -\frac{1}{2\sigma^2} \|y - DS\|^2 \right\}} \qquad (10)$$

According to Bayes' rule, the posterior over **S** is given by(11),

$$
\begin{aligned}
&p(\mathbf{S}, \alpha, \sigma^2 | y) \\
&= p(\alpha, \sigma^2 | y) p(\mathbf{S}|y, \alpha, \sigma^2) \\
&= p(\alpha, \sigma^2 | y) \frac{p(y|S, \alpha, \sigma^2) p(S|\alpha, \sigma^2)}{p(y|\alpha, \sigma^2)} \\
&\propto p(\alpha, \sigma^2 | y) p(y|S, \alpha, \sigma^2) p(S|\alpha) \\
&= p(\alpha, \sigma^2 | y)(2\pi)^{-(M+1)/2} |\Delta|^{-1/2} e^{\left\{ -\frac{1}{2}(\mathbf{S}-u)^T \Delta^{-1}(\mathbf{S}-u) \right\}}
\end{aligned} \qquad (11)
$$

Where, the posterior covariance and mean are respectively $u = \alpha_0 \Delta D^T y$, $\Delta = (\alpha_0 D^T D + A)^{-1}$.

$$A = diag(\alpha_1, \alpha_2, \ldots, \alpha_M).$$

In (11), When $\mathbf{S} = \mu$, $p(s|y, \alpha, \sigma^2)$ has the biggest value, as to $p(\alpha, \sigma^2 | y)$, shown by (10)

$$p(\alpha, \sigma^2 | y) \propto p(y|\alpha, \sigma^2) p(\alpha) p(\sigma^2) \qquad (12)$$

We can estimate α, α_0 with the method of maximum likelihood for the suitable value, obtain the maximum $p(y|\alpha, \sigma^2)$, and finally recover sparse coefficients **S** from the maximum posteriori estimate value.

3 EXPLOITING FSR COMPRESSIVE SENSING ON PIPELINE LEAK SIGNAL

We applied the FSR Compressive Sensing method for multi-group negative pressure and optical fiber pipeline leak signals got from an oil field in north China, as shown in Figure1, by experimental validation, results and conclusions are given at last.

(a) The optical fiber pipeline leak signal

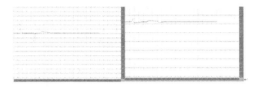

(b) The negetive pressure pipeline leak signal

Figure 1. The pipeline leak signal.

Compare the commonly used Compressive Sensing method, such as, Basic Pursuit (BP) , Fast Bayesian Matching Pursuit (FBMP),Compress Sampling Matching Pursuit (CoSaMP). We obtain the Signal to Noise Ratio (SNR), Mean Square Error as well as the ERP as shown (13)-(15), measuring the similarity between reconstruction results and the original signal .

$$ ERP = \sum_{i=1}^{N}\left[\hat{f}(i) \right]^2 \Big/ \sum_{i=1}^{N}\left[f(i) \right]^2 \qquad (13) $$

$$ MSE = \sqrt{\sum_{i=1}^{N}\left[f(i) - \hat{f}(i) \right]^2} \Big/ \sqrt{\sum_{i=1}^{N}\left[f(i) \right]^2} \qquad (14) $$

$$ SNR = 10\lg\left[\sum_{i=1}^{N}\left| f(i) \right|^2 \Big/ \sum_{i=1}^{N}\left| f(i) - \hat{f}(i) \right|^2 \right] \qquad (15) $$

Where, $f(i)$ is the original data, $\hat{f}(i)$ is the reconstructed data, and N is the sample length.

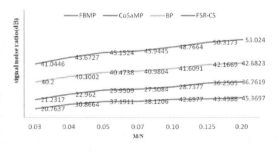

Figure 2. Compression performance evaluation of pipeline leak signals with compressive sensing.

4 CONCLUSIONS AND OUTLOOK

In this study, a new FS Compressive Sensing (FSCS) method for mutation smooth signal reconsturction was developed with a new model for fast sparse processing based on the sparse signal's characteristics in wavelet space. Combined with the developed FSCS, a reconstruction algorithm was developed to reconstruct the data measured with low sampling rate, which is different from the traditional Nyquist sampling theroy in principal. It was demonstated that, with the desirable recovery accuracy and signal to nosie ratio of recovered signal, the burden of the bandwidth of the industrial monitoring networks, e.g., for pipeline leakgage monitoring, can be significantly reduced with the developed FSCS method. Besides, the analysis time ofthe compressed measurement data can be reduced with lower hardware cost. With the compression ratio of 30:1, the parameter value of ERP is less than 1×10^{-3}, which ensures the positioning accuracy of pipleline leakage detection. Similarly, the developed FSCS method can also be applied to other industrial monitoring networks, for example, it can be used to compress power quality sigal or electrical power diagnosis signal for high utilizationof communication bandwidth with low sampling frequency.

ACKNOWLEDGMENTS

This study was supported by: Beijing Higher Education Young Elite Teacher Project,YETP1767.

REFERENCES

[1] Donoho D L. Compressed sensing[J]. IEEE Transactions on Information Theory, 2006, 52(4): 1289–1306.
[2] Tingting Li;, "Super-Resolution using Regularized Orthogonal Matching Pursuit based on compressed sensing theory in the wavelet domain," IEEE International Symposium on Computational Intelligence in Robotics and Automation(CIRA) ,2009,12(15): 234–239.
[3] Justin Romberg. Imaging via compressive sampling[J]. IEEE Signal Processing Magazine,2008,25(2),14–20.
[4] L. Chai, B. Hu and P. Jiang, Distributed state estimation based on quantized observations in a bandwidth constrained sensor network[C], Proc. 7thWorld Congress on Intelligent Control and Automation, 2008,2411–2415.
[5] JIN TAO, QUE PEIWEN, LI LIANG.Research on MFL Inspection Large Capacity and Hi-Fi Data Compression, [J] Journal of basic science and engineering , 2005,13(2):154–161.
[6] Do, T.T.; Lu Gan; Nguyen, N.H.; Tran, T.D.; , "Fast and Efficient Compressive Sensing Using Structurally

Random Matrices," IEEE Transactions on Signal Processing, 2012,60(1), 139–154.

[7] Shoa, A.; Shirani , S. ; "Tree structure search for matching pursuit," IEEE International Conference on Image Processing(ICIP), 2005,3(9),908–911.

[8] Dan Su, Xuewei Wang, HongjieWan ,"TMP compressed sensing in pipeline leak detection,"Journal of Information and Computational Science, 2012,18(9),5731–5740.

[9] WEI MAOAN, JIN SHIJIU, LIYINGYING.Study On wavelet compression technique for magnetic flux leakage image of oil and gas pipeline defects. [J] Opto-electronic engineering, 2004,31(4):57–60.

[10] K E Abhulimen, A A Susu. Modeling Complex Pipeline Network Leak Detection Systems[J].Process Safety and Environmental Protection, 2007,85(6): 579–598.

[11] XU QUANSHENG, LIN SEN. Compression technique of pipeline magnetic leakage signal[J]. Journal of Shenyang university of technology. 2008,30(5):555–558.

[12] YAN JUN,QUE PEIWEN. Implementation of Data Compression in Ultrasonic Pipeline Inspection [J]. Computer Measurement & Control,2007,15(6):717–719.

[13] LIU LINLIN, WANG LIN, WANG XUEWEI. Compression algorithm of pipeline leak detected data based on SPIHT coding [J]. Journal of Liao Ning Technical University (Natural Science Edition), 2011,30 (2):280–283.

[14] TONG YUN, HUANG SONGLING, ZHAO WEI. Oil and gas pipeline EMAT data compression algo-rithm.JT singhua Univ(Sci & Tech) , 2010, 50(10): 1619–1622.

[15] LI LIN,YANG HONG-GENG,ZHAO YAN-FEN. Short duration power quality disturbance signal detection method based on two dimensional discrete static wavelet transform[J]. Power System Technology,2007,31(11): 49–53.

[16] Brechet, L.; Lucas, M.-F.; Doncarli, C.; Farina, D.; "Compression of Biomedical Signals With Mother Wavelet Optimization and Best-Basis Wavelet Packet Selection," IEEE Transactions on Biomedical Engineering, 2007,54(12),2186–2192.

[17] AI QINGSONG;LIU QUAN;YUAN TINGTING; Gestures recognition based on Wavelet and LLE[J]. Australasian Physical and Engineering Sciences in Medicine, 2013, 36(2),167–176.

[18] Flad, Heinz-Jürgen. Hackbusch, Wolfgang; Wavelet approximation of correlated wave functions. I. Basics[J]. Journal of Chemical Physics, 2002,116(22),9641–9657.

[19] Hongya Ge; Kirsteins, I.P.; Xiaoli Wang; "Does canon-ical correlation analysis provide reliable informa-tion on data correlation in array processing?," IEEE International Conference on Acoustics, Speech and Signal Processing(CASSP),2009, 4(19),2113–2116.

Network Security and Communication Engineering – Chan (Ed.)
© 2015 Taylor & Francis Group, London, ISBN: 978-1-138-02821-0

Laser refractive surgery: New materials and methods and optimal market strategies

Vitaliy Kalashnikov & Jose de Jesus Rocha Salazar
UANL, Economy Faculty, Monterrey, Mexico

M.D. Mariya Kalashnikova
D.I.F., Monterrey, Mexico

M.D. Oksana Beda
SumDU, Sumy, Ucraine

ABSTRACT: The market of medical laser surgery is not widely studied and analyzed in Mexico. Still, different kinds of special services are offered by medical institutions with the appropriate capacity and technology. Examples are as follows: plastic surgery, refractive surgery, radiology services, microbiology, oncology, neurosurgery etc. In this research we studied refractive surgery market and the impact of new laser technologies and materials which became available recently, especially in Ophthalmology. According to a study by the INEGI, the visual and motor disabilities are the most frequent among Mexican population. Refractive surgery serves visual disabilities such as nearsightedness, farsightedness and astigmatism; performance of this type of surgery requires specialized equipment which is costly, only a few ophthalmologists can afford it. We try to establish a dynamic economical model to analyze the effects of financial and technical progress.

KEYWORDS: Test theories and applications, control theories and technologies, laser technologies, medical care, medical insurance.

1 INTRODUCTION

This paper will analyze the market of refractive surgery and considervarious factors. First, we will consider the existence of a single service provider (monopoly) and the problem of optimal control in continuous case where the optimum production of refractive surgery will be sought and the production of substitute goods will be considered(control variables). Next, we will introduce a state owned medical service which provides maximization of the social welfare. At last, we will raise a problem of oligopoly in which we consider the presence of n service providers who seek optimal production that maximizes their profits in a period time. We do this in order to prove or disapprove the hypothesis that the price is higher in oligopoly and monopoly, which is subject to the restriction in the flow of replacement demand. And at last, we compare the obtained results with that a public company would be interested in maximizing social welfare. The paper is organized as follows. After introduction we present some theoretical antecedents and frameworks of our models. Next, we work out three mathematical models: Monopoly, Social Welfare and Oligopoly Competition. Conclusions, Acknowledges and Reference List will be the last part of this paper. Due to the volume restrictions, we omit certain too lengthy as well as preliminary proofs, which will appear elsewhere.

2 THEORETICAL FRAMEWORK AND ANTECEDENTS

Examinations of mixed oligopolies, in which social surplus-maximizing public firms compete with profit-maximizing private firms, have become increasingly popular in recent years. For pioneering works on mixed oligopolies, see Merrill and Schneider (1966), Harris and Wiens (1980), and Bös (1986, 1991). Excellent surveys can be found in Vickers and Yarrow (1988), De Fraja and Delbono (1990), Nett (1993). The interest in mixed oligopolies is high because of their importance in economies of Europe (Germany, England and others), Canada and Japan (see Matsushima and Matsumura, 2003) for analysis of "herd behaviour" by private firms in many branches of the economy in Japan). There are examples of mixed oligopolies in United States such as the packaging and overnight-delivery industries. Mixed oligopolies are also common in the East European and former Soviet Union transitional economies, in which competition among public and private firms existed or still exists in many industries such as banking, house loan, airline, telecommunication, natural gas, electric power, hospital, health care, railways and others. These situations have been investigated in different ways. Many works have analyzed Cournot and Stackelberg models with the role of each firm assigned exogenously. However, it is reasonable to assume that each firm decides what actions to take, and when to take them. DeFraja and Delbono (1989) are pioneers in these investigations. They showed that in simultaneous-move games, privatization of the public firm may improve welfare. In 1989 Matsumura showed that under certain conditions, the partial privatization of the public firm improved welfare.

The optimal control theory has developed considerably over recent years. The optimal control theory allows us to solve dynamic problems of varied nature, where the evolution of a system that depends on the timing may be controlled partially by the decisions of an agent (see Kalish, S. (1983) and Mate, K. (1982)).

3 MATHEMATICAL MODEL

3.1 Oligopoly competition

Let us introduce the most important functions in order to build the model:

$R(t) \in C2$ – quantity of persons who may require laser surgery at time t.

$S(t) \in C2$ – quantity of persons who pay for laser surgery at time t.

$P(t)$ – the price for refractive surgery at time t.

The dependence of the $P(t)$ on $R(t)$ and $S(t)$ is typical for microeconomic theory.

In this model we assume that persons who need surgery may also replace it by substituting non-laser technologies. Thus, we have the following relations:

$$P(t) = P[R(t),S(t)] \text{ with } \frac{\partial P}{\partial R} < 0, \frac{\partial P}{\partial S} < 0.$$

Let us consider n competing producers at the laser refractive surgery market. Thus we have total demand $R(t)$ represented by $R(t) = \sum_{i=1}^{n} R_i(t)$.

Every company is choosing its level of production R_i which maximizes its returns in finite horizon, given the production levels of the competitors.

As always, we take into consideration customer flow restriction and initial point for maximization strategy.

$$\underset{\{R_i(t)\}}{Max} \int_0^T e^{-rt}[P(R,S)R_i - CR_i]dt$$

Such that $\dot{S}(t) = \beta S(t) - R(t)$ and $S(0) = So$.

General Solutions will be:

$$R_{OLI} = \frac{n(P-C)\beta}{-\dfrac{\partial P}{\partial S} - \beta \dfrac{\partial P}{\partial R}} \text{ and } S_{OLI} = \frac{n(P-C)}{-\dfrac{\partial P}{\partial S} - \beta \dfrac{\partial P}{\partial R}}.$$

Here we have P= POLI as price of the laser refractive surgery for oligopoly case, ROLI represents the number of operated patients and SOLI represents the number of persons using technologies of substitution.

Now we consider a special case. If we assume the linear relation between G and R, as well as linear dependence of H with respect to S, we may obtain the following relation:

$$P(R,S) = \alpha_0 + \alpha_1 R + \alpha_2 S$$

Solving this, we will get equilibrium relations as follows:

$$P_M = \frac{a_0 + C}{2} ; P_{OLI} = \frac{a_0 + nC}{n+1} ; P_{PS} = C.$$

Due to space restriction of the NSCE2014 we will present mathematical representation for special cases as monopoly and social surplus somewhere else.

4 CONCLUSIONS

Industry Refractive Surgery provides an important need in the population, a need that is considered to be vital for most.

It has been theoretically shown that in dynamic oligopoly competition, which is subject to the restriction of the flow of the substitute, the prices are higher than that charged by a company interested in maximizing the social welfare.

Current refractive surgery industry has an oligopolistic behavior due to the existing barriers to entry.

The prices currently charged in the refractive surgery industry are very high, and this has forced consumers to opt for eyeglasses. It is noted that 43.24% of the population is visually impaired, but only about 0.3 % of the population undergo refractive surgery.

ACKNOWLEDGMENTS

This research activity was supported by the National Council of Science and Technology (CONACYT) of Mexico as part of the project CB-2011-01-169765; PROMEP 103.5/11/4330, and PAICYT 464-10. The authors also acknowledge their profound gratitude to referees, whose valuable comments and suggestions have helped greatly to improve the presentation.

REFERENCES

[1] Bös, D. (1986). Public Enterprise Economics. North-Holland: Amsterdam.

[2] Bös, D. (1986). Privatization: A Theoretical Treatment. Clarendon Press: Oxford.

[3] Mate, K. (1982). "Optimal Advertising Strategies of Competing Firms Marketing New Products," Working Paper, University of Washington.

[4] Kalish, S. (1983). "Monopolist Pricing with Dynamic Demand and Production Cost," Marketing Science, 2 (Spring), 135–159.

[5] De Fraja, G., and Delbono, F. (1990). Game theoretic models of mixed oligopoly. Journal of Economic Surveys, 4, pp. 1–17.

[6] Vickers, J., Yarrow, G. (1998). Privatization-An Economic Analysis. MIT Press: Cambridge.

[7] Matsumura, T. (2003). Stackelberg mixed duopoly with a foreign competitor. Bulletin of Economic Research, 55, pp. 275–287.

[8] [Nett, L. (1993). Mixed oligopoly with homogeneous goods. Annals of Public and Cooperative Economics, 64, pp. 367–393.

[9] Harris, R.G., and Wiens, E.G. (1980). Government enterprise: An instrument for the internal regulation of industry. Canadian Journal of Economics, 13, pp. 125–132.

Network Security and Communication Engineering – Chan (Ed.)
© 2015 Taylor & Francis Group, London, ISBN: 978-1-138-02821-0

Application of integrated network storage architecture in digital campus construction

H.Y. Ge & J.T. Zhang
Xuchang Vocational Technical College, Xuchang, China

ABSTRACT: The common network storage technology is viewed, then FC SAN and IP SAN is introduced; finally a whole application programs of digital campus integrated network storage based on FC SAN and IP SAN technology for the current digital campus construction and development is proposed in this study, to meet challenges and requirements of data security faced by digital campus, with reliable storage and efficient use.

KEYWORDS: Storage technology, network storage, integrated architecture, digital campus.

1 INTRODUCTION

Data of government and enterprises information systems are increased explosively with the rapid development and application of information technology and network technology, large amounts of data are to be integrated, shared, and analyzed, data center is established in many organizations, the hinge is data storage platform of massive, high-speed, safe, and reasonable architecture [1].

2 STORAGE TECHNOLOGY

Storage device is only as a subsidiary device of server in traditional direct attached storage (DAS) technology and directly to server by point to point through IDE, parallel SCSI, SATA and other interfaces. It depends on server, stack of hardware, and without any storage operating system [2].

Network attached storage (NAS) is a technology that data centers of large, centralized management is integrated by distributed and independent data to facilitate access to different hosts and by servers [2]. Generally, NAS has its own node in a LAN, and server intervention is not need. Accessing to data on network is allowed by user, and data on network are managed, centralized and processed by NAS, and load is unloaded from application server.

Storage Area Network (SAN) technology is also emerged as an alternative to DAS technology. With the processing speed of computer becoming faster and faster, the data amount growing bigger, storage subsystems growing, tape libraries, RAID (cheap redundant disk array) and other SCSI devices requiring more and more large space, the problem of SCSI cable length is brought.

In order to meet these needs, fiber channel (FC) protocol came into being that can provide gigabit serial network access for storage devices, a new technology that combined storage applications and network technology: storage area network (SAN) emerged with FC protocol has been gradually accepted by market [2].

Future storage technology will be more diversified, and develop imperceptibly to fusion on this basis, and the current ascendant cloud computing is a product that fused storage and running [4].

3 FC SAN TECHNOLOGY

FC SAN is a network constructed by storage devices and system components [3]. FC SAN not only provides high-performance connection and data backup speed to data devices, but also adds redundant connections to storage system, and provides support for high availability cluster systems.

FC SAN network is often called as FC + LAN dual-network structure, as shown in Figure 1, and is constituted by fiber channel FC and Ethernet LAN networks. LAN Local Area Network (first network) is responsible for data transmission and exchange between client and server; fiber channel network (second network) is only responsible for high-bandwidth transmission of actual data, so characteristics of the two networks are used fully, and maximized efficiently.

4 IP SAN TECHNOLOGY

IP SAN is a storage technology developed on the basis of FC SAN, adopted iSCSI (Internet SCSI) standard of internet engineering task force (IETF), and is used for SCSI data blocks mapped into Ethernet packets [5]. IP protocol defines rules and methods for storing data on level of TCP / IP network sending, receiving block (data block). Initiator (the sender) encapsulates SCSI commands and data into TCP / IP packets and forwards by network, target (the receiver) receives TCP / IP packet and reverts to SCSI commands and data and executed, then the returned SCSI commands and data are encapsulated into TCP / IP packet and re-transmitted to the sender.

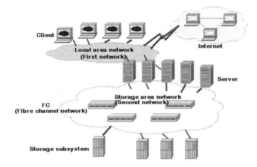

Figure 1. FC SAN dual-network structure figure.

Ethernet switch existed is replaced by expensive fabric switch and only dedicated by FC SAN in IP SAN; client's iSCSI card or Initiator is replaced by host HBA card; and cost-effective storage device with iSCSI interface is replaced by optical disk array. Meanwhile, IP SAN can directly use the existing lines on transmission media, and provides faster scheme than FC SAN.

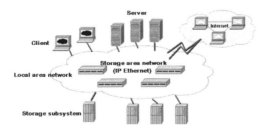

Figure 2. IP SAN network structure Figure.

IP SAN generally adopts single network structure; storage area networks can share the same IP network with business network, as shown in Figure 2.

5 DIGITAL CAMPUS NETWORK STORAGE SOLUTIONS OF INTEGRATED ARCHITECTURE BASED ON FC SAN AND IP SAN

5.1 *Requirements and characteristics of the campus network storage*

At present, information teaching is carried out vigorously in colleges and universities, and more and more variety applications and users are on digital campus network, thus the contradiction between storage, application of large-capacity data and data security, transmission is more and more prominent. Therefore, a good campus network storage system solution is an important content in constructing, developing and planning of digital campus [6]. Taking our college as an example, the campus network data storage requirements include the following aspects.

School Web Site: Including WWW master server, each secondary institution, department, and colleges' website and dedicated website, nearly up to 40, the large data need support of backed services.

Large-scale mail systems: Mail users are divided into teacher and student, and network storage space.

Campus Network Accounting System: Importance and security of user data are very high, needed to backup real-time, saved requirement of logs is high.

Video-on-demand systems, courses, high-quality courses and online courses, etc, amounts data of these applications developed based on multimedia is very large and need vast storage capacity.

Some important software systems (such as asset management systems, educational management system and recruitment and employment systems, etc.) contain a large important data.

Large databases extensive use of SQL SERVER, ORACLE and others. Frequency of reading and writing data is high, concurrency requirements (such as evaluation teaching and selective course system, etc.) are high, and storage requirements are much higher.

Campus card system requires data absolutely accurate and secure; stability, reliability, parallel processing capability and network performance demanding are very high.

Moreover, Library for data extension request of e-books and the future of digital campus construction, and etc.

In short, the network storage access requirements of our colleges basically represent variety needs and characteristics of universities to data storage and application.

5.2 *Storage solutions analysis of existed network and improvement ideas*

Currently, traditional DAS has been replaced gradually by NAS and SAN, SAN has FC SAN and IP SAN.

FC SAN technology is adopted in network storage systems of safe, efficient, high stability, and mainly in the field of finance, telecommunications, avionics systems, and FC SAN is adopted in network storage systems of Fudan University, East China University of Science and Technology and many other universities. However, shortcomings such as high cost, application compatibility with heterogeneous network and storage silos formed easily for coverage are not more than 50 km, especially in remote disaster recovery system.

Lower input costs, low-cost of management and maintenance, no distance limitations, flexible to set up, high scalability of pure IP SAN are superiorities, as well as widely popular in some small and medium agencies. Noise of collisions, low transmission rate, complexity of IP network environment and security of IP SAN are shortcomings.

In order to address the problems: How to integrate advantages of the two solutions, how to solve SAN scheme design of security, stability of data, and larger difference of transmission rate requirements effectively, and how to control cost and other issues, we proposed network storage solutions combining FC SAN and IP SAN on ideas of data of hierarchical storage and application.

5.3 Solution design

Storage architecture combining FC SAN and IP SAN is adopted based on the above analysis, solutions as follows:

1 Main storage parts
   One intelligent switch, two storage servers, backup software, one fiber disk array, two SCSI disk array. In this scenario, intelligent switch is the main equipment that can achieve storage subsystem function. Virtualized storage management services, data backup and copy services, data snapshot services and so on are integrated in the intelligent switch. Function provided is equal to sum function of optical switches and virtual storage management server. It can avoid "bottleneck" effectively generated by using virtual storage management server.

2 Hierarchical Storage
   Campus network billing system, important software management system, campus card and other core applications are connected by FC SAN. Primary storage and applications are achieved in high performance, and high availability.

   The existing storage devices are integrated, application data that "read" access is large and "write" is small, such as fine courses, including those read-only library information, VOD is stored on those equipments (original SCSI disk arrays and IDE disk array). As can be seen from topology, hierarchical

storage approach can fully utilize equipment performance and organize data scientifically and rationally.

3 Asynchronous mirroring
   Offsite storage server connected by intelligent switches can provide local asynchronous mirroring, without adding any other components.

4 Scalable
   Intelligent switch used has FC ports and IP ports, and is very convenient to add network device, insert corresponding port and setup simply.

5 Existing investments protection
   Investment of college is protected in designing campus network. All devices are used rationally and play their role fully from topology.

Network storage topology combined by FC SAN and IP SAN is shown in Figure 3.

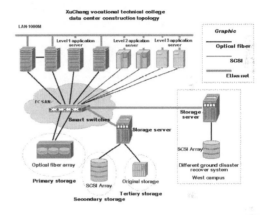

Figure 3. Network storage topology combined by FC SAN and IP SAN.

6 APPLICATION EFFECT

The solution solved storage, data replication, I/O efficiency, network performance and other problems in building centralized data center, and has characteristics of less investment, easy to maintain, high efficiency, safety and reliability, leading technology, and fully use of the existing storage resources and reduction the TCO (Total cost). The project as an important part of a brand demonstration project in Henan and digital campus project demonstration agencies, has a strong reference value to other universities and enterprises of similar projects.

REFERENCES

[1] Changdong, Fu. 2004, (04). Research and development of network-storage architecture. Journal of Chinese Computer Systems:24–26.

[2] Changsheng, Xie. 2003, (02). A storage network system that merges NFS and SAN. Computer Science:89–105.

[3] Chao, Li. 2006, (03). Performance Analysis and Monitoring of FC-SAN System for Massive Digital Resource Management. Tsinghua University:256–271.

[4] Li, Liu. 2011, (06). Strategies of campus network data storage based on NAS and SAN technology. Journal of Fuyang Teachers College(Natural Sciences):78–81.

[5] Yuntang, Wang. 2004. Cluster System Using Fibre Channel as Interconnection Network Analysis. Proceedings of Asia-pacific Optical Communications 2004–5626volume:212–218.

Network Security and Communication Engineering – Chan (Ed.)
© *2015 Taylor & Francis Group, London, ISBN: 978-1-138-02821-0*

Analysis and research on information management of SME based on BPR method

Z.X. Lu, Y. Han & R.M. Wang
Oxbridge College, Kunming University of Science and Technology

ABSTRACT: SMEs are an important part of Chinese economy, the construction of SME informatization is not only the inherent requirement of the development and growth, but also the process of restructuring and reengineering by the use of information technology. The internal demands of SME informatization construction are to be bigger and stronger. Business process of reengineering is important for SEMs developing and managing. In this paper, analysis and research are done on the process of restructuring and reengineering of SMEs informatization, and then the principle and recommendation of informatization construction of SMEs were put forward.

1 THE CONCEPT OF BUSINESS PROCESS REENGINEERING

BPR (Business Process Reengineering) is the fundamental thinking and thorough reconstruction for enterprise Business Process. Companies can be significantly improved through business process reengineering in terms of cost, quality, service and speed etc, and make its utmost to adapt to changes in customer, competition and the characteristics of modern enterprise management environment. In fact, many enterprises have realized the promotion enterprise's management through the implementation of BPR; BPR plays an important part in improving the core competitiveness in the domestic and foreign well-known enterprises. BPR has become a revolution of business management.

BPR is the process of reengineering; its main technology is simplification and optimization process. The main idea of simplification BPR is: the process to simplify the process of strategic streamlined dispersed; to correct the wrong process function; to remove the redundant execution.

2 THE NECESSITY OF BPR IMPLEMENTATION OF CHINA'S SMES IN THE ENTERPRISE INFORMATION

2.1 *The background information of software design*

Large companies can co-operate with developed software company to match the business needs of the information process, but SMEs due to limited funds, tend to buy ready-made direct management information system software. As the current management of SMEs, the design requirements of existing business processes for small and medium corresponding transformation and restructuring. And advanced information technology software should be established.

2.2 *The purpose of the SME informatization construction*

The purpose of SMEs through information technology is to achieve business-related project management, business process automation and information technology, and strengthen enterprise management. The ultimate goal of SME information for businesses is to construct and put it into practice. Then continuously optimize their business processes, so that the whole business activities tend can get improved and the competitiveness and efficiency of enterprises can be enhanced.

2.3 *SME information system*

SME information construction is the transition from single applications to integrated applications. In the 2012 SME information survey in China, 50% of enterprises in the single application stage, 45% in the partially integrated application stage, 5% in the fully integrated application stage. SME information is not purely a technology introduced in the application, but rather to achieve its management system to integrate systems engineering.

3 CONTENTS OF THE SME INFORMATIZATION PROCESS REENGINEERING

SME information construction should focus on business objectives with a clear corporate strategic positioning and meeting customers' demand, and management of business processes should achieve reintegration and reconstruction by enterprise information. Process reengineering includes process changes, restructuring weaknesses, human resource restructuring to new corporate culture instead of the old.

3.1 Restructuring process

Because of the short-term development of SMEs, and week management foundation, most enterprises have not established a comprehensive enterprise management system and business process system. The restructuring of the process of information plays an important role in promoting corporate restructuring management standardization and re-sort the information needs of the company. In order to establish an efficient information system, a reasonable process is necessary. Process reengineering is the basic condition for SME informatization construction.

3.2 The transformation of the weak links

As bucket theory in management, management and development of SMEs also depends on its "short board". Enterprises will be eliminated in the competition as the company's weaknesses, which is particularly important for the competition. Companies should be targeted to find out its weak links in production and management processes, and then make adjustments and improvements. At last the company should combine project construction with rebuilding of the weak link.

3.3 The integration of human resources

Integration of Human resource and innovation of organizational structure are the basis for SMEs to achieve process optimization, but also the carrier to achieve strategic objectives and structure of the core competitiveness. In traditional organizational structure, departments and hierarchical division of the enterprise formed a "pyramid" type of organization hierarchy and commanded control system. The responsibility was not clear, the information transfer was not smooth, and the working efficiency was not high. SME information construction process should be re-organized, which is subject to the company's overall strategic goals. The boundaries of existing departments should be broken, and job

responsibilities re-defined, in order to flatten the organization instead of the pyramid organization, to achieve the integration of human resources.

3.4 Reengineering of the corporate culture

The process of SME information construction reengineering is not a simple process of reorganization, but a system of engineering IT-based applications, it's also to create a new corporate culture. Corporate culture based on operational business processes are likely to deviate from corporate strategic direction without guide. Lack of corporate culture support, the combat effectiveness of the team is difficult to continue. SME information construction process reengineering should be the corporate culture as a management tool for process reengineering, and the existing concept of corporate culture in business management processes to implement the whole process. By reshaping the core values of corporate culture, enhancing the cohesion and influence of enterprises, and strengthening coordination of forces external to the organization, we can enable enterprises to successfully implement BPR, and realize truly sustainable development.

4 THE PRINCIPLES OF THE SME INFORMATIZATION PROCESS REENGINEERING

4.1 Based on business objectives, focus on the overall flow of the best

SME information is information system implemented according to systems thinking, focusing on the global optimum rather than a process or part of the best. SEM failed to highlight the overall business strategy and goals, and work is divided into a number of tasks prior to the construction of SME information. Process design information after the organizational structure should be recycled, and with job standards, record production, quality and other aspects of running the system for effective integration of the report, and then structure flat, smooth communication. The recording is complete, the process clear, management efficient, business objectives achieved.

4.2 The process has good maneuverability

SME informatization process reengineering should design, plan, and highlight the management and business process reengineering in accordance with the principles of software technology suitable for the process, and not only considering software technology itself. The effectiveness of certain SME information was not significant, the crux lies precisely in process design operability. SME information construction must adhere to management and business process optimization, business operations to achieve cost reduction and management efficiencies.

4.3 Give full play to the role of each employee in the whole business process

Employees of SMEs is implementation of enterprise information ultimate power lies, without the full participation of all employees, any changes are unlikely to succeed. SMEs in the recycling process of implementation of information flow, whether it is training and reserve in advance, or something in support of participation, as well as improvement and feedback afterwards, the participation of all employees to ensure the smooth implementation of information technology has played a significant role.

4.4 Focusing on systems fully integrated application

The current stage of development of SME information is transited from single application to integration applications, and they are not yet fully integrated into the application stage. SME informatization construction mode is usually called "stop-gap", beginning from a certain computer information systems, each department is only required for the selection of software products in the field, and did not conduct a comprehensive information system between system integration. If SME do not carry out comprehensive system integration, it will inevitably lead to a different sector enterprises and information redundancy and waste. In the planning stage of Information Systems Process Reengineering should establish standard information resources to ensure data consistency, to achieve effectiveness and efficiency of the process. Information technology systems integration should be implemented and fully integrated as early as possible.

4.5 Establish control procedures to make decisions in a business process

Before entering the fully integrated application stage, SME informatization constructions performer, monitoring and decision-makers are strictly separated. And there's no time or the obligation to monitor the processes and decisions. Information construction process should be abandoned pyramid management structure, the magnitude of the compression levels of management and administration; with a flat

organizational structure constructed rapid response capability level employees can achieve their own decisions and control. Practice has proved that once employees become self-managed and self-makers, organizational structure, and its low efficiency and bureaucracy accompanied by a pyramid will disappear.

4.6 Put the information-processing work into the real work to produce this information

Due to the small size and limited personnel of SMEs, information process of reengineering work should be integrated into a real work. The employee is responsible for collecting the information and no longer need another group of people to coordinate and process information, so it cuts the amount of information from external contact and reduce the error rate of processing information, thereby it improves the management and production efficiency.

5 DEADLINE

Business Process Reengineering is the indispensable step for SME informatization. It is necessary to carry out business process reengineering to achieve the desired results.

REFERENCES

[1] Jaeger p, lin t,grimes j. Cloud computing and information policy: computing in a policy cloud [d]. 2008(5):76.
[2] Zhangz, matthew kol,peih,et al. a framework of erp systems implementation success in china:an empirical study [j]. Int. J. Production economics 98. 2005(11):52–53,56–80, 122–123.
[3] Danma, abraham seidmann. The prieing strategy analysis for the oftware-as-a-serviee, business model [j] .grid eeonomiesand business models, 2008(8):103–112s.
[4] Gian paolo carraro, fred chong. Software as a serviee(saas): anenterprise pers peetive [eb/ol], Http://msdnz.mierosoft.eom/enus/arehiteeture/aa905332.as p.2009-10-05.
[5] Sal inesi, c,bouzid,r. And elfassy,e. An experience of reuse based requirements engineering in erp implementat ion projects[c].l 1th ieee internat ional enterprise distributed object computing conference,2007: 379.
[6] Aeger p, lin t,grimes j. Cloud comptıng and information policy: computing in a policy cloud [d]. 2008(5):76.

Network Security and Communication Engineering – Chan (Ed.)
© *2015 Taylor & Francis Group, London, ISBN: 978-1-138-02821-0*

FPGA-based Kalman filtering for motor control

A.M. Romanov & B.V. Slaschov
Moscow State Technical University of Radio Engineering, Electronics and Automation, Russian Federation, Moscow

ABSTRACT: The article presents a new FPGA-based coprocessor with original architecture, that directly operates with matrix variables and is designed specially for Kalman filtering in motor control applications. Simulation and implementation results showed that the new coprocessor achieves higher computing performance per one clock period, having same or lesser FPGA resource utilization than analogues.

KEYWORDS: DSP, FPGA, Kalman filter, Matrix operations, Motor control.

1 INTRODUCTION

Kalman filter is one of the most powerful engineering instruments for variables estimation. Developed in 1960[1], this algorithm is widely used in different areas of technology. Nowadays there are many variations of this filter. All of them require calculation of different matrix operations and have quite high demands for computation resources. Since most DSP operate only on scalar (not matrix) variables, real-time performance, that is strongly needed in motor control applications, is difficult to achieve with cost-effective chips. FPGA-based Kalman filter implementations that use soft-processors have the same problem. Like DSP they do not use all of FPGA advantages such as parallel mathematical and memory access operations. This article presents a new FPGA-based coprocessor, that directly operates with matrix variables and is designed specially for Kalman filtering in motor control applications.

2 REQUIREMENTS FOR MATRIX COPROCESSOR

Analysis of Kalman filter usage in motor control algorithms shows that they operate on matrices with size 5x5 or less. Usually they are based on one of three most popular Kalman filter modifications: linear (KF), extended (EKF) and unscented (UKF) Kalman filters [2,3,4]. This allows to define a set of matrix operations, required for Kalman filtering in motor control: matrix addition, scalar multiplication (matrix by number), matrix by matrix multiplication, matrix inversion, Cholesky decomposition, matrix transposition, forming diagonal and triangular matrices.

There are lots of FPGA intellectual property (IP) cores for matrix processing, but most of them are developed to operate with matrices of high and very high degree. As a result they use huge amount of resources[5,6,7] and require expensive FPGA, that makes such solutions not competitive on the market. Also those IP cores usually implement only one matrix operation, for example matrix inversion or matrix multiplication. The only found matrix DSP that performs a full set of operations required for a Kalman filter is the uMP matrix processor introduced by a Canadian company Ukalta. Unfortunately, its development was not completed, and Ukalta stopped its existence. This is why it was decided to develop a new special matrix coprocessor for Kalman filtering in motor control applications.

In modern highly integrated motion control systems the central processing unit based on FPGA or DSP is often placed near power transistors and motor. This means it should be able to work in relatively high temperatures, that cause high timing delays inside chip. As a result it is difficult to achieve central processing unit operation on frequencies higher than 100 Mhz.

Yet, input signals of a Kalman filter in motion control usually come from an analog-to-digital converter with a sample rate of 100 kHz or less. That is why operating filter with higher sample rate is not necessary.

The goal of this project was to develop a matrix soft-coprocessor, that would be able to compute a 5th degree Kalman filter with sampling rate up to 100 kHz on a low-cost FPGA clocked with frequency that is not higher than 100 Mhz.

To reduce FPGA resource usage and make the coprocessor more task-oriented it was decided to implement it with a modular architecture, which allows to synthesize into FPGA only those matrix and scalar operations, that are required in the specific solution.

3 COPROCESSOR ARCHITECTURE

A structural diagram of the new matrix coprocessor architecture is shown in Fig. 1. For storing matrices the coprocessor uses special matrix memory that consists of two identical modules of two-port memory and a memory commutator. Matrix memory has 4 independent access lines. Generally, lines A and B are used to access memory module AB, and C,D – to access module CD. Operands of matrix operation are obtained throught access lines A,B, and the results are written using access lines C,D. This allows to effectively use all memory access lines during calculations and obtain maximum performance. The memory commutator is used to reconnect C and D lines to module AB, and A,B lines – to CD. Thus, the results of the previous operation can be used as operands for the next operation without coping them to the other memory module.

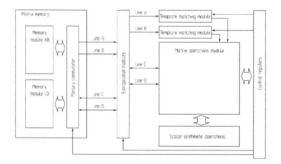

Figure 1. Matix coprocessor architecture.

Before reaching mathematical operations the lines A,B,C,D of matrix memory pass through transposition modules, which can switch indices of matrix elements in the requested memory module. This allows processing matrix transposition in parallel with any other operation, increasing calculation time by only one FPGA clock period, regardless of the matrix size.

The same approach is used for forming unity, zero, diagonal and triangle matrices. Lines A, B pass through a template matching module, which can change input matrix elements' values from the memory module to a fixed constant, depending on accessed matrix element indices and predefined template. Currently there are implemented templates for unity, zero, diagonal, upper and lower triangle matrices.

The core of the matrix coprocessor is a matrix mathematical operations module, which consists of separate operations modules, which are parallely connected to memory access lines with logical OR. Depending on selected configuration the coprocessor may include the following operations :

1 matrix addition;
2 multiplication of matrix by number ;
3 linear combination of two matrices;
4 element by element matrix multiplication;
5 square root of each element of matrix;
6 matrix by matrix multiplication;
7 elementwise matrix inversion;
8 inversion of custom size square matrix with calculation of the matrix determinant;
9 fast 2×2 matrix inversion with calculation matrix determinant;
10 Cholesky decomposition.

Each of these operations can be executed both on a complete matrix from the memory and on a part of it. For example it is possible to take the first row of one matrix variable and multiply it by the second column of another variable from matrix memory. The main advantage is that sub-matrix extraction operation don't even require additional computing resources.

In compliance with the modular architecture, the matrix operations do not include any scalar operations (number by number arithmetic operations) and use them as external IP cores. The full set of scalar operations required for the maximal matrix coprocessor configuration consists of 2 pipelined adders, 2 pipelined multipliers, 1 division module and 1 square root module.

Separation of matrix and scalar operations has two advantages. First, it makes it possible to use arithmetic IP cores optimized for specific FPGA family. Second, it makes matrix operations number format independent. For example, originally the coprocessor was designed to operate only with fixed-point variables, but it is possible to adapt it for floating point variables just by changing fixed-point arithmetic operations IP cores to floating-point ones.

Downloading/uploading data into matrix memory and operations choice is carried out through control registers.

4 SIMULATION AND IMPLEMENTATION RESULTS

In its maximum configuration the coprocessor with matrix memory size of 6 matrices 4×4×32 bit requires 3900 LUT, 2 block RAM modules and two hardware multipliers 18×18 for implementation on FPGA Spartan 3E. This can be compared with systems, based on 32-bit soft-processors MicroBlaze, aeMB, Mico32. Unlike these solutions the developed coprocessor is fully scalable by changing matrix memory capacity, matrix elements size and the set of matrix operations. Also matrix operations included in the developed coprocessor are fully pipelined and use four memory access lines in parallel. As a result the developed coprocessor has higher computing performance per one FPGA clock period with same or lesser FPGA resource usage.

Using the developed coprocessor, a flux linkage estimator for an asynchronous motor, based on an extended Kalman filter, was implemented. To simplify comparison with existing solutions filter equations were taken the same as in [8]: the process model was chosen as a non-linear system of 5th degree with two observed states.

In [8] the Kalman filter is implemented on a 32-bit processor Freescale MPC555 with frequency 40MHz achieved a sample rate of 1kHz. The same Kalman filter is implemented on the developed coprocessor and FPGA Xilinx Spartan 6 with frequency of 100MHz achieved a sample rate of 100 kHz. Thus performance was increased up to 40 times.

5 CONCLUSION

A new FPGA-based soft-coprocessor, that directly operates with matrix variables and is designed specially for Kalman filtering in motor control applications was presented. It has an original modular architecture that allows achieving higher computing performance per one FPGA clock period, having same or lesser FPGA resource utilization than analogues. Experiments showed that new coprocessor is able to increase performance in the non-linear Kalman filtering task up to 40 times compared to a known DSP-based solution.

REFERENCES

[1] Kalman R. E. A New Approach to Linear Filtering and Prediction Problems // Journal of Basic Engineering 82 (1), 1960: pp. 35–45.

[2] Simon D. Kalman Filtering // Embedded Systems Programming, vol. 14, no. 6, pp. 72–79, June 2001.

[3] Simon D. Using Nonlinear Kalman Filtering to Estimate Signals // Embedded Systems Design, vol. 19, no. 7, pp. 38–53, July 2006.

[4] Julier S.J., Ulhmann J.K. A new method for nonlinear transformation of means and covariance's in filters and estimators // IEEE Transactions on Automatic Control. – 2000. – Vol. 45. – P. 472–478.

[5] Karkooti, M.; Cavallaro, J.R.; Dick, C., FPGA Implementation of Matrix Inversion Using QRD-RLS Algorithm // Signals, Systems and Computers, 2005. Conference Record of the Thirty-Ninth Asilomar Conference on , vol., no., pp.1625–1629, October 28 - November 1, 2005.

[6] Nirav Dave, Kermin Fleming, Myron King, Michael Pellauer, Muralidaran Vijayaraghavan Hardware Acceleration of Matrix Multiplication on a Xilinx FPGA // Proceedings of the 5th IEEE/ACM International Conference on Formal Methods and Models for Codesign (MEMOCODE '07), 2007, pp. 97–100.

[7] Ganchosov P.N., Kuzmanov G.K., Kabakchiev H., Behar V., Romansky R.P., Gaydadjiev G.N. FPGA Implementation of Modified Gram-Schmidt QR-Decomposition // Proc. 3rd HiPEAC Workshop on Reconfigurable Computing, Paphos, Cyprus, Jan 2009, pp. 41–51.

[8] Extended Kalman Filter - Flux and Speed Observer — Information on www.mathworks.in/matlabcentral/fx_files/10439/1/content/mpc555_mtrctrl/help/EKF/EKF.pdf

Network Security and Communication Engineering – Chan (Ed.)
© *2015 Taylor & Francis Group, London, ISBN: 978-1-138-02821-0*

Using narrow-band communications technologies, increase of radio resource utilization

S.Q. Liu, Y. G. Li, C.X. Yang & J.D. Zhou
Room 2003, Fuhaiguojigang Building, No.17 Daliushu Road, Haidian District, Beijing China

X.Z. Hou & H.L. Sun
No. 80 of The Middle Section of Mount Huangshan Road, Yubei District, Chongqing City

ABSTRACT: With the expanding volume of rapid information transmission, frequency resources become more and more valuable. How to make the best use of frequency band, to be as much as possible to increase the amount of available channels , and to increase traffic capacity , have become a focus of global wireless communications research. Communication need, is one aspect of speed and capacity. The greater the bandwidth is, the higher the rate of communication is and accordingly, the communication capacity is lower on the premise to ensure the communication speed to minimize bandwidth, into the direction of development of communication technology. At the international level, narrow-band communication technology is increasingly used, This is an example for 470MHz wireless communication devices: This band of wireless devices, primarily transceiver and radios after the 25kHz/12.5kHz extension which has been used for decades, in recent years in the implementation of the 6.25 kHz communication technology, has achieved good results. This paper describes a "narrow-band carrier" communication technology in order to fully utilize the limited radio resources and significantly reduce the possibility of interference.

KEYWORDS: Narrowband communication, Bandwidth, Channel capacity, Anti-interference, Interference probability.

1 INTRODUCTION

In the information society, in order to occupy an important position in the field of communications wireless communications, radio resources are finite, and overlapping frequency signal will make communication parties interfere with each other, resulting in communications which are carried out properly, In this case, parties can only send the signal over the spaces so that the parties can at a certain signal "fit" the aim that signals will be delivered successfully and in this way, also causes a flood of signals in the air, seriously interfere with proper communication equipment.

In the communication, if the bandwidth is greater and speed can be greater; most communication applications have fixed speed, Bandwidth is not the higher the better, instead, the greater the bandwidth is, the higher probability to be interfered, and anti-interference means more needs.

At present, Radio Management Committee of the people's Republic of China approval metering wireless communication band is 470-510MHz, the communication bandwidth is 200KHz; the State Grid

Corporation of China interoperability standards for 471-486MHz. According to the regulations of radio frequency that is made by the people's Republic of China, other users in this area are radio station, transceiver and so on. The normal communication of the intelligent electric meter will be interfered when these equipment are used.

In terms of wireless communication in smart metering, what we urgently need is to find a communicating technology, which not only fix the communication rate, but also can be more anti-interference.

2 THE MAIN SOURCE OF INTERFERENCE OF INTELLIGENT ELECTRIC METER COMMUNICATION FREQUENCY

The main source of interference is transceiver and radio station, the power is from 0.5W to 30W. The channel bandwidth of the transceiver, according to the different brands, is 25/12.5 KHz and 25/12.5/6.25KHz. And for the radio station, the bandwidth is 25/20/12.5KHz.

The device such as the transceiver and the radio station, the power they cost is 10 times or even 100 times more than the micro power wireless communication module, and if the channel frequency is in the channel of the micro-power wireless communication module, the module cannot communicate.

3 THE SOLUTION FOR ANTI-INTERFERENCE OF INTELLIGENT ELECTRIC METER

As we all know, We use different frequencies to prevent interference in a wireless communication device. If the low-power device is in communication, there is a signal in a communications channel beyond a certain value(Signal strength > communications devices receive signal - 20dBm), then the communications will be disturbed. The data for smart meter is "character", and the error data is difficult to restore. —So, if there is a interference signal above a certain strength in the communication channels, the communications will be disturbed and unable to communicate properly. The signal strength of interference sources such as the transceiver and the radio station is far more than a smart meter communication module, and communication module is unable to communicate within a certain distance(This distance will be in 1.5~15km), In other words, a higher power radio station which launches in 10km, may affect the module's work.

To solve the problem of anti-interference of intelligent electric meter, we come up with the following way.

Module will "listen" if there are signals with the same frequency before communication, if there is no one, it will wait for a random time, and once again, if there is no signals with the same frequency, communication will be started.

Repeated listening, if there are still interference signals with the same frequency, module will be "hopping" operation, and another channel group is entered into, with the purpose of trying communication.

As we know, there is working frequency, modulation frequency deviation and "bandwidth" for the frequency parameter of wireless communication technology. In figure 1,the smaller of the bandwidth is, the more number of working channels there is, in the same working frequency range, If the signal frequency into the bandwidth, interference is possible. Small bandwidth can reduce the interference probability.

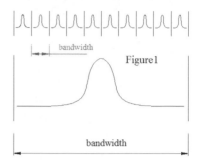

Figure 1.

4 FEASIBILITY FOR THE NARROWBAND COMMUNICATION TECHNOLOGY

The range of communication bandwidth decides the channel capacity, and is the biggest amount of the channel signal that can be transmitted in a unit of time, which is the transmission capacity of channel.

According to the Nyquist stability criterion, if the interval is

$$\pi / \omega \ (\omega = 2\pi f) \tag{1}$$

the narrowband signal transmits through by the ideal communication channel, and the interference will not be generated between the front and rear symbol. Therefore, for the maximum data of binary data signal:

The relationship between Rmax and B communication channel bandwidth(Unit:Hz) is:

$$Rmax=2B \ (bit/s) \tag{2}$$

For binary data, if the channel bandwidth B= f=20,000Hz, the maximum rate of data transmission is 40,000bps.

That is to say that, when the channel bandwidth B is 20kHz, the maximum transfer rate without interference of the channel is 2B, and the highest system efficiency (transmission rate within per unit band) is 40kHz.

If the rate of sender transmission side is twice times more than the bandwidth of the baseband channel, the code will be interfered

According to the rule for the State Grid Corporation of China, the transmission rate of communication module is 10kbps, and to calculate by Nyquist criterion, the communication bandwidth must be more than 5kHz. Considering of the channel power distribution, in order to ensure the communication rate, the optimal bandwidth is set to 20kHz.

In narrowband communication, we need to consider the energy distribution, and other parameter. And, it will be harder for narrowband to design the hardware than the broadband because of the energy distribution. In order to guarantee the quality of communication, more than 99% power needs to fall into the channel which means that the adjacent channel power needs to be less than -60dB.

By the Shannon's theorem: when the channel is with random thermal noise transport signal, the relationship between the data transmission rate Rmax, channel bandwidth B and Sinad S/N is:

$$Rmax=B*log2(1+S/N) \qquad (3)$$

On the basis of this relationship, Bandwidth (B) of the optimal value is 20kHz, Sinad is 12(general requirements for communication equipment designed), and there goes to: Rmax=B*log2(1+S/N) =20000*log2 (1+12) =20000*3.7004=74kbps. That is to say that, restricted by hardware, and when the Sinad is 12, the communication speed can be achieved at 74kbps, which can meet the requirements of 10kbps completely.

Now, there is the chip enough to support narrowband communication technology for sale, such as CC1020, CC1121from TI etc. And there is the corresponding radio-frequency circuit. The hardware design, especially the design of filter is not difficult.

5 THE MATHEMATICAL MODEL FOR INTERFERENCE PROBABILITY

To calculate the probability of interference, the method is:

A: The number of channels to determine the number of users, different circumstances, a combination of user using channel number;

B: In the above case, a combinational number of user using a certain channel;

C: The above combinational number, the numerical value of B/A, which is the probability of user using a certain channel;

D: It will be interfered with each other when two users use a same channel at the same time; the square of C is the probability of interference;

E: Frequency hopping to another channel, that has a probability of user use, is the same value as C. That is to say: third-party C is the probability that frequency hopping still interferes.

The distribution of user and the channel are as following:

A number of combinations of all users are concentrated in the same channel as it is shown in Table 1

Table 1. Figures in the table stand for the number of users.

CH1	CH2	CH3	CH4	...	CH k-1	CH k
m				...		
	m			...		
		m		...		
			m	...		
...	...	...	...	...	...	...
					m	
				...		m

The combinational number in this situation is

$$C (K, 1). \qquad (4)$$

This is the combinational number that all of the users are distributed in two channels (Table 2)

Table 2. Figures in the table stand for the number of users.

CH 1	CH 2	CH 3	CH 4	...	CH k-1	CH k
m-1	1			...		
m-1		1		...		
m-1			1	...		
...	...	...	...	...	...	...
m-1					1	
m-1						1
m-2	2					
m-2		2				
...	...	...	...	...	...	...
m-2					2	
m-2						2
...	...	...	...	...	...	...
1	m-1			...		
...	...	...	...	...	...	...
1				...		m-1
	m-1	1		...		
...	...	...	...	...	...	...
					1	m-1

The combinational number in this situation is:

$$C (K, 2)*C(m-1,1) \qquad (5)$$

This is the combinational number that all of the users are distributed in three channels (Table 3)

The combinational number in this situation is:

$$C(k,3)*C(m-1,2) \qquad (6)$$

Table 3. Figures in the table stand for the number of users.

CH 1	CH 2	CH 3	CH 4	...	CH k-1	CH k
m-2	1	1		...		
m-2	1		1			
...	...	...	...	...	...	...
m-3	2	1				
m-3	2		1			
...	...	...	...	...	...	...
m-n	n-1	1				
...	...	...	...	...	...	...
			1		1	m-2

There are two kinds of situations that each user has its own channel, when m<k, the combination goes as it is shown in Table 4.

Table 4. Figures in the table stand for the number of users.

CH 1	CH 2	CH 3	...	CH m	CH m+1	...	CH k
1	1	1	1	1		...	
1	1	1	1		1	...	
...	...	...	...	...	...	...	...
			1	1	1	...	1

The combinational number in this situation is:

$$C(k,m)*C(m-1,m-1) \tag{7}$$

Users distributed across all channels, scilicet when m>k, the combination goes as it is shown in Table 5. The combinational number in this situation is:

$$C(k,k)*C(m-1,k-1) \tag{8}$$

Table 5. Figures in the table stand for the number of users.

CH 1	CH 2	CH 3	CH 4	...	CH k-1	CH k
m-(k+1)	1	1	1	...	1	1
m-(k+2)	2	1	1	...	1	1
m-(k+2)	1	2	1		1	1
...	...	...	...	...	...	...
1	1	1	1	...	1	m-(k+1)

Based on the two kinds of situation above, when users are distributed in every channel, the combinational number is

$$C(k,n)*C(m-1,n) \tag{9}$$

among that, n is the smaller than anyone between m and k-1.

The statistics of distribution over all users, the combinational number of all distribution is:

$$Zm=C(k,1)+C(k,2)*C(m-1,1)+C(k,3)* C(m-1,2)+...+C(k,n)*C(m-1,n). \tag{10}$$

among that, n is the smaller than anyone between m and k-1.

The distribution that user falls into certain signal channel is (take CH 1 as example):

When all of the users are distributed in one same channels, the combinational number of channel 1 is (Table 6):

Table 6. Figures in the table stand for the number of users.

CH 1	CH 2	CH 3	CH 4	...	CH k-1	CH k
m				...		

In this situation, the situation that users having communication within some channel is unique, just like Ch1.

When all of the users are distributed in channel 2, the combinational number of users that channel 1 has is (Table 7):

Table 7. Figures in the table stand for the number of users.

CH 1	CH 2	CH 3	CH 4	...	CH k-1	CH k
m-2	1			...		
m-2		1		...		
...	...			...	...	...
m-2						1
...				...	...	...
m-3	2					
m-3		2				
...	...			...	...	...
m-n					n-1	
...	...	...	...	...	...	...
1						m-2

In this situation, the combinational number of user communication that channel 1 has is:

$$C(k-1,1)*C(m-1,1) \tag{11}$$

There are two kinds of situations that each user has its own channel, one of it: m<k, the combinational number of users that channel 1 has is (Table 8)

560

Table 8. Figures in the table stand for the number of users.

CH 1	CH 2	CH 3	...	CH m	CH m+1	...	CH k
1	1	1	1	1		...	
1	1	1	1		1	...	
...	...	...	...	...	...	...	...
1				1	1	...	1

The combinational number in this situation is:

$$C(k-1,m-1)*C(m-1,m-1). \tag{12}$$

Users distributed across all channels, scilicet m>k, the combinational number of users that channel 1 has is (Table 9):

Table 9.

	CH 1	CH 2	CH 3	CH 4	...	CH k-1	CH k
m-(k+1)	1	1	1	...	1	1	
m-(k+2)	2	1	1	...	1	1	
m-(k+2)	1	2	1	...	1	1	
...	...	...	...	...	...	...	
1	1	1	1	...	1	m-(k+1)	

The combinational number in this situation is:

$$C(k-1,k-1)*C(m-1,k-1). \tag{13}$$

The statistics of distribution over all users, the combinational number of user in some channel is:

$$Y_m = 1+C(k-1,1)*C(m-1,1)+C(k-1,2)*C(m-1,2)$$
$$+...+C(k-1,n)*C(m-1,n). \tag{14}$$

Among that, n is the smaller than anyone between m-1 and k-1.

Therefore, the probability of communicating in some channel is:

$$X_m = Y_m/Z_m. \tag{15}$$

The probability that two channels can be the same which (interfered by each other) is:

$$A = Xm^2. \tag{16}$$

The probability that after a frequency hopping, channels will be interfered by each other is as follows:

$$AJ = Xm^3 \tag{17}$$

(In this case the source of interference is another communication apparatus in general)

6 TO CALCULATE THE PROBABILITY OF TECHNOLOGY INTERFERENCE WHEN BROADBAND AND NARROW BAND GET COMMUNICATION

The frequency range of the smart meter communication module is allocated for 471-486MHz now,.and total 15MHz, and 200 kHz for communication bandwidth, 75 channels can be used.

If the communication bandwidth is reduced to 20 kHz, the number of communication channel that can be used will increase to 750. The possibility of mutual interference will be reduced. Calculated by the probability theory, the probability will be much smaller when 750 channels are interfered by the same external frequency single than that 75 channels are interfered by each other, the probability of interference can be calculated as follows:

When the bandwidth is 200 kHz, k=75; when the bandwidth is 20 kHz, k=750

The value in the following table is calculated by the probability of interference (Table 10):

Table 10.

	Wide band 200kHz(k=75)			Narrow band 20kHz(k=750)		
The number of users (m)	Probability of the channel	interference probability Without hopping	interference probability after hopping	Probability of the channel	interference probability Without hopping	interference probability after hopping
2	1.33%	0.018%	0.00024%	0.13%	0.00018%	0.0000%
5	6.33%	0.40%	0.025%	0.66%	0.0044%	0.0000%
10	11.90%	1.42%	0.17%	1.32%	0.017%	0.0002%
20	21.28%	4.53%	0.96%	2.60%	0.068%	0.0018%
50	40.32%	16.26%	6.56%	6.26%	0.39%	0.025%
100	57.47%	33.03%	18.98%	11.78%	1.39%	0.16%
200	72.99%	53.28%	38.89%	21.07%	4.44%	0.94%
300	80.21%	64.34%	51.61%	28.60%	8.17%	2.34%

It can be seen from the values calculated above that when the user number adds up to 50, the contrast of interference probability is as follows (Table 11):

Table 11.

	Probability of the channel	interference probability Without hopping	interference probability after hopping
Bandwidth 200kHz	40.32%	16.26%	6.56%
Bandwidth 20kHz	6.26%	0.39%	0.16%

We can see, that when a plurality of different users use a communication equipment with same frequency band at the same time, without the frequency hopping, the interference probability for bandwidth 20kHz will be ten times smaller than the bandwidth 200kHz, even smaller than the value when broadband use the frequency hopping.

7 CONCLUSION

For the communication of the smart meter, narrowband with 20 kHz can completely satisfy the communication requirements, and get more available channels.

The above mathematical model analysis shows, that when we use 20 kHz to communicate, frequency hopping is not required, and the probability of interference will be much smaller than 200kHz.

The design of hardware has been a mature solution, and there is not much difference between the cost and broadband.

Advice:

At present, required by the communication parameters of smart meter communications, "narrowband communications technology" can completely achieved. Anti interference is much better than the current use of wideband communications, technology implementation is no longer difficult, and the cost is controlled.

We consider communication as anti jamming effect, channel capacity, the direction of technical development, cost etc. We recommend that smart metering communication terminal use the "narrow-band communication technology", to achieve better communication effect, and ensure that the simultaneous use of conditions in the more users can be normal communication.

REFERENCES

[1] Q/GDW 1374.3–2013 《Power user electric energy data acquire system technical specification Part 3:communication unit》.
[2] ZigBee wireless standard (ISBN:953-7044-03-3).
[3] Response & Stability; Cambridge University Press (ISBN 0-521-31994-3).
[4] R. J. V. dos Santos, Generalization of Shannon's theorem for Tsallis entropy, J. Math. Phys. 38, 4104 (1997).

Network Security and Communication Engineering – Chan (Ed.)
© 2015 Taylor & Francis Group, London, ISBN: 978-1-138-02821-0

Non-uniformity spatial sensitivity of an electrostatic sensor

M. Abdalla, J. Zhang, R. Cheng & D. Xu
Teesside University, Middlesbrough, UK

ABSTRACT: The solids flow rate and concentration measurements of gas–solids flow often have to face non-even distribution of solids in gas. The uniform spatial sensitivity of a sensor is essentially and ideally required for such applications. However, there are not many meters that inherently have such property. The circular electrostatic meter is a popular choice for gas–solids flow measurement with its high sensitivity to flow, robust construction, low cost for installation and non-intrusive nature. However it does not have a uniform sensitivity. This paper analyses the spatial sensitivity of circular electrostatic meter in general with experimental results to support. The paper not only studies the spatial sensitivity in term of response to a single charged particle, but also the spatial sensitivity of the sensor to a roping flow stream, which is practically more useful. The results from both experiments and simulation reveal that the spatial sensitivity is more complex than it is believed at present. The ultimate aim of this study is to achieve a uniform spatial sensitivity for such sensors through signal analysis and manipulation, this method is proposed in this paper.

KEYWORDS: Spatial sensitivity, Electrostatic sensor, Pneumatic conveying.

1   INTRODUCTION

Coal fired plant typically burn pulverised fuel (coal), where coal is pulverised, mixed with air, and pneumatically conveyed to the furnace burners. Accurate flow rate measurement is essential to efficiency of combustion and vital for reducing greenhouse gas and pollution. However due to complex flow regimes (profiles) and non-ideal characteristics of sensors, the accurate measurement of such flow has yet to be achieved. This paper discusses one of problems regarding the characteristics of electrostatic sensors, which is their "spatial sensitivity".

Provided that a suitable meter could be made, the possibility of measuring and balancing the flow between conveyors should lead to important improvements of the industry in terms of energy saving and pollution reduction. Non-uniform spatial sensitivity presents a significant drawback and fundamental challenge to researchers. The problem can be exasperated in large diameter pipes, which are often used in coal fired power stations. The focus of this paper is addressing and finding solution to this problem.

2   SPATIAL SENSITIVITY OF CIRCULAR ELECTROSTATIC METERS

The spatial sensitivity is defined as the induced charge on the electrode to a single charged particle at a different location. The study will be carried out firstly by investigating the response of the electrode to a charged particle moving axially and radially, and then the response of the electrode to a flow stream at different radial position will be studied, and compared with experimental results.

In principle, in order that the indication of meter is unaffected by solids distribution, there is a possible way to achieve it by compensate the non-uniform sensitivity of sensor and make the overall sensitivity nearly uniform. This paper proposes a method to achieve a uniform spatial sensitivity by using frequency spectrum of the signal induced by the naturally charged pneumatically conveyed solids. The study is still under way; the findings from the recent investigation are included.

3   CHARGE DETECTION

Theoretically, when a ring shaped sensing electrode reaches an electrostatic equilibrium for a stationary charge particle, the relationship between the induced charge, position of the charged particle and the source charge carried by this particle are governed by the following set of Gaussian functions:

$$\nabla . (\varepsilon . \nabla \Phi) = - \rho. \tag{1}$$

$$
\begin{aligned}
\phi(r,\theta,z)\,|(r,\theta,z)\,\varepsilon\,\Gamma p &= 0 \\
\phi(r,\theta,z)\,|(r,\theta,z)\,\varepsilon\,\Gamma s &= 0 \cdot \\
\phi(r,\theta,z)\,|(r,\theta,z)\,\varepsilon\,\Gamma e &= cons,
\end{aligned}
\tag{2}
$$

The electric field intensity E and charge distribution on the surface of the electrode can be found based on Equation 3, 4 with the flux Φ is found from Equation 1 and 2.

$$E = -\nabla \Phi. \tag{3}$$

$$\nabla^2 \cdot \Phi = -\rho / \varepsilon 0. \tag{4}$$

∇ is Laplace's operator and Φ, ρ and $\varepsilon 0$ represent the electric potential, charge density and the permittivity of free space respectively. Besides Γp, Γs and Γe represent the pipeline boundaries, shield and electrode respectively. The induced charge density σ on an electrode can be represented as:

$$\sigma = D = \varepsilon 0 \ E = \varepsilon 0. \nabla \phi. \tag{5}$$

Electric displacement vector, near the inner wall of electrode is represented by D and E is the electrostatic filed intensity. For given boundary conditions, σ the surface charge density on the electrode surface.

$$q = \int_s \sigma. \ ds. \tag{6}$$

In Equation 6, s represents the inner surface area of the electrode. This equation can be solved numerically, for example through Finite Element Analysis. The simulation results in this paper were obtained by using ANSYS fluent. When the sensor's structural parameters along with unity charge position are known, the induced charge q can be solved with Equation 6.

The location of the spatial sensitivity can be represented by the three dimentions (x, y, z) in the sensing field of the theelectrostatic sensor. The spatial sensitivity of a given electrode to a single charged particle is defined as:

$$S(x, y, z) = \left| \frac{Q}{q(\text{x}, \text{y}, \text{z})} \right| \tag{7}$$

where $q(x, y, z)$ stands for charge and Q refers to the charge induced on the electrodes due to $q$in the location (x,y,z) [1].

The simulation set-up is depictedin Figure 1, where a metallic ring-shaped electrode is mounted flush with the inner surface of the earthed conveyor. The electrode is insulated from the conducting conveyor wall and exposed to the flowing air-solid mixture. The charge and potential induced on the electrode for a charged particle in the vicinity of the electrode is calculated using Finite Element Simulation.

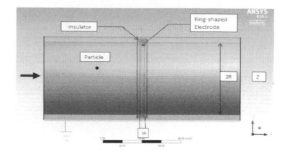

Figure 1. The ring-shaped electrode installation by ANSYS Fluent software.

4 RESULT

The Finite Element analysis using ANSYS Fluent has been conducted to investigate the spatial sensitivity of a ring-shaped electrostatic meter to a single charged particle. In accordance with the dimensions of the meters used in the Teesside Pneumatic conveying rig, the diameter of the electrode and pipe is set as 40mm, the thickness of insulator is 2 mm, and the width of electrode is 2 mm which is flush with inner pipe wall.

The spatial scnsitivity has been defined in Equation 7. In order to simplify the analysis, the cylindrical coordinates system is used. Hence the spatial sensitivity is obtained by calculating, the induced charge ratio on the electrode to the source charge carried by a single particle at a given location (r, θ, z), where (z) represents axial dimension, (r) represent radial dimension, (θ) is the angle between a fixed diameter and the radius where the particle locates (See Figure 2). Due to the axial symmetrical property of the ring-shaped electrode, (θ) doesn't affect the induced charge, therefore this variable isn't included in the results.

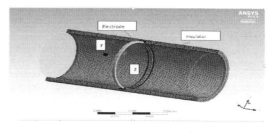

Figure 2. Single particle at a given location.

The simulations were conducted in the following two cases:

1 Calculation of induced charge and variation on the electrode while the charged particle move along pipe axis (z).
2 Calculation of the induced charge and its variation when the location of the single charged particle changes radially (r).

Based on the above simulations, further work will be conducted to reveal the response of the electrode to a stream of particles while they move simultaneously radially, from which the response of the electrode to a flow stream can be obtained.

4.1 z (axis)

As seen in Figure 3, the vertical coordinate depicts the induced charge on the electrode, the horizontal axis represent the axial location of the particle. The simulation was conducted at several points with pixel length 0.02 m along the pipe central line.

Clearly it can be seen that the spatial sensitivity is the highest at the $(z-0)$, i.e. at the central point. It decays with (z) stretched toward both sides.

The simulation results agree with the results published [1, 5] and it can be explained theoretically. It proves ANSYS Fluent is a valid tool.

4.2 r (radial)

Figure 4 provides the results for induced charge measurement in radial axis at several points with pixel 0.005 m along the pipe central radial. At different radial positions (r), the sensitivity is found to be different.

With an increase in radial coordinate, it results in an increase of sensitivity and it illustrates that the particles are adjacent to pipe wall are more sensitive in a sensor.

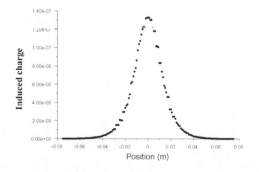

Figure 3. The charge induced curve from a single particle to an electrode at (r =0).

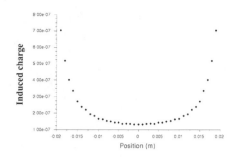

Figure 4. The sensitivity of the electrode in radial axis at (z =0).

5 CURRENT METHOD AVAILABLE

Based on Cheng's model, it was suggested that a uniform sensitivity can be achieved through compensation using two parallel electrodes with different geometric widths. The apparatus comprises two electrodes, the narrow and the wide electrodes. With their different characteristics, by properly weighting two signals from the electrodes, the sensitivity can be improved and the near uniform sensitivity can be achieved, which is illustrated in Figure 5 [4].

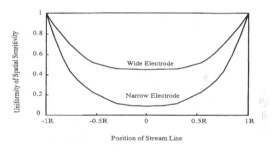

Figure 5. Effect of signal scaling on the combined signals.

However, this method was based on the sensitivity near the cross section of the electrode. The experiments with roping flow stream revealed that the spatial sensitivity changes the overall shape with axial locations. Due to this fact, the above method has its limitation.

6 EXPERIMENTAL RESULTS TO ROPING FLOW STREAM

Ignoring the effect of permittivity of particles, according to superposition theorem, the response of the above electrode to a stream of particles in parallel with the pipe axis can be found as the sum of response of each individual particle at different axial location for a given radial position of the stream. Following this principle, the response of the electrode to the above stream of charged particle can be experimentally obtained if a roping stream of flow can be created.

This has been done in a 14" pneumatic conveying system; the results are shown in Figure 6.

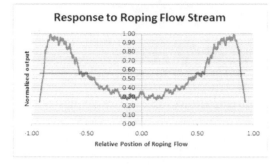

Response to Roping Flow Stream

Figure 6. The response of the electrode to a stream of particles in parallel with the pipe axis.

In this figure, the relative position of roping represents the radial position of the flow stream; the unit of this position is R, the radius. It is clearly that the induced signal on the electrode increases with the "stream" move from the central position towards pipe wall, which is similar to the response to a single charged particle. However when the "stream" approaches the pipe wall further, at about 0.7-0.8R, the above trend is reversed. It is due to reduce sensing zone along the pipe axis when the flow "stream" is too close to the inner pipe wall.

Further simulations are required to work out analytical expressions of spatial sensitivity to a roping flow stream in the sensing volume. However from the above figures, the overall spatial sensitivity can be outlined. The sensing volume can be defined based on these types of simulations, i.e. the response to a single charged particle and the response to a roping flow stream.

7 PROPOSED METHOD FOR SPATIAL SENSITIVITY COMPENSATION

Based on the model for a single charged particle, the frequency response of a ring-shaped electrode has been studied. The experimental results shown in Figure 7 obtained on a 14" pneumatic conveying rig verified the theoretical prediction [2]. In the frequency domain shows that the stream in the centre of pipe contributes to low frequency components with lower amplitude, whereas the stream closer to the pipe wall generates high frequency components with higher amplitudes as shown in Figure 7.

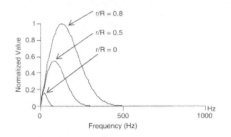

Figure 7. The spatial sensitivity of the meter in frequency domain.

From this figure, it can be seen that the weighted spectrum can be obtained to achieve a uniform spatial sensitivity. For example, if the spectrum can be split into a series of components with regard to radial position, for the same component corresponding to r/R=0 (central position), it should have a higher weight compared to that at r/R=0.8. As discussed in the experimental results, more work needs to be done to the spectrum when the flow stream is closer to the pipe wall. This is an on-going project; the further findings will be published in the future.

8 CONCLUSION

This paper presents the spatial sensitivity of ring-shaped electrostatic meters, showing that the response to a flow stream is quite different to a single charged particle. A method to achieve uniform spatial sensitivity has been proposed based on the findings. Further research through simulation, modelling, and experiments are essential to test the viability of this method, if the proposed method is proven to be achievable measurement accuracy can be significantly improved and the installation requirements of the costs can be further reduced.

REFERENCES

[1] C. Xu, J. Li, H. Gao,S. Wang, Investigations into sensing characteristics of electrostatic sensor arrays through computational modelling and practical experimentation, Journal of Electrostatics, 70 (2012) 1: 60–71.
[2] J. Zhang,J. Coulthard, Theoretical and experimental studies of the spatial sensitivity of an electrostatic pulverised fuel meter, Journal of Electrostatics, 63, (2005) 12: 1133–1149.
[3] J. Zhang, A Study of an Electrostatic Flow Meter, PhD thesis, Teesside University,Middlesbrough, UK, (2002).
[4] J. Coulthard,R. Cheng,Flow metering, ABB, Kent-Taylor Limited, US6305231 B, (1995).
[5] R. Cheng, A study of electrostatic pulverised fuel meters', Ph. D Thesis, University of Teesside, Middlesbrough, UK, (1996).

Intelligent Computing and Communication System

A study on 2D dimensional transmission based on color LED matrix

Soon-Ho Jung, Min-Woo Lee & Jae-Sang Cha
GraduateSchool of NID Fusion Technology, Seoul National University of Science and Technology Seoul, Korea

Goo-Man Park
Department of Electronics and IT Media Engineering, Seoul National University of Science and Technology, Seoul, Korea

Kae-Won Choi
Department of Computer Science and Engineering, Seoul National University of Science and Technology Seoul, Korea

ABSTRACT: Recently, development of LED lighting communication technology has been increasing. LED lighting and broadcasting communication technology is being studied which can be used in a wide variety of fields. In this paper, we proposed a study on 2D dimensional transmission based on color LED matrix. This is a new data transmission concept system of convergence LED and ID (Identification) using the color code. Also, we described enough utility value of the color LED matrix based on the 2D dimensional transmission technology.

1 INTRODUCTION

In this paper, we proposed the 2D dimensional transmission via color LED patterns using LED-ID communication technology. Color LED matrix that generates picture, pattern, language and specific identification code can perform the role of LED-ID Tag. It is necessary to design control boards to control a LED matrix. Thus, we proposed a 2D dimensional transmission using color LED matrix.

2 2D DEMENSIONAL TRANSMISSION THEORY USING LED-ID TECHNOLOGY

According to Kato, H. 2D barcodes for mobile phones, the LED-ID communication technology is in the spotlight as next generation lighting. It is new optical-wireless communication technology with information transfer capability. It also can install the LED lighting. It is called dubiquitous information and communication technology. LED-ID communication technique that is mainly using light which is recognizable for the human eyes. It is possible to recognize range of the transmission data. Also, it is eco-friendly technology, because it has none of electromagnetic waves harmless to human.

According to Jang, Y. M. Cooperative MAC protocol for LED-ID systems, LED-ID communication technology has a lot of advantages as follows. Firstly, LED-ID is harmless to human body unlike the RFID. Secondly, it is not necessary to get the frequency permission. Thirdly, there is no frequency interference. Fourthly, LED light (visible light, infrared, ultraviolet, etc.) band may be used widely. Finally, it is possible to send high-speed multi-media data. Therefore, LED-ID communication technology can be applied to LED light-based facility such as Home-appliances, mobile communications equipment, lighting applications, advertising sector, medical sector, environmental sector, Agriculture and Fisheries. Figure 1 shows a concept of LED-ID communication technology.

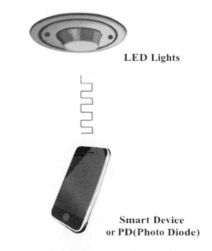

Figure 1. Concept of LED-ID communication technology.

Color code is the name of the first ever "3D Barcode" utilized by a server-based content delivery system known as a ColorZip.

At the moment, color code is a closed and managed source code symbology designed for use by Brand Managers and Marketing Communication professionals through ColorZip Media - a flexible, real-time, interactive content management system that provides up-to-the-minute CRM data to the consumer.

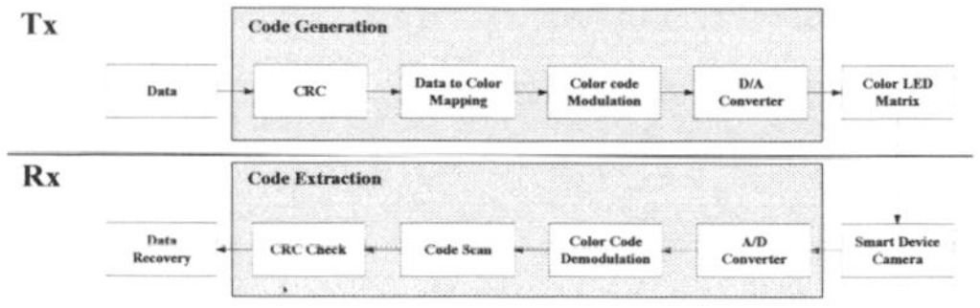

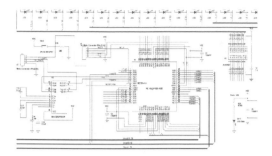

Figure 2. Diagram of 2D dimensional transmission using LED-ID technology.

IEEE 802.15.7a SG-OCC scientists came up with a new kind of code based on color patterns, which can be easily incorporated into company logos or other graphic designs. Colors are much easier for scanners to read, and they help users home in the content. Color code has the added advantage of simplicity. Figure 2 shows a diagram of 2D dimensional transmission technology. By entering the data in the transmission unit, it is displayed on the matrix via the code generation process. Smart device receives the color code data to recover the data via the code extraction process in receiver.

3 DESIGN OF COLOR LED MATRIX FOR 2D DIMENSIONAL TRANSMISSION

LED-ID communication technology is composed of information code including information data. It is possible to express various code expression.

Color LED matrix control transceiver module operated in the DC Power 3 ~ 5V. It was designed to use a rechargeable internal paper battery. In addition, it is possible to control LED lights using current driver.

Color LED matrix is composed of very small LED(0.4mm).LED control section is possible to control a circuit of LED module using current Driver (like Darlington pair) by switching Anode or Cathode of LED arrays

Figure 3 shows output of a control signal from a8-bit microprocessor. MPU can transmit and receive information using UART. We can change LED module's display pattern by transmitting information from PC using UART.

Figure 3. MPU, UART driver circuit.

4 IMPLEMENTATION OF COLOR LED MATRIX FOR 2D DIMENSIONAL TRANSMISSION

In this section, we studied content on color code for sending the 2D dimensional data. A color code can send 2D dimensional data according to various color change using RGB 3-color.

Figure 4. Implementation of Color LED matrix.

A color LED matrix can transmit information to use minimum 256 LED piece. A LED Driver is connected with PC through RS-232C protocol via program. Data to wish transmission should be sent to fit LED matrix. Transmitted data is displayed through 1024 LED piece. In order to send the color code information, color LED matrix circuit s composed of 6 by 16 of LED matrix 4 pieces. Also, it is composed of 1024 LED pieces.

Figure 5. Experiment of LED-ID communication technology.

Figure 5. shows experiments of 2D dimensional transmission using smart device. We confirmed that the data was correctly sent and received.

5 CONCLUSIONS

In this paper, we demonstrated usefulness of 2D dimensional transmission technique via the color LED matrix experiments. Also, we were verified that it is possible to transfer the information using LED-ID communication technology and color code.

We learned a variety of information transmission techniques and circuit-based techniques about LED-ID via the color LED matrix. We demonstrated the usefulness of the 2D dimensional transmission technique applied color code. Also, we are scheduled to implement a miniaturized circuit using color code. So, we consistently make progress in a research which can transmit information.

REFERENCES

[1] Le,N. T. et al. 2011. Cooperative MAC protocol for LED-ID systems.*In Proc. Int. Conf. on ICT Convergence (ICTC)*:144–150.
[2] Huynh,V. V. et al. 2012. Collision reduction using modified Q-Algorithm with moving readers in LED-ID system.*J.KICS*37A(5): 358–366.
[3] Komine, T.&Nakagawa, M. 2004.Fundamental Analysis for Visible-Light Communication System using LED Lights. *IEEE Trans on Consumer Electronics*50(1): 100–107.
[4] Yucek, T.&Arslan, H.2009. A survey of spectrum sensing algorithms for cognitive radio applications. *IEEECommun.Surv.& Tutorials*11(1): 116–130.
[5] Park,I. H. et al. 2010.Scalable optical relay for LED-ID system.*International Conference on Technology Convergence (ICTC)*: 415–420.
[6] Kato, H.&Tan,K.T. 2005. 2d barcodes for mobile phones.*IEEE International Conference on Mobile Technology, Applications and Systems*:8.
[7] Rekimoto, J.&Ayatsuka, Y. J.2000. Cybercode : Designing augmented reality environments with visual tags.*Proceedings of ACM International Conference on Designing Augmented Reality Environments*: 1–10.

Network Security and Communication Engineering – Chan (Ed.)
© 2015 Taylor & Francis Group, London, ISBN: 978-1-138-02821-0

The implementation of an intelligent user interface for OpenATNA system

T.Y. Lin, C.Y. Yang & C.T. Liu
Graduate Institute of Biomedical Informatics of Taipei Medical University, Taipei, Taiwan

ABSTRACT: ATNA is an Integrating the Healthcare Enterprise (IHE) security profile representing Audit Trail and Node Authentication (ATNA). OpenATNA is an implementation of an Audit Record Repository of the IHE ATNA profile supporting RFC 3881 audit messages over BSD Syslog as well as RFC 5424-5426 (UDP and TLS). The authors in this paper implemented the ATNA profile based on the OpenATNA and developed intelligent user interfaces for the OpenATNA to facilitate users query to the Audit Message.

1 INTRODUCTION

The development of network services leads to the transformation of health care, for example, medical records have evolved from paper to the electronic ones. However, patient's health information is exposed to an open network. As a result, protecting patient privacy and ensuring information security has become an importance issue.

In order to solve the issue, the Integrating the Healthcare Enterprise (IHE) has not only proposed the Cross Enterprise Document Sharing profile for exchange of electronic medical documents, but also the security requirements of information exchange under the IHE Infrastructure scheme. However, the OpenATNA user interface does not support the auto complete function, the system manager is hard to find the events which happened in our affinity domain. The aim of this research is develop an intelligent user interface for the OpenATNA for users to search information more conveniently.

2 METHODSTAKE IT INFRASTRUCTURE

2.1 *ATNA profile function and operation*

The IHE is an international project for the integration of medical information system. According to the medical applications of different areas and their characteristics, the IHE established related technical Frameworks. For example, the IT infrastructure specifies the integrating regulation of exchanging the information of patient between hospitals. It also proposed XDS (Cross Enterprise Document Sharing) profile for exchanging medical documents between different hospitals or institutions.

There are several actors in the XDS profile, such as Document Source, Document Repository, Document Registry, Patient Identity Source and the Document Consumer. The provider of documents, which are usually generated by different departments of the hospital. Document Repository is the management system of documents, which is responsible for storing and managing medical records. Each document includes metadata, which is corresponding to the description of the document. Document Registry is the index center for documents, where all the metadata are up-loaded to register and storage. Patient Identity Source is a system to check patients' identity. Patients' data will be added and sent to the Document Registry. Document Consumer is the user system of documents, which is the searching and retrieval for medical records. To enable all the actors in XDS to satisfy the information security during exchanging EMR, the Audit Trail and Node Authentication (ATNA) was proposed.

ATNA is one of the standard of IHE ITI-TF (IT Infrastructure Technical Framework). ITI-TF standardize the way all the IHE system audit log. It also expressly provided that each of the Actors in the XDS specification systems audit record formats, using RFC-3881 specification (Security Audit and Access Accountability Message XML Data Format Definition for Healthcare Applications), UDP (RFC-5426) and TLS (RFC-5425) transmission specification for logging transactions. In the RFC-3881 specification, transaction audit schema explicitly defines all the schema Description, Optionality and Format/Values, which in some Format/Values also defines optional attributes. ATNA standard specification mainly by the Audit Repository as the central management and control, audit user access to protected health information (Protected Health Information; PHI). All IHE Actors must be time synchronized by ITI-1 transaction before starting Record Audit Event transaction by ITI-20 transaction, will be based on the corresponding audit log canonical format store to Audit Repository. ATNA providing audit records for all Actors in the overall architecture and node controls, including service starts, stops, each transaction record, certificate authentication and management, the main control scope to the point of view of the user provides "What patient's PHI (Protected Health Information) can be obtained?". To the patient's PHI (Protected Health Information) point of view, compared with "These PHI which allows users to access?". In System perspective include the "user authentication failure message which needs to be reported?"and "validation failure message which nodes need to be reported?".

ATNA a complete specification of the construction of the foundation of information security methods, including not only user identity identification, authentication, authorization , access control approach to planning, but also provides a node-based audit system mechanism way, and through this transaction record, providing future information security audit personnel-related transactions.

IHE XDS integration framework was shown in Figure 1, transaction standards used are as shown in table 1. PIX Source used ITI-8 transaction to upload the patients metadata to Registry registered and stored. Registry used ITI-42 transaction to Repository to query medical documents. Document Source and Image Source used ITI-41 transaction to store medical documents to Repository. Document Consumer and Image Consumer used ITI-18 transaction to Registry to query the position of medical documents in Repository, and used ITI-43 transaction to Repository to download medical documents.

ATNA plays a role in audit trail the relationship of Actors and Transactions shown in Figure 2.When Actor generate transaction, each executing a transaction will be sent a message to ATNA.ATNA will be time synchronized by ITI-1 transaction, then Record Audit Event transaction used ITI-20 transaction, will be based on the corresponding audit log canonical format store to Audit Repository. The Audit Repository divided into many of the Table, each message is stored among the various table.

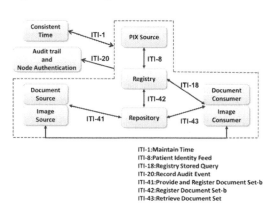

Figure 1.   IHE XDS Integration framework.

Table 1.   Used of transaction standards.

Transaction	Function
ITI-1	Maintain Time
ITI-8	Patient Identity Feed
ITI-18	Registry Stored Query
ITI-19	Authenticate Node
ITI-20	Record Audit Event
ITI-41	Provide And Register Document Set
ITI-42	Register Document Set
ITI-43	Retrieve Document Set

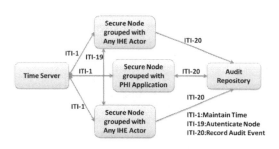

Figure 2.   Audit trail and node authentication diagram.

2.2 OpenATNA system features and defections

OpenATNA is an implementation of an Audit Record Repository, supporting RFC 3881 audit messages over BSD Syslog as well as RFC 5424-5426 (UDP and TLS). (website: https://www.projects.open-healthtools.org/sf/projects/openatna/) The features of OpenATNA list in Table2.

Table 2.   The feature of OpenATNA.

1	Optimizes the database so that entities such as active participants, audit sources, participant objects and codes are not duplicated for every message received.
2	Performs validation of messages as they arrive.
3	Optionally logs errors as well as audit messages.
4	Supports TLS and UDP.
5	Supports Java NIO for demanding environments with many concurrent clients.
6	Supports multiple databases (Postgres, MySQL, Oracle, Derby have been tested).
7	Can archive to file and reload messages in a database neutral way, simplifying management of long term audit trails.
8	Extensible API that allows you to insert your own business logic into the message processing chain.

2.3 Design of OpenATNA database

In this research we used the Oracle database. The Audit Message of ATNA is mainly divided into Event Identification, Active Participant, Audit Source Identification, and Participant Object Identification.

3 OPENATNA INTELLIGENT QUERY INTERFACE

In this research we used the Oracle database to implement ATNA Audit Messages, which include Event Identification, Active Participant, Audit-Source Identification, and Participant Object-Identification. The OpenATNA user interface does not support the auto complete function, the system manager is hard to find the events which happened in our affinity domain, therefore, this study designed an intelligent interface that allows users to query Audit Messages easily.

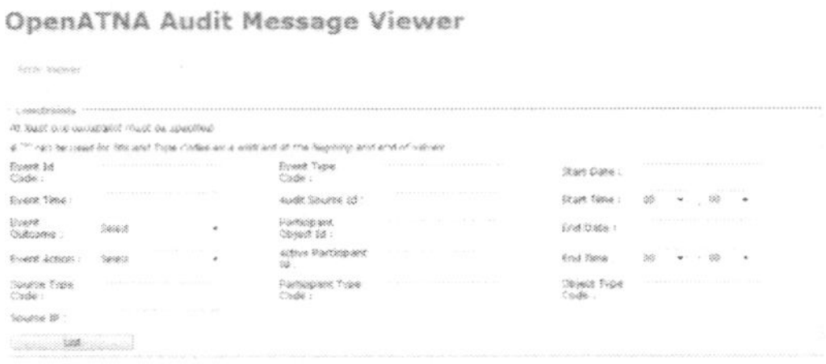

Figure 3. OpenATNA user interface - audit message viewer.

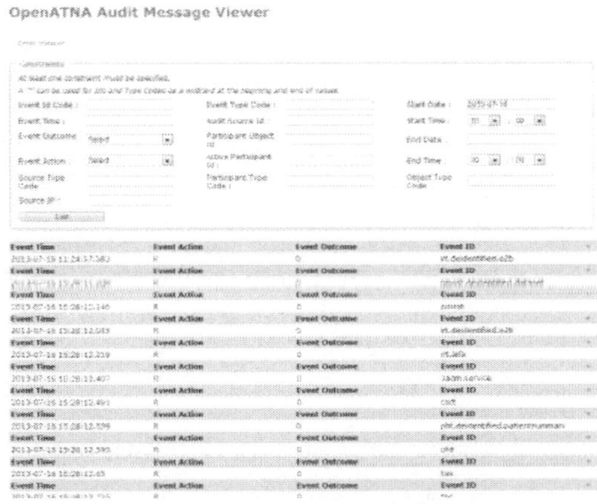

Figure 4. Sample of audit message viewer.

There is a user interface for OpenATNA.

However, the OpenATNA user interface does not support the auto complete function, the system manager is hard to find the events which happened in our affinity domain. Therefore, this study was proposed to design an intelligent user interface with AJAX techniques that allows managers more easily to query Audit Message. The system could auto pop out the possible message events simultaneously during user keying. Audit Message of ATNA is mainly divided into Event Identification, Active Participant, Audit Source Identification, and Participant Object Identification. The structure of the four messages can be converted to database schema. So we can easily formulate query conditions using these four schema. The interface allow users to specify the conditions. For example, input the date you want to query in the Start Date, the system performs the query and display all the Audit Messages matched with the condition, from the start date to current date. Event Outcome represents the success or failure of the event. There are four value to select. The value are 0, 4, 8, 12, represent success, minor failure, serious failure and severe failure, respectively. User can query all messages of success or failure by given the value in this field. Event Action represents the type of actions, the value can be C, R, U, D, E, representing Create, Read, Update, Delete, Execute, respectively. A user can query the actions of messages by typing the value into this field.

4 CONCLUSION AND DISCUSSION

With the advance in network and information technology, information security and privacy protection become more and more important. Under XDS electronic medical records across agencies based medical information exchange platform, ATNA (Audit Trail and Node Authentication) safety code for electronic medical records can make electronic medical records in the inter institutional exchange more secure. Our OpenATNA intelligent user interface shown in this paper can be convenient for the user to query message of whether the process of electronic medical records exchange is success and normative.

ACKNOWLEDGEMENTS

This project has been partially supported by Taiwan's Ministry of Science and Technology (MOST 103-2622-E-038-001-CC2).

REFERENCES

[1] IHEInternational Homepagehttp://www.ihe.net/.
[2] IHE IT Infrastructure (ITI) Technical Framework Volume 1 Revision 10.1 – Final Text October 25, 2013, 9 Audit Trail and Node Authentication (ATNA).

[3] IHE IT Infrastructure (ITI) Technical Framework Volume 2a Revision 10.0 – Final Text September 27, 2013, 3.20 Record Audit Event.

[4] RFC-3881(Security Audit and Access Accountability Message XML Data Definition for Healthcare Applications). http://tools.ietf.org/html/rfc3881#section-5.

[5] Bill Gregg, Horacio D_Agostino, and Eduardo Gonzalez Toledo, *Creating an IHE ATNA-Based Audit Repository.*

[6] Berthold B. Wein, Marco Eichelberg, Alexander Ihls, Eric Poiseau, *IHE "Integrating the Healthcare Enterprise"-an update for Information Technology Infrastructure for 2005.*

[7] Chia-Hung Hsiao, Chung-Yueh Lien, Mei-Chun Kang, Ming-Fang Kuo, *The analysis of IHE system frameworks.*

[8] DICOM PS 3.15 2011 - Security and System Management Profiles, A.5 Audit Trail Message Format Profile. http://www.dabsoft.ch/dicom/15/A.5/.

[9] Chung-Yueh Lien, Chia-Hung Hsiao, Tsung-Lung Yang, Tsair Kao, *Design and Implementation of IHE XDS-Compliant New Generation PortableElectronic Health Records.*

[10] IHE Wiki-Audit Trail and Node Authentication http://wiki.ihe.net/index.php?title=Audit_Trail_and_Node_Authentication/.

[11] IHE - Privacy and Security Profiles - Audit Trail and Node Authentication. http://healthcaresecprivacy.blogspot.tw/2011/05/ihe-privacy-and-security-profiles-audit.html.

[12] openATNAhttps://www.projects.openhealthtools.org/sf/projects/openatna/.

[13] Sheng-Chi Tseng, Der-Ming Liou, *The implementation of IHE ATNA for the EHR system.*

Network Security and Communication Engineering – Chan (Ed.)
© 2015 Taylor & Francis Group, London, ISBN: 978-1-138-02821-0

Design and implementation of campus intelligent power consumption monitoring system based on internet of things

T.J. Wang & P.J. Zhao
XinYang College of Agriculture and Forestry, Xinyang, China

ABSTRACT: IOT technology is combined for designing and developing campus intelligent power consumption monitoring system based on IOT under the general environment of constructing conservation-oriented campus at present in China. The system includes two parts of intelligent air-conditioning energy conservation monitoring and dormitory high-power electric appliance monitoring. Different control methods are respectively proposed aiming at respective features in the two parts. Wireless sensor network and insertion technology are combined for introducing realization mode of the system in details from the perspective of hierarchy structure of IOT. SQL Server database management system is applied for providing scientific decision-making basis for conservation-oriented campus construction.

1 GENERAL INSTRUCTIONS

Although great achievements have been made in total consumption with rapid development of economy in China, such achievements are made by sacrificing the environment and consuming a lot of resources as price. Problems in China, such as insufficient resources and ecological vulnerability, are further exposed. Therefore, it is urgent to accelerate the construction of a resource-saving and environment-friendly society[1]. University is regarded as an important part of the whole social system, which also acts as a high energy consumer. Energy consumption accounts for one tenth of total national social energy consumption. Therefore, energy saving in university can make great contribution to constructing conservation-oriented society[2].

IOT is regarded as the third wave of world information industry following computer, Internet and mobile communication network, which has become a global research hotspot[3]. At present, concept for IOT that is generally accepted is shown as follows: connection among people, between people and object as well as among objects can be realized anytime and anywhere according to agreed protocol through radio frequency identification, infrared sensor, global positioning system, laser scanner and other information sensing device, thereby conducting information exchange and communication, and realizing intelligent identification, positioning, tracking, monitoring and management. IOT belongs to a vast network system. In the paper, characteristics of power consumption in universities are combined on the basis of IOT

technology[4]. Power consumption condition of air conditioning and high-power electric appliances in universities are systematically analyzed as a whole. A set of power saving electricity system suitable for universities is designed and developed, and the system is described in details.

2 GENERAL DESIGN OF SYSTEM

Air conditioning temperature in public room: Since public rooms on campus are either used as large conference room or as large lecture hall, participants pass in and out frequently, temperature fluctuation is high, and the temperature control has hysteretic nature. If air conditioning minimum temperature is still restricted strictly in accordance with energy saving standards, human comfort will be seriously affected, and it is very important to select appropriate control strategies and methods. A good control policy can not only achieve energy-saving purpose, but also does not affect human comfort. Air conditioning air-conditioning energy-saving monitoring system is used for monitoring air conditioning power consumption uniformly in the entire campus; air conditioning temperature of the whole campus is supervised, thereby greatly lowering energy consumption of air conditioner.

Bedroom high power monitoring system: fire disasters are frequently caused due to illegal use of high power electric appliances in school dormitory in recent years, thereby causing personnel and property damage. Many schools organize students to carry out regular checks on the dormitories, high-power electric appliances, such as heat faster, etc. They are collected, however fire disaster accidents still occur frequently in universities. In view of the situation, the system is used for monitoring power consumption condition of student dormitories according to nature of electric appliances from the student power consumption terminals. Once high power resistance load connected to the circuit is discovered, power will be cut on the dormitory immediately, thereby fundamentally eliminating the hidden danger. Division of IOT system structure is combined. Feasibility analysis is conducted on IOT application system development. Campus energy-saving monitoring system vertical system structure is shown in the following Figure 1.

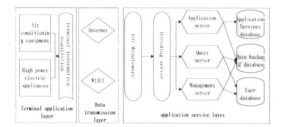

Figure 1. Campus energy-saving monitoring system vertical system structure.

The entire campus energy-saving monitoring system is divided into three levels of terminal application layer, data transmission layer and application service layer, wherein, terminal application layer includes real-time information of various teaching devices and acceptance of remote control commands; data transmission layer is mainly used for sensor data exchange and real-time communication, wireless transmission network on campus mainly adopts ZigBee communication protocol; application service layer is mainly used for storing and managing various resource information, including various database servers, application servers and various application service windows based on network. The terminal application layer and data layer belong to hardware-building scope, and application service layer belongs to software development scope according to vertical perspective of system development. However, all layers are necessarily linked, thereby forming organic service integrity.

3 INTELLIGENT CONTROL STRATEGY

3.1 *Control plan design of air conditioning energy-saving monitoring system*

People pass in and out large public places frequently, thereby leading to larger temperature fluctuations. Temperature control has hysteretic nature. A good control strategy can not only achieve energy-saving

purpose, but also does not affect human comfort. Control strategy of combining expert control and fuzzy PID control box is adopted for solving the problem.

Feature recognition and signal processing module is the core of expert control. It is used for realizing system information extraction and processing and providing basis for real-time inference of inference engine. In the monitoring system, the module is mainly used for solving actual quantity of personnel in a public places, providing basis for the inference engine to infer the best real-time temperature, and providing judgment basis for automatically adjusting air conditioning air direction. Fuzzy control refers to computer intelligent control based on fuzzy set theory, fuzzy linguistic variables and fuzzy logic reasoning. In the system, the minimum (maximum) temperature can be determined according to actual number of people, human comfort and optimum energy-saving temperature in the room. Since large spaces are characterized by high personnel flow and rapid change, fuzzy control capable of absorbing PID robustness is adopted.

3.2 *Control plan design of high power electric appliance monitoring system*

Real-time power consumption monitoring is conducted on student dormitory throughout the campus. If sudden change of power in some dormitory is discovered within short time, or total power exceeds the regulated value, five-minute power-cut measure is adopted. Power can be restored automatically five minutes later. Dormitory number of each automatic power cut should be recorded in database. Dormitory with power-cut record frequency exceeding the regulated value should be monitored, spot checks should be adopted in necessary condition, thereby fully guaranteeing the safety of power consumption and preventing hidden danger.

4 HARDWARE DESIGN OF SYSTEM

4.1 *Embedded hardware platform construction*

Embedded system is usually composed of embedded operating system, application software, peripherals, embedded microprocessor, etc, which has characteristics of high reliability, low cost, high real-time performance, high integration, low power consumption, support of Ethernet technology, easy access to third-party applications, etc. The system has been widely applied in all aspects of industrial control field. Embedded microprocessor is the core of embedded system. The system adopts ARM9 processor based on ARM920T core. System hardware structure composition is shown in the Figure 2.

Figure 2. Hardware structure of campus intelligent power consumption monitoring system.

4.2 Design and implementation of IOT intermediate layer

In the paper, IOT intermediate layer refers to network transmission layer. Wireless transmission technology is utilized; ZigBee is synonymous to IEEOE802.15.4 protocol. Technology regulated according to the protocol belongs to a short-distance and low energy consumption wireless communication technology. It is characterized by short transmission distance, low complexity, self-organization, low energy consumption, high data transmission speed, which is mainly suitable for automatic control box remote control field. Various devices can be embedded. ZigBee as the wireless transmission tool has the greatest characteristic that it can form network automatically, all modules can automatically complete networking after being electrified[5-6]. The system has automatic repair function. Communication of the whole network is not affected even when some modules in the network are damaged.

4.3 Design and implementation of terminal application layer

Embedded air conditioning temperature remote intelligent control device based on IOT is designed. The design of the device is described with energy-saving control of one air conditioner as example only here in order to facilitate description. In the device, DSI8B20 temperature sensor is adopted for collecting site temperature, ZigBee is adopted as wireless transmission module, S3C2440A is adopted as processor, and temperature controller analog remote controller is responsible for sending out instruction for realizing real-time control on indoor temperature. Temperature is controlled within set scope, thereby effectively lowering air conditioner energy consumption and achieving the energy-saving purpose. The intelligent control module consists of dual-loop control system composed of expert control box fuzzy control. Fuzzy control belongs to outer loop for controlling dynamic balance between air-conditioner real-time temperature and optimum energy-saving temperature. Expert control belongs to inner loop. Actual people quantity in large public space is combined for feeding back and adjusting real-time temperature. Both parts act

together for achieving the purpose of saving energy and not affecting human comfort. The hardware design structure is shown in the following Figure 3.

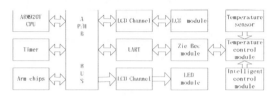

Figure 3. The hardware design structure of Energy saving device of air conditioner.

Currently popular method combining IOT technology, embedding technology and wireless sensor network technology is adopted for IOT high power electricity consumption intelligent monitoring device. Power detection ammeter is carefully selected for automatically testing high power electricity consumption condition. The system has remote access and control functions. Principle of current transformer is utilized in the device. Voltage sensed by current is collected. Voltage is converted into digital signal for processing through A/D conversion. One resister with suitable resistance is added in the circuit in the overload protection aspect. Voltage on both ends of the resister is converted into digital signal through A/D conversion, which is processed by SCM. Current flowing through the resister can be obtained. One threshold value should be set. When the current is larger than the threshold value, SCM can control the relay to act. Electric shock can be disconnected, thereby realizing overload protection.

5 DESIGN OF SYSTEM SOFTWARE

In the paper, Red Hat Linux5.0 operating system is installed on virtual machine VM Ware6.5. Embedded development environment is constructed on the basis. Arm-linux-gcc cross compiler tools are adopted for cross-compiling on the program. QT3 embedded GUI development tools are adopted for graphical interface program development. Communication between host machine and development board is completed through serial port and USB interface. Linux system transplantation, drivers and graphical interface application procedure development are critical for software development of the system.

5.1 Linux system transplantation

Linux system transplantation mainly includes three parts: Bootloader preparation and transplantation, Linux configuration and construction as well as Linux file creation. Cross-compiling environment and necessary toolkit should be prepared for correct installation before system transplantation. Bootloader (vivi) of Samsung Company is adopted as Bootloader in the transplantation. (1) Command #make menu-config is executed under the catalogue storing vivi, the required vivi can be generated; (2) Default kernel configuration file is adopted for configuring Linux kernel, #make zImage is executed for compiling the kernel and producing zImage file; (3) Creation file of Linux is placed under root_qtppia directory for execution. Appropriate image file is produced. The vivi, zImage, and image files are respectively downloaded to development board N and Flash, and the system is restarted.

5.2 Drivers transplantation

In the paper, dynamic link method is combined with the kernel. Drivers of the system mainly include Linux equipment driver, A / D driver under Linux, and driver of DS18B20 temperature sensor. Existing modules can be used for two former drivers, which are not described in details. DS18B20 driver is mainly reset, it is required that CPU 500 should pull the data wire downwards for 500 microseconds and then release it. When a signal is received by DS18B20, it should wait for about 16–60 seconds. Low pulse can be sent, and reset is successful when main CPU receives the signal.

5.3 GUI application procedure

QTDesigner under Qtopia3.0 development platform is adopted as graphical interface procedure development tool for the entire system. Relation between Qt signals and slots is utilized. Button is regarded as procedure trigger signal. Functions for realizing corresponding functions are regarded as slot functions in order to establish connection relation between signals and slots.

6 DATABASE DESIGN

Database design is regarded as an essential part in campus energy-saving construction, which plays an important role in the whole system. It can provide scientific decision-making basis for energy-saving construction of the whole campus. Database design quality can directly affect system development, subsequent maintenance, etc. After demand of system is repeatedly analyzed, SQL Server 2005 is selected as database management system of IOT campus energy-saving monitoring system.

Various factors and demands are combined. ADO link mode is adopted for data communication and transmission between database and host computer. Monitoring data are counted, summarized and stored day by day according to principles of different categories and different items in order to facilitate statistic analysis of campus energy consumption condition. Storage includes two parts of real-time data storage and historical data storage. Data receiving and storage can be completed through XTTECT software, detection data can be transmitted to receiving software through wireless transmission network. After the receiving software receives data, the trigger can send out real-time data storage procedure once every 5s, thereby real-time data table can be updated once. Historical data storage procedure can be sent once every 1 minute. Data can be stored in historical data table. Since the data storage procedure is slightly changed due to different testing contents, others are not changed basically.

7 CONCLUSION

IOT technology is combined for designing and developing a set of intelligent campus power consumption energy-saving monitoring system under current major environment of energy conservation and exhaust reduction. Construction of IOT campus energy-saving monitoring system is described in details, and intelligent control method is combined in the design. Since individual ability is limited, some aspects still should be improved and researched mainly in the following two aspects in the future: (1) hardware design is not perfect enough, which should be further improved; (2) The system database design is not complete enough, some functions are not completed, and construction should be further strengthened.

REFERENCES

[1] Guohui Kou. The explorat ion of approaches on energy cons ervat ion in chinese colleges[J]. Anhui Building. 2011,5:57–58.
[2] Zhengjin Shi, Yaocheng Wang. Measures and ways to the construction of a conservation oriented university[J]. Energy Conservation.2006,5:58–59.
[3] Yu Geng. Present situation and countermeasures of energy savingwork in colleges and universities[J]. China Electronic Education. 2007,4:39–41.
[4] Hongxu Wang, Yubao Sun. Discussion on the Development Future of the Internet of Things in University[J]. Modern computer. 2010,1:20–31.
[5] Geer D. Users make a Beeline for ZigBee sensor technology[J]. Computer. 2005,38(12):16–19.
[6] Jaafar A. A proposed the Oretical Model for Development and Implementation of Capital Projects[J]. Project Management Journal.2000,31(1):44–52.

Network Security and Communication Engineering – Chan (Ed.)
© 2015 Taylor & Francis Group, London, ISBN: 978-1-138-02821-0

Building the smart emergency warning system with FM-RDS protocol

C.H. Chang & C.H. Chang
Department of Health Care Administration, Chang Jung Christian University, Tainan, Taiwan

M.C. Liu
Department of Information Management, National Kaohsiung First University of Science and Technology, Kaohsiung, Taiwan

ABSTRACT: In recent years, radio broadcast application is no longer able to meet our needs for real time information. This proposal aims to show this important issue: How to deliver the real time emergency warning information to every vehicle driver on the road such as traffic accident or traffic disaster? It is very important from the user-point of view that nobody could get enough real-time warning information to make the decision of driving. Here, we propose two types of information for vehicle driver: 1. Radio Data System Traffic Message Channel (RDS-TMC) information from global FM station (Police Radio Station in Taiwan broadcasts RDS-TMC). 2. Real time states of traffic flow information from local area RDS transponder. This paper has attempted to show the connection between vehicle and infrastructure through the innovative design of RDS Network (RDS-NT) architecture and RDS coding structure. Besides, we also try to solve the limitation of no real-time warning information about the traffic accident or traffic disaster. Finally, this paper would explain the smart emergency warning system of RDS network about how to work by three devices of hardware and three protocols of software.

KEYWORDS: RDS traffic message channel, RDS network, RDS coding structure, Smart emergency Warning system.

1 INTRODUCTION

Frequency modulation (FM) has a long history of its application and is widely used in radio broadcast. To transmit stereo music, FM is improved by stereo multiplexing which carries both left-side and right-side audio channel content. With the digital world, Radio Data System (RDS) starts enabling FM to carry text information such as traffic, weather, and radio station information which can be displayed on the end-user's device interface [1]. Currently, the global brand mobile phones (Apple's 5th iPod Nano, HTC, Sony Ericsson, Nokia, Motorola, and Samsung) and consumer mobile devices will have an integrated FM-RDS module [8]. The RDS is an add-on data transmission service with subcarrier and uses frequency from 87.5Mhz to 108 MHz. The goal of RDS is to increase the digital wireless application of system functionality. The RDS also offers a standard altering of Emergency Warning System (EWS) with this 9A group defined by RDS standard document. The special EWS RDS receivers can be alerted in the case of emergencies or disasters. This means fast and also cost-efficient warning for the most people when dangers like tsunami waves, industry accidents with hazardous emissions etc. arise [6]. This proposal aims to solve two important issues: (1) How to deliver the real time emergency event information to the driver? and (2) How to get the traffic flow information in time form dynamic traffic block? In this paper, there are two basics: (1) RDS Network (RDS-NT) architecture (three hardware devices) and (2) RDS coding structure (three software protocols). In advance, this paper will claim to develop the new future application.

2 RELATED WORK

2.1 *Radio data system for digital application*

The Clock synchronization is critical for Wireless Sensor Networks (WSNs) due to the collaborative information computing. In particular, it would be interesting to know how to introduce a new clock synchronization methodology by RDS of FM radio stations. The system could receive pulse from any FM broadcasts as clock information by the new FM digital data receiver. The research presented that RDS clock is very stable and easy to calibrate the clocks of large-scale city-wide sensor networks [2]. The RDS wireless network is a strong technology to deliver the demand response messages and includes two features: low cost implementation and ubiquitous coverage. In addition, security concerns arise due to the wireless nature of the communication channel [4]. This research used the simulation method to evaluate the performance with three protocols and proved the long distance transaction with all data correctness [3].

2.2 Radio data system for Traffic Message Channel (TMC)

In Taiwan, Institute of Transportation (IOT) started the RDS-TMC program and spent more money and time installing eight RDS encoders at the Police Radio Station from 2008. In the same period, IOT built the Taiwan Location Table and Taiwan Event Table (such as traffic block, traffic accident, and traffic maintain) followed ISO 14819-2 and 14819-3. Currently, every driver could install any Navigator embedded RDS-TMC function and receive the real time traffic information [7, 8]. In 2011, Taiwan Telematics Industry Association (TTIA) gathered IOT, global vehicle corporations (BMW, Mercedes-Benz, and Volkswagen), e-Map corporations (PAPAGO), and Navigator corporations (Mio) to cooperate the Taiwan RDS-TMC certification.

3 FM-RDS NETWORK DESCRIPTION

In the paper, we propose two types of information for vehicle driver: 1. RDS TMC information form global FM station (TMC is broadcasted by Police Radio in Taiwan); 2. Real time states of traffic flow and light from local area RDS transponder. Besides, the RDS transponder should have itself identification ID number to match with local location table and the RDS vehicle terminal also should have two-way transmission to communicate with vehicle and infrastructure. In this RDS network, every vehicle driver should install one RDS vehicle terminal to read the real time traffic information and feedback its own vehicle driving information to the most closed vehicle and the most closed RDS transponder. The RDS network will update the newest traffic information to every vehicle under covered. Additionally, this network can collect the information from navigator to collocate and share with each vehicle driver.

V2V: Vehicle to Vehicle
V2I: Vehicle to Infrastructure

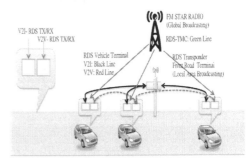

Figure 1. RDS network architecture.

3.1 RDS Algorithm

RDS data were formatted in 16 groups and divided into type A and type B. These groups contained different data like program type or enhanced other networks information data. Every group contains 104 bits and separates 4 blocks. A block contains 26 bits and is divided into the information word and the check word with the offset. The information word contains 16 bits and carries data. The check word with the offset is for error correction and synchronization. In this paper, we addressed the first 6 bits of information word as the local area location table and the last 10 bits of information word as the continuous states of traffic flow and light. We could use the 9B, 10A, 11B, 12A and 12B groups of open data application (ODA) with RDS local algorithm. Every group has 32 bits to be written such as TMC algorithm [6, 8].

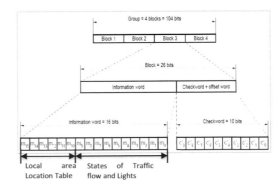

Figure 2. Coding structure of RDS network.

3.2 RDS network architecture

In this paper, we launched the RDS network architecture with the use of three devices of hardware

582

and collected the real time traffic flow information. Besides, we also designed the network architecture with the use of three protocols to send or read data with each other by RDS local algorithm.

3.2.1 Hardware

A. RDS Terminal: This device looks like navigator and uses the touch panel to operate. It concludes the two-way transmission mode to send and receive the traffic data.
B. RDS Transponder: This device looks like Wi-Fi access point and installs outside near the crossroad. It collects all message when each vehicle drives into its local area in 3 km. Additionally, the back-edge server computes all information and shares back to every RDS terminal.
C. RDS Back-edge server: This device works as cloud data server to manage every RDS transponder by IPV6 network and computes every traffic event as RDS network code to deliver.

3.2.2 Software

A. RDS-NT Protocol #1: This algorithm is used among RDS terminals that we call it protocol #1.
B. RDS-NT Protocol #2: This algorithm is used between RDS terminal and RDS transponder that we call it protocol #2.
C. RDS-NT Protocol #3: This algorithm is used between RDS transponder and RDS back server that we call it protocol #3.

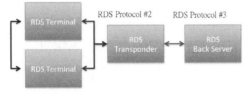

Figure 3. Hardware and software operation of RDS network.

3.3 Emergency warning system for vehicle driver

The primary consiaderation of RDS network is implemented for the real time need of vehicle driver to make the decision of choosing driving pathway. At present, there is no solution to answer which pathway would be quick even the navigator also could not response the precise map guiding. Specially, the emergency warning information of traffic accident and traffic disaster could not send into the vehicle in time. That is why this paper would like to study the RDS and innovatively design the devices and protocols to create the network.

4 SYSTEM LIMITATION

In the process of research, the first limitation of RDS-NT is data transmission capacity that each RDS raw protocol only contains 37 bits of defined user data. The second limitation, the transmission distance of RDS-NT is depending on the RDS transponder's power and antenna type. In a word, the data transmission capacity is lower than GSM network or Wi-Fi network. But the transmission distance is longer than GSM network of Wi-Fi network.

5 CONCLUSIONS

This paper has attempted to show the basics of RDS Network (RDS-NT) architecture (three hardware devices) and RDS coding structure (three software protocols) on how to cooperate with vehicle and infrastructure. In this paper, we introduced the RDS application such as RDS-TMC which is more mature application worldwide and the RDS coding structure to act as the new network communication interface within vehicle and infrastructure. Finally, this paper developed the new solution of RDS network by implementing the new hardware and software. In the near future, the smart phones embedded FM-RDS will develop various applications such as disaster inform, advertisement push and others.

ACKNOWLEDGMENT

This work was supported in part by the National Science Council (NSC), Taiwan, under the Grant NSC100-2221-E-309-005 and by the Institute of Transportation, Taiwan, under Grant MOTC-IOT-101-IBA012. The authors would like to thank the patrons for their research operation support.

REFERENCES

[1] Ge, Liang, Tan, EK, & Kelly, Joe. 2012. Introduction to FM-Stereo-RDS Modulation. http://www1.verigy.com/cntrprod/groups/public/documents/file/fm-stereo-rds.pdf.
[2] Liqun Li, Guoliang Xing, Limin Sun, Wei Huangfu, Ruogu Zhou & Hongsong Zhu. 2011. Demo: a sensor network time synchronization protocol based on fm radio data system. Mobile systems, applications, and services (MobiSys '11) 9: 367–368.

[3] Liqun Li, Guoliang Xing, Limin Sun, Wei Huangfu, Ruogu Zhou & Hongsong Zhu. 2011. Exploiting FM radio data system for adaptive clock calibration in sensor networks. Mobile systems, applications, and services (MobiSys '11) 9: 169–182.

[4] Monageng Kgwadi & Thomas Kunz. 2010. Securing RDS broadcast messages for smart grid applications. Wireless Communications and Mobile Computing Conference (IWCMC '10) 6: 1177–1181.

[5] Radio Data System. http://en.wikipedia.org/wiki/Radio_Data_System, accessed: July 15, 2012.

[6] RDS Basics. http://www.2wcom.com/fileadmin/redaktion/dokumente/Company/RDS_Basics.pdf, accessed: July 15, 2012.

[7] RDS-TMC_Introduction. http://e-iot.iot.gov.tw/EIntegration_new/RDS/RDS-TMC_Introduction.doc, accessed: July 15, 2012.

[8] Traffic Message Channel. http ://en.wikipedia.org/wiki/Traffic_Message_Channel, accessed: July 15, 2012.

Design and realization of unmanned mobile platform control system

Y.K. Li & Y.CH. Wang
Department of UAV Engineering, Ordnance Engineering College, Shijiazhuang, China

ABSTRACT: Unmanned Mobile Platform (UMP) is an intelligent and automated transport vehicle which can implement kinds of tasks by equipped different appurtenances. A four-wheel car with front wheels steer and rear wheels drive is selected as the moving mechanism. The paper researches and designs a control system to make the platform follow the path's guide-line with the master control chip MC9S12DG128. Platform control system is divided into longitude and lateral parts based on system model. Because it's a nonlinear and time-variation system, the effect of normal PID controller is not good. The paper combines with fuzzy-controller and PID controller to adjust the PID parameters on-line, and improves the system fault tolerance and robustness. Both the model and controller are developed in the Matlab/Simulink environment, and they are proved to be feasible.

1   INTRODUCTION

Unmanned Mobile Platform (UMP) is an intelligent and automated transport vehicle which can implement kinds of tasks by equipped different appurtenances. A four-wheel car with front wheels steer and rear wheels drive is selected as the moving mechanism. The paper researches and designs a control system to make the platform follow the path's guide-line with the master control chip MC9S12DG128.

UMP control system is the collection of environment awareness, planning decision and automatic drive, and other functions in an integrated system, involved in real-time embedded systems, computer control, image processing and target recognition, automatic control and power transmission, and other disciplines. It is a typical combination of new and high technology.

The UMP party needs a control system and is suitable for the mobile platform, in the process of independent travels down the line should meet the requirements of high speed, be stable, at the same time should also have strong robustness. Path tracking control is the purpose of coordinating the driving speed and direction, the platform moving platform on the premise of accurate path tracking, improve its average driving speed.UMP party longitudinal control design by using PID control algorithm, lateral control using Fuzzy - PID algorithm.

The UMP is mainly composed of path identification, intelligent decision-making, actuators and condition monitoring subsystem, its structure as shown in Figure 1.

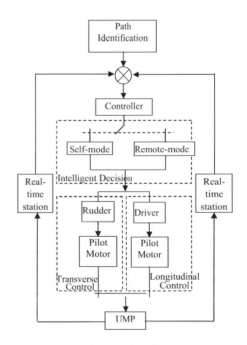

Figure 1.   Structure diagram of platformControl system.

2   UMP KINEMATICAL MODEL

The UMP Kinematical model is shown as figure 2.

For the sake of convenience in the research, the paper makes following assumptions:

1 The whole system is on the flat surface to exercise;
2 Bodywork concerning its lengthways stalk line is symmetric;
3 UMP party wheels and the surface are rigid;
4 Car wheel and circulate surface always keep contact with each other orderly;
5 The car Wheel in the surface of the rolling operation, has no side.

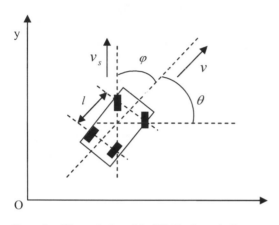

Figure 2.   Kinematical model of UMP schematic diagram.

Build BMP model of the ambulation terrace is as follows:

$$\dot{x} = v_s \cos\varphi \cos\theta$$
$$\dot{y} = v_s \cos\varphi \sin\theta$$

The kinetics equation of vehicle is as follows:

$$\begin{cases} x(k+1) = x(k) + v(k)\cos\theta(k)T_s \\ y(k+1) = y(k) + v(k)\sin\theta(k)T_s \\ \theta(k+1) = \theta(k) + \omega(k)T_s \end{cases}$$

3   THE RUDDER PILOT MOTORCONTROL STRATEGY

In UMP party, by the steering gear to drive the front wheel steering, and control platform of steering, the output of the steering angle is equal to the front wheel steering angle, this is a typical servo system. In this system, the control object is the servo steering gear, control variable is a duty ratio changing PWM wave signal, object variables of path information, collected is a CCD camera.

For steering gear control, considering the time-delay for mechanical structure characteristics, even with the steering gear will have obvious to delay, delay time in the big angle between 0.1 s to 0.3 s, when the speed 2.0 m/s, delay the process of the platform has 0.2 m to 0.6 m. So in the turn, although the controller to decision,

but due to mechanical delay, not immediately transferred to the angle of hope, deviating from the road. In addition, the steering gear steering system is nonlinear, time-varying and model uncertainty of complex systems. If we use the traditional PID control, the control effect is poorer. The PID parameter is not set to be very difficult. It is difficult to get the expected control effect. Fuzzy control does not need accurate mathematical model of controlled object and control flexibility and robustness is strong, the lag of time-varying, nonlinear and complex system has obvious advantages, but it does not have integral link, and in general it is difficult to completely eliminate the steady-state error in fuzzy control system, and in the case of not enough variables classification, often near the equilibrium point will exist small oscillation phenomena. Therefore, this paper proposes an adaptive setting Fuzzy-PID control strategy.

3.1   The determination of the typical PID parameters

According to the driving rules, the algorithm is designed to make UMP always stay within the rules of the runway, and the whole process of exercise to high speed, smooth, smooth. Therefore, for different road conditions give the corresponding control law.

From the system's stability, response speed, overshoot and steady state accuracy and so on various aspects to consider, the parameters of the PID function as follows: K_p's role is to accelerate the response speed of system, improve the adjustment precision of the system; K_i is used to eliminate the system steady-state error; K_d improve the dynamic performance of system.

In this paper, through a large number of tests, according to the role of each parameter and K_p, K_i, and K_d, along with the change of error was obtained under the three conditions suitable PID parameters, the experimental data are as follows:

1 On the straight road, systems take $k_p = 2$, $k_i = 4$, $k_d = 2$, the test environment for a length of 5 m long straights, the average speed of 2 m/s;
2 On the big detours road, systems take $k_p = 8$, $k_i = 0$, $k_d = 1$, teste nvironment for a radius of 50 cm 3/4 of the circle, the average speed of 1.8 m/s;
3 On the "S" type curve road, systems take $k_p - 4, k_i - 2, k_d = 2$, test environment for six 90° arc form "S" type curve road, the average speed of 1.8 m/S.

From the above test data it shows that the structure of the UMP party and characteristic parameters are changing with load and the influence of interference factors, using traditional PID control, K_p, K_i, and K_d is a fixed value, under different road conditions, the control effect will be difficult to meet the requirements. So, there is a need for K_p, K_i, and K_d online adaptive setting.

3.2 The pilot motor Fuzzy-PID control

Because the UMP party within the error range can be regarded as a left and right sides is symmetrical, select deviation $|E|$ and deviation rate $|EC|$, common as input language variable, variables can not only improve the classification accuracy, and does not increase the complexity of the algorithm, the $|E|$ theory area then takes 7 language value $(0,28)$ for the (VZ,VS,S,VM,M,B,VB). In the meantime, in consideration of the examination of the system time the partition is only very short for 20 m and $|EC|$ orgy areas can choose for the $(0,5)$ and take 5 language value (VZ,Z,S,M,B).Output the quantity as K_p, K_i, and K_d, convenient for single chip microcomputer processing, $|E|$ and $|EC|$ and K_p, K_i, and K_d are linear membership functions, as shown in Figure 3.

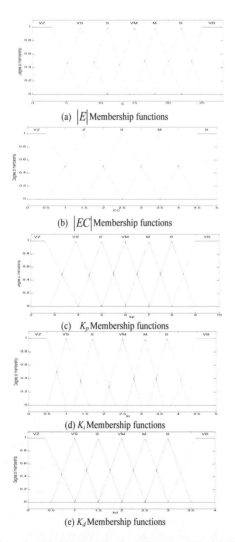

(a) $|E|$ Membership functions

(b) $|EC|$ Membership functions

(c) K_p Membership functions

(d) K_i Membership functions

(e) K_d Membership functions

Figure 3.   Membership functions diagram of all variable.

Reference times of test data, on the basis of chapter 3.1 described in the role of PID parameters, combined with theoretical analysis can induce deviation $|E|$, deviation rate $|EC|$ and K_p, K_i, and K_d three PID controller parameters, which shows the following:

1 When $|E|$ is bigger, to make the system has good tracking performance quickly, it should take bigger K_p with smaller K_d at the same time to avoid a larger response overshoot, dealing with the integral limit, usually take K_i=0;

2 When the $|E|$ is placed in medium etc. size, to make the system response has smaller overshoot, K_p should get smaller, the size of the K_i and K_d wants moderate, to ensure that the system response speed, which had a greater influence on the K_d's value of system response;

3 When $|E|$ is small, to make the system has good steady-state performance, K_p and K_i shall obtain bigger, at the same time to avoid system oscillation, and consider the anti-jamming performance of the system, when $|EC|$ is small, the K_p value to get bigger, vice versa.

According to above-mentioned adjust regulation, we can build up the fuzzy rule form of K_p, K_i, and K_d parameters, as shown in Table 1~3:

Table 1.   Fuzzy control rules of K_p.

EC	E						
	VZ	VS	S	VM	M	B	VB
VZ	VZ	VZ	VS	S	VM	VM	B
Z	VZ	VS	S	S	VM	M	B
S	VZ	VS	S	VM	M	M	B
M	VS	VS	VM	M	M	B	VB
B	VS	VM	M	M	B	B	VB

Table 2.   Fuzzy control rules of K_i.

EC	E						
	VZ	VS	S	VM	M	B	VB
VZ	VB	B	M	VM	VM	S	S
Z	B	B	M	S	S	S	VS
S	B	M	VM	S	S	VS	VZ
M	M	VM	VM	VM	VS	VS	VZ
B	VM	VM	VM	S	VZ	VZ	VZ

Table 3.   Fuzzy control rules of K_d.

EC	E						
	VZ	VS	S	VM	M	B	VB
VZ	VB	B	M	M	M	S	VM
Z	B	B	M	M	M	S	S
S	B	M	VM	M	M	S	S
M	M	VM	VM	VM	VM	S	VZ
B	S	VM	VM	S	VM	VS	VZ

There are many kinds of fuzzy reasoning method, combining with chosen MC9S12DG128 microcontroller, this system adopts the method of MAX - MIN. Namely every rule in the strength of the former pieces minimum values, when the same thing the rules of the strength is not at the same time, fuzzy output take its maximum value.

Above is fuzzy quantity is obtained by fuzzy reasoning, and used to actually control the amount must be clear, so you need to transform fuzzy quantity into clear, this is the fuzzy calculation to complete the task. Commonly used the fuzzy calculation of maximum membership degree method, the median method and weighted average method. The control requirements of this system in order to achieve high adopts weighted average method, computation formula is as follows:

$$k_m = \frac{\sum\limits_{i=1}^{n} u_i A(u_i)}{\sum\limits_{i=1}^{n} A(u_i)} (m = p, i, d)$$

Among them, $k_m (m = p, i.d)$ is the final output accurate quantity, u_i for fuzzy output, corresponding to the membership degree of $A(u_i)$ to u_i. Using MATLAB simulation analysis was carried out on the control rules, the result is shown in figure 4.

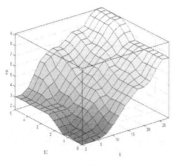

(a) k_p

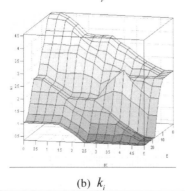

(b) k_i

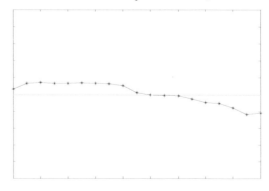

(c) k_d

Figure 4. Analytic result of fuzzy control rules by MATLAB.

4 EXPERIMENT RESULT

To test and verify the effectiveness of the proposed control strategy, this paper introduced the UMP party path tracking under the condition of indoor experiments. Experiments, laying a simulated runway indoors, with fine sand leakage way to write down the UMP party center position on the road, each the size of the deviation distance is measured with a ruler, sub-millimeter accuracy level measurements, along the path tracking direction, 20 cm per interval measurement a point.

On the straight road, the average speed of 2 m/s, error curve as shown in figure 5, the path of biggest position deviation is less than 2 cm, visible path tracking effect is good. "S" type road for 50 cm by 6 section of the radius of curvature of the bend, the circular arc of the Angle is 90°, the average speed of 1.8 m/s, the tracking error curve is shown in figure 6.Path biggest position deviation is less than 11 cm, the edge distance is far less than the runway (30 cm), conform to the requirements of the error. Curve can be found on the amount of error is bigger than on the straights, because the existence of pre at distance, turn ahead

Figure 5. Tracking errors on the straight road.

of the mobile platform, copying a faster route, this is the human and the real driving process is consistent.

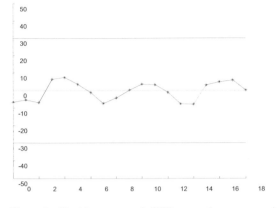

Figure 6.   Tracking error on the "S" type road.

5   CONCLUSION

This paper mainly studied the UMP party control system control strategy, the design of the control system can be divided into vertical and horizontal two parts. Fuzzy-PID control algorithm is mainly studied, using the method of dynamic characteristic parameters setting of PID parameters, and the simulation in MATLAB environment. Because the platform of the steering system is a nonlinear, time-varying, uncertain complex system model, the different shape of the track its affected by objective factors such as mechanical properties and road friction force is very big, it is almost impossible to establish accurate mathematical model, so that part of the controller used the Fuzzy-PID control algorithm, experiments show that the algorithm can adjust the control parameters online, get ideal control effect.

REFERENCES

[1] Chad Karl Tobler. Development of An Autonomous Navigation Technology TestVehicle. 2004.
[2] MC9S12DG128 Device User Guide. Freescale Semiconductor, Inc.2005.
[3] Wayne Wolf, Burak Ozer and Tiehan Lu. Smart Cameras as Embedded System [J]. IEEE computer, 22(5), 48–55, 2002.
[4] Arnold S. Berger. Embedded Systems Design-An Introduction to Processes [J]. Tools and Techniques, CMP Books, Nov.2001.
[5] D.Lewis. Fundamentals of Embedded Software: Where C and Assembly meet. Prentice Hall, Feb.2002.
[6] Jiang G Y, Chio T Y, Hong S K. Lane and obstacle detection based on fast inverse perspective mapping algorithm.2000 IEEE International Conference Systems, man and Cybernetics [C].USA, Nashville, 2000.2969–2974.
[7] Morteza Majdi, Roohollah Barzamini.AGV Path Planning in UnknownEnvironment Using Fuzzy Inference Systems.IEEE, 2006.
[8] H. Marbera, J. Quinonero, M. Izquierdo, A. ASkarmeta,"i-Fork: a Flexible AGV System using Topological andGrid Maps," International Conference on Robotics andAutomation, Taiwan, 2003.
[9] Miro Samek. Practical StateCharts in C/C++ - Quantum Programming for Embedded Systems. CMP Books, July, 2002.
[10] Andrew J. Hoffman. Object Oriented Programmable Integrated. Naval Postgraduate School, December 2006.

Network Security and Communication Engineering – Chan (Ed.)
© *2015 Taylor & Francis Group, London, ISBN: 978-1-138-02821-0*

Term-based character feature planning for automatic story generation

H.T. Wang
China National Institute of Standardization, Beijing, China

F. Zhu
School of Computer Science and Engineering, Jiangsu University of Science and Technology, Jiangsu, China

F. Zhang
China National Institute of Standardization, Beijing, China

Y. Gao
Academy of Mathematics and Systems Science, Chinese Academy of Sciences, Beijing, China

Y.H. Cheng
China National Institute of Standardization, Beijing, China

ABSTRACT: After the first automatic story generation system "TALE-SPIN" was developed in 1976, numerous scientists have been attracted to such field. How to improve the effect of stories is always the focus of this study. In this paper, we put forward our approach of Character Feature Planning Algorithm, which is capable of generating feature description dynamically. It could significantly ameliorate believability of stories.

KEYWORDS: Automatic story generation, character feature planning, multi-agent system.

1 INTRODUCTION

Automatic story generation is a promising research area of computer science, sociology, psychology, linguistics, etc. It is not a process of mechanical assembly, but an intelligent and artistic creation indeed. This research area comprises several special issues in computer science, such as Logic, Virtual Reality, Nature Language Understanding and Generation, Knowledge Engineering, etc. It is not only a research subject, but also could be widely used for many Intelligent Tutoring and Entertainment Industry purposes. The current research on characters is to study how to create agents that appear reactive, goal-directed, emotional, moderately intelligent, and capable of using natural language.

In this paper, we put forward our approach of Character Feature Planning Algorithm (CFPA), which is capable of generating feature description dynamically, and significantly ameliorate believability of stories. In the next section, we show some grand system of story generation. The section 3 is the overview of our approach. And then, we'll present the Character Feature Knowledge Acquisition in section 4. In section5 and 6, we'll get the Character Feature Planning Algorithm, and give an example. The last section is our conclusion.

2 RELATED RESEARCHES

In 1976, James R. Meehan developed the first automatic story generation system – "TALE-SPIN" [1]. Planning was used to generate fables about animals with simple drives and goals. The system's memorable, amusing errors revealed how difficult it is to automatically generate interesting stories.

From then on, many scientists and research groups devoted themselves to this field.

One of the most famous systems is Oz project, developed at Carnegie Mellon University, which studies the construction of artistically effective simulated world [5] which includes several agents. Oz system includes a simulated physical world, several characters, an interactor, a theory of presentation, and a drama manager. It also has three primary research focuses: characters, presentation, and drama. Oz system allows users to create and present interactive dramas in two different presentation models: textual and animated.

Another is "Virtual Storyteller", a part of the AVEIRO project, which is a multi-agent system for story creation led by Mariet Theune in Parlevink Research group at University of Twente [4][7]. It built virtual environments inhabited by autonomous embodied agents. The plot development is mainly based on character interaction. The current version aims at story presentation by a traditional storyteller: an embodied, talking agent that tells a story using appropriate prosody, gestures etc. The agents are used to perform specific tasks (tutoring, reception, navigation). Some of them are embodied and capable of conversing face-to-face with visitors of the environment.

The Virtual Theater project, directed by Barbara Hayes-Roth, as a part of the Adaptive Intelligent Systems (AIS) project at Stanford University[6], provides a multimedia environment, in which users can play all of the creative roles, and perform plays and stories in an improvisational theater. Currently, this research focuses on building individual characters that can act according to the direction from the user or the environment in ways that are consistent with their unique emotions, moods, and personalities.

The Mimesis system is developed by R. Michael Young with his group at North Carolina State University[8]. It defines architecture for building intelligent interactive narrative worlds that run across a range of platforms (3D graphical game engines, PDAs, cell phones, web browsers). The goal is to build a system capable of creating structured interaction within virtual worlds that achieve the same kind of cognitive and affective responses to interactive stories as that seen in the participants of conventional narrative media. The approach taken in the design of the Mimesis architecture is to exploit a well-founded, declarative model of action and intention, in combination with new computational models of narrative structure.

In University of Teesside, Marc Cavazza and his research group implemented Interactive Storytelling system[2][3], which is a character-based system, and creates dynamic narratives with which the user can interact. Stories that emerged from the roles played by the virtual actors have a well-defined storyline, from which many variants can unfold based on characters' interaction of user intervention. The emphasis of this work is on the relations between narrative descriptions and the dynamic generation of virtual actors' behaviors. It shows that dynamic interaction between artificial actors can be a powerful drive for story generation, even in the absence of a centralized plot representation.

Most of story creation systems, however, could only generate stories of some static styles with predefined story characters (agents). These systems are obviously inflexible to expand.

Feature design is an indispensable component of story generation system. There must be characters and features about character in story. Feature design could influence the effect and believability of a story. Spectators won't bear a bald uninteresting story in which characters have same appearance and properties.

Character features is inconstant, and will change dynamically and temporally. It is influenced by emotion, mental state, health, identity, occupation, and experience. Hypothetically, when a little boy, a story character, grew up, he would have some beard on his jaw. Moreover, when he became old, the beard would become white from black. When audiences saw a girl cried with tears in her eyes, it may imply that something bad happened to her. Character features also indicate the difference between characters, which could help spectators distinguish protagonist with deuteragonist.

For a general purposed story generation system, it is unforeseen how the characters would be. So, it is impossible to preset the features of characters. The only effective way to set character features is by using a dynamic generation algorithm.

In our research, we put forward a character feature planning algorithm, and develop an all-purpose story generation system, in which story character features are run-time assigned, and can change dynamically.

3 OVERVIEW OF OUR APPROACH

Our Story Generation Frame (SGF) is capable of creating emotional stories according to uses' requirements and constrains. The whole process is mainly based on multi-agents interaction, and the storyline which is a sequence of story characters' actions. Stories are created by a goal-driven partial-order planning algorithm. A story character is an agent who inherits one or more pre-defined roles such as teacher, doctor, manager, etc.

Since it is not necessary for SGF to ask users to predetermine all the possible story characters, SGF has the ability of building characters dynamically with roles. Whereas roles haven't been assigned any features, we define a special agent to provide feature descriptions to SGF, which has an interface with SGF. So, SGF will take charge of arranging these descriptions in proper order, and representing in proper form.

Figure 1 shows the architecture of our approach and the relation with SGF.

It mainly includes three parts:

a. Character Feature Knowledge Base (CFKB), consisting of various feature rules, and a set of axioms of these rules.

b. Character Feature Planning Agent (CFPA), which is responsible for selecting proper rules from CFKB according to the initial requirement, where Character Feature Planning Algorithm is implemented.

c. The Interface Module (IM), which take charge of receiving the initial conditions for planning, and outputting the planning result of character features. It is a component of CFPA.

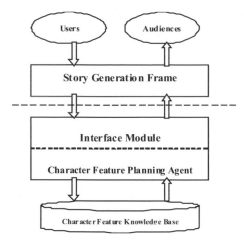

Figure 1.   System architecture.

4   CHARACTER FEATURE KNOWLEDGE ACQUISITION

4.1   *Representing character feature knowledge*

Every piece of character feature knowledge can be regarded as a rule. Such rule contains a set of preconditions and some purposed descriptions of features.

The syntax of rule is like "$P => PAV_1 \wedge PAV_2 ... \wedge PAV_n$", where:

P: first-order Well-Formed Formula of proposition

PAV_k (1<=k<=n): Formalized Feature Description (FFD), like "PAV (x, y, z, w)"

A rule means that after P is (are) satisfied, PAV_1, PAV_2, ..., PAV_n are activated.

As for "PAV" in rules, it is separated into four parts by commas. The first part is the parameter that indicates the character to deal with, denoted as c. The second part is about which part of character is to be described, denoted as p. The third part is the attribute of p, denoted as a. The last part is the value of a, denoted as v.

E.g., rules are like the following:

$\forall$x: Person $Gender$(x, Female)$\wedge Crying$(x) => PAV(x, eyes, color, red) $\wedge PAV$(x, face, color, red)

$\forall$x: Boy fat(x) => PAV(x, face, shape, round) $\wedge PAV$(x, eyes, size, big)

The system-defined predicates are listed in table1. Figure 2 shows predefined tree-like role structure. Figure 3 is partial definition about story character in Automatic Generation Frame, which can be seen as a category, and every real character in story is an instance of it. It is like the idea of OO (Object-Oriented) someway.

Before talking about the knowledge acquisition algorithm, we introduce some definitions.

Definition 1

CFKB is defined as a tuple of <R, A, $\Re$>, where R is a set of feature rules.

A is a set of axioms about R, in the form of Well-Formed Formula (WFF) of first-order propositions.

$\Re \subseteq R*R$

Definition 2

For any different d, d' of FFD,

d and d' are equivalent, denoted as $d \equiv d'$, iff p_d is same to $p_{d'}$, a_d is same to $a_{d'}$, and v_d is same to $v_{d'}$.

d and d' are different, denoted as $d \neq d'$, iff p_d is not same to $p_{d'}$, called equivalent.

d and d' are similar, denoted as $d \propto d'$, iff p_d is same to $p_{d'}$, a_d is same to $a_{d'}$, but v_d is not same to $v_{d'}$.

This definition shows some possible relations between two descriptions.

Definition 3

For any different r, $r' \in R$, d and d' are FFD

r, r' are equivalent, denoted as $(r, r') \in \Re_E$, iff $P_r \leftrightarrow P_{r'}$, $\forall d$ in r, $\exists d'$ in r', such that $d \equiv d'$, concurrently, $\forall d'$ in r', $\exists d$ in r, such that $d' \equiv d$

r conflicts with r', denoted as $(r, r') \in \Re_C$, iff $P_r \rightarrow P_{r'}$, $\forall d$ in r, $\exists d'$ in r', such that $d \equiv d'$

r implies r', denoted as $(r, r') \in \Re_I$, iff $P_{r'} \rightarrow P_r$, and $\forall d'$ in r', $\exists d$ in r, such that $d \equiv d'$

And we also setup some theorems for Â.

Theorem 1:

$\Re_E$ is reflexive, transitive, and symmetric.

Theorem 2:

$\Re_C$ is symmetric.

Theorem 3:

$\Re_I$ is reflexive, transitive.

Definition 4

$\Re =_{df} \Re_E \cup \Re_C \cup \Re_I$

Additionally, the axioms in CFKB are WFF propositions, such as:

$\forall$x (x: Human) ISA(x, BlindMan) => Unavailable (x, eyes), which means any person who is blind , doesn't use the descriptions about his(her) eyes.

4.2   *Organizing character feature knowledge base*

According to the described character, the feature rules can be organized in terms of roles, age, and gender. Thus, the rules about upper roles in role-tree are able to be inherited by lower roles. For example, the rules about parent can be used in father, or mother, such as wrinkle on face, white hair, etc.

Otherwise, the feature rules can also be arranged with general feature rules and special feature rules. General feature rules are those that describe characters who are in an ordinary environment, with ordinary mental state

Table 1. Predefined predicates.

Predicate Form	Parameter Specification	Interpretation
Gender(x, y)	x: character; y: female or male	whether x's gender is y
ISA(x, y)	x: character; y: name of role, such as police, teacher, doctor, etc	whether x is an instance of y
Age(x, y)	x: character; y: a point or interval of positive integer	whether x's age is y(or in y)
Temper(x, y)	x: character; y: type of temper, such as hot temper, mild temper, etc	whether x's temper is y
Health(x, y)	x: character; y: healthy, weak, medium	whether x is y
Action(x, y)	x: character; y: action defined in roles, such as laughing, crying, running, etc	whether x is doing y
Emotion(x, y)	x: character; y: state of mood, such as sad, glad, etc	whether x feels y
Disposition(x, y)	x: character; y: one's usual mood, such as brave, cowardly, etc	whether x's disposition is y
Occupation(x, y)	x: character; y: name of occupation	whether x's occupation is y
Experience(x, y)	x: character; y: something one had done	whether x has done y
FaceShape(x, y)	x: character; y: shape of one's face	whether shape of x's face is y
Skin(x, y)	x: character; y: state of one's skin	whether x's skin is y
Visage(x, y)	x: character; y: description of one's face	whether the statement of x's face is y
SocialClass(x, y)	x: character; y: name of social class	whether x belongs to y
EconomyState(x, y)	x: character; Y: statement of one's economy	whether economy state of x is y
Marriage(x)	x: character	whether x is married

Family_Role
　Senior
　　Grandparent
　　　Grandfather
　　　Grandmother
　　Parent
　　　Father
　　　Mother
　　　Uncle
　　　Aunt
　……

School_Role
　Teacher
　Student
　……

Medical_Role
　Doctor
　Nurse
　Patient
　……

Sports_Role
　Coach
　Spectator
　Athlete
　Referee
　Member of cheering squad
　……

Figure 2. Some of predefined roles.

Defcategory StoryCharacter
{
　Slot: family name
　　: Type character string
　Slot: first name
　　: Type character string
　Slot: gender
　　: Type character string
　　: domain female, male
　Slot: age
　　: Type positive integer
　Slot: temper
　　: Type character string
　Slot: inherited roles
　　: Type array of defined role
　Slot: face
　　: Type character string
　Slot: eyes
　　: Type character string
　Slot: eyebrows
　　: Type character string
　Slot: hair
　　: Type character string
　Slot: voice
　　: Type character string
　Slot: gait
　　: Type character string
　Slot: clothing
　　: Type character string
　……
}

Figure 3. Part of definition about story character.

and ordinary physical state, for instance, a student with a schoolbag on shoulders, a doctor in white gown.

Correspondingly, special feature rules are those that describe characters who are in special environment, or with special mental state, or special physical state, e.g. a crying child with tears in eyes, a drunkard shambling.

4.3 *Character feature knowledge acquisition*

We extract pieces of description knowledge from Internet with a key-word driven extracting program, and formalize them into rules.

The whole feature description knowledge acquisition process is arranged as follows:

Step 1 Collect the key words about human feature including appearance and clothes, such as *eye, nose, face, wear,* etc;

Step 2 Extract the context texts which contain any of these key words;

Step 3 For every context:

Step 3.1 mark the description of the feature, and replace the embodied person with a parameter;

Step 3.2 Analyze implicit and explicit necessary assumptions in the description which are the primary conditions to adopt it;

Step 3.3 Rewrite the description text as a rule;

Step 3.4 Verify the validity;

Step 3.5 Add it into CFKB if it hasn't existed in CFKB;

Step 3.6 Explore the axiom(s) about how to use the descriptions.

We will skip the testing for validity of CFKB here, because it is not the point we want to discuss in this paper.

5 CHARACTER FEATURE PLANNING ALGORITHM

The character feature planning process is encapsulated in CFPA. CFPA receives the initial condition(s) and planning constrain(s) from the SGF, selects proper rules from CFKB according to the inputting condition(s), organizes the descriptions of selected rules, and outputs them to SGF.

Briefly, we denote a feature rule as r, the description (PAV) in r as d, the set of selected rules as SR, the result set of description as RD.

The CFPA is as following:

Step 1 Set SR and RD NULL;

Step 2 While there is any available rule in CFKB; otherwise go to Step 3;

Step 2.1 Get the next rule r whose pre-conditions are satisfied by initial conditions;

Step 2.2 If there is $r' \in SR$, $(r, r') \in \Re_I$, $SR = SR \cup \{r\} - \{r'\}$, goto Step 2;

Step 2.3 If there is no $r' \in SR$, $(r, r') \in \Re_E$, and $(r, r') \in \Re_C$, $SR = SR \cup \{r\}$, goto Step 2;

Step 2.4 For all r in SR, for all d_i in r, if d_i doesn't conflict with any of A (axioms in CFKB), and $\forall d' \in RD$, $d_i \neq d'$, then $RD = RD \cup \{d_i\}$, else continue;

Step 3 Return RD to SGF;

6 STRATEGIES OF FEATURE PLANNING

When the feature planning algorithm gets the result descriptions of character, it doesn't need to use all of them because every character should be distinct.

Therefore, we introduce some strategies to resolve this trouble.

a. For every story character c, if c is a protagonist, c should be described as particular as possible.

b. For any minor character mc, and any protagonist pc, pc's descriptions should be more than mc's as possible.

c. For every different story character c and c', it is prohibited modifying c and c' with the same descriptions.

d. For all feature rules, special feature rules are assigned with higher priority than general rules.

e. For all feature rules, rules that are about occupation, social position are assigned with higher priority than others.

7 AN EXAMPLE

Here we show an example to illustrate our approach.

Suppose SGF establishes a goal that a girl is sad and is crying. Meanwhile, there are some rules in CFKB including:

$\forall$x: Person *Crying*(x) => PAV(x, eyes, color, red)

$\forall$x: Girl *Crying*(x) => PAV(x, voice, tone, sobbing) $\wedge PAV$(x, shoulder, state, trembling)

CFPA firstly adds both of them to SR, and then tests whether every of these "PAVs" is conflicted with axioms. If not, it means the girl has the ability of seeing, it will add "PAV(x, eyes, color, red)" to RD. So do the rest two of "PAVs".

At last, SGF will use them to embellish the crying girl with red eyes, sobbing, and trembling shoulders. Thus, readers will easily imagine a vivid crying girl while reading this part of story.

8 CONCLUSION

In our current research, the semantic analysis of feature rules has not been utilized. As a result, the algorithm mentioned above may conclude that "*PAV*(x, hands, size, big)" and "*PAV*(x, hands, size, very big)" are different. So, semantic analysis will be envisaged in our future work.

All in all, automatic story generation is a challenging work, in which many fields are syncretized together. Stories are works of art as well as intelligent products. A good story should be at least plot coherent and character believable. Our Character Feature Planning Algorithm is able to significantly improve the believability of an intelligent story. Furthermore, another obvious quality of it is easily to shift textual stories to animations.

ACKNOWLEDGMENT

This work is partially supported by grants (522013Y-3062, 522014B-0053, 522014Y-3354, 20110252) from China National Institute of Standardization. The authors also gratefully acknowledge the helpful comments and suggestions of the reviewers, which have improved the presentation.

REFERENCES

[1] Meehan J. TALE-SPIN, An Interactive Program that Writes Stories. In Proceedings of the Fifth International Joint Conference on Artificial Intelligence.

[2] Cavazza, M., Charles, F., and Mead, S. J. Narrative Representations and Causality in Character-Based Interactive Storytelling. In Proceedings of CAST2001, (Bonn, Germany), 139–142.

[3] Charles, F., Mead, S. J., Cavazza, M. Character-driven Story Generation in Interactive Storytelling. In VSMM, (Berkeley, USA, October 2001), 17–24.

[4] Heylen, D., Nijholt, A., and Poel, M. Embodied Agents in Virtual Environments : The AVEIRO Project. In Proceedings of the European Symposium on Intelligent Technologies, Hybrid Systems & their implementation on Smart Adaptive Systems (Eunite 2001), 110–111.

[5] Mateas, M. An Oz-Centric Review of Interactive Drama and Believable Agents. Technical Report CMU-CS-97-156, School of Computer Science, Carnegie Mellon University, Pittsburgh, PA. (June 1997).

[6] Riedl, M. O., Young, R. M. Character-Focused Narrative Generation for Execution in Virtual Worlds. In Proceedings of the International Conference on Virtual Storytelling, (2003), 47–56.

[7] Theune, M., Rensen, S., Akker, R., Heylen, D., and Nijholt, A. Emotional characters for automatic plot creation. In Proceedings of Technologies for Interactive Digital Storytelling and Entertainment, Lecture Notes in Computer Science 3105, Springer-Verlag Berlin Heidelberg, (2004), 95–100.

[8] Young, R. M. Notes on the Use of Plan Structures in the Creation of Interactive Plot. In Workshop Notes of the AAAI-99 Fall Symposium on Narrative Intelligence, (Cape Cod, MA, 1999), 164–167.

[9] Liu, H., Singh, P. MAKEBELIEVE: Using Commonsense to Generate Stories. In Proceedings of the Eighteenth National Conference on Artificial Intelligence, AAAI 2002, (Edmonton, Alberta, Canada, July 28 - August 1, 2002), 957–958.

[10] Rousseau, D., Hayes-Roth, B. Improvisational Synthetic Actors with Flexible Personalities. Stanford Knowledge Systems Laboratory Report, (1997), KSL-97-10.

[11] Willmott, S., Constantinescu, I., and Calisti, M. Multilingual Agents: Ontologies, Languages and Abstractions. In Proceedings of the Workshop on Ontologies in Agent Systems. 5th International Conference on Autonomous Agents. (Montreal. Canada. 2001).

Network Security and Communication Engineering – Chan (Ed.)
© 2015 Taylor & Francis Group, London, ISBN: 978-1-138-02821-0

Research on key technology of three-dimensional talking-head driven by the Chinese text

L.J. Shi
College of Electronic Information and Engineering Changchun University, Changchun, China
College of Opto-electronic Engineering Changchun University of Science and Technology, Changchun, China

P. Feng
College of Computer Science and Technology Changchun University, Changchun, China

J. Zhao
College of Computer Science and Technology Changchun University, Changchun, China
College of Instrumentation and Electrical Engineering, Jilin University, Changchun, China

J.J. Tao
College of Electronic Information and Engineering Changchun University of Science and Technology, Changchun, China

ABSTRACT: Visual model which suits hearing-impaired children's language rehabilitation is researched. Standard mouth shape database is built and the text analysis is done to the input Chinese. By calculating and experiments, Chinese visemes are divided into 16 categories. The calculated amount of mouth shape composition is lessened. The relationship between the text and three-dimensional mouth shape is built by using the phoneme and viseme. Interpolation is done between the key frame mouth shapes to form the intermediate frame. The mouth moving animation which is dealt through joint and interpolation fusion has the smooth and natural transition. Then the real and natural mouth shape which is synchronous with pronunciation can be formed and the pronouncing movement of virtual mouth can change when Chinese text is input in the three-dimensional visual model.

1   INTRODUCTION

Visual pronunciation is the multimedia technology which comprehensively considers the movements of each organ during the pronunciation. When the environmental noise is loud or the listener has hearing impairment, signal to noise ratio can be increased about 8-12dB if a visual pronunciation model, a head portrait which can show the movement of each facial organ, is given when the acoustic information is given.

In terms of visual pronunciation composition, since F. Parke successfully synthetized the first parameterized human face model all over the world in the 1970s, many research organizations have done research on the three-dimensional model of the real human face[1,2]. The Biology and Calculation Center of MIT in America, University of Cambridge, University of Geneva in Switzerland, and other organizations have done much representative work. In 2001, the first virtual host Ananova was born in the world, which was produced by Associated Press of England.

Then a virtual singer Alana was born in China and Vivian was born in America. In China, there are also some research groups doing the research of visual pronunciation[3,4], for example, vision and calculation group in Microsoft Research Asia, pattern recognition lab of automation department of Chinese Academy of Sciences, Zhejiang University, University of Science and Technology of China, Harbin Institute of Technology, etc. Among them, Zhejiang University has achieved great success in terms of modeling and computer animation, and computer research institution of Chinese Academy of Sciences has got important improvement in terms of human facial expression analysis. Tsinghua University and System House of Intel Company jointly built a large-scale bimodule voice library of stand Chinese. They recorded the audio and video data of people speaking in the natural scene, which makes the teaching system aiming at the disabled more authentic and amiable. The Computer Department of Harbin Institute of Technology has successfully solved the technical problems such as the synthesis of certain human face picture in the advanced behaviors, expression synthesis, lip movement synthesis, etc.

At present some achievements has got in terms of visual pronunciation and it has also been applied in many fields. However, there is not much study of visual pronunciation aiming at rehabilitation of hearing-impaired children. So the research about application of three-dimensional visual pronunciation to language rehabilitation training is done in the paper. Three-dimensional visual pronunciation can generate the standard mouth shape synchronizing with pronunciation, the right position of the tongue, palate and other vocal organs and the movement change of virtual mouth cavity, which can help with correct the pronunciation of hearing-impaired children. In this way, the synthetic virtual teacher can be used to have the one-to-one teaching model. As long as the students input or choose the text information, they can learn by themselves repeatedly anywhere at any time in the system. This not only relives the teachers' working capacity and pressure, but also strengthens the initiative of the students by the multimedia communicative way which is composed of text, voice and video.

2 THE DESIGN SCHEME OF THE SYSTEM

In the study, the basic mouth shape database is built according to the features of hearing-impaired children. The method based on parameter control is adopted to do the research of synthetic techniques of three-dimensional visual pronunciation. This makes the model generate the standard mouth shape synchronizing with pronunciation, the right positions of tongue, palate and other vocal organs and the movement change of virtual mouth cavity when the Chinese text is input. It aims to help correct the pronunciation of hearing-impaired children.

2.1 *Building the standard mouth shape database*

Facial animation parameters (FAP) in MPEG-4 is used in the study. The lip shape FAP is combined with the common FAP parameters. In the facial animation based on MPEG-4, FAP is the quantization parameter of facial feature point moving relative to neutral resting position. In order to drive the

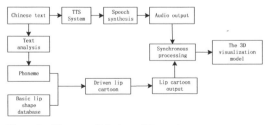

Figure.1　The overall design of the system.

Chinese conversation of three-dimensional virtual head portrait, viseme FAP parameters which adapt to the Chinese pronunciation features needs to be got. As to the extraction method of FAP in MPEG-4, three-dimensional facial movement capture technology is planned to be used in the study. Through the displacement of human face gauge point, the value of FAP is analyzed and calculated[5,6].

The pronunciation process of people's speaking is finished with the assistance of tongue, palate, lip and other organs. In order to get the authentic three-dimensional conversation head portrait, the pronunciation information of the tongue and palate is also got in the study. Palatal sensor technology is used for obtaining the tongue and palate pronunciation information. 126 gold-plating palatal sensors are equally distributed in the net-frame structure. When the tongue contacts with these electrodes, the collecting equipment for the special purpose can get the contact state of tongue and artificial palate and get the palate measuring data which shows the position of tongue on the palate when speaking. By using the data processing technology, the palate measuring data and pronunciation data are combined. After the test, the input rate of palate data is 100/s.

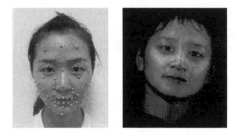

Figure.2　T Establish the standard database.

2.2 *Text Analysis*

Text analysis is a relatively mature technology. The common practice is that the system can separate any input text into the individual characters. Generally one Chinese character is one syllable, and then it is converted into Chinese Pinyin. Two parts of 19 initial consonants and 39 simple or compound vowels are composed of Chinese Pinyin and vowels can be divided into the simple vowel, compound vowel, nasal vowels, etc.

In order to reduce the calculative amount of key frame abstraction, the corresponding mouth shapes are classified according to the feature of each mouth shape in the Chinese phoneme. In the study, for getting the real mouth shape, they are divided into 16 groups of visual voice phonemes according to the difference of mouth shape amplification and contraction proportion of pronouncing each phoneme[7], which is seen in Chart One.

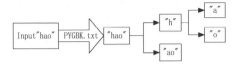

Figure 3.   Text analysis.

3   STUDY ON THE DRIVING FUNCTION AND ALGORITHM

In the study, key frame method and image mosaic method are mixed together. By using the key frame in the mouth shape database and the input text, the interpolation is done between the key frames to form

Table 1.   16 groups of visual speech phoneme.

No.	Chinese Visual phoneme	Mouth shape features
F1	a	The length of the same almost The width increase 50%
F2	b p m	The length of the same almost The width increase 25%
F3	d t l	The length increase 25% The width increase 25%
F4	e	The length increase 50% The width increase 15%
F5	f	The length increase 15% The width increase 15%
F6	g k h	The length increase 25% The width increase 15%
F7	i y	The length increase 50% The width increase 15%
F8	j q x	The length increase 15% The width increase 25%
F9	o	The length Smaller 50% The width increase 25%
F10	r	The length Smaller 50% The width of the same almost
F11	u v	The length Smaller 25% The width of the same almost
F12	z c s	The length increase 50% The width of the same almost
F13	zh ch sh	The length increase 30% The width increase 25%
F14	n	The length increase 10% The width increase 10%
F15	ing	The length increase 10% The width of the same almost
F16	No pronunciation	Lips closed

the transition frame. Then each transition frame is output successively between two key frames according to the time of interpolation while outputting the image sequence.

First, MPGE-4 facial animation parameter FAP standard is used to confirm and define the parameters[8]. According to

$$F_p = \sum_i \frac{\Delta_{pi}}{\left[M_{RC}(p_i) \times w_{RC} + M_{RBF}(p_i) \times w_{RBF} \right] \times U \times n_p} \quad (1)$$

the value of FAP acting on the feature point P is calculated. In the formula, shows displacement component in the given movement direction of one certain point in the effecting area of feature point P, is the amount of grid points which are affected by the feature point P in the area; U is the collected facial animation unit FAPU. Raised cosine function $M_{RC}(p_i) = 1 + \cos(\pi d_p / d_{max})$, Radial basis function $M_{RBF}(p_i) = \exp(-d_p^2 / (2d_{max}^2))$. d_p is p to the feature point distance, d_{max} is the most long distance influence region distance feature point the point, w_{RC} is the deformation weight raised cosine function influence, w_{RBF} is a deformation weight P point affected by radial basis function. For a set of input FAP set, Driving method according to MPEG- 4 animation, It can be calculated by the displacement of a certain time each vertex 3D face model, and finally obtains the key frame.

As to the synthesis of visual pronunciation, the corresponding FAP parameters can be generated through joint and fusion based on the Chinese syllable mouth shape database built in the study. Each syllable in the mouth shape database corresponds to one animation frame set represented by FAP set. Therefore when the given text is used as input, the FAP set corresponding to Chinese characters are matched one by one, and the linear interpolation between FAP frames are done in the bordering syllables, that is [9].

$$F(t) = \frac{t_1 - t}{t_1 - t_0} F(t_0) + \frac{t - t_0}{t_1 - t_0} F(t_1) \quad (2)$$

During the transition period of the mouth shape, the vision weight of the prior phone in the generated mouth shape decreases little by little while that of the latter phone increases little by little. Finally the transition of mouth shape from the prior phone to the latter one is finished. Because the transition mouth shape between two neighboring phones is the mouth shape between the corresponding mouth shapes of two phones, all possible transition mouth shapes between two phones can be generated by linear interpolation while people are speaking without expressions. The expression of linear interpolation is seen as follows[10]:

$$Vis_{itp} = (1-\alpha)Vis_{f1} + \alpha Vis_{f2} \qquad (3)$$

In it, Vis is the apparent place of interpolation frame, Vis_{f1} is the apparent place of the prior key frame, Vis_{f2} is the apparent place of the latter key frame, α is the parameter of control interpolation.

In order to reduce the influence of coarticulation to mouth shape, the No.n pronunciation is made close to the prior or latter mouth shape to some extent. In the continuous articulation sentences, except the fact that the vowel has the great influence on the prior consonant, generally speaking, the mouth shape of the prior voice has a bigger influence on that of the latter voice while the latter voice has smaller influence on that of the prior one. Therefore, only the difference of different rhyme of initial consonants is considered in terms of influence of the latter voice to the prior one. The influence of the prior voice to the latter one is adjusted according to the parameter of the mouth shape of each prior voice to the mouth shape of the latter voice. The specific processing method is as follows: increase A times of the differential value which is the same as the parameter value of prior voice to all FAP parameter value of the latter voice[11], that is:

$$Fap'(n) = Fap(n) + \alpha \left(Fap(n\text{-}1)\text{-}Fap(n) \right) \qquad (4)$$

4 CONCLUSION

In the study, basic mouth shape database is built. The method based on key frame and transition frame interpolation is used for driven algorithm. After inputting the text, it can make the three dimensions output the corresponding mouth shape sequence. Through the control of pronunciation feature points, it can smoothly reflect the articulation movement of the input text. When doing research on coarticulation phenomenon, the mutual influence of articulation mouth shape of each character in the continuous flow of speech is relatively great. Aiming at the features of Chinese mandarin pronunciation and according to the difference of pronunciation and mouth shape influence, the vowels and consonants in the Pinyin are concluded and classified. The coarticulation model is improved to make the three-dimensional visual model generate the authentic mouth shape synchronizing with the pronunciation while inputting the Chinese text, then it can help with the voice training of hearing-impaired children.

ACKNOWLEDGMENTS

This work was financially supported in part by China Postdoctoral Science Foundation under Grant (2013M541298); The Fund of Ministry of Education of China(Z2012062); Jilin Province Science and Technology Development Plan under Grant (20110 1088,20130522162JH); JiLin Province Education Development Plan under Grant (2013278, 2013285, 2014281,2011223). Changchun Administration of Science & Technology Plan under Grant (13GH03).

REFERENCES

[1] MA Yan-hua, GENG Guo-hua, ZHOU Quan-ming et al. 2009. Localization and extraction methods of facial feature points[J], Computer engineering and Application.(18).

[2] T.Wakasugi, M.Nishiura, K.Fukui. Robust Lip. 2004. Contour Extraction using Separability of Multi-Dimensional Distributions[J]. in Proc. 6th IEEE Int. Conf. Automatic Face and Gesture Recognition, p. 415–420.

[3] WANG Zhi-ming. 2002. CAI Lian-hong, WU Zhi-Yong. Study of Text to Visual Speech in Chinese[J], MINI-MICRO SYSTEM, Vol. 23 No.4: 475–477.

[4] ZHAO Jian-guo. 2005. Facial animation new Technology to promote the development of animation industry in China[J], Chinese Intellectual Property News.

[5] Shi Li juan, An Zhi yong, Zhao, Jian, Wang Lirong, et al. 2011. The 3D Face Feature Extraction and Driving Method on Pronunciation Rehabilitation for Impaired Hearing[c]. In Proceeding of the 2011 International Workshop on Assistive Engineering and Information Technologies 2011. p. 547–550.

[6] ZHAO Jian, SHI Li-juan, DU Qin-sheng et al. 2013, Special Human-Computer Interaction Model Oriented to Hearing-impaired Children[J], Journal of Convergence Information Technology, 8(3): 503–511.

[7] TAO Jing-jing, 2013. Research on 3D Text-Visual Synthesis Technology,[D] Changchun University of Science and Technology.

[8] FENG Zhe, SUN Ji-gui, ZHANG Chang-sheng et al. 2007. Research Advance of Chinese Speech Synthesis[J], Journal of Jilin University (Information Science Edition) Vol.25 No.2:198–206.

[9] LI Gang WANG Meng-Jun LIN Ling Mandarin, 2007, Chinese Visual speech Database for Speech-impaired People[J], Chinese Journal of Biomedical Engineering,Vol.26 No.3:355–360.

[10] Wang Zhiming, Tao Jianhua. 2006. A Review of Text to Visual Speech Synthesis[J], Journal of Computer Research and Development 43(1):145–152.

[11] ZHOU Wei, 2008, Research on Chinese speech synchronized realistic 3D facial animation[D], University of Science and Technology of China.

Network Security and Communication Engineering – Chan (Ed.)
© 2015 Taylor & Francis Group, London, ISBN: 978-1-138-02821-0

An oriented random walk model for simulation test in the indoor navigation for visually impaired users

Z. Yu, X. B. Zhang, K. Huang & W. Wang

Zhejiang Provincial Key Laboratory of Service Robot, College of Computer Science, Zhejiang University, Hangzhou, China

ABSTRACT: Position Detection is an important application in Wireless Sensor Network and has received much research attention. Recently, it has been used for indoor navigation for people with visual impairment. However, experiment and evaluation of these systems usually incur a high cost in that users with disability must be engaged. To address this issue, this paper proposes an Oriented Random Walk Model for the simulation test before sensors are deployed. Cost can be reduced and more efficient deployment can be deployed.

1 INTRODUCTION

Wireless Sensor Network (WSN) has played an important role in monitoring the environmental conditions, such as temperature, humidity, pressure, and so on. It has narrowed the gap between real world and digital world. With the development of Radio-Frequency Identification (RFID), WSN can also detect the position and movement of an object easily, which improves our daily life very well, especially for the people with disabilities.

It has been widely recognized that people should share equal rights in accessing to the technologies and contents in this information age. Information accessibility for people with disability is thus an important topic both in academia and in industry. WSN and RFID suggest a method that helps people with visual impairment traveling in a building without any difficulty as other people. With a perfect position detection result and the description of the indoor environment, we can support accurate in door navigation to visually impaired people for going to the destination in a building, which advance the travelling will of them to get in touch with more people.

Our team has joined into a group aiming to improve the life of people with disability. There are some prototype equipments have been developed. This wearable equipment is similar to a hat, with a RFID for position detection, and the communication equipment for receiving the requirement and supplying the indoor navigation information by sound for visually impaired people. Also, we map the wireless sensors in a visually impaired library as the real experiment place, and a lot of systems are developed for indoor navigation system, emergency processing system and so on.

Unfortunately, WSNs present several failures and bugs in real exoteric environment without being well-organized or well-tested. Several problems cannot be easily detected with real test or simulation test. Unlike other applications of WSNs, the real tests of position detecting and indoor navigation system on visually impaired people cost a lot, especially the cost for hiring volunteers. Because the volunteers should be the people who cannot see clearly, so the protection for their safeties needs the most attentions. So we should do a lot of simulation tests to avoid the risk of accidents in real tests, which means that the design of simulation test affect the practicability of the developed equipment very much.

Because the movement for visually impaired people in this library with guild information is not similar to the normal random walk, which has been discussed very well before, so we introduce a new Oriented Random Walk model to simulate the movements, and we use the result of this model to do a better simulation test on our WSN and indoor navigation system.

2 RELATED WORK

Position and movement Detection has become an important application in WSN research area, and the demand of indoor positioning is increasing significantly (Keyong, L. et al. 2012). Varied solution has been fully discussed for a long time, such as infrared-based positioning system (Want, R. et al. 1992), ultrasound-based system (Priyantha, N. et al. 2000) and so on. Also, there are many types of signal measurements, such as the most basic Received Signal Strength Indication (RSSI) is widely used in majority of such applications (Guvenc, I. et al. 2007, Patwari, N. 2007).

Random walk model is a mathematical formalization for generating a random path as people walking in this space, which has been proposed for a long time (Kaye, B. H. 1989) and has been changed a lot to satisfy the environment requirements, such as walking on graphs (Göbel, F. et al. 1974), walking on grids (Doerr, E. et al. 2009) and so on. Also, there are several applications in WSN which are based on random walk model (Jin, Z. Y. et al. 2013).

3 ORIENTED RANDOM WALK MODEL

In this section, we introduce the Oriented Random Walk model for our simulation tests. At first, we need to discuss the challenges of the simulation tests

3.1 Environment Description

Unlike the normal random walk model, there are a lot of special features for the walking space in a blind library, such as in a floor as Figure 1:

- There are several Rooms and Passages in this floor. And there are several floor entrances for going upstairs or downstairs;
- There is a special room named *Touch Area*, where there are many models of the newest technology products, such as airplane, satellite, rocket and so on. This area aims to supply an interaction method for visually impaired people to understand these products by the feelings and the introduction information given from the speakers;
- All the wireless sensors are located in the middle of the passages, the top of the doors of rooms, and the top of floor entrance. There is no sensor in any room, except the Touch Area.

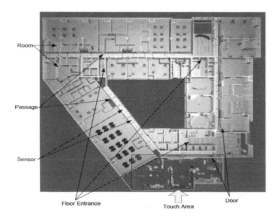

Figure 1.   Real test environment: A floor plan of blind library.

3.2 Characteristic movement

Considering the situation that there are many visually impaired people traveling in this floor and each one gain the indoor navigation information from communication equipment, many features for their movements can be summarized also:

- Each person can only enter this floor from any entrance, similar when they want to leave;
- Each person can stay in a room or Touch Area for a long time, while cannot stop at any point in passages. Rooms or Touch Area are the only destination choices for their oriented travelling;
- When people walking in the passages, they have special destination to visit. Because their move speeds is lower than other people, and their going with an accurate indoor navigation, we can assume that they will go to the destination with the shortest path;
- After visiting a special destination choice, people would like to randomly select a unvisited destination as next movements, or go to the nearest entrance to leave this floor;
- For there is no sensor in normal room, we assume that people stay in the position of the door, until they leave this room;
- For there are 2 doors in the Touch Area, we assume that when people enjoy the sights of these models, they would like to go to another door to leave Touch Area with skimming a lot of models;
- For there are many sensors in Touch Area, which are located nearby each model, we assume that each people would randomly select a unvisited model to enjoy, and stay a short time behind the sensor near this model.

Using these patterns, we can model the movements much more exactly.

3.3 Oriented random walk model

The most important difference between our walk model and normal random walk model is that: In Oriented Random Walk model, people randomly select a destination, and walk straight to this destination, not as people randomly select a direction to go in each step without any long-term destination as normal random walk model, which makes our model can simulate the real movements much better.

We assume that all people are not in this floor at the beginning, and all people leave this floor in the end, and the global controller of our model is that:

Algorithm 1: Global controller

1. Initialize all people's label as *Unvisited;*
2. While any people's label is not *Visited*
3. For the people labeled as *Unvisited*, randomly change their label into *Visiting* and select an entrance as the beginning position to begin the travelling;
4. For all the people labeled as *Visiting*, randomly change their label into *Visited* and select an entrance as the destination to finish the traveling;
5. For all the people labeled as *Visiting*, randomly select a unvisited Room or Touch area as the destination and begin to walk to this target;
6. End;

In our model, destination position may be long distance away from people's position now (all positions are represented by the nearest sensors' position).To make an exact indoor navigation, we must tell the nearest next sensor as the *ToPoint* for this step. We can prepare a WSN connection map, which shows whether 2 sensors can be connected strictly without crossing any other sensors or walls. Also we can get the shortest *path distances* between each 2 sensors according to the connection map, instead of the real Euclidean distance. With this information, we can use the following algorithm to find out the next *ToPoint*:

Algorithm 2: Find the to point

INPUT: Present Position, Destination Position
OUTPUT: ToPoint sensor ID;
1. Find the nearest sensor ID a,b with Present Position and Destination Position;
2. For all the sensor c connected to sensor a
3. If the path distance between a and c add the distance between c and b equals to the distance between a and b, return c as the result;
4. End;

Using Algorithm 2, we can show a wonderful movement simulation result for people traveling among the normal rooms, while cannot gain a good performance in Touch Area, whose connection map is as Figure 2:

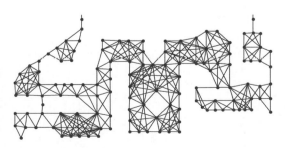

Figure 2.   Connection map in Touch Area.

From this map, we can see that if we use the shortest path to travel around this area, all the people would leave this area with a lot of models unvisited by any one, which is not similar to the real movements in a library or museum. So we use another algorithm to find the ToPoint in Touch Area:

Using the Algorithm 3, we can let the people traveling in Touch Area more randomly and diversified, which is much similar to the real situation.

Combining all these 3 algorithms, we can generate a good movement results as the dataset for simulation test.

Algorithm 3: Find the to point in touch area

INPUT: Present Position, Destination Position
OUTPUT: ToPoint sensor ID;
1. Find the nearest sensor ID a,b with Present Position and Destination Position;
2. Randomly generate a integer;
3. If the integer is odd, randomly select a unvisited sensor connected to a as the result;
4. Otherwise using the Algorithm 2 to calculate the result;

4   EXPERIMENTS & RESULTS

In this section, we do some experiments on Oriented Random Walk model and the simulation test demo.

4.1   *Experiment on oriented random walk*

Dataset: The real WSNs include 403 sensors located in this floor, and there are 32 normal rooms, 36 doors and 8 entrances. According to the position of each sensors and the floor plan, we manually generate the connection map between each nearby sensors, and calculate the path distance between any 2 sensors. Using this data set, we generate 50 people traveling in them, 2 examples are as follows:

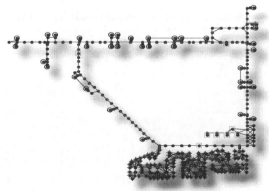

Figure 3.   Oriented random walk path example 1.

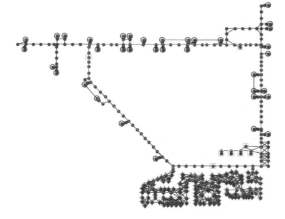

Figure 4.   Oriented random walk path example 2.

Figure 6.   Demo project example 2.

From Figure 3 and 4, we can see that our model can generate different walk to simulate the real movements. Also these 2 simulated persons come into this floor through different entrances. These Figures show the traveling position in 1 hour only, so only the second person leave this floor while the first person haven't finish his travel. All of them go to several rooms (nodes in circles) as the destinations, and the paths between normal rooms are mostly the shortest paths which are similar to each other. While the traveling paths in the Touch Area (nodes at the bottom area) are distinctly different, which are more like the real persons traveling in Touch Area.

4.2 *Simulation test demo*

Using the walk path generated by our proposed model, we can do better simulation test. So in this section, we introduce the simulation test demo.

We are using the Unity 4 to build the demo program, using the map model given by the library to visualize the floor, as follows:

Figure 5.   Demo project example 1.

When we have uploaded the walk paths for all simulated person, we can run the simulation and do the detection for each person. People's action analysis can be done at the same time.

5   CONCLUSION AND FUTURE WORK

In this paper, we discuss the different requirements and features for simulating the visually impaired people traveling in the visually impaired library with an exactly indoor navigation. According to the special features in the real environment and the patterns for people's movements, we propose an Oriented Random Walk model to simulate their traveling situation. The experiment results show that our propose model can simulate the real situation much better. Also we develop a demo project to do further analysis.

At the same time, we find several shortages in our model waiting for further discussion, such as we can add the Collision Detection for person walking in the passages, and more diversified features in the environment influencing the path selection as the toilet selection by the gender of each person.

ACKNOWLEDGMENT

This work is supported by National Key Technology R&D Program (Grant No. 2012BAI34B01).

REFERENCES

[1] Doerr, E. & Friedrich, T. 2009. Deterministic Random Walks on the Two-Dimensional Grid. Combinatorics, Probability and Computing, vol. 18, no. 1-2, pp. 123–144.

[2] Göbel, F. & Jagers, A. A. 1974. Random walks on graphs. Stochastic Processes and their Applications, vol. 2, no. 4, pp. 311–336.

[3] Guvenc, I. & Chong C.C. & Watanabe, F. Mar. 2007. NLOS Identification and Mitigation for UWB Localization Systems, Proc. WirelessComm. and Networking Conf, pp. 1571–1576.

[4] Jin, Z. Y. & Shi, D. X. & Wu, Q. Y. & Yan, H. N. 2013. Random Walk Based Location Prediction in Wireless Sensor Networks. International Journal of distributed sensor networks, vol. 2013, pp. 1–9.

[5] Kaye, B. H. 1989. A random walk through fractal dimensions, New York: VCH Publishers.

[6] Keyong, L. & Dong, G. & Yingwei, L. & Loannis, C. P. Dec. 2012. Position and Movement Detection of Wireless Sensor Network Devices Relative to a Landmark Graph, IEEE Trans. Mobile Computing, vol. 11, no. 12, pp. 1970–1982.

[7] Priyantha, N.B. & Chakraborty, A. & Balakrishnan, H. 2000. The Cricket Location-Support System, Proc. Sixth Ann. Int'l Mobile Computing and Networking, pp. 32–43.

[8] Patwari, N. & Kasera, S. 2007. Robust Location Distinction Using Temporal Link Signatures. Proc. ACM MobiCom, pp. 111–122.

[9] Want, R. & Hopper, A. & Falcao, V. & Gibbons, J. Jan. 1992.The Active Badge Location System, ACM Trans. Information Systems, vol. 10, pp. 91–102.

Network Security and Communication Engineering – Chan (Ed.)
© 2015 Taylor & Francis Group, London, ISBN: 978-1-138-02821-0

The design of LED display based on CPLD dot matrix

Y. Xiao

Shandong Polytechnic, Jinan China

ABSTRACT: This paper introduces the design of practical and powerful large LED display systems based on CPLD, which takes AT89S52 as core and uses separate scan mode.

KEYWORDS: MCU, Graph Matrix Display Screen, CPLD.

1   INTRODUCTION

Information technology promotes the development of display technology. As high-tech products, large LED dot matrix display systems have come into people's daily life.

Compared with traditional displays, LED displays have higher brightness, better dynamic image display effect, lower energy consumption, longer working life, a larger variety of display contents, more flexible display method, and higher cost performance. With these advantages, LED displays have come into all walks of life.

With better display effect, larger variety of display contents, lower power consumption, simpler structure, and easier operation, LED dot matrix displays controlled by SCM have high market potential and have already been widely used in such fields as banking, securities, films, televisions, sports, and traffic.

2   THE OVERALL BLOCK DIAGRAM OF THE SYSTEM

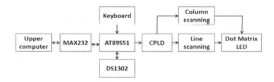

Figure 1.   The overall block diagram of the system.

The entire display system can be divided into eight portions, ① CPU main circuit, serial data transfer and timing ② control section, ③ CPLD ranks decoding scanning part, ④ real-time clock control circuit, ⑤ and PC serial communications section, ⑥ buttons the control circuit part, ⑦ transistor drive circuit portion, ⑧ dot matrix display section.

3   EACH UNIT CIRCUIT DESIGN

3.1   *Scan control logic design*

To transmit a larger amount of data, large LED dot matrix displays have to have a faster refresh rate, or the display would shake and flicker. To solve this problem, there are many ways. For example, to increase the rate of data transmission by using PC DMA controller, toadopt parallel data transmission, to employ multi-CPU control, etc.

CPU is used to control; complex programmable logic device (CPLD) is used for line scan and column control module; VHDL language is used to design control logic; the scanning system uses time-serial data transmission column, line scan mode. This kind of design makes best of the advantages of CPLD and the EDA design of digital circuits. The hardware circuit design is simple and reliable, with high stability. As a result, the entire display system works smoothly. CPU controls scanning frequency so that the screen does not flicker.

This design employs EPM7128SQC160-10 chip which has 160 I/O ports. Through simple programming, only one CPLD chip is enough to simulate 8 pieces of 74LS595 and one piece of 74LS154. The internal logic is as shown in Figure 2.

The column data transfer control circuit works like this. Display data from SCM is serial input to CPLD. VHDL language is used to design the bus read logic. The CPLD chip completes eight pieces 8-bit 3-state serial input and parallel output. Serial is transformed to parallel by latched shift register. The CPLD's 64 I/O ports output data to control LED dot matrix's 64 columns.

The column data transfer control circuit works like this–display data from SCM is serial input to CPLD. VHDL language is used to design the bus read logic. The CPLD chip completes eight pieces 8-bit 3-state serial input and parallel output. Serial is transformed to parallel by latched shift register.

The CPLD's 64 I/O ports output data to control LED dot matrix's 64 columns.

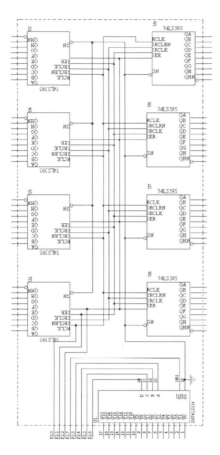

Figure 2.   CPLD internal logic diagram.

Strobe control signal output, RCK, SCK, SCLR was sent by the MCU control signals and RCK to latch control signal. SCK is a clock control signal, and SCLR clears signal for each piece of RCK, SCK, SCLR connected together. Si scan data output is connected to the microcontroller, and the first piece of 74LS595 (implemented by CPLD) shift output is the end of the second piece of data input 74LS595 Si2. Eight were cascaded to achieve a scan a byte namely: eight parallel output and serial shift function.

Line scan control is the use of CPLD programming four - sixteen line decoder functions to achieve 74LS154, as shown, AA, BB, CC, DD control data output line connected to the microcontroller in Figure 2, the line scan H1-H16 as CPLD output control signal, then the row driver circuitry.

3.2   Line drive circuit design

The system is a part of the monochrome dot matrix displaying LED dot matrix module, and 8 × 8 dot matrix module is connected to a 16 × 64 matrix. Because a line scanning tube simultaneously controlling multiple LED line-off, so it carries a large current. In each of the light emitting diode flowing current of 10mA calculated lattice screen a 64; each line scan tubes withstand the current is 10mA × 64 = 0.64A, for which we choose Darlington transistor structure to ensure a line drive capability.

3.3   Scanning frequency control

Because the human eye's persistence of vision phenomenon, a bright LED light tube in a second if more than 24 times, then do not feel the human eye blinks. Thus when one screen to screen 25 times per second continuous frequency cycle, the impression is stable. For this purpose as long as the use of CPU controls the row decoder implemented by the CPLD decoding speed, to ensure each second decoder 16 × 25 = 400 times, can guarantee the stability of the screen. Design, we make AT89C51 timer / counter T0 in Mode 2 (auto reload mode). At this setting M1M0 10, mode 2 in the 16 split into two parts counter, where TL0 is 8 counters; TH0 used to store and maintain the initial count. When TL0 count in the overflow flag is set to 1 while TF0, will automatically reload the initial value TH0 TL0, so in the initialization process, the software only once initial value. Its period is:

$$T = (2^8 - TH0\ \text{initial}) \times \text{Clock cycle} \times 12$$

When using 12MHz crystal oscillator, count rate of about 1MHz, the period of the input pulse interval 1uS, by calculating, TH0 the initial value of 243, that is 0XF3 (OF3H). Interrupt the row decoder controlled scanning frequency, you can ensure flicker-free screen display.

3.4   Design of the real time control circuit

The design also provides real time display, using the US DALLAS's chip DS1302, It has high performance, low power consumption and has a internal RAM of real-time clock chip.Itcancost years, months, days, weeks, hours, minutes and seconds and has a leap year compensation, Three-wire synchronous communication interfaces with the CPU, and once the burst mode is used to transfer clock signal or more bytes of data RAM, battery backed clock chip. You only need to set the initial time. DS1302 using SCM control, the use of split screen display year, within the months, days, minutes, seconds, can be adjusted.

3.5 MCU and PC communication unit design

3.5.1 The communication unit circuit

The unit is mainly composed of two parts: Upper Computer, MCU interface with cable and the middle level conversion circuit. The system uses a three-wire, monopolizing the CPU serial port. The computer's RS-232 signal level microcontroller serial signal level is inconsistent, and it must be level conversion between the two, in this use of integrated level converter chip MAX232 and level converter chip. It only uses a single + 5V power supply, which is connected with four 1μF electrolytic capacitors to complete the conversion and RS-232 level between TTL level. Its circuit diagram is shown in Figure 3. Conversion is complete serial signal TXD, RXD direct and 89C51 serial port connection.

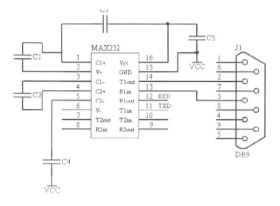

Figure 3. Level conversion circuit.

3.5.2 Communication protocol

In this system the PC undertakes the task master and the microcontroller to accept the PC command and control display information according to the instructions, and then modifies it the displays. We use RS-232 serial asynchronous communication, a start bit, 8 data bits, 1 stop bit, no parity, baud rate, 9600b / s, to transfer data adopting ASCII mode. PC transfers control information and data, and the control information includes modify display parameters, and modifying display mode. Meanwhile PC function displays the contents transferred to the microcontroller. Lower machine work as received instruction. If the master sends the wrong instruction,it will be without any control, display Error promptly, and automatically return after 1 second.

3.6 Keyboard control circuit design

The system set up three keys (k1, k2, k3) were adjusted to control the time, date, and time scrolling display, the system default state is the display time. First, the keyboard scan to determines if k1 key is pressed, Ifk1 key is pressed, pressing the enter date only displays the status. Double-click it to enter the order when scrolling displays status information, and if you do not press it , which is determined by whether the key pressedis k2 setting state k2 key is pressed then enter the time and date. K3 key works when the adjustment time, the time and date of the completion accumulates until the adjustment is up to date and need.

4 SOFTWARE DESIGN

The modular software design and programming methods, use C51 programming. The entire software system is simple and clear, but also has good scalability. The entire software system, includes the main program, the ranks of the control logic CPLD program, the clock control subroutine and PC serial communication routines four modules. The main program is responsible for handling the keyboard, display refresh, information calls and transmission control.

5 SUMMARY

The 16 × 64 dot matrix switching system with a key. The screen brightness is continuously adjustable vertically and horizontally. Scrolling display of information and stored information timed cycle, which use DS1302 real-time time display and other functions can be displayed directly on the PC serial port information control updates with refresh speed, high brightness, low power consumption. And unit-level display unit display module can receive data transmitting down the information and commands from the controller or the information, and these data and commands information are then transmitted without any change to the next-stage module unit when the display panel can be expanded to display more of the display unit, when a plurality of display units cascading only requires a corresponding change in the internal logic circuit and changes CPLD data which the number of bytes per row can be transferred.

REFERENCES

[1] S.W.Wang, Xin.Yu, S.J.Wu:"Electronic Innovations Design and Practice" Beijing: National Defense Industry Press.
[2] M.Q.Lin, W.W.Ma: "VHDL digital control system design paradigm", Beijing: Electronic Industry Press.
[3] Song.Pan, J.Y.Huang: "EDA technology Practical Guide", Beijing: Science Press.
[4] Y.P.Liao, R.Q.Lu: "CPLD digital circuit design - using the MAX + plus II" Beijing: Tsinghua University Press.

Network Security and Communication Engineering – Chan (Ed.)
© 2015 Taylor & Francis Group, London, ISBN: 978-1-138-02821-0

Literature semantic retrieve based on domain ontology

X.Q Zhang & H.M Gao
East China University of Science and Technology, Shanghai, China

ABSTRACT: The traditional literature retrieval system is usually based on keyword-matching model, and it can only extract keywords in user's query, but lacks ability to understand implicit knowledge and cannot achieve associated expression on semantic level. In order to implement retrieval on semantic level, literature domain ontology with the characteristics of literature resources is presented in this paper. And then a set of inference rules based on the literature domain ontology are defined with OWL description language and SWRL rules. With Jena inference tool, some implicit correlation information is obtained by these rules and related concept expansion. It helps users to hit literature effectively and quickly. At last, a literature semantic retrieval system prototype is developed to verify the advantage of the literature semantic retrieval system based on domain ontology.

KEYWORDS: Literature semantic retrieval, domain ontology, semantic reasoning, literature retrieval system.

1 INTRODUCTION

The traditional literature retrieval platforms are usually based on the method of keyword matching. It has a lot of defects when we process the document resources, such as honest expression, island vocabulary, expression difference etc. Ontology, as the key technology of semantic web, has concept hierarchy structure and logical reasoning ability. It has been widely used in natural language processing, software agents, medical modeling, information retrieval and other fields.

Castells, et al [1] proposed a model for the exploitation of ontology based knowledge library to improve search over large document repositories. Semantic search is combined with conventional keyword based retrieval to achieve tolerance to knowledge base incompleteness. Huang, et al [2] presented a rough ontology which was based on a semantic information retrieval system named as ROSRS. The experiment results had shown that ROSRS could retrieve information semantically not only from precise ontology but also from rough ontology. Furthermore, the recall ratio and precision ratio of retrieval results were improved availably. Li, et al [3] proposed a framework on fuzzy ontology based knowledge reasoning. By reducing the fuzzy ontology, checking consistency and repairing inconsistency, the efficiency and accuracy of the fuzzy ontology based knowledge reasoning could be improved effectively. Ma, et al [4] proposed a rule based ontology reasoning method with expanding 5 groups reasoning rules on Jena reasoning engine, and demonstrated this method can greatly widen the solution space and increase opportunities for product

innovation by an experiment. Oaman, N.A, et al [5] proposed a semantic indexing and retrieval approach which includes ontology based information retrieval framework. The proposed ontology based retrieval model is based on an adaptation of the classic vector space model, which includes document annotation, weighting and concept based ranking algorithm.

In this paper, based on the above methods, the literature ontology and the semantic dictionary ontology are presented for literature retrieval. With the ontology, the main concepts of the literature in a special field and the relationships of document metadata are considered together. The literature concept hierarchy structure and the domain semantic dictionary are constructed to help optimize the literature retrieval result. Also the ontology reasoning mechanism is applied to generate inference rules to help find implicit knowledge in literature, and push more interesting information to retrieval user. This method can improve the accuracy and breadth of literature retrieval compared with the traditional keywords matching retrieval method.

2 RELATIVE THEORY OF ONTOLOGY

2.1 *The related concept of ontology*

Ontology is an explicit formal specification of a shared conceptual model [6]. According to the degree of dependence on domain, ontology can be divided into top-level ontology, domain ontology, task ontology and application ontology. Among them, the domain ontology built with concepts in a special field is always used to describe the specified domain knowledge. Domain ontology is expressed by four tuples:

$$O = \{C, I, R, Ax\}.$$

Here, C, I, R, A_x represent a set of concepts, instances, relations and axioms respectively.

To construct ontology, a lot of methods are presented according to different fields, such as seven-step method, life-cycle method, skel eton method, IDEF-5 method and so on [7]. The seven-step method is always used to build domain ontology. Its main steps are:

Step1: Analyzing requirements make these questions clear: the application domain, range of the ontology and the objective of the ontology construction.

Step2: Find the reusable ontology. It is the most effective and fast method to build the ontology.

Step3: Abstracting the concepts of ontology into classes.

Step4: Define classes and their hierarchies. The hierarchy of classes is defined according to the top-down rule or the bottom-up rule or their combination.

Step5: Define the attributes of classes. The attribute includes object attribute and data attribute.

Step6: Define the restrictions of attributes.

Step7: Adde instances for the classes.

2.2 Ontology reasoning

Ontology reasoning is to obtain the implicit knowledge from the known definitions and declarations. It is different from blind reasoning, but organizes reasoning mechanism according to the specific demands and applications. For the user of ontology, the main purpose of ontology reasoning is to obtain the implicit knowledge and solve more problems.

Jena, as a general ontology reasoning tool, contains a series of inference rules named the general rules. In addition, according to the individual demands in different fields, user can also define their own inference rules to create the specific reasoning mechanism with Jena.

3 THE LITERATURE DOMAIN ONTOLOGY AND ONTOLOGY REASONING

3.1 The literature domain ontology

According to the characteristics of the academic literature, a literature domain ontology which contains the literature ontology and semantic dictionary ontology is presented in this paper. The literature ontology is used to describe the important metadata and their relations in literature. The semantic dictionary ontology is used to extract the main concepts

in professional field and find the semantic relations of concepts, such as synonymy, hyponymy, near-synonymy, abbreviations etc. And finally a conceptual hierarchy tree is constructed. The main purpose of literature domain ontology is to achieve semantic reasoning and conceptual expansion of the query content.

3.1.1 The literature ontology

Referencing the seven-step method described above, the main building processes of the literature ontology are summed as follows:

Step1: Define the requirements and ranges of the literature ontology. Here, the literature ontology is built on the basis of academic thesis.

Step2: Define the essential classes and the hierarchies of these classes. Five conceptual entities in academic papers which are *author, paper, journal, conference* and *organization*, are defined. Here, the class of *paper* includes *journal_paper, degree_thesis*, and *conference_paper*. The hierarchies of the literature ontology are shown in figure 1.

Step3: Define the attributes of classes and add the instances. The attribute includes object attribute and data attribute. The object attributes are shown in table 1. Among them, *learn_from* and *teach, citing* and *cited* are two pairs of reciprocal and symmetric attributes. Meanwhile, the data attributes of classes are also defined. But due to the space limitations of this article, only the data attributes of *paper* are listed in table 2. Finally, by web crawler tool, the academic papers or theses are added to the literature ontology as instances.

Step4: Store the literature ontology. The literature ontology is saved as *paper.owl* and also stored in SQL Server database.

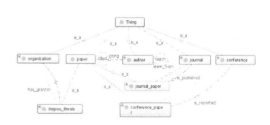

Figure 1.   Hierarchy of literature ontology.

3.1.2 The Semantic dictionary ontology

In this paper, the semantic dictionary ontology is presented to help implement conceptual expansion. Usually, when users begin a retrieve, they will use different query keywords according to their own preferences even if they express the same meaning.

This will lead to different retrieval results in traditional keywords-matching retrieval pattern. But with semantic dictionary ontology, the same retrieval result can be obtained.

In order to construct the semantic dictionary ontology, the main step is to obtain the term in some professional field and their relations. For example, if we want to construct the semantic dictionary ontology in information security field, the main concepts and its hyponym, synonym, abbreviations, English or Chinese term are extracted from the professional references, such as *The Principle and Application of Network Security* [8], *Chinese Classified Subject Thesaurus*, the standards *GB/T20274* and *GB/T25069*. Then the conceptual hierarchies, semantic relationships of concepts and attributes of concepts are defined. The specific description of the class is shown in table 3. Finally, the top-down rule is adopted to establish the hierarchy tree.

Table 1. Description of object attributes.

Basic Attributes	Attribute Name	Domain	Range
	creatte	author	Paper
	is_published	paper	Journal
	is_reported	paper	Conference
	has_grantor	paper	Organization
Inferential Attributes	learn_from(teach)	author	Author
	citing(cited)	paper	Paper
	both_citing	paper	Paper
	both_cited	paper	Paper
	same_tutor	author	Author
	sametutor_paper	paper	Paper
	sameorg_tutor	author	Author
	sameorg_tutor_paper	paper	Paper

Table 2. Data attributes of *paper* attributes.

Attribute Name	Domain	Range
title	paper	String
abstract	paper	String
keywords	paper	String
degree_level	paper	String
discipline	paper	String
grantor	paper	String
tutor	paper	String
CLC	paper	String
year	paper	String
fund	paper	String

Table 3. Description of the class (Term).

Attribute Name	Attribute Description	Domain	Range
term	Chinese of the termT	term	String
english	English of the term	term	String
synonym	Synonym of the term	term	String
hypernym	Hypernym of the term	term	String
hyponym	Hyponym of the term	term	String

3.2 User-defined inference rules

Usually, there are a lot of implicit relations existed in field literature, such as the citing relations, the teacher-student relationship, and the relationship between colleague, etc. According to data structure of literature and query experience of users, these relations are summarized as some inference rules. These rules are described with OWL description language and SWRL rules and saved as *paper.rules*, as follows.

rule1:(?x fa:citing ?z) (?y fa:citing ?z)
notEqual(?x,?y) ->(?x fa:both_citing ?y)
rule2:(?x fa:cited ?z) (?y fa:cited ?z)
notEqual(?x,?y) -> (?x fa:both_cited ?y)
rule3:(?x fa:learn_from ?z) (?y fa:learn_from ?z)
notEqual(?x,?y) -> (?x fa:same_tutor ?y)
rule4:(?x fa:create ?a) (?y fa:create ?b)
(?x fa:same_tutor ?y) notEqual(?a,?b) ->
(?a fa:sametutor_paper ?b)
rule5:(?x fa:teach ?a) (?y fa:teach ?b)
(?x fa:author_organization ?z) (?y fa:author_organization ?z) notEqual(?x,?y) ->
(?x fa:sameorg_tutor ?y)
rule6:(?x fa:create ?a) (?y fa:create ?b)
(?x fa:sameorg_tutor ?y) notEqual(?a,?b) -> (?a fa:sameorg_tutor_paper ?b)

3.3 Framework of literature semantic retrieval based on domain ontology

A framework of literature semantic retrieval system based on domain ontology is shown in figure 2. When user inputs the query keyword, a set of relation concepts will be obtained from literature domain ontology and user-defined inference rules. With these extension concepts, more accurate and broader information can be obtained and more feedback can be got from users.

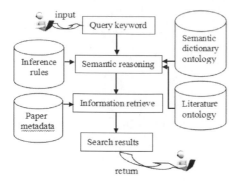

Figure 2.   Framework of literature semantic retrieve system.

4   EXPERIMENTAL AND ANALYSES

We collected 112 degree theses, 156 journal papers and 68 conference papers in information security field by web crawler tool as experiment samples. There are three designed experiments to verify the effective of the presented approach. The literature semantic retrieval system is developed with JAVA language, and the database is SQL SEVER 2008.

4.1   The test of ontology reasoning

This experimental system is used to show the application of ontology reasoning. In figure 3, when users retrieve *Xueqin Zhang*, her papers and related authors and tutors are returned by the third and fifth inference rule. The reasoning results are shown in the red strings. When we click the blue authors-strings, the related papers will be gotten.

From the experiment, the information of students and colleagues in the same field are obtained by ontology reasoning. Different from traditional simple keyword-matching retrieval system, this implicit information are directly obtained in our retrieval system in a query.

Figure 3.   The application of user-defined rules.

4.2   The test of conceptual expansion

The experiment is used to check the effect of the semantic dictionary ontology. When user inputs *intrusion detection*, the retrieval system can not only find the papers containing *intrusion detection*, but also those papers containing *distributed intrusion detection* and *DIDS* as shown in figure 4. Here, *distributed intrusion detection* is the hyponym of *intrusion detection*, and *DIDS* is the abbreviations of *distributed intrusion detection* which are defined in semantic dictionary ontology.

The ontology is also designed to support the Chinese-English conversion query in the semantic retrieve. It means when Chinese word is inputted, the equivalent English and Chinese papers can be gotten

Figure 4.   The expansion of synonym and hyponym.

together, vice versa.

4.3   The analyses of semantic expansion

We use all of the experiment samples to analyze the advantage of the literature semantic retrieval based on domain ontology as shown in table 5. Here, *NK* represents the number of papers with a specified query keyword in experiment samples. Correspondingly, *NS, NH* and *NC* represent the number of papers with the specified query keyword's synonym, hyponym and equivalent Chinese term respectively. *KMM* means the number of papers retrieved by keywords-matching method, and *LSM*

Table 5.   Retrieval results on the related papers.

Query keywords	NK	NS	NH	NC	KMM	LSM
cryptography	10	13	24	10	**10**	**57**
information hiding	10	6	19	12	**10**	**47**
intrusion detection	26	0	7	30	**26**	**63**
biometrics identification	10	5	8	13	**10**	**36**

means the number of papers retrieved by our literature semantic method. From table 5, it can be seen that the proposed semantic retrieval system can feedback results not only having accuracy, but also having more breadth.

5 CONCLUSIONS

In this paper, according to the characteristics of the academic literature, literature ontology and semantic dictionary ontology are presented to build the literature domain ontology. With the semantic dictionary ontology, a lot of conceptual expansions are implemented, and by analyzing the literature ontology and users' interests, ontology reasoning is adopted to get user-defined inference rules. Based on these methods, a literature semantic retrieval framework is presented and a web retrieval system is developed. The experiments on information security academic literature verified this framework that can not only make retrieve accurately, but also enhance the retrieve breadth. Considering Chinese user, this system makes the retrieve more convenient compared with the traditional keywords-matching retrieve pattern.

REFERENCES

[1] Castells P, Fernandez M, Vallet D. 2007. An Adaptation of the Vector-Space Model for Ontology-Based Information Retrieval. Knowledge and Data Engineering19(2):261−272.
[2] Huang Yinghui, Li Guanyu, Li Qiangqiang. 2013. Rough Ontology Based Semantic Information Retrieval. Computational Intelligence and Design 1(6):63−67.
[3] Li Guanyu, Ma Di, Loua, V. 2012. Fuzzy ontology based knowledge reasoning framework design. Software Engineering and Service Science(ICSESS), Beijing,22−24 June 2012:345−350.
[4] Ma Jianhong, Zhang Quan, Zhang Jinling, et al. 2012. A rule-based ontology reasoning for scientific effects retrieval. Management of Innovation and Technology(ICMIT), Sanur Bali, 11−13 June 2012:232−237.
[5] Osman, N.A, Noah, S.A.M, Omar, N. 2010. Semantic Search in Digital Library-semantic technology. Information Technology 3:1504−1507.
[6] Studer R, Benjamins V R, Fensel D. 1998. Knowledge Engineering, Principles and Methods. Data and Knowledgeing 25(122):161−197.
[7] USEHOLD M. 1996. Ontologies principles, methods and applications. Knowledge Engineering Renew 6(11):2−3.
[8] Zhang Shiyong. 2003. The principle and application of network security. Beijing: Science press.

Network Security and Communication Engineering – Chan (Ed.)
© *2015 Taylor & Francis Group, London, ISBN: 978-1-138-02821-0*

About single biometric network

S. Mahmudova & T. Kazimov
Institute of Information Technology of ANAS, Baku, Azerbaijan

ABSTRACT: In this paper, we propose biometric network in history, the essence of the problem and identification. The system finds facial images in auto mode via video camera, encodes them, and identifies them by using the images available in a database. The joint use of integrated or distributed biometric database is one of the main goals. For this purpose, the used data is very important to be in accordance with international quality standards.

1 INTRODUCTION

The researches on the biometric features, such as human face, fingerprints, hand shape, sound parameters, iris and etc. And the development of new biometric identification systems are of great importance. The use of computer search engines for the human face recognition has become widespread in modern times. Note that the key data used in the scientific and practical issues includes the images of the studied objects. Availability of different types of pattern recognition systems expands the scope of the solved issues [1].

High quality identification maximum accuracy can be achieved by using biometric features, such as human face, fingerprints, and iris. Dimensions of the data bases and the quality of the stored are very important for biometric systems. The joint use of integrated or distributed biometric database is one of the main goals. For this purpose, the used data is very important to be in accordance with international quality standards. Thus, the two-dimensional images, including the three-dimensional images which are appropriate to be created [1].

In some works, a new biometric-based user authentication mechanism is proposed in heterogeneous wireless sensor networks [2].

In law enforcement bodies, the databases with a large number of human photo portraits are used. The photos images of criminals are compared with the photos in such databases and found. In modern times, false documents, and photos taken by mobile phones, the photographs taken from social networks are widely spread. Identification based on the facial image has developed rapidly, and connected to the information infrastructure of law enforcement agencies all over the world. The efficiency of the recognition algorithms has significantly increased in recent years and approved by experiments. In a network, the success of the algorithm depends on the length of the key that user uses [4].

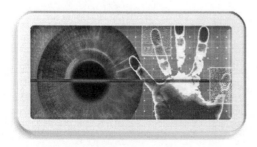

Figure 1. Fingerprints and eye-socket of the people.

Figure 2. Database and network environment.

The system finds facial images in auto mode via video camera, encodes them, and identifies them by using the images available in a database. In this case, the algorithms intended for three-dimensional images are used more commonly for recognition. The system forms electronic or print reports, which contain full facial information in the database, including personal information.

Figure 4.   mobile devices.

Figure 3.   Taken by a video camera and image.

The system includes integrated interfaces, which realize information exchange with image databases of law enforcement agencies and other departments. Using three-dimensional models improves the quality of recognition significantly. In this case, the quality of the cameras is of great importance [1].

In some papers, the security breaches are mitigated, and a new biometric-based user authentication scheme is proposed without using password for WSN [3].

The followings shall be taken into account to solve a number of issues:

• Improve the quality of the obtained images before their automatic evaluation, search, and sending to recognition servers;
• Maintain all the characteristics the Database Management System shall be realized based on the mobile facilities.

Biological neural networks are commonly used for the combined application of distributed biometric databases. Biological neural network is composed of chemical neuron groups or connected neurons. A neuron can be connected to many other neurons, and the total number of neurons in the network and their connections can be overwhelming. The location of neurons connection is called synapse. Transmission of the impulses is carried out by the mediator or electricity through transmission of ions from on cage into another.

Some researchers try to imitate some features of the biological neural networks in their studies in the cognitive field of artificial intelligence and modeling. Artificial neural networks are successfully applied in the development of program agents or various robots for the recognition of images and speech in the field of artificial intelligence. Currently, the majority of artificial neural networks used in the field of artificial intelligence is developed based on the statistical methods, optimization and control theory. Some works explore the problem of look-alike faces and their effect on human performance and automatic face recognition algorithms [5].

Figure 5.   The simple neural network: green- input layer, blue - hidden layer, and purple - output layer.

Let's inform some of the scientists involved in biometric network. Bain is one of them. As for Bain, any action leads to the activation of a certain neurons collection. Connection between these neurons is improved during the repetition of their performance. According to his theory, this repetition leads to the formation of the memory. At that time, his theory was regarded with suspicion by the international scientific society, because the neural connections in the brain are excessive. The human brain has an extremely complex structure, and it is able to function simultaneously in multiple issues [4].

The theory of another scientist, James, is similar to the Bain's theory. However, at the same time, James assumed that the formation of memory is realized due to the transmission of the impulses between the neurons, without requiring connections of neurons for each act of remembering or action.

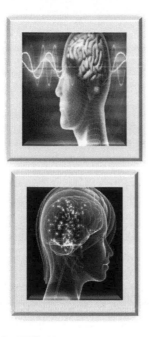

Figure 6. Man Wolfpack.

In 1943, Mc.Culloch and Pitts developed computer model for neural networks based on mathematical algorithms [6]. They called the model "threshold logic".

In late 1940s, Canadian psychologist and physiologist Donald Hebb,basedon the mechanism of neural plasticity, suggested the hypothesis of training

interpretation known as Hebb theorem. Hebb theory is viewed as a typical self-learning, which studies the issue directly without intervention of the expert system [7].

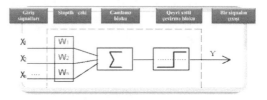

Figure 7. Neural network processes.

Neural Networks Researches stopped after the publication of computer learning by Minsk and Papert in 1969. They found two main problems related to computing machines, so they were processing neural networks. The first problem was that logical computation was not possible in one-layer neural networks. The second major problem was that computers did not have enough computational power to process a large volume of calculations necessary for the efficient neural networks. Neural Networks Researches were delayed until computers achieved massive computing power. One of the major achievements was the development of a method of backward propagation of errors, thus, the logical computation solved the problem. Summation block (denoted by net) is the sum of the multiplied appropriate weight coefficients (w) of total input signals (x).

$$net = \sum_{j=1}^{n} W_j x_j$$

Multi-layer neural networks "Cognition" developed by K.Fukusima in 1975, was one of the first papers in this field. The actual structure of the network and methods used for Cognitron to provide relative weights of relations was replaced by another strategy, so that each of the strategies had its pros and cons. The networks were able to distribute information in only one direction or to drop the information from another end. However, all nodes did not activate, and the network was not finite. To achieve two-way transfer of information between neurons nodes that was possible in Hopfield network (1982), and it was first included in the hybrid networks for specific purposes. In the mid-1980s, parallel distributed processing algorithm known as connectives was developed (connectives is based on network theories, which are difficultly organized and self-organizing systems). Rumelhart and McClelland (1986) made full use of connectives for computer modeling of neural processes.

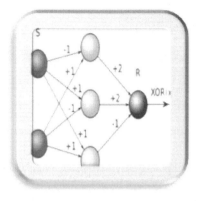

Figure 8. Hopfield network.

The spread of the method based on backward propagation of errors caused great enthusiasm, as well as much controversy in the scientific community. Thus, there was a doubt whether this study can be realized in the brain or not, because backward transmission mechanism of the signal was not known then. In addition, in 2006, a number of training algorithms of neural networks were proposed. These algorithms can be used with and without output signals to comprehend the main features of sensor signals distribution, which move in each layers of neural network, and to study the intermediate presentations.

Algorithm for parallel distributed data processing was known as connectives in the mid-1980s. Rumelhart and McClelland (1986) made full use of connectives for computer modeling of neural processes.

Neural networks used in the field of artificial intelligence are traditionally viewed as the simplified models of neurons of the brain. Thus, artificial neural networks of any size reflect the real structure of the brain.

Theoretically, the main subject of researches is the complexity in neurobiology and its properties, so they should have the separate neurons.

2 CONCLUSIONS

In this paper, we have proposed history single biometric network. And it is widely used in neural networks. Neural networks used in the field of artificial intelligence are traditionally viewed as the simplified models of neurons of the brain. Thus, artificial neural networks of any size reflect the real structure of the brain. Theoretically, the main subject of researches is the complexity in neurobiology and its properties, so they should have the separate neurons.

REFERENCES

[1] CarevNikolay, Povishenie effektivnosti raboti sotrudnikov QUVD q. Moskvi putem vnedreniya peredovix biometriceskix texnoloqiy. Rukovoditel napravleniya, 2011.
[2] Ashok Kumar Das, Bezawada Bruhadeshwar. A Biometric-Based User Authentication Scheme for Heterogeneous Wireless Sensor Networks / 2013 27th International Conference on Advanced Information Networking and Applications Workshops. Barcelona, Spain, March 25-March 28, 2013.
[3] Eun-Jun Yoon, Kee-Young Yoo, A New Biometric-based User Authentication Scheme without Using Password for Wireless Sensor Networks / 2011 IEEE 20th International Workshops on Enabling Technologies: Infrastructure for Collaborative Enterprises. Paris, France June 27-June 29, 2011.
[4] V.V. Satyanrayanarayana Tallapragada, E.G. Rajan. Multilevel Network Security Based on Iris Biometric / 2010 International Conference on Advances in Computer EngineeringBangalore, India, June 20-June 21, 2010.
[5] Hemank Lamba, Ankit Sarkar, Mayank Vatsa, Richa Singh, Afzel Noore. Face recognition for look-alikes: A preliminary study / 2011 International Joint Conference on Biometrics. October 11–13, 2011, West Virginia Univeristy, USA.
[6] Grove, A.T. 1980. Geomorphic evolution of the Sahara and the Nile. In M.A.J. Williams & H. Faure (eds), *The Sahara and the Nile*: 21–35. Rotterdam: Balkema.
[7] https://ru.wikipedia.org/wiki/Mak-Kallok, Uoppen.
[8] http://en.wikipedia.org/wiki/Donald_O._Hebb.

Network Security and Communication Engineering – Chan (Ed.)
© 2015 Taylor & Francis Group, London, ISBN: 978-1-138-02821-0

Reliability analysis of communication subsystem in the train control system

L.G. Zeng

College of Mathematics, Physics and Information Engineering, Zhejiang Normal University, Jinhua, China

ABSTRACT: The train communication system is an important part of the train control system, and its reliability is important for the high speed train. In this paper, we researched the communication subsystem's reliability of the train control system and established the dynamic fault tree of this subsystem using fault tree analysis method and Markov model and combining the simulation platform for experiment. We have obtained the fault probability of this communication system and provided reliable basis for diagnosing whether the system has fault.

KEYWORDS: The train control system, Fault tree, Fault probability.

1 INTRODUCTION

With the implementation of high-technology in railway system, especially the development of high speed trains, people can enjoy a more convenient means of transportation now. In recent years, we have mastered the 200 km/h railway technology and some train's speed can reach 350km/h. Some high-speed lines were put into operation, so that we can set out in the morning and arrive at night. But how to ensure the safety of the high-speed trains has become a research focus and difficulty, such as timely response, short interval, reliability, drive steady, and arriving on time etc. So, the train control system must ensure the safety of train operation. We mainly adopted the train control system named CTCS-3 and CTCS-4.

The reliability of the train control system needs to be completely ensured in the whole life cycle of the vehicle operation. The reliability can be estimated and got by some engineering means, test tools and simulation means etc., so we can know the TCS's ability to fulfill the required function in specific environment in limited time. Of course, how to estimate the system reliability is an important way to find the defect in the system design phase so that we can perfect the system. Currently, the traditional qualitative method includes: the failure model and the impact analysis (FEMA), Hazard and operability analysis (HAZOPS) and the qualitative fault tree analysis (FTA). The traditional quantitative analysis method includes: event tree analysis ETA), Markov model and quantitative fault tree analysis based on the probability calculation and statistical analysis method and so on. Because of the importance of the train control system and the complexity of the MIMO, we adapted the dynamic FAT to analyze the communication system of the train control system.

2 RELIABILITY ANALYSIS THEORY

2.1 Dynamic fault tree (DFAT)

The dynamic fault tree introduces dynamic logical gate based on the static fault tree analysis. It can analyze the dynamic events with the rules of fault and time sequence. After combining the advantages of the fault tree analysis method and the Markov model, we put the dynamic fault tree into smaller modules firstly, so we can get the static/dynamic subtree, and calculate subtree with BDD and Markov process method. Because the BDD was introduced into the FAT analysis method, the qualitative and quantitative analysis efficiency of the fault tree was improved immensely. The large scale fault tree is also efficiently and automatically analyzed and processed. BDD is a non-orthogonal decomposition that FAT based on the Shannon decomposition. We convert the fault tree into BDD by using *ite (if–then–else)* method

$$ite(F, G, H) = (F \wedge G) \vee (F \wedge H) \qquad (1)$$

2.2 Reliability index

1 Failure distribution function (F (t))
That the research object loses its function was called failure, and any object failure exists randomly. There is a random parameter—T from proper work to failure. The distribution function F(t) about T is called the failure distribution function. It denotes the probability that working time is not more than t from start to failure.

$$F(t) = P(T \leq t) \qquad (2)$$

2 Reliability function

The reliability function shows the probability that the research object can complete specific function in specific circumstance during specific time; it is a function about time.

$$R(t) = P(T \geq t) = 1 - F(t) = \int_0^\infty f(t)dt \qquad (3)$$

3 Failure rate function

It denotes the probability of the research object within the per unit time after the research object normality works to time t, it is showed by $\lambda(t)$.

3 STRUCTURE TRAIN AND GROUND COMMUNICATION SYSTEM FAT

3.1 *The communication system composition*

The TCS that we used is the China train control system—CTCS. The CTCS includes 5 technical grades to meet the demand of different routes, but the CTCS–3 was used generally. The TCS includes some function models. Because of the paper length limit, we will only study the communication between train and ground here.

The GSM for Railways—GSM-R is a private digit mobile communication system which is only for rail communication. The GSM-R mainly comprises of wireless trains dispatch communication, marshalling dispatch communication, zone maintenance operation communication, emergency communication and tunnel communication etc. It could compete transferring the train auto control information and detection information and provide automatic addressing and passenger train service. Its composition is shown in figure 1.

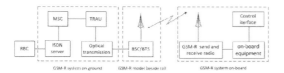

Figure 1. GSM-R communication system schematic diagram.

3.2 *The FAT construction and analysis*

Based on the principle of fault tree, we established the top event. The train with ground communication module is as shown in figure 1. The fault tree chooses train and ground communication failure as the top event. If any unit or any wireless channel

fails, this communication system would fail. So the intermediate events and bottom events are shown in the table 1:

Table 1. The FAT events.

Event number	Event name	property
T	Train—ground communication system fault	Top event
G1	GSM-R system on-broad fault	intermediate events
G2	GSM - R system on ground fault	intermediate
G3	GSM-R network fault	intermediate
G4	Wireless communication mode fault	intermediate
G5	Wireless channel fault	intermediate events
E1	GSM-R radio-station fault	Bottom even
E2	GSM-R model 1 on ground fault	Bottom event
E3	GSM-R model 2 on ground fault	Bottom event
E4	GSM-R model beside rail fault	Bottom even
E5	Wireless communication model 1 fault	Bottom even
E6	Wireless communication model 2 fault	Bottom even
E7	Wireless channel 1 fault	Bottom even
E8	Wireless channel 2 fault	Bottom even
E9	Wireless channel 3 fault	Bottom even

The dynamic fault tree of the train-ground communication system is shown in figure 2.

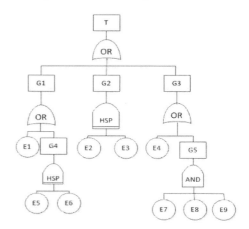

Figure 2. Train - ground communication system dynamic fault tree.

The fault tree in figure 1 is divided into several modules. There is no need to divide the module into smaller module; we can directly analyze the module using Markov model. We get 6 sub-trees through module decomposition: T, G1, G5, G3, G4 and G2 (G4 and G2 are dynamic FAT). We use the depth-first traversal to visit the FAT and the result is shown in table 2.

Table 2. Result based on depth-first traversal.

No. step	1	2	3	4	5	6	7	8	9	10	11
visited event	T	G1	E1	G4	E5	E6	G4	G1	G2	E2	E3
No. step	12	13	14	15	16	17	18	19	20	21	
visited event	G2	G3	E4	G5	E7	E8	E9	G5	G3	T	

4 SIMULATION RESULT

Based on the technical standard of CTCS-S, the fault probability of each bottom event was shown in the table 3:

Table 3. Fault probability for each bottom event.

Event label	Event name	Fault probability	MTBF
E1	GSM-R radio-station fault	1.5×10^{-8}	6.9×10^7
E2	GSM-R model 1 on ground fault	2.0×10^{-4}	4.8×10^3
E3	GSM-R model 2 on ground fault	2.0×10^{-4}	4.8×10^3
E4	GSM-R model beside rail fault	1.2×10^{-6}	8.3×10^5
E5	Wireless communication model 1 fault	1.8×10^{-4}	5.6×10^3
E6	Wireless communication model 2 fault	1.8×10^{-4}	5.6×10^3
E7	Wireless channel 1 fault	1.5×10^{-5}	6.7×10^4
E8	Wireless channel 2 fault	1.5×10^{-5}	6.7×10^4
E9	Wireless channel 3 fault	1.5×10^{-5}	6.7×10^4

So by using the analysis method of Markov chain and combining the simulation platform, the FAT G1, G2, G3, G4 are simulated. The top event probability of this 4 sub-trees are shown in the table 4:

Table 4. FAT sub tree event probability.

FAT subtree(G_i)	Top event probability $P(G_i)$	FAT subtree(G_i)	Top event probability $P(G_i)$
G1	4.8×10^{-8}	G2	4.5×10^{-8}
G3	1.2×10^{-6}	G4	3.4×10^{-8}

As shown in Fig. 1, the fault tree top event $T = G_1 \vee G_2 \vee G_3$. The probability of the whole communication system fault tree top P (T) is shown below:

$$P (T) = P (G1) + P (G2) + P (G3) = 1.293 \times 10^{-6}$$

The average time between failures of the system is: MTBF$=2.4 \times 10^3$

5 CONCLUSION

In this paper, we have studied the reliability of the communication system included in the train control system. Firstly, we used the dynamic fault tree analytical method to structure FAT, and then analyzed the simulation data based on the simulation platform. Secondly, we have obtained the fault probability of the communication system. With these completed, we can master the reliability of the communication system which ensures the reliability of the train control system by providing the solution to study the reliability of the train control system.

ACKNOWLEDGEMENT

This work was financially supported by the Education Department of Zhejiang Province (Y201328293) and Opening Fund of Top Key Discipline of Computer Software and Theory in Zhejiang Provincial Colleges at Zhejiang Normal University.

REFERENCE

[1] Y Dutuit, A Rauzy Approximate estimation of system reliability via fault trees [J]. Reliability Engineering and System Safety 2005, (87): 163–172.
[2] Barua, A., Sinha, P., Khorasani, K., Tafazoli, S.Control Appli-cations, 2005.CCA 2005.@roceedings of 2005 IEEE Conference on.2005.
[3] Hongli Zhao, Tianhua Xu, Tao Tang. Towards modeling and evaluation of availability of communication based train control (CBTC) system. Communications Technology and Applications, IEEE, 2009, pp860–863.
[4] Li Zhu, Yu, F.R, Bing Ning. Availability Improvement for WLAN-Based Train-Ground Communication Systems in Communication-Based Train Control (CBTC). IEEE.2010.page(s): 1–5.
[5] John G.Proakis.Digital Communication (4th edition) [M].USA: The McGraw-Hill Companies, 2001.

Network Security and Communication Engineering – Chan (Ed.)
© *2015 Taylor & Francis Group, London, ISBN: 978-1-138-02821-0*

Group management application for android social network service

S.S. Lee

Department of Computer Science and Engineering, Jeonju University, Wansan-Gu, Jeonju, Jonbuk, South Korea

ABSTRACT: This study proposes a Social Network Service (SNS) which is used exclusively for groups. In order to develop a system, this study implements functions such as group service, chatting, location-tracking, transmission, drawing and editing, management of photo gallery, and file transmission and combines the whole module to design a group management application system for Android.

KEYWORDS: Social network system, Google map, GPS, Smart Phone App, Software engineering.

1   INTRODUCTION

Most people are engaged in a variety of social activities such as gatherings in small groups, with colleagues, and club meetings. Modern people make a phone call or use group Kakao Talk or text message to make plans for get-togethers, but it is not easy to adjust the schedule among group members as there are occasional constraints in communication. This study aims to create a "social network service (SNS) available and exclusive for groups," which is necessary for group gatherings. The following functions are required to complete this system: Chatting service of group members, Photo gallery of group members, Contact information of group members, Detailed schedule management of group members, Current location information that shows the movements of group members, and Transmission of current location information of group members. The purpose of this study is as follows: this system is an application for Android as well as a program for smart phones targeting the user group that is growing in number with the rapid increasing users of smart phones. Currently, social network services of messengers such as Facebook, Twitter and Kakao Talk are wielding their power. The intention of this study is to open up a new market for social network . It is necessary to enable many users to use this application and provide more stable service in terms or functionality and performance. This paper is organized in the following structure. Chapter 2 presents an analysis of functions and requirements proposed by this study along with the system structure map. Chapter 3 introduces the database communication and XML Parsing method using JSP as the implementation technology. Chapter 4 draws the use case diagram and specification sheet, and introduces the sequence diagram and class diagram. Finally, the paper presents a conclusion and discusses problems for future improvement.

2   ANALYSIS OF CUSTOMER DEMAND

First, the following scenario is set up in order to explain the theoretical background of this study. Let's see an example of the scenario. Office worker *K* regularly meets with five of his friends from high school. Setting the date and place or communicating about their gatherings is inconvenient for them. The busiest member *L* sets the time and sends texts to all the other members, but *P* and *M* are not available on that day. They have to contact with each other to adjust the time, which requires a great deal of time and is highly inconvenient. Then *K* hears about an application that can manage small groups, and he installs it and creates a group. After that he invites his five friends on the contact list to join via Kakao Talk (or text) message. Once the installation is completed, they undergo the process of authorization via cell phone and then complete the group joining process. They check other members on the [Contact List] menu and make phone calls or send texts and Kakao Talk messages. *L* posts his selca on the Photo Gallery, which sends real-time push notification alarm to other friends. The friends who check the photo can exchange thoughts about it by posting comments below. They register the date and time for their gatherings on the [Schedule] menu, and members can receive push notification *X* hours before the fixed time for their gathering. Sometimes it is difficult to describe the precise location for the gathering; in that case, specific location information can be sent to a friend using the 'Send Location' function so that the friends can find their way to the place.

When this application is runnig, you can be authorized with your cell phone number. The group list appears after authorization. The invited friends receive the invitation message via text or Kakao Talk. The invited users select the link on the message and move on to the Market's application download page. After installation, they undergo the process of authorization via cell phone number and are automatically added to the group. Their group name and cell phone numbers are encrypted and saved on the database.

Figure 1. Diagram of group manager.

The group network configuration and friend registration process are designed as follows. Friends are registered whenever new group members are added, and you can check your friends on the [Contact List] menu. You can call, text or send a Kakao Talk message to the registered friends. It is so designed that data is transmitted for the user to use other services such as chatting, photo gallery or schedule through the group network. Group members can use real-time chatting service. If one group member enters something on the chatting screen, it is sent to the database server and notifies other group members through push notification alarm. If you select the alarm, the application is automatically running and switched to the chatting screen. You can select Alarm ON or OFF on the [Setting] menu.

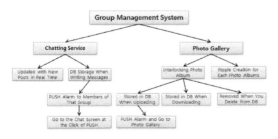

Figure 2. Diagram of chatter and photo manager.

The current location of group members can also be identified on a real-time basis. The members' consent to provide location information is obtained initially when the application is installed. The GPS information of the group members is transmitted to the database server with a certain time period, and if you select the location identification menu, the latest GPS on the database server is loaded and marked on the map. Sending a specific location means sending specific location information on the map to other group members instead of sending your own location. A few structure maps of the main system are shown in Figure 1 to Figure 2.

3   IMPLEMENTATION TECHNOLOGY

3.1   *Communication method of database using JSP*

First, it is important to use various communication modules efficiently as this is an application for group management. In particular, the communication technique of database using JSP which is shown in (Figure 4) can be described as follows. Sending data to the DB on an Android application requires the following process. This is an example of a method that is called when a friend is invited. Three things are received as parameters: requestURL indicates the address of jsp, groupnumber indicates group num-

Table 1.   Implementation example of DB communication.

```
public InputStream requestWriteAddressInvite
                    (String requestURL,
String groupnumber, String phonenumber) {
try {
    HttpClient client = new DefaultHttpClient();
    List<NameValuePair> dataList =
        new ArrayList<Name ValuePair>();
    dataList.add(new BasicNameValuePair
    ("groupnumber", groupnumber));
    dataList.add(new BasicNameValuePair(
    "phonenumber", phonenumber));
    requestURL = requestURL + "?"
    + URLEncodedUtils.format (dataList,
    "UTF-8");
    HttpGet get = new HttpGet(requestURL);
    HttpResponse response =
client.execute(get);
    HttpEntity entity = response.getEntity();
    InputStream is = entity.getContent();
/* Some Code is Omitted */
```

ber, and phonenumber indicates cell phone number A custom list called dataList is created, and the data is saved. Then the data is attached to the URL in the form of UTF-8, and this URL is given to the HttpGet format as a parameter. Then HttpResponse is created and the execution (get) of the HttpClient object is called, after which the transmission is completed with the parameter attached to the relevant URL. The following <Table 1> is an example of implementing the above mentioned process.

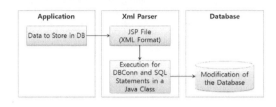

Figure 3. Sequence diagram between application and DB communication.

After it is completed as prepareStatement, executeUpdate() is running to change the DB content. <Table 3> is an example of implementing this process.

Table 2. Implementation example of XML Parser.

<?xml version="1.0" encoding="utf-8" standalone="no" ?>
<%@page import="com.parser.*"%>
<%@page import="java.io.*, java.sql.*, java.util.*, java.net.*, java.io.BufferedReader" contentType="text/xml; charset=utf-8"%>
<addressInvite>
<% String getGroupnumber = request. getParameter("groupnumber");
String getPhonenumber = request. getParameter("phonenumber");
Address obj = new Address();
 getPhonenumber = obj. extractNumber(getPhonenumber);
 obj.writeAddressInvite (getGroupnumber, getPhonenumber); %>
</addressInvite>

Figure 4. Sequence diagram of XML parsing.

3.2 XML Parsing method of database using JSP

First, for operation on the application, this study explains XML Parsing of database using JSP asit is

shown in (Figure 5). The method for parsing on the application is shown in <Table 4>. The address and parameter are entered using XmlPullParser, and the parser is set up. Then, while () is performed and the TAG information and main text are called. The called information is put into the array list called userList for use. The following process must be carried out for the system to operate on XML. The first thing is to operate on readUserList.jsp. Count, or group number, must be received through the parameter. A file in the XML format is created, the database connected, and the completed SQL statement is executed to save the called results on rs. rs.next() is looped, saved on retResult, and printed on screen. It is a JSP file, but all are in the XML format.

Table 3. Implementation example of database in java class.

```
public Address() { try {
Class.forName("org.gjt.mm.mysql.Driver");
        con = DriverManager.getConnection(url,
userid, passwd);
            } catch (Exception e) {
            e.printStackTrace(); }}
 public void writeAddressInvite(String groupnumber,
 String phonenumber) {
        try {
        String sql = "INSERT INTO
GROUPINVITE
                (GROUPNUMBER,
PHONENUMBER) VALUES(?, ?)";
        stmt = con.prepareStatement(sql);
        stmt.setString(1, groupnumber);
        stmt.setString(2, phonenumber);
        stmt.executeUpdate();
        } catch (Exception e) {
        e.printStackTrace();
        }
        finally {

        try { stmt.close(); con.close();}
        catch (SQLException se) {
se.printStackTrace();
}}}
```

3.3 Communication of SQLite database

Saving data on SQLite is as follows. When we create an app program, it is sometimes necessary to use the DB of the application itself, and this is when the current group number is used. If the CUR GROUP exists, it must be DROPPED, and the new CUR GROUP table is created. After that it is INSERTED to save the group number. This is coded as which is shown in <Table 5>.

The determined location information is sent to the database server through write GPS.

Table 4. Implementation example of XML parsing.

```
/* Some Code is Omitted */
StrictMode.ThreadPolicy policy =
 new StrictMode.ThreadPolicy.Builder()
                      .permitAll().build();
StrictMode.setThreadPolicy(policy);
String xml = "http://ADDRESS/jspXmlParsing/
readAddressUserList.jsp?count ="+ groupNumber;
URL URL = new URL(xml);
InputStream in = URL.openStream();
XmlPullParserFactory factory =
XmlPullParserFactory.newInstance();
XmlPullParser parser = factory.newPullParser();
parser.setInput(in, "utf-8");
int eventType = parser.getEventType();
while (eventType != XmlPullParser.END_
DOCUMENT) {
        switch (eventType) {
        case XmlPullParser.START_DOCUMENT:
        break;
        case XmlPullParser.END_DOCUMENT:
        break;
        case XmlPullParser.START_TAG:
            if (parser.getName().equals("name")) {
                inname = true;
                } else if (parser.getName().
equals("number")) {
                innumber = true;
                } else if (parser.getName().
equals("date")) {
        indate = true; } break;
        case XmlPullParser.END_TAG:
        break;
        case XmlPullParser.TEXT:
        if (inname) {        Name = parser.
getText(); inname = false;
        } else if (innumber) { Number = parser.
getText(); innumber = false;
        } else if (indate) {
        Date = parser.getText();
        indate = false;
        userList.add(new UserData(Name, Number,
Date));
        } break; }
        eventType = parser.next();}/* Some Code is
Omitted */
```

Table 5. Example of data in SQLite.

```
SQLiteDatabase db;
SQLiteHelper mDBHelper = new
SQLiteHelper(ACTIVITY-NAME.this);
db = mDBHelper.getWritableDatabase();
String groupNum = groupList.get(position).
getGroupcount();
db.execSQL("DROP TABLE IF EXISTS
CURGROUP");
db.execSQL("CREATE TABLE
CURGROUP(NUMBER TEXT)");
db.execSQL("INSERT INTO CURGROUP VALUES("
+ groupNum + ")");
```

Scenario Name	Group Management
Participating Actor	Driven by User
Starting Condition	Begins when the user runs application
Flow of Event	* Normal Flow ① If there is a group to which the user is invited on the DB, it shows group name, number of members, inviter and message in a list-view style, and marks [Join] and [Reject]. ② Receive nickname when joining, and complete joining process by clicking on [Joining Process Complete] button through redundancy check. When rejected, it is removed from the list. ③ Groups joined by the user are printed in a list-view style on screen. ④ Move to creation screen by clicking on [Create Group] button. ⑤ Receive group name, group description and nickname. ⑥ Group name must undergo the process of [Redundancy Check] to create a new group. ⑦ After the group creation is complete, the groups joined are printed in a list-view style.
Ending Condition	Ends when the user ends application or logs into a group.

4 UNDERSTANDING OF CODES USING DIAGRAM

4.1 Usecase specification

Usecase specifications of friend management and chatting service are omitted.

4.2 Usecase specification of photo gallery

Usecase specifications of schedule management, location tracking, and location transmission are omitted.

5 CONCLUSION

This study designs a group management application system for Android, implements functions such as group service, chatting, location-tracking, transmission, drawing and editing, management of photo gallery and file transmission, and combines the whole module. This is uploaded on Google Play (formerly Market) as "Talk Talk Among Ourselves" and is actually used by many users. As a messenger application,

Scenario Name	Photo Gallery
Participating Actor	Driven by User
Starting Condition	Begins when user selects [Photo Gallery] menu.
* Flow of Event · Print a list view under the photos by bringing the comment data of application. * Selective Flow · If a comment is written by selecting [Post], data is saved on DB and 5 to 7 of the normal flow are operated.	* Normal Flow ① Bring the number of photo galleries from DB and compare with the number of photo galleries in the application. ② Bring the amount of the number in the application subtracted from the number on DB, and save them on the application. ③ Bring nine of the application data per screen and print in the grid view method. ④ The selected photo is enlarged, and the user can write a comment below. ⑤ Bring the number of comments in the Photo Gallery selected on DB and compare with the number of comments in the Photo Gallery selected in the application. ⑥ Bring the amount of the number in the application subtracted from the number on DB, and save them on the application.
Ending Condition	Ends when the user ends application or selects a different menu.

it must secure higher speed in terms of performance, and prevent loss or leakage of data by thoroughly managing the database. Moreover, the program must have constant version upgrade with regard to the feedback of users.

ACKNOWLEDGEMENTS

This is short for 'self camera'. The term was first used in the following sentence from the Dong-A Ilbo article on June 18, 2004 in Korean: "The growing trend of selca is one example that shows how much people are interested in 'being photographed well'.

REFERENCES

[1] Ian Sommerville, Software Engineering, 9th Ed., ISBN 978-0-13-703515-1, 2011.
[2] Open API. http://en.wikipedia.org/wiki/Application_ programming_ interface
[3] http://developer.android.com/reference/android/pack-age-summary.html
[4] Maser Peiravian and Xingquan Zhu, "Machine Learning for Android Malware Detection Using Permission and API Calls," 2013 IEEE 25th International Conference on Tools with Artificial Intelligence (ICTAI), Herndon, VA, USA, (2013) 300–305.

Network Security and Communication Engineering – Chan (Ed.)
© *2015 Taylor & Francis Group, London, ISBN: 978-1-138-02821-0*

Description of multi-ethnic facial features based on natural linguistic interpretation

X.D. Duan
Dalian Key Lab of Digital Technology for National Culture, Dalian Nationalities University, Dalian, China

Z.D. Li
Institute of System Science, Northeastern University, Shenyang, China

ABSTRACT: In this paper, the semantic membership function is established to describe facial features based on AFS. Meanwhile, according to the similarity of the same ethnic and difference of various ethnic, the data mining method is proposed to extract the individual semantic. The method proposed has fewer rules and is easy to understand. Lastly, there is an experiment of Uygur, Tibetan and Zhuang. The result shows that there is a balance between semantic interpretation and accuracy for individual attribute.

1   INTRODUCTION

China is a multi-ethnic country, different factors like region, culture, and living style lead to distinctive facial features of people from different nations. For example, it's common for people from Zhuang to have wide forehead and short eyebrow, Tibetan ladder-shaped face, and Uighurs straight nose and thick eyebrows, and etc. [1-2]. The facial feature semantics are defined by the respective sections such as eyes, eyebrow, and features of a face. Meanwhile, it shows that utilizing the semantics of one's face can directly deliver the difference among nations.

The distinction of facial feature and section semantic is derived from face image recognition technology, which can be divided into tensor space and facial feature location method. The classical methods of PCA and LDA extracting image features and practicing recognition through analyzing the latitude variation of face image in tensor space, though with better precision, omit the geometrical characteristics of human face, and are complicated in calculation. Facial feature location method analyzes geometrical morphology and changes rule by detecting the facial features and utilizing topology of each feature. However its weakness lies on its less explanation on feature data. It is of higher superiority as for fuzzy rule-based method on tackling with the above problem, its models are widely used on data analysis because of its priority on explaining the data attribute, and it is easy to be understood. But on the contrary, excessive rules would decrease the interpretability on data; meanwhile they increase the difficulty in model establishment and the liability to fitting phenomenon, which would affect the

accuracy. Axiomatic Fuzzy Set[3-5] is a new method proposed on base of fuzzy rule, which can establish the fuzzy assembly according to the original assignment and fuzzy semantics. In the framework of AFS theory, based on the subjective and objective uncertainty, the information included in the data store is transferred from subset membership to logic operator. All the above will be helpful to meet human recognition by making description on data.

This paper utilizes human facial feature to establish ethnic facial feature data assembly, and it founds the individual and ethnic faces semantics set. At the same time, this paper proposes one mining algorithm between the classes based on the individual difference of ethnics and another mining algorithm in the class on the basis of the individual similarity of the same ethnic. By using the above methods, this paper is trying to extract different facial semantics feature of ethnics, establish face semantic expression sets of different ethnics, describe ethnic human facial features, and finally distinguish individual ethnic attribute on the basis of ethnic human facial semantics sets.

2   RELATED WORKS

2.1   *AFS algebra*

Definition 1[3-5]: let *M* be one human face semantics set, and *EM* be any composite human face definition. Then *EM* can be expressed as follows

$$EM = \{\Sigma_{i\in I} Ai \mid A_i \subseteq M, i \in I, I \text{ is a non-empty indexing set }\} \qquad (1)$$

Suppose $\sum_{i\in I} A_i, \sum_{i\in J} B_j \in EM$, and $(M,\wedge,\vee)$ is completely distributive lattice, "x" represents the relationship of "and" or "or", we can get the following

$$\sum_{i\in I} A_i \vee \sum_{j\in J} B_j = \sum_{k\in I \sqcup J} C_k \qquad (2)$$

$$\sum_{i\in I} A_i \wedge \sum_{j\in J} B_j = \sum_{i\in I, j\in J} A_i \cup B_j \qquad (3)$$

In formula (2), (3) "∪" indicates the relationship between I and J which is neither mixed nor aggregated. When $u\in I$, $C_u=A_u$ will be concluded; when $u\in J$, $C_u=B_u$.

2.2 AFS structure

AFS structure is another important subject in AFS theory. AFS structure (M,τ,X) is defined on the basis of face sample space X and semantics set M. By comparing each factor's belonging degree in semantics, fuzzy set is expressed as binary relation on X.

Definition 2[3-5]: Suppose X as facial feature data set, then $X=\{x_1, x_2, \ldots, x_n\}$, $M=\{m_1, m_2, \ldots, m_p\}$. $\forall m\in M$, m is the composite semantics composed by some simple semantics on X, then the relationship of any three factors $x_1, x_2, x_3 \in X$ on AFS structure (M,τ,X) can be expressed as follows:

AX1: $\forall(x_1, x_2)\in X\times X$, $\tau(x_1, x_2)\subseteq \tau(x_1, x_1)$ \qquad (4)

AX2: $\forall(x_1, x_2), (x_2, x_3)\in X\times X$,

$\tau(x_1, x_2)\cap\tau(x_2, x_3)\subseteq \tau(x_1, x_3)$ \qquad (5)

X is called universe of discourse; M is called a concept set and τ is called a structure.

2.3 Coherence membership function of AFS

Theorem 1[6]. Let (Ω, F, P) be a probability measure space and M be a set of fuzzy terms on Ω, Let ρ_γ be the weight function for a fuzzy tern $\gamma \in M$, Let $X \subseteq \Omega$ be a finite set of observed samples from the probability space (Ω, F, P). If for any $m \in M$ and any $x \in \Omega$, $\{m\}\geq(x) \in F$. Then the following assertions hold.

$\{\mu\xi(x)\mid \xi\in EM\}$ is a set of coherence membership functions of $(EM, \wedge, \vee)$, provided that the membership function for each fuzzy set $\xi = \sum_{i\in I}(\prod_{m\in A_i} m) \in EM$ has two situations, which are discrete and continuous. And, they are defined as follows

$$\mu_\xi(x) = \sup_{i\in I} \inf_{\gamma\in A_i} \frac{\sum_{u\in A_i(x)} \rho_\gamma(u)N_u}{\sum_{u\in X} \rho_\gamma(u)N_u}, \forall x \in X \qquad (6)$$

3 FACIAL SEMANTICS DESCRIPTION

According to AFS framework, this paper proposes a new approach, which establishes the facial semantic using the features difference of individual facial features. So, this section has two parts, which extracts individual semantic and ethnic semantic.

3.1 Procedure A: Select individual facial semantics

In this section, let X be a set of training samples, in which it has C ethnic, and the number of each class is n_i. $F=\{f_1,f_2,\ldots,f_s\}$ is a set of features, and each feature has l semantic. $M=\{m_j \mid 1\leq j\leq sl\}$ is a set of fuzzy terms, where $m_1,m_2,\ldots,m_j$ are fuzzy terms associated with the feature f_j in F. The training samples are labeled and belong to c classes. We can obtain a sample semantic set for each individual x_i according to (6). The fuzzy description of x_i is described as following. $M\{xi\}=\{\sum m_j\mid M$ is non-empty set$\}$. Then, utilizing the set M, we can get a complex semantic set for each x_i, as following. $EM\{x_i\}=\{\sum A_i \mid A_i=\prod m_i, m_i\subseteq M\}$

The A_i is one complex semantic to describe the x_i of EM, and it is not unique. So, we can change a best semantic for x_i in EM. We set an evaluation function as following

$$V_{x_i}^{C_i}(A_i) = \mu_{x_i}(A_i) - \mu_{\overline{x_i}}(A_i) \qquad (7)$$

$\overline{x_i}$ is a set, in which the ethnic of any sample is different from x_i. $\mu_{A_i}(\overline{x_i})$ means that it is the maximum membership degree of x_i in the semantic Ai.

$$EM^* select \{x_i\}=\{A_i\mid \max\{V_{x_i}^{C_i}(A_k)\}, A_k \subseteq EM\} \qquad (8)$$

In this section, we can obtain a better semantic set for each individual, but the complexity of semantic rules will be large. In order to search the best semantic for each class, we give an evaluation index for the semantic set in Procedure B.

3.2 Procedure B: Select ethnic facial semantics

Now, we has gained the semantic $EM^*select\{x_i\}$, which is a description of each individual x_i. The semantic description of Ci class will be found in ξ_{Ci}.

$$\xi_{Ci}=\{\sum_{k\in Ci} A_i \mid A_i \subseteq EM^*_{Ci}\{X\}\} \qquad (9)$$

Due to the complexity of semantic set ξ_{Ci}, in this section, we still build a criterion to extract the best semantic A_i for each class from ξ_{Ci}.

$$V_{x_i}^{c_i} = \left\{A_k \mid \max\left\{\sum \mu_{A_k}(x_i)\right\}, A_k \in \zeta_{C_i}, i \in n_i\right\} \qquad (10)$$

According to (10), we need only the semantic description corresponding to maximum value of $V_{x_i}^{c_i}$.

4 EXPERIMENTAL RESULTS

This paper uses female facial feature data respectively from Uyghur, Tibet and Zhuang nationalities to procede class test. Extract each ethnic famale facial feature data 70 groups, choose randomly 40 groups to act as training samples, and leave the remaining 30 groups data to procede class test for 5 times and calculate the average recognition rate. Meanwhile it uses the methods of PCA and LDA in the same way to procede class test.

4.1 Multi-ethnic facial feature and semantics

Ethnic facial semantics data set, which is established in this paper, needs to position the facial feature. Therefore, the method of ASM based on global feature is used to position the human facial features.

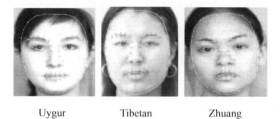

| Uygur | Tibetan | Zhuang |

Figure 1. Ethnic facial feature position.

Utilizing facial feature points got in the process of feature positioning, this paper established facial features from ten ethnic groups, each of which is standardized on the basis of the distance between eyes. The features established are demonstrated in the following charge.

This paper established facial features from ten ethnic groups, each of which is standardized on the basis of the distance between eyes. The features established are demonstrated in the following charge.

Table 1. Facial features and semantic range.

Number	Facial features	Semantic range
f_1	Left eye width	$m_1 \sim m_6$
f_2	Nose width	$m_7 \sim m_{12}$
f_3	Mouth width	$m_{13} \sim m_{18}$
f_4	Eyebrow span	$m_{19} \sim m_{24}$
f_5	Nasal height	$m_{25} \sim m_{32}$
f_6	upper left cheek	$m_{33} \sim m_{36}$
f_7	upper right cheek	$m_{37} \sim m_{42}$
f_8	lower left cheek	$m_{43} \sim m_{48}$
f_9	lower right cheek	$m_{49} \sim m_{54}$
f_{10}	Lower jaw width	$m_{55} \sim m_{60}$

4.2 Individual face semantics mining

According to the facial features of Table 1, we can extract the parsimonious description of individual with the Procedure A in section 3. For example, We mine the semantic of training sample 1. The result is as follow

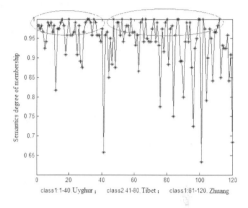

Figure 2. Non-parsimonious semantics membership.

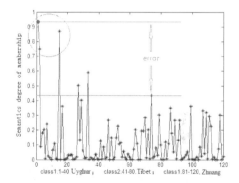

Figure 3. Parsimonious semantics membership.

Figure 2 shows, the compound semantics sets composed by sample 1 single-attribute semantics will describe well sample 1 and faces within the same ethnic group. However, the well description of sample 1 on most non-homogeneous samples shows the similar semantics existing in sample 1 to non-homogeneous sample, and the similarity should be deleted.

Figure 3 shows that, based on inbetween mining method, this paper chooses the semantics sets which both can better describe itself and get less similarities with the non-homogeneous ones. The semantics sets achieved in this process can better distinguish themselves from other non-homogeneous ones, and at the same time, they pormise the similarity within the same samples, which evate the isolated sample.

633

4.3 Facial semantic description

According to the method proposed in this paper, we can get the semantic of each ethnic. Then, we can gain the membership degree distribution as following using the method, we can also get the semantic of each ethnic. The semantic interpretation is simple and understandable.

The facial features of **Uygur** are described as following

ξ_{C1}="$m_{39}m_{45}$"," *the Uygur which has medium upper right cheek and medium lower left cheek*"

The facial features of **Tibetan** are described as following

ξ_{C2}="$m_{36}m_{50}$","*the Tibetan which has not narrow left right cheek and not wide lower left cheek*"

The facial features of **Zhuang** are described as following

ξ_{C3}="$m_{41}m_{48}$","*the **Zhuang** which has narrow left right cheek **and** narrow lower left cheek*"

4.4 Ethnic classification based on facial semantic sets

This paper uses female facial feature data respectively from Uyghur, Tibet and Zhuang nationalities to procede class test. Extract each ethnic famale facial feature data 70 groups, choose randomly 40 groups to act as training samples, and leave the remaining 30 groups data to procede class test for 5 times and calculate the average recognition rate. Meanwhile it uses the methods of PCA and LDA in the same way to procede class test.

Table 2. Ethnic semantic ruels and accuracy.

Method	Average rules	Average accuracy
Our method	2.4	67.55
PCA	—	65.33
LDA	—	66.22
DecisionTable	9.2	64
JRip	4.8	66.22
PART	7.8	65.34

Compared with the tensor space algorithm of PCA and LDA, our method has a higher accuracy. In addition, our methods have a better interpratation and fewer rules than fuzzy rule algorithm of DecisionTable, JRip and PART.

5 CONCLUSION

This paper is an expanded application in facial feature position method. Through the distribution of ethnic facial feature, this paper established membership function to explain ethnic face semantics, which can not only depict facial features of different ethnics, but can judge the ethnic attribute of different individuals efficiently. This method, as a new method to procede ethnic facial feature analysis, is of great significance. Meanwhile, the fusion of different ethnics brings about various patterns of manifestation of ethnic facial features. The different ethnic facial feature semantics expression proposed in this paper will also provide efficient evidence to analyze facial features of people from different ethnics. This is also one of the aspects that this paper will do research.

ACKNOWLEDGEMENTS

This work is supported by National Nature Science Foundation of China (NO. 61370146) and Liaoning Science & Technology of Liaoning Province of China (No. 2013405003, 2013020026).

REFERENCES

[1] DUAN Xiaodong, WANG Cunrui, LI Zhijie, ZHANG Qingling. The Ethnic Facial Research Based on Haarlike Features. Microelectronics & Computer, 2011, 28(7): 17–20 (in Chinese).
[2] DUAN Xiaodong, WANG Cunrui, Liu Xiangdong, LIU Hui. Minorities Features Extraction and Recognition of Human Faces. Computer Science, 2010,37 (8): 276–279, 301 (in Chinese).
[3] X.D. Liu, The fuzzy sets and systems based on AFS structure, EI algebra and EII algebra, Fuzzy Sets and Systems, 1998,95(2): 179–188.
[4] X.D. Liu. The fuzzy theory based on AFS algebras and AFS structure, Journal of Mathematical Analysis and Applications , 1998, 217(2): 459–478.
[5] Liu X D. A New Mathematical Axiomatic System of Fuzzy Sets and Systems. Journal of Fuzzy mathematics, 1995, 3: 559–560.
[6] Xiaodong Liu, Witold Pedrycz. Axiomatic fuzzy set theory and its applications, Berlin:Springer, 2008.
[7] LI Ze-dong, DUAN Xiao-dong, ZHANG Qing-ling. A Novel Survey Based on Multiethnic Facial Semantic Web, TELKOMNIKA Indonesian Journal of Electrical Engineering, 2013, 11(9), 5076–5083.
[8] J. Casillas, O. Cordón, F. Herrera, L. Magdalena, Interpretability Issues in Fuzzy Modeling, Springer, Berlin, 2003.
[9] X.Z. Wang, J.H. Zhai, S.X. Lu. Induction of multiple fuzzy decision trees based on rough set technique, Journal of Information Sciences, 2008, 178(16), pp: 3188–3202.

Network Security and Communication Engineering – Chan (Ed.)
© 2015 Taylor & Francis Group, London, ISBN: 978-1-138-02821-0

The research on the space–time statistic characteristics of moving Sonar reverberation and its simulation method

J.G. Liu & L. Zhu

School of Marine Science and Technology, Northwestern Polytechnical University, Xi'an, China

ABSTRACT: when the active sonar platform is moving, the Doppler frequency aliasing of reverberation is related to the spatial cone angle. The traditional method to describe this space–time coupling property is to generate samples of the reverberation data at first, and then estimate the space–time covariance matrix. A new method is presented: directly deduce the reverberation covariance matrix with the relationship between scattering units with the array sensors, and then obtain the space–time samples of reverberation. This method can obtain reverberation covariance matrix on any simulation condition and does not need to calculate covariance matrix frequently at the same time, so the simulation of space–time reverberation is move rapid. The theoretical results have been verified by the computer simulations.

KEYWORDS: Space–time coupling property, reverberation, reverberation covariance matrix, space–time samples.

1 INTRODUCTION

When the sonar platform is moving, there are relative motions between the platform with different scattering units, and the relative speeds of different spatial cone angles are different [1-3]. So the Doppler frequency aliasing of the reverberation from different incident angles is different too. The difference is shown by the frequency expansion of reverberation which is called the space–time coupling property. Most of the methods to describe the space–time characteristics of reverberation are based on the Klemm clutter model[4,5]. Klemm analyzed the relationship between scattering cells with array antenna. Then it is assumed that echoes phases obey uniform distribution and the amplitudes obey the Rayleigh distribution. Different phases are introduced to scattering echoes. This model to describe the space–time coupling characteristics of the clutter in airborne radar has been proven by many tests[6].

However, the reverberation at any time is the sum of the echoes from the scattering units of all the volume elements and surface units within some spherical shell and rang ring, so the calculation of the numerical simulation is very large. The clutter covariance matrix of reverberation is unknown at the space–time adaptive processing. The common method is to simulate reverberation samples with the Klemm clutter model, and then obtain the estimation of the covariance matrix. This method always requires enough reverberation samples to ensure the IID properties of the covariance matrix.

In this paper, the geometric relationship between the clutter cell and the array antenna are also used as the Klemm clutter model, and then the covariance matrix of reverberation are deduced. The covariance matrix is decomposed and the correlation processing is operated with uncorrelated random normal processes, and finally the simulation space–time reverberation samples are obtained. With this method, the covariance matrix of reverberation is obtained, and the simulation of space–time reverberation moves rapid.

2 THE ANALYSIS OF THE SPACE–TIME CHARACTERISTICS OF REVERBERATION

Assume that any spatial stationary discrete array is shown in Figure 1, the number of the sensors is N, choose the origin of coordinate as the reference point, position vectors of sensors are $r_n=(x_n, y_n, z_n)^T$, n=0, 1, 2, … , N-1, and its speed is v_i. The position vector of i-th scattering unit in the space is u_i. The relative time delay is

$$\tau_{ni} = \frac{\vec{u}_i^T \vec{r}_n}{C} \tag{1}$$

On the condition of far filed and narrowband, the echo of the i-th scattering unit received by the m-th array sensor is

$$v_{mi}(t) = d_T(\theta,\phi)b(t)u(t-\tau-\tau_{ni})^* \tag{2}$$
$$\exp(j2\pi f_d(t-\tau-\tau_{mi}))$$

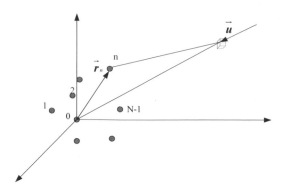

Figure 1. The space–time echo model of the array element moving point.

u(t) is the emission signal, $d_T(\theta,\phi)$ is the emitted direction function of the array, f_d is the Doppler shift caused by the relative movement between the m-th array sensor and the scattering unit i-th.

According to the volume, seabed and surface reverberation, divide the spherical shell or the distance ring in some distance into small volume units or surface units. Assume that the scattering intensity of all scattering units obey the complex Gauss distribution, and the phase is random; the speed of random motion obeys Maxwell distribution.

$$W_v(|\mathbf{v}|) = \frac{2}{\pi\sigma_v^3}|\mathbf{v}|^2 \exp\left(-|\mathbf{v}|^2 \Big/ \sigma_v^2\right) \qquad (3)$$

The angles of the speed vectors of the scattering units obey uniform distribution; there is no relation between the scattering intensity and the motion of each scattering unit. Then the normalized motion spectrum D(f) caused by the random motion of the scattering units can be deduced.

Suppose further that within one period of the signal, the relative displacement of the scattering unit is so small that the displacement can be ignored, and the motion of the array can be regard as the reverse motion of the scattering units. Assume that at the distance r, the reverberation scattering channel is WSSUS, and then the reverberation cross spectral density matrix between the n-th and the m-th sensor can be obtained.

$$\mathbf{R}_{mn}^i(r,f) = \sigma_i^2(r)\mathbf{Y}_{mn}^i(f,r)*|u(f)|^2 * D(f) \qquad (4)$$

While * is the convolution, u(f) is the signal's envelope spectrum, $\sigma^2(r)$ is the reverberation power caused by all scattering units within some spherical shell or the ring at the distance r; $\mathbf{Y}_{mn}^i(f,r)$ is a K×K dimensional matrix, and it is the envelope spectral density matrix caused by the relative motions between the scattering units with the array:

$$\mathbf{Y}_{mn}^i(f,r) = \mathbf{a}_m^i\left(\mathbf{a}_n^i\right)^{\mathrm{H}} \qquad (5)$$

$\mathbf{a}_m^i$ is the time orientation vector of the m-th array sensor related to the small volume unit or the surface unit i:

$$\mathbf{a}_m^i = \mathrm{FT}\left[e^{j2\pi f_0 \tau_m(i)} \exp(j2\pi f_d(i)\frac{k}{f_s})\right] \qquad (6)$$

$$k = [0,1,...K-1]^{\mathrm{T}}$$

In the formula (4), only $Y_{mn}^i(f,r)$ is related to the small volume unit or the surface unit i, and it can be gotten by the integration within the spherical shell or the ring. As for the distance r, the reverberation cross spectral density matrix received by the nth and the m-th array sensor:

$$\mathbf{R}_{mn}(r,f) = \sigma^2(r)\mathbf{Y}_{mn}(f,r)*|u(f)|^2 * D(f) \qquad (7)$$

$\sigma^2(r)$ is the reverberation power caused by all the scattering units in some spherical zone or the ring at the distance r, and it can be got by the reverberation attenuation curve. $\mathbf{Y}_{mn}(f,r)$ is:

$$\mathbf{Y}_{mn}(f,r) = \int_\Omega \mathbf{Y}_{mn}^i(f,r)\,\mathrm{d}\Omega \qquad (8)$$

Ω is the integral interval(the spherical zone or the ring at the distance r).

3 THE SIMULATION METHOD

Assume that the width of the signal pulse is τ, choose a section of data from the received data, suppose that time domain sampling points is K, x_{nk} (n=1, 2, …, N, k=1, 2, …, K) is the k-th sample of the n-th array sensor, and then the space–time sample vector can be expressed as $x=[\mathbf{x}_{s,1}^{\mathrm{T}}, \mathbf{x}_{s,2}^{\mathrm{T}}, …, \mathbf{x}_{s,K}^{\mathrm{T}}]^{\mathrm{T}}$ (NK×1), while $\mathbf{x}_{s,k}=[x_{1k}, x_{1k}, …, x_{Nk}]$ (N×1) is the k(k=1, 2, …, K)) spatial data snapshots, and T is the transpose operation of vectors or matrixes. The space–time sample vector $\mathbf{X}$ is formed by the sequentially arrange of all the K spatial data snapshots. The covariance matrix of the array is:

$$\mathbf{R}_x = \mathrm{E}\left\{\mathbf{x}(n)\mathbf{x}^H(n)\right\} \qquad (9)$$

In the traditional STAP method, the covariance matrix of the clutter or reverberation is unknown. In order to guarantee the IID characteristics of the reverberation covariance matrix, enough samples are required to be estimated.

$$\hat{\mathbf{R}}_x = \frac{1}{N}\sum_{n=0}^{N-1}\mathbf{x}(n)\mathbf{x}^H(n) \qquad (10)$$

In this paper the cross spectral density matrix in the previous section is used to deduce the covariance matrix of the reverberation space–time data samples directly, and then the covariance matrix is decomposed. With the correlation processing with random samples which obey normal distribution, the space–time data samples of reverberation will be obtained. The covariance matrix of the reverberation $\mathbf{R}_x$ is a NM×NM matrix. Use the block matrix properties to obtain as follows:

$$\mathbf{R}_x = E\left\{\mathbf{x}(n)\mathbf{x}^H(n)\right\} =$$

$$\begin{pmatrix} \mathbf{Q}_{00} & \mathbf{Q}_{01} & \mathbf{Q}_{0,N-1} \\ \mathbf{Q}_{10} & \mathbf{Q}_{11} & \mathbf{Q}_{1,N-1} \\ \\ \mathbf{Q}_{N-1,1} & \mathbf{Q}_{N-1,1} & \mathbf{Q}_{N-1,N-1} \end{pmatrix} \qquad (11)$$

While $\mathbf{Q}_{ij}$ is

$$\mathbf{Q}_{ij} = E\left\{\mathbf{x}_i(n)\mathbf{x}_j^H(n)\right\} \qquad (12)$$

$\mathbf{Q}_{ij}$ is M×M dimensional matrix, and it is the correlation matrix between the sensors $\mathbf{x}_i$ with $\mathbf{x}_j$. Calculate cross spectral density matrix $\mathbf{R}_{ij}(r,f)$ at the distance r, do the inverse Fourier transform and then obtain the time-varying covariance matrix of the oceanic reverberation. Obviously, $\mathbf{R}_x(r)_{NK\times NK}$ is a Hermit matrix, and there exists a NK×NK dimensional matrix:

$$\mathbf{U}^H\mathbf{R}_x(r)_{NK\times NK}\mathbf{U} = diag(\lambda_1,\lambda_2,...\lambda_{NK}) = \Lambda \qquad (13)$$

Let $\Lambda' = diag(\sqrt{\lambda_1},\sqrt{\lambda_2},...\sqrt{\lambda_{NK}})$, and the decomposition of $\mathbf{R}_x(r)_{NK\times NK}$ is:

$$\mathbf{R}_x(r)_{NK\times NK} = \mathbf{U}\Lambda\cdot(\mathbf{U}\Lambda)^H = \mathbf{V}\mathbf{V}^H \qquad (14)$$

While $\mathbf{V} = \mathbf{U}\Lambda$. Assume NK is uncorrected random samples which obey to normal distribution, the space–time data sample of reverberation is a NK×1 vector

$$\mathbf{x}(t) = \mathbf{N}(t)\cdot\mathbf{V} \qquad (15)$$

With the formula (15), the space–time samples of locally stationary reverberation at the distance r can be obtained. Change the integral interval, which means changing the spherical shell and the distance ring, calculate the cross spectral density matrix of the array and the space–time covariance matrix $\mathbf{R}_{ij}(r,f)$,

then use the method of matrix decomposition as above, and the space–time samples of reverberation at different distance r are obtained.

4 COMPUTER SIMULATION

In order to verify the numerical simulation method and the reverberation space–time characteristics of the moving sonar, computer simulations are presented. Assume the center frequency of the emission signal is 2.5kHz, and the monostatic array is the ULA with 16 sensors. The space between sensors is half of the wavelength of the center frequency. The array ULA is placed on the X-axis, the sonar platform moves along the Y-axis, whose speed is 10m/s. The vertical beam angle of the array is (-20°~20°), the horizontal beam angle is (-60°~60°).

Figure 2. The power spectral of the reverberation space–time.

Use formula (7) to get cross the spectral density matrix of the array on each element, then use IFFT, and then each block matrix $\mathbf{Q}_{ij}$ of the covariance matrix $\mathbf{R}_x(r)_{NK\times NK}$ are presented. With conventional space–time processing methods, the power spectrum of the space–time reverberation are obtained with the covariance matrix $\mathbf{R}_x(r)_{NK\times NK}$, which is shown as Figure 3. Under the condition of the known covariance matrix of reverberation, the MV spectrum is also obtained with the minimum variance spectrum (MV) method:

$$P(\alpha, f_d) = \frac{1}{\mathbf{S}^H(\alpha, f_d)\mathbf{R}_x^{-1}(r)\mathbf{S}(\alpha, f_d)} \qquad (16)$$

While $\mathbf{S}(\alpha, f_d)$ is the space–time steering vector. The MV spectrum of reverberation is shown as Figure 4. As for the forward looking array, the distribution in the space–time plane is a semi-ring. The result of simulation is consistent with the theoretical analysis. The strong reverberation appears in the region of space where the launching main beam is mapping and side lobe area reverberation is weak but also distributed. In Figure 4, the distribution of MV spectrum is fluctuant, which is caused by the reason that the echoes of different scattering units are suppressed by different phases.

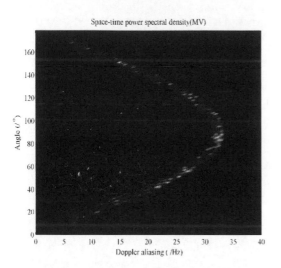

Figure 3.   The MV spectral of the reverberation.

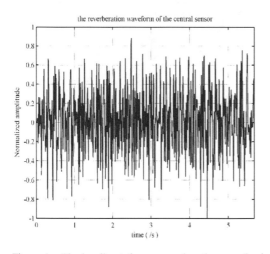

Figure 4.   The locally stationary reverberation samples in the center array at the 2000m distance.

With the above method in this paper, the space–time samples of reverberation at the distance of r can be obtained. In Figure 5, reverberation samples of the central sensor are presented, while the distance is 2000 meters.

5   CONCLUSION

The conventional methods to model the space–time reverberation of the moving platform are based on the Klemm clutter model. They produce reverberation samples, and then estimate the reverberation covariance matrix with those data samples. These methods always need a great amount of calculation, and enough reverberation data samples are needed to guarantee the IID properties of the covariance matrix. A new method is presented: using the geometric relationship between the scattering unit and the array antenna to deduce the cross spectral density matrix between the sensors. Then get the reverberation covariance matrix by the calculation of the block matrix. Decompose the covariance matrix, and then the locally stationary reverberation space–time samples at different distance can be obtained with the correlation operations with the uncorrelated random processes. With this method, the covariance matrix of reverberation is obtained firstly, and then reverberation data samples can be presented quickly. It can not only get the reverberation covariance matrix directly, but also is unnecessary to calculate the covariance matrix and decompose the matrix frequently, so the calculation amount of reverberation simulation is greatly decreased. The theoretical results are verified by Computer simulations.

ACKNOWLEDGEMENTS

The research work was supported by National Natural Science Foundation of China under Grant No. 61371152 and the Basic Research Foundation of North-western Polytechnical University under Grant No. JC20120220.

REFERENCES

[1] Brennan L E, Reed I S. Theory of Adaptive Radar. IEEE Transactions on Aerospace and Electronic System, 1973, 9(2): 237–252.
[2] Klemm R. Adaptive airborne MTI: an auxiliary channel approach. IEE Proc. F&H, 1987, 134(3): 269–276.
[3] Klemm R. Principles of space-time adaptive processing. Boston: the Institution of Electrical Engineers, 2002.
[4] Melvin W L. A STAP overview. IEEE A&E Systems Magazine, 2004, 19(1): 19–35.
[5] Yunhan D. Approximate Invariance of the Inverse of the Covariance Matrix and Its Applications. International Conference on Radar, 2006.
[6] Klemm, R.; Nickel, U.; Adaptive monopluse with STAP. International Conference on Radar, 2006.

Network Security and Communication Engineering – Chan (Ed.)
© 2015 Taylor & Francis Group, London, ISBN: 978-1-138-02821-0

Intellectual feedback in adaptive control systems

S. Manko
119454, Russia, Moscow, prosp. Vernadskogo 78, aud. 232

S. Diane
117036, Russia, Moscow, prosp. 60-letiya Oktyabrya 20, kv. 46

A. Panin
141014, Russia, Mytyshy, ul. Semashko 26/1, kv. 196

ABSTRACT: The paper presents information processing methodology for autonomous robot control based on intellectual feedback. The proposed approach is aimed at robot external situations analysis using neural networks and classification trees. We describe how a learning process may be implemented in terms of robot knowledge base formation and demonstrate the application of our method for the task of mobile robot control.

KEYWORDS: Intellectual feedback, Neural networks, Classification trees, Machine learning.

1 INTRODUCTION

The main goal of the conducted research is to increase the autonomy and reliability of robot control under conditions of uncertainties in the environment. Such conditions are typical for reconnaissance tasks, search and rescue operations, military missions, and others.

Providing a substantial degree of autonomy requires incorporation of artificial intelligence methods into robot control system in order to accumulate and effectively analyze measurements of sensory system, thus synthesizing robot's control policy and adjusting it towards more effective functioning.

Embedding an intellectual feedback into the robot control loop can be a solution to increasing robot autonomy. Through intelligent sensory information processing the key features of an environment may be extracted and used for situation analysis. The decision-making process then may be described as a situational control.

The identification of environmental situations can be performed using classification and regression trees (CART) [1]. This method provides accurate and rapid-learning solutions for low complexity problems. However, it should be understood that the dimensionality of the data generated in the robot sensory subsystem may in practice be too large for effective processing by means of this technique.

One possible solution to this problem may be the use of neural networks. The results of recent research in this area [2] strongly confirm perspectives of the neural network approach to hierarchical pattern recognition in visual, acoustic and textual data. Thus, it appears to be a proper technique for a robot control system to reduce the dimensionality of gathered information.

2 CONTROL WITH INTELLECTUAL FEEDBACK

The design of intellectual feedback (IF) for complex dynamic object (CDO) control, in particular the robot control problem, suggests the presence of two major subsystems: the informational-measuring subsystem (IMS), and intellectual information processing subsystem (IIPS).

Informational-measuring subsystem provides an accumulation of diverse sensory and command information for complex dynamic object control. Intellectual information processing subsystem performs an analysis of the gathered data, shaping a continual formation of knowledge about the patterns in changing environmental situations and the state of the controlled CDO itself.

Intellectual feedbacks are able to provide a number of advantages in comparison to feedback without intelligent information processing:

- an ability to predict the state of CDO due to accumulation and generalization of the rules of its interaction with the environment;
- an ability to partially compensate for missing information about the state of CDO or conditions of the environment;

• an ability to adapt the control policy to changing characteristics of CDO or varying conditions of the environment.

Fig. 1 shows a block diagram of a control circuit with intellectual feedback. The IF circuit is formed by IMS, and IIPS containing the sensory measurements database (SMDB) and the knowledge base (KB) about CDO functioning principles and appropriate control rules.

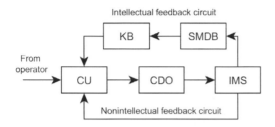

Figure 1. Structure of the control system with an intellectual feedback loop.

SMDB is formed by the accumulation of retrospective data about the measured parameters of the CDO and the environment. The knowledge base includes a set of criteria for CDO state evaluation or feature extraction from the surrounding environment. It can also contain a mathematical model of interaction between CDO and its environment, usable in forecasting of the changes in the situation. In the simplest case KB only contains situational control rules with no account for environment model.

Embedding IF into the control system structure can increase its decision-making efficiency. Knowledge gained in IF loop may be used to optimize the non-intellectual control system part, leading to control unit (CU) indirect adaptation. Alternatively, the high-level features of the environment state, discovered in the IF circuit, (both immediate and predicted) can be used directly to organize the situational or forecasting control of the CDO.

3 SITUATION CONTROL BASED ON CLASSIFICATION TREES

A fundamental approach to intellectual CDO control was proposed within the framework of the theory of situational control [4]. Based on the key provisions of this theory, each class of situations, occurrence of which is considered probable during the operation of the system, is associated with an adequate control action. In this case, the situation, identified by the informational-measurement system, can be assigned to a certain class, for which a known control policy exists.

Classification of situations can be effectively carried out using classification trees [2]. The construction of a classification tree is associated with the analysis of a large number of sample observations. Input data for this process can be represented as pairs $\mathbf{S}_i = \{\mathbf{X}_i, y_i\}$,

where $\mathbf{X}_i = \{x_1, x_2, ..., x_m\}$, are the input parameters based on which the classification should be performed and y_i is the required classification label.

It should be noted that the generation of sample set in case of active learning takes place continually during system runtime on the basis of collected data about outcomes of various situations. In other words, the set of sample labels for the input vectors can be generated through retrospective analysis of the consequences of various control actions on the CDO.

The CART composition algorithm is based on the dichotomization procedure. For a given set of training samples $\mathbf{S}$, an entropy value is given by the formula:

$$H(\mathbf{S}) = -\sum_{y \in D(y)} P(\mathbf{S}, y) \log_2 P(\mathbf{S}, y),$$

where $D(y)$ is a set of discrete values of the parameter y; $P(\mathbf{S},y)$ is the statistical probability of observing a value y on the set of examples $\mathbf{S}$. Let us define a logical predicate d on the set $\mathbf{S}_x = \{\mathbf{X}_1, ..., \mathbf{X}_n\}$. Then the set of examples can be divided into two subsets: a subset on which d is true and a subset, where d is false. For these two subsets in the same way as for the original set, the values of entropy are computed using the same formula:

$$H(\mathbf{S}^d) = -\sum_{y \in D(y)} P(\mathbf{S}^d, y) \log_2 P(\mathbf{S}^d, y),$$

$$H(\mathbf{S}^{\bar{d}}) = -\sum_{y \in D(y)} P(\mathbf{S}^{\bar{d}}, y) \log_2 P(\mathbf{S}^{\bar{d}}, y).$$

In accordance to the property of dichotomization, the obtained subsets will satisfy the following criterion: $P(\mathbf{S}^d | \mathbf{S}) H(\mathbf{S}^d) + P(\mathbf{S}^{\bar{d}} | \mathbf{S}) H(\mathbf{S}^{\bar{d}}) \le H(\mathbf{S})$. In other words after dichotomy of the set $\mathbf{S}$ the verity or the falsity of parameter y will be more evident than in the original set.

The described dichotomization process may be iteratively applied to every subset derived from it, ending up with the subsets of minimal entropy. In this way due to the fact that every consequent split will decrease summarized set entropy, a repetitive dichotomization will converge to decomposition with minimal total entropy. However, the greedy algorithm described here does not guarantee the tree to have an optimal number of nodes.

4 PREPROCESSING OF SENSORY INFORMATION USING NEURAL NETWORKS

The use of classification trees in robot control is complicated by the nature of data generated in the informational-measuring subsystem. In real world problems, associated with image processing, the high dimensionality of the input vector can dramatically degrade performance of the CART algorithm. This problem can be solved with neural networks technology, well known to be successful at tasks of feature extraction and dimensionality reduction.

The requirement for high efficiency of image preprocessing implies the possibility of multi-stage image analysis for feature identification. Such an analysis can be performed using neural networks with hierarchical connectivity (HCNN) [2]. HCNN architecture is based on a multilayer feedforward neural network (MFNN) with some specified number of layers L and a number of neurons in each of these layers $n_1, \dots, n_L$.

The output of each neuron is given by the formula:

$$Y = \sigma\left(\sum_{i=0}^{N_w} x_i w_i\right),$$

where x_i is the input values of the neuron; w_i are weighting coefficients of neuron connections with the input values; N_w is a number of inputs of the neuron; σ is a sigmoid function.

The output values of neurons in each layer are arranged in two-dimensional information images. Each neuron in a layer with an index $i > 1$ is connected to a subset of values from the data of the image belonging to the layer of $i - 1$. The neurons of the first layer are connected directly to informational image values.

It is important to note that in the proposed structure adjacent layers are not fully connected, which otherwise would increase the computational complexity. In addition, the use of local receptive fields at each level of the neural network allows organizing a multi-stage aggregation of information from small areas of the input image to the larger ones. Each neuron becomes a detector revealing a specific combination of features in the underlying layer. For example in task of visual image analysis successive layers of neurons can perform the following functions: identification of small segments at different angles; identification of object parts; identification of objects and complex scenes.

Along with the problem of setting the topology of connections of the neural network an important role belongs to the search of its weighting coefficients. We use an unsupervised learning algorithm based on biologically inspired principles of detector selectivity formation [4]. This algorithm allows the following requirements to be fulfilled:

- the expectation of the output value of each neuron in time should be close to a specified constant of temporal activation frequency: this requirement does not allow the neurons to stay in a passive or an active state only;
- the variance of the output value of neurons along in the coordinates of the layer should be close to the maximum; this requirement is needed to avoid situations in which all neurons layer will be active at once not allowing to identify actual features.

The generalized learning algorithm for one layer of HCNN can be described as follows:

1 initialization of weights to small random values;
2 calculation of the neurons activations in every layer;
3 local contrast normalization in each layer;
4 learning of the neurons by the rules of detector selectivity formation;
5 if the criterion of completion of training is not performed, transition to the step 2, otherwise termination of the algorithm.

A criterion of completion of the learning process is lowering the rate of weight coefficients change to a small value. Training procedure of HCNN is achieved by separate training of every layer, starting from the layers closest to network inputs.

5 VIRTUAL MODELING ENVIRONMENT

One possible approach to research and debugging of the intellectual control algorithms for a mobile robot is to simulate its typical functioning environment including the robot itself. Thus, we developed a virtual environment by means of an object-oriented programming language C#, Box2DX solid-state physics library and Windows GDI graphics. The fragment of the simulation of the four-wheeled mobile robot in the developed software is shown in Fig. 2, a.

An informational-measurement system of the virtual robot includes range-finders and a vision sensor (camera). Both of them are used to organize the intellectual feedback.

As it was mentioned earlier the two components of the IF-loop involved in the robot learning process are sensor measurement database and knowledge base. SMDB accumulates video frames for consequent knowledge extraction, whereas the KB is formed from the preprocessing HCNN and a classification tree for situation analysis, both of which learn in an unsupervised manner.

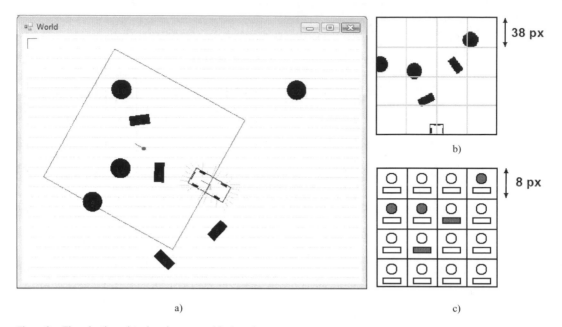

a)   b)   c)

Figure 2. The robot in a virtual environment with obstacles.

Objects that surround the robot have approximately equal probabilities to occur in every position near the robot, and hence the same features can be extracted at different regions of robot's field of view. Thus the original visual image may be split into several regions for separate analysis with the use of a smaller HCNN than it would be needed for a whole image. The size of original visual image is of 150×150 pixels size (Fig. 2, b). The size of each of the regions is 38 × 38 pixels (resulting from 4 × 4 grid split).

A two-layer HCNN is used first to learn on sub-images from the regions and then to reduce their dimension. First layer of the network contains 33 × 33 neurons with 6 × 6 sized overlapping receptive fields. Second layer has 8 × 8 neurons with 10 × 10 receptive fields.

Ultimately, preprocessing of the visual image obtained from robot VS, using an array of 16 hierarchically connected networks allows us to get a feature vector of local environment of the robot (Fig. 2, c) smaller by dimensions than the original video frame (32 × 32 compared to 150 × 150).

Combining two channels of intellectual feedback allows robot to discover regularities in the environment. A fact of collision detected via range-finders can be used as a label for environment features vector. Hence a sample database for an automated supervised learning can be formed during the functioning of the robot.

After the finishing of the learning process, it is possible to predict whether the situation at time t

leads to collision at time $t + \Delta t$. The resulting knowledge may become a basis for predictive control of the robot. In particular, when classifying the current situation as leading to a collision, speed of the robot should be reduced.

6 CONCLUSION

The rate of development of adaptive information processing and robot control algorithms is increasing every year. At the same time it becomes more obvious that the construction of effective intelligent control systems based on a single technology is impossible.

Conversely, the use of several techniques within a concept of situational control can produce a solution, capable of boosting system effectiveness and reliability in its application to the modern robotics problems. Our study attempts to combine the methods of classification trees and neural networks in intellectual feedback loop to optimize the control system of autonomous mobile robot.

The set of high-level features extracted with the neural network from visual image is classified using a CART algorithm thus allowing autonomous robot to predict collisions with objects in the surrounding environment and providing decision support when controlling the robot manually.

ACKNOWLEDGEMENT

This work was financially supported by the Ministry of Education and Science of the Russian Federation under grant № 2014/112 (project № 994).

REFERENCES

[1] J. Quinlan: Induction of decision trees. Machine Learning, V. 1, (1986), 81–106.

[2] Q. Le et al.: Building High-level Features Using Large Scale Unsupervised Learning. In: Proceedings of the 29th International Conference on Machine Learning, ICML 2012, Edinburgh, Scotland, UK.

[3] D. A. Pospelov: Situational Control: Theory and Practice. Batelle Memorial Institute, (Michael N. Golovin and Nne Grunewald, Trans.), Columbus, Ohio, USA, (1986).

[4] E. Bienenstock, L. Cooper, P. Munro et al.: Theory for the development of neuron selectivity: orientation specificity and binocular interaction in visual cortex. The Journal of Neuroscience V. 2, No. 1, (1982), 32–48.

Network Security and Communication Engineering – Chan (Ed.)
© 2015 Taylor & Francis Group, London, ISBN: 978-1-138-02821-0

Pattern recognition with applications to pre-diagnosis of pathologies in the vocal tract

Regiane Denise Solgon Bassi
University of São Paulo, São Carlos, São Paulo, Brazil
North of São Paulo University, São José do Rio Preto, São Paulo, Brazil

Henrique Dezani
São Paulo State University, São José do Rio Preto, São Paulo, Brazil
Faculty of Technology, São José do Rio Preto, São Paulo, Brazil

Kátia Cristina Silva Paulo
University of São Paulo, São Carlos, São Paulo, Brazil
North of São Paulo University, São José do Rio Preto, São Paulo, Brazil

Rodrigo Capobianco Guido
São Paulo State University, São José do Rio Preto, São Paulo, Brazil

Ivan Nunes da Silva
University of São Paulo, São Carlos, São Paulo, Brazil

Norian Marranghello
São Paulo State University, São José do Rio Preto, São Paulo, Brazil

ABSTRACT: For the detection of laryngeal pathologies, in general medical examinations, for example laryngoscopy and stroboscopy, are adopted. Besides being considered invasive and uncomfortable procedures, they are made only by medical request when the diseases are already on advanced levels. In order to perform a computational pre-diagnosis of such conditions, this paper presents a non-invasive technique in which three classifiers are tested and compared: Euclidian distance, RBF Neural Network with the Gaussian kernel, and RBF Neural Network with the modified Gaussian kernel. Based on a database of normal and pathological voices, tests that demonstrate the effectiveness of the proposed technique, which can be implemented in real-time, were performed.

KEYWORDS: Signal processing, Larynx pathologies, Euclidian Distance, RBF Neural Networks.

1 INTRODUCTION

Based on works such as the one by GUIDO [1], BARBON JR [2], and SOUZA [3], it is clear that the voice provides all the information necessary for their analysis and detection of pathologies in the larynx, by loading extremely important characteristics of the vocal tract. The most common disorders, such as laryngeal edema, generally caused by the use of cigarettes, and the nodules, generated by overexertion of the voice, among other, affect the dynamics of the speech production system as discussed in SCALASSARA [4], generating small variations in acoustic parameters of the vocal tract, when observed over time phonation.

2 BRIEF REVIEW OF IMPORTANT CONCEPTS

In this section, the review is not extensive and is limited to the essential concepts required to understand the motivation for the proposed approach. The human speech apparatus can be divided into three main anatomical subsystems: respiratory, laryngeal and articulatory. The respiratory tract is essential to the act of speech as it is like an air pump that provides energy to the aerodynamic and laryngeal articulation. The larynx is a cartilaginous tube of irregular shape that connects the pharynx with the trachea. It is located in the upper neck, being a hollow chamber where the voice is produced.

According to KENT [10], the articulation of the soft palate is the organ that determines whether a sound is oral or nasal, and the passage of air. When the air from the lungs passes through the vocal cords, and our neural commands that produce muscular adjustments make pressures, voice production occurs. In the airway, the vocal folds control its flow, making this excitation signal to be periodic, vibrating at a certain frequency.

3 CLASSIFIER I: TECHNIQUE BASED ON EUCLIDEAN DISTANCE

The training and testing of the proposed system, based on Euclidean distance classifier, occur in

---

Algorithm 3.1: Training Steps Classifier System I

- **STEP T_1:** extract the raw data from all the voice files used in the experiment, which are stored in WAVE format, sampled at 22050Hz, 16-bit;
- **STEP T_2:** for each voice signal, coming from normal and pathological individuals, produce the feature vector representing the raw data. This vector should be composed of T = 42 values, so that:
  - **average** energies of J windows of the signal under analysis, filtered with a low-pass filter of order designed for critical band **Bark scale 1**. Each window will have size that equals 1024 samples, and the subsequent window overlaps the previous one in 50%;
  - Repeat to for the **Bark scale 2** critical band;
  - (. . .)
  - Repeat to the **Bark scale 21** critical band, which corresponds to the maximum possible bandwidth in this case in view of the sampling rate of the speech signals;
  - normalized **variance** of J windows of the signal under analysis, filtered with a low-pass filter of order 1 designed for critical band **Bark scale 1**. Each window will have size that equals 1024 samples, and the subsequent window overlaps the previous one in 50%;
  - Repeat for the **Bark scale 2** critical band; (. . .)
  - Repeat for the **Bark scale 21** critical band, which corresponds to the maximum possible bandwidth in this case in view of the sampling rate of the speech signals;
- **STEP T_3:** separate the feature vectors of normal voices and pathological ones;
- **STEP T_4:** isolate, randomly, the training and testing subsets;

Agorithm 3.2: Steps Test / Use Classifier System I

STEP U_1: take the feature vectors, derived from normal and pathological voices, which were isolated for testing, in step T_4 of the training procedure previously described;
STEP U_2: for each testing vector, $\vec{a_i}$, with size 42, measure the Euclidean distance between itself and all training vectors, $\vec{b_j}$.
STEP U_3: every time an $\vec{a_i}$ vector is compared with all $\vec{b_j}$, the lowest value is stored. If the vector $\vec{b_j}$ that generated the lowest value is normal, then $\vec{a_i}$ is considered as normal, otherwise as pathological.

---

accordance with the algorithms 3.1 and 3.2. In this classifier, the term "training" is being used only for allowing an analogy with the other classifiers adopted (II and III), as the classifier is not trained but comparisons with template models are used instead.

4 CLASSIFIER II: RBF NEURAL NETWORK WITH GAUSSIAN KERNEL

Training and testing procedures for this classifier occur as follows.

Algorithm 3.3: Training Steps Classifier System II

- **STEP T_5:** Set the RBF network to have T = 42 entries, R neurons in the hidden layer, where R is the number of training cases, and 1 neuron in the final layer;
- **STEP T_6:** using Gaussian kernels, adjust the outputs of the hidden layer so that each i-th neuron shows maximum output, equal to 1, for the i-th training case. For other examples, the i-th neuron will present an output value between 0 and 1. Simply adjust the center of the respective Gaussian function, i.e., μi , to the point that is identical to the feature vector being analysed. This will make the natural exponential kernel function results in the value 1, which is its maximum. Particularly, the Gaussian used corresponds to the natural exponential of the negative of the Euclidean distance between the input vector and the mean vector μi ;
- **STEP T_7:** establish a linear system of R equations and R unknowns. Each equation is formed based on the output values of each neuron in the hidden layer, drilling through each training case. The variables correspond to existing weights between the hidden layer and the final layer. The use of the values 1 or −1 are used to match the equations, representing, respectively, normal voices or pathological voices;
- **STEP T_8:** solve the system, by triangularization or any other method, finding the values of the weights.

Algorithm 3.4: Steps Test/Use Classifier System II

- **STEP U_1:** for each testing vector, apply the corresponding features in the input of the RBF network previously trained. Collect the data present at the output of the hidden layer neurons, making a linear combination of them with the weights determined during the training phase.
- **STEP U_3:** The value from the previous linear combination is interpreted to provide the system response to the input vector. If this value is closer to 1 than to −1, the corresponding voice will be considered normal, otherwise it will be considered pathological.

---

5 CLASSIFIER III: RBF NEURAL NETWORK AND MODIFIED GAUSSIAN KERNEL

The training and testing procedures for this classifier are the same of classifier II, detailed above, but changing the kernel function. Particularly, the variance in commonly used Gaussian kernel was replaced by quasi-variance, in order to apply an unbiased classifier. The modified kernel function is therefore the following

$$e^{-\left(\dfrac{\left(\frac{x-\sum x_i}{N}\right)^2}{\frac{2\sigma^2}{N-1}}\right)} \tag{1}$$

Therefore, the partially modified kernel is given by:

$$e^{-\left(\frac{1}{2\sigma^2}\right)} \tag{2}$$

being,

$$= -(N^{-2} - N^{-3})(x^2 N^2 + (\sum x_i)^2) - N^{-1} 2x \sum x_i (1 + N^{-1}) \tag{3}$$

From t, we have:

$$= (x - \mu)^2 - \frac{(x-\mu)^2}{N} \tag{4}$$

Substituting (8) into (6) we have:

$$e^{-\frac{((x-\mu)^2) - \frac{(x-\mu)^2}{N}}{2\sigma^2}} \tag{5}$$

6 IMPLEMENTATION / METHODS AND MATERIALS USED

In phases, training and testing / implementation, the algorithms were implemented in C/C++ language under LINUX environment Conectiva 7.0 and Intel Core i5 processor, on a machine with 4 Gb of RAM. The compiler used was the GNU g++, associated with kwrite text editor. To develop this project, all signs of voices used correspond to various combinations of parameters extracted from the voiced phonemes / a /, sustained for five seconds on average, which were extracted from the database of voices from the Hospital das Clínicas, USP, Ribeirão Preto, Brazil. This database has 118 voices with normal features and 33 others belonging to individuals with the following conditions in the larynx: lump / cyst / calluses on the vocal folds, forming a kind of pathology analysis, and Reinke's edema, forming another type of pathology sought. All subjects were previously examined by medical professionals to confirm their healthy state or pathological one. The database is separated and labeled with the name of each examined subject.

7 TEST PROCEDURES AND RESULTS WITH THE USE OF CLASSIFIER I - EUCLIDEAN DISTANCE

Following the battery of tests, and the results obtained, lead to the conclusions of this work. In particular, tests were carried out in order to try:

- 10% of the base for training and the remaining 90% for testing;
- 20% of the base for training and the remaining 80% for testing; (. . .)
- 90% of the base for training and the remaining 10% for testing.

In each case above, 100 random combinations were used to separate the amount of signals intended for training and the amount allocated for testing, with mutually exclusive sets. The results presented below, expressed as confusion matrices, correspond to averages of each 100 tests. Noting the above results, it can be seen that the number of voices correctly classified as being normal is quite relevant even for training with a small number of voices. Where normal voices are seen as pathological, the risk posed by prediagnosis system is not significant, considering that an additional medical examination is able to dispel possible doubts or point out errors in ratings. A negative point that can be observed in the results is that there are a considerable number of pathological voices classified as normal. Such errors are more serious and constitute a disadvantage of this classifier I.

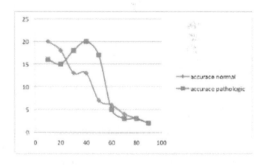

Figure 1.   Accuracy of normal and pathological voices.

8 TEST PROCEDURES AND RESULTS WITH THE USE SORTER II - RBF NEURAL NETWORK AND GAUSSIAN KERNEL

In this case, 50 random combinations were used to separate the amount of signals intended for training and the amount allocated for testing, mutually exclusive. The results presented below are expressed in the form of confusion matrices, pointing out the best and worst cases.

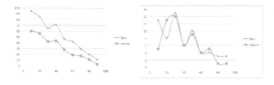

Figure 2. – [left]: Accuracy of normal voices in the best and the worse case; [right]: Accuracy of pathological voices in the best and the worse case.

Now looking at the previous results, it is still possible to observe that the number of voices correctly classified as being normal is quite relevant even for training with a small number of voices. Although smaller in magnitude, a negative point is still observed: there are a considerable number of pathological voices classified as normal.

9 TEST PROCEDURES AND RESULTS WITH THE USE OF CLASSIFIER III - RBF NEURAL NETWORK AND MODIFIED GAUSSIAN KERNEL

For this classifier, the results are the following, as expressed in figures 3.

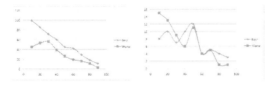

Figure 3. [left]: Accuracy of normal voices in the best and the worse case; [right]: Accuracy of pathological voices in the best and the worse case.

10 CONCLUSIONS

After we have analyzed the results, a very important observation is that, for all three tested classifiers, I, II and III, the largest and smallest total number of hits were considered as best and worst cases, respectively. Thus, there is no advantage when the classifier III, for example, is compared with classifier II, that is, the total number of hits is similar. However, when comparing the false negatives, i.e., pathological voices misrecognized as being normal, one sees advantage in using the Classifier III compared with classifier II, specially when more than 50% of the base is used for training. This fact is relevant, considering that false negatives are a major concern of clinicians.

REFERENCES

[1] GUIDO, R. C. ; PEREIRA, J ; SLAETS, J . Introduction to the special issue on Emergent Applications of Fractals and Wavelets in Biology and Biomedicine. Applied Mathematics and Computation, v. 207, p. 3–4, 2009.
[2] BARBON JR, S. Dynamic time warping baseado na transformada wavelet. 2007. 113p. Dissertação (Mestrado). Instituto de Física de São Carlos da Universidade de São Paulo, São Carlos, 2007.
[3] SOUZA, L.M. Detecção inteligente de patologias na laringe baseada em máquinas de vetores de suporte e na transformada wavelet. 2010. 102p. Dissertação (Mestrado em Ciências). Instituto de Física de São Carlos, Universidade de São Paulo, São Carlos, 2010.
[4] SCALASSARA, P. R. Utilização de Medidas de Previsibilidade em Sinais de Voz para Discriminação dc Patologias de Laringe. 2009. 267 f. Tese (Doutorado). Escola de Engenharia de São Carlos, Universidade de São Paulo, São Carlos, 2009.
[5] BAKEN R. J. Clinical Measurement of Speech and Voice. Allyn and Bacon, a division of Simon and Schuster, Inc., Needham Heights, MA. 1987.
[6] CHILDERS D. G., Bae K. S. Detection of Laryngeal Function Using Speech and Electroglottographic Data, IEEE Transactions on Biomedical Engineering, Vol. 39, Issue 1, Pages 19–25, Janeiro 1992.
[7] FROHLICH M., Michaelis D., Werner S. H. Acoustic breathiness measures in the description of pathologic voices, Procedings of the 1998 IEEE International.
[8] DAJER, M. E. Análise de sinais de voz por padrões visuais de dinâmica vocal. 2010. pp. 154. Tese (Doutorado). Escola de Engenharia de São Carlos, Universidade de São Paulo, São Carlos.
[9] MARTINEZ C. E., Rufiner H. L. Acoustic Analysis of speech for detection of Laryngeal Pathologies, Information Technology Applications in Biomedicine. Proceedings. 2000 IEEE EMBS International Conference on, 2000.
[10] KENT R. D. and Read C. The Acoustics Analysis of Speech. Singular Publishing Group, Inc., San Diego, CA.1992.
[11] PARRAGA, A. Aplicação da Transformada Wavelet-Packet na Análise e Classificação de Sinais de Vozes Patológi-cas..2002.163p. Dissertação (Mestrado). Departamento de Engenharia Elétrica, Universidade Federal do Rio Grande do Sul. Porto Alegre, 2002.
[12] FONSECA, E. S.; GUIDO, R. C.; SCALASSARA, P. R.;MACIEL, C.D.; PEREIRA, J. C.; Speech processing: Wavelet Time-frequency Analysis and Least Squares Support Vector Machines for the Identification of Voice Disorders. EE/UCLA - School of Engineering and Applied Sciences, University of California at Los Angeles, CA, USA, 2006.
[13] MUSSOI, F. L. R.; Resposta em Frequência: filtros passivos. 2.ed. Florianópolis, Julho 2004.
[14] HAYKIN, S.; VEEN, B. V. Sinais e sistemas. Porto Alegre: Bookman, 2002.
[15] SILVA I. N. DA,; SPATTI, D. H. ; FLAUZINO, R. A. Redes Neurais Artificiais para engenharia e ciências aplicadas. São Paulo, Artliber, 2010.

Network Security and Communication Engineering – Chan (Ed.)
© 2015 Taylor & Francis Group, London, ISBN: 978-1-138-02821-0

Smart fire detection system based on the IEEE 802.15.4 standard for smart buildings

K.B. Deve, A. Kumar & G.P. Hancke
Electrical, Electronics, and Computer Engineering Department, University of Pretoria, South Africa

ABSTRACT: A smart fire detection system for smart buildings has been developed in compliance with the IEEE802.15.4 standard. With the use of this system evacuation procedures have been designed according to the level of the fire hazard. The wireless sensor network is implemented using the Global System for Mobile (GSM) communication so as to detect fires effectively and reduce false positives. In the proposed system a WSN was set up with temperature and smoke sensors. In addition, Short Message Service (SMS) capability via GSM was implemented for the occupants to interact with the fire detection system and aid in the detection of false positives. The fire detection system detects fires effectively and reduces false positives as predicted.

KEYWORDS: Wireless sensor networks, Smart buildings, Sensor array, Global system for mobile, False positives.

1 INTRODUCTION

Smart building is the term commonly used to define a building equipped with an ambient intelligence system, which can react to predefined conditions in real time [1]. These are context-aware systems in the sense that they can collect data from various sensors embedded in the home environment, extract information, and act accordingly. In addition, home automation appliances controlled by building automation networks (BAN) is often referred to as a smart buildings or intelligent buildings. According to building automation standards, fire safety is a major aspect of building design [2]-[5]. The use of fire detection systems has immensely promoted the awareness and safety of people and their property against harmful fires in buildings. The most of them are the common causes of structural fires in residential areas. This paper comprises the detection of fire in a kitchen, which automatically qualifies the fire detection system to be a life safety system. Kitchens are a work environment for people and preservation of life surpasses all other criteria [6].

Positioning of these fire detectors is another crucial factor when setting up a fire detection system. It is an obvious fact that these heat and smoke detectors can only detect fires that are remote to the position of the detector itself [7]-[14], hence it is of great importance to have them positioned in the right locations to ensure effective detection.

The smart fire detection system has been developed in compliance with the IEEE 802.15.4 standard. The developed system is suitable for SMS capability via GSM.

2 SYSTEM OVERVIEW

The combination of WSN and GSM communication was the basis of this design and this was done by putting together control variables and thresholds that would act as indicators when a fire is present and when a false positive occurs. The Fig. 1 below shows the basic concept for the smart fire detection system, which includes a processing device, a GSM communication device and a set of sensor nodes in a wireless sensor network (WSN).

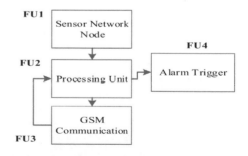

Figure 1. Block diagram of the developed system.

The functional units FU1 comprised sensors that detect the characteristics of a fire and FU3, a GSM communication link to mobile devices. The processing unit FU2 processes the information gathered from FU1 and FU2 and determines an outcome, i.e. whether there is a fire or not.

2.1 Wireless sensor node

The functional unit FU1 was the wireless sensor node as explained before and comprised ZigBee wireless devices that use the Zigbee protocol, MQ-2 smoke sensors and LM35 temperature sensors. The technical specifications of the ZigBee devices include a peak transmit current of 40mA and a peak receive current of 50mA at a voltage (Vcc) of 3.3V and has an operating temperature of maximum 85°C. They are low power devices that can be used for continuous transmission and reception [15]-[17]. These were the reasons for this choice and in addition they are small size and low cost [3].

The system detected two kinds of temperature instances, the first was checking whether a certain threshold has been reached and if so it gives a warning and the other was checking if a high rate of change in temperature has occurred and again it gives a warning signal. For the detection of smoke only the checking of a specific threshold was used as a trigger mechanism.

The design was specific for a kitchen and for simplicity it was divided into two sections. Each section has one sensor node that has a smoke and temperature sensor connected to it and another sensor node with just a temperature sensor on it. These nodes send their readings to the central node that is connected to the processing unit. This was in a star network as described earlier. The GSM communication was implemented via a universal serial bus (USB) 3G modem connected directly to the raspberry pi (RPI). The main aim of the GSM communication is to reduce false positives by giving the occupants in the kitchen the opportunity to vote and disable the alarm if there is no fire and as an added process to the system is the activation and deactivation of alarm via sms. The control of this 3G modem is via AT commands which are sent to the modem via a Python script. The photograph of the developed sensor node is shown in Fig. 2.

Figure 2.   The photograph of the developed sensor node.

2.2 Software design

For this system to have worked efficiently certain algorithms were developed to aid in the detection process and reduction of the detection of false positives. The overall design of the fire detection system encompasses the combination of the GSM communication and the WSN so as to effectively detect fire hazards. The written code starts by initializing and importing the required modules for the Python script to run efficiently. These modules include ZigBee, serial, time and general purpose input/output (GPIO). The next stage was the checking for the sms (short message service) that activates the alarm and in this case the word "Activate" was used as the trigger word to activate the alarm via sms. The system would then read the temperature values from the ZigBee's and check the thresholds, then loop back after going through all the sensors.

3 EXPERIMENTAL EVALUATION

First of all we tested the communication module, i.e. the ZigBee module. The sampling rate of the ZigBee was set to 1000ms. Each ZigBee device has a 64 bit assigned identity that was used in its identification; the 64 bit identities were used to locate where ZigBee sent information at and what time via a Python script. We have developed the temperature sensor module successfully. The temperature sensor module responds linearly and the output voltage is proportional to the temperature reading in millivolts.

The smoke sensors module was tested for the different concentrations of smoke in the ppm range. The smoke sensor output voltage rose to 0.7 V and kept on rising with the increase in smoke intensity, Table I shows the results of the experiment at the different intensities of smoke. In the testing of the GSM communication system, via AT commands, each AT command sent requires a time delay so as to receive a response. The delay was set at 0.5s. Table II shows the results for the time intervals between the sending and receiving of sms's.

After the testing of the separate sensor modules and the sensor network, we have tested the fire

Table 1.   Smoke sensor output (ppm).

Vout (V)	R_S (Ω)	R_L (Ω)	R_S/R_o	Smoke (ppm)
0.3	166k	10k	10	–
0.7	47.1k	10k	2.83	≈200
0.8	41.2k	10k	2.48	≈500
0.9	36.6k	10k	2.20	≈600
1.0	33k	10k	1.98	≈800

Table 2. Sms communication response time.

	No of AT commands	Time taken(s)
Sending out sms	5	≈13.5
Receiving sms	5	≈16
Voting system	–	120

detection criteria of the whole system and how the system responded to false positives. The set thresholds for smoke and temperature were calibrated for the previous experiments and set as designed for in the detailed design of this project. The results shown in Table III are for the detection system without the GSM communication. The test was performed 10 times for a 'fire' condition and a 'no fire' condition and showed that the system has a positive prediction value of 66.66% and a negative prediction value of 75%.

The warning sms's were received as expected and the triggering of the alarm via sms worked as expected, including the voting system as well, but the quantification of this could not be verified because the voting system is mainly subjective to the occupants in the building.

4 RESULTS AND DISCUSSION

In this article, we present a low-cost and energy-effic ient prototype of a smart fire detection system. The complete system was meant to effectively detect fire and reduce false positives and also integrate a voting system. The sensor modules of the temperature sensor and smoke sensor have been successfully used in the smart fire detection system. These systems put together detected fire and gave a warning to show the presence of a fire in the form of a buzzer. The system sent out sms's (short message services) to pre-warn occupants of the presence of a fire hazard, and this was the first means of reducing false positives, because the sms would prompt the occupants to vote whether there is a fire or not. This pre-warning process took approximately 13.5 s (for sending out ansms). The detection system could also be activated and deactivated by a sms. Thisa response time was approximately 16 s. This shows that the sms system is an efficient method of switching on and off the alarm. The system was designed to reduce false positives in a kitchen. The reduction of false positives is based on the elements that cause false positives in a kitchen, which are variations in temperature above the normal or excess smoke due to the cooking process or by any other appliance in a kitchen. The fire detection system did not detect specific appliances and whether it caused the false positive, but looked at the overall effect of the appliance and sees if there is an increase in temperature or smokes that is above normal.

5 CONCLUSION

In this paper we present a low-cost and energy-efficient prototype of a smart fire detection system. The developed system is suitable for reducing false positives using the wisdom of crowds. Overall most of these specifications for the system were met. On the other hand, there was some difficulty due to the testing environment, but overall the system did what it was designed to do.

In further research, the addition of a GUI (graphical user interface) for the setting of customised threshold can be done and also the capability for occupants to check the number of sms's (short message service) left in the system and be able to recharge the SIM easily. And also we will focus on the potential alternative communication technology (such as DASH 7), which allows better signal propagation (penetration) through walls, windows, doors, etc.

Table 3. Testing results of the fire detection system.

		Condition		
		Fire	No Fire	
System Response	Fire	True positive 8	False positive 4	Positive prediction value = 66.6%
	No Fire	False negative 2	True negative 6	Negative prediction value = 75%
		Sensitivity = 80%	Specificity = 60%	

REFERENCES

[1] T. A. Nguyen and M. Aiello, "Energy intelligent buildings based on user activity: A survey," Energy and Buildings, vol. 56, pp. 244–257, 2013.

[2] J. Y. Hong, E. H. Suh, and S. J. kim, "Context aware systems: a literature review and classification," Expert Systems with Applications, vol. 36, pp. 8509–8522, 2009.

[3] A. Kumar and G. P. Hancke, "Energy efficient environment monitoring system based on the IEEE 802.15.4 standard for low cost requirements", IEEE Sensors Journal, vol. 14, no. 8, pp. 2557–2566, Aug. 2014.

[4] A. Kumar and G. P. Hancke, "An energy-efficient smart comfort sensing system based on the IEEE 1451 standards for green buildings, IEEE Sensors Journal, vol. 14, No. 12, pp. 4245–4252, Dec. 2014.

[5] A. Kumar, I. P. Singh, and S. K. Sud, "An approach towards development of PMV based thermal comfort smart sensor, International Journal of Smart Sensing and Intelligent System, vol. 3, no. 4, pp. 621–642, Dec. 2010.

[6] Fire Protection Association of South Africa, 2013. [Online]. Available: http://www.fpasa.co.za/journals/fire-statistics.html.

[7] R. W. Bukowski, Techniques For Fire Detection, Center for Fire Research, National Bureau of Standards, 1987, p. 15.

[8] Oxford University Press, 2013.[Online]. Available: .

[9] T. J. Ohlemiller, "Smoldering and Combustion," in SFPE Handbook of Fire Protection Engineering, pp. 2/171–179.

[10] G. W. Muholland, "Smoke Production and Properties," in SFPE Handbook of Fire Protection Engineering, pp. 2/217-2/227.

[11] Fairfax County Fire and Rescue Department, "Smoke Alarms obey the   law, save your life," 2003. [Online]. Available:www.fairfaxcounty.gov/fr/prevention/fmpublications/obeythelaw.pdf.

[12] Central Station Alarm Association, "A Practical Guide to Fire Alarm Systems," 2011.

[13] M. Ahrens, "False Alarms and Unwanted Activations from U.S. Experience with Smoke Alarms and other Fire Detection/Alarm Equipment," National Fire Protection Association, Quincy, MA, November 2004.

[14] H. Soliman, K. Sudan and A. Mishra, "A smart forest-fire early detection sensory system: Another approach of utilizing wireless sensor and neural networks," Sensors, 2010 IEEE , pp. 1900–1904, 2010.

[15] A. Kumar, I. P. Singh, and S. K. Sud, "Energy efficient and low cost indoor environment monitoring system based on IEEE 1451 standards, IEEE Sensors Journal, vol. 11, no. 10, pp. 2598–2610, Oct 2011.

[16] NSC Data Sheet, LM35CZ: Precision Centigrade Temperature Sensor, National Semiconductor Corporation. Santa Clara, CA. [Online]. Available:http://www.national.com, (visited on 20 Nov 2012).

[17] Smoke Sensor: MQ-2, [Online].http://www.pololu.com/product/1480.

Network Security and Communication Engineering – Chan (Ed.)
© *2015 Taylor & Francis Group, London, ISBN: 978-1-138-02821-0*

Research on MDA and SOA integrated software craftsmanship

Q.H. Wang
College of Computer Science and Technology, Institute of Computer Application Technology, Inner Mongolia University for the Nationalities, Tongliao, China

L.S. Liu
Automation Workstation People's Laborational Army Troops, Tonghua, China

J. Chen, J. Lian & Z.L. Pei
College of Computer Science and Technology, Department of Information Center, The Affiliated Hospital, Inner Mongolia University for the Nationalities, Tongliao, China

ABSTRACT: Traditional hand workshop software development mode brings low software developing consistency, limited software size and uncontrollability of software cost. The reasonable and effective integration scheme is insufficient in the face of the more and more demanding of software platform integration. This paper gives a profound analysis of the very nature of software platform integration and craftsmanship, explores the invariant nature of craftsmanship model and puts forward the guiding architecture of software integration and craftsmanship.

1 INTRODUCTION

Throughout the development of worldwide software industry, the popular situation of new technology of software development has been rare. Software is still a high labor-intensive hand workshop industry. Its productivity and quality rely heavily on the individual quality of software developers. It does not form the scaled effect nor facilitates the advantage exploitability of teamwork. High system coupling is always the key factor which restricts the flexibility of software. Generally large-scale software consists of multiple decentralized, functional independence subsystems. Effective system integration proposes higher demands to software design.

2 IMPORTANT COMPONENT OF SOFTWARE CRAFTSMANSHIP—MDA

So-called software craftsmanship is meant to convert the original requests to highest quality software system problems that can solve business issues at the lowest cost. Traditional software development mode is a function prototype development that focuses on the business after specifying requirements. Everything starts from scratch. It is a hand workshop development mode that repairer first made wheels. For the purpose of avoiding the drawbacks of traditional software development mode, the solution is to adopt the MDA (Model Driven Architecture) mode that focuses on the general development. By means of abstracting business and systems, MDA builds kinds of models that can describe and implement business and system. It defines and manages the software process.

To realize the code automation, a series of standards and constraints are needed. The code automatic generation is not a free and optional process. The standards and constraints should be put on the invariant which is an abstract macro model to a system. The model is the core of MDA which meaning that the whole software process of MDA is based on model-driven. The models of MDA are divided into PIM (Platform Independent Model) and PSM (Platform Specific Model) according to the difference of abstracting levels. PIM (Platform Independent Model) surveys the software from the software system perspective which is the highest level model abstract and not relying on the generalized specific technology realization.

MDA is model driven that every phases in software process there are isomorphic component models and abstract levels are different in different stage models. Standard map mechanism of the whole system is scheduled and realized in the development stage. By means of mapping, the models of upper stage is transformed to the models of next stage. This PIM (Platform Independent Model) of different abstract levels is transformed to PSM (Platform Specific Model) which is relevant to the specific realizing technologies. Ultimately, codes relying on the specific realizing technologies are transformed by these PSM (Platform Specific Model) which drive the whole software development process.

3 SOFTWARE INTEGRATION IMPORTANT COMPONENT—SOA

Software integration usually can be divided into identification integration, data integration, application integration, process integration and platform integration by integration levels. Identification be integrated in the means of portal integration. Data integration is usually taken by ETL among multi isomorphic or isomeric data source. Application integration is taken by function between two systems or by the means of API. Process integration is based on work flow to realize the business process integration. The highest level of integration is platform integration which is to realize the application integration through unified integration platform middleware. It can eliminate the dispersion of business systems which is the source of isolated island of information.

Good solution of application integration is SOA (Service-Orient Architecture). SOA is not a software programming language nor a generalized specific technology realization. It is a new software architecture model that follows the open architecture standards. The original requirement forced the emergency of SOA is to solve the business integration of different business applications in the mode of B/S. From the view of service, development is based on service and the service result. The tendency is that application software taken from the service perspective is a brand new way of thinking.

4 BLUEPRINT OF INTEGRATED SOFTWARE CRAFTSMANSHIP PLATFORM

The strategy of software platform is an excellent scheme meeting the requirement of communication integration and craftsmanship. Integrated software craftsmanship platform ought to be a communication platform which is fault tolerance, standard and integrating enterprise technology assets. This paper puts forward an innovational idea that adopts the large granularity software life cycle management to the platform level management. The platform contains the whole process of system software life cycle model management, which makes a standard software process quantized controlled from the product project and approval to the software craftsmanship development process and the whole process of utterly deployment and launch. Development on platform is a synthesis solution that can help enterprises avoid software risks, control software cost, realize quick development of software craftsmanship and embody technology vane.

Integrated software craftsmanship platform defines two key objects: high integration and software craftsmanship. It is for achieving the high integration goal applying SOA, while for achieving the software craftsmanship goal by applying MDA throughout code automatic generation to realize quick software development. A reasonable design method is to take the platform into multi sub platforms. Three basic platforms as cores of the platform include model virtual sub platform, integrated development environmental sub platform and synthesized management sub platform. The model virtual sub platform is the core that takes charge of the dispatch and control of business system basic logic. Integrated development environmental sub platform is the visual development window of software craftsmanship which focus on code automatic generation and accelerating the development of project. Synthesized management sub platform is the periphery management tools set of platform that implement the visual synthesized management function through kinds of management tools. The blueprint of integrated software craftsmanship platform is shown in Fig.1.

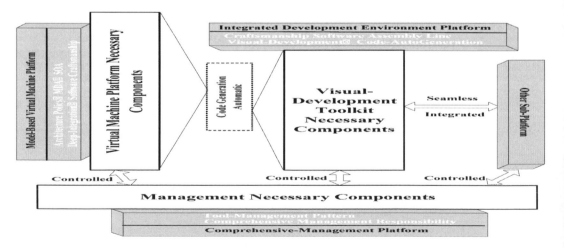

Figure 1. Blueprint of integrated software craftsmanship platform.

5 GUIDING ARCHITECTURE SCHEME OF VISUAL MACHINE PLATFORM

Visual machine platform involves two designing measures: platform independent abstract designing and platform specific abstract designing that are successive relationship instead of parallel relationship. Platform independent abstract designing focuses on the whole platform and models the system elements independent of specific platform, which is a set of PIM (Platform Independent Model). Platform specific abstract designing relies on elements modeling referring to specific technology details, which is a set of PSM (Platform Specific Model). The evolution process from platform independent abstract designing to platform specific abstract designing is in essence a process of modeling generalization. This generalized relationship can be understood as a relationship of definition and realization. Then PIM (Platform Independent Model) is the definition while PSM (Platform Specific Model) is the realization of definition. The definition is invariable while the ways of realization can be various. It is as shown in Fig.2.

Figure 2. Two designing measures of platform.

The realization of software craftsmanship platform relies on two key designing points.

Implementing environmental platform of target software: the design of model visual machine sub platform. The design ought to model the business system and has the capacity of cutting software system and recombining the system which is the basis of software craftsmanship realization. Considering the target of craftsmanship, the core architecture mode adopts MDA. Based on the demands of software craftsmanship, the inner support system modeling and flow modeling cater to the original need of craftsmanship basic design.

Craftsmanship flow line: the design of integrated development environmental platform. Only with the support of model visual machine platform craftsmanship, it cannot be a software craftsmanship platform. It only states that the software has the character of craftsmanship. Referring to the craftsmanship, there must be craftsmanship flow line, which is not only the integrated development environmental sub platform but also the custom interface of software craftsmanship.

Throughout unifying models, sub platforms achieve seamless joint. The craftsmanship goal relies on good design between two sub platforms. It is as shown in Fig. 3.

Model visual machine platform is the core of integrated software craftsmanship platform which follows the principle of layered architecture design. Its core architecture model adopts MDA+SOA scheme. The whole architecture focuses on the modeling of system implementing flow. It effectively cut off and recombine dynamically the target business systems that run on it, which embodies high flexibility and craftsmanship

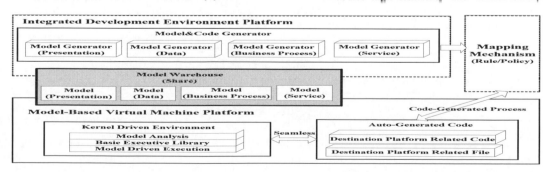

Figure 3. Seamless integrated design of platform.

character. Model visual machine platform is high integrated and adopts the component-oriented development mode. Directing at kernel programming, it can exert the highest value of codes which embodies high code reusability and platform universality.

The architecture focuses on loose coupling design. Development process pays attention to the design results and codes realization reflects clearly the design concept. The recommending model visual machine platform architecture is as shown in Fig.4.

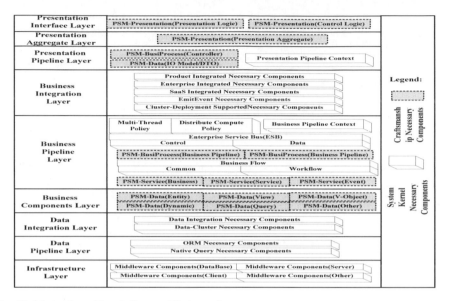

Figure 4. Model visual machine platform architecture scheme.

6   CONCLUSION

Integration and software craftsmanship is the motivation and tendency of modern software which can effectively resolve the difficulties in modern software domain. Development model based on platform is the only choice of integration and software craftsmanship. This paper discusses the craftsmanship model of integration software platform based on the MDA and SOA technologies and puts forward the corresponding guiding scheme. The most important feature of software craftsmanship is its process of pipeline whose focus is the invariant of software. It will be the next stage work of this paper to study the common model of software process from the theory perspective.

ACKNOWLEDGMENTS

This work was financially supported by the National Natural Science Foundation of China (61163034, 61373067), the Inner Mongolia talent development fund(2011), the Grassland Excellent Talents Project of Inner Mongolia Autonomous Region(2013), Supported By Program for Young Talents of Science and Technology in Universities of Inner Mongolia Autonomous Region(NJYT-14-A09), the Inner Mongolia Natural Science Foundation(2013MS0911, 2013MS0910), the 321 Talents Project the two level of Inner Mongolia Autonomous Region(2010), and the science research project of Inner Mongolia University for the Nationalities(NMD1316, NMD1229). The Corresponding author is Z.L. Pei, and the Corresponding email is zhilipei@sina.com.

REFERENCES

[1] Mak JKH, Choy CST, Lun DPK. 2004. Precise modeling of design patterns in UML. *Proc.of the 26th Int'l Conf.on Software Engineering(ICSE04)* .

[2] Than Xuede, Miao Huaikou, Liu Ling. 2004. Formalizing the semantics of UML statecharts with Z. *The Fourth International Conference on Computer and Information Technology (CIT 2004).*

[3] Shane Sendall, Wojtek Kozaczynski. 2003. Model Transformation: The Heart and Soulof Model-Driven Software Development. *IEEE Software* .

[4] Jean Bezivin,, Slimane *Hammoudi*, DenivaldoLope. 2004. Applying MDA approach for web service platform. *Proc of the8th IEEE Int'1Enterprise Distributed Object Computing Conf* .

[5] Tae Young Kim, Sunjae Lee, Kwangsoo Kim, Cheol-Han Kim. 2006. A modeling framework for agile and interoperable virtual enterprises. *Computers in Industry.*

Network Security and Communication Engineering – Chan (Ed.)
© 2015 Taylor & Francis Group, London, ISBN: 978-1-138-02821-0

Program engineering, development stages, problems, and perspectives in Azerbaijan

T.A. Bayramova

Azerbaijan National Academy of Sciences, Institute of Information Technology, Baku, Azerbaijan

ABSTRACT: The establishment and development stages of program engineering in Azerbaijan are provided. As a result of continuing and sustainable development policy of Azerbaijan in current stage, the development of information–communication systems is described. The works to be carried out for program engineering to gain a specific importance in the economy and the problems in this sphere are specified.

KEYWORDS: Program engineering, Brainware, Software, Information society, Critical systems.

The development of information-communication technology and systems performing based on this caused the difficulties in the process of development, accompaniment and improvement of software. The increase of the capacity of program complexes and the complexity in terms of architecture and technological solutions necessitated the engagement of quite a large specialist collectives for the development of such programs.

The specified factor required the regimentation of all specialists' peformance participating in systematization of requests, development, accompaniment of software and its adaption to user requests. Program engineering mainly considers the complex solution of such issues.

In Azerbaijan, the science of cybernetics and programming has commenced to be shaped starting from 60's. (For comparison, it's worth to mention that this sphere has been developed starting grm 1985 in India, the country which is considered as a leader on export of program products in the world). During USSR period, as in other republics, computer sciences have been developing in Azerbaijan. Thousands of specialists worked on programming problems. Those specialists possessed quite large knowledge in science spheres such as applied mathematics and cybernetics. Although during Soviet rule programmer position existed as a specialty in state system, program engineering mentality has not been shaped yet. After the collapse of USSR, the majority of specialists became unemployed. By that time, very few programmer collectives existed. In the end of 80's and in the beginning of 90's a "trend", a spurt happened in programming in the world [1]. We must regretfully mention that, because Azerbaijan has missed that "trend" for the known reasons, it has no capacity to compete with leading countries on this sphere at current time.

After Azerbaijan gained its independence once again, dynamic development in economic and socio-economic sectors has been achieved owing to successful external and internal policy conducted, reforms and large-scale project carried out in republic. The international prestige, defence power, the increase of economic potential of republic and improvement of well-being of population have created a strong basis for Azerbaijan government to stand abreast the development of countries in theglobalized world.

In current stage, being one of priority trends of continuing and sustainable development of Azerbaijan, information-communication techno-logies (ICT) has rapidly penetrated all spheres of socio-economic system and daily spheres of people and become an inseperable part of development of the economy in Azerbaijan. It is known that economic development of Azerbaijan is based on oil sector. It is not a coincidence that a great attention and concern is directed to ICT sector and the president of the country has proclaimed ICT sector as a second priority sphere after energy sector. Hence, as an alternative sphere to oil sector in the country, the development of ICT sector is specifically considered. The State Program on communication and information technologies development in Azerbaijan (Electron Azerbaijan) was adopted for 2005-2008 years. The measures considered in "Electron Azerbaijan" Government Program covered the initial works to be carried out for the provision of transition to information society. 2013 year was proclaimed as "Information-communication technologies year" in our country with the decree of Azerbaijan President [3]. Considering the development and implication of new technologies in different spheres of our alife in 2013, the President of Republic of Azerbaijan has adopted the decree on the establishment of Information Technologies University for the purpose of strenghtening the cadre potential.

According to research of UN Department of Economic and Social Affairs (UN DESA), the diagram has been prepared indicating the rating of e-goverment ratings in 2012. In this diagram, Azerbaijan occupies 96th place (Figure 1).

In global information society the knowledge becomes a commodity, as a result knowledge economy is shaped. The establishment of knowledge-based economy without educated people possessing high motivation and professional knowledge is very difficult. The educational and scientific potential in information technologies sphere is highly sufficient in our country. In high technologies sphere, it is desirable that Azerbaijan would export their own products. By directing the income gained from oil sector to human resources development, Azerbaijan can occupy fitting place in ICT sector among development countries. By ICT products, hardware, software and, brainware as a bit larger definition products are considered. While system software and hardware products are supervised by well-known and major companies, our state can gain profits in the competitive market by developing software, brainware products [2].

It must be mentioned that there is a reason for Azerbaijan becoming the regional center in high technologies sphere. In the report prepared by International Trade Center (jointly with UN and World trade organisation) it is mentioned that Azerbaijan possessed a potential for becoming a leader in Transcaucasus cybernetics market on ICT. Owing to cadre potential, the pespectivity of information technologies (IT) market and advantegeous geographical location, Azerbaijan can be in the center of all regional projects [4].

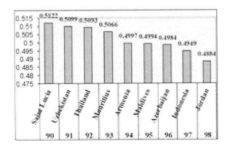

Figure 1. The diagram indicating the development rating of some e-governments.

Considering all these factors we can state that it will be reasonable to talk about shaping of "electron economy" in Azerbaijan in near future and program engineering will gain a specific importance in economy in relation with rapid development of ICT.

Program engineering is not only concerned with development of software. Alongside with technical sides of the analysis and documentation of requirements directed to program, projecting, development of program code, testing and attestation, operation and accompaniment stages, each program covers economic aspects of those software. Only a particular part of programs are carried out in accordance with budget facilities. In others, departures from budget, postponement problems occur. Some programs are not able to be terminated due to spending of the budget prior to deadline or to low-quality. Thus, skilled experts are required for the accurate estimation of prime cost of software, the provision of efficient performance of the system to be developed, program engineering management sphere for accurate selection of the collective carrying out the project. Information security imposes specific duties on program engineers. The most important of those is the production of products developed and tested by our local programmers in information systems serving to national interests.

Let us mention the several important issues standing in front of Azerbaijan for occupying the place abreast the developed countries in program engineering:

1 The absence of state program directed to program products development in Azerbaijan;
2 The few number of experts attained the international certificates (In Azerbaijan, there is no enterprise that attained CMM certificate, for comparison, lets mention that in India, more than 300 enterprises attained CMM certificate, 27 of them have 5th level CMM certificate);
3 The absence of management systems of production quality known by programmers and experts trained in accordance with world standards in project management;
4 The absence of standards in this sphere in the country.

The absence of large-scale program projects in Azerbaijan does not allow the program engineering to shape as an industry. As a result of the cheap assessment of programmer labor, low level of education, absence of educational specialties for the training of IT-specialists, we have to face the cadre migration.

Nowadays, one of the most important issues is the creation of productive environment for development of program engineering, the increase of interests of IT specialists in this sector, Internet-forums, electronic notifications, periodicals, practise in program engineering based on exhibitions and seminars, the exchange of information and knowledge. The preparation of books and other publications in Azerbaijani language is also important.

Refering to world standards, the adoption of modern standards in progrem engineering at state level, the approval of appropriate education and scientific specialties, the attraction of IT experts for education abroad for the increase of professional skills is important.

The main objective of information society must constitute of the creation of relations with foreign organisations specialising in program engineering, the participation in the development of modern international certification systems, the assistance to certification of Azerbaijan software companies in accordance with ISO and CMM standards and etc. For the purpose of discussion and investigation of solution trends of these issues, it will be useful to the realization of round tables with the participation of IT-specialists, programmers and etc.

REFERENCES

[1] Vasenin V.A. 2012. Modernizasiya ekonomiki i novie aspekti injenerii proqramm. *Proqramnaya injeneriya*,Moskva.
[2] T.H.Kazimov, T.A.Bayramova. 2013. Proqram muhendisliyi. İnformasiya Texnologiyaları, Azerbaijan: Baku.
[3] http://www.president.az/articles/564?locale=az
[4] http://www.mincom.gov.az/layiheler/elektron-hokumet/

Network Security and Communication Engineering – Chan (Ed.)
© *2015 Taylor & Francis Group, London, ISBN: 978-1-138-02821-0*

The study of diminishing marginal utility in university teaching

M.Z. Cui
College of Economy and Management, Shang Hai Ocean University, Shang Hai, China

X. Yang
Department of Marketing, Hang Seng Management College, Siu Lek Yuen, N.T., Hong Kong

ABSTRACT: The effect of diminishing marginal utility always has obvious influence in our daily life, which is also penetrating in the process of university teaching. This paper clarifies the characteristics of the effect of diminishing marginal utility in teaching and explains why the effect of diminishing marginal utility occurs in university teaching. Then this study analyses the teaching performance with the influence of the effect of diminishing marginal utility. Finally, this paper gives some suggestions to cope with the diminishing marginal utility in university teaching.

KEYWORDS: Margin diminishing, Marginal utility, University teaching.

1 INTRODUCTION

In the economic field, margin means that the utility each unit of new incremental resource leads to. For example, marginal cost is defined as the every new incremental cost when produce one product, and marginal profit is gained when produce one product.

The effect of diminishing marginal utility is very important in the method marginal analysis. In the consuming process, although the whole utility is increased with the increase of products consumption, the marginal efficiency will decrease and even to a negative influence. In this situation, the whole utility will decrease with the increase of the consumed products. Students always have a high motivation to learn in the beginning of the semester. However, they will feel bored or dissatisfied when the learning contents become more and more. This phenomenon will influence the whole utility of learning and teaching. This phenomenon is called the effect of diminishing marginal utility in teaching.

2 THE EFFECT OF DIMINISHING MARGINAL UTILITY

Classroom teaching is a perfect combination of interaction between teachers and students. In this process, teachers provide knowledge, information, train their ability and provide spiritual cultivation. At the same time, students accept knowledge, receive information, obtain skills, digest, and absorb spiritual cultivation from teachers. In this interactive process, there also exists the phenomenon of diminishing marginal utility. If we carry out in-depth analysis and discussion on this phenomenon, it will be helpful to improve college teachers' classroom teaching ability and teaching skills.

For example, a class of students can be regarded as a spiritual consumption. In fact, teaching is to provide students information with a spirit of servicing. In this process, the size of students' spirit service acceptance is the size of the utility. With the increase of the number of students attending a lecture, the marginal utility reduces gradually. If the number is too large, it may cause the students' reverse psychology and bring the total utility reduce.

The effect of diminishing marginal utility is treated as the basic research method in economics. It is not only applied in academic fields, but also in economic society. It is very useful to bring the effect of diminishing marginal utility into the field of teaching management, because teachers' teaching performance has great influence on students' learning process. This effect is more important to the education in China than in other countries, because the government of China is lack of resources and capital in the education budget. It will be very meaningful if we can know and handle this effect in education field.

3 THE ROLE OF DIMINISHING MARGINAL UTILITY ON TEACHING MANAGEMENT

If we want to know the effect of diminishing marginal utility on teaching management, we must know the essence of this effect first. Normally, if the function of $y=f(x)$ could be derivate, then the function of $f'(x)$ is called marginal function. In teaching filed, independent variable x can be treated as the investment of teaching resource, such as the construction of basic facilities, investment of teaching members, logistic guarantee and so on. The performance of teaching can be treated as a dependent variable, including the increase of students' ability and knowledge after learning in class. Therefore, $f'(x)$ can be treated as the increase of students' ability and knowledge when each unit of incremental teaching resources is invested in teaching.

According to the effect of diminishing marginal utility, the $f'(x)$ is decreased which means that if the condition of teaching is controlled, the speed of increase of students' ability and knowledge is reduced when each new unit of teaching resource be invested. This effect is neither subjective nor can be controlled by ourselves.

There are several reasons that lead to the diminishing marginal utility in teaching management:

Firstly, this effect can be explained with the physiological and psychological reasons. The degree of satisfaction and reaction to repeated stimulus will reduce with the increase of investment of teaching resource.

Secondly, we can use the resource's multiple usages to understand the diminishing marginal utility in teaching management. Students always consider the first unit of teaching contents as the most important thing, and the second unit as the relatively less important thing. In this situation, teaching service's marginal utility will decrease with the decrease of usage's importance.

Thirdly, there exists the inconsistency matching between information and time. The most obvious characteristics of different levels of education in China are less learning time, more learning contents and classes. In the beginning, students are always interest in new knowledge with good learning performance such as quick and thorough understanding and high motivation etc. However, students always feel that they are receiving relatively less knowledge in the same time and are less interested in study with time goes on. This phenomenon is strongly supported by the diminishing marginal utility in teaching management.

There is no now doubt about the existence of the diminishing marginal utility in teaching management. This urges us to think about this question-how to arrange or use the limited teaching resources to achieve the maximal teaching result.

4 THE PERFORMANCE OF DIMINISHING MARGINAL UTILITY IN TEACHING MANAGEMENT

There are many kinds of performance of diminishing marginal utility in teaching management. They especially occur in the following parts:

First, the effect of diminishing marginal utility occurs in students' knowledge, ability and accomplishments. For students, they hope that they can get greater utility in class or learning from teachers. If they can get the teaching resource with high quality and suitable training, they will show a high level of satisfaction to knowledge receiving, leading to a high level of motivation to study harder. However, teaching contents are boring and training ways are rigid as well as unsatisfying teaching performance. One of the important reasons which lead to the negative consequence is the diminishing marginal utility in students' consumption of teaching service from teachers.

Second, the diminishing marginal utility occurs with the praise from teachers. Praise from teachers will lead to positive motivation in student, which makes students feel satisfied. Students, who get the praise for the first time, will have a high marginal utility that leads to a high level of motivation. However, if they are praised many times in a period, the diminishing marginal utility will occur with the incremental of praise, leading to a low level of motivation or even lead to students' ignoring of the praise. Moreover, students who do not get praise at all may dislike other students' getting praise from teachers.

Third, the diminishing marginal utility occurs in criticizing. Criticizing is a motivation with negative consequences. It will bring unsatisfied feelings to students who are criticized. This psychological pain motives students to study harder. Therefore, appropriate criticizing is also a good way to stimulate students' motivation of learning.

Students who are criticized for the first time will have the most impressive memory and the strongest psychological pain, which will lead to students' determination of improving himself and working harder. However, students will have less psychological pain when they are criticized for the second time. With the increase of criticizing times, students will be adjusted to the criticizing, and the criticizing will lead to low level of motivation. Thus the diminishing marginal utility will occur in criticizing.

Finally, the diminishing marginal utility occurs in teaching scenarios. Students will consume various resources in specific teaching scenarios. Moreover, the teaching scenarios will also produce satisfaction or utility for the students. Teaching scenarios include hardware and software. Hardware incudes teaching environments and teaching facilities; software consists of teaching policies and rules, teaching methods, teachers' personalities, studying atmosphere and moral standards etc. If the students receive a relatively less active teaching scenario with poor quality, the diminishing marginal utility will occur again in students' learning.

5 METHODS TO COPE WITH THE DIMINISHING MARGINAL UTILITY IN TEACHING PERFORMANCE

First, teachers should improve their comprehensive quality.

They should:

a. renew their understanding of the concepts and theories of teaching as well as their professional skills
b. understand students' cognitive situation
c. prepare lessons adequately and grasp the teaching contents
d. know what to teach and how to teach.
e. learn how to strengthen the knowledge renewal and accumulation in the teaching practice.

Second, there should be a rich variety of teaching forms. In response to the teaching material and contents, teaching methods should be timely updated. Different types of teaching plans should be designed. A variety of teaching methods and teaching programmers are alternately combined, such as lecture, pronunciation, polemics, barrier method, investigation method. They all enable students to accept different teaching methods and experience different types of teaching plan. All these methods will decrease the diminishing marginal utility effect in the classroom teaching.

Third, there should be a comprehensive use of modern teaching means. We should use comprehensive teaching means, such as teaching tools, teaching plans, the slide projector, videos and sound design. These will give students different types of stimulation and adjust teaching content and teaching density. Then we could improve the teaching efficiency and the teaching of the marginal utility.

Fourth, praise should be combined with criticism, and criticism should be used carefully. We should pay attention to students' individual difference in the recognition of a student. For the same praise, students of different learning attitudes, achievements, and psychological expectations often have different reactions. Some students may get great encouragement, while others may be indifferent. We use praise to complement one's measures to local conditions and distinguish them from each other. We should pay attention to the time and appropriateness of praising. We should make praise a strong attraction to every student and try to reduce the marginal utility. Because students react strongly to critics for the first time, therefore college teachers should be cautious in using criticism. Otherwise, we will seriously damage the students' self-esteem and self-confidence and lead to a counter effect.

6 SUMMARY

In short, we should pay enough attention to the law of diminishing marginal utility and its performance in school education. By doing so, we will be able to take appropriate measures against its negative influence and achieve better education results.

REFERENCES

[1] Cui Maozhong; WEN Yanping. Application Research of Marginal Ideas in School Teaching [J].The Guide of Science & Education, 2011.12.
[2] Cai Wei. Marginal utility of classroom teaching[J]. Journal of Hebei Normal University(Educational Science Edition), 2010.02.
[3] Wen Shaoting, LI Xiaodong. Analysis of the law of diminishing marginal utility in teaching application[J]. Exam weekly, 2012.14.
[4] Yu Jiatai. The revelation of diminishing marginal utility in the classroom teaching[J]. Shanghai Research on Education.
[5] Wu Qing. Analysis of the marginal concept, improve their teaching methods[J]. Technology and Market, 2009.09.
[6] Liu Fuying. Discussion of diminishing marginal utility of classroom teaching phenomenon[J]. Journal of Zhuzhou Teachers College, 2002.03.
[7] Bao Shihong, Jiang Shihong. The rediscover of the law of diminishing marginal utility[J]. Value Engineering, 2012.06.
[8] LI Zhi Feng. The law of diminishing marginal utility and moral education innovation[J]. Journal of Chongqing University of Science and Technology(Social Sciences Edition), 2012.01.

Network Security and Communication Engineering – Chan (Ed.)
© *2015 Taylor & Francis Group, London, ISBN: 978-1-138-02821-0*

Optimization of cold-test parameters on Helix geometry for S-Band TWT

J.Y. Kou
NASA Center for Aerospace Device Research and Education, North Carolina Central University
Beijing Vacuum Electronics Research Institute, Beijing, China

Y. Tang & B. Vlahovic
NASA Center for Aerospace Device Research and Education, North Carolina Central University

H.F. Wang
Beijing Vacuum Electronics Research Institute, Beijing, China

ABSTRACT: The main structure parameters of TWT SWS (slow wave structure) are important. The pitch and radius of helix, shell inner radius and other parameters are decisive. But the geometry of helix could also affect the cold-test parameters of TWT. Cold-test parameters are composed of dispersion, interaction impedance on-axis of helix SWS, and attenuation. There is an optimum of cold-test parameters on helix width and thickness for BWI efficiency in S-Band TWT. After optimization of helix width and thickness, the BWI efficiency of S-band TWT could rise by 1-2 percentage points, i.e. about 5%.

KEYWORDS: TWT (traveling-wave tube), Helix, SWS (Slow-Wave Structure), Dispersion, Interaction Impedance, Optimization, BWI efficiency.

1 INTRODUCTION

The frequency, power and voltage of TWT are determined by the main structure parameters of TWT SWS. i.e, pitch and radius of helix, shell inner radius and other parameters for helix TWT. But the helix geometry w also affects the cold-test parameters of TWT SWS. Helix geometry refers to width and thickness of helix. Cold-test parameters of TWT SWS are composed of dispersion, interaction impedance on-axis of helix SWS, and attenuation. The cold-test parameters could be calculated by general EM software[1][2], but also could be calculated by special software[4].

There are different dispersion, interaction impedance on-axis of helix SWS, and attenuation on different width and thickness of helix for S-Band helix TWT SWS. The BWI efficiency of TWT would be higher on condition that the dispersion is appropriate, the interaction impedance is greater, and the attenuation is less. There is an optimal point of BWI efficiency for cold-test parameters on helix width and thickness in S-Band TWT.

2 CALCULATION OF COLD-TEST PARAMETERS FOR S-BAND TWT

The method for calculating the dispersion is similar to experimental methods. With quasi-periodic boundary conditions frequency-phase characteristics are determined by measuring the resonant frequencies in a section of circuit shorted at both ends. The dispersion is expressed as phase velocity. The function to calculate phase velocity is defined as the following.

$$v_p = 2\pi f L / \varphi \qquad (1)$$

The calculation of interaction impedance is also similar the experiment method. With the quasi- periodic boundary conditions using perturbation theory. With a cylindrical dielectric rod placed on the central axis of helix. the resonance frequency shift is calculated, and then interaction impedance could be calculated by the following equation.

$$K_c = \frac{2}{\omega\beta(\varepsilon_r - 1)\varepsilon_0\pi b^2} \frac{v_p}{v_g} \frac{\Delta f}{f} \qquad (2)$$

Attenuation could be calculated by the following equation[5].

$$d = 20\lg e^{\alpha l} = 8.686\alpha l = 8.686\frac{P_L}{2Wv_g} \cdot L \qquad (3)$$

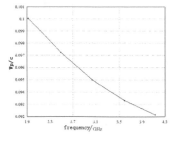

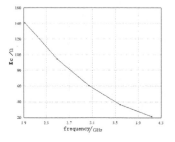

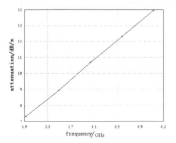

Figure 1a. Phase velocity of helix SWS.

Figure 1b. Interaction impedance of helix SWS.

Figure 1c. Attenuation of helix SWS.

where α is attenuating constant, i.e. the ratio of attenuating power to the total power in periodical length, W is the total power in cavity, v_g is group velocity, L is the period of SWS.

Fig.1 are the cold-test parameters of a S-band TWT helical SWS calculated by [4]. Fig. 1a is phase velocity, and Fig. 1b and 1c are interaction impedance and attenuation respectively.

3 CALCULATION OF LARGE-SIGNAL FOR S-BAND TWT

After the cold-test parameters of TWT helix SWS is obtained, the BWI efficiency could be calculated by large-signal program. There are some large-signal programs for TWT nowadays. Microwave-tube simulator suite [4] is usually used for calculating BWI efficiency in China.

4 OPTIMIZATION OF COLD-TEST PARAMETERS AND BWI CALCULATION FOR S-BAND TWT

With the variations of axial width and thickness of the helix, there is a slight effect in the phase velocity, but more effect in the interaction impedance and attenuation. The relation of phase velocity and helix width is present at Fig. 2a; and the relation of interaction impedance and helix width is present at Fig. 2b. The helix width is normalized by the original width. The relation of phase velocity and helix thickness is present at Fig. 3a; and the relation of interaction impedance and helix thickness is present at Fig. 3b. The normalized thickness refers to the thickness of helix tape normalized by the original, where the inter radius of helix is fixed. The relation of attenuation and helix width, thickness is upward direction, and not present.

There are different cold-tested parameters of helix SWS, on different width and thickness of helix for S-Band Helix TWT SWS. And different BWI

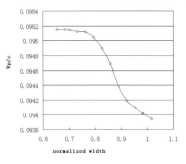

Figure 2a. The relation of phase velocity and helix width.

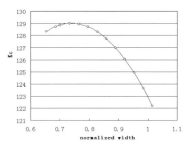

Figure 2b. The relation of interaction impedance and helix width.

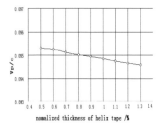

Figure 3a. The relation of phase velocity and helix thicknes.

666

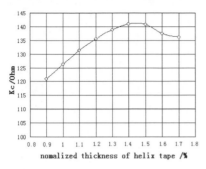

Figure 3b. The relation of interaction impedance and helix thickness.

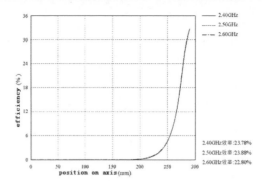

Figure 4b. Large-signal result after optimizing helix geometry.

efficiency is corresponding to different phase velocity, interaction impedance and attenuation. The BWI efficiency would be more high on condition that the dispersion is appropriate, the interaction impedance is greater, and the attenuation is less. There is an optimal point of helix width and thickness for BWI efficiency in S-Band TWT.

Fig. 4a is the original large-signal result, and Fig. 4b is the large-signal result for helix width and thickness optimized of helix. It can be seen from the figure that the BWI efficiency of S-band TWT rise by 1-2 percentage points after optimizing width and thickness of helix.

5 CONCLUDING REMARKS

The main structure parameters of TWT SWS determine The frequency, power and voltage of TWT. The geometry of helix could also affect the cold-test parameters of TWT.

There is an optimum of cold-test parameters on helix width and thickness for BWI efficiency in S-Band TWT. After optimization of helix width and thickness, the BWI efficiency of S-band TWT could rise by 1-2 percentage points, i. e. about 5%.

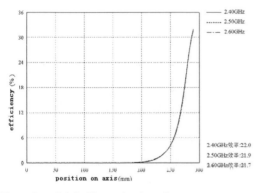

Figure 4a. Original large-signal result.

REFERENCES

[1] Kory C L,Dayton J A. Accurate cold-test model of helical TWT slow-wave circuits[J].IEEE Trans Electron Devices,1998,45(4);966–971.

[2] Kory C L. Three-dimensional simulation of helical traveling-wave tube cold-test characteristics using MAFIA[J]. IEEE Trans Electron Devices,1996, 43(8);1317–1319.

[3] Li Shi, et al, Effect of Helical Slow-wave structure Parameter Variations on TWT Cold-Test Characteristics, proceding of IVESC2004,Beijing.

[4] Bin Li, Zhonghai Yang, et al, Theory and Design of Microwave-Tube Simulator Suite, IEEE transactions on electron devices, Vol. 56, NO. 5, May 2009.

[5] Jianyong Kou, Research on Efficiency Improvement of CCTWT, doctoral dissert. of Eng ,BVERI, China, Apr. 2007.

Author index

/